PERIODIC TABLE OF THE ELEMENTS

I A	II A												III A	IV A	V A	VI A	VII A	Noble gases
1 H 1.0																		2 He 4.0
3 Li 6.9	4 Be 9.0												5 B 10.8	6 C 12.0	7 N 14.0	8 O 16.0	9 F 19.0	10 Ne 20.2
11 Na 23.0	12 Mg 24.3	III B	IV B	V B	VI B	VII B		VIIIB		I B	II B		13 Al 27.0	14 Si 28.1	15 P 31.0	16 S 32.1	17 Cl 35.5	18 Ar 39.9
19 K 39.1	20 Ca 40.1	21 Sc 45.0	22 Ti 47.9	23 V 50.9	24 Cr 52.0	25 Mn 54.9	26 Fe 55.8	27 Co 58.9	28 Ni 58.7	29 Cu 63.5	30 Zn 65.4	31 Ga 69.7	32 Ge 72.6	33 As 74.9	34 Se 79.0	35 Br 79.9	36 Kr 83.8
37 Rb 85.5	38 Sr 87.6	39 Y 88.9	40 Zr 91.2	41 Nb 92.9	42 Mo 95.9	43 Tc (98)‡	44 Ru 101.1	45 Rh 102.9	46 Pd 106.4	47 Ag 107.9	48 Cd 112.4	49 In 114.8	50 Sn 118.7	51 Sb 121.8	52 Te 127.6	53 I 126.9	54 Xe 131.3
55 Cs 132.9	56 Ba 137.3	57 La* 138.9	72 Hf 178.5	73 Ta 180.9	74 W 183.9	75 Re 186.2	76 Os 190.2	77 Ir 192.2	78 Pt 195.1	79 Au 197.0	80 Hg 200.6	81 Tl 204.4	82 Pb 207.2	83 Bi 209.0	84 Po (210)	85 At (210)	86 Rn (222)
87 Fr (223)	88 Ra (226)	89 Ac† (227)	104 Unq (261)	105 Unp (262)	106 Unh (263)	107 Uns (262)	108 —	109 Une (266)									

Atomic number — 6
Symbol — C
Atomic weight — 12.0

58 Ce 140.1	59 Pr 140.9	60 Nd 144.2	61 Pm (147)	62 Sm 150.4	63 Eu 152.0	64 Gd 157.3	65 Tb 158.9	66 Dy 162.5	67 Ho 164.9	68 Er 167.3	69 Tm 168.9	70 Yb 173.0	71 Lu 175.0
†	90 Th 232.0	91 Pa (231)	92 U 238.0	93 Np (237)	94 Pu (242)	95 Am (243)	96 Cm (247)	97 Bk (247)	98 Cf (249)	99 Es (254)	100 Fm (253)	101 Md (256)	102 No (254)	103 Lr (256)

‡ Parentheses around atomic weight indicate that weight given is that of the most stable known isotope.

CHEMISTRY AND LIFE

An Introduction to General, Organic, and Biological Chemistry

THIRD EDITION

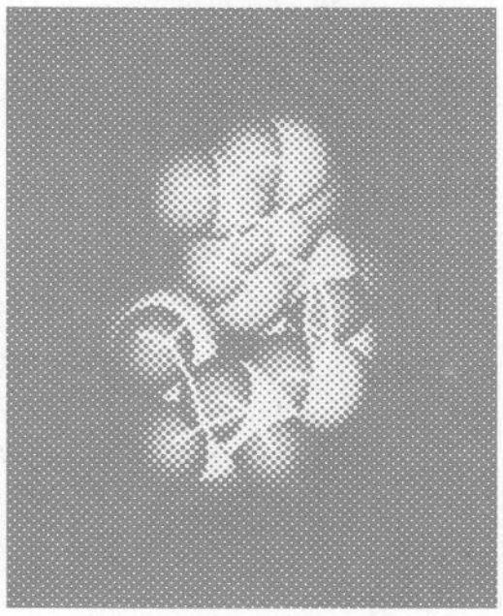

John W. Hill

University of Wisconsin
River Falls, Wisconsin

Dorothy M. Feigl

Saint Mary's College
Notre Dame, Indiana

MACMILLAN PUBLISHING COMPANY
New York
COLLIER MACMILLAN PUBLISHERS
London

Cover photograph: Computer graphics model of an adenosine triphosphate (ATP) molecule and its exterior surface. The ATP molecule is central to life. A part of the energy we obtain from food is stored in our cells as ATP. These molecules provide the energy for the mechanical work of muscles, the transport of substances across cell membranes, and the synthesis of other molecules needed in the cells. ATP thus serves as the link between the energy-producing reactions and the energy-requiring reactions of our cells. ATP is often called the energy currency of the cell. Computer graphics by Helga Thorvaldsdottir, University of North Carolina at Chapel Hill Department of Computer Science. The spheres were generated using a program by Thomas Porter, National Institutes of Health, and Conrad Huang and Thomas Ferrin, University of California—San Francisco. The UNC Molecular Graphics facility is a National Institutes of Health Research Resource.

Copyright © 1978, 1983, 1987, by Macmillan Publishing Company, a division of Macmillan, Inc.

Printed in the United States of America

Library of Congress Cataloging-in-Publication Data
Hill, John William, 1933–
    Chemistry and life.

    Includes index.
    1. Chemistry. 2. Biological chemistry. I. Feigl,
Dorothy M. II. Title.
QD31.2.H56 1987   540   86-23911
ISBN 0-02-354970-X

Macmillan Publishing Company
866 Third Avenue, New York, New York 10022

Collier Macmillan Canada, Inc.

Printing: 1 2 3 4 5 6 7 8   Year: 7 8 9 0 1 2 3 4 5

ISBN 0-02-354970-X

Contents

Preface

Our world has been transformed by science and technology. The impact of science on the quality of human life is profound. Yet, to beginning students, the scientific disciplines that daily influence their lives often seem mysterious and incomprehensible. Those of us who enjoy the study of science, however, find it a fascinating and rewarding experience precisely because it *can* provide reasonable explanations for seemingly mysterious phenomena.

Chemistry and Life has been written in that spirit. Apparently obscure phenomena are explained in an informal, readable style. We assume that the student has little or no chemistry background and clearly explain each new concept as it is introduced. Chemical principles and biological applications are carefully integrated throughout the text, with liberal use of drawings, diagrams, and photographs.

Our selection of topics and choice of examples make the text especially appropriate for students in health and life sciences, but it is also suitable for anyone seeking to become a better-informed citizen of our technological society. The text provides ample material for a full year's course. The twelve Special Topics cover optional material for added flexibility. They may be omitted or assigned as outside reading without loss of continuity. If the length of your course is one semester (or one or two quarters), we recommend *General, Organic, and Biological Chemistry: Foundations of Life,* by Feigl and Hill (New York: Macmillan Publishing Company).

CHANGES IN THE THIRD EDITION

In response to the suggestions of users of the second edition and our own experience, we have made the following modifications in the third edition.

- A new chapter on bioenergetics has been added (Chapter 25, Energy and Life).

- The number of problems at the ends of chapters has been almost tripled, and the selection of worked-out examples within chapters has been expanded. Answers to odd-numbered numerical end-of-chapter problems are provided in Appendix IV, as are answers to a selection of other odd-numbered end-of-chapter problems.
- A discussion of significant figures (Appendix III) has been added, and answers to problems are presented in the appropriate number of significant figures.
- The coverage of some fundamental concepts in chemistry has been expanded, including: the introduction to energy (Chapter 1), atoms (Chapter 2), atomic weight (Chapter 3), electronegativity (Chapter 4), activation energy and Le Chatelier's principle (Chapter 5), the ideal gas law (Chapter 6), ratio concentrations, strengths of drugs, colligative properties, and osmolarity (Chapter 9), pH with other than unit values (Chapter 11), esters of inorganic acids (Chapter 17), chemistry of the nervous system (Special Topic G), and bulk properties of polymers (Chapter 19), physical properties of carbohydrates (Chapter 20), physical properties of amino acids (Chapter 22), induced-fit model of enzyme action (Chapter 24), thermodynamic laws, free energy, coupled reactions, and synthesis of ATP (Chapter 25), and regulating steps in glycolysis and the Krebs cycle (Chapter 27).
- New health-related topics have been added, including, among others: medical imaging techniques, including X ray, CAT, PET, magnetic resonance, and ultrasonography (Chapter 3), gases used in medicine (Chapter 6), benzoyl peroxide as antiacne medicine (Chapter 8), hypertension (Chapter 12), fluorinated compounds as blood substitutes (Special Topic E), increased emphasis on *how* drugs act (Special Topic F), biomedical polymers (Chapter 19), cellulose as dietary fiber (Chapter 20), physiologically active peptides (Chapter 22), oncogenes (Chapter 23), drugs as enzyme inhibitors (Chapter 24), muscle action (Chapter 25), autoimmune diseases, AIDS, and monoclonal antibodies (Chapter 26), lactose intolerance (Special Topic L), and lipoproteins and cardiovascular disease (Chapter 28).

SUPPLEMENTS

Study Guide. The *Study Guide* contains a key word list, chapter summary, study hints, additional worked-out examples and practice problems, and a self-test for each chapter. The text itself includes many learning aids, but the *Study Guide* should help your students learn the material more effectively and more efficiently.

Laboratory Manual. A laboratory manual, also published by Macmillan Publishing Company, complements the text very well. *Chemistry and Life in the Laboratory: Experiments in General, Organic, and Biological Chemistry,* by

Heasley, Christensen, and Heasley, also uses examples from students' lives to communicate the science of chemistry. The manual covers the same general topics as the textbook, instructions are clear and thorough, and the experiments are well written and imaginative. All the experiments work! There are 33 experiments to choose from and all have been thoroughly tested over many years with students in the laboratory. Special care has been taken in the second edition to reduce the cost of chemicals required in the experiments and to make the experiments even safer. An *Instructor's Manual* is available from the publisher.

Instructor's Manual and Solutions Manual.   The *Instructor's Manual and Solutions Manual* to accompany the text is available from the publisher for adopters of the text. It offers chapter overviews, scheduling suggestions, teaching tips, and answers to all end-of-chapter problems in the text.

Test Bank.   A selection of multiple-choice test questions on $5\frac{1}{2}$-by-$8\frac{1}{2}$-inch cards is available to instructors who adopt the text.

Transparency Masters.   A set of transparency masters of illustrations and tables from the text is available from the publisher.

ACKNOWLEDGMENTS

In preparing this edition, we have had the benefit of thoughtful reviews by a number of teachers. For generously sharing insights garnered from their own classroom experience, we thank: John J. Houser, The University of Akron; Richard P. McKee, Pasadena City College; Louis E. Perlgut, California State University, Long Beach; J. W. Phillips, University of South Alabama; Norman C. Rose, Portland State University; Paul G. Seybold, Wright State University; and Charles R. Willms, Southwest Texas State University.

No book—or other educational device—can replace a good teacher; thus we have designed this book as an aid to the classroom teacher. The only valid test of this or any text is in a classroom. We would greatly appreciate receiving comments and suggestions based on your experience with this book.

To the Student

What is chemistry?

Chemistry is such a broad, all-encompassing area of study that people almost despair in trying to define it. Indeed, some have taken a cop-out approach by defining chemistry as "what chemists do." But that won't do; it's much too narrow a view.

Chemistry is what we all do. We bathe, clean, and cook. We put chemicals on our faces, hands, and hair. Collectively, we use more than 10 000 consumer chemical products in our homes. Professionals in the health and life sciences use thousands of additional chemicals as drugs, antiseptics, or reagents for diagnostic tests.

Your body itself is a remarkable chemical factory. You eat and breathe, taking in raw materials for the factory. You convert these supplies into an unbelievable array of products, some incredibly complex. This chemical factory—your body—also generates its own energy. It detects its own malfunctions and can regenerate and repair some of its component parts. It senses changes in its environment and adapts to these changes. With the aid of a neighboring facility, this fabulous factory can create other factories much like itself.

Everything you do involves chemistry. You read this sentence; light energy is converted to chemical energy. You think; protein molecules are synthesized and stored in your brain. All of us are chemists.

Chemistry affects society as well as individuals. Chemistry is the language—and the principal tool—of the biological sciences, the health sciences, and the agricultural and earth sciences.

Chemistry has illuminated all of the natural world, from the tiny atomic nucleus to the immense cosmos. We believe that a knowledge of chemistry can help you. We have written this book in the firm belief that beginning chemistry can be related immediately to problems and opportunities in the life and health sciences. And we believe that this can make the study of chemistry interesting and exciting, especially to nonchemists.

For example, an "ion" is more than a chemical abstraction. Mercury ions in the wrong place can kill you, but calcium ions in the right place can keep you from bleeding to death. "$P \times V = c$" is an equation, but it is also the basis for the respiratory therapy that has saved untold lives in hospitals. "Hydrogen bonding" is a chemical phenomenon, but it also accounts for the fact that a dog has puppies while a cat has kittens and a human has human offspring. Hundreds of similar fundamental and interesting applications of chemistry to life can be cited.

A knowledge of chemistry has already had a profound effect on the quality of life. Its impact on the future will be even more dramatic. At present we can control diabetes, cure some forms of cancer, and prevent some forms of mental retardation because of our understanding of the chemistry of the body. We can't *cure* diabetes or cure *all* forms of cancer or *all* mental retardation, because our knowledge is still limited. So learn as much as you can. Your work will be enhanced and your life enriched by your greater understanding.

Be prepared. Something good might happen to you—and to others because of you.

1

Matter and Measurement

This book is called *Chemistry and Life*. What does chemistry have to do with life? What is chemistry? For that matter, what is life?

The last question is more than rhetorical. Progress in science, technology, and medicine has blurred the distinction between life and death. Is someone whose heart has stopped beating dead? Is someone whose vital functions are being maintained by machine alive? We won't even attempt in this book to supply a definitive answer to the question "What is life?" We'll simply note its critical significance for our society.

How about the first question, "What does chemistry have to do with life?" A chemist would say, "Just about everything." The human body, for example, is the most extraordinarily complicated, most elegantly designed, and most efficiently operated chemical laboratory there is. Our attempts to answer that first question will fill most of this text.

That leaves the middle question, "What is chemistry?" And that is the subject of this first chapter. We shall see how science in general and chemistry in particular have developed from earlier human endeavors. Our study will include a consideration of the methods of science and the manner of its progress. Finally, we shall also develop some basic concepts necessary to our study of chemistry and its relationship to life.

1.1 SCIENCE AND THE HUMAN CONDITION

We are taught in elementary school that people have three basic needs: food, clothing, and shelter. Certainly those three things—if adequate in quantity and

quality—are enough to keep us alive. Most of us, however, would also agree to adding two more requirements for the good life: reasonable health and some chance for happiness.

In early human societies, nearly all human efforts were directed toward the hunting and gathering of food, the making of clothing, and the provision of shelter. Our early ancestors had no knowledge of the biological and chemical bases of illness, and they could do little about their health except to pray and make sacrifices to their gods. With the coming of civilization, some people gained enough leisure to turn their thoughts to the human condition and to the natural world around them. Over the centuries what we now call science grew out of their speculations. As this scientific study of the material universe progressed, the responsibility for adding to the growing body of knowledge was divided among various disciplines. Among these disciplines was chemistry.

Modern chemistry's roots are firmly planted in alchemy, a kind of mystical chemistry that flourished in Europe during the Middle Ages (Figure 1.1). And modern chemists have inherited from the alchemists an abiding interest in those aspects of their study that relate to human health and to the quality of life.

Consider, for example, the fact that alchemists not only searched for a "philosopher's stone" that would turn cheaper metals into gold but also sought an "elixir" that would confer immortality on those exposed to it. Alchemists never achieved their primary goals, but they did discover many new chemical substances. As early as the ninth century, a Persian alchemist, Al-Razi, described the use of plaster of Paris for casts to set broken bones. And alchemists perfected techniques such as distillation and extraction that are still useful in our time.

It was a Swiss physician, Theophrastus Bombastus von Hohenheim (1493–1541), who urged alchemists to turn away from their attempts to make gold and to seek instead medicines with which to treat disease (Figure 1.2). Possessed of a monstrous ego, von Hohenheim (who preferred the self-chosen name Paracelsus) alienated many of his contemporaries. His followers, however, were numerous enough to ally forever the science of chemistry with the art of medicine.

By the seventeenth century, a changed attitude, characterized by a reliance on experimentation, had been adopted by astronomers, physicists, physiologists, and philosophers. It was this change in orientation that signaled the emergence of chemistry from alchemy. The English philosopher Sir Francis Bacon (1561–1626) had visions of these new scientific methods endowing human life with new inventions and wealth (Figure 1.3).

By the middle of the twentieth century, it appeared that science and its application in technology had made the dreams of Bacon and von Hohenheim a reality. Many diseases had been virtually eliminated. The coordination of chemistry and medicine had also made the difference between operations in which four strong men were employed to hold the patient down and ones in which the patient was painlessly anesthetized. Fertilizers, pesticides, and scientific breeding had made food more abundant. Nutritionists were applying their science to designing diets that would produce healthier, stronger people. New materials were being developed to improve our clothing and shelter. Industry was offering an almost endless variety of products at relatively low cost to the average consumer.

Indeed, it seemed that, despite its sometimes less than honorable intentions, science could do no wrong. For example, during World War I, when the German armies' supply of ammonia (which they needed to make nitrate explosives) was cut off, the Haber process provided them with an alternate supply (Figure 1.4). Fritz Haber's work probably lengthened the war, but it is far more significant for its influence on modern agriculture. Ammonia and nitrates are the stuff of which fertilizers are made, and fertilizers are essential to modern high-yield farming. In fact, most of the ammonia made by the Haber process today goes into fertilizer.

Much of the technology of the early twentieth century grew out of scientific discoveries. New technological developments were used in turn by scientists as tools to make new discoveries. These developments in science and technology became hallmarks of the modern world. Hardly anyone questioned that science in the early part of the twentieth century had significantly improved the human condition.

FIGURE 1.3
Sir Francis Bacon, English philosopher and Lord Chancellor to James I. (Duplication courtesy of the Smithsonian Institution, Washington, D.C.)

1.2 PROBLEMS IN PARADISE

If during the first half of the twentieth century science was viewed as humankind's savior, during the latter half it is sometimes viewed as quite the opposite. Those anesthetics that made surgery painless for the patient have caused female anesthesiologists, surgeons, and surgical nurses to suffer a high percentage of miscarriages compared with other health personnel. Fertilizer runoff from farms has polluted streams, and insecticide residues have had a devastating effect on wildlife. Some of those industrial workers making modern products for our use have died from diseases caused by the chemicals they worked with, and chemical waste dumps seem to threaten the health of us all.

One solution to these problems would be simply to throw out science. But do we really wish to return to surgery without anesthetics? Most of us don't. We need scientists, for it is they who will search for safer anesthetics, for approaches to increased agricultural production compatible with the natural environment, and for analytical techniques that will ensure healthful working conditions for industrial personnel.

FIGURE 1.4
Fritz Haber, the German chemist who invented a process for manufacturing ammonia. (Courtesy of Encyclopaedia Britannica, Inc., Chicago.)

The simple fact is that chemistry and its products, both good and bad, are so intimately involved in determining the quality of life that to ignore the subject is to court disaster. It will take an educated, informed society to ensure that science is used for the human good.

1.3. THE WAY SCIENCE WORKS—SOMETIMES

Textbooks often define science as a "body of knowledge," and it is frequently taught as a finished work rather than an ever-changing approach to learning. Science is organized into concepts. For example, even though we will often speak of atoms as if they were readily observed, the atomic concept is merely a convenience that successfully describes many observable facts in a metaphorical way. It is not the "body of facts" that characterizes science but the *organization* given to those facts. To be useful, concepts must have predictive value. If the atomic theory is to be useful, it should enable a scientist to predict how matter will behave.

The most distinguishing characteristic of science is its use of processes or methods. The making of observations and the cataloguing of facts are bare, though necessary, beginnings to these intellectual processes. Scientists must be able to make careful measurements, but they must also be able to grasp the central theme of these observations. They must recognize the variables and be able to note the effect of changing one variable at a time. Scientists must be able to sort out the useful aspects of information and ignore irrelevancies. Perhaps basic to these intellectual processes is the ability to formulate testable hypotheses. Even an educated guess is of little value to scientists unless an experiment can be devised to test the guess. In fact, if a hypothesis cannot be tested, the question is generally considered to lie outside the realm of science.

Science is not totally different from other disciplines. For example, creativity is central to both science and the humanities. Science does not involve cold logic to the exclusion of other human characteristics. Albert Einstein recognized that there was no *logical* path to some of the laws that he formulated. Even he relied on intuition based on experience and understanding.

It is important that you realize there is no single "scientific method" that, when followed, produces guaranteed results. Scientists observe, gather facts, and make hypotheses, but somewhere along the way they test their hunches and their organization of facts by experimenting. Scientists, like other human beings, use intuition and may generalize from a limited number of facts. Sometimes they are wrong. One of the strengths of science lies in the fact that results of experiments are published in scientific journals. These results are read—and often checked—by other scientists in all parts of the world. To become an accepted part of the "body of knowledge," the results must be reproducible. Scientists also extend each other's work, sometimes to the point that we see a "bandwagon" effect. One breakthrough sometimes results in the unleashing of vast quantities of new data and leads to the development of new concepts. For example, early in the nineteenth century it was thought that certain chemical substances, called *organic* compounds, could be produced only by living tissue, such as someone's liver or the leaf of a plant. These substances were in contrast to other materials, labeled *inorganic*, which could be prepared by a chemist in a laboratory. In 1828, a German chemist named Friedrich Wöhler (1800–1882) succeeded in making an organic compound from an inorganic one in the laboratory. The belief that such a compound could not be prepared in this manner was so strong that Wöhler did the same thing over and over again to assure himself that he had really done the "impossible." When he finally published his work, other chemists quickly repeated it and then proceeded to make hundreds of thousands of organic compounds. That bandwagon is still rolling today, with chemists making hormones, vitamins, and even genes in the laboratory.

Thus, contrary to an often-expressed popular notion, scientific knowledge is not absolute. Science is cumulative, and the "body of knowledge" is dynamic and constantly changing. Old concepts or even old "facts" are discarded as new tools, new questions, and new techniques reveal new data or generate new concepts. To truly understand what science is, one has to observe what the whole, worldwide community of scientists has done over a period of several years rather than look over the shoulder of a single scientist for a few days.

1.4 WHAT IS CHEMISTRY? SOME FUNDAMENTAL CONCEPTS

Nowadays, chemistry is often defined as a study of matter and the changes that it undergoes. Changes in matter are accompanied by changes in energy. Since the entire universe is made up of nothing more than matter and energy, the field of chemistry extends from atoms to stars, from rocks to living organisms. Matter and energy are such fundamental concepts that definitions are difficult. **Matter** is the stuff that makes up all material things. It occupies space and has mass. Wood, sand, water, and people have mass and occupy space. So does air, although one usually needs a flat tire or a 50 km/hr wind to emphasize the point. **Mass** is a measure of the quantity of matter. **Weight,** on the other hand, measures a force. On Earth, it measures the force of attraction between the Earth and the mass in question. For most of its history, the human race was restricted to the surface of the Earth. The terms "mass" and "weight" can be and are used interchangeably in such restricted circumstances. When the exploration of space began, however, it became apparent to most people that mass and weight are not the same thing. The mass of an astronaut on the

moon is the same as his mass on Earth. The matter that makes up the astronaut does not change. His weight on the moon, however, is one-sixth that on Earth because the moon's pull is only one-sixth as strong as the Earth's (Figure 1.5). Weight varies with gravity; mass does not.

Matter comes in three familiar states: solid, liquid, and gas. **Solid** objects generally maintain their shape and volume regardless of their location. Many solids are crystalline in nature. Liquids assume the shape of their containers (except for a generally flat surface at the top). Like solids, however, liquids maintain a fairly constant volume. Unlike solids, they flow rather readily. Gases maintain neither shape nor volume. Rather, they expand to fill completely whatever container one puts them in and can be easily compressed. For example, enough air for many minutes of breathing can be compressed into a steel tank for underwater diving. We will consider the states of matter in considerably more detail in later chapters.

Energy is often defined as the capacity for doing work. Work is related to the force needed to move a mass through a distance. (By this definition, play involving exercise is called work.) Energy is the basis for change in the material world. When something moves or breaks or cools or shines or grows or decays, energy is involved. Energy can be classified as either potential energy or kinetic energy. A system has **potential energy** by virtue of its position or composition, that is, by being in a particular arrangement or place. An example of such a system is a boulder on a cliff (Figure 1.6). Simply because it is at the top of

FIGURE 1.6
A boulder on a cliff represents potential energy. A falling boulder represents kinetic energy. The falling boulder can do work (smash a house, for example). The boulder on the cliff has a potential for doing the same thing.

the cliff and not at the bottom, the boulder has potential energy, which it can use to do work. The boulder can fall from the cliff and destroy a house at the base (thereby doing work), whereas a boulder sitting at the base of the cliff cannot do the same thing.

Kinetic energy (K.E.) is energy in motion. The boulder that was sitting at the base of the cliff could destroy the house if it were rolling along the ground at a good clip. Its ability to do work in this case would depend on its motion. If you were really determined to destroy the house using kinetic energy, your best bet would be to use a boulder rather than a stone. Given the choice, it would also be better to get the boulder moving at a brisk rather than a leisurely pace. That is just another way of saying that kinetic energy depends on mass and velocity. The bigger an object is and the faster it is moving, the more kinetic energy it has and the more work it can do. Mathematically, the kinetic energy is one-half the mass times the square of the velocity.

$$K.E. = \tfrac{1}{2}mv^2$$

Energy is often classified in other ways. These classifications are based on some characteristic of the energy being considered, for example, its source. It is easier to discuss some of these types of energy by indicating their significance in the overall pattern of energy flow on Earth.

The source of nearly all our energy is the sun. Solar energy radiates through space as light. A small portion of this **radiant energy** reaches Earth, where part of it is converted to **thermal (heat) energy.** This heat causes water to evaporate and then rise to form clouds. The water in the clouds has potential energy. As the water falls through the air and then flows in rivers, the potential energy is converted to kinetic energy.

The kinetic energy of the flowing stream can be used to turn a turbine, which converts a part of the stream's energy to **electrical energy.** The electricity thus produced can be transported by wires to homes and factories, where it is converted to light energy or to heat or to still other forms of energy.

Some of the solar energy striking the Earth is absorbed by green plants, which use a complicated chemical process called photosynthesis to convert **radiant** (solar) **energy** into **chemical energy.** The chemical energy stored by plants—now and in ages past—is used by humankind for food and fuel. Nearly all the vast quantities of energy used in our modern civilization come ultimately from the sun by way of green plants. Plants of the current age are harvested in forestry and agriculture. Those of ancient ages are reaped as fossil fuels—coal, oil, and gas.

Another form of energy is **nuclear energy.** This type of energy was stored in the Earth's crust when the solar system was formed some 4 or 5 billion years ago. We recover it for use or misuse when we mine uranium and build nuclear reactors or atomic bombs.

example
1.1

What kind of energy does each of the following possess?

a. a thrown softball
b. a softball resting at the edge of a table
c. gasoline

a. A thrown softball has kinetic energy; it is moving through the air.
b. A softball at the edge of a table has potential energy by virtue of its position; it could release energy by falling.
c. Gasoline contains chemical energy (a form of potential energy); the energy is released when the gasoline burns.

example
1.2

What energy changes occur during each of the following events?

a. Fuel oil is burned.
b. A softball falls off the edge of a table.
c. Sunlight falls on your back.

a. Chemical energy is converted to thermal energy.
b. Potential energy is converted to kinetic energy.
c. Radiant energy is converted to thermal energy.

But chemistry is not concerned only with matter and its changes. It is also concerned with the energy transformations that accompany these changes. To deal with energy transformations, chemistry often borrows fundamental concepts from its neighboring discipline, physics. One such concept is **force.** A force is a push or a pull that sets an object in motion, or stops a moving object, or holds an object in place. **Gravity** is a force. Objects—including us—are held to the surface of the Earth by gravity, the attraction of its mass for our mass. The weight of an object is the force of gravity that exists between it and the Earth.

Electrical forces are extremely important in chemistry. Particles of matter bear two types of electrical charges, called positive ($+$) and negative ($-$). No one can really tell exactly what an electrical charge is. We simply accept the fact that a particle with a "charge" can exert a force, that is, can push or pull another particle that also has a "charge" on it. The particles do not have to be touching one another to attract or repel. For this reason, we say charged particles have force fields about them. Even at a distance they attract and repel one another, although these forces get weaker as the particles get farther apart. Particles with **like** charges (both positive or both negative) repel one another. Those with **unlike** charges (one positive and one negative) attract one another (Figure 1.7).

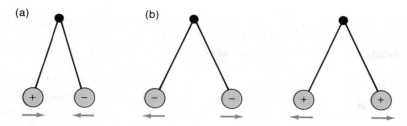

FIGURE 1.7 *(a) Particles with unlike charges attract one another. (b) Those with like charges repel one another.*

This phenomenon of charged substances is not unfamiliar to you. Anyone who has pulled clothes from an automatic dryer on a cold winter's day has probably seen what commercials like to call "static cling," pieces of clothing sticking to one another. The "cling" is due to the attraction of *unlike* charges. If, on the other hand, you brushed your hair vigorously on this same cold day, it might have become "unmanageable" (another great commercial term). Instead of lying flat against your head, it may have stuck out, each strand seemingly trying to get away from all the other strands. And that is exactly what was happening. The strands had *like* charges on them and were repelling one another.

1.5 MEASUREMENT: THE MODERN METRIC SYSTEM

Accurate measurement of such quantities as mass (weight), volume, time, and temperature is essential to the compilation of dependable scientific "facts." Such facts may be used by a chemist interested in basic research, but similar information is of critical importance in every science-related field. Certainly we are all aware that measurements of both temperature and blood pressure are routinely made in medicine. It is also true that modern medical diagnosis depends on a whole battery of other measurements, including careful chemical analyses of blood and urine.

The system of measurement used by most scientists is an updated metric plan called the International System of Measurements, or SI (from the French Système International). Indeed, SI has been adopted worldwide for everyday use. Even the United States is committed to change to SI, but conversion is voluntary and has proceeded rather slowly so far.

The beauty of SI is that it is based on the decimal system. This makes conversion from one unit to another rather simple. The SI has only a few basic units. For example, the basic unit of length is the **meter** (m), a distance only slightly greater than a yard. The SI unit of mass is the **kilogram** (kg), a quantity slightly greater than two pounds. All other units for length, mass, and volume can be derived from these basic units. For example, area can be measured in square meters (m^2) and volume in cubic meters (m^3).

A disadvantage of the basic SI units is that they are often of awkward magnitude. We seldom work with kilogram quantities in the laboratory. A cubic meter of liquid would fill a very large test tube. More convenient units can be derived by the use of (or, in the case of the kilogram, the deletion of) prefixes (Table 1.1). For example, in the laboratory we may work with grams (g) or milligrams (mg) of material. The prefix **milli-** means 1/1000 or 0.001. Thus, one milligram equals 1/1000 gram or 0.001 gram. The relationship between milligrams and grams is given by

$$1 \text{ mg} = 0.001 \text{ g}$$

or

$$1000 \text{ mg} = 1 \text{ g}$$

You should learn the more common prefixes (marked with an asterisk in Table 1.1) right away.

For volume we may choose cubic centimeters (cm^3) or cubic decimeters (dm^3) as more convenient units. The cubic decimeter has a special name, the **liter** (L). From that is derived the milliliter (mL), a unit that is the same as the cubic centimeter.

TABLE 1.1 *Approved Numerical Prefixes*

Exponential Expression	Decimal Equivalent	Prefix	Phonic	Symbol
10^{12}	1 000 000 000 000	Tera-	ter' a	T
10^{9}	1 000 000 000	Giga-	ji' ga	G
10^{6}	1 000 000	Mega-	meg' a	M*
10^{3}	1 000	Kilo-	kil' o	k*
10^{2}	100	Hecto-	hek' to	h
10	10	Deka-	dek' a	da
10^{-1}	0.1	Deci-	des' i	d
10^{-2}	0.01	Centi-	sen' ti	c*
10^{-3}	0.001	Milli-	mil' i	m*
10^{-6}	0.000 001	Micro-	mi' kro	μ*
10^{-9}	0.000 000 001	Nano-	nan' o	n
10^{-12}	0.000 000 000 001	Pico-	pe' ko	p
10^{-15}	0.000 000 000 000 001	Femto-	fem' to	f
10^{-18}	0.000 000 000 000 000 001	Atto-	at' to	a

* Designates most commonly used units.

$$1 \text{ dm}^3 = 1 \text{ L}$$
$$1 \text{ L} = 1000 \text{ mL}$$
$$1 \text{ mL} = 1 \text{ cm}^3$$

Conversions within the SI system are much easier than those using the more familiar units of pounds, feet, and pints. This is best shown by examples.

Convert 0.742 kg to grams.

$$0.742 \text{ kg} \times \frac{1000 \text{ g}}{1 \text{ kg}} = 742 \text{ g}$$

Note: If you are not familiar with the use of conversion factors or would like to review the mathematics of conversions, see Special Topic A immediately following this chapter.

example 1.3

Convert 0.742 lb to ounces.

$$0.742 \text{ lb} \times \frac{16 \text{ oz}}{1 \text{ lb}} = 11.9 \text{ oz}$$

example 1.4

Convert 1247 mm to meters.

$$1247 \text{ mm} \times \frac{1 \text{ m}}{1000 \text{ mm}} = 1.247 \text{ m}$$

example 1.5

Convert 1247 in. to yards.

$$1247 \text{ in.} \times \frac{1 \text{ yd}}{36 \text{ in.}} = 34.64 \text{ yd}$$

example 1.6

In conversions involving pounds, ounces, inches, and yards, you multiply and divide by numbers such as 16 or 36. In metric conversions, you multiply by 10 or 100 or 1000, and so on; you need only to shift the decimal point.

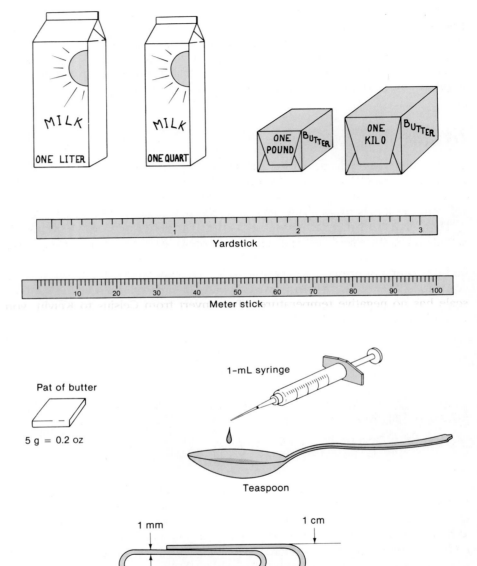

FIGURE 1.8   A comparison of common and metric units.

Conversions between systems are seldom necessary. If you need to learn how to do this, such conversions are discussed in Special Topic A following this chapter. What you do need is some idea of the relative size of a few such units. Some comparisons are shown in Figure 1.8. Others easily remembered are that the U.S. ten-cent coin, the dime, is about 1 mm thick and the U.S. five-cent piece, the nickel, just happens to weigh about 5 g.

1.6 MEASURING ENERGY: TEMPERATURE AND HEAT

The SI unit for temperature is the **kelvin** (K), but for much of their work scientists still use the **Celsius scale.** On this scale, the freezing temperature of water is defined as 0 °C and the boiling point as 100 °C. The scale between these two reference points is divided into 100 equal divisions, each a Celsius degree.

The kelvin and the Celsius degree are the same size. The Kelvin scale is called an **absolute scale** because its zero point is the coldest temperature possible, or absolute zero. This fact was determined by theoretical considerations and has been confirmed by experiment. The zero point on the Kelvin scale is equal to -273 °C. The freezing point of water is 273 K. Note that the absolute scale has no negative temperatures. To convert from Celsius to Kelvin, you merely add 273 to the Celsius temperature.

$$K = °C + 273$$

What is the boiling point of water on the Kelvin scale? The boiling point of water is 100 °C.

$$100 °C + 273 = 373 K$$

example 1.7

In the United States, weather reports and cooking recipes still use the Fahrenheit scale. This scale defines the freezing temperature of water as 32 °F and the boiling point of water as 212 °F. Exact conversions between °F and °C are seldom necessary. Figure 1.9 compares the Fahrenheit and Celsius scales. Some additional temperature equivalents and formulas for converting one scale to the other are found in Appendix I.

Scientists often need to measure amounts of heat. You should not confuse heat with temperature. **Heat** is a form of energy. **Temperature** is a measure of intensity, that is, of how energetic each particle of the sample is. A glass of water at 70 °C contains less heat (thermal energy) than a bathtub of water at 60 °C. The particles of water in the glass are more energetic, on the average, than those in the tub, but there is far more water in the tub and its total heat

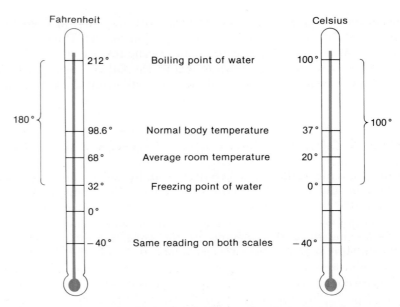

FIGURE 1.9   *A comparison of the Fahrenheit and Celsius temperature scales.*

content is greater. The SI unit of energy is the **joule** (J) but we will use the more familiar calorie (cal).

$$1.00 \text{ cal} = 4.18 \text{ J}$$
$$1000 \text{ cal} = 1.000 \text{ kcal} = 4184 \text{ J}$$

A **calorie** is the amount of heat required to raise the temperature of 1 g of water 1 °C. A more precise definition than this exists, but for our purposes the simpler version will do.

Different substances have varied capacities for heat. The **specific heat** of a substance is the amount of heat required to raise the temperature of 1 g of the substance by 1 °C. The definition of the calorie indicates the heat capacity of water is 1 cal/g °C.

example
1.8

How many calories would it take to raise the temperature of 250 g of water from 25 °C to 100 °C?

The specific heat of water is

$$\frac{1 \text{ cal}}{(1 \text{ g})(1 \text{ °C})}$$

The temperature change is 100 °C − 25 °C, or 75 °C, and the amount of water is 250 g.

$$250 \text{ g} \times 75 \text{ °C} \times \frac{1 \text{ cal}}{(1 \text{ g})(1 \text{ °C})} = 19\ 000 \text{ cal*}$$

For the measurement of the energy content of foods, the **large Calorie** (Cal) (note the capital C), or **kilocalorie** (kcal), is sometimes used. A dieter may be aware that a banana split contains 1500 Cal. If the same dieter realized that that meant 1 500 000 calories, giving up the banana split might be easier.

If a hot-water bag contains 1 kg of water at 65 °C, how much heat will it have supplied to someone's aching muscles by the time it has cooled to 20 °C?

example 1.9

$$1 \text{ kg} = 1000 \text{ g}$$

$$1000 \text{ g} \times (65 - 20) \text{ °C} \times \frac{1 \text{ cal}}{(1 \text{ g})(1 \text{ °C})} = 1000 \text{ g} \times 45 \text{ °C} \times \frac{1 \text{ cal}}{(1 \text{ g})(1 \text{ °C})}$$

$$= 45\ 000 \text{ cal or } 45 \text{ kcal}$$

Calculations such as the above can also be made using an equation,

Heat absorbed or released = mass × specific heat × ΔT

where ΔT is the change in temperature. The setup and mathematics are exactly the same.

1.7 DENSITY

An important property of matter, particularly in scientific work, is **density.** When one speaks of lead as "heavy" or aluminum as "light," one is referring to the density of these metals. This term is defined as the amount of mass (or weight) per unit of volume.

$$D = \frac{M}{V}$$

---

* Answers are given with the proper number of significant figures. If you do not understand significant figures, please see Appendix III.

Rearrangement of this equation gives

$$M = D \times V \quad \text{and} \quad V = \frac{M}{D}$$

These equations are useful for making calculations. Density values are usually reported in grams per milliliter (g/mL) or grams per cubic centimeter (g/cm^3).

example 1.10

What is the density of iron if 156 g of iron occupies a volume of 20 cm^3?

$$\frac{156 \text{ g}}{20 \text{ cm}^3} = 7.8 \text{ g/cm}^3$$

example 1.11

How much will a liter of gasoline weigh if its density is 0.66 g/mL?

$$\frac{0.66 \text{ g}}{1 \text{ mL}} \times 1 \text{ L} \times \frac{1000 \text{ mL}}{1 \text{ L}} = 660 \text{ g}$$

example 1.12

What volume is occupied by 461 g of mercury? The density of mercury is 13.6 g/mL.

$$461 \text{ g} \times \frac{1 \text{ mL}}{13.6 \text{ g}} = 33.9 \text{ mL}$$

The density of water is 1 g/mL, or 1 g/cm^3 (remember that 1 mL = 1 cm^3). This nice round number for the density of water is not an accident. The metric system was originally set up in such a way as to ensure that this was the case.

A term related to density is **specific gravity.** Specific gravity is the ratio of the mass of any substance to the mass of an equal volume of water. The specific gravity of water itself, therefore, is 1. Mercury (the silvery liquid in thermometers) has a specific gravity of 13.6. That means it has a density 13.6 times as great as water's. The specific gravity of alcohol is 0.8; thus, alcohol is less dense than water. The specific gravity of human fat is 0.903. If mixed with water, the fat would float on top. The higher the proportion of body fat in a person, the more buoyant that person is in water (Figure 1.10). Because it is the ratio of two values, specific gravity is a number without units, whereas

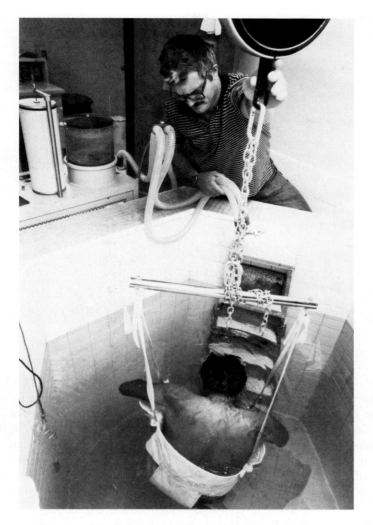

FIGURE 1.10 *Percent of body fat can be determined by weighing a person who is completely submerged in water. The calculation must include a correction for the volume of air in the lungs. (Photo courtesy of The National Institutes of Health.)*

density is reported in units of mass per volume. And because the density of water is 1 g/mL, the specific gravity of a substance is numerically the same as its density.

$$\text{Specific gravity of a substance} = \frac{\text{density of the substance}}{\text{density of water}}$$

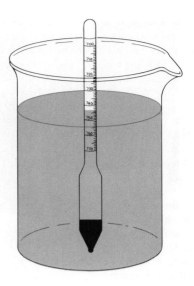

FIGURE 1.11
A hydrometer. The one
shown here measures
specific gravities over a
range of 0.700 to 0.770.

example	The density of chloroform is 1.5 g/mL. What is its specific gravity?
1.13	

$$\frac{1.5 \text{ g/mL}}{1.0 \text{ g/mL}} = 1.5$$

Specific gravity of a liquid is frequently measured by a device called a **hydrometer.** The hydrometer is placed in the liquid. How far it dips down into the liquid is determined by the density of the liquid (Figure 1.11). The stem of the hydrometer is marked in such a way that the specific gravity can be read directly at the surface of the liquid. Hydrometers can be used to measure the strength of the "battery acid" in your car, sugar content in maple syrup, dissolved solids in urine, alcohol content of wine, and many other properties of solutions that are related to specific gravity.

For those who are interested and those whose work might require it, measurement is discussed further in Appendix I, and a variety of conversion tables are collected there for convenient reference.

PROBLEMS

1. List three characteristics of science. Which one serves best to distinguish science from other disciplines?
2. Define each of the following terms:
   a. chemistry   **b.** matter
   c. mass
   d. weight
   e. energy
   f. heat
   g. temperature
   h. calorie
   i. density
   j. specific gravity

3. What is a hypothesis? What kind of hypothesis is necessary in science?

4. Explain the difference between mass and weight.

5. Two samples are weighed under identical conditions in a laboratory. Sample A weighs 1 lb and Sample B weighs 2 lb. Does B have twice the mass of A?

6. What has changed when a person completes a successful diet: the person's weight or the person's mass?

7. Sample A, which is on the moon, has exactly the same mass as Sample B, which is on Earth. Do the two samples weigh the same?

8. How do gases, liquids, and solids differ in their properties?

9. Which is the most compressible form of water: steam, ice, or liquid?

10. What is the difference between kinetic energy and potential energy?

11. Which has the greater kinetic energy: a sprinter or a long-distance runner (assume the two weigh the same)?

12. Which has the greater kinetic energy: a cannonball or a bullet, both fired at the same speed?

13. Which has the greater kinetic energy: a bicyclist traveling at 15 mi/hr or an automobile traveling at 50 mi/hr?

14. Which has the greater kinetic energy: a 110-kg football tackle moving slowly across the field or an 80-kg halfback racing quickly down the field?

15. Which has the greater potential energy: a diver on the 1-m board or that diver on the 10-m platform?

16. Which has the greater potential energy: an elevator stopped at the twentieth floor or one stopped at the twelfth floor?

17. Which has the greater potential energy: a roller coaster as it starts to climb the first hill or as it reaches the top of the hill?

18. Describe what happens to two particles with like charges when they are brought close together. What happens to particles with unlike charges when they are brought close together?

19. What are the basic units of length, volume, and mass in the SI system? What derived units are more often employed in the laboratory?

20. How many millimeters and how many centimeters are there in 1 m?

21. For each of the following, indicate which is the larger unit:

   a. mm or cm   **b.** kg or g   **c.** dL or μL

22. For each of the following, indicate which is the larger unit:

   a. L or cm^3 **b.** cm^3 or mL

23. For each of the following, indicate which is the larger unit:

   a. in. or m   **b.** lb or g   **c.** L or gal

24. For each of the following, indicate which is the larger unit:

   a. °C or °F   **b.** cal or Cal

25. How many meters are there in each of the following?

   a. 50 km   **b.** 25 cm

26. How many millimeters are there in each of the following?

   a. 1.5 m   **b.** 16 cm

27. How many deciliters are there in each of the following?

   a. 1.0 L   **b.** 20 mL

28. How many liters are there in each of the following?

   a. 2056 mL   **b.** 47 kL

29. Make the following conversions:

   a. 15 000 mg to g   **b.** 0.086 g to mg

30. Make the following conversions:

   a. 0.149 L to mL   **b.** 47 mL to L

31. Make the following conversions:

   a. 1.5 L to mL   **b.** 18 mL to L

32. How many milliliters are there in 1 cm^3? In 15 cm^3?

33. Order the temperatures from coldest to hottest: 0 K, 0 °C, 0 °F.

34. Convert the following to kelvins:

   a. 37 °C   **b.** −100 °C   **c.** 273 °C

35. Which is hotter: 100 °C or 100 °F?

36. Convert the following to degrees Celsius:

   a. 298 K   **c.** 373 K

   b. 233 K   **d.** 273 K

37. How many calories are there in each of the following?

   a. 2.75 kcal   **b.** 0.74 Cal

38. How many calories are required to raise the temperature of 50 g of water from 20 °C to 50 °C?

39. How many calories are required to raise the temperature of 13 g of water from 15 °C to 95 °C?

40. How much heat would be released by 2.0 kg of water cooling from 90 °C to 20 °C?

41. How much water can be heated from 20 °C to 50 °C by 800 cal of energy?

42. One gram of fat provides 9 kcal of energy. A nutrition article says margarine, a fat, provides 100 kcal per

tablespoon (T). How many grams of margarine are there in 1.0 T of margarine?

43. What is the density of a salt solution if 50 cm³ of the solution has a mass of 57 g?

44. What is the density of a liquid that has a mass of 60.2 g and a volume of 25.0 mL?

45. A 1.0-L container of carbon tetrachloride weighs 1.6 kg. What is the density of carbon tetrachloride?

46. A piece of tin has a volume of 16.4 cm³. The density of tin is 5.75 g/cm³. What is the mass of the tin?

47. What is the mass of 50.0 mL of mercury? The density of mercury is 13.6 g/mL.

48. If the density of a normal urine sample is 1.02 g/mL, what is its specific gravity?

49. What is the density of a urine sample if 150 mL has a mass of 157 g?

50. A 59.0-g piece of lead has a volume of 5.20 cm³. Calculate the density of lead.

51. The density of ethyl alcohol at 20 °C is 0.789 g/mL. What is the mass of a 50.0-mL sample?

52. What volume is occupied by 253 g of bromoform? The density of bromoform is 2.90 g/mL.

53. In Problem 42, you calculated the mass of a tablespoon (T) of margarine. If 1.0 T has a volume of 15 mL, what is the density of the margarine?

54. A tall, layered drink called a pousse-café is made of four different liqueurs. The liqueurs (20.0 mL of each) are carefully poured into a single container so that they do not mix. The names, colors, and masses (per 20.0 mL) of the liqueurs follow.

Liqueur	Color	Mass (g)
Cassis (Black currant)	dark red	23.2
Crème de menthe	green	21.8
Irish Mist	amber	20.4
Triple Sec	white	21.2

Calculate the density of each liqueur and make a diagram of the drink, indicating the color of each layer.

A

Unit Conversions

Sometimes we need to convert a measurement from one kind of unit to another. A powerful method for doing these unit conversions is the **factor-label method.** (The approach is also often called **dimensional analysis,** although the units employed are not always dimensions.) Whatever we call it, the method employs the units, such as L, cm/ft, mi/hr, or g/cm^3, as aids in setting up and solving problems. The general approach is to multiply the known quantity (and its units!) by one or more conversion factors so that the answer is obtained in the desired units.

Known quantity and unit(s) × conversion factor(s) = answer (in desired units)

The method is best learned by practice. We urge you to learn it now to save yourself a lot of time and wasted effort later.

A.1 CONVERSIONS WITHIN A SYSTEM

Quantities can be expressed in a variety of units. For example, you can buy beverages by the 12-oz can or by the pint, quart, gallon, or liter. If you wish to compare prices, you must be able to convert from one unit to another. Such a conversion changes the numbers and units, but it does not change the quantity. Your actual weight, for example, remains unchanged whether it is expressed in pounds, ounces, or kilograms.

You know that multiplying a number by 1 doesn't change its value. Multiplying by a fraction equal to 1 also leaves the value unchanged. A fraction is equal to 1 when the numerator is equal to the denominator. For example, we know that

$$1 \text{ ft} = 12 \text{ in.}$$

Therefore, the fraction

$$\frac{1 \text{ ft}}{12 \text{ in.}} = 1$$

Similarly

$$\frac{12 \text{ in.}}{1 \text{ ft}} = 1$$

If we wish to convert from inches to feet, we can do so by choosing one of the fractions as a **conversion factor.** Which one do we choose? The one that gives us an answer with the right units! Let's illustrate by an example.

example A.1

My bed is 72 in. long. What is its length in feet? You know the answer, of course, but let's show how the answer is obtained by using unit conversions. We need to multiply 72 in. by one of the above fractions. Which one? The known quantity and unit is 72 in.

$$72 \text{ in.} \times \text{conversion factor} = ? \text{ ft}$$

For the conversion factor, choose the fraction that, when inserted in the equation, cancels the unit *inch* and leaves the unit *feet*.

$$72 \text{ in.} \times \frac{1 \text{ ft}}{12 \text{ in.}} = 6.0 \text{ ft}$$

Now let's try the other conversion factor.

$$72 \text{ in.} \times \frac{12 \text{ in.}}{1 \text{ ft}} = 860 \text{ in.}^2/\text{ft}$$

Absurd! How can a bed be 860 in.^{2}/ft? You should have no difficulty choosing between the two possible answers.

Please note that conversion factors are not usually given in a problem. Rather, they must be obtained from tables such as those in Appendix I. For the following examples, please refer to those tables for conversion factors that you don't know already.

It is possible (and frequently necessary) to manipulate units in the denominator as well as in the numerator of a problem. Just remember to use conversion factors in such a way that the unwanted units cancel.

At a track meet a runner completes the mile in 4.0 min. How fast is this in miles per hour?

The time for 1 mi is 4.0 min, or 1 mi per 4.0 min, or

$$\frac{1 \text{ mi}}{4.0 \text{ min}}$$

This is the known quantity and unit. The unit desired is miles per hour.

$$\frac{1 \text{ mi}}{4.0 \text{ min}} \times \text{conversion factor} = ? \frac{\text{mi}}{\text{hr}}$$

To change minutes to hours, we need the equivalence

$$1 \text{ hr} = 60 \text{ min}$$

Now arrange this equivalence as the fractions

$$\frac{1 \text{ hr}}{60 \text{ min}} \text{ and } \frac{60 \text{ min}}{1 \text{ hr}}$$

Choose for your conversion factor the fraction that cancels the unit *minute* and leaves the unit *hour* in the denominator.

$$\frac{1 \text{ mi}}{4.0 \text{ min}} \times \frac{60 \text{ min}}{1 \text{ hr}} = \frac{15 \text{ mi}}{1 \text{ hr}} \text{ (or 15 mi/hr)}$$

example A.2

Problems may involve the use of several conversion factors.

If your heart beats at a rate of 72 times per minute and your lifetime will be 70 years, how many times will your heart beat during your lifetime?

Two equivalences are given in the problem.

$$72 \text{ beats} = 1 \text{ min}$$
$$1 \text{ lifetime} = 70 \text{ years}$$

example A.3

Three others that you will need you can recall from memory.

$$1 \text{ year} = 365 \text{ days}$$
$$1 \text{ day} = 24 \text{ hr}$$
$$1 \text{ hr} = 60 \text{ min}$$

Start now with the factor 72 beats/1 min (the known quantity and unit) and apply conversion factors as needed to get an answer in beats/lifetime (the desired unit).

$$\frac{72 \text{ beats}}{1 \text{ min}} \times \frac{60 \text{ min}}{1 \text{ hr}} \times \frac{24 \text{ hr}}{1 \text{ day}} \times \frac{365 \text{ days}}{1 \text{ year}} \times \frac{70 \text{ years}}{1 \text{ lifetime}}$$
$$= 2\,600\,000\,000 \text{ beats/lifetime}$$

Most problems that you will have to deal with will be much simpler than example A.3.

Now let's do some conversions within the metric system. They are much easier.

example A.4

How many milliliters are there in a 2.0-L bottle of soda pop?

From memory or from the tables in Appendix I you find that

$$1 \text{ L} = 1000 \text{ mL}$$

$$2.0 \text{ L} \times \frac{1000 \text{ mL}}{1 \text{ L}} = 2000 \text{ mL}$$

Notice that we picked the conversion factor that allowed us to cancel liters and obtain an answer in the desired unit, milliliters.

example A.5

In the United States, the usual soda pop can holds 360 mL. How many such cans could be filled from one 2.0-L bottle?

The problem tells us that

$$1 \text{ can} = 360 \text{ mL}$$

Using that equivalence, we can calculate the answer.

$$2.0 \text{ L} \times \frac{1000 \text{ mL}}{1 \text{ L}} \times \frac{1 \text{ can}}{360 \text{ mL}} = 5.6 \text{ cans}$$

How many 325-mg aspirin tablets can be made from 875 g of aspirin? **example**
   The problem tells us that **A.6**

$$1 \text{ tablet} = 325 \text{ mg}$$

From memory or the tables, we have

$$1 \text{ g} = 1000 \text{ mg}$$

$$875 \text{ g} \times \frac{1000 \text{ mg}}{1 \text{ g}} \times \frac{1 \text{ tablet}}{325 \text{ mg}} = 2690 \text{ tablets}$$

A.2 CONVERSIONS BETWEEN SYSTEMS

To convert from one system of measurement to another, you need a conversion table such as that in Appendix I. Let's plunge right in and work some examples.

You know that your weight is 140 lb, but the job application form asks for **example**
your weight in kilograms. What is it? **A.7**
   From the table we find

$$1.00 \text{ lb} = 0.454 \text{ kg}$$

$$140 \text{ lb} \times \frac{0.454 \text{ kg}}{1.00 \text{ lb}} = 63.6 \text{ kg}$$

A recipe calls for 750 mL of milk, but your measuring cup is calibrated in fluid **example**
ounces. How many ounces of milk will you need? **A.8**

$$750 \text{ mL} \times \frac{1.00 \text{ fl oz}}{29.6 \text{ mL}} = 25.3 \text{ fl oz}$$

A doctor puts you on a diet of 5500 kilojoules (kJ) per day. How many kilo- **example**
calories (food "Calories") is that? **A.9**

$$5500 \text{ kJ} \times \frac{1000 \text{ J}}{1 \text{ kJ}} \times \frac{1.000 \text{ kcal}}{4184 \text{ J}} = 1315 \text{ kcal}$$

example
A.10

A sprinter runs the 100-m dash in 11 s. What is her speed in kilometers per hour? The given speed is 100 m per 11 s.

$$\frac{100 \text{ m}}{11 \text{ s}}$$

The conversion factors that we need are found in the tables or recalled from memory.

$$\frac{100 \text{ m}}{11 \text{ s}} \times \frac{1 \text{ km}}{1000 \text{ m}} \times \frac{60 \text{ s}}{1 \text{ min}} \times \frac{60 \text{ min}}{1 \text{ hr}} = 33 \text{ km/hr}$$

Note that the first conversion factor changes meters to kilometers. It takes two factors to change seconds to hours. Note also that we could have first applied the factors that convert seconds to hours and then converted meters to kilometers. The answer would be the same.

A.3 DENSITY PROBLEMS

Almost any type of problem can be solved using dimensional analysis. Let's try some involving density.

example
A.11

A bottle filled to the 25-mL mark contains 75 g of bromine. What is the density of bromine?

If you know that density is ordinarily reported in grams per milliliter (Chapter 1), then you know how to set up the problem. Arrange the data so the answer will come out in grams per milliliter. The density of bromine is

$$\frac{75 \text{ g}}{25 \text{ mL}} = 3.0 \text{ g/mL}$$

example
A.12

The density of alcohol is 0.79 g/mL. What is the mass (weight) of 150 mL of alcohol?

We know that mass is measured in grams, so we set up the problem this way.

$$150 \text{ mL} \times \frac{0.79 \text{ g}}{1.0 \text{ mL}} = 120 \text{ g}$$

What volume is occupied by 451 g of mercury? The density of mercury is 13.6 g/mL.

example A.13

This time we want an answer in units of volume, so we set it up this way.

$$451 \text{ g} \times \frac{1.00 \text{ mL}}{13.6 \text{ g}} = 33.2 \text{ mL}$$

You can always be reasonably confident of your answer if it has the proper units.

We will use unit conversions to solve other problems as we proceed through subsequent chapters. For now, though, you should improve your proficiency by working through the following problems. Answers are given in Appendix IV.

PROBLEMS

1. How many meters are there in each of the following?
   a. 3.4 km
   b. 570 cm
   c. 4.76 in.
   d. 100 yd
2. How many liters are there in each of the following?
   a. 3300 mL
   b. 51 dL
   c. 51 qt
   d. 81 gal
3. Make the following metric conversions:
   a. 1.29 kg to g
   b. 1575 mg to g
   c. 0.421 g to mg
   d. 255 cm to m
   e. 1.83 cm to mm
   f. 4.22 L to mL
4. Make the following conversions. (You may refer to the conversion tables in Appendix I when necessary.)
   a. 16.4 in. to cm
   b. 4.17 qt to L
   c. 1.61 kg to lb
   d. 9.34 g to oz
   e. 2.05 fl oz to mL
   f. 105 lb to g
5. Football player William "The Refrigerator" Perry weighs 310 lb. What is his weight (mass) in kilograms?
6. Basketball player Manute Bol is 7.5 ft tall. What is his height in meters? In centimeters?
7. How many pounds are there in 144 kg?
8. How many centimeters are there in 18.4 in.?
9. State your weight in pounds. Calculate your weight (mass) in kilograms.
10. State your height in feet. Calculate your height in centimeters.
11. An aspirin tablet has a mass of 325 mg. What is its mass in grams? In ounces?
12. How would you describe a young man who is 160 cm tall and who has a mass of 90 kg?
13. Convert 12 km/min to miles per minute.
14. Convert 5.8 lb/gal to kilograms per liter.
15. Convert 45 mi/hr to kilometers per hour.
16. The speed limit in the Saskatchewan countryside is 88 km/hr. What is the speed in miles per hour?
17. The speed of light is 186 000 mi/s. What is the speed in meters per second?
18. The density of gasoline is 5.9 lb/gal. What is the density in grams per milliliter?
19. Milk costs $2.09 per gal or 59¢ per L. Which is cheaper?
20. One gram of fat provides 9 kcal of energy. How many kilocalories are furnished by 1 lb of fat?
21. According to the *Guinness Book of Records*, the heaviest recorded man in Great Britain weighed 53 stone, 8 lb (1 stone = 14 lb). What was his weight in pounds?
22. When a service station in Toronto converted to the metric system, the price of gasoline changed from 95.9¢ per imperial gal to 21.1¢ per L. Did the price increase, decrease, or remain unchanged? (There are 4.545 L in 1.000 imperial gal.)

23. A slug expends about 0.10 mL of mucus per 1.0 m crawled. (It is necessary for the slug to generate mucus continually in order to move.) Calculate the slug's progress in miles per gallon.

24. The food Calorie is actually 1000 cal (1 kcal). In SI, we measure our food energy intake in kilojoules (1.00 kcal = 4.18 kJ). How many kilojoules of energy are there in a can of pop that has 165 Cal (165 kcal)?

Note: Use unit conversions (rather than the formulas in Chapter 1) to solve Problems 25–26.

25. The density of iron is 7.86 g/cm^3. What is the mass of a piece of iron that has a volume of 80.5 cm^3?

26. What volume is occupied by 55.5 g of chloroform? The density of chloroform is 1.48 g/mL.

2

Atoms

A tom. The word is so familiar that it seems it must always have been with us. Yet the birth of the concept was a difficult one, spanning more than two millennia. It took another century or so for our idea of the atom to mature—and what started as a hard sphere grew to a complicated, fuzzy cloud of matter. In this chapter we will look at the development of the concept of the atom. We will examine the parts from which atoms are made. Then we will consider how the parts may be arranged to make up the hundred-odd kinds of atoms of which the entire universe is made.

For all practical purposes, atoms are eternal. We can't get rid of those that we don't want. We can bury them under the ground or throw them into the sea. We can combine them in different ways. Today, we can even do what a tenth-century alchemist could only dream of doing: we can change one kind of atom into another. But they still won't go away. We should keep that in mind when we consider solutions to pollution problems.

A knowledge of the structure of atoms is of utmost importance. Most of the analyses for components of blood and urine—analyses upon which many medical diagnoses are based—depend on a knowledge of the structure of atoms and the way that structure changes when energy is absorbed. Learn all you can about atoms. They are what you and all the universe are made of.

2.1 ATOMS: THE GREEK IDEA

In the fifth century B.C., Leucippus and his pupil Democritus (Figure 2.1) strolled upon an Aegean beach. Leucippus is said to have wondered aloud whether the water of the sea was continuous, as it appeared to be, or whether it might be

FIGURE 2.1
Democritus. (Duplication courtesy of the Smithsonian Institution, Washington, D.C.)

composed of tiny, separate particles like the grains of sand on the beach. From a distance, the sand appeared continuous, but closer inspection revealed the separate grains. Leucippus could divide the water into drops and each drop into smaller drops. Was there any reason this process could not be continued indefinitely, yielding ever smaller drops of water? This idea of endless divisibility was the prevailing view of the Greek philosophers of that time, but Leucippus, on the basis of intuition alone, concluded that there must be a limit to divisibility—that there must be ultimate particles that could not be further subdivided.

Democritus, who lived from about 470 to 380 B.C., gave these ultimate particles names. He called them *atomos*, meaning "indivisible." It is from this name that we get our modern word *atom*.

Today we know that Democritus was right, although at the time his was a minority view. The view that matter was continuous, rather than atomistic, prevailed for 2000 years. We approach the question differently today: we base our ideas on experiments. We gather evidence from experiments to substantiate or negate our ideas. The ancient Greeks almost never performed experiments. They preferred to reason from what they called "first principles."

> "To understand the very large, we must understand the very small."
>
> Democritus

2.2 THE FRENCH: SCIENTIFIC AND POLITICAL REVOLUTION

Modern science developed about 300 years ago. At that time, scientists began to make careful observations and accurate measurements.

Antoine Laurent Lavoisier (Figure 2.2), a Frenchman (1743–1794), perhaps did more than anyone else to establish chemistry as a quantitative science. He found that when a chemical reaction was carried out in a closed system, the total mass of the system was not changed. Perhaps the most important chemical reaction that Lavoisier performed was the decomposition of the red oxide of mercury to form metallic mercury and a gas that he named oxygen. The reaction had been carried out before—by Karl Wilhelm Scheele, a Swedish apothecary (1742–1786), and by Joseph Priestley, a Unitarian minister who later fled England and settled in America—but Lavoisier was the first to weigh all the substances present before and after the reaction. Lavoisier was also the first to interpret the reaction correctly.

Lavoisier performed many quantitative experiments. He found that when coal was burned, it united with oxygen to form carbon dioxide. He also experi-

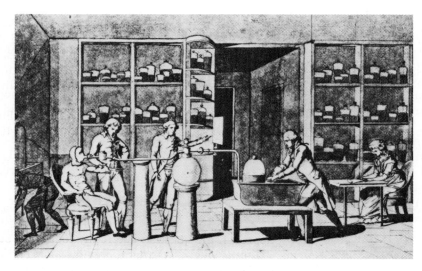

FIGURE 2.2   *Lavoisier in his laboratory. This sketch, adapted from one made by his wife, shows him studying respiration. (Reprinted with permission from Smith, Henry D.,* Torchbearers of Chemistry, *New York: Academic Press, 1949. Copyright* © *1949 by Academic Press, Inc.)*

mented with animals. When a guinea pig breathed, oxygen was consumed and carbon dioxide was formed. Lavoisier therefore concluded that respiration was related to combustion. In each of these reactions he found that matter was conserved.

Lavoisier performed many quantitative experiments. He found that when mulating a scientific law.* His **law of conservation of mass** holds that matter is neither created nor destroyed during a chemical change. In other words, if one weighed all the products of a reaction—solids, liquids, and gases—the total would be the same as the mass of all the original substances (called reactants, or starting materials).

Scientists by this time were almost universally using Robert Boyle's operational definition of an element, put forth over a century before. Boyle, an Englishman, in his book *The Sceptical Chymist*, published in 1661, said that supposed **elements** must be tested to see if they really were simple. If a substance could be broken down into simpler substances, it was not an element. The simpler substances might be elements and would be so regarded until such time (if it ever came) as they in turn could be broken down into still simpler

---

* **Scientific ''laws''** summarize experimental data. For example, Lavoisier found that in each of the reactions he carried out, the total mass of products was equal to the total mass of reactants. The law of conservation of mass simply summarizes these findings. A modification of this law was made necessary when Albert Einstein pointed out the relationship between matter and energy. We shall consider this modification later. For the moment, we shall work with Lavoisier's law.

substances. On the other hand, two or more elements might combine to form a complex substance, called a **compound.**

Using Boyle's definition, Lavoisier included a table of elements in his book, *Elementary Treatise on Chemistry.* His table included some substances we now know to be compounds. Lavoisier was the first to use modern, somewhat systematic names for the chemical elements. He is often called the "father of chemistry," and his book is regarded as the first chemistry textbook. Incidentally, Lavoisier lost his head (on the guillotine) during the French Revolution, but not because of his chemical research. In those days no one was a full-time chemist. Lavoisier had another job on the side. He was a tax collector for Louis XVI, and it was in this capacity that he incurred the wrath of the French peasantry.

In 1799, Joseph Louis Proust showed that a substance called copper carbonate, whether prepared in the laboratory or obtained from natural sources, contained the same three elements—copper, carbon, and oxygen—and always in the same proportion by mass—5.3 parts of copper to 4 parts of oxygen to 1 part of carbon (Figure 2.3). To summarize this and numerous other experiments, Proust formulated a new law. A compound, he said, always contains elements in certain definite proportions, and in no other combinations. This generalization he called the **law of definite proportions** (sometimes referred to as the **law of constant composition**).

Proust, like Lavoisier, was a member of the French nobility, but he was working in Spain, temporarily safe from the ravages of the French Revolution. His laboratory was destroyed, however, and he was reduced to poverty when the French troops of Napoleon Bonaparte occupied Madrid in 1808.

One of the earliest illustrations of the law of definite proportions is found in the work of a Swedish chemist, J. J. Berzelius (1779–1848). He heated 10 g of lead with varying amounts of sulfur to form lead sulfide. Since lead is a soft,

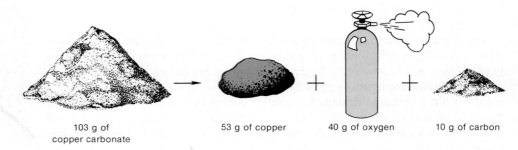

103 g of
copper carbonate   53 g of copper   40 g of oxygen   10 g of carbon

FIGURE 2.3   Whether synthesized in the laboratory or obtained from various natural sources, copper carbonate always has the same composition. Analysis of this compound led Proust to formulate the law of definite proportions.

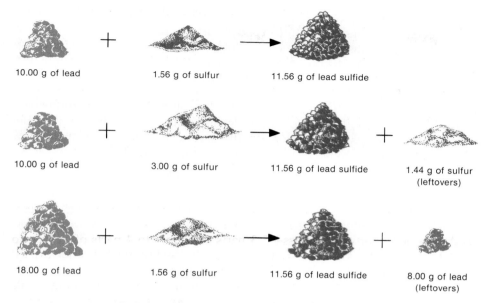

FIGURE 2.4   *The law of definite proportions.*

grayish metal, sulfur is a yellow solid, and lead sulfide is a shiny black solid, it was easy to tell when all the lead had reacted. Excess sulfur was easily washed away by carbon disulfide, a liquid that dissolves sulfur but not lead sulfide. As long as he used at least 1.56 g of sulfur, he got exactly 11.56 g of lead sulfide. Any sulfur in excess of 1.56 g was left over, unreacted. If he used more than 10 g of lead with 1.56 g of sulfur, he got 11.56 g of lead sulfide with lead left over. These reactions are illustrated in Figure 2.4 and explained in Figure 2.7.

2.3 DALTON'S ATOMIC THEORY

Lavoisier's law of conservation of mass and Proust's law of definite proportions were repeatedly confirmed by experiment. This led to attempts to formulate theories that would account for these laws. In science, a **theory** is a model that consistently explains observations.

John Dalton (Figure 2.5), an English schoolteacher, was one of those who proposed a model to explain the accumulating experimental data. As he developed the details of his model, he uncovered a third "law" that his theory would have to explain. Proust had previously observed that a compound contains elements in certain definite proportions and only those proportions. Dalton's new law stated that elements might combine in more than one set of proportions. If they did, though, each different combination would produce a

FIGURE 2.5
John Dalton. (Duplication courtesy of the Smithsonian Institution, Washington, D.C.)

different compound. For example, carbon combined with oxygen in a proportion of 3 parts by mass of the former to 8 parts by mass of the latter to form carbon dioxide, a gas familiar as a product of respiration and of the burning of wood or coal. But Dalton found that 3 parts by mass of carbon would also combine with 4 parts by mass of oxygen to form a poisonous gas that we know today as carbon monoxide. After observing similar multiple combinations for other sets of elements, Dalton put forth his **law of multiple proportions.**

In the same year, he set down the details of his **atomic theory,** a model that offered a logical explanation for the several laws we have mentioned. The most important points of Dalton's atomic theory are these:

1. All elements are made up of small, indestructible, and indivisible particles called **atoms.**
2. All atoms of a given element are identical, but the atoms of one element differ from the atoms of any other element.
3. Atoms of different elements can form combinations to give compounds.
4. A chemical reaction involves a change not in the atoms themselves, but in the way atoms are combined to form compounds.

Dalton's reasoning went something like this. If matter is continuous, why should 1 part by mass of x always combine with 3 parts by mass of y? Why shouldn't 1 part of x also combine with 2.9 parts of y? Or 3.1 parts of y? On the other hand, if matter is atomistic, atoms are indivisible, and only whole atoms should combine (Figure 2.6). If an atom of y has a mass three times that of an atom of x, then the compound formed by the combination of an x and a y would have to consist of 1 part by mass of x and 3 parts by mass of y. An analogy may clarify this point. If you have mashed potatoes and gravy (call them continuous foods), you can have a lot of mashed potatoes and a little bit of gravy, or vice versa. There are no set amounts that go together. On the other hand, if you have a hot dog sandwich (an atomistic food), you need a bun for every hot dog. If the hot dogs have a mass twice that of the buns, then the law of definite proportions simply notes that hot dog sandwiches always consist

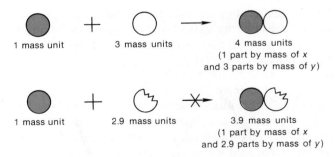

FIGURE 2.6   According to the atomic theory, only whole atoms may combine. The second scheme is impossible.

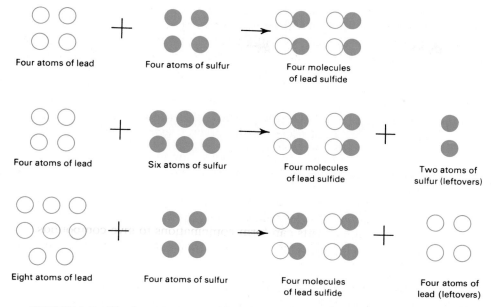

FIGURE 2.7 *The law of definite proportions. Berzelius's experiment interpreted in terms of Dalton's atomic theory.*

of 2 parts by mass of meat and 1 part by mass of bread. If you have 27 hot dogs and only 25 buns, you can only make 25 hot dog sandwiches. The other 2 hot dogs are simply left over (see Berzelius's experiment, in Figure 2.4, and then consider Figure 2.7). Dalton concluded that matter must be atomistic in order for the law of definite proportions to hold.

Dorothy's delectable bologna sandwiches are made with 1 slice of bologna, 2 slices of bread, and 4 pickle slices. How many sandwiches can be made with 10 slices of bologna, 21 slices of bread and 43 pickle slices?

example 2.1

$$10 \text{ slices bologna} \times \frac{1 \text{ sandwich}}{1 \text{ slice bologna}} = 10 \text{ sandwiches}$$

$$21 \text{ slices bread} \times \frac{1 \text{ sandwich}}{2 \text{ slices bread}} = 10 \text{ sandwiches } (+1 \text{ slice bread})$$

$$43 \text{ pickle slices} \times \frac{1 \text{ sandwich}}{4 \text{ pickle slices}} = 10 \text{ sandwiches } (+3 \text{ pickle slices})$$

There are enough parts for 10 sandwiches, with 1 slice of bread and 3 pickle slices left over.

example 2.2

One unit of hydrogen sulfide (called a molecule) is made from 1 sulfur atom and 2 hydrogen atoms. How many hydrogen sulfide molecules can be made from 45 sulfur atoms and 97 hydrogen atoms?

$$45 \text{ sulfur atoms} \times \frac{1 \text{ molecule}}{1 \text{ sulfur atom}} = 45 \text{ molecules}$$

$$97 \text{ hydrogen atoms} \times \frac{1 \text{ molecule}}{2 \text{ hydrogen atoms}} = 48 \text{ molecules } (+1 \text{ hydrogen atom})$$

Since the molecules cannot be made without sulfur atoms, only 45 hydrogen sulfide molecules can be made. That requires only

$$45 \text{ molecules} \times \frac{2 \text{ hydrogen atoms}}{1 \text{ molecule}} = 90 \text{ hydrogen atoms}$$

The other 7 hydrogen atoms are simply left over.

example 2.3

An 18.0-g sample of water always consists of 2.0 g of hydrogen and 16.0 g of oxygen. What weight of water can be made from 64.0 g of oxygen and 40 g of hydrogen?

$$64.0 \text{ g oxygen} \times \frac{18.0 \text{ g water}}{16.0 \text{ g oxygen}} = 72.0 \text{ g water}$$

$$40 \text{ g hydrogen} \times \frac{18.0 \text{ g water}}{2.0 \text{ g hydrogen}} = 360 \text{ g water}$$

We have enough hydrogen to make 360 g water. However, water cannot be made without oxygen, so only 72.0 g of water can be made. That requires only

$$72.0 \text{ g water} \times \frac{2.0 \text{ g hydrogen}}{18.0 \text{ g water}} = 8.0 \text{ g hydrogen}$$

The remaining 32 g of hydrogen are simply left over, unreacted.

The atomic theory also explained the law of multiple proportions. The difference between carbon dioxide and carbon monoxide is that, in the former compound, one atom of carbon always combines with two atoms of oxygen and, in the latter, one atom of carbon always combines with only one atom of oxygen. Figure 2.8 shows how oxygen and nitrogen combine.

Finally, the law of conservation of mass can also be understood in terms of the atomic theory. When a reaction occurs and the reactants change to pro-

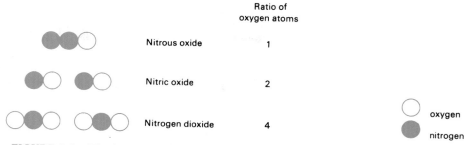

Ratio of
oxygen atoms

Nitrous oxide   1

Nitric oxide   2

Nitrogen dioxide   4

○ oxygen

● nitrogen

FIGURE 2.8   *The law of multiple proportions. The amount of nitrogen is the same in each sample (two atoms). The oxygen ratio for the three compounds is 1:2:4.*

ducts, this change simply involves a reordering of atoms. Matter is neither lost nor gained; it is simply rearranged (Figure 2.9).

As a most important part of this theory, Dalton set up a table of relative masses for the various kinds of atoms. Many of his entries in the table are wrong, mainly because he assumed each unit of the compound water was made of one hydrogen atom and one oxygen atom. (We now know that each unit of water consists of *two* hydrogen atoms and one oxygen atom.) Atoms are extremely tiny, and it was impossible to determine actual atomic masses in Dalton's time. Indirect measurements, however, could indicate their relative masses. These relative masses are usually expressed in terms of **atomic mass units** (amu). This unit of mass was simply invented for atoms and molecules. Hydrogen atoms have a mass of 1 amu. Since oxygen atoms have a mass 16 times as great, the mass of the oxygen atom is 16 amu.

Dalton also used various kinds of circles (similar to those in Figures 2.6 through 2.9) to represent the different kinds of atoms. Jöns Jakob Berzelius (Figure 2.10), who corrected many of Dalton's inaccurate relative masses, also proposed modern chemical symbols (next section).

Despite its inaccuracies, Dalton's atomic theory was a great success. Why? Because it served—and still serves—to *explain* a large body of experimental data. It also successfully predicts how matter will behave under a wide variety of circumstances. Dalton arrived at his atomic theory by reasoning from experimental data, and with modest modification the theory has stood the test

FIGURE 2.10
Jöns Jakob Berzelius, a Swedish chemist, invented modern chemical symbols. (Reprinted with permission from Weeks, Mary E., **Discovery of the Elements,** Easton, Pa.: *Chemical Education Publishing, 1968. Copyright © 1968 by the Chemical Education Publishing Company.*)

One atom of carbon   Two atoms of oxygen   One molecule of carbon dioxide

FIGURE 2.9   *The law of conservation of mass. In a chemical reaction, atoms are merely rearranged (not created or destroyed); thus, matter is conserved.*

of time and the assault of modern, highly sophisticated instrumentation. Formulation of so successful a theory was quite a triumph for a Quaker schoolteacher in the year 1803!

2.4 LEUCIPPUS REVISITED: ELEMENTS, COMPOUNDS, AND MIXTURES

Now back to Leucippus's musings by the seashore. We now know that if we kept dividing those drops of water into smaller drops, we would ultimately obtain a small particle—called a **molecule**—that would still be water. (A **molecule** is just a tightly bound collection of atoms that act as a unit.) If we divided that particle still further, we would obtain two atoms of hydrogen and one atom of oxygen. And if we divided those.... But that's a story for another time. Dalton regarded the atom as indivisible.

We define an **element** in terms of the atoms of which it is composed. In an element, all the atoms are of the same kind. Dalton would have said that "the same kind" meant "the same mass." However, this is one aspect of the atomic theory that had to be revised slightly. We will consider precisely what "the same kind" means a bit later in this chapter.

An element and the atoms of the element are represented by a **symbol.** Over a hundred elements are known today, each represented by a symbol of one or two letters derived from the name of the element (or, sometimes, the Latin name of the element). The first letter of the symbol is always capital: the second is always lowercase. (It does make a difference. For example, Hf is the symbol for hafnium, an element; HF is the formula for hydrogen fluoride, a compound. Similarly, Co is cobalt, an element; CO is carbon monoxide, a compound. More about compounds shortly.) Symbols for a few of the more important elements are given in Table 2.1.

A **compound** is made up of two or more elements, *always in fixed proportions.* For example, when the compound water is broken down, it is always found to be composed of 89% oxygen by mass and 11% hydrogen by mass. Water is made up of molecules that contain two hydrogen atoms and one oxygen atom. Each hydrogen atom has a mass of 1, and each oxygen atom has a mass of 16.

Two hydrogen atoms have a mass of   2
One oxygen atom has a mass of   16
One water molecule has a mass of $\overline{18}$

$$\text{Percentage of hydrogen in water} = \frac{2}{18} \times 100 = 11\%$$

$$\text{Percentage of oxygen in water} = \frac{16}{18} \times 100 = 89\%$$

Any size sample of water will be 89% (or $\frac{16}{18}$) oxygen and 11% (or $\frac{2}{18}$) hydrogen.

TABLE 2.1   *Some Elements and Their Symbols*

Element	Symbol	Element	Symbol
Hydrogen	H	Iron (ferrum)	Fe
Helium	He	Cobalt	Co
Carbon	C	Copper (cuprum)	Cu
Nitrogen	N	Zinc	Zn
Oxygen	O	Bromine	Br
Fluorine	F	Silver (argentum)	Ag
Sodium (natrium)	Na	Tin (stannum)	Sn
Magnesium	Mg	Iodine	I
Aluminum	Al	Barium	Ba
Silicon	Si	Mercury (hydrargyrum)	Hg
Phosphorus	P	Gold (aurum)	Au
Sulfur	S	Lead (plumbum)	Pb
Chlorine	Cl	Radium	Ra
Potassium (kalium)	K	Uranium	U
Calcium	Ca	Plutonium	Pu

How many grams of oxygen are there in 100 g of water?

$$100 \text{ g of water} \times 0.89 = 89 \text{ g of oxygen}$$

or

$$100 \text{ g of water} \times \frac{16 \text{ g of oxygen}}{18 \text{ g of water}} = 89 \text{ g of oxygen}$$

example 2.4

How many pounds of hydrogen are there in 36 lb of water?

$$36 \text{ lb of water} \times 0.11 = 4.0 \text{ lb of hydrogen}$$

or

$$36 \text{ lb of water} \times \frac{2.0 \text{ lb of hydrogen}}{18 \text{ lb of water}} = 4.0 \text{ lb of hydrogen}$$

example 2.5

How much water could be made from 32 g of oxygen?

$$32 \text{ g of oxygen} \times \frac{18 \text{ g of water}}{16 \text{ g of oxygen}} = 36 \text{ g of water}$$

example 2.6

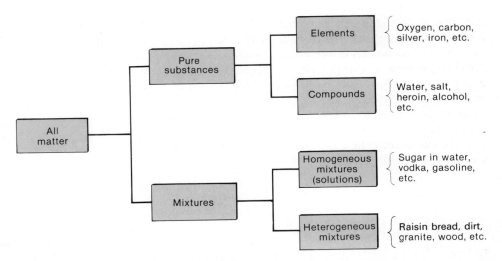

FIGURE 2.11   *A scheme for classifying matter.*

If the compound sulfur trioxide has a mass ratio of 2 parts of sulfur to 3 parts of oxygen, how much oxygen is there in 45 g of sulfur trioxide?

Out of a total of 5 parts (3 + 2) of sulfur trioxide, $\frac{2}{5}$ is sulfur and $\frac{3}{5}$ is oxygen. Therefore,

$$45 \text{ g of sulfur trioxide} \times \frac{3 \text{ g of oxygen}}{5 \text{ g of sulfur trioxide}} = 27 \text{ g of oxygen}$$

The law of definite proportions distinguishes compounds from **mixtures.***
One can *mix* lead with sulfur in any proportions. However, if sulfur and lead are heated together, they form a compound, lead sulfide, in which the elements are always found in the same definite proportions (87% lead, 13% sulfur). Any extra lead or sulfur is simply left over, unreacted (see Figure 2.4).

Matter is characterized by its properties. This simply means that we know we have water and not, say, gasoline because water has certain characteristics or properties that distinguish it from gasoline. For example, it is hard to start a fire by dousing something with water and then lighting it with a match. **Chemical properties** describe how one substance reacts with other

---

* Mixtures have variable compositions. Figure 2.11 presents a scheme for classifying matter. **Homogeneous** substances are the same throughout. The composition of **heterogeneous** substances varies from one part to another.

substances; to demonstrate a chemical property, a substance must undergo a change in composition. **Physical properties** involve no change in composition. Such characteristics as color, hardness, density, and melting point are physical properties. Some physical properties of sulfur, for example, are that it is a brittle yellow solid with a density of 2.07 g/cm³ at room temperature. If you perform an experiment to determine the density of sulfur, you will still have sulfur when you are finished. Some chemical properties of sulfur are that it reacts with oxygen, with carbon, and with iron. These reactions yield, in turn, sulfur dioxide, carbon disulfide, and iron sulfide—all new substances with unique properties.

FIGURE 2.12
Michael Faraday, a British scientist who helped prove that matter is electrical in nature. (Reprinted with permission from Smith, Henry D., Torchbearers of Chemistry, *New York: Academic Press, 1949. Copyright © 1949 by Academic Press, Inc.)*

2.5 THE DIVISIBLE ATOM

Dalton, who set forth his atomic theory in 1803, regarded the atom as hard and indivisible. It wasn't long, however, before that simple picture underwent significant modification. Indeed, even a few years prior to the publication of Dalton's theory, evidence that suggested a more complicated structure began to accumulate. In 1800, two English chemists, William Nicholson and Anthony Carlisle, using the recently invented electric battery, passed an electric current through water and decomposed it into hydrogen and oxygen. Clearly matter could somehow interact with electricity. Dalton's model of the atom failed to show how. Evidence of the electrical nature of the atom poured out of nineteenth-century laboratories, particularly that of the English scientist Michael Faraday (Figure 2.12). Each new experiment provided information about the structure of atoms. There was, in fact, so much data that we shall limit ourselves to a consideration of just a few critical experiments.

Of major significance was the experimental work of William Crookes (Figure 2.13) and Joseph John Thomson (Figure 2.14). In 1875, Crookes carried out studies in an evacuated glass tube. Into the tube were inserted two metal disks that were connected to a source of electric current. The two disks were called **electrodes,** and when the system was connected to the voltage source, one disk became positively charged and the other became negatively charged. The positive electrode was called the **anode;** the negative one, the **cathode.** Crookes's tube was designed so that the air could be removed from it and the interior maintained under a vacuum. When this was done, a beam of current passed from the cathode to the anode in a straight line through the vacuum. The beam was termed a **cathode ray** (Figure 2.15).

But just what were these cathode rays? J. J. Thomson provided the answer in 1897. He showed that the cathode rays were deflected in an electric field. To one side of the Crookes tube was placed a metal plate that carried a positive charge, and to the opposite side was placed a plate with a negative charge.

FIGURE 2.13
Sir William Crookes, a British scientist who invented the cathode-ray tube. (Reprinted with permission from Weeks, Mary E., Discovery of the Elements, *Easton, Pa.: Chemical Education Publishing, 1968. Copyright © 1968 by the Chemical Education Publishing Company.)*

FIGURE 2.14
J. J. Thomson, a British scientist who determined the mass-to-charge ratio of the electron. (Reprinted from Thomson, George, The Electron, *Oak Ridge, Tenn.: U. S. Department of Energy, 1972.)*

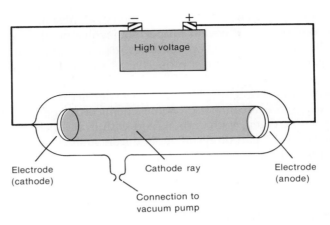

FIGURE 2.15 *A simple gas discharge tube.*

As the ray traveled from the cathode to the anode, it was attracted by the positive plate and repelled by the negative plate (Figure 2.16). Thus, the rays must be composed of negatively charged *particles*. (It was known that light waves were not deflected by electric fields.) The name **electron** was given to those negatively charged particles. For his work, Thomson won the Nobel Prize in physics in 1906.

The properties of the cathode ray were the same regardless of the material from which the cathode was made. Cathodes made from different elements emitted streams of particles with identical mass and charge. Thus,

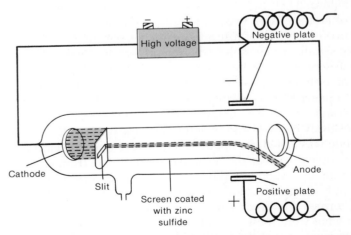

FIGURE 2.16 *Thomson's apparatus showing deflection of an electron beam in an electric field. The screen is coated with zinc sulfide, a substance that glows when struck by electrons.*

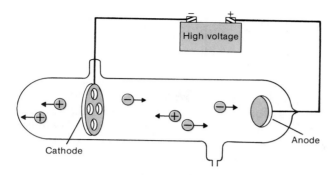

FIGURE 2.17   *Goldstein's apparatus for study of positive particles.*

electrons came to be regarded as constituents of all matter. Note that Dalton's model of the atom could not account for these experimental results. Indivisible atoms could not give off particles.

In 1886, a German scientist named Eugen Goldstein conducted some experiments in equipment similar to Crookes's. Goldstein, however, worked with small amounts of various gases in the tube and used a cathode disk that had perforations in it (Figure 2.17). He found that, at the same time that electrons left the cathode and sped off toward the anode, positive rays were also formed within his tube and were attracted toward the cathode. These positive rays raced toward the cathode, passed through the holes in the cathode, and eventually struck the end of the tube, causing it to glow. Although all of the negatively charged particles were identical, the positively charged particles differed depending on what gas was in the tube. For example, helium produced more massive positively charged particles than did hydrogen. The lightest positive particles found were those from hydrogen, and they were given a special name, **protons.** The charges on the proton and the electron are the same size (although opposite in sign), but the proton was found to be 1837 times more massive than the electron. Thus, the smallest positive particle isolated was many time heavier than the negatively charged electron.

Atoms, then, are not indivisible after all. Negatively charged electrons can be plucked from every kind of atom.

2.6 SERENDIPITY IN SCIENCE: X RAYS AND RADIOACTIVITY

Other discoveries at the turn of the century helped to reveal more of the secrets of the atom. Two important discoveries can be described as happy accidents. Something unexpected happens, and, as a result, whole new areas of study are opened. Such accidents are of no value unless they happen to a trained observer who can grasp their significance. It is the combination of the right scientist and the right accident that qualifies an event as a case of scientific *serendipity*.

Two such happy accidents occurred in the waning years of the nineteenth century. In 1895 a German scientist, Wilhelm Konrad Roentgen, was working in a darkroom, studying certain substances that glowed while exposed to cathode rays. To his surprise, he noted this glow in a chemically treated piece of paper held some distance from the cathode-ray tube. The paper even glowed if the tube was located in the next room. Roentgen had discovered a new type of ray, one that could travel through walls! The rays were given off from the anode whenever the cathode-ray tube was operating. With seeming lack of imagination, Roentgen called the mysterious, penetrating rays **X rays.** X rays are related to light, unlike cathode rays, which are streams of charged particles.

As is often the case when an exciting new discovery is made, many other scientists began to study the new phenomenon.* One, Antoine Henri Becquerel, a Frenchman, had been studying fluorescence, a phenomenon in which certain chemicals, when exposed to strong sunlight, continued to glow even when taken into a dark room. Was this phenomenon related to X rays? he wondered. Becquerel wrapped photographic film in black paper, placed a few crystals of the fluorescing chemical on top of the paper, and then placed it in strong sunlight. He reasoned that if the glow were ordinary light, it would not pass through the black paper. If, on the other hand, the glow were similar to X rays, it would pass through the black paper and fog the film.

Before Becquerel's studies of fluorescence had progressed very far, he made an important discovery. He was testing a crystal of a uranium compound. After being placed in sunlight, it fogged the covered photographic film, suggesting that here, perhaps, was a compound that emitted X rays when it fluoresced. During several cloudy days when work in sunlight was impossible, Becquerel prepared samples and placed them in a drawer. To his great surprise, the photographic film was exposed even though the uranium compound had not been exposed to sunlight. Further experiments showed that, in fact, this radiation had no connection with fluorescence but was a characteristic of the element uranium.

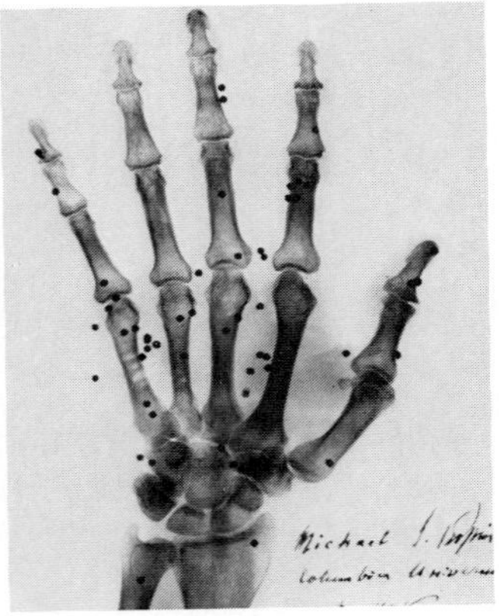

FIGURE 2.18
An early example of the use of X rays in medicine. Professor Michael Pupin, of Columbia University, made this X ray in 1896 to aid in the removal by surgery of gunshot pellets (dark spots) from the hand of a patient. (Courtesy of Burndy Library, Norwalk, Conn.)

* The medical community immediately recognized the significance of the penetrating X rays. The X-ray picture shown in Figure 2.18 was taken in February 1896, within two months of the publication of Roentgen's discovery. The round black dots are gunshot pellets whose positions were established for removal by surgery through this picture.

Unfortunately, details of the X-ray apparatus and commercial versions of the equipment were readily available to the public, who were as fascinated by the phenomenon as were scientists. (Roentgen himself would not patent his discovery and, thus, refused to reap its commercial rewards.) While X rays made major contributions to medicine and to basic science, their indiscriminate use by people (including many scientists and physicians) resulted in a number of cases of severe burns and some deaths. It took some time to fully appreciate that rays that could pass through the body could also inflict biological damage as they made their way through.

As happened with the discovery of X rays, many other scientists began to study this new type of radiation. One, Marie Sklodowska Curie (Figure 2.19), gave this new phenomenon the name **radioactivity.** Marie Sklodowska was born in Poland in 1867. She came to Paris to work for her doctor's degree in mathematics and physics. There she met and married Pierre Curie, a French physicist of some note. She, often with the help of her husband, discovered a number of new radioactive elements, including radium. Pierre Curie was killed in a traffic accident 3 years after he, Marie, and Becquerel were awarded the 1903 Nobel Prize in physics. Marie Curie continued to work with radioactivity, winning a second Nobel Prize (for chemistry) in 1911. She died in 1934 of pernicious anemia, perhaps brought on by deprivation, hard work, and long exposure to radiation from the materials she had studied.

FIGURE 2.19
Marie Curie in her
laboratory. (Culver Pictures.)

2.7 A NEW ATOM FOR A NEW CENTURY

It was soon realized that the radiation emanating from uranium, radium, and other radioactive elements was of three types. When this radiation was passed through a strong magnetic field, one portion was deflected in one direction, another portion was deflected in the opposite direction, and a third was not deflected at all (Figure 2.20). These portions were named **alpha (α), beta (β),** and **gamma (γ)** rays, respectively, by Ernest Rutherford, a New Zealander working at McGill University, in Montreal. The alpha rays were shown to have a mass four times that of the hydrogen atom and a positive charge twice that

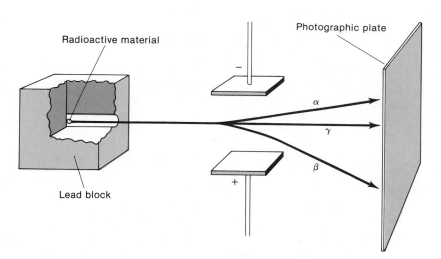

FIGURE 2.20 Behavior of radioactive rays in an electric field.

TABLE 2.2 *Types of Radioactivity*

Name	Symbol	Mass (in atomic mass units)	Charge
Alpha	α	4	$2+$
Beta	β	$\frac{1}{1837}$	$1-$
Gamma	γ	0	0

of the proton (or twice the magnitude of the electron's negative charge). Beta rays were shown to be identical to cathode rays; that is, they were streams of electrons. The charge on an electron or beta particle is assigned a value of $1-$. Therefore, the charge on an alpha particle is $2+$. Gamma rays were shown to be very much like X rays but even more penetrating. Gamma rays have no mass and no charge. These properties are summarized in Table 2.2.

Rutherford soon used the positively charged alpha particles to make an important discovery. He placed some highly radioactive material in a lead-lined box that had a tiny hole. Most of the alpha particles emitted by the material in the box were absorbed by the lead, but those escaping through the hole formed a narrow stream of very high-energy particles. This apparatus, then, could be aimed like a gun at some "target" (Figure 2.21).

One target that Rutherford selected was a thin piece of gold foil. Rutherford expected most of the positively charged alpha particles to be deflected only slightly by the positive charges in the atoms of the gold foil. What he

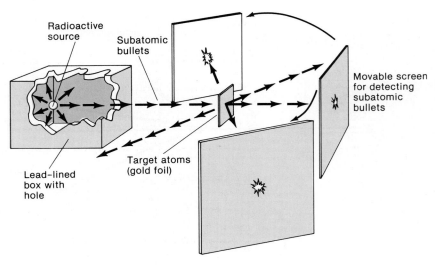

FIGURE 2.21 *Rutherford's gold foil experiment.*

found was that most of the alpha particles went through the foil, which was about 2000 atoms thick, without being deflected at all. This result is understandable if one assumes that the positive charges in the atom are actually rather thinly spread and no match at all for a high-energy alpha bullet. The tiny electrons in the atom would not be expected to stop the much more massive alpha particle either. What really amazed Rutherford was the fact that a few of the particles were deflected very sharply, some even bouncing back in the direction from which they came.

On the basis of these results Rutherford proposed an arrangement of the positive and negative parts of the atom (Figure 2.22). To account for the very strong deflections of some of the alpha particles, he pictured all the positive charge and almost all of the mass concentrated in a very small space. When an alpha particle approached this concentrated, positively charged matter (called the **nucleus** of the atom), it was strongly repelled and, thus, strongly deflected. Since very few alpha particles were deflected, Rutherford concluded that the nucleus must occupy only a fraction of the total volume of the atom. Most of the particles passed right through the atom, because most of the atom was empty space. But this space was not completely empty: it was here that Rutherford placed the negatively charged electrons. He concluded that electrons were so small that their presence in no way interfered with the passage of the alpha particles through the atom, a situation analogous to that of a mouse trying to stop the charge of an enraged bull elephant.

Rutherford's nuclear theory of the atom, set forth in 1911, was revolutionary indeed. He postulated that all the positive charge and virtually all the mass

FIGURE 2.22
Model explaining the results of Rutherford's experiment.

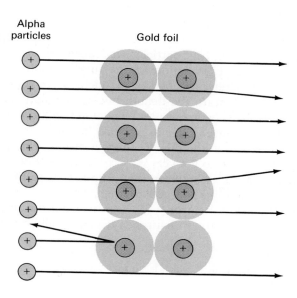

were concentrated in the tiny, tiny nucleus. The negatively charged electrons, he said, had virtually no mass yet occupied nearly all the volume of the atom. Rutherford's model can be pictured as follows: a BB (the nucleus) in the middle of a volume of space equivalent to that inside the New Orleans Superdome (the atom) surrounded by some mosquitoes (electrons) that flit here and there throughout the stadium.

2.8 THE STRUCTURE OF THE NUCLEUS

In 1914, Rutherford suggested that the smallest positive-ray particle (i.e., that which is formed when there is hydrogen gas in the Goldstein apparatus) is the unit of positive charge in the nucleus. This particle, called a **proton,** has a charge equal in magnitude to that of the electron and has nearly the same mass as the hydrogen atom. Rutherford's suggestion was that protons constitute the positively charged matter in all atoms.

We can now define an **element** more precisely. All the atoms of an element have the same number of protons in the nucleus; that is, they have the same **atomic number.**

The hydrogen atom's nucleus consists of one proton, and the nuclei of larger atoms contain a number of protons. Except for hydrogen atoms, though, atomic nuclei are heavier than would be indicated by the number of positive charges (number of protons). For example, the helium nucleus has a charge of 2 + (and, therefore, it contains two protons, according to Rutherford's theory), but its mass is four times that of hydrogen. This excess mass puzzled scientists until 1932, when the English physicist James Chadwick discovered a particle with about the same mass as a proton, but with *no* electrical charge. This particle was called the **neutron,** and its existence made possible an explanation of the unexpectedly high mass of the helium nucleus. Whereas the hydrogen nucleus contains only a proton of 1 atomic mass unit (amu), the helium nucleus contains not only two protons (2 amu) but also two neutrons (2 amu), giving a total mass to the nucleus of 4 amu.*

One can even offer a rationale for the presence of neutrons in the helium nucleus and in the nuceli of bigger atoms. The nucleus is very small, and two protons have to get very close together within it. Perhaps the neutrons act as a sort of buffer between the protons to ease the repulsive forces experienced by the two like-charged protons. Hydrogen, with only one proton, doesn't really need a neutron—a neutron would not hurt the hydrogen nucleus, but the single proton in hydrogen doesn't really require buffering of any sort.

---

* The alpha particle, which has a mass of 4 and a charge of 2 +, is identical to the nucleus of a helium atom.

TABLE 2.3   *Subatomic Particles*

Particle	Symbol	Mass (in atomic mass units)	Charge	Location in Atom
Proton	p^+	1	$1+$	Nucleus
Neutron	n	1	0	Nucleus
Electron	e^-	$\frac{1}{1837}$	$1-$	Outside nucleus

With the discovery of the neutron, the list of "building blocks" we will need for "constructing" atoms is complete. The properties of these particles are summarized in Table 2.3.

All atoms, except hydrogen (which has only a proton for a nucleus), have a tiny nucleus composed of protons and neutrons, and they have electrons somewhere outside the nucleus.

2.9 ELECTRON STRUCTURE: THE BOHR MODEL

First, an aside. Light, such as that emitted by the sun or an incandescent bulb, is a form of pure energy. When white light from an incandescent lamp is passed through a prism, it is separated into a continuous spectrum, or rainbow, of colors (Figure 2.23). When sunlight passes through a raindrop, the same phenomenon occurs. The different colors of light represent different amounts of energy. Blue light packs more energy than does red light of the

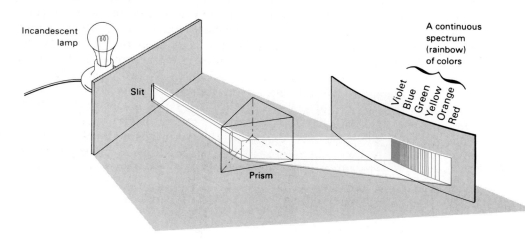

FIGURE 2.23   *White light is separated into a continuous spectrum.*

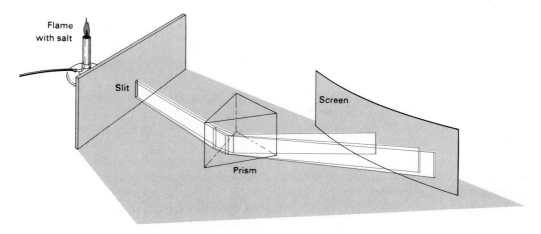

FIGURE 2.24 A line spectrum.

same intensity. White light is simply a combination of all the various colors of different energies.

Now back to atoms. If the light from a flame in which a certain element is being heated is passed through a prism, only narrow colored lines are obtained (Figure 2.24). Each line corresponds to light of definite energy. The pattern of colored lines emitted by each element (called its line spectrum) is characteristic of that element and can be used to identify it (Figure 2.25). Consider, for example, the fact that the only thing reaching the Earth from the stars or the planets is light. Until recently, everything we knew about these heavenly bodies had to be deduced from our examination of this light. Scientists have used line spectra to determine the chemical makeup of the stars and the atmospheres of the planets.

FIGURE 2.25 Characteristic line spectra of some elements.

In a crude way, the color of a flame can be used to identify an element. Sodium salts give a persistent yellow flame; potassium salts, a fleeting lavender; lithium salts, a brilliant red. The colors in fireworks displays are due to the characteristic energies of the light given off by specific elements. Sodium compounds are used for yellow fireworks, strontium salts for red, and barium compounds for green. Copper salts produce an ordinary blue. No substance produces a brilliant blue.

When last we saw the atom, it included electrons flitting about the nucleus like mosquitoes. Like anything in motion, electrons have kinetic energy. And because they are outside the nucleus, they have a potential energy. Electrons, in this respect, are like rocks on a cliff. The rocks can fall and give up energy; if the electrons were to fall toward the nucleus, they too would give up energy.

How does all this concern the line spectra of the elements? Well, the first satisfactory explanation of line spectra was set forth by Niels Bohr, a Danish physicist, in 1913. He made the revolutionary suggestion that electrons cannot have just any amount of energy but can have only certain specified values; that is to say, the total energy of an electron is **quantized.** An electron, by absorbing a quantum (a packet of specific size) of energy (for example, when atoms of the element are heated in a flame), is elevated to a higher energy level. By giving up a quantum of energy, the electron can return to a lower energy level. The energy given off, having only certain specified values, shows up as a line spectrum (Figure 2.26). An electron moves instantaneously from one energy level to another.

An analogy would be that of a person on a ladder. The person can stand on the first rung, the second rung, the third rung, and so on but is unable to stand between rungs. As the person goes from one rung to another, the potential energy (energy due to position) changes by definite amounts, or quanta. For an electron, its total energy (both potential and kinetic) is changed as it moves from one energy level to another.

Bohr's model of the atom was based on the laws of planetary motion that had been set down by the Danish astronomer Johannes Kepler three centuries before. He imagined the electrons to be orbiting about the nucleus much as planets orbit the sun (Figure 2.27). Different energy levels can be pictured as different orbits. The electron in the hydrogen atom is usually in the first energy level (the innermost orbit). Given the choice, electrons usually remain in their lowest possible energy levels (those nearest the nucleus); atoms whose electrons are in this situation are said to be in their **ground state.** When a flame supplies energy to an atom (hydrogen, for example) and its electron jumps from the first level to the second or a higher level, the atom is said to be **excited.** An atom

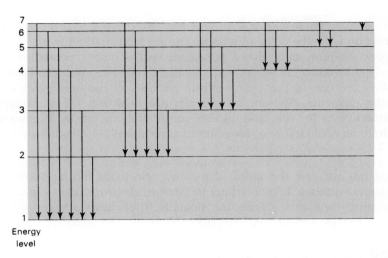

FIGURE 2.26 *Possible electron shifts between energy levels in atoms to produce the lines found in spectra. Not all the lines are in the visible portion of the spectrum.*

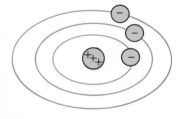

FIGURE 2.27
Bohr visualized the atom as planetary electrons circling a nuclear sun.

with its electron in one of these upper levels or in an excited state eventually emits energy (light) as the electron jumps back down to one of the lower levels and ultimately reaches the ground state.

Atoms larger than hydrogen have more than one electron, and Bohr was also able to deduce that the various energy levels of an atom could only handle a certain number of electrons at one time. We shall simply state Bohr's findings in this regard. The maximum number of electrons that can be in a given level is indicated by the formula $2n^2$, where n is equal to the energy level being considered. For the first energy level ($n = 1$), the maximum population is $2(1)^2$, or 2. For the second energy level ($n = 2$), the maximum number of electrons is $2(2)^2$, or 8. For the third level, the maximum is $2(3)^2$, or 18.

Imagine building up atoms by adding one electron to the proper energy level, as protons are added to the nucleus, keeping in mind that electrons will go to the lowest energy level available. For hydrogen (H), with a nucleus of only 1 proton, the 1 electron goes into the first energy level. For helium (He), with a nucleus of 2 protons (and 2 neutrons), 2 electrons go into the first energy level. According to Bohr, 2 electrons is the maximum population of the first energy level; thus, that level is filled in the helium atom. With lithium (Li), which has 3 electrons, the extra electron goes into the second energy level. This process of adding electrons is continued until the second energy level is filled with 8 electrons, as in the neon (Ne) atom, which has 2 of its 10 electrons in the first energy level and the remaining 8 in the second energy level. The sodium (Na) atom has 11 electrons. Of these, 2 are in the first energy level, the second level is filled with 8 electrons, and the remaining electron is in the third energy level.

We can use a modified Bohr diagram to indicate this **electron configuration** (or arrangement).

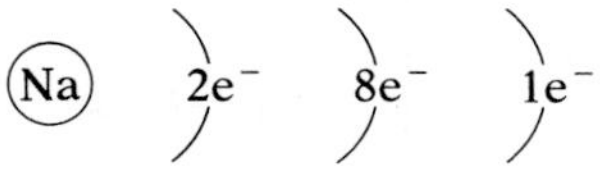

(Na)   2e⁻   8e⁻   1e⁻

The circle with the symbol indicates the sodium nucleus. The arcs represent the energy levels, with the one closest to the nucleus being the first energy level, the next the second, and so on.

We could now continue to add electrons to the third energy level until we get to argon (Ar).

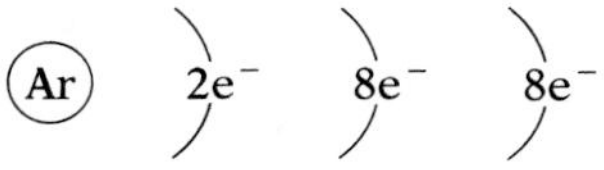

(Ar)   2e⁻   8e⁻   8e⁻

Even though the third energy level can have up to 18 electrons, it is temporarily filled with only 8. The next element, potassium (K), has the electron configuration

(K)   2e⁻   8e⁻   8e⁻   1e⁻

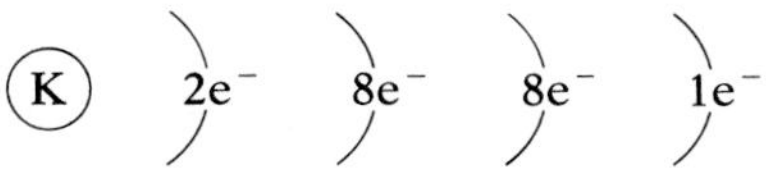

Similarly, calcium (Ca) has 2 electrons in the fourth energy level.

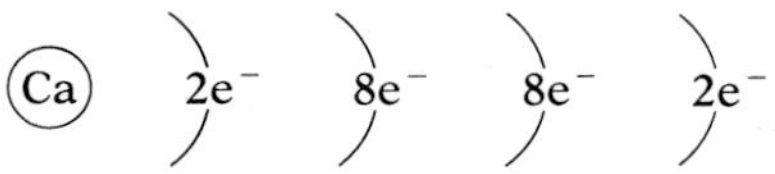

(Ca)   2e⁻   8e⁻   8e⁻   2e⁻

After calcium, things get a bit more complicated as the third energy level resumes filling.

The buildup of atoms to calcium is diagrammed in Figure 2.28. The entire topic of atomic structure is discussed in more detail in Special Topic B. Using rules outlined there, we could (in our imagination, at least) draw Bohr diagrams for all the known elements. For now, though, let us practice with some simple examples.

Draw a Bohr diagram for fluorine (F).

Fluorine is element number 9; it has nine electrons. Two of these go into the first energy level, the remaining seven into the second:

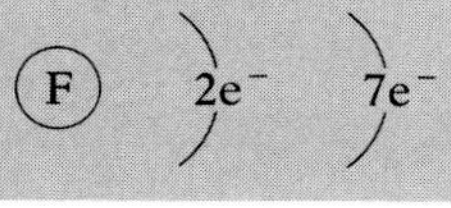

(F)   2e⁻   7e⁻

example 2.8

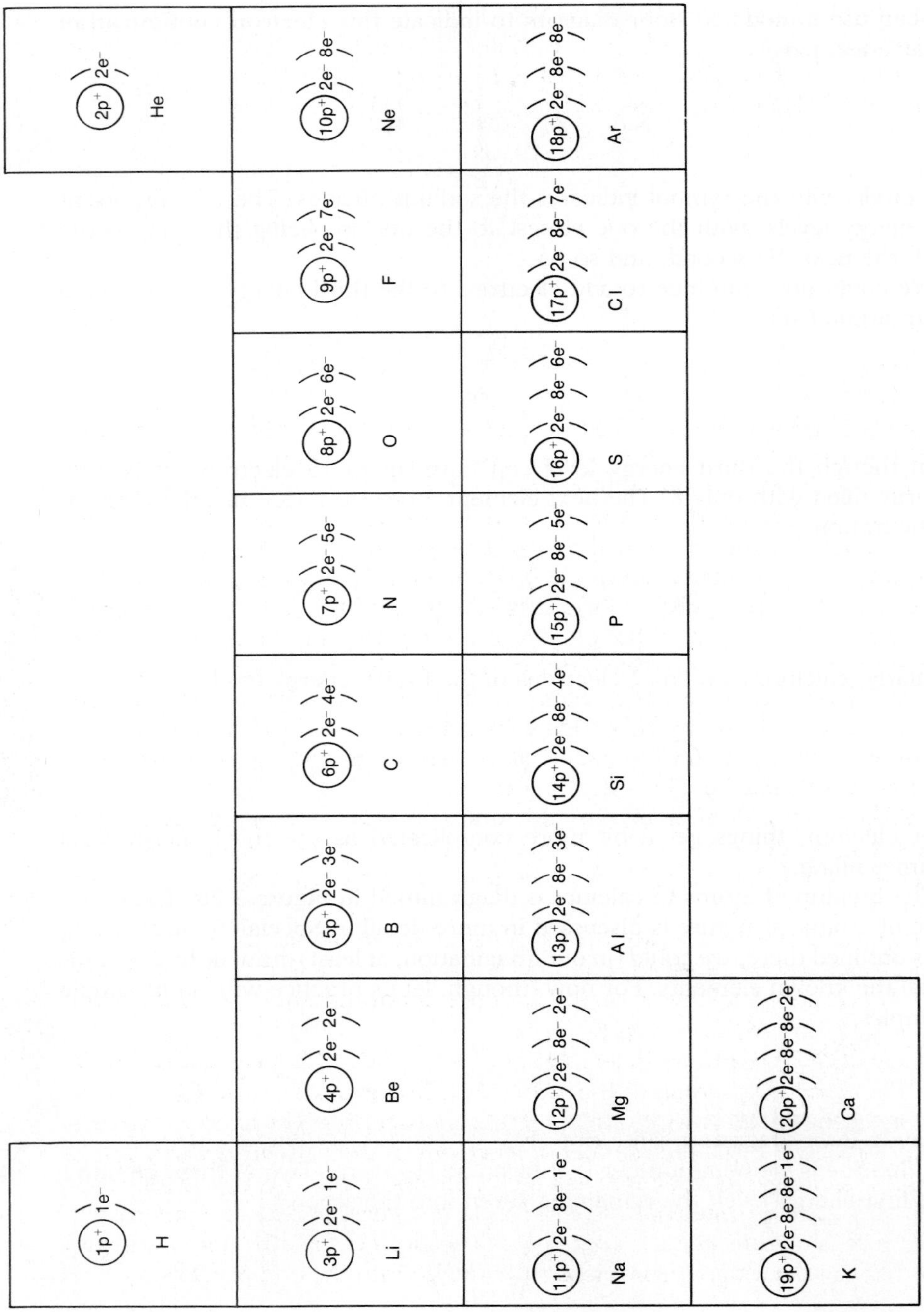

FIGURE 2.28 Bohr diagrams of the first 20 elements.

example
2.9

Draw a Bohr diagram for sodium (Na).

Sodium is element number 11; it has eleven electrons. Two of these can go into the first energy level. Of the remaining, eight go into the second energy level. That leaves one electron for the third energy level:

$$\text{(Na)} \quad 2e^- \quad 8e^- \quad 1e^-$$

2.10 MENDELÉEV AND THE PERIODIC TABLE

Let's go back now to the 1800s. New elements were discovered with surprising frequency. By 1830 there were 55 known elements, all with seemingly different properties and with no apparent order in these properties. Several attempts were made to arrange the elements in some sort of systematic fashion. The most successful arrangement was that of Dmitri Ivanovich Mendeléev (1834–1907), a Russian chemist (Figure 2.29). He published a **periodic table** of the elements in 1869. His table had the elements arranged, primarily, in order of increasing atomic weight, although in a few cases he placed a slightly heavier element before a lighter one. He did this to get elements with similar chemical properties in the same row. For example, he placed tellurium (which has an atomic weight of 128) ahead of iodine (which has an atomic weight of 127) because the former resembled sulfur and selenium in its properties, whereas the latter was similar to chlorine and bromine.

Mendeléev left a number of gaps in his table. Instead of looking upon those blank spaces as defects, he boldly predicted the existence of elements as yet undiscovered. Furthermore, he even predicted the properties of some of these missing elements.

In succeeding years many of the gaps were filled in by the discovery of new elements. The properties were often quite close to those Mendeléev had predicted. It was the predictive value of this great innovation that led to the wide acceptance of Mendeléev's chart.

A copy of the modern periodic table appears in the inside front cover of this book. The vertical columns of the table are called **groups** or **families.** Each group includes elements with similar chemical properties. The elements within a group also have the same number of electrons in their *highest* energy level. This is not a coincidence, as we shall see in Section 2.11. The horizontal rows of the table are called **periods.** For elements in a given period, the outer electrons are in the same main energy level. (The situation is a bit more complicated than this, but there is no need for us to go into the complications.)

FIGURE 2.29
Dmitri Mendeléev, a Russian chemist who invented the periodic table of the elements. (Reprinted with permission from Weeks, Mary E., Discovery of the Elements, *Easton, Pa.: Chemical Education Publishing, 1968. Copyright © 1968 by the Chemical Education Publishing Company.)*

Certain groupings of elements in the periodic table are designated by special names. The heavy stepped diagonal line on the table divides the elements into two major classes. Those to the left of the line are called **metals,** and those to the right, **nonmetals.** Group IA elements are known as **alkali metals;** Group IIA are **alkaline earth metals;** Group VIIA, **halogens.** The group at the extreme right of the table contains the **noble gases.** Group B elements are called **transition metals.**

You'll find the periodic table an invaluable aid in your study of chemistry. As you gain experience, you will be able to use the table to determine the electron configuration of the atoms of an element (Section 2.11 and Special Topic B). The periodic table also summarizes much of what we know about the properties of the elements and their compounds—as we shall see in the chapters that follow.

2.11 ENERGY LEVELS AND THE PERIODIC TABLE

The chemical properties of atoms depend mainly on the number of electrons in the outermost energy level. Sodium (Na) is the element immediately after neon; therefore, it has one more electron than neon and must place this electron into the third energy level. Sodium, with one electron in its outermost level (the third) is similar in properties to lithium, which also has only one electron in its outermost level (the second). Helium and neon, having filled outer energy levels, are similar to each other in that both are inert, that is, in that they tend not to undergo reactions. Apparently, not only do similar chemical properties occur in atoms of similar electron configuration (or arrangement), but certain configurations appear to be more stable (less reactive) than others. This point is explored in detail in Chapter 4.

In Mendeléev's periodic table, the elements were arranged by atomic weight for the most part, and this arrangement revealed the periodicity of chemical properties. However, it is the number of electrons that determines chemical properties and that should and now does determine the order of elements in the modern periodic table. In the modern table, the elements are arranged according to **atomic number.** This number indicates both the number of protons and the number of electrons in an electrically neutral atom of an element. The modern table, arranged in order of increasing atomic number, and Mendeléev's table, arranged in order of increasing atomic weight, parallel one another because an increase in atomic number is generally accompanied by an increase in atomic weight. In only a few cases (noted by Mendeléev) do the masses fall out of order. Atomic weights do not increase in precisely the same order as atomic numbers because both protons and neutrons contribute to the mass of an atom. It is possible for an atom of lower atomic number to have more neutrons than one with a higher atomic number. Thus, it is

TABLE 2.4 *Composition of the Commonest Isotopes of the First 10 Elements*

	H	He	Li	Be	B	C	N	O	F	Ne
Number of protons	1	2	3	4	5	6	7	8	9	10
Number of electrons	1	2	3	4	5	6	7	8	9	10
Number of neutrons*	0	2	4	5	6	6	7	8	10	10

* Note that, although the number of neutrons generally increases, it does not follow the same regular increase as the number of protons and electrons.

possible for an atom with a lower atomic number to have a mass greater than an atom with a higher atomic number.

Atoms of most elements have varying numbers of neutrons in the nucleus. For example, all carbon atoms have six protons in their nucleus (or else they would not be carbon atoms), but some have six neutrons, some have seven, and others have eight. Atoms of the same element that have different masses are called **isotopes.** The different masses are due to varying numbers of neutrons (Table 2.4). Isotopes are discussed in more detail in the next chapter.

As you might suspect, the Bohr theory had a number of flaws. However, we are rapidly moving into rather abstract chemistry, which bases many of its conclusions on intricate mathematics. We shall just say that, among other things, Bohr's theory could not satisfactorily account for the spectra of atoms more complicated than hydrogen. Thus, physicists and chemists went on to develop still more sophisticated—and more nearly accurate—models for the atom.

For most purposes in this text, we will use the Bohr model of the atom. We'll even find that Dalton's model sometimes proves to be the best way to describe certain phenomena (the behavior of gases, for example). For those who are interested or whose work might require it, atomic structure is discussed in more detail in Special Topic B.

PROBLEMS

1. What is the distinction between the atomistic and the continuous view of matter?
2. Outline the main points of Dalton's atomic theory.
3. Which of the following facts is in conflict with Dalton's atomic theory? Give a reason for your answer.
   a. An atom of calcium has a mass of 40 amu, and an atom of carbon has a mass of 12 amu.
   b. An atom of nitrogen has a mass of 14 amu, and an atom of carbon has a mass of 14 amu.
   c. An atom of hydrogen has a mass of 1 amu, and another atom of hydrogen has a mass of 2 amu.
4. State the law of conservation of mass.
5. What is a chemical compound?
6. What is an element?

7. What is an atom?
8. What is a molecule?
9. What is the difference between a pure substance and a mixture?
10. State the law of definite proportions.
11. Name two compounds that illustrate the law of multiple proportions.
12. Explain, in terms of Dalton's atomic theory, the law of conservation of mass, the law of definite proportions, and the law of multiple proportions. Illustrate each law by an example.
13. What is a scientific "law"? How does a scientific law differ from a governmental law?
14. What is a theory?
15. Discuss evidence indicating the electrical nature of matter.
16. Without consulting Table 2.1, give names for the elements with the following symbols:

    a. He   f. S   k. Sn
    b. N   g. K   l. Ba
    c. F   h. Fe   m. Au
    d. Mg   i. Cu   n. Ra
    e. Si   j. Br   o. Pu

17. Without consulting Table 2.1, give symbols for the following elements:

    a. hydrogen   f. phosphorus   k. silver
    b. carbon   g. chlorine   l. iodine
    c. oxygen   h. calcium   m. mercury
    d. sodium   i. cobalt   n. lead
       aluminum   j. zinc   o. uranium

18. Identify each of the following as a pure substance or a mixture:

    a. carbon dioxide   f. soup
    b. oxygen   g. sodium chloride (salt)
    c. smog   h. a carrot
    d. gasoline   i. 24-karat gold
    e. mercury

19. Which of the symbols or formulas represent elements and which represent compounds?

    a. H   d. C   g. CO_2
    b. He   e. CO   h. Cl_2
    c. HF   f. Ca   i. NaBr

20. Distinguish between chemical and physical properties.
21. For each of the following, indicate whether a physical or a chemical change is described:
    a. Sheep are sheared, and the wool is spun into yarn.

    b. Silkworms feed on mulberry leaves and produce silk.
    c. Because a lawn is watered and fertilized, it grows thicker.
    d. An overgrown lawn is manicured by mowing it with a lawn mower.
    e. Ice cubes form when a tray filled with water is placed in a freezer.
    f. Milk that has been left outside a refrigerator for many hours turns sour.
22. What is a cathode ray?
23. What is radioactivity?
24. What are alpha particles? What are beta particles? What are gamma rays?
25. How do gamma rays differ from X rays? How are the two kinds of rays similar?
26. What evidence is there that electrons are particles?
27. How did the discovery of radioactivity contradict Dalton's atomic theory?
28. Give the distinguishing characteristics of the proton, the neutron, and the electron.
29. What is the nucleus? Describe the nuclear atom.
30. What are isotopes?
31. The table below describes four atoms:

	Atom A	Atom B	Atom C	Atom D
Number of protons	10	11	11	10
Number of neutrons	11	10	11	10
Number of electrons	10	11	11	10

Are atoms A and B isotopes? Are atoms A and C isotopes? A and D? B and C?
32. What are the masses of the atoms in Problem 31?
33. Use the periodic table to determine the number of protons in atoms of the following elements:
    a. helium (He)   d. oxygen (O)
    b. potassium (K)   e. magnesium (Mg)
    c. chlorine (Cl)   f. sulfur (S)
34. How many electrons are there in the neutral atoms of the elements listed in Problem 33?
35. Without referring to Table 2.4, indicate how many electrons and how many protons there are in a neutral atom of each of these elements. (You may use the periodic table.)
    a. calcium (Ca)   b. fluorine (F)

c. beryllium (Be)   **e.** argon (Ar)
d. sodium (Na)   **f.** nitrogen (N)

36. How was the nuclear model of the atom refined by Bohr?

37. According to Bohr, what is the maximum number of electrons in the fourth energy level ($n = 4$)?

38. If the third energy level of electrons contains 2 electrons, what is the total number of electrons in the atom?

39. When light is emitted by an atom, what change has occurred within the atom?

40. Which atom absorbed more energy: one in which an electron moved from the second energy level to the third level or an otherwise identical atom in which an electron moved from the first to the third energy level?

41. Consulting only the periodic table, draw Bohr diagrams for these elements:
a. helium (He)   **d.** sulfur (S)
b. magnesium (Mg)   **e.** oxygen (O)
c. carbon (C)   **f.** silicon (Si)

42. What similarity in electron structure is shared by lithium, sodium, and potassium? By beryllium, magnesium, and calcium?

43. How many electrons are there in the outermost energy level of an atom of each of the following elements?
a. C   **b.** Ne   **c.** F   **d.** Al   **e.** Mg

44. The following Bohr diagram is supposed to represent the neutral atoms of an element. The diagram is incorrectly drawn. Identify the error.

$$\boxed{3p^+} \quad)\;2e^- \quad)\;2e^-$$

45. Referring only to the periodic table, indicate what similarity in electron structure is shared by fluorine (F) and chlorine (Cl) and by carbon (C) and silicon (Si). What is the difference in their electron structures? What is the difference in the electron configurations of oxygen (O) and fluorine (F)?

46. Where are the nonmetals and the metals located on the periodic table?

47. Identify the groups of elements in the periodic table that are classified as halogens, alkali metals, noble gases, alkaline earth metals, transition metals.

48. Which of the following are metals and which are nonmetals?
a. C   **b.** Fe   **c.** Na   **d.** Cl   **e.** Mg

49. Which of the following elements are in the same group? Which are in the same period?
a. Mg   **b.** Al   **c.** Ca   **d.** Cl   **e.** Sr

50. Which of the following is an alkali metal? (You may consult the periodic table.)
a. N   **b.** Na   **c.** Ne   **d.** Ni   **e.** No

51. Which of the following is a transition metal? (You may consult the periodic table.)
a. C   **b.** Ca   **c.** Cl   **d.** Cr   **e.** Cs

52. Which of the following is a halogen?
a. B   **b.** Ba   **c.** Be   **d.** Bk   **e.** Br

53. Which of the following is an alkaline earth metal?
a. S   **b.** Sc   **c.** Se   **d.** Si   **e.** Sr

54. Elements are defined on a theoretical basis as being composed of atoms that have the same atomic number. On the basis of this theory, would you think it possible that someone might discover a new element that would fit between magnesium (atomic number of 12) and aluminum (atomic number of 13)?

55. When 18 g of water is decomposed by electrolysis, 16 g of oxygen and 2 g of hydrogen are formed. According to the law of definite proportions, how much hydrogen would be formed by the electrolysis of 180 g of water? Hydrogen, from the decomposition of water, has been promoted as the fuel of the future. How many metric tons (t) of water would have to be electrolyzed for 2 t of hydrogen to be produced? How much would have to be electrolyzed for 1 t of hydrogen to be produced?

56. Carbon can burn in air to form either carbon monoxide or carbon dioxide, depending on the amount of oxygen present during combustion. Using the mass ratios given on page 34, calculate how much carbon would have to be burned for the following to be produced:
a. 70 g of carbon monoxide
b. 1100 g of carbon dioxide

57. The gas silane can be decomposed to yield silicon and hydrogen in a ratio of 7 parts of silicon by mass to 1 part of hydrogen by mass. If the relative mass of silicon atoms is 28 when the mass of hydrogen is taken to be 1, how many hydrogen atoms are combined with each silicon atom in silane?

58. When we burn a 10-kg piece of wood, only 0.05 kg of ash is left. Explain this apparent contradiction of the law of conservation of mass.

59. When 3 g of carbon is burned in 8 g of oxygen, 11 g of carbon dioxide is formed. What weight of carbon dioxide would be formed if 3 g of carbon were burned in 50 g of oxygen? What law does this illustrate?

60. Jan Baptista van Helmont, a Flemish alchemist (1579–1644), performed an experiment in which he planted a young willow tree in a weighed bucket of soil. After 5 years, he found that the tree had gained 75 kg, yet the soil had lost only 57 g (0.057 kg). Van Helmont had added only water to the system. He concluded that the substance of the tree had come from water. Criticize van Helmont's conclusion.

Energy Sublevels and Orbitals

The Bohr theory of electrons in energy levels has been replaced for many purposes by quantum mechanical models. These more sophisticated models explain more data in greater detail than the simpler Bohr model. This presents us with something of a quandary. We can more accurately interpret the nature of matter only by using a model that is more difficult to understand. Fortunately, however, we need not understand the laborious mathematics to make use of some of the results obtained from quantum mechanical calculations. The quantum mechanical model can make any chemistry course more meaningful by providing a deeper understanding of the structure of atoms and molecules.

B.1 ORBITALS: ELECTRON CHARGE CLOUD MODELS

Quantum mechanics is a highly mathematical discipline that treats electrons as waves and bases their location on probabilities.

It was the young French physicist Louis de Broglie who first suggested (in 1923) that the electron had wavelike properties. In other words, de Broglie said that a beam of electrons should behave like a beam of light. Since Thomson had "proved" (in 1897) that electrons were negatively charged particles, this suggestion that they be treated as waves was hard to accept. Nevertheless,

FIGURE B.1
Erwin Schrödinger, an Austrian physicist, developed mathematical equations for the structure of atoms in which electrons were treated as waves. One such equation is the following.

$$E(\Psi) = \frac{-h^2}{8\pi^2 m} \nabla^2(\Psi) + V(\Psi)$$

(Courtesy of the Nobel Institute, Stockholm.)

de Broglie's theory was experimentally verified within a few years. Electron microscopes, which make use of the wave nature of electrons, are now important pieces of equipment in many scientific laboratories.

A model of the atom, based on this wave nature of the electron, was developed in the late 1920s, principally by the Austrian physicist Erwin Schrödinger (Figure B.1). Elaborate equations that describe the properties of electron waves in atoms are fundamental to the modern picture of the atom. Solutions to these equations are called **wave functions.** These wave functions can be manipulated to provide a measure of the probability that an electron is located in a given volume of space. The definite planetary orbits of the Bohr atom are abandoned* in favor of three-dimensional charge clouds—volumes of space in which the electrons move. These clouds are referred to as **orbitals** to distinguish them from Bohr's orbits.

The calculations involved in the solution of wave equations are complicated and time-consuming. We don't need to be able to do these elaborate calculations, however, to use some of the results of quantum mechanics. For example, a plot of the probability distribution of electrons in the lowest energy level always has a spherical symmetry. To understand this statement, suppose you had a camera that could photograph electrons (there is no such thing, but we are just supposing) and you left the shutter open while the electron zipped about the nucleus. When you developed the picture, you would have a record of where the electron had been. Doing the same thing with an electric fan that was turned on would give you a picture in which the blades of the fan looked like a solid disk of material. The blades move so rapidly that their photographic image is blurred. Back to the electrons. This imagined picture of electrons in the first energy level would look like a fuzzy ball (that's what **spherical symmetry** means). The quantum mechanical orbital corresponds to this fuzzy ball (Figure B.2). Orbitals such as this are designated 1s, and, like Bohr's lowest energy level, the 1s orbital can contain only two electrons.

Quantum mechanical calculations split the second lowest energy level (which, like Bohr's, can hold eight electrons) into four different orbitals. The first two electrons in the second energy level also have spherical symmetry. The orbital for these two electrons is designated 2s (2 because it is in the second energy level, s because it is spherical, like the 1s orbital). The next six electrons of the second energy level go into three orbitals that are dumbbell

---

* It is interesting to note, however, that the figure given by Bohr for the radius of the first orbit, 52.9 picometers (pm), is the same as the "most probable" distance from the nucleus of an electron in the first energy level—as revealed by wave mechanical calculations. Here is a point worth commenting on. A new theory replaces an old one when the new theory explains something the old one can't. However, the new theory must also answer all those questions the old theory was able to handle, and the new answers must be at least as good as the old.

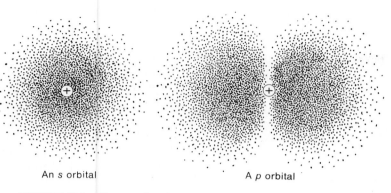

An s orbital   A p orbital

FIGURE B.2   Charge cloud representations of atomic orbitals.

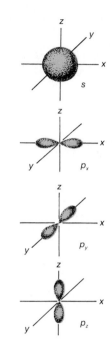

FIGURE B.3
Outline representations of atomic orbitals. The three p *orbitals differ only in their orientation in space.*

shaped (Figure B.3) and differ from one another only in the direction in which they point. These are designated the $2p$ orbitals. Again, the 2 indicates that these orbitals are in the second energy level, and the p means that they have the dumbbell shape. To distinguish the different p orbitals, they are sometimes referred to as the $2p_x$, $2p_y$, and $2p_z$ orbitals.

An orbital can be occupied by 1 electron or by 2 paired electrons, but no more. Thus, the 2 electrons of the lowest energy level can be accommodated by the $1s$ orbital. The second level requires four different orbitals to accommodate its 8 electrons. The third level (with 18 electrons) requires nine orbitals: a $3s$ orbital, three $3p$ orbitals, and a new set of orbitals, the $3d$, of which there are five. We shall not bother to draw all of these orbitals. The point is that each higher energy level has more orbitals than the one below it, to accommodate the larger number of electrons in it.

Strictly speaking, these various orbitals would have to be infinite in size to account for an electron with 100% certainty. Quantum mechanics states that there is at least a tiny probability that any electron can be anywhere in the universe. For convenience in representing orbitals, however, an outline drawing is used to indicate a 95% probability that the electron is located within the volume so outlined (see Figure B.3).

B.2 ELECTRON CONFIGURATIONS

In the second energy level, the p electrons are in a slightly higher energy **sublevel** than the s electrons. In building up the electron configuration of atoms of the various elements, the lower sublevels are filled first. Let's illustrate by "building" a few atoms.

A hydrogen atom (H), with atomic number 1, has only one electron. That single electron goes into the first energy level, which has only one kind of orbital, and s. We write the electron configuration of hydrogen, then, as $1s^1$. The superscript (1 in this case) gives the number of electrons in the orbital. Similarly, the two electrons of helium (He) are in the first energy level; helium's electron structure is $1s^2$. Lithium (Li) has three electrons. Two go into the first energy level; the third must be placed in the s orbital of the second energy level. The electron structure for lithium is written $1s^2 2s^1$. Beryllium (Be) has four electrons; its electron configuration is $1s^2 2s^2$. Boron (B) has five electrons. Two are placed in the $1s$ orbital, two in the $2s$, and the one remaining is placed in one of the three $2p$ orbitals. The electron structure of boron, then, is $1s^2 2s^2 2p^1$.

The three p orbitals (that is, the p sublevel) hold a maximum of six electrons (two per orbital). The elements boron, carbon, nitrogen, oxygen, fluorine, and neon correspond to the filling of the $2p$ sublevel with one through six electrons, respectively. The electron configuration of neon, then, is $1s^2 2s^2 2p^6$. This corresponds to a filled p sublevel—and, of course, to a completely filled second energy level.

example B.1	Write out the electron configuration, using sublevel notation, for nitrogen (N). Nitrogen has seven electrons. Place them in the lowest unfilled energy sublevels. Two go into the $1s$ orbital and two into the $2s$ orbital. That leaves three electrons to be placed in the p sublevel. The electron configuration is $1s^2 2s^2 2p^3$.

example B.2	Write out the electron configuration for fluorine (F). Fluorine has nine electrons. Place two in the $1s$ orbital and two in the $2s$ orbital. That leaves five electrons for the three p orbitals (the p sublevel). The configuration is $1s^2 2s^2 2p^5$.

We could continue to build up electron configurations by following a simple procedure. First, find the atomic number for the element (use the periodic table), then place that number of electrons in the lowest possible energy sublevels. Keep in mind that the maximum number of electrons in an s sublevel is two and in a p sublevel is six.

Write out the electron configuration for sulfur (S).

Sulfur atoms have 16 electrons each. The electron configuration is $1s^2 2s^2 2p^6 3s^2 3p^4$. Note that the total of the superscripts is 16, and that we have not exceeded the maximum capacity for any sublevel.

example
B.3

Next let's write the electron configuration for argon (Ar). Argon has 18 electrons; its configuration is $1s^2 2s^2 2p^6 3s^2 3p^6$. Note that the highest occupied energy sublevel, the $3p$, is filled. Recall, however, from Chapter 2, that the third energy level has a maximum of 18 electrons. We might reasonably expect that potassium (K), with 19 electrons, would have the argon configuration plus the 19th electron in a $3d$ sublevel. It doesn't. Rather, the potassium atom has the electron structure $1s^2 2s^2 2p^6 3s^2 3p^6 4s^1$. The $4s$ sublevel fills before the $3d$. Calcium, with 20 electrons, has the structure $1s^2 2s^2 2p^6 3s^2 3p^6 4s^2$. With Scandium (Sc), the $3d$ sublevel begins to fill. The filling of the $3d$ sublevel corresponds to the **transition elements** of the periodic table.

We will not further pursue here the topic of electron configurations. For those who wish to go further, an "order-of-filling" chart is given in Figure B.4. Using this chart, and the additional information that the maximum capacity for a d sublevel is 10 electrons and that of an f sublevel is 14 electrons, you can write out the electron structure of nearly any element. (There are, as you might expect, a few exceptions that do not follow these rules exactly.) We will deal in this book, for the most part, with the first 20 elements. The electron structure of atoms of these elements is given, for convenience, in Table B.1.

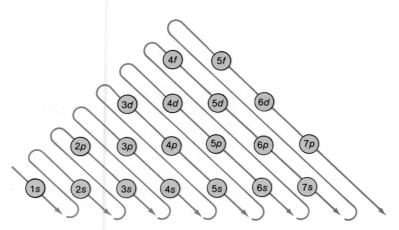

FIGURE B.4   *An order-of-filling chart for determining the electron configurations of atoms.*

TABLE B.1   Electron Structures for Atoms of the First 20 Elements

Name	Atomic Number	Electronic Structure
Hydrogen	1	$1s^1$
Helium	2	$1s^2$
Lithium	3	$1s^2 2s^1$
Beryllium	4	$1s^2 2s^2$
Boron	5	$1s^2 2s^2 2p^1$
Carbon	6	$1s^2 2s^2 2p^2$
Nitrogen	7	$1s^2 2s^2 2p^3$
Oxygen	8	$1s^2 2s^2 2p^4$
Fluorine	9	$1s^2 2s^2 2p^5$
Neon	10	$1s^2 2s^2 2p^6$
Sodium	11	$1s^2 2s^2 2p^6 3s^1$
Magnesium	12	$1s^2 2s^2 2p^6 3s^2$
Aluminum	13	$1s^2 2s^2 2p^6 3s^2 3p^1$
Silicon	14	$1s^2 2s^2 2p^6 3s^2 3p^2$
Phosphorus	15	$1s^2 2s^2 2p^6 3s^2 3p^3$
Sulfur	16	$1s^2 2s^2 2p^6 3s^2 3p^4$
Chlorine	17	$1s^2 2s^2 2p^6 3s^2 3p^5$
Argon	18	$1s^2 2s^2 2p^6 3s^2 3p^6$
Potassium	19	$1s^2 2s^2 2p^6 3s^2 3p^6 4s^1$
Calcium	20	$1s^2 2s^2 2p^6 3s^2 3p^6 4s^2$

B.3 ELECTRON CONFIGURATION AND THE PERIODIC TABLE

Recall from Chapter 2 that the number of electrons in the highest energy level is the same for each element in a group in the periodic table. We can now extend that idea to electron configurations using sublevel notations.

All the elements in Group IA (the alkali metals) have their single outermost electron in an s orbital. The outer electron in lithium (Li) is denoted $2s^1$, that in sodium (Na) as $3s^1$, that in potassium (K) as $4s^1$, and so on. We can say, then, that all Group IA metals have the outer electron structure ns^1, where n denotes the main energy level. Similarly, all the Group IIA elements have the outer electron structure ns^2, that is, the two outermost electrons in any Group IIA atom are in an s orbital. All Group VIIA elements have their seven outermost electrons in the ns^2np^5 configuration. These electron configurations correlate with and help to explain the similar chemical properties of elements within a group.

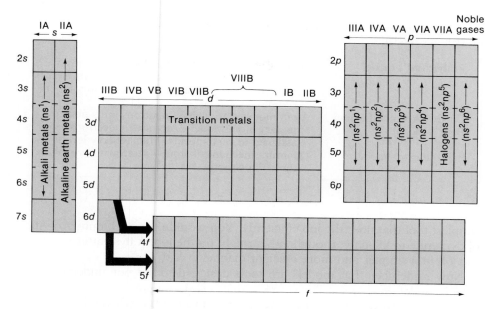

Write out the sublevel notation for the electrons in the highest main energy level for strontium (Sr) and arsenic (As).

Strontium is in Group IIA and the fifth period of the periodic chart. Its outer electron configuration is $5s^2$. (Each period of the periodic table corresponds to the filling, or at least partial filling, of a main energy level.) Arsenic is in Group VA and the fourth period. Its five outer electrons have the configuration denoted by $4s^2 4p^3$.

example B.4

Figure B.5 relates the sublevel configurations to the various groups of the periodic table.

B.4 WHICH MODEL TO USE

Table B.2 summarizes the various models of the atom that we have encountered. Which model should we use? Let's use an analogy to help us decide. How do you choose a tool to do a job around the house? Generally, you choose the

TABLE B.2   *Models of the Atom*

Dalton's model	Billiard ball model—Atom a hard, indestructible particle with a particular weight
Rutherford's model	Nuclear atom model—Atom composed of positive parts (protons) in tiny nucleus and negative parts (electrons) situated throughout its mostly empty volume
Chadwick's modification	Atom modified to include neutrons in its nucleus
Bohr's model	Planetary atom model—Atom with electrons assigned to very specific orbits (energy levels)
Schrödinger's model	Quantum mechanical atom—Atom with orbitals in three dimensions

simplest tool that does the job efficiently. You wouldn't use a hammer or stick of dynamite to kill a fly, though either might do the job. A flyswatter would do the job in a simpler and more efficient manner.

Now remember that models are tools designed to aid our understanding. Which model do we use? The simplest that will do the job.

For our purposes, we will use whatever model is most helpful for understanding a particular concept. In trying to evaluate the behavior of gases, chemists often use Dalton's hard, indivisible (billiard ball) model. In discussing how atoms combine to form molecules, we will use the Bohr model. To explain the shape of molecules (very important in explaining the action of drugs, for instance), we will use an extension of the Bohr theory called the valence shell electron pair repulsion (VSEPR) theory, for the most part. Occasionally we will employ the orbital atom theory. Don't let this inconsistent use disturb you. Remember that a model or a theory is used to explain phenomena. When a better model or theory is invented, the old is sometimes discarded or pushed into a secondary role, but occasionally it can be used to clarify some point that a newer, more complicated model only obscures.

▬▬▬▬▬▬ PROBLEMS ▬▬▬▬▬▬

1. Using sublevel notation and referring only to the periodic table, write out the electron configuration of each of the following:
   a. Al   d. Cl   g. Na
   b. B   e. He   h. O
   c. C   f. Li   i. Si

2. Using sublevel notation and referring only to the periodic table, write out the electron configuration of each of the following:

   a. Ar   d. H   g. Ne
   b. Be   e. K   h. P
   c. Ca   f. Mg

3. Using sublevel notation and referring only to the periodic table, write out the electron configuration for the highest main energy level for each of the following:

   a. Ba   c. Rb   e. As
   b. Br   d. Se   f. Sn

4. Using sublevel notation and referring only to the periodic table, write out the electron configuration for the highest main energy level for each of the following:
   a. Ga
   b. Cs
   c. Te
   d. Sb
   e. I
   f. Sr

5. Give the symbols for the elements whose atoms have the following electron configurations. You may refer to the periodic table.
   a. $1s^2 2s^2 2p^6$
   b. $1s^2 2s^2 2p^6 3s^2 3p^2$
   c. $1s^2 2s^2 2p^6 3s^1$
   d. $1s^2 2s^2$
   e. $1s^2 2s^2 2p^3$
   f. $1s^2 2s^2 2p^6 3s^2 3p^1$

6. In the sublevel notation $2p^6$, how many electrons are described in the notation? What is the general shape of the orbitals described in the notation? How many orbitals are included in the notation?

7. None of the following electron configurations is reasonable. In each case explain why.
   a. $1s^2 2s^2 3s^2$
   b. $1s^2 2s^2 2p^2 3s^1$
   c. $1s^2 2s^2 2p^6 2d^5$

8. Without referring to the periodic table, give the atomic number of each element in Problem 5.

9. What is the maximum number of electrons that can go into a p orbital? A p sublevel?

10. Which member of each set of orbitals is higher in energy?
    a. $2s$ or $3s$
    b. $2s$ or $2p$
    c. $3s$ or $2p$
    d. $3s$ or $3p$

3

The Atomic Nucleus

I n Chapter 2, we took a brief look at the atomic nucleus and then turned our attention to electrons. For a major portion of this book, we will take special note of the outermost electrons. It is the interaction of these outer electrons that holds atoms together to form the compounds that make up our world. For the moment, however, let us turn our gaze inward to that tiny speck of matter called the nucleus. Atomic nuclei are 10 000 times smaller than whole atoms, yet it is the nucleus that holds the power that has become the symbol of our age.

The nuclear age is a paradox. Nuclear power unleashed in wrath can destroy cities and perhaps civilizations. Controlled nuclear power can provide energy necessary to maintain our civilization. Even here, however, there is a paradox. The controlled, peaceful use of nuclear power is not without its own potential dangers.

As citizens of the nuclear age we have difficult decisions to make. Nuclear bombs may kill, but nuclear medicine saves lives. Our knowledge of nuclear science gives us both power and responsibility. How we exercise that responsibility will determine how future generations remember us—even whether there will *be* future generations to remember us.

3.1 A PARTIAL PARTS LIST FOR THE ATOMIC NUCLEUS

We saw in Chapter 2 that atomic nuclei are made up of protons and neutrons. Although the nuclear physicists have extended this parts list to more than a hundred particles, most of these have a transitory existence and are of little

TABLE 3.1 *Subatomic Particles*

Particle	Symbol	Approximate Mass	Charge
Proton	p^+	1	$1+$
Neutron	n	1	0
Electron	e^-	0	$1-$

interest to a chemist. We shall take the oversimplified, but useful, view that atomic nuclei are made up of protons and neutrons—and some sort of force that holds them together.

A proton and neutron have virtually the same mass, 1.007 276 amu and 1.008 665 amu, respectively. That's equivalent to saying that two different people weigh 100.7 kg and 100.9 kg. The difference is so small it can be ignored. Thus, for many purposes, we assume the masses of the proton and the neutron to be the same, 1 amu. The proton has a charge equal in magnitude but opposite in sign to that of an electron. This charge on a proton is written as $1+$. The electron has a charge of $1-$ and a mass of 0.000 549 amu. The electrons in an atom contribute so little of its total mass that they are usually disregarded and are treated as if their mass were 0. The subatomic particles of greatest interest to us are summarized in Table 3.1.

The number of protons in the nucleus of an atom of any element is the atomic number of that element. This number determines the kind of atom, that is, the identity of the element. For example, an atom with 26 protons (one whose atomic number is 26) is an atom of iron (Fe). An atom with 92 protons is an atom of uranium (U). (You are not expected to memorize the atomic number of every element. But you should be able to find this information in the periodic table.) The number of neutrons in the nucleus of atoms of a given element may vary. Most hydrogen (H) atoms have a nucleus consisting of a single proton and no neutrons. About 1 hydrogen atom in 5000, however, does have a neutron as well as a proton in the nucleus. Both kinds are hydrogen atoms (any atom with atomic number 1, that is, with one proton, is a hydrogen atom). Atoms that have this sort of relationship—the same number of protons but differing numbers of neutrons—are called **isotopes.** There is a third, very rare isotope of hydrogen, called tritium, which has two neutrons and one proton in the nucleus. Most, but not all, elements exist in nature in isotopic forms. An interesting and easy-to-remember example is the element tin (Sn), which exists in 10 isotopic forms. (Tin . . . ten isotopes.)

Isotopes are of little importance in ordinary chemical reactions. That simply means that the existence of different isotopes of any element has essentially no effect on the course of a chemical reaction and little effect on the rate of a chemical reaction. Both light (ordinary) hydrogen atoms and heavy hydrogen (deuterium) atoms react with oxygen to form water. Since deuterium atoms are about twice as heavy as ordinary hydrogen atoms, compounds formed from the

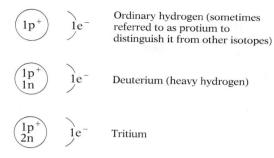

Ordinary hydrogen (sometimes referred to as protium to distinguish it from other isotopes)

Deuterium (heavy hydrogen)

Tritium

two isotopes have different physical properties, but such differences are usually slight. *Chemical reactions* involve the outer electrons of an atom. Differences in the number of neutrons buried deep in the heart of the atom would not be expected to have a major effect. However, in *nuclear reactions*, reactions that do involve the heart of the atom, isotopes are of utmost importance, as we shall see later in this chapter.

3.2 NUCLEAR ARITHMETIC

Before we talk about nuclear reactions, it will be necessary to discuss a special symbolism used in writing such reactions. To represent different isotopes, symbols with subscripted and superscripted numbers are used. In the generalized symbol

$$_Z^A X$$

the Z represents the **atomic number,** that is, the number of protons. The A represents the **mass number,** the number of protons plus the number of neutrons. From the symbol

$$_{17}^{35}Cl$$

we know that the number of protons is 17. The number of neutrons is 18 $(35 - 17)$.

The Z number is not really necessary, since the elemental symbol establishes the number of protons in the nucleus being considered. However, it is convenient to have this information at one's fingertips while writing nuclear equations.

Write the symbol for an isotope with a mass number of 58 and an atomic number of 27.

In the general symbol $_Z^A X$, A is the mass number, which is 58 in this case.

$$_Z^{58}X$$

example 3.1

Z is the atomic number of 27.

$$^{58}_{27}\text{X}$$

From the periodic table we can determine that the element with atomic number 27 is cobalt.

$$^{58}_{27}\text{Co}$$

example 3.2

Indicate the number of protons, neutrons, and electrons in a neutral atom of the isotope $^{235}_{92}\text{U}$.

The atomic number gives the number of protons and electrons in a neutral atom of the isotope.

$$\text{Atomic number} = \text{protons} = \text{electrons} = 92$$

The number of neutrons is obtained by subtracting the atomic number from the mass number.

Mass number	235
Atomic number	− 92
Number of neutrons	143

example 3.3

Which of the following are isotopes of the same element? We are using the letter X as the symbol for all so that the symbol will not identify the elements.

$$^{16}_{8}\text{X} \qquad ^{16}_{7}\text{X} \qquad ^{14}_{7}\text{X} \qquad ^{14}_{6}\text{X} \qquad ^{12}_{6}\text{X}$$

$^{16}_{7}\text{X}$ and $^{14}_{7}\text{X}$ are isotopes of the element nitrogen (N). $^{14}_{6}\text{X}$ and $^{12}_{6}\text{X}$ are isotopes of the element carbon (C).

Which of the original five atoms have identical mass numbers?

$^{16}_{8}\text{X}$ and $^{16}_{7}\text{X}$ have the same mass number. The first is an isotope of oxygen, and the second is an isotope of nitrogen. $^{14}_{7}\text{X}$ and $^{14}_{6}\text{X}$ have the same mass number. The first is an isotope of nitrogen, and the second is an isotope of carbon.

Which of the original five atoms have the same number of neutrons?

$^{16}_{8}\text{X}$ (16 − 8 = 8 neutrons) and $^{14}_{6}\text{X}$ (14 − 6 = 8 neutrons) have the same number of neutrons.

A specific isotope can be designated by its nuclear symbol (for example, $^{12}_{6}\text{C}$) or by the element name followed by the mass number (for example, carbon-12).

The atoms of the carbon-12 isotope are defined as having a mass of 12 amu. The relative masses of all other atoms are determined by comparing them to this standard. An atom of the isotope $_1^1H$ has one-twelfth the mass of an atom of $_6^{12}C$; thus, atoms of $_1^1H$ have an atomic mass of 1 amu. Atoms of deuterium ($_1^2H$) have one-sixth the mass of atoms of $_6^{12}C$; the atomic mass of deuterium atoms is 2 amu.

The **atomic weight** of an element, as listed in the periodic table, is related to the atomic mass in the following way. The atomic weight is the average mass of the atoms in a representative sample of an element. Each atom in the sample has a particular atomic mass. For example, in a sample of the element bromine, about half the atoms have an atomic mass of 79 amu, and half have an atomic mass of 81 amu. The atomic weight of bromine is about 80 amu, the average of 79 amu and 81 amu. (The atomic weight of bromine is listed in the periodic table as 79.9 amu, indicating that there is slightly more of the bromine-79 isotope than of the bromine-81.) In another example, about three-quarters of the atoms of the element chlorine have an atomic mass of 35 amu, and about one-quarter have an atomic mass of 37 amu. The atomic weight of chlorine is 35.5 amu; the average mass in this case is much closer to that of the chlorine-35 isotope because there is more of that isotope in the sample (Figure 3.1).

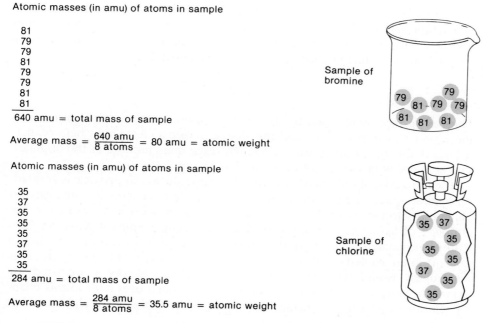

Atomic masses (in amu) of atoms in sample

81
79
79
81
79
79
81
81
___
640 amu = total mass of sample

Average mass = $\frac{640 \text{ amu}}{8 \text{ atoms}}$ = 80 amu = atomic weight

Sample of bromine

Atomic masses (in amu) of atoms in sample

35
37
35
35
35
37
35
35
___
284 amu = total mass of sample

Average mass = $\frac{284 \text{ amu}}{8 \text{ atoms}}$ = 35.5 amu = atomic weight

Sample of chlorine

FIGURE 3.1   Atomic weights of elements are averages of isotopic masses.

3.3 NATURAL RADIOACTIVITY

Some nuclei are unstable as they occur in nature. Radium atoms with a mass number of 226, for example, break down spontaneously, giving off **alpha** particles. Since alpha particles are identical to helium nuclei, this process may be summarized by the equation

$$^{226}_{88}\text{Ra} \longrightarrow \, ^{4}_{2}\text{He} + \, ^{222}_{86}\text{Rn}$$

The new element, with two fewer protons, is identified by its atomic number (86) as radon (Rn). Note that the mass number of the reactant must equal the sum of the mass numbers of the products. The same is true for the atomic numbers. The use of the symbol $^{4}_{2}\text{He}$ for the alpha particle is preferred to the symbol α because the former allows one to check the balance of mass and atomic numbers more readily.

Thorium nuclei are also unstable. The isotope with a mass number of 234 decomposes by **beta decay.** Since a beta particle is identical with an electron, this process may be written

$$^{234}_{90}\text{Th} \longrightarrow \, ^{0}_{-1}\text{e} + \, ^{234}_{91}\text{Pa}$$

The new element, with one more proton, is identified by its atomic number as protactinium (Pa). How can the original nucleus, which contains only protons and neutrons, emit an electron? One can envision one of the neutrons in the original nucleus changing into a proton and an electron.

$$^{1}_{0}\text{n} \longrightarrow \, ^{1}_{1}\text{p} + \, ^{0}_{-1}\text{e}$$

The new proton is retained by the nucleus (therefore the atomic number of the product increases by 1), and the almost massless electron or beta particle is kicked out (the product nucleus has the same mass number as the original). The beta decay of carbon-14 is pictured in Figure 3.4(a), p. 82.

The third type of radioactivity is called **gamma decay.** No particle is emitted; remember, gamma rays are related to light. Gamma emission involves no change in atomic number or mass number. The process is analogous to the emission of light from an atom. In that case, light is emitted when an electron in an excited state drops to a lower energy level. In gamma emission a nucleus in an excited state drops to a lower energy state. This type of decay is particularly useful when radioisotopes (radioactive isotopes) are used for diagnostic purposes in medicine. We will discuss such uses later in this chapter. Gamma emission frequently accompanies the emission of alpha or beta particles.

TABLE 3.2   *Radioactive Decay and Nuclear Change*

Type of Radiation	Symbol	Mass Number	Charge	Change in Mass Number	Change in Atomic Number
Alpha	α	4	2+	Decreases by 4	Decreases by 2
Beta	β	0	1−	No change	Increases by 1
Gamma	γ	0	0	No change	No change

The major types of radioactive decay and the ensuing nuclear changes are summarized in Table 3.2.

Plutonium-239 emits an alpha particle when it decays. What new element is formed?

example 3.4

$$^{239}_{94}\text{Pu} \longrightarrow {}^{4}_{2}\text{He} + ?$$

Mass and charge are conserved. The new element must have a mass of $239 - 4 = 235$ and a charge of $94 - 2 = 92$. The nuclear charge (atomic number) of 92 identifies the element as uranium (U).

$$^{239}_{94}\text{Pu} \longrightarrow {}^{4}_{2}\text{He} + {}^{235}_{92}\text{U}$$

Protactinium-234 undergoes beta decay. What new element is formed?

example 3.5

$$^{234}_{91}\text{Pa} \longrightarrow {}_{1}{}^{0}_{-}\text{e} + ?$$

The new element still has a mass number of 234. It must have a nuclear charge of 92 in order for the charge to be the same on each side of the equation. The nuclear charge identifies the new atom as another isotope of uranium (U).

$$^{234}_{91}\text{Pa} \longrightarrow {}_{1}{}^{0}_{-}\text{e} + {}^{234}_{92}\text{U}$$

3.4 HALF-LIFE

Thus far we have discussed radioactivity as applied to single atoms. In the laboratory one generally deals with great numbers of atoms—numbers far larger than the number of people on Earth. If we could see the nucleus of an individual atom, we could note its composition and tell *whether* it would undergo radioactive decay. Previous observations have indicated that certain combinations of protons and neutrons are unstable. We could not, however, determine *when* the atom would undergo a change. Radioactivity is a random process, a process generally independent of outside influence.

With a large number of atoms, the process becomes more predictable. The situation is similar to that encountered by life insurance companies. It is impossible for such companies to predict how long an individual will live and the precise date of death. However, when dealing with larger populations, these same companies can predict with great accuracy how many people will die within a particular period. Because of this ability, they can profitably maintain their business. The same sort of thing can be done with large numbers of radioactive atoms. And even 1 microgram (μg) contains an enormous number of atoms (perhaps a thousand million million atoms).

It is possible to characterize a radioactive isotope by a quantity called its **half-life.** The half-life of a radioactive isotope is that period of time in which one-half of the radioactive atoms present undergo decay. Suppose, for example, that one had 16 billions atoms of the radioactive iodine isotope, $^{131}_{53}$I. The half-life of this isotope is 8 days. This means that in 8 days one-half, or 8 billion, of the atoms will have undergone radioactive decay. In another 8 days, half the remaining 8 billion atoms will have decayed. That is, after two half-lives, 4 billion atoms (one-quarter of the original atoms) would remain unchanged. Two half-lives do not make a whole! This concept of half-life is shown graphically in Figure 3.2.

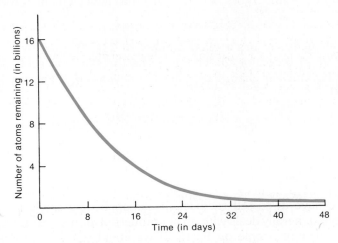

FIGURE 3.2   The radioactive decay of $^{131}_{53}$I.

What fraction of radioactive atoms remains after five half-lives?

example
3.6

After one half-life, $\frac{1}{2}$ of the original atoms remain. After two half-lives, $\frac{1}{2} \times \frac{1}{2}$ ($\frac{1}{2}$ of $\frac{1}{2}$) $= \frac{1}{4}$ of the original atoms remain. After three half-lives, $\frac{1}{2} \times \frac{1}{2} \times \frac{1}{2} = 1/(2^3) = \frac{1}{8}$ of the original atoms remain. After four half-lives, $\frac{1}{2} \times \frac{1}{2} \times \frac{1}{2} \times \frac{1}{2} = 1/(2^4) = \frac{1}{16}$ of the original atoms remain. After five half-lives, $\frac{1}{2} \times \frac{1}{2} \times \frac{1}{2} \times \frac{1}{2} \times \frac{1}{2} = 1/(2^5) = \frac{1}{32}$ of the original atoms remain. To determine what fraction of the original sample remains after a certain number *(n)* of half-lives, calculate the following value.

$$\frac{1}{2^n}$$

If you started with 16 million radioactive atoms, how many would you have left after four half-lives?

example
3.7

Since $n = 4$, the fraction remaining would be $1/(2^4)$, which is $\frac{1}{16}$, and $\frac{1}{16} \times$ 16 million atoms $= 1$ million. A million of the original atoms would be left.

It is impossible to say exactly when *all* the $^{131}_{53}\text{I}$ will have decayed. For practical purposes, one may assume that nearly all the radioactivity will be gone after about 10 half-lives. For the iodine sample considered here, 10 half-lives would be 80 days, at which time only a tiny fraction of the original atoms would still be present.

Much of the concern over nuclear power centers on the long half-lives of some isotopes that can be released in a nuclear accident or that are simply left as by-products of the normal operation of a nuclear reactor. In the first instance, the fear is that large areas could be rendered uninhabitable for thousands of years if an explosion released long-lived radioisotopes into the atmosphere. Even with no accidents, normal operation of a reactor produces nuclear wastes, which must be safely stored for thousands of years.

The half-life of an isotope may be very long, as with $^{238}_{92}\text{U}$ (whose half-life is 4.5 billion years), or very short, as with $^{9}_{5}\text{B}$ (whose half-life is 8×10^{-19} s).

3.5 ARTIFICIAL TRANSMUTATION

The forms of radioactivity encountered thus far occur in nature. Other nuclear reactions may be brought about by bombardment of stable nuclei with alpha particles, neutrons, or other subatomic particles. These particles, given sufficient energy, penetrate the formerly stable nucleus and bring about some form of radioactive emission. Like natural radioactivity, this sort of nuclear change brings about a **transmutation:** one element is changed into another. Because the change would not have occurred naturally, the process is called **artificial transmutation.**

Ernest Rutherford, a few years after his famous gold foil experiment (Chapter 2), studied the bombardment of a variety of light elements with alpha particles. One such experiment, in which he bombarded nitrogen, resulted in the production of protons.

$$^{14}_{7}N + {}^{4}_{2}He \longrightarrow {}^{17}_{8}O + {}^{1}_{1}H$$

(Recall that the hydrogen nucleus is a proton; hence the alternative symbol ${}^{1}_{1}H$ for the proton.) Notice that the sum of the mass numbers on the left equals the sum of the mass numbers on the right. The atomic numbers are also balanced. Rutherford had postulated the existence of protons in nuclei in 1914. An experiment published in 1919 gave the first empirical verification of the existence of these fundamental particles. Eugen Goldstein had earlier produced protons in his discharge tube experiments (Chapter 2). He obtained these particles from hydrogen gas in the tube by knocking an electron away from the hydrogen atom. The significance of Rutherford's experiment lay in the fact that he obtained protons from the nucleus of an atom other than hydrogen, thus establishing their nature as fundamental particles. By **fundamental particles** we mean basic units from which more complicated structures (such as the nitrogen nucleus) can be fashioned. Rutherford's experiment was the first induced nuclear reaction.

A great many transmutations were performed during the 1920s. And in the 1930s one such reaction led to the discovery of another fundamental particle. James Chadwick, in 1932, bombarded beryllium with alpha particles.

$$^{9}_{4}Be + {}^{4}_{2}He \longrightarrow {}^{12}_{6}C + {}^{1}_{0}n$$

Among the products was the neutron.

example 3.8

When potassium-39 is bombarded with neutrons, chlorine-36 is produced. What other particle is emitted?

$$^{39}_{19}K + {}^{1}_{0}n \longrightarrow {}^{36}_{17}Cl + ?$$

To balance the equation, a particle with a mass of 4 and an atomic number of 2 is required. That's an alpha particle.

$$^{39}_{19}K + {}^{1}_{0}n \longrightarrow {}^{36}_{17}Cl + {}^{4}_{2}He$$

3.6 INDUCED RADIOACTIVITY

The first artificial nuclear reactions produced isotopes already known to occur in nature. This was perhaps fortuitous, for it was inevitable that an unstable nucleus would be produced sooner or later. Irène Curie (daughter of the 1903 Nobel Prize winners) and her husband, Frédéric Joliot (Figure 3.3), were studying the bombardment of aluminum with alpha particles. Neutrons were produced, leaving behind an isotope of phosphorus.

$$^{27}_{13}\text{Al} + {}^{4}_{2}\text{He} \longrightarrow {}^{30}_{15}\text{P} + {}^{1}_{0}\text{n}$$

Much to their surprise, the target continued to emit particles after the bombardment was halted. The isotope of phosphorus was radioactive, emitting particles equal in mass to the electron but opposite in charge. These particles are called **positrons.** The reaction they were observing is written

$$^{30}_{15}\text{P} \longrightarrow {}^{0}_{1+}\text{e} + {}^{30}_{14}\text{Si}$$

Once again the question arises: Where does this particle come from if the nucleus contains only protons and neutrons? Previously we accounted for a beta particle (an electron) popping out of a nucleus by saying a neutron changed into a proton and an electron. Perhaps a similar happening can account for

FIGURE 3.3   Frédéric and Irène Joloit-Curie discovered artificially induced radioactivity in 1934. (Reprinted with permission from Weeks, Mary E., Discovery of the Elements, *Easton, Pa.: Chemical Education Publishing, 1968. Copyright © 1968 by the Chemical Education Publishing Company.)*

the appearance of a positron. Imagine a proton changing to a neutron and a positron (a proton is the same as a hydrogen nucleus).

$$_1^1H \longrightarrow \, _{+1}^0e + \, _0^1n$$

Everything balances rather nicely in this equation. When the positron is emitted, the original radioactive nucleus suddenly has one less proton and one more neutron than before. Therefore, the mass of the product nucleus is the same, but its atomic number is one less than that of the original nucleus (Figure 3.4(b).

This work won the Joliot-Curies a Nobel Prize of their own in 1935. (The Joliot-Curies adopted the combined surname to perpetuate the Curie name. Marie and Pierre Curie had two daughters, but no son.)

example 3.9

Carbon-10, a radioactive isotope, emits a positron when it decays. Write an equation for this process.

$$_6^{10}C \longrightarrow \, _{+1}^0e + ?$$

To balance the equation, a particle with a mass of 10 and an atomic number of 5 (boron) is required.

$$_6^{10}C \longrightarrow \, _{+1}^0e + \, _5^{10}B$$

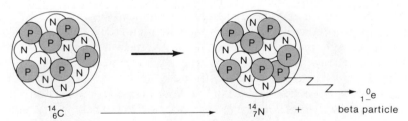

(a) Nuclear changes accompanying beta decay.

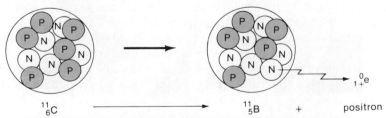

(b) Nuclear change accompanying positron emission.

FIGURE 3.4   Nuclear emission of (a) a beta particle, $_{-1}^0e$, and (b) a positron, $_{+1}^0e$.

3.7 MEASURING RADIATION: THE UNITS

There are many ways of measuring radiation. A chemist or physicist might wish to know how frequently nuclear disintegrations are occurring within a particular sample. Such nuclear activity is measured in **disintegrations per second** or in **curies** (Ci), named in honor of Marie Curie, the discoverer of radium.

$$1 \text{ Ci} = 3.7 \times 10^{10} \text{ disintegrations/second (the activity of 1 g of radium)}$$

A source used in the externally applied cobalt radiation treatment of cancer, for example, might be rated at 3000 Ci. For less active samples, metric prefixes are used.

$$1 \text{ millicurie (mCi)} = 0.001 \text{ Ci} = 3.7 \times 10^7 \text{ disintegrations/second}$$

A dose of radioactive isotope whose activity is 1 mCi might be taken internally by an adult about to undergo certain types of diagnostic scanning procedures.

$$1 \text{ microcurie } (\mu\text{Ci}) = 0.000\ 001 \text{ Ci} = 3.7 \times 10^4 \text{ disintegrations/second}$$

A child would be subjected to a smaller dose, perhaps 10 to 50 μCi for some kinds of diagnostic scanning. Obviously, a sample whose activity is measured in microcuries represents far less potential danger than a sample with activity measured in curies.

Most of us are more interested in our *exposure* to radiation than in simply counting disintegrations in some sample of radioactive material. The unit of exposure to gamma rays or X rays is the **roentgen** (R). Named after Wilhelm Roentgen, who discovered X rays in 1895, a roentgen is defined as the amount of gamma or X rays required to produce ions* carrying a total of 2.1 billion units of electrical charge (+ or −) in 1 cm³ of dry air at 0 °C and normal atmospheric pressure. The definition is a classic of scientific terminology, very precise but probably meaningless to most people. (If you read it over again, you will find in the definition about a half-dozen terms that were introduced in the first two chapters of this book.) The roentgen is a measure of the energy of the beam of rays, indicating how badly an air sample would be disrupted by a particular source of X rays or gamma rays. It is still not quite the unit that relates directly to us. The degree of ionization in tissue is not quite the same as that in air.

What we are really interested in is damage to tissue. The SI unit for absorbed dose of ionizing radiation is the **Gray,** named for Harold Gray, a British scientist. In the United States, a smaller unit, the rad, is frequently used.

$$1 \text{ Gray (gy)} = 100 \text{ rads}$$

---

* **Ions** are atoms with net electrical charges. Ions are discussed more fully in Chapter 4.

Rad is the acronym for radiation absorbed dose. An exposure to 1 rad means that each gram of absorbing tissue has absorbed 100 **ergs** of radiation energy.

$$1 \text{ erg} = 10^{-7} \text{ J} = 2.39 \times 10^{-8} \text{ cal}$$

$$1 \text{ rad} = 100 \text{ ergs absorbed} = 2.39 \times 10^{-6} \text{ cal absorbed}$$

In terms of the amount of heat energy absorbed, the rad is an extremely small unit (1 000 000 rads would be only 2.39 cal). The "absorption" of a diet cola provides perhaps a thousand times that much energy to the body. But it isn't the heat that is dangerous. It is the formation of highly reactive ions (hence the term **ionizing radiation**) and other molecular fragments within the cell that makes radiation so hazardous to us and other living things. In fact, it is estimated that about 500 rads would kill most of us. A single dose of 1000 rads would kill nearly any mammal. For comparison, 1 rad is about the average yearly radiation dose absorbed by an individual from medical and dental X rays.

There is one other unit for measuring radiation that is worth mentioning. This is the **rem,** which is an acronym for roentgen equivalent in man. It is a measure of the biological damage produced by a particular dose of radiation. Again, there is a precise definition for the rem, but it will probably be useful for us to define the term more generally. A value in rems tells you how the biological damage resulting from exposure to some radiation compares with the damage that would have resulted from a set standard. For example, if a dose of a particular radiation is rated at 4.5 rems, that means that the dose caused the same biological damage as would 4.5 rads of the standard radiation.

These terms are often confusing, particularly the roentgen, the rad, and the rem. It is not critical for you to know how many disintegrations occur in 30 min in a 2.7-Ci sample of radium. What you should know is that the number of curies tells you how active the sample is; the number of roentgens, how much energy a particular dose transfers to the air; the number of rads, how much energy a dose transfers to tissue; and the number of rems, how much biological damage the dose can accomplish. In all instances, it can be said that, the larger the number, the more potential danger the sample represents.

For our purposes, we may consider a roentgen and a rad to be about equal. When absorbed in muscle tissue, 1 R will generate about 1 rad of energy.

3.8 RADIATION DETECTORS

Generally, we can't see, hear, feel, taste, or smell radiation. Then how do we know that it is there? Most of us have seen or heard of **Geiger counters.** These are instruments that extend our senses so that we can see or hear the effects

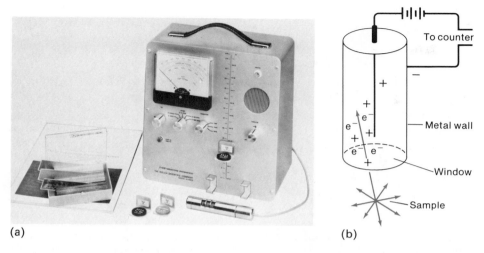

(a)   (b)

FIGURE 3.5   *The Geiger counter. (a) Photograph. (Photo courtesy of Sargent-Welch Scientific Company, Skokie, Ill.) (b) Schematic diagram.*

of radiation. A Geiger counter has a tube with a window at one end (Figure 3.5). The tube is filled with a gas at low pressure. A wire with a high positive potential (a large excess of positive charge) extends down the center of the tube. Radiation entering the tube will produce ions by knocking electrons off the gas molecules, and the electrons will be attracted to the positively charged wire in the center of the tube. Positive ions will move to the metal wall. A small electrical current, which can be amplified and used to produce a meter reading or a clicking sound or to cause a light to flash, will flow between the two electrodes. The meter can be calibrated in counts per minute or in other convenient units.

Another type of instrument, called a **scintillation counter,** makes use of the fact that tiny flashes of light are produced when radiation strikes certain materials called **phosphors.** (Zinc sulfide is a phosphor. This substance was used to detect radiation in early experiments carried out at the turn of the century.) These phosphors have the ability to convert the kinetic energy of a moving particle to visible light. A device called a **photomultiplier** converts the light output of a scintillator into a pulse of electricity that can be counted automatically.

Many variations of these basic radiation detectors are available for special purposes. Their functions, though, are the same in principle as the functions of the Geiger counter and the scintillation counter.

Individuals who work with radioactive materials wear personal detectors, simple film badges (Figure 3.6) worn on their pockets or at their waists or as rings. These are not sophisticated electronic devices; instead, they react to

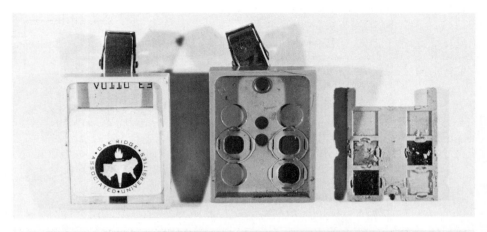

FIGURE 3.6 *Film badges are worn by people working around radioactive materials. (Photo courtesy of Oak Ridge Associated Universities, U.S. Department of Energy, Oak Ridge, Tenn., 1985.)*

radiation as film does to light. A certain dose of radiation will cause the film in the badge to become exposed, alerting the wearer to the potential danger. Individuals working in nuclear power plants or in hospital radiology laboratories use such devices.

3.9 INTO EACH LIFE SOME RADIATION . . .

Just sitting there you are getting zapped by all kinds of radiation. There is radiation from natural sources striking us at all times. This **background radiation** is made up of cosmic rays and of radiation from naturally occurring radioisotopes buried in the Earth's crust (or in materials derived from this source—the bricks of a house, for example). Background radiation varies from place to place. People who live at high altitudes receive more cosmic radiation. Since air absorbs some of the rays, those who live at low altitudes receive less radiation from this source.

About 50 of the 350 or so natural isotopes are radioactive. The most important of these are potassium-40 ($^{40}_{19}$K), thorium-232 ($^{232}_{90}$Th), and the isotopes of uranium. Gamma rays from potassium-40 and thorium-232 are more significant sources of *external* background radiation, that is, of radiation outside our bodies. Carbon-14 ($^{14}_{6}$C) and tritium ($^{3}_{1}$H), formed in the upper atmosphere by cosmic rays, contribute little to our external exposure, but they are important

TABLE 3.3   The Natural Background Radiation

Type	Source	Dose Rate (in millrads per year)	Varies With
External	Cosmic rays	30–60	Latitude and altitude
	Soil potassium-40, thorium, uranium	30–100	Location (mineral deposits) and dwelling (least in tents, greatest in stone buildings)
Internal	Thorium, uranium, decay products	40–400	Location and water supply
	Potassium-40	20	Not very variable
	Carbon-14	2	Not very variable
	Tritium	2	Not very variable
Total		100–600	

Adapted from Frigerio, Norman A., *Your Body and Radiation*, Oak Ridge, Tenn.: U.S. Department of Energy, 1967.

sources of *internal* background radiation. Carbon and hydrogen are important elements in our bodies, and the radioactive isotopes are incorporated in body tissue along with the more common nonradioactive isotopes. Potassium-40 contributes to internal background radiation also (Table 3.3).

We should comment a bit on one characteristic of the various kinds of radiation. It is termed **penetrating power.** All other things being equal, the more massive the particle, the less its penetrating power. Of alpha, beta, and gamma rays, alpha rays are the least penetrating. These are streams of helium nuclei, each particle with a mass number of 4. Beta rays are more penetrating than alpha rays. The electrons that make up the stream of beta particles are assigned a mass number of 0. These particles are not really massless, but they are very much lighter than alpha particles. Gamma rays are a form of high-energy radiation and are truly massless. These are the most penetrating of the three radiations.

It may seem contrary to common sense that the biggest particles make the least headway. Consider that penetrating power reflects the ability of the radiation to make its way through a sample of matter. It is as if you were trying to roll some rocks through a field of boulders (Figure 3.7). The alpha particle acts as if it were a boulder itself. Because of its size, it can't get very far before it bumps into and is stopped by the other boulders. The beta particle acts like a small stone. It can sneak between boulders and perhaps ricochet off one or another until it has made its way farther into the field. The gamma ray can be compared to an insect that can get through the smallest openings, and although it may brush against some of the boulders, it can, in general, make its way through most of the field without being stopped.

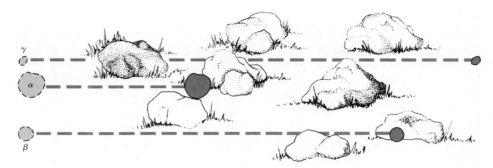

FIGURE 3.7   *Shooting radioactive particles through matter is like rolling rocks through a field of boulders—the larger rocks are more quickly stopped.*

At the beginning of this discussion, we said that, all other things being equal, this is how things worked. But all other things are not always equal. The faster a particle moves or the more energetic the radiation is, the more penetrating power it has. Cosmic rays with very high energy have fantastic penetrating power. Particles that would ordinarily be stopped by the skin can pass right through the body when traveling at the speeds attained by some cosmic ray particles.

Let's get back to particles of equal energy. Which kind does the most damage? In general, the big ones. The big ones don't travel very far, so they dump all of their energy into a small volume of tissue. The small ones have longer paths, and they release their energy all along this path. The effect is diluted over this larger area. Apparently the body can repair some low-level damage. But if a critical molecule comes under massive assault, it may be irreparably damaged, and the body will suffer a permanent effect.

Now we can return to our consideration of background radiation. Radiation from uranium and thorium penetrates air and skin so little that it presents no external threat. Uranium and thorium present serious problems, however, when swallowed or inhaled: they break down to other radioactive elements, including radium. Radium resembles calcium in its chemical properties (if you check the periodic table, you will find that both are Group IIA elements), and it readily replaces calcium in our teeth and bones. Indirectly, then, uranium and thorium contribute to our internal background radiation. The same lack of penetrating power that makes elements relatively innocuous on the outside makes them more significant on the inside. If their radiation is not penetrating, then all of it is trapped within the body, where it destroys vital components of our cells.

Failure to distinguish between damage done by a radiation source external to the body and one that is internal frequently leads to very different assessments of potential danger. In 1979, radioactive material was released during an emergency shutdown of a nuclear generator at Three Mile Island in Pennsylvania. The danger associated with this release was disputed by two different groups. On the one hand, it was argued that the people in neighboring communities were exposed to very little radiation, not much more than the normal background radiation (that which comes from natural sources like radioisotopes in the Earth's crust). On the opposite side of the argument was a group that maintained that the released radioactive isotopes would ultimately enter the body through food or inhaled air. The radioisotopes would then concentrate within the body, these people argued, and remain to do long-term damage to various organs. Only the future can tell us which group is correct in its assessment.

So you see that we are being constantly bombarded by radiation. In fact, we ourselves are radioactive. Since people have been exposed to background radiation throughout the course of human evolution, there is evidently little permanent damage to our bodies (or else, whatever damage is done we have come to accept as the normal condition of the body). Some are becoming concerned, though, about additional radiation exposure from medical sources, atmospheric testing of nuclear weapons, and nuclear power plants. Most of us get a maximum of 150 mrems of background radiation per year. We may get an additional 20 to 50 mrems from medical X rays. Fallout varies a great deal. France, China, and India conducted atmospheric tests of nuclear weapons in 1974. We probably get at most 1.5 mrems from these tests. Radiation released from nuclear power plants each year exposes us to about 0.85 mrems of radiation (less than 0.6% of background). If we increase our dependence on nuclear power, this last category could increase somewhat.

3.10 RADIATION AND LIVING THINGS

High doses of radiation (on the order of 2000 rads) cause gross destruction of tissue. A person exposed to about 2000 rads passes into shock and dies in a few hours. With exposures of 500 to 1000 rads, the person usually survives long enough to exhibit several phases of the acute radiation syndrome.

Phase 1: A short latent period (a few hours) when no effects are observed.
Phase 2: A period of nausea and vomiting (which usually ends in 24 hours) and a drop in white blood cell count.
Phase 3: A period of few symptoms. A low-grade fever may persist.
Phase 4: The last period. Loss of hair begins rather abruptly anywhere from about the 17th through the 21st day. Increased discomfort sets in with loss of appetite and diarrhea. The body temperature rises and the patient complains of pain in the throat and gums. Emaciation sets in, the general condition deteriorates, and the patient dies.

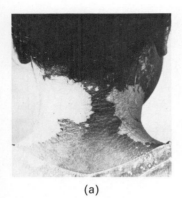

(a)

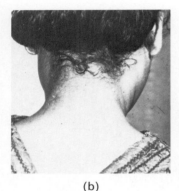

(b)

FIGURE 3.8
Burns from beta rays cause changes in skin pigmentation. (a) The skin 1 month after exposure. (b) The skin upon full recovery, 1 year after exposure. (Reprinted from Frigerio, Norman A., Your Body and Radiation, Oak Ridge, Tenn.: U.S. Department of Energy, 1967.)

TABLE 3.4 $LD_{50}/30$ days Values for Various Species

Organism	LD_{50} (in rems)
Dog	310
Pig	375
Human	400–450 (estimated)
Monkey	600
Rat	790
Yeast	10 000
Bacterium	100 000
Virus	1 000 000

Adapted from Williams, K., Smith, C. L., and Chalke, H. D., *Radiation and Health*, Boston: Little, Brown, 1962.

With smaller doses, recovery can take place (Figure 3.8). Such recovery may not be complete. Malignancy may show up, even years later. And genetic effects may appear in succeeding generations.

Some cells are more susceptible to radiation than others. Those that are being constantly and rapidly replaced are affected most. These include the blood cells and those organs responsible for producing blood cells (such as the bone marrow), the intestinal mucosa, germ cells, and embryonic cells. Damage to reproductive cells will show up as abnormalities in the descendants of affected persons.

The effect of radiation varies greatly from species to species. It may even vary a great deal among organisms within a species. To take this variability into account, research workers use a term known as the **$LD_{50}/30$** days. This indicates the dosage required to kill 50% of the individuals in a large group within 30 days. Primitive organisms such as viruses, bacteria, and yeasts have much greater ability to withstand radiation than do mammals (Table 3.4). These doses are whole-body exposures. Certain parts of the body can be exposed to much higher doses without death of the subject resulting. For example, while exposure of bone marrow to 500 to 1000 rads will lead to death, a 100 000-rad dose to the arm might be survived. This is not to say that no damage occurs. Indeed, it is very likely that the arm would have to be amputated. However, the functioning of the arm is not as critical to the overall health of the body as the functioning of the bone marrow.

3.11 RADIATION AND CANCER

Cancer is not one disease but many. Some forms are particulary susceptible to radiation therapy. Radiation is carefully aimed at the cancerous tissue, and exposure of normal cells is minimized. If the cancer cells are killed by the destructive effects of the radiation, the malignancy is halted. But persons undergoing radiation therapy often get quite sick from the treatment. Nausea and vomiting are the usual symptoms of radiation sickness. (Remember that intestinal mucosa is particularly susceptible to radiation.) Thus, the aim of radiation therapy is to destroy the cancerous cells before too much damage is done to the healthy cells. Radiation is most lethal to rapidly reproducing cells, and this is precisely the characteristic of cancer cells that allows the therapy to be successfully applied.

For many years radium salts were used for the radiation treatment of cancer. Radium-226 is an alpha and gamma emitter.

$$^{226}_{88}\text{Ra} \longrightarrow \, ^{222}_{86}\text{Rn} + \, ^{4}_{2}\text{He} + \, ^{0}_{0}\gamma$$

The radon (Rn) product is a radioactive gas. To prevent its escape, the radium was sealed in tiny hollow needles made of gold or platinum. These could be inserted directly into a tumor and left there to irradiate the surrounding tumor tissue until the desired dosage had been administered. Unfortunately, the needles were so small that they were sometimes lost. Frantic efforts to find them were not always successful.

During recent years, cobalt-60 has come into widespread use. It is easily made by neutron bombardment of ordinary cobalt-59 in certain facilities that specialize in the synthesis of radioisotopes.

$$^{59}_{27}\text{Co} + \, ^{1}_{0}\text{n} \longrightarrow \, ^{60}_{27}\text{Co}$$

The cobalt-60 emits beta rays and strong gamma rays, but in medical practice the betas are screened out. Cobalt-60 teletherapy units with intensities of over 1000 Ci are now quite common (Figure 3.9). The radioactive source has a limited life. The half-life of cobalt-60 is 5.3 years. Over a period of time the source will weaken beyond useful limits and have to be replaced. Nevertheless, cobalt-60 is cheaper than radium. Its shorter half-life also means that a given radiation dosage can be achieved with a smaller sample.

Radiation therapy is not without its costs. In addition to sickness, the radiation can cause mutations. Occupational exposure of health professionals may lead to increased birth defects in their children. Indeed, radiation is known to cause some forms of leukemia, a disease of the blood-forming organs.

People working with radioactive materials can do several things to protect themselves. The simplest is to move away from the source, for intensity of radiation decreases as the square of the distance increases. If the intensity at 1 m is 100 units, it will be only 25 units at 2 m ($1/(2^2)$ or $\frac{1}{4}$ as much). At 3 m, the intensity will be only 11 units ($1/(3^2)$ or $\frac{1}{9}$ as much), and at 10 m, the radiation intensity will be only 1 unit ($1/(10^2)$ or $\frac{1}{100}$ as much).

Workers can also be protected by shielding. A sheet of paper will stop most alpha particles. A block of wood or a thin sheet of aluminum will stop beta particles. However, it takes several meters of concrete or several centimeters of lead to stop gamma rays.

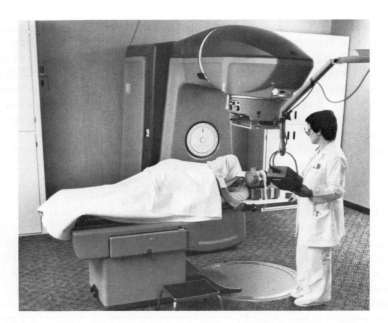

FIGURE 3.9 *A cobalt-60 unit for radiation therapy. (Photo by Raymond C. Carballada, Department of Medical Photography, Geisinger Medical Center.)*

You may recall Rutherford's experiment that led to his proposing the nuclear atom. He fired alpha particles at a sheet of gold foil. Most went straight through. Wait a minute, you say. How can this be so if a sheet of plain old paper will stop most alpha particles? A piece of gold foil about 2000 atoms thick, believe it or not, is very, very thin (about 0.002 mm). An ordinary sheet of paper is perhaps a hundred times as thick as that and will stop all but the most energetic alpha particles.

Hospital patients who have ingested therapeutic doses of radioisotopes (but not those who have been treated with an external source) must themselves be regarded as sources of radiation while the radioisotope maintains significant activity. Health personnel who can expect to be exposed to many such patients during their careers must exercise caution in order to avoid exposing themselves to a damaging dose of radiation over a long period.

3.12 NUCLEAR SLEUTHS: RADIOACTIVE TRACERS

Scientists in a wide variety of fields make use of radioisotopes as **tracers** in physical, chemical, and biological systems. Isotopes of a given element, whether radioactive or not, behave nearly identically in chemical and physical processes.

Since radioactive isotopes are easily detected, it is relatively simple to trace their movement, even through a complicated system.

As a simple example, let us consider the flow of a liquid through a pipe. Suppose a pipe buried beneath a concrete floor sprang a leak. One could locate the leak by digging up extensive areas of the floor. Or one could add a small amount of radioactive material and trace the flow of liquid with a Geiger counter. Once the leak had been located, only a small area of the floor would need to be dug up for the leak to be repaired. Short-lived isotopes, which disappear soon after doing their job, are usually employed for such purposes.

In a similar manner, one could trace the uptake of phosphorus by a plant. The plant might be fed some fertilizer containing radioactive phosphorus. A simple method of detection would involve placing the plant on photographic film. Radiation from the phosphorus isotopes would expose the film, much as light would. (The film badges previously mentioned in connection with the detection of radiation are based on this principle. In fact, the original discovery of radioactivity by Becquerel resulted from an unexpectedly exposed film.) The picture produced by the radioactive plant would be called an **autoradiograph,** and it would show the distribution of phosphorus in the plant (Figure 3.10).

FIGURE 3.10
Autoradiograph showing the uptake of phosphorus in a plant. (Courtesy of U.S. Department of Energy.)

Radioisotopes can also be used to trace the pathways of complicated chemical and biochemical reactions. Perhaps their most important applications, however, have been in the field of medicine.

3.13 NUCLEAR TOOLS IN MEDICAL DIAGNOSIS

We have spoken of the therapeutic uses of radioisotopes, uses intended to treat or cure a disease. Radioisotopes are also used for diagnostic purposes, to help provide information about the type or extent of illness.

Radioactive iodine-131 ($^{131}_{53}$I) is used to determine the size, shape, and activity of the thyroid gland as well as to treat cancer located in this gland and to control a hyperactive thyroid. One merely needs to drink some potassium iodide containing iodine-131. The body concentrates iodide in the thyroid. Large doses are used for treatment of thyroid cancer; the radiation from the isotope concentrates in the thyroid cancer cells even if the cancer has spread to other parts of the body. For diagnostic purposes, however, only a small amount is needed. Again the material is concentrated in the thyroid. A detector is set up so that readings are translated into a permanent visual record showing the differential uptake of the isotope. The "picture" that results is referred to as a **photoscan** (Figure 3.11).

Technetium-99^{m} is used in a variety of diagnostic tests. The *m* stands for *metastable*, which means that this isotope will give up some energy to become a more stable version of the same isotope (same atomic number, same atomic weight). The energy it gives up is the gamma ray needed to detect the isotope.

$$^{99}_{43}\text{Tc}^{m} \longrightarrow {}^{99}_{43}\text{Tc} + \gamma$$

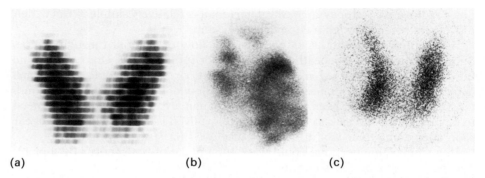

(a)   (b)   (c)

FIGURE 3.11   A linear photoscanner produced these pictures of (a) a normal thyroid, (b) a multinodal goiter, and (c) a thyroid adenoma. (Photos by Joseph J. Mentrikoski, Department of Medical Photography, Geisinger Medical Center.)

Notice that the decay of technetium-99^m produces no alpha or beta particles, which could cause unnecessary damage to the body. Technetium-99^m also has a short half-life (about 6 hr), which means that the radioactivity does not linger in the body long after the scan has been completed. With this short a half-life, use of the isotope must be carefully planned. In fact, the isotope

TABLE 3.5   Some Radioisotopes and Their Application in Medicine

Isotope	Name	Radiation	Uses
^{51}Cr	Chromium-51	γ	Determination of volume of red blood and total blood volume
^{57}Co	Cobalt-57	γ	Determination of uptake of vitamin B_{12}
^{60}Co	Cobalt-60	β, γ	Radiation treatment of cancer
^{153}Gd	Gadolinium-153	γ	Determination of bone density
^{131}I	Iodine-131	β, γ	Detection of thyroid malfunction; measurement of liver activity and fat metabolism; treatment of thyroid cancer
^{59}Fe	Iron-59	β, γ	Measurement of rate of formation and lifetime of red blood cells
^{32}P	Phosphorus-32	β	Detection of skin cancer or cancer of tissue exposed by surgery
^{226}Ra	Radium-226	α, γ	Radiation therapy for cancer
^{24}Na	Sodium-24	β, γ	Detection of constrictions and obstructions in the circulatory system
^{99}Tc^m	Technetium-99^m	γ	Imaging of brain, thyroid, liver, kidney, lung, and cardiovascular system
^{3}H	Tritium	β	Determination of total body water

itself is not what is purchased. Technetium-99^m is formed by the decay of molybdenum-99:

$$_{42}^{99}\text{Mo} \longrightarrow {}_{43}^{99}\text{Tc}^m + {}_{-1}^{0}\text{e} + \gamma$$

A container of this molybdenum isotope is obtained, and the decay product, technetium-99^m, is "milked" from the container as needed.

The radioisotope most widely used in medicine is gadolinium-153. This isotope is used to determine bone mineralization. Its popularity is an indication of the large number of people, mainly women, who suffer from osteoporosis (reduction in the quantity of bone) as they grow older. Gadolinium-153 gives off two characteristic radiations, a gamma ray and an X ray. A scanning device compares these radiations after they pass through bone. Bone densities are then determined by differences in absorption of the rays.

Table 3.5 lists a variety of radioisotopes used in medicine.

3.14 MEDICAL IMAGING

Medical imaging provides a means of looking at internal organs without resorting to surgery. The history of medical imaging dates back to the discovery of X rays at the turn of the century. As noted earlier in this chapter, the penetrating X rays were almost immediately used to visualize skeletal structure. Radiation from an external X-ray source passes through the body (except where it is absorbed by more dense structures, such as bone) and exposes a film, thus providing a picture that distinguishes the more dense structures from less dense tissue. Softer tissue can be visualized by introducing material that absorbs X rays into the area to be studied. For example, compounds of barium are used to visualize portions of the digestive tract (Figure 3.12).

As indicated in the previous section, radioisotopes can also be used to visualize internal organs. In this technique, the source of the radiation is inside the body, and the radiation (usually gamma rays) is detected as it emerges from the body.

Recently both X-ray technology and nuclear imaging have been coupled with computer technology to provide versatile and powerful imaging techniques. In computed tomography (referred to as CT or, sometimes, CAT scanning), many X-ray readings are obtained, processed by a computer, and then displayed. The resulting pictures present cross-sectional slices of a portion of the body. A series of these pictures gives a three-dimensional view of organs such as the brain.

Positron emission tomography (PET) can be used to measure dynamic processes occurring in the body, such as blood flow or the rate at which oxygen or glucose is being metabolized. PET scans are being used to pinpoint the area of brain damage that triggers severe epileptic seizures. Compounds incorporating positron-emitting isotopes, such as carbon-11, are inhaled or injected prior to

FIGURE 3.12
Barium sulfate, BaSO$_4$, is insoluble in water and opaque to X rays. When swallowed, this salt can be used to outline the stomach for X-ray photographs. (Courtesy of B. Levin, Michael Reese Hospital and Medical Center, Chicago, Ill.)

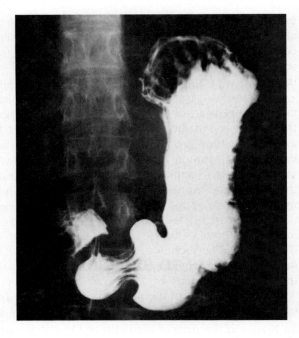

the scan. Before the emitted positron can travel very far in the body, it encounters an electron (in any ordinary matter there are numerous electrons), and two gamma rays are produced.

$$^{11}_{6}C \longrightarrow {}^{11}_{5}B + {}^{0}_{1+}e$$
$$_{1+}^{0}e + {}^{0}_{1-}e \longrightarrow 2\gamma$$

These exit from the body in exactly opposite directions. Detectors positioned on opposite sides of the patient record the gamma rays. If the recorders are set so that two simultaneous gamma rays must be "seen," gamma rays resulting from natural background radiation are ignored. A computer is then used to calculate the point within the body at which the annihilation of the positron and electron occurs, and an image of that area is produced.

Both X rays and nuclear radiation are ionizing radiations, which means there is always some tissue damage involved. Modern techniques keep this damage to an absolute minimum. Other imaging techniques use nonionizing radiation.

Most of you are familiar with ultrasonography. In this technique, high frequency sound waves are bounced off tissue, and the echo of the sound wave is recorded. Once again, a computer is used to process the data and produce an image of the tissue. (The technique is related to sonar detection of submarines.) Because no ionizing radiation is involved, ultrasonography has been extensively used in obstetrics for following fetal development. A 1984 conference on the use of ultrasonography (sponsored by the National Institutes of Health) concluded, however, that even this technique should not be used casually. The conference recommended that an ultrasonogram be obtained only if there is a sound medical reason for doing so (for example, to evaluate fetal growth in mothers suffering from diabetes or hypertension).

The newest imaging technique is magnetic resonance imaging (MRI). No radioactivity is involved, but the technique depends on a property of nuclei that we have not discussed before. Some nuclei behave as if they were little magnets. If they are placed in a strong magnetic field (for example, between the poles of a much more powerful magnet), the nuclei line up in a certain manner. When supplied with the right amount of energy, the nuclei absorb the energy and flip over in the magnetic field. The energy required to flip the nuclei is provided by nonionizing radiation. The absorption of the energy by the nuclei can be detected by appropriate equipment, and an image of the tissue in which the nuclei reside is produced. (Again, a computer is used to process the image.) The technique is still largely experimental but has demonstrated its potential for providing not only images of organs but also information about the metabolic activity in particular tissues.

The cost of most of the newer, computer-based technologies is high, ranging into the millions of dollars for a single installation (Figure 3.13) and thousands

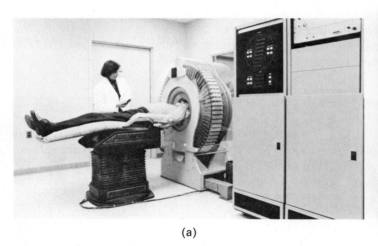
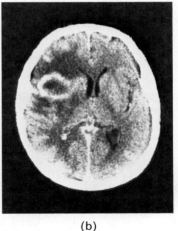

(a)   (b)

FIGURE 3.13   Modern computer technology used for medical diagnosis. (a) Patient in position for positron emission tomography (PET), a technique that uses radioisotopes to scan internal organs. (Courtesy of Brookhaven National Laboratory and New York University Medical Center.) (b) Image of section of brain created by computed tomography (CT), a scanning technique that uses X rays rather than radioisotopes, shows tumor in left temporoparietal area (white circle). (Courtesy of the National Institute of Neurological and Communicative Disorders and Stroke.)

of dollars for a single scan. The great advantage of the techniques is that they provide information that could otherwise be obtained only by subjecting the individual to the risks of surgery. In some instances, the information provided by the newer techniques is unavailable by any other route.

3.15 RADIOISOTOPES AND ANIMALS AND VEGETABLES

Radioisotopes are also used in veterinary medicine for both therapeutic and diagnostic purposes. Figure 3.14 shows one example.

The irradiation of foodstuffs as a method of preservation is being intensively studied. Such methods can be used to prevent food spoilage because they destroy microorganisms that cause such spoilage. The food so treated shows little change in taste or appearance (Figure 3.15). The process is used at present in several countries. In the United States, radiation is used only to preserve spices. The finding of residues of chemical fumigants in foods in 1984 spurred increased interest in radiation as a method of food preservation. Extensive testing is currently being carried out to establish that the products of such treatment are not harmful to human beings.

In the meantime, radioactive tracers are put to good use in agricultural research. They are used to study the effectiveness of fertilizers and weed killers, to compare the nutritional value of various feeds, and to determine the best methods for controlling insects. The purposeful mutation of plants by irradiation has produced new and improved strains of commercially valuable crop plants ranging from tobacco to peanuts.

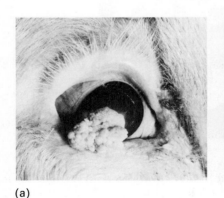
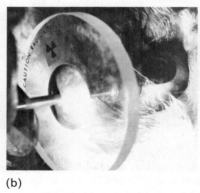
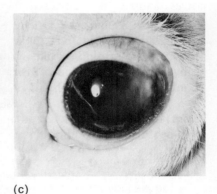

(a)   (b)   (c)

FIGURE 3.14   *Treatment of eye cancer in a cow. (a) The eye after the outer portion of the cancer has been removed surgically. (b) The inner portion of the cancer being treated with beta radiation from strontium-90. (c) The eye after successful treatment. (Reprinted from Frigerio, Norman A.,* Your Body and Radiation, *Oak Ridge, Tenn.: U.S. Department of Energy, 1967.)*

FIGURE 3.15
Gamma radiation prevents
sprouting in potatoes. Both
the potatoes were stored for
16 months. The bottom one
was irradiated; the top one
was not. (Reprinted from
Pizer, Vernon, Preserving
Food With Atomic Energy
Oak Ridge, Tenn.: U.S.
Department of Energy, 1970.)

3.16 THE NUCLEAR AGE

As we have seen, nuclear chemistry is extremely important in medicine. It has also made substantial contributions to industry and agriculture. Hardly a single facet of our lives has been left untouched by developments in nuclear science. In our daily lives, however, we don't hear or read much about nuclear medicine. Rather, the news is filled with stories about nuclear bombs and nuclear power plants. These applications of nuclear science are discussed in Special Topic C.

PROBLEMS

1. Define each of the following terms:
   a. isotope
   b. deuterium
   c. alpha particle
   d. beta particle
   e. gamma ray
   f. half-life
   g. positron
   h. curie (Ci)
   i. roentgen (R)
   j. rad
   k. rem
   l. Geiger counter
   m. scintillation counter
   n. background radiation
   o. LD_{50}/30 days
   p. radioisotope
   q. radioactive tracer
   r. artificial transmutation

2. Draw the nuclear symbols for protium, deuterium, and tritium (hydrogen-1, hydrogen-2, and hydrogen-3, respectively).

3. Draw nuclear symbols for the following isotopes. You may refer to the periodic table.
   a. an element with a mass number of 8 and an atomic number of 5
   b. an element with $Z = 35$ and $A = 83$
   c. an element with 53 protons and 72 neutrons

4. Indicate the number of protons and neutrons in atoms of the following isotopes. You may refer to the periodic table.
   a. $^{62}_{30}Zn$
   b. $^{241}_{94}Pu$
   c. $^{99m}_{43}Tc$
   d. $^{81m}_{36}Kr$
   e. gallium-69
   f. molybdenum-98
   g. molybdenum-99
   h. technetium-98

5. Which of the following sets represents isotopes?
   a. $^{70}_{34}X$, $^{70}_{33}X$ **c.** $^{186}_{74}X$, $^{186}_{74}X$ **e.** $^{22}_{11}X$, $^{44}_{22}X$
   b. $^{57}_{28}X$, $^{66}_{28}X$ **d.** $^{8}_{2}X$, $^{6}_{4}X$

6. The two principal isotopes of lithium are lithium-6 and lithium-7. The atomic weight of lithium is 6.9 amu. Which is the predominant isotope of lithium?

7. Out of every five atoms of boron, one has a mass of 10 and four have a mass of 11. What is the atomic weight of boron? Use the periodic table only to check your answer.

8. How are X rays and gamma rays similar? How are they different?

9. In each case, describe the changes in atomic mass and atomic number that occur when a nucleus emits each of the following:
   a. a beta particle   **d.** a proton
   b. an alpha particle   **e.** a positron
   c. a neutron   **f.** a gamma ray

10. Complete the following equations:
    a. $^{179}_{79}Au \longrightarrow {}^{175}_{77}Ir + ?$
    b. $^{23}_{10}Ne \longrightarrow {}^{23}_{11}Na + ?$
    c. $^{10}_{5}B + {}^{1}_{0}n \longrightarrow {}^{4}_{2}He + ?$
    d. $^{12}_{6}C + {}^{2}_{1}H \longrightarrow {}^{13}_{6}C + ?$
    e. $^{121}_{51}Sb + ? \longrightarrow {}^{121}_{52}Te + {}^{1}_{0}n$
    f. $^{154}_{62}Sm + {}^{1}_{0}n \longrightarrow 2\,{}^{1}_{0}n + ?$

11. When magnesium-24 is bombarded with a neutron, a proton is ejected. What new element is formed? (Hint: Write a balanced nuclear equation.)

12. A radioactive isotope decays to give an alpha particle and bismuth-211. What was the original element?

13. Lead-209 undergoes beta decay. Write a balanced equation for this reaction.

14. C. E. Bemis and colleagues at Oak Ridge National Laboratory confirmed the synthesis of element 104, the half-life of which was only 4.5 s. Only 3000 atoms of the element were created in the tests. How many atoms were left after 4.5 s? After a total of 9.0 s?

15. Krypton-81^m is used for lung ventilation studies. Its half-life is 13 s. How long will it take the activity of this isotope to reach one-quarter of its original value?

16. Explain how radioisotopes can be used for therapeutic purposes.

17. Describe the use of a radioisotope as a diagnostic tool in medicine.

18. The activity of a radiation source is 500 Ci. The activity of another source is 10 mCi. To be used properly, one source is taken into the body and the other remains outside the body during treatment. Which is likely to be the internal source and which the external?

19. An iodine-131 sample with an activity of 150 mCi is taken internally by patient A. Patient B is given an internal dose of 15 μCi. In which patient was the iodine-131 being used to treat a malignancy, and in which patient was the isotope used for imaging the thyroid gland?

20. About 2 mCi of thallium-201 is given by intravenous administration for imaging the heart. It is estimated that the total body radiation dose in humans is about 0.07 rad per mCi of thallium-201. How does the radiation dose from this procedure compare to the lethal dose for humans?

21. List two ways in which workers can protect themselves from the radioactive materials with which they work.

22. What are some of the characteristics that make technetium-99^m such a useful radioisotope for diagnostic purposes?

23. Plutonium is especially hazardous when inhaled or ingested because it emits alpha particles. Why would alpha particles cause more damage to tissue than beta particles?

24. What form of radiation is detected in CT (CAT) scans? What form is detected in PET scans?

25. What is the advantage of using nonionizing radiation for medical imaging? Name two imaging techniques that do not use ionizing radiation.

26. Radium-223 nuclei usually decay by alpha emission. For every billion alpha decays, one atom emits a carbon-14 nucleus. Write a balanced nuclear equation for each type of emission.

27. To make element 106, a 0.25-mg sample of $^{249}_{98}Cf$ was used as the target. Four neutrons were emitted to yield a nucleus with 106 protons and a mass of 263 amu. What was the bombarding particle?

$$^{249}_{98}Cf + ? \longrightarrow 4\,{}^{1}_{0}n + {}^{263}_{106}X$$

28. One atom of element 109 with a mass number of 266 was produced in 1982 by bombarding a target

of bismuth-209 with iron-58 nuclei for 1 week. How many neutrons were released in the process?

$$^{209}_{83}Bi + {}^{58}_{26}Fe \longrightarrow {}^{266}_{109}X + ? \, {}^{1}_{0}n$$

29. Element 109 undergoes alpha emission to form element 107, which in turn also emits an alpha particle. What is the atomic number and mass number of the isotope formed by these two steps? Write balanced nuclear equations for the two reactions.
30. Describe how a film badge works.
31. What is ionizing radiation?
32. A patient is given a 128-μCi dose of iodine 131, which has a half-life of 8 days. How much remains in her body after 32 days?
33. Tritium (hydrogen-3) is a beta emitter. What new isotope is formed?
34. When nitrogen-14 is bombarded with alpha particles, oxygen-17 is formed. What other product is formed?
35. When aluminum-27 is bombarded with alpha particles, a neutron is ejected. What new isotope is formed?
36. How can an atomic nucleus emit a beta particle (an electron) when there are no electrons in the nucleus?
37. A radioisotope placed near a Geiger counter gives a reading of 80 counts per second. Eight hours later, the reading is 10 counts per second. What is the half-life of the isotope?
38. A patient is given a 128-μCi dosage of iodine-123, which has a half-life of 13 hr. How much will remain in her body after 26 hr? After 52 hr? After 104 hr?
39. Why is iodine-123, with a half-life of 13 hr, more desirable for diagnostic work than iodine-131, which has a half-life of 8 days (192 hr)?
40. Why aren't alpha emitters used in diagnostic work?
41. What is the main source of radiation to which people in developed countries are exposed?

C

Nuclear Power

In Chapter 3, we discussed the nature of the atomic nucleus, a variety of nuclear reactions, and some applications of radioactive isotopes. Now let's turn our attention to some additional nuclear reactions—ones that unleash vast quantities of energy. This energy can be released almost instantaneously, as in nuclear bombs, or in a controlled fashion in a nuclear power plant. In either case, new radioactive materials are formed that can cause a variety of problems. This special topic will focus on some of the processes of nuclear energy and on some of the problems that arise.

C.1 EINSTEIN'S EQUATION

Let us return to our study of the history of radioactivity. Perhaps the most noted development of our age was a theoretical one, worked out with a pencil and notepad. These are not the tools one usually associates with a scientist. Albert Einstein (Figure C.1) may well be the best-known scientist of all time, yet his achievements are those of the mind, not the laboratory.

By 1905, Einstein had worked out his special theory of relativity. In doing this he derived a relationship between matter and energy. The now-famous equation is written

$$E = mc^2$$

where E represents energy, m represents mass, and c is the speed of light. According to Einstein, energy and mass are different aspects of the same thing.

FIGURE C.1
Albert Einstein during the period of his most important discoveries. (Photo by Martin Hohlig. Courtesy Vanity Fair. *Copyright © 1923 (renewed 1951) by The Condé Nast Publications Inc.)*

The energy equivalent of 1 g of matter is 9×10^{13} J, an enormous amount of energy—enough to heat the average home for a thousand years.

Einstein's reasoning was not verified until 40 years later. The verification shook the world.

C.2 NUCLEAR FISSION

FIGURE C.2
Enrico Fermi. (Courtesy of Mrs. Laura Fermi.)

A group of physicists headed by the Italian Enrico Fermi (Figure C.2) was first to study the bombardment of uranium with neutrons, but they failed to interpret their work properly. The German chemists Otto Hahn and Fritz Strassman first correctly interpreted the experiments after discovering that the reaction products were much lighter elements than uranium and included barium (Ba), lanthanum (La), and cerium (Ce). The uranium atom had been split! In this reaction, it wasn't a matter of a small piece (an alpha or a beta particle, for example) being chipped from the original nucleus. Here the nucleus was cleaved, split into two major fragments. Lise Meitner (Figure C.3) and Otto Frisch calculated the energy associated with the **fission** of uranium and found it to be several times greater than that of any previously known nuclear reac-

FIGURE C.3   *Lise Meitner and Otto Hahn in Hahn's laboratory.*

tion. In addition, the splitting of the nucleus into two major fragments was accompanied by the release of some smaller fragments, more neutrons. The neutrons thus produced could split other uranium atoms, yielding enormous amounts of energy in a **chain reaction** (Figure C.4).

In 1938, Nazi Germany invaded and annexed Austria. Lise Meitner, an Austrian Jew, fled to Sweden. There she released the news of these momentous discoveries. This news was carried to the United States by Niels Bohr, the Danish physicist who is so well known for his quantum theory of the electron structure of atoms. Fermi, who had fled Italy to the United States because his wife was Jewish, prevailed on Einstein to sign a letter to President Franklin D. Roosevelt indicating the importance of the discovery.

The United States government launched a massive research project for the study of atomic energy and designated it the Manhattan Project. Uranium had to be collected and the isotopes separated, for only the relatively rare uranium-235 isotope is fissionable. The neutrons, it was found, had to be slowed by graphite to increase the probability of their hitting a uranium nucleus.

Fermi and his group achieved the first sustained nuclear reaction on 2 December 1942 under the bleachers at Stagg Field of the University of Chicago.

Separation of the uranium-235 isotope from the more abundant uranium-238 proceeded very slowly at a top secret installation at Oak Ridge, Tennessee.

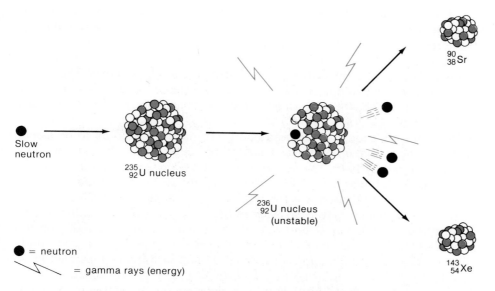

Slow neutron

$^{235}_{92}$U nucleus

$^{236}_{92}$U nucleus (unstable)

$^{90}_{38}$Sr

$^{143}_{54}$Xe

● = neutron

⋀⋀⋀ = gamma rays (energy)

FIGURE C.4 *The splitting of a uranium atom. The neutrons produced in this fission can split other uranium atoms, thus sustaining a chain reaction. The splitting of one uranium-235 atom yields 8.9 × 10^{-18} kwh of energy. Fission of a mole of uranium-235 (6.02 × 10^{23} atoms) produces 5 300 000 kwh of energy.*

This separation could not be done by chemical reaction, for the isotopes behave identically chemically. Separation was accomplished by conversion of all the uranium to volatile uranium hexafluoride. Molecules of the latter containing the uranium-235 isotope are slightly lighter and move slightly more rapidly than molecules containing the uranium-238 isotope. The vapors of uranium hexafluoride were allowed to pass through thousands of consecutive pinholes, a process in which the molecules that contained uranium-235 gradually outdistanced the others. The scientists eventually obtained 15 kg of the separated uranium-235 isotope, enough to make a small explosive device.

While the tedious work of separating uranium isotopes was under way at Oak Ridge, other workers, led by Glenn T. Seaborg, approached the problem of obtaining fissionable material by another route. It was known that uranium-238 would not fission when bombarded by neutrons. However, it had been determined that when uranium-238 was bombarded by neutrons (and certain other particles), a new element named neptunium (Np) was formed, and this product quickly decayed to another new element, plutonium (Pu).

$$^{238}_{92}U + ^{1}_{0}n \longrightarrow ^{239}_{92}U$$
$$^{239}_{92}U \longrightarrow ^{239}_{93}Np + ^{0}_{-1}e$$
$$^{239}_{93}Np \longrightarrow ^{239}_{94}Pu + ^{0}_{-1}e$$

The isotope plutonium-239 was found to be fissionable and, thus, was suitable material for the making of a bomb. A series of large reactors were built near Hanford, Washington, for the making of plutonium.

Before a fissionable material can sustain a chain reaction, a certain minimum amount, called the **critical mass,** must be brought together. There must be enough fissionable nuclei that the neutrons released in one fission process will have a good chance of finding another fissionable nucleus before escaping from the mass.

By July 1945 enough plutonium had been made for a bomb to be assembled. This first atomic bomb was tested in the desert near Alamogordo, New Mexico, on 16 July 1945. The heat from the explosion vaporized the 30-m steel tower on which it was placed and melted the sand for several hectares around the site. The light released was the brightest anyone had ever seen.

Some of the scientists were so awed by the force of the blast that they argued against its use on Japan. A few, led by Leo Szilard, argued for a demonstration of its power at an uninhabited site. But fear of a well-publicized ''dud'' and the desire to avoid millions of casualties in an invasion of Japan led President Harry S Truman to order the dropping of the bombs on Japanese cities. A uranium bomb called ''Little Boy'' was dropped on Hiroshima on 6 August 1945, causing over 100 000 casualties (Figures C.5 and C.6). Three days later, a plutonium bomb called ''Fat Man'' was dropped on Nagasaki with comparable results. World War II ended with the surrender of Japan on 14 August 1945.

FIGURE C.5 *Nuclear bomb of the type exploded over Hiroshima. The diameter of the bomb is 71 cm, and its length is 305 cm. It weighs 4.1 t and has explosive power equivalent to about 18 000 t of TNT. (Smithsonian Institution, Washington, D.C.)*

FIGURE C.6
The now-familiar mushroom cloud that follows an atomic explosion. (Smithsonian Institution, Washington, D.C.)

C.3 RADIOACTIVE FALLOUT

When the atomic bomb explodes, it produces a fantastic amount of heat, devastating shock waves, and deadly gamma radiation. Even then it is not through, for the products of uranium fission are radioactive. These materials may rain upon parts of the Earth even thousands of miles away, days and weeks later.

A typical fission reaction (see Figure C.4) might be

$$^{235}_{92}U + ^{1}_{0}n \longrightarrow ^{90}_{38}Sr + ^{143}_{54}Xe + 3\,^{1}_{0}n$$

The neutrons may strike additional uranium atoms, carrying on the chain reaction. The strontium (Sr) and xenon (Xe) atoms are radioactive and are a part of the fallout.

The uranium atom can split in 40 or more ways, producing 80 or 90 primary radioactive products. Some of these produce radioactive daughter isotopes. For example, xenon-143 undergoes beta decay, with a half-life of 1 s.

$$^{143}_{54}Xe \longrightarrow ^{0}_{-1}e + ^{143}_{55}Cs$$

The cesium isotope is also radioactive—as are its daughters and their daughters. Still other radioisotopes, such as carbon-14 and tritium, are formed by the impact of neutrons produced in the explosion on molecules of the atmosphere. Thus, fallout is exceedingly complex. We will consider only a few of the more important isotopes.

Of all the isotopes, strontium-90 presents the greatest hazard to people. This isotope has a half-life of 28 years. Strontium-90 reaches us primarily through milk and vegetables (Figure C.7). Because of its similarity to calcium (both are Group IIA elements), strontium-90 is incorporated into bone. There it remains a source of internal radiation for many years.

Although strontium-90 is a greater long-term hazard, iodine-131 may present a greater threat immediately after a nuclear explosion. The half-life of iodine-131 is only 8 days, but it is produced in relatively large amounts. Iodine-131 is efficiently carried through the food chain. In the body it is concentrated in a small area, the thyroid gland. It is precisely this characteristic that makes this isotope so useful for diagnostic scanning. However, for a healthy individual, the incorporation of radioactive iodine in the thyroid gland offers no useful information, only damaging side effects.

Another important isotope in fallout is cesium-137. Cesium is similar to potassium (both are Group IA elements), and it mimics potassium in the body. Cesium-137 is a gamma emitter and has a half-life of 30 years. It is less of a threat than strontium-90, though, because it is removed from the body more readily. We get cesium-137 through vegetables, milk, and meat.

Concern over radiation damage from nuclear fallout led most nations to halt atmospheric testing of nuclear weapons. Only France, the People's Republic of China, and India have continued above-ground tests.

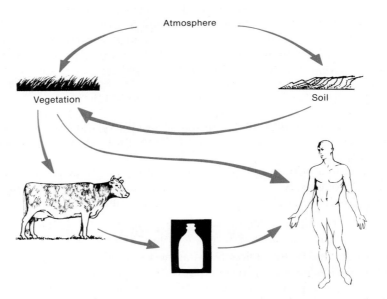

FIGURE C.7   Pathways of strontium-90 from fallout. (Reprinted from Comar, C.L., Fallout, Oak Ridge, Tenn.: U.S. Department of Energy, 1966.)

C.4 NUCLEAR POWER PLANTS

Nuclear power was envisioned by some as a fulfillment of the biblical prophecy of a fiery end to our world. Indeed, as the cold war between the United States and the Union of Soviet Socialist Republics intensified through the 1950s, it was difficult to see how nuclear war could be avoided. If it came, it could very well be the end of civilization.

Others saw nuclear power as a source of unlimited energy. Predictions were common in the later 1940s that electricity from nuclear plants would be so cheap that it would not have to be metered. Lights in public buildings could be left on continuously; switches to turn them on and off would not be needed! Obviously, such predictions have not come true. Let's take a look at nuclear power plants to see why such facilities have not yet lived up to their promise of plentiful power.

Nuclear power plants use the same fission reactions as nuclear bombs. The reaction is controlled, though, by the insertion of boron steel or cadmium control rods. Boron (B) and cadmium (Cd) absorb neutrons readily, thus preventing the neutrons from participating in the chain reaction. These rods are installed as the reactor is built. Removing them partway starts the chain reaction; the reaction is stopped if the rods are pushed all the way in.

The tremendous heat of the nuclear reaction is used to produce steam. The steam is used to turn turbines, which generate electricity (Figure C.8). Nuclear power plants have one great advantage over coal- and oil-burning plants—they do not pollute the air with soot, fly ash, sulfur dioxide, and other noxious chemicals. However, nuclear plants offer some disadvantages as well. First, the reactor requires heavy shielding to protect operating personnel from radiation. Second, fissionable fuel is rare and expensive. The supply of high-grade (thus, easily obtained) uranium ore will run out, it is estimated, by the year 2000. Third, the radioactive fission products present a serious disposal problem. Putting them in deep wells or mines or burying them at sea is like sweeping them under the rug. Do we have the right to leave our descendants with a problem that they will have to contend with for 10 000 years? Fourth, the waste heat from the generating plants heats up the environment. This effect is known as thermal pollution. Fifth, no matter how carefully constructed, nuclear plants release some radioactivity into the environment. Although proponents of nuclear power say that the amount is negligible, others say that *any* increase in radiation exposure is dangerous. Sixth, there is the possibility of a major accident (although nuclear explosions are *not* possible) at a reactor site. Such an accident could release large amounts of radioactivity to the surrounding areas, as occurred in 1986 at Chernobyl, the Ukraine.

There is considerable controversy over most of these points, and scientists stand on each side. While they may be able to agree on the results of laboratory experiments, scientists obviously do not agree on what is best for society.

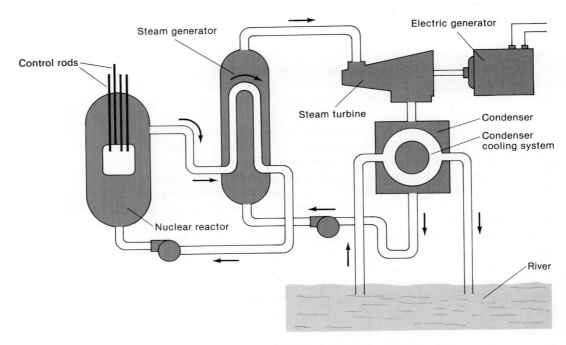

FIGURE C.8 Diagram of a nuclear power plant.

C.5 BINDING ENERGY

Nuclear reactions involve tremendous energy changes. Where do these enormous amounts of atomic energy come from? Let's consider as an example the building of a helium nucleus from its parts. If two protons and two neutrons are put together to form a helium nucleus, it should have a mass of 4.031 882 amu.

Two protons weigh 2 × 1.007 276 = 2.014 552 amu
Two neutrons weigh 2 × 1.008 665 = 2.017 330 amu
Total mass = 4.031 882 amu

The actual mass of the helium nucleus is 4.001 506 amu. The missing mass, called the **mass defect** or **binding energy** (Figure C.9), is 0.030 376 amu. Binding energy is energy that the nucleus *doesn't* have. It is the energy given up when the nucleus forms. The larger the binding energy, the more stable the nucleus.

Remembering Einstein's $E = mc^2$, we see that this small amount of mass is equivalent to a large amount of energy. If one calculates the binding energy for all the elements, as was done above for helium, and plots a graph using

FIGURE C.9

Binding energy or mass defect. A helium nucleus has less *mass than the two protons and two neutrons from which it is made.*

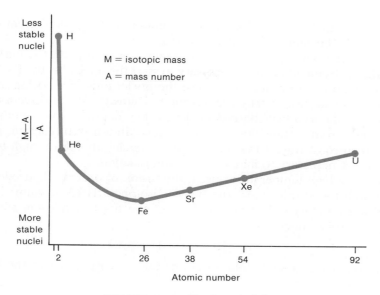

FIGURE C.10   *Nuclear stability.*

the resulting values, one obtains a curve like that in Figure C.10. The most stable nuclei, those with the highest binding energy, are in the vicinity of iron (atomic number = 26). The splitting of very large nuclei, like those of uranium atoms, produces smaller daughter nuclei, which are more stable; consequently, this splitting releases a lot of energy.

One can also see from the graph that even more energy would be produced if very small nuclei (such as hydrogen nuclei) were combined to form larger, more stable nuclei. This sort of nuclear **fusion** occurs in the explosion of a hydrogen bomb.

C.6 THERMONUCLEAR REACTIONS

The source of nearly all our energy on Earth is the thermonuclear reaction taking place in the sun. Intense temperatures in the center of the sun cause nuclei to fuse, releasing tremendous amounts of energy. The principal net reaction is believed to be the fusion of four hydrogen nuclei to produce one helium nucleus.

$$4\,^1_1\text{H} \longrightarrow \,^4_2\text{He} + 2\,^0_{+1}\text{e}$$

Upon fusion, 1 g of hydrogen releases an amount of energy equivalent to that released by the burning of 17 000 kg of coal.

The hydrogen bomb makes use of a uranium or plutonium (fission) bomb to provide the tremendous heat necessary to start the nuclei fusing. Extremely high temperatures are required because the nuclei must be moving at great speeds in order to fuse. If they are moving too slowly, the repulsive force experienced by the two like-charged nuclei as they approach one another will prevent them from getting close enough to fuse. In a fission reaction, neutral particles (neutrons) trigger the splitting of the nuclei; therefore, high temperatures are not necessary to initiate the fission reaction.

Even at the high temperatures used, the fusion of ordinary hydrogen ($_1^1H$) occurs much too slowly; thus, the heavier isotopes $_1^2H$ (deuterium) and $_1^3H$ (tritium) are employed. The intense heat of the fission explosion starts the fusion of hydrogen nuclei.

$$_1^2H + {_1^3H} \longrightarrow {_2^4He} + {_0^1n}$$

The neutron released splits lithium atoms (also incorporated in the bomb), forming more tritium.

$$_3^6Li + {_0^1n} \longrightarrow {_2^4He} + {_1^3H}$$

To date, the fusion reactions are useful only for the making of bombs. Research is progressing in the control of nuclear fusion. Controlled fusion would have several advantages over the nuclear fission reactors. The principal fuel, deuterium ($_1^2H$), is plentiful and is easily obtained by the fractional electrolysis (the splitting apart by means of electricity) of water, even though only 1 hydrogen atom in 5000 is the deuterium isotope. (We have oceans full of water.) The problem of radioactive wastes would be minimized. The end product, helium, is stable and biologically inert. Escape of tritium might be a problem, because this hydrogen isotope would be readily incorporated into organisms. Tritium ($_1^3H$) undergoes beta decay with a half-life of 12.3 years. And there is one other problem associated with any production and use of energy: an unavoidable loss of part of the energy as heat. We would still have to be concerned with thermal pollution.

Great technical difficulties must be overcome before a controlled fusion reaction can be used in the production of energy. Temperatures of 50 000 000 °C must be attained. No material on Earth can withstand more than a few thousand degrees. At a temperature of 50 000 000 °C, no molecule can hold together, nor can the atoms from which molecules are made. All atoms are stripped of their electrons. The nuclei and free electrons form a mixture called **plasma** (no relation to blood plasma). This plasma can be contained by a strong magnetic field. Scientists in several nations are closing in on the conditions necessary for a controlled fusion reaction, but even when that reaction is attained in laboratories, a lot more work will be needed before it becomes a practical source of energy.

C.7 THE NUCLEAR AGE REVISITED

We live in an age in which fantastic forces have been unleashed. The threat of nuclear war has been a constant specter in our time. Nuclear bombs have been used to destroy cities—and men, women, and children. Science and scientists have been very much involved in it all.

Still, who can believe that the world would be a better place had we not discovered the secrets of the atomic nucleus? For one thing, more lives have been saved through nuclear medicine than have been destroyed by nuclear bombs. And no nuclear bombs have been used in warfare since 1945. Perhaps the terror of nuclear holocaust has done more to prevent World War III than anything else. Nuclear power has not realized the potential once expected. Yet it remains, despite the problems involved, one of our best hopes for a plentiful supply of energy until well into the twenty-first century.

PROBLEMS

1. Compare nuclear fission and nuclear fusion. Why is energy liberated in both processes?
2. Did President Harry S Truman make the right decision when he decided to drop the nuclear bombs on Japanese cities? Would your answer be the same if you were living in 1945 and had relatives among the troops preparing for the invasion of Japan? If you were an inhabitant of one of the cities bombed?
3. Why are such high temperatures required to cause hydrogen nuclei to fuse?
4. What conditions limit the use of power from nuclear fission?
5. What conditions limit the use of power from nuclear fusion?
6. Discuss the impact of nuclear science on the following:
   a. war and peace   **c.** our energy needs
   b. medicine
7. Silicon for electronic devices can be doped by a process called neutron transmutation doping. Pure silicon is composed of three isotopes with mass numbers of 28, 29, and 30. Silicon-30, which makes up about 3% of all silicon atoms, can be converted to phosphorus-31 by capturing a neutron followed by radioactive decay. What isotope is formed when ^{30}Si captures a neutron? What particle is given off when that isotope undergoes radioactive decay to form ^{31}P? Write balanced nuclear equations for both processes.
8. Uranium has a density of 19 g/cm^3. What volume is occupied by a critical mass of 8 kg of uranium?
9. What is a chain reaction? Why does a chain reaction occur when uranium-235 undergoes fission?
10. A uranium-235 nucleus absorbs a neutron and splits into a krypton-91 isotope and a barium-142 isotope. How many neutrons are released? Write a balanced nuclear equation for the process.
11. How does the mass of a helium-4 nucleus compare to the mass of the two protons and two neutrons from which it is made? What is the difference in mass called?
12. Which process would release energy from thorium (Th)—fission or fusion?
13. Strontium-90 has a half-life of 28 years. If a sample of radioactive waste containing this isotope has an activity of 256 mCi, how many years will it take for the activity to drop to 1 mCi?
14. The half-life of cesium-137 is 30 years. If 100 mg of this isotope is released in a nuclear accident, how much will remain after 60 years? After 120 years?

Chemical Bonds

I n Chapter 2, we discussed the structure of atoms. There are only about 100 kinds of atoms. From these, as far as we know, the entire universe is made. Although there are relatively few kinds of atoms, the world is made of millions upon millions of different kinds of materials. In this chapter, we consider how atoms combine to form some of the chemical compounds that make up this myriad of materials.

The forces that hold atoms or ions together to form molecules or crystals are called **chemical bonds.** The type of bonding within molecules also determines the forces between molecules. The state of matter—whether it is solid, liquid, or gas—depends on bonding forces. Bonds even determine the physical shapes of molecules, that is, whether they are spherical or flat, rigid or wobbly.

All this is of great interest to chemists, but what significance can it have for life processes? Let us offer a few pertinent examples.

Carbon has a unique ability to form huge molecules by bonding to itself and to other elements. Those molecules are used to build carbohydrates, fats, proteins, and people.

The bonds of the ozone molecule make life on Earth possible. This substance will break apart when struck by sunlight. Its sacrifice stops the sun's ultraviolet rays from burning us all to a crisp.

The bonds of certain chemicals store the energy we need for breathing and for maintaining our heartbeat.

Because carbon monoxide bonds more tightly than oxygen to hemoglobin in our blood, carbon monoxide can kill us.

The action of some drugs depends on the shape (determined by bonding) of the drug molecule. Knowledge of molecular structure has enabled chemists to design not only drugs but also synthetic fabrics, plastics, insecticides, and a thousand other compounds with specific properties.

And if you aren't convinced yet that bonding is important to life, here is one last example. The DNA molecule is the chemical basis of heredity. It contains two strands of bonded atoms, and these strands are joined to each other by another special kind of bonding. Within this molecule is stored the genetic information that one generation passes on to the next (Chapter 23). Whether an organism is fish, fowl, hippopotamus, or human is determined by the bonding in DNA. As techniques of genetic engineering improve, we may literally be able to custom-tailor genes. Our ability to rearrange bonds in DNA molecules already gives us some control over the structure of living matter. Surely that's reason enough to learn as much about chemical bonding as we can.

4.1 DEDUCTION: STABLE ELECTRON CONFIGURATIONS

In our discussion of the atom and its structure (Chapter 2), we followed the historical development of some of the more important concepts. Some of the nuclear concepts (Chapter 3) were approached in the same manner. We could continue to look at chemistry in this manner, but that would require several volumes of print—if we got very far—and more of your time, perhaps, than you would care to spend. We won't abandon the historical approach entirely, but we will emphasize that other aspect of scientific enterprise: deduction.

The art of deduction works something like this.

Fact	*Theory*	*Deduction*
The noble gases, such as helium, neon, and argon, are inert (i.e., they undergo few, if any, chemical reactions).	The inertness of the noble gases is due to their electron structure (each has a filled outermost energy level).	If other elements could alter their electron structure to become more like noble gases, they would become less reactive.

To illustrate, let's look at an atom of the element sodium (Na). It has 11 electrons, 2 in the first energy level, 8 in the second, and 1 in the third. If it could get rid of an electron, it would have the same electron structure as an atom of the inert gas neon (Ne).

$$\left(\begin{array}{c}11p^+\\12n\end{array}\right)\ 2e^-8e^-1e^- \longrightarrow \left(\begin{array}{c}11p^+\\12n\end{array}\right)\ 2e^-8e^- + 1e^-$$

Na   Na$^+$

Recall that neon has the structure

$$\left(\begin{array}{c}10p^+\\10n\end{array}\right)\ 2e^-8e^-$$

Ne

Let us immediately emphasize that the structure represented by Na$^+$ and the neon atom (Ne) are not identical. The electron arrangement is the same, but the nuclei—and resulting charges—are not. As long as sodium keeps its 11 protons, it is still a form of sodium, but it is sodium *ion*, not sodium *atom*. **Ions** are charged structures, that is, structures in which the number of electrons is *not* equal to the number of protons. The sodium, having lost an electron, becomes positively charged. It has 11 protons (11+) and only 10 electrons (10−). Positively charged ions are called **cations.** The sodium ion (Na$^+$) is a cation.

If a chlorine atom (Cl) could gain an electron, it would have the same structure as argon (Ar).

$$\left(\begin{array}{c}17p^+\\18n\end{array}\right)\ 2e^-8e^-7e^- + 1e^- \longrightarrow \left(\begin{array}{c}17p^+\\18n\end{array}\right)\ 2e^-8e^-8e^-$$

Cl   Cl$^-$

The structure of the argon atom is

$$\left(\begin{array}{c}18p^+\\22n\end{array}\right)\ 2e^-8e^-8e^-$$

Ar

The chlorine atom, having gained an electron, becomes negatively charged. It has 17 protons (17+) and 18 electrons (18−). It is written Cl$^-$ and is called *chloride ion.* (More about names later.) Negatively charged ions are called **anions.** The chloride ion is an anion.

What is the charge on an aluminum atom that has lost three electrons?
   The neutral aluminum has 13 electrons and 13 protons (its atomic number is 13). The ion would have 13 protons (13+) and 10 electrons (10−). The net charge on the aluminum ion would be 3+. The symbol is Al^{3+}.

example 4.1

example
4.2

What is the charge on a sulfur atom that has gained two electrons?

The atomic number of sulfur is 16. Therefore, it has 16 protons (16+) and, if it gains 2 electrons, 18 electrons (18−). The net charge is 2−. The symbol is S^{2-}. This is called *sulfide ion.*

In forming ions, the nuclei of sodium and chlorine and the *inner* levels (i.e., lower energy levels) of electrons do not change. Therefore, it is convenient to let the nucleus *and* the inner levels be represented by the *symbol* alone. Electrons in the outer, or **valence,** level are represented by dots. Thus, the energy level diagrams for the ionization of sodium and chlorine are reduced to

$$Na \cdot \longrightarrow Na^+ + 1e^-$$

and

$$\cdot \overset{..}{\underset{..}{Cl}} : + 1e^- \longrightarrow : \overset{..}{\underset{..}{Cl}} : ^-$$

Representations of this sort are called **electron dot** symbols.

4.2 SYMBOLISM IN CHEMISTRY

The mystery of chemistry to the nonchemist is probably due in large part to chemists' use of symbolism. Chemists find it convenient to represent the sodium atom as Na· rather than as the more complex energy level diagram. Therefore, we do just that. And it is easier to write the electron dot symbol for chlorine than the energy level diagram. Chemists usually use the shorter form.

$\left(\begin{array}{c} 11p^+ \\ 12n \end{array} \right)$ $2e^- 8e^- 1e^-$ is represented by Na·

$\left(\begin{array}{c} 17p^+ \\ 18n \end{array} \right)$ $2e^- 8e^- 7e^-$ is represented by $: \overset{..}{\underset{..}{Cl}} \cdot$

It is very easy to write electron dot formulas for elements in the first three periods (horizontal rows) of the periodic chart. The number of electrons in the outer level is equal to the **group number.** Aluminum (Al) is in Group IIIA; therefore, it has three outer electrons. Sulfur (S) is in Group VIA; thus, it has six outer electrons. This generalization works fairly well for elements in A subgroups even beyond the first three periods. Thus, iodine (I), in Group VIIA, has seven outer electrons. The B subgroups are not quite so regular, and we shall

TABLE 4.1   *Electron Dot Formulas for Selected Elements*

IA	IIA	IIIA	IVA	VA	VIA	VIIA	Noble Gases
H·							He:
Li·	·Be·	·B·	·C·	:N·	:O·	:F:	:Ne:
Na·	·Mg·	·Al·	·Si·	:P·	:S·	:Cl:	:Ar:
K·	·Ca·					:Br:	:Kr:
Rb·	·Sr·					:I:	:Xe:
Cs·	·Ba·						

simply ignore them for the moment. Table 4.1 gives the electron dot formulas for selected elements. Notice that there is a pattern to the way in which the dots are drawn. For elements with four or fewer outer electrons, the electrons are isolated from one another. With the appearance of the fifth electron, a pairing up begins. This is a useful convention—as we shall see shortly.

It doesn't matter whether you draw lithium in any of the following ways. They are all correct.

Li   Li·   Li   ·Li

And magnesium can be drawn

Mg·   Mg   ·Mg   ·Mg·   Mg

As long as you show two outer electrons, you're right. As we get into a discussion of compounds, you will see that occasionally sticking to one choice of structures will simplify writing symbols for compounds.

Symbolism is a convenient, shorthand way of conveying a lot of information in compact form. It is the chemist's most efficient and economical form of communication. Learning this symbolism is much like learning a foreign language. Once you master a certain basic "vocabulary," the rest is easier.

4.3 SODIUM REACTS WITH CHLORINE: THE FACTS

Sodium is a highly reactive metal. It is soft enough to be cut with a knife. When freshly cut, it is bright and silvery, but it dulls rapidly as a result of reacting with oxygen or water in the air. In fact, it reacts so readily in air that it is usually stored under oil or kerosene. Sodium reacts violently with water

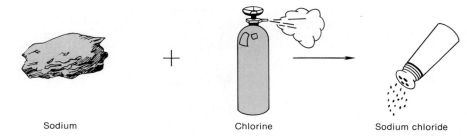

Sodium   Chlorine   Sodium chloride

FIGURE 4.1   Sodium, a soft silvery metal, reacts with chlorine, a greenish gas, to form sodium chloride (ordinary table salt).

in larger amounts, the reaction producing so much heat energy that any un-reacted sodium melts. A small piece of sodium will form a spherical bead after melting and race around on the surface of the water as it continues to react.

Chlorine is a greenish yellow gas. It is familiar as a disinfectant for swimming pools and city water supplies. (The actual substance added may be a compound that reacts with water to form chlorine.) Who hasn't been swimming in a pool having so much chlorine that it could be tasted? Chlorine is quite irritating to the respiratory tract. Indeed, chlorine was used as a poison gas in World War I.

If a piece of sodium is dropped into a flask containing chlorine gas, a violent reaction ensues. A white solid is formed that is quite stable. It is a familiar compound—sodium chloride, or table salt (Figure 4.1).

4.4 THE SODIUM-CHLORINE REACTION: THEORY

A sodium atom forms a less reactive species, a sodium ion, by *losing* an electron. A chlorine atom becomes a less reactive chloride ion by *gaining* an electron. A chlorine atom can't just pluck an electron from nowhere, nor can a sodium atom kick out an electron unless something else is willing to take it on. What happens when sodium atoms come into contact with chlorine atoms?* The obvious. A sodium atom transfers an electron to a chlorine atom.

---

* Actually, the greenish yellow gas is composed of chlorine molecules, each molecule consisting of two atoms. More about that later.

FIGURE 4.2
*The arrangement of ions in
a sodium chloride crystal.*

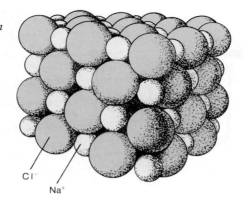

Cl⁻
Na⁺

$$\left(\begin{array}{c}11p^+\\12n\end{array}\right) \;2e^-\;8e^-\;1e^- \;+\; \left(\begin{array}{c}17p^+\\18n\end{array}\right)\;2e^-\;8e^-\;7e^-$$

$$\downarrow$$

$$\left(\begin{array}{c}11p^+\\12n\end{array}\right) \;2e^-\;8e^- \;+\; \left(\begin{array}{c}17p^+\\18n\end{array}\right)\;2e^-\;8e^-\;8e^-$$

In the abbreviated electron dot form, this reaction is written

$$\text{Na·} + \cdot \overset{\cdot\cdot}{\underset{\cdot\cdot}{\text{Cl}}}\text{:} \longrightarrow \text{Na}^+ + \text{:}\overset{\cdot\cdot}{\underset{\cdot\cdot}{\text{Cl}}}\text{:}^-$$

The sodium ion (Na^+) and the chloride ion (Cl^-) not only have electron structures like those of two noble gases (neon and argon, respectively) but have
opposite charges. Opposite charges attract. Remember that, for even a small
amount of salt, there are billions and billions of particles. These arrange themselves in an orderly fashion (Figure 4.2). These arrangements are repeated in
all directions—above and below, left and right, top and bottom—to make up
a **crystal** of sodium chloride. Each sodium ion attracts (and is attracted by)
six chloride ions (the ones to the front and back, the top and the bottom, and
both sides). Each chloride ion attracts (and is attracted by) the surrounding
six sodium ions. The forces holding the crystal together (the attractive forces
between positive and negative charges) are called **ionic bonds.**

Models. Scientists sometimes use different models to represent the
same system. The model employed in Figure 4.2 is a space-filling model
showing the relative sizes of the sodium and chloride ions. Sometimes
a ball-and-stick model is employed to better show the geometry of

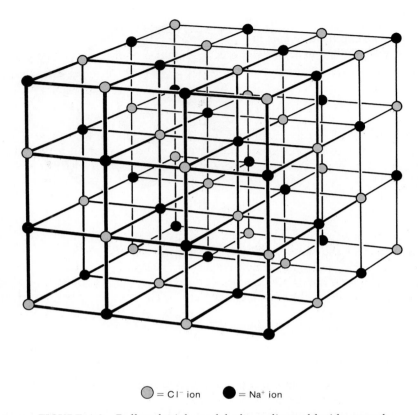

$\bigcirc$ = Cl$^-$ ion $\bullet$ = Na$^+$ ion

FIGURE 4.3   *Ball-and-stick model of a sodium chloride crystal.*

the crystal (Figure 4.3). From this model it is easy to see the cubic arrangement of the ions. One can also see that for each sodium ion there is one chloride ion; thus, the ratio of ions is one to one, and the simplest formula for the compound is NaCl. The symbols Na and Cl, written together, stand for the compound sodium chloride. The formula is also used to represent one sodium ion and one chloride ion.

4.5 IONIC BONDS: SOME GENERAL CONSIDERATIONS

Potassium (K), a metal similar to sodium, can also react with chlorine to yield a compound called potassium chloride (KCl).

$$K \cdot + \cdot \overset{\cdot \cdot}{\underset{\cdot \cdot}{Cl}} : \longrightarrow K^+ + : \overset{\cdot \cdot}{\underset{\cdot \cdot}{Cl}} :^-$$

And potassium reacts with bromine, a reddish brown liquid that is chemically similar to chlorine, to form a stable white crystalline solid called potassium bromide (KBr).

$$K \cdot + \cdot \overset{\cdot \cdot}{\underset{\cdot \cdot}{Br}} : \longrightarrow K^+ + : \overset{\cdot \cdot}{\underset{\cdot \cdot}{Br}} :^-$$

Sodium can also form a compound with bromine: sodium bromide. Magnesium, a metal harder and less reactive than sodium, reacts with oxygen, a colorless gas, to form another stable white crystalline solid called magnesium oxide (MgO).

$$\overset{\cdot}{Mg} \cdot + \cdot \overset{\cdot \cdot}{O} : \longrightarrow Mg^{2+} + \overset{\cdot \cdot}{\underset{\cdot \cdot}{O}} :^{2-}$$

Magnesium must give up two electrons and oxygen must gain two electrons for each to have the same configuration as the noble gas neon.

An atom such as oxygen, which needs two electrons, may react with potassium atoms, which have only one electron each to give. In this case, two atoms of potassium are needed for each oxygen atom. The product is potassium oxide (K_2O).

$$\begin{matrix} K \cdot \\ + \cdot \overset{\cdot \cdot}{O} : \longrightarrow \\ K \cdot \end{matrix} \quad \begin{matrix} K^+ \\ + : \overset{\cdot \cdot}{\underset{\cdot \cdot}{O}} :^{2-} \\ K^+ \end{matrix}$$

By this process, each potassium atom achieves the argon configuration. Oxygen again assumes the neon configuration.

One last example is the reaction of magnesium and nitrogen to give magnesium nitride (Mg_3N_2).

$$\begin{matrix} \overset{\cdot}{Mg} \cdot \\ \overset{\cdot}{Mg} \cdot + \\ \overset{\cdot}{Mg} \cdot \end{matrix} \quad \begin{matrix} \cdot \overset{\cdot}{N} \cdot \\ \cdot \overset{\cdot}{\underset{\cdot}{N}} \cdot \end{matrix} \longrightarrow \begin{matrix} Mg^{2+} \\ Mg^{2+} + \\ Mg^{2+} \end{matrix} \quad \begin{matrix} : \overset{\cdot \cdot}{\underset{\cdot \cdot}{N}} :^{3-} \\ : \overset{\cdot \cdot}{\underset{\cdot \cdot}{N}} :^{3-} \end{matrix}$$

Each of three magnesium atoms gives up two electrons (a total of six), and each of two nitrogen atoms acquires three (a total of six). Notice that the total positive and negative charges on the products are equal (6+ and 6−).

Generally speaking, those elements on the left side of the periodic table (especially those on the far left) react with those elements on the far right (excluding the noble gases) to form stable crystalline solids. The theory is that the elements on the left (called **metals**) tend to give up electrons to the elements on the right **(nonmetals).** The crystalline solids are held together by the attraction of oppositely charged ions. This attraction is called an **ionic bond.**

4.6 NAMES OF SIMPLE IONS AND IONIC COMPOUNDS

Names of simple positive ions are derived from those of the parent elements by addition of the word *ion*. A sodium atom (Na), upon losing an electron, becomes a *sodium ion* (Na^+). A magnesium atom (Mg), upon losing two electrons, becomes a *magnesium ion* (Mg^{2+}). Names of simple negative ions are derived from those of the parent elements by change of the usual ending to *-ide* and addition of the word ion. A chlor*ine* atom (Cl), upon gaining an electron, becomes a chlor*ide* ion (Cl^-). A sulf*ur* atom (S), upon gaining two electrons, becomes a sulf*ide* ion (S^{2-}).

Names and symbols for several important simple ions are given in Table 4.2. Note that the charge on an ion of a Group IA element is $1+$ (usually written simply as $+$). The charge on an ion of a Group IIA element is $2+$, and that on an ion of a Group IIIA element is $3+$. You can calculate the charge on the negative ions in the table by subtracting 8 from the group number. For example, the charge on the oxide ion (oxygen is in Group VIA) is $6 - 8 = -2$. The charge

TABLE 4.2   *Symbols and Names for Some Simple Ions*

Group	Element	Name of Ion	Symbol for Ion
IA	Hydrogen	Hydrogen ion*	H^+
	Lithium	Lithium ion	Li^+
	Sodium	Sodium ion	Na^+
	Potassium	Potassium ion	K^+
IIA	Magnesium	Magnesium ion	Mg^{2+}
	Calcium	Calcium ion	Ca^{2+}
IIIA	Aluminum	Aluminum ion	Al^{3+}
VA	Nitrogen	Nitride ion	N^{3-}
VIA	Oxygen	Oxide ion	O^{2-}
	Sulfur	Sulfide ion	S^{2-}
VIIA	Chlorine	Chloride ion	Cl^-
	Bromine	Bromide ion	Br^-
	Iodine	Iodide ion	I^-
IB	Copper	Copper(I) ion (cuprous ion)	Cu^+
		Copper(II) ion (cupric ion)	Cu^{2+}
	Silver	Silver ion	Ag^+
IIB	Zinc	Zinc ion	Zn^{2+}
VIIIB	Iron	Iron(II) ion (ferrous ion)	Fe^{2+}
		Iron(III) ion (ferric ion)	Fe^{3+}

* Discussed in Chapter 10.

on a nitride ion (nitrogen is in Group VA) is $5 - 8 = -3$. The periodic relationship of these simple ions is shown in Figure 4.4.

There is no simple way to determine the most likely charge on ions formed from elements in B subgroups. Indeed, you may have noticed that these can form ions with different charges. In such cases, chemists use Roman numerals with the names to indicate the charge. Thus, **iron(II) ion** means Fe^{2+}, and **iron(III) ion** means Fe^{3+}. An older terminology called Fe^{2+} **ferrous ion** and Fe^{3+} **ferric ion.** See similar names for the two copper ions in Table 4.2.

Compounds such as sodium chloride, potassium bromide, magnesium oxide, and potassium oxide are called **ionic compounds.** The constituent units of these compounds are charged particles—ions. Yet the compound as a whole is electrically neutral. One can use this principle of electrical neutrality to determine the combining ratio of ions. Potassium ions (K^+) would combine with bromide ions (Br^-) in a ratio of one to one. The formula KBr expresses this ratio and represents the compound potassium bromide. The combining ratio ($1:1$) and the ionic charges are understood.

Let's try another example. One calcium ion (Ca^{2+}) combines with *two* chloride ions (Cl^-). This ratio is expressed in the formula $CaCl_2$. In this formula, the ionic charges are understood. As with a coefficient of 1 in algebra, a subscript of 1 is understood where no other number appears, so, in the formula $CaCl_2$, the subscript 1 for calcium ion is understood, but the 2 for chloride is explicitly written (not Ca_1Cl_2 but $CaCl_2$). Thus, $CaCl_2$ not only gives us the combining ratio ($1:2$) but stands for the compound calcium chloride. It is a shorthand way of writing ($1\ Ca^{2+}$) and ($2\ Cl^-$).

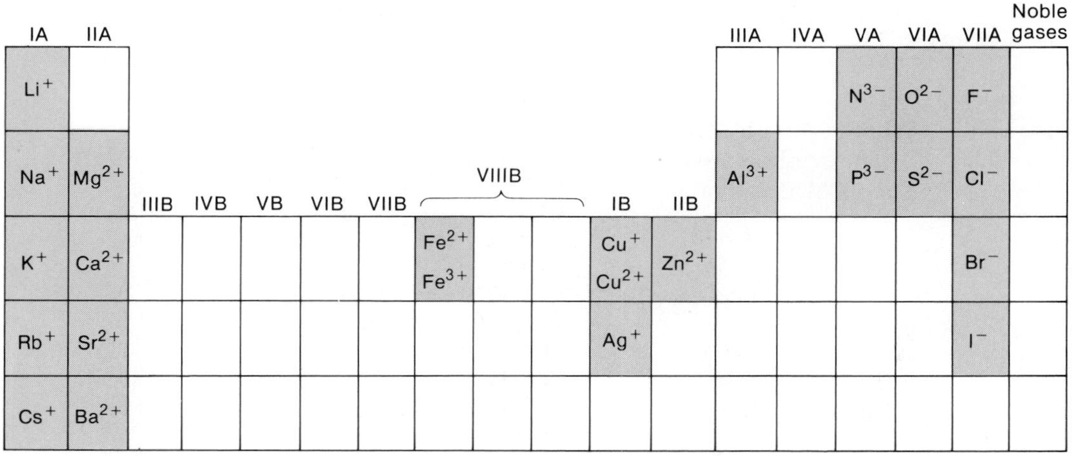

IA	IIA											IB	IIB	IIIA	IVA	VA	VIA	VIIA	Noble gases
Li^+																N^{3-}	O^{2-}	F^-	
Na^+	Mg^{2+}	IIIB	IVB	VB	VIB	VIIB		VIIIB						Al^{3+}		P^{3-}	S^{2-}	Cl^-	
K^+	Ca^{2+}						Fe^{2+} Fe^{3+}			Cu^+ Cu^{2+}	Zn^{2+}							Br^-	
Rb^+	Sr^{2+}									Ag^+								I^-	
Cs^+	Ba^{2+}																		

FIGURE 4.4   The periodic relationships of some simple ions.

You can use the charges on the ions in Table 4.2 to determine formulas for compounds of these elements. You can use the periodic table to predict the charge on ions formed from subgroup A elements, with Group IVA being the dividing line between positive and negative ions.

**example
4.3**

What is the formula for sodium sulfide?

First, write the symbols for the ions (positive ion first). Sodium is in Group IA; therefore, its charge is $1+$. Sulfur is in Group VIA, and its charge is $2-$ $(6-8)$. The symbols are Na^+ and S^{2-}. The smallest number into which both charges can be evenly divided, that is, the **least common multiple** (LCM), is 2. The least common multiple simply indicates the smallest number of electrons that can be evenly exchanged between the two elements. The subscript for each symbol can be determined by division of its charge (without the plus or minus) into the least common multiple. This step determines how many atoms of each element are needed to supply or accept the smallest common number of electrons. For Na^+,

$$\frac{2 \text{ (LCM)}}{1 \text{ (charge)}} = 2$$

For S^{2-},

$$\frac{2 \text{ (LCM)}}{2 \text{ (charge)}} = 1$$

Thus, we have the formula Na_2S_1 ($2\,Na^+$ and $1\,S^{2-}$), or Na_2S.

**example
4.4**

What is the formula for aluminium oxide?

The symbols are Al^{3+} and O^{2-} (Al is in Group IIIA and O is in Group VIA). The LCM is 6. For Al^{3+},

$$\frac{6}{3} = 2$$

For O^{2-},

$$\frac{6}{2} = 3$$

The formula is, therefore, Al_2O_3 ($2\,Al^{3+}$ and $3\,O^{2-}$).

What is the formula for calcium sulfide?
    The symbols are Ca^{2+} and S^{2-}. The LCM is 2. For Ca^{2+},

example
4.5

$$\frac{2}{2} = 1$$

For S^{2-},

$$\frac{2}{2} = 1$$

The formula is, therefore, CaS (Ca^{2+} and S^{2-}).

    Naming these ionic compounds is simple. Write the name of the positive ion first and then the name of the negative ion. (The word *ion* is not used. It is understood in each case.)

What is the name for the compound Na_2S?
    Find the constituent ions in Table 4.2. They are sodium ion (Na^+) and sulfide ion (S^{2-}). The compound is sodium sulfide.

example
4.6

What is the name for the compound FeS?
    There are two kinds of iron ions. Since sulfur exists as the S^{2-} ion and one iron ion is combined with it, the iron ion in this compound must be Fe^{2+}. The name of FeS is iron(II) sulfide.

example
4.7

What is the name of the compound $FeCl_3$?
    Since the charge on the chloride ion is 1 − and three of these ions are combined with one iron ion, the iron ion must be Fe^{3+}. The name of $FeCl_3$ is iron(III) chloride.

example
4.8

    Ionic compounds generally exist as crystalline solids. However, many of them are soluble in water. Ionic compounds are found dissolved in all natural waters—including the water in the cells of our bodies—where they are involved in such critical functions as the transmission of nerve impulses.

We cannot emphasize too strongly the difference between ions and the atoms from which they are made. They are as different as a whole peach (an atom) and a peach pit (an ion). The names and symbols may look a lot alike, but the substances themselves are quite different. Unfortunately, the situation is confused because people talk about needing "iron" to perk up "tired blood" and "calcium" for healthy teeth and bones. What they really mean is iron(II) ions (Fe^{2+}) and calcium ions (Ca^{2+}). No one would think of eating iron nails to get "iron." Nor would they eat highly reactive calcium metal. Although careful distinction is not always made in everyday life, we will try to use precise terminology here.

4.7 COVALENT BONDS: SHARED PAIRS OF ELECTRONS

You might expect a hydrogen atom, with its one electron, to tend to acquire an electron and assume the helium structure. Indeed, hydrogen atoms do just that in the presence of atoms of a reactive metal such as lithium, an element that readily gives up electrons.

$$Li\cdot + H\cdot \longrightarrow Li^+ + H\mathbf{:}^-$$

But what if there are no other kinds of atoms around? What if there are only hydrogen atoms (as in a sample of the pure element)? One hydrogen atom can hardly grab an electron from another, for among hydrogen atoms all have an equal attraction for electrons. (Even more important, perhaps, hydrogen atoms do not have a tendency to lose electrons at all, for the result would be a highly reactive bare proton—the hydrogen nucleus.) Hydrogen atoms can compromise, however, by *sharing* a pair of electrons.

$$H\cdot + \cdot H \longrightarrow H\mathbf{:}H$$

It is as if the two hydrogen atoms, in approaching one another, get their electron clouds or orbitals so thoroughly enmeshed that they can't easily pull them apart again.

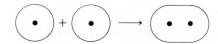

Although there is evidence that the bonding electrons can be found anywhere in the orbital, the greatest probability is that they will be found between the two nuclei. The electron dot representation usually used, H:H, is therefore a fairly good picture. (If we were to attribute human qualities to hydrogen atoms, we would suggest that they are a bit stupid. Each one looks around, sees two electrons, and happily decides that they are its very own electrons and that, therefore, it has an arrangement like that of helium, one of the noble gases.) This combination of hydrogen atoms is called a *hydrogen molecule.* The bond formed by a shared pair of electrons is called a **covalent bond.**

A chlorine atom will pick up an extra electron from anything willing to give one up. But, again, what if the only thing around is another chlorine atom? Chlorine atoms too can attain a more stable arrangement by sharing a pair of electrons.

$$:\ddot{\text{C}}\text{l}\cdot + \cdot\ddot{\text{C}}\text{l}: \longrightarrow :\ddot{\text{C}}\text{l}:\ddot{\text{C}}\text{l}:$$

Each chlorine atom in the chlorine molecule has eight electrons around it, an arrangement like that of the noble gas argon. This stable **octet** of electrons is the arrangement characteristic of all the noble gases except helium. Covalently bonded atoms that we shall consider, except hydrogen, follow the **octet rule;** that is, they seek an arrangement that will surround them with eight electrons. The shared pair of electrons in the chlorine molecule also creates a covalent bond.

For simplicity, the hydrogen molecule is often represented as H_2 and the chlorine molecule as Cl_2. The subscripts indicate two atoms *per molecule.* In each case, the covalent bond between the atoms is understood. Sometimes the covalent bond is indicated by a dash, H—H and Cl—Cl. Unshared valence (outer) electrons often are not shown.

This sharing of electrons is not limited to one pair of electrons. Consider, for example, the nitrogen atom. Its electron dot symbol is

$$:\dot{\ddot{\text{N}}}\cdot$$

Now we know this atom will be reactive after all we've learned about the octet rule. It has only five electrons in its outermost level. It could share a pair of electrons with another nitrogen atom and would then look like this:

$$:\dot{\ddot{\text{N}}}:\dot{\ddot{\text{N}}}: \quad \text{(Incorrect structure)}$$

Each atom in this arrangement has only six electrons around it. And six is not eight. Each has two electrons hanging out there without partners, so, to solve the dilemma, each nitrogen shares two additional pairs of electrons, for a total of three pairs.

$$:\text{N}:::\text{N}: \text{ (or } :\text{N}\vdots\text{N}: \text{ or } \text{N}{\equiv}\text{N})$$

In drawing the nitrogen molecule (N_2), we have simply drawn all the electrons that are being shared in the space between the two atoms. Each nitrogen has now satisfied the octet rule. A molecule in which three pairs of electrons (a total of six individual electrons) are being shared is said to contain a **triple bond.** Note that we could have drawn the *unshared* pair of electrons above or below the atomic symbol. Such a drawing would represent the same molecule.

4.8 UNEQUAL SHARING: POLAR COVALENT BONDS

So far we have seen that atoms combine in two different ways. Some that are quite different in electronic structure (from opposite ends of the periodic table) react by the complete transfer of an electron from one atom to another (ionic bond formation). Atoms that are identical combine by sharing one or more pairs of electrons (covalent bond formation). Now, let's look at some "in-betweeners."

Hydrogen and chlorine react to form a colorless gas called hydrogen chloride. This reaction may be represented schematically by

$$\text{H·} + \text{·}\overset{..}{\underset{..}{\text{Cl}}}\text{:} \longrightarrow \text{H:}\overset{..}{\underset{..}{\text{Cl}}}\text{: (or H—Cl)}$$

Both hydrogen and chlorine want an electron, so they compromise by sharing and forming a covalent bond. Since both hydrogen and chlorine actually consist of diatomic molecules, the reaction is more accurately represented by the scheme

$$\text{H:H} + \text{:}\overset{..}{\underset{..}{\text{Cl}}}\text{:}\overset{..}{\underset{..}{\text{Cl}}}\text{:} \longrightarrow 2\,\text{H:}\overset{..}{\underset{..}{\text{Cl}}}\text{:}$$

One might reasonably ask why the hydrogen molecule and the chlorine molecule react at all. Have we not just explained that they themselves were formed to provide a more stable arrangement of electrons? Yes, indeed, we did say that. But there is stable, and there is more stable. The chlorine molecule represents a more stable arrangement than two separate chlorine atoms. But, given the opportunity, a chlorine atom would rather form a bond with a hydrogen atom than with another chlorine atom. Why? Tune in again (in Chapter 5) for the answer.

For the sake of convenience and simplicity, the reaction of hydrogen (molecule) and chlorine (molecule) to form hydrogen chloride can be reduced to

$$\text{H}_2 + \text{Cl}_2 \longrightarrow 2\,\text{HCl}$$

Molecules of hydrogen chloride consist of one atom of hydrogen and one atom of chlorine. These unlike atoms share a pair of electrons. Sharing does not mean sharing equally, though. A wealth of evidence points to the fact that chlorine atoms have a greater attraction for a shared pair of electrons than do hydrogen atoms. Chlorine is said to be more **electronegative** than hydrogen.

Thus, shared electrons spend more time near the chlorine atom than they do near the hydrogen atom (Figure 4.5). If you think of an orbital as a fuzzy-looking cloud, then the cloud is denser near the chlorine. Since the electrons are more often found in the vicinity of the chlorine nucleus, that end of the molecule is more negative than the other. The hydrogen end is slightly positive. Such a covalent bond, in which the electron sharing is not equal, is called a **polar covalent bond,** or simply a **polar bond.** Covalent bonds in which electrons are equally shared (as in Cl_2 or H_2) are referred to as **nonpolar covalent bonds** to distinguish them from polar bonds. The polar covalent bond is *not* an ionic bond. In the latter, one atom completely loses an electron. In the former, the atom at the positive end of the bond (hydrogen in HCl) still has some share in the bonding pair of electrons. To distinguish this arrangement from that in an ionic bond, the following notation is used:

$$\overset{\delta+}{H}{-}\overset{\delta-}{Cl}$$

The line between the two atoms represents the covalent bond, a pair of shared electrons. The $\delta+$ and $\delta-$ (read "delta plus" and "delta minus") signify which end is partially positive and which is partially negative (the word *partially* is used to distinguish this charge from the full charge an ion has). This unequal sharing of electrons has a marked effect on the properties of a compound (Figure 4.6).

The gas hydrogen chloride dissolves readily in water. The aqueous solution formed is called hydrochloric acid (sometimes muriatic acid). This acid is used for, among other things, cleaning toilet bowls and removing excess mortar from new brick buildings. Hydrochloric acid is also the well-known "stomach acid." Acids are defined and further discussed in Chapter 10.

4.9 ELECTRONEGATIVITY

Before we look at the structure of any more molecules, we should consider in more detail the concept of **electronegativity.** When we speak of the electronegativity of an atom, we describe its tendency to attract electrons to itself. The atoms to the right in the periodic table are, in general, more electronegative than those to the left. The ones on the right are precisely those atoms that, in forming ions, tend to gain electrons and form negative ions. The ones on the left, the metals, tend to give up electrons, to become positive ions. The more electronegative an atom, the greater its tendency to pull the electrons in the bond toward its end of the bond when it is involved in covalent bonds.

The most electronegative element in the periodic table is fluorine, in the upper right-hand corner of the table. You can use this fact as a guide. Within

FIGURE 4.5
Chlorine hogs the electron blanket, leaving hydrogen partially, but positively, exposed. To hydrogen's pleas for more cover, chlorine's answer is partially negative.

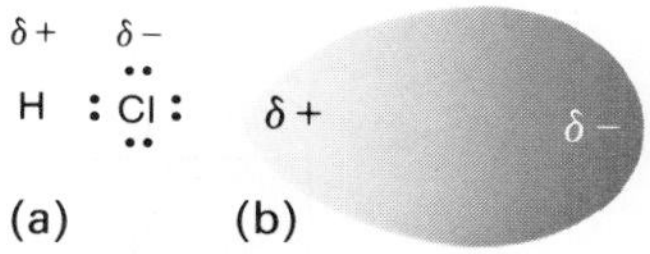

$\delta+$ $\delta-$

H :Cl: $\delta+$ $\delta-$

(a) (b)

FIGURE 4.6
Representation of the polar hydrogen chloride molecule. (a) The electron dot formula, with the shared electron pair shown nearer the chlorine atom. The symbols $\delta+$ and $\delta-$ indicate partial positive and partial negative charges, respectively. (b) A diagram depicting the unequal distribution of electron density in the hydrogen chloride molecule.

a period (a row) of the table, elements become more electronegative toward the right. Oxygen is more electronegative than nitrogen. Within a group (column) of the table, elements become less electronegative toward the bottom. Chlorine is less electronegative than fluorine. The comparison is not quite so straightforward when one is considering two elements that are in neither the same period nor the same group.

Hydrogen is difficult to place (see the unique position assigned to it in the periodic table, on the inside front cover). It could be placed in Group IA, because, like lithium, it has one electron in its outer level, but it really isn't much like the Group IA metals. Or it could be placed in Group VIIA, because, like fluorine, it is just one electron short of looking like a noble gas. However, hydrogen isn't nearly as electronegative as fluorine. If you look at a number of versions of the periodic table, you'll see that chemists still haven't decided what to do with it. In this book, we have taken the coward's way out. We've placed it at the top middle, away from everything (Figure 4.7). Its electronegativity fits that position. It will take electrons from an atom that gives them up readily (lithium, for example), and it will shift electrons to an electronegative element (fluorine, for example).

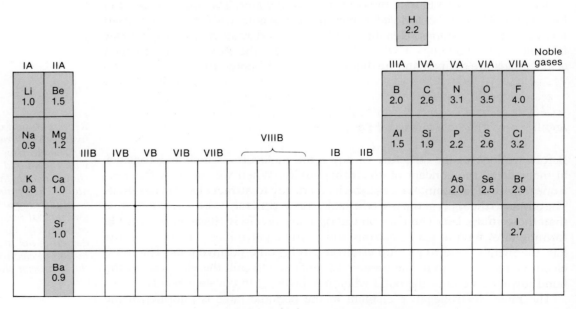

FIGURE 4.7   Relative electronegativities of some common elements.

To consider covalent bonding in any detail, you must have an understanding of this concept of electronegativity.

4.10 WATER: A BENT MOLECULE

Water is one of the most familiar chemical substances. The electrolysis experiment of Nicholson and Carlisle (Chapter 2) and ample evidence since their time indicate that the molecular formula for water is H_2O. When we considered oxygen previously, we noted that it tended to share two pairs of electrons, so that it could be surrounded by a total of eight. But a hydrogen atom only tends to share one pair of electrons. Therefore, an oxygen atom must bond with two hydrogen atoms.

This arrangement completes the valence energy level of oxygen, which now has the neon structure. It also completes the outer energy level of the hydrogen atoms, each of which now has the helium structure.

We have chosen to represent the water molecule with the hydrogen atoms arranged at an angle rather than on a straight line with the oxygen atom. This arrangement is necessary to explain the polar nature of the water molecule. When placed between two charged plates, water molecules align themselves as dipoles. By a **dipole** we mean a molecule that has a positive end and a negative end.

It is evident that hydrogen-oxygen bonds ought to be polar, for hydrogen and oxygen would not be expected to share a pair of electrons equally. Oxygen is a strongly electronegative atom. However, if the atoms were in a straight row (that is, in a linear arrangement), the two polar bonds would cancel one another, and there would be no net dipole.

$$\overset{\delta^+}{H}\!-\!\overset{\delta^-}{O}\!-\!\overset{\delta^+}{H} \qquad \text{(Incorrect structure)}$$

To act like dipoles, the atoms must be arranged on an angle. Precise (but indirect) measurements with sophisticated instruments show that the bond angle (that is, the angle between the two bonds) is 104.5°. Thus, the charge distribution is as follows:

The molecules would align themselves between charged plates in the following way:

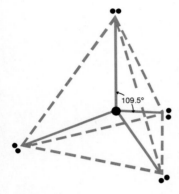

109.5°

FIGURE 4.8

The four electron pairs around a central atom point toward the corners of a regular tetrahedron. Each angle is 109.5°.

There are several theories of chemical bonding that account for the shape of the water molecule. One of the simplest and most satisfying is the valence shell electron pair repulsion (VSEPR) theory. According to this theory, the arrangement of the bonds in water can be explained as follows. There are four pairs of electrons surrounding the oxygen atom in the water molecule. Since all electrons bear a like (negative) charge, it is reasonable to expect them to get as far apart as possible. If each pair of electrons is represented as a line extending out from the oxygen atom, the farthest apart these lines can get is 109.5°. Further, if the ends of these lines were all connected, the connecting lines would inscribe a regular tetrahedron (Figure 4.8).

The predicted tetrahedral angle of 109.5° is not far from the actual value of 104.5° for water. The disagreement is accounted for by the theory that in water the **nonbonding pairs** of electrons, those that are not shared with hydrogen atoms, occupy a greater volume than the bonding pairs because the nonbonding pairs are not squeezed between *two* nuclei. Thus, the nonbonding pairs push the bonding pairs a little bit closer together.

We will take a look at the unique properties of water in Chapter 7.

The **tetrahedral** arrangement is not limited to molecules. The farthest apart any four equivalent things can get, if they are tied to a common center, is 109.5°. Four balloons tied together would automatically take up the tetrahedral shape.

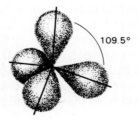

109.5°

Two balloons so tied, as you might expect, would point in opposite directions. To put it more elegantly, they would take up positions such that the angle between them would be 180°, a **linear** arrangement.

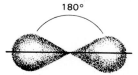

Three balloons would all lie in the same plane and assume positions such that the angles between them would be 120°, a **trigonal** arrangement.

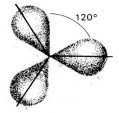

4.11 AMMONIA: A PYRAMIDAL MOLECULE

An atom of the element nitrogen (N) has five electrons in its valence energy level. It can assume the neon configuration by sharing three pairs of electrons with *three* hydrogen atoms and forming the compound ammonia.

$$\left.\begin{array}{ccc} & \cdot\ddot{N}\cdot & \\ & \cdot & \\ H\cdot & H\cdot & H\cdot \end{array}\right\} \quad H\!:\!\ddot{N}\!:\!H \quad \left(\text{or } H\!-\!\overset{\cdot\cdot}{\underset{|}{N}}\!-\!H\right)$$

There are four pairs of electrons on the central nitrogen atom in the ammonia molecule. Using the electron pair repulsion theory, one would expect a tetrahedral arrangement of the four pairs and bond angles of 109.5°. The actual bond angles are 107°, quite close to the theoretical value. Presumably the unshared pair occupies a greater volume than the shared pairs occupy, pushing the latter slightly closer together. The bond angles in ammonia are larger (closer to 109.5°) than the bond angle in water because there is only one unshared pair of electrons doing the pushing in ammonia while there are two in water.

In ammonia, the bond arrangement is that of a tripod with a hydrogen atom at the end of each "leg" and the nitrogen atom with its unshared pair of electrons sitting at the top (Figure 4.9). Ammonia is also a polar molecule, with the top of the tripod being the end with the partial negative charge.

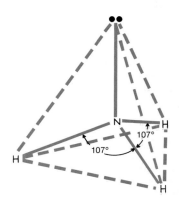

FIGURE 4.9
The pyramidal ammonia molecule.

Ammonia (NH_3) is a gas at room temperature. Vast quantities of it are compressed into tanks and used as fertilizer. Ammonia is quite soluble in water. It forms an aqueous solution that is basic, that is, one which will neutralize acids (Chapter 10). Such aqueous preparations are familiar household cleansing solutions.

4.12 METHANE: A TETRAHEDRAL MOLECULE

An atom of carbon (C) has four electrons in its valence energy level. It can assume the neon configuration by sharing pairs of electrons with four hydrogen atoms, forming the compound methane.

$$\cdot \overset{\cdot}{\underset{\cdot}{C}} \cdot \quad \left. \begin{array}{cccc} H\cdot & H\cdot & H\cdot & H\cdot \end{array} \right\} \quad \begin{array}{c} H \\ H\!:\!\overset{\cdot\cdot}{\underset{\cdot\cdot}{C}}\!:\!H \\ H \end{array}$$

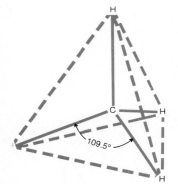

FIGURE 4.10
The tetrahedral methane molecule.

There are four pairs of electrons on the central carbon in methane. By the VSEPR theory, one would expect a tetrahedral arrangement and bond angles of 109.5°. The actual bond angles are 109.5°, in perfect agreement with theory (Figure 4.10). All four electron pairs are shared with hydrogen atoms; thus, all four pairs occupy identical volumes. The arrangement of the four bonds would make methane a nonpolar molecule; any polarity in the individual bonds would mutually cancel out. The situation here is analogous to that pointed out for a "straight" arrangement of the water molecule (p. 133). In fact, carbon and hydrogen have very similar electronegativities, so the individual bonds themselves would not be expected to be very polar.

Methane is represented by the formula CH_4. It is the principal component of natural gas, used as a fuel. Methane burns with a hot flame, and if sufficient oxygen is present, the principal products are the relatively innocuous carbon dioxide and water.

Methane is produced by the decay of plant and animal material. It is often seen bubbling to the surface of swamps, hence its common name "marsh gas." Bacterial metabolism in the intestinal tract also produces methane, making it a component of intestinal gas.

4.13 SOME RULES FOR BUILDING MOLECULES

We have seen in the preceding section how different atoms form different numbers of bonds. Hydrogen forms one bond, chlorine one, oxygen two, nitrogen three, and carbon four. The number of covalent bonds that an atom can form

is called its **valence.** Recall that the lowest energy level of an atom can be occupied by at most two electrons. These may be either shared or unshared. The
hydrogen atom has one electron. It can form *one* bond by sharing this electron
with another atom. The other atom also furnishes one electron to form a shared
pair with the hydrogen electron. Hydrogen is said to be **univalent.**

The helium (He) atom has a filled innermost energy level. It cannot share
electrons with other atoms; hence its valence is zero.

The second energy level can hold a maximum of eight electrons (that is, four
pairs of electrons). In the element neon (Ne), this level is filled, and neon forms
no bonds. Its valence is zero (or, one could say, it has *no* valence).

The fluorine atom has seven electrons in its outer, or valence, energy level.
It can therefore form one bond by sharing one electron with another atom.
The other atom furnishes one electron to form a shared pair. Thus, fluorine,
like hydrogen, is univalent, having three unshared pairs of electrons and the
capacity to form one bond involving one shared pair.

Similar considerations reveal that oxygen is **bivalent,** nitrogen is **trivalent,**
and carbon is **tetravalent.** These guidelines for building molecules are summarized and illustrated in Table 4.3 and Figure 4.11.

TABLE 4.3   *Valence*

Electron Dot Picture	Valence Bond Picture	Number of Bonds (Valence)	Representative Molecules
H·	H—	1	H—H H—Cl
He:	none	0	none
·C·	—C—	4	H—C—H (formaldehyde, CO_2)
·N·	—N—	3	H—N—H N≡N
·O:	—O—	2	H—O—H (formaldehyde)
·F:	—F	1	H—F F—F
·Cl:	—Cl	1	H—Cl Cl—Cl

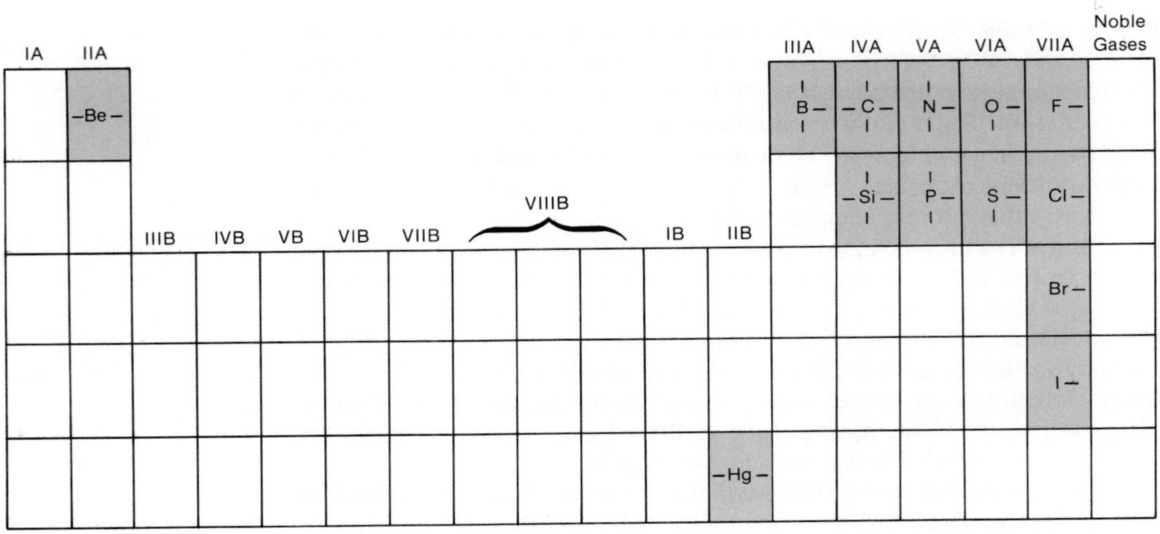

FIGURE 4.11 Covalent bonding of representative elements of the periodic table.

You will notice that the table does not include all of the elements in which the second energy level is not yet filled. (Lithium is one that's missing.) That's because these elements have a tendency to form ionic bonds rather than the covalent bonds in which we are presently interested. The electron dot symbols of beryllium and boron are

$$\cdot Be \cdot \qquad \cdot \overset{\cdot}{B} \cdot$$

These two elements can use their unshared electrons to form two and three covalent bonds, respectively. When they do so, however, they do not achieve a noble gas configuration. Therefore, the compounds thus formed are ordinarily quite reactive.

4.14 THE CONSTRUCTION OF MOLECULES

There is a chemical substance called hydrogen peroxide. Chemical analysis shows that each molecule of hydrogen peroxide is composed of two hydrogen atoms and two oxygen atoms. This information is indicated in the formula H_2O_2. What is the structure of the hydrogen peroxide molecule? The parts list is as follows:

Two H— Two $: \overset{|}{\underset{\cdot\cdot}{O}} —$

For the sake of conserving ink, we will omit drawing the nonbonding electrons, that is, the unshared pairs. No connections can be made *through* hydrogen, for each hydrogen has only one connector. To be in between you have to be able to hold on with two hands (or bonds) at least. The only kind of molecule that one can make using two atoms of hydrogen and two of oxygen is shown in arrangement I.

$$\underset{\text{I}}{\overset{\displaystyle H}{\underset{\displaystyle H}{\ddot{O}-\ddot{O}}}} \qquad \underset{\text{II}}{\overset{\displaystyle H}{\underset{\displaystyle H}{\ddot{O}-\ddot{O}}}}$$

Arrangement II, which may look like a different molecule, really isn't. Arrangement I can be converted into arrangement II—and vice versa—by rotation (spinning) about the O—O bond.

$$\overset{\displaystyle H}{\underset{\displaystyle H}{\ddot{O}-\ddot{O}}}$$

We did not offer

$$H-\ddot{O}-\ddot{O}-H$$

as a possibility. The molecule is really bent (for precisely the same reason that the water molecule is bent). The unshared electron pairs are shown for clarity. You might see the structure drawn in this straight form, but that is because it is easier to type the straight formula. And this formula does show which atoms are attached to which. However, the presumption is that we all understand the molecule is really bent.

Let's build another molecule. There is a chemical substance called ethane that, like methane, is a constituent of natural gas. Chemical analysis indicates that the molecular formula for ethane is C_2H_6. By the rules of molecular architecture, the molecule can be put together as follows. The parts list includes

$$\text{Six } H- \qquad \text{Two}-\overset{\displaystyle |}{\underset{\displaystyle |}{C}}-$$

The only way six atoms of hydrogen and two of carbon can be put together according to the rules is

$$H-\overset{\displaystyle H}{\underset{\displaystyle H}{C}}-\overset{\displaystyle H}{\underset{\displaystyle H}{C}}-H$$

The molecule is really three-dimensional (carbon's bonds are arranged in the tetrahedral form). The restrictions imposed by the two-dimensional nature of the printed page have caused us to draw it flat here.

Methane (CH_4) and ethane (C_2H_6) are merely the first members of a series of similar compounds called *alkanes* or *saturated hydrocarbons*. Other members of the series are discussed in Chapter 14.

Following are a few other examples of molecule building.

---

example 4.9

If the **molecular formula** of a substance is HCN, what is its **structural formula** (the formula that shows the attachment of the atoms)?

We have

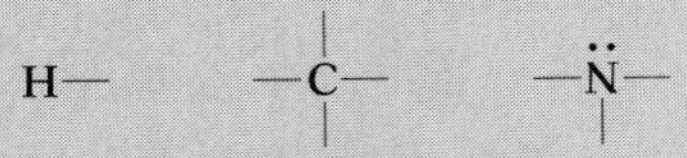

$$H— \qquad —\overset{\displaystyle |}{\underset{\displaystyle |}{C}}— \qquad —\overset{\displaystyle ..}{N}—$$

The carbon needs to form four bonds, the hydrogen needs to form one bond, and the nitrogen needs to form three bonds. The only way this combination can be put together is

$$H—C\equiv N:$$

Note that the carbon has formed four bonds (a single and a triple), the nitrogen has formed three, and the hydrogen has formed one.

---

example 4.10

If the molecular formula of a substance is $COCl_2$, what is its structural formula?

The parts are

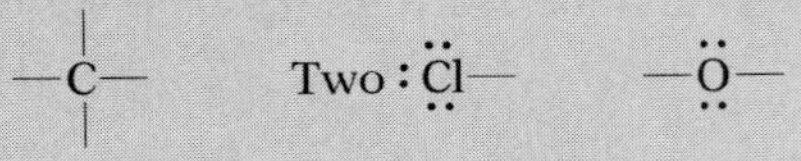

$$—\overset{\displaystyle |}{\underset{\displaystyle |}{C}}— \qquad Two : \overset{\displaystyle ..}{\underset{\displaystyle ..}{Cl}}— \qquad —\overset{\displaystyle ..}{\underset{\displaystyle ..}{O}}—$$

The formula is

$$:\overset{..}{\underset{..}{Cl}}—C—\overset{..}{\underset{..}{Cl}}: \\ \quad \overset{\displaystyle \|}{\underset{\displaystyle :O:}{}}$$

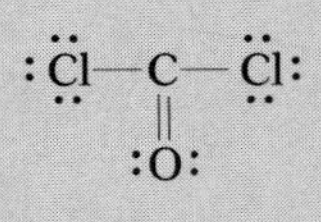

If you attempt to attach the chlorine atoms to the oxygen rather than the carbon, you'll find yourself with some leftover connectors. Try it. (The compound is carbonyl chloride, alias phosgene, a poisonous gas.)

Extremely complicated molecules can be constructed by these rules. Problem 36 at the end of the chapter will give you some practice in drawing a few simpler ones.

4.15 POLYATOMIC IONS

There are many familiar substances (for example, sodium hydroxide, also called **lye**) that have both ionic and covalent bonds. As might be expected, sodium hydroxide contains sodium ions (Na^+). The "hydroxide" part contains an oxygen atom covalently bonded to hydrogen. There is an "extra" electron in this unit that gives it a negative charge.

$$\cdot \ddot{O} \cdot + \cdot H \quad \cdot \ddot{O} : H \left.\begin{array}{c} \\ + \\ e^- \end{array}\right\} \quad [:\ddot{O}:H]^-$$

Hydroxide ion

There are many such groups of atoms in nature. They hang together through most chemical reactions. For example, sodium hydroxide will react with aluminum chloride to form aluminum hydroxide. A list of the more common polyatomic ions is given in Table 4.4, which you can use, in combination with Table 4.2, to write formulas for compounds containing polyatomic ions.

TABLE 4.4   *Some Common Polyatomic Ions*

Charge	Name	Formula
1+	Ammonium ion	NH_4^+
	Hydronium ion*	H_3O^+
1−	Hydrogen carbonate (bicarbonate) ion	HCO_3^-
	Hydrogen sulfate (bisulfate) ion	HSO_4^-
	Acetate ion	$CH_3CO_2^-$ (or $C_2H_3O_2^-$)
	Nitrite ion	NO_2^-
	Nitrate ion	NO_3^-
	Cyanide ion	CN^-
	Hydroxide ion	OH^-
	Dihydrogen phosphate ion	$H_2PO_4^-$
	Permanganate ion	MnO_4^-
2−	Carbonate ion	CO_3^{2-}
	Sulfate ion	SO_4^{2-}
	Monohydrogen phosphate ion	HPO_4^{2-}
	Oxalate ion	$C_2O_4^{2-}$
	Dichromate ion	$Cr_2O_7^{2-}$
3−	Phosphate ion	PO_4^{3-}

*Discussed in Chapter 10.

example
4.11

What is the formula for sodium sulfate?
    First, write the formulas for the ions.

$$Na^+ \qquad SO_4{}^{2-}$$

The least common multiple is 2. The formula is therefore Na_2SO_4.

example
4.12

What is the formula for ammonium sulfide?
    The ions are

$$NH_4{}^+ \qquad S^{2-}$$

The least common multiple is 2. The formula is $(NH_4)_2S$. The parentheses are necessary to indicate that the entire ammonium unit is taken twice.

example
4.13

What is the formula for ammonium nitrate?
    The ions are

$$NH_4{}^+ \qquad NO_3{}^-$$

The least common multiple is 1. The formula for ammonium nitrate is NH_4NO_3.

example
4.14

What is the name for the compound NaCN?
    The ions are

$$Na^+ \qquad CN^-$$

The name is sodium cyanide.

example
4.15

What is the name for KH_2PO_4?
    The ions are

$$K^+ \qquad H_2PO_4{}^-$$

The name is potassium dihydrogen phosphate.

What is the name for $(NH_4)_2HPO_4$?
    The ions are

$$NH_4^+ \qquad HPO_4^{2-}$$

The name is ammonium monohydrogen phosphate.

example
4.16

4.16 BONDING AND THE SHAPE OF MOLECULES

If a molecule consists of a central atom surrounded by two, three, or four other atoms, it is possible to predict its shape from simple rules. These rules are derived from the VSEPR rules of electron pair repulsion. Table 4.5 summarizes and illustrates these rules. Study the table, then draw electron dot structures for each of the example molecules.

What is the shape of each of the following molecules?

a. BH_3 b. PH_3

example
4.17

First draw electron dot structures for each.

```
      H                H
      ··               ··
      B:H             :P:H
      ··               ··
      H                H
```

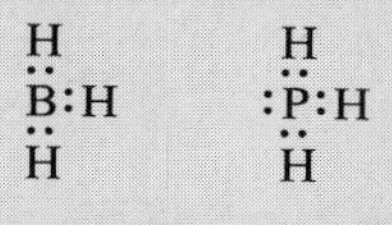

(Boron has only three electrons; BH_3 does not follow the octet rule.) The BH_3 molecule has three bonds and no unshared pairs. The three bonds get as far apart as possible; the molecule is trigonal. The PH_3 molecule has three bonds and one unshared pair; the molecule is pyramidal.

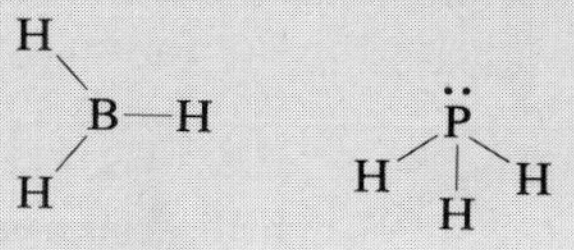

```
   H                    ··
    \                   P
     B—H            H  /|  H
    /                   |
   H                    H
```

TABLE 4.5   *Bonding and the Shape of Molecules*

Number of Bonds	Number of Unshared Pairs	Shape	Examples
2	0	Linear	$BeCl_2$, $HgCl_2$
3	0	Trigonal	BF_3
4	0	Tetrahedral	CH_4, $SiCl_4$
3	1	Pyramidal	NH_3, PCl_3
2	2	Bent	H_2O, H_2S, SCl_2

PROBLEMS

1. What is the structural difference between a sodium atom and a sodium ion? How does sodium metal differ from sodium ions (in sodium chloride, for example) in properties?
2. What is the structural difference between a sodium ion and a neon atom? What is similar?
3. What is the structural difference among chlorine atoms, chlorine molecules, and chloride ions? How do their properties differ?
4. Write electron dot symbols for each of the following elements. You may use the periodic table.
   a. sodium   f. oxygen
   b. fluorine   g. aluminum
   c. carbon   h. potassium
   d. magnesium   i. chlorine
   e. nitrogen
5. Indicate the charges on simple ions formed from the following elements:
   a. Group IIIA   c. Group IA
   b. Group VIA   d. Group VIIA
6. Using electron dot formulas, draw the formation of ion from atom for each of the following:
   a. barium   c. aluminum
   b. bromine   d. sulfur
7. Identify the ions of Problem 6 as cations or anions.
8. Draw the formulas (e.g., Cl^-) for the following ions:
   a. lithium ion   c. calcium ion
   b. iodide ion
9. Draw the formulas for the following ions:
   a. ammonium ion   c. phosphate ion
   b. hydrogen carbonate ion
10. Name the following ions:
    a. K^+ b. O^{2-} c. Al^{3+}
11. Name the following ions:
    a. CO_3^{2-} c. NO_3^-
    b. HPO_4^{2-} d. OH^-
12. Write correct formulas for the following compounds:
    a. lithium fluoride   c. aluminum bromide
    b. calcium iodide   d. aluminum sulfide
13. Write correct formulas for the following compounds:
    a. magnesium sulfate   c. sodium cyanide
    b. potassium nitrate   d. calcium oxalate
14. Write correct formulas for the following compounds:
    a. ferrous sulfate

b. ammonium phosphate
c. magnesium phosphate
d. calcium monohydrogen phosphate
15. Give correct names for the compounds represented by the following formulas:
    a. NaBr   b. $CaCl_2$ c. Al_2O_3
16. Give correct names for the compounds represented by the following formulas:
    a. KNO_2 d. $NaNO_3$ g. $KHCO_3$
    b. LiCN   e. $CaSO_4$ h. $Al(OH)_3$
    c. NH_4I f. $NaHSO_4$ i. Na_2CO_3
17. Give correct names for the compounds represented by the following formulas.
    a. $Mg(CH_3CO_2)_2$ f. $Ca(H_2PO_4)_2$
    b. $Al(C_2H_3O_2)_3$ g. $Mg(HCO_3)_2$
    c. $(NH_4)_3PO_4$ h. $Ca(HSO_4)_2$
    d. $(NH_4)_2C_2O_4$ i. NH_4NO_2
    e. Na_2HPO_4
18. Fill in this table assuming that elements X, Y, and Z are all in A subgroups in the periodic table.

	Element X	Element Y	Element Z
Group Number	IA	___	___
Electron Dot Formula	___	$\cdot \dot{Y} \cdot$	___
Charge on Ion	___	___	$2-$

19. Consider the hypothetical elements X, Y, and Z with electron dot formulas:

$$:\overset{..}{\underset{..}{X}}\cdot \qquad :\dot{Y}\cdot \qquad :\overset{.}{\underset{.}{Z}}\cdot$$

    a. To which group in the periodic table would each belong?
    b. Write the electron dot formula for the simplest compound of each with hydrogen.
    c. Write electron dot formulas for the ions formed when X and Y react with sodium.
20. Draw electron dot formulas for the following covalent molecules:
    a. CH_4O c. CH_5N
    b. NOH_3 d. N_2H_4
21. Draw electron dot formulas for the following covalent molecules:

   a. NF_3 **c.** C_2H_2
   b. C_2H_4 **d.** CH_2O

22. There are two different covalent molecules with the formula C_2H_6O. Draw electron dot formulas for the two molecules.

23. Shared pairs of electrons can be represented by dashed lines. The electron dot formula for the hydrogen molecule, H:H, can be translated to H—H. Draw dashed-line formulas for the molecules in Problems 20–22.

24. Name the following covalent compounds:
   a. CS_2 **c.** N_2S_4
   b. CBr_4 **d.** PBr_3

25. If atoms of the two elements in each set below were joined by a covalent bond, which atom would more strongly attract the electrons in the bond? You may refer to Figure 4.7.
   a. N and S   **b.** B and Cl   **c.** As and F

26. *Without* referring to Figure 4.7 but using the periodic table, indicate which element in each set is more electronegative.
   a. Br or F   **b.** Br or Se   **c.** Cl or As

27. Solutions of iodine chloride (ICl) are used as disinfectants. Are the molecules of ICl ionic, polar covalent, or nonpolar covalent?

28. Potassium is a soft silvery metal that reacts violently with water and ignites spontaneously in air. Your doctor recommends you take a potassium supplement. Would you take potassium metal? What would you take?

29. Is there any such thing as a sodium chloride molecule? Explain.

30. The compound $Ca(NO_3)_2$ is composed of Ca^{2+} and NO_3^- ions. Write the formulas for the ions in each of the following:
   a. NaBr   **c.** $MgCO_3$ **e.** KH_2PO_4
   b. $CaCl_2$ **d.** $NaHCO_3$ **f.** $Mg_3(PO_4)_2$

31. How many covalent bonds do each of the following usually form? You may refer to the periodic table.
   a. H   **c.** C   **e.** O
   b. F   **d.** N   **f.** Br

32. Why does neon not form bonds?

33. Classify the following bonds as ionic or covalent. For those bonds that are covalent, indicate whether they are polar or nonpolar.
   a. KF   **d.** IBr   **g.** MgS
   b. NO   **e.** CaO   **h.** NaBr
   c. Br_2 **f.** F_2 **i.** HCl

34. Give examples of univalent, bivalent, trivalent, and tetravalent elements.

35. Use the VSEPR theory to predict the shape for these molecules.
   a. hydrogen sulfide (H_2S)
   b. silane (SiH_4)
   c. beryllium chloride ($BeCl_2$)
   d. boron fluoride (BF_3)

36. Use the valence rules to construct structural formulas for the following compounds:
   a. CH_3F **d.** C_2H_5I **g.** CH_2O_2
   b. HCOF   **e.** C_2H_7N
   c. NH_2Cl **f.** C_2H_3N

5

Chemical Reactions

The complex chemical processes in the living cell involve changes in energy as well as changes in chemical composition. Some reactions provide the energy that keeps the cell alive and well. Other reactions, vital to life processes, require an input of energy.

Chemical reactions proceed at various rates. Some are explosively fast; others, exceedingly slow. Rates are affected by a number of factors. Perhaps the most important of these factors, for living organisms, are complex molecules called enzymes that accelerate reaction rates enormously. Yet, strange as it may seem, the enzymes are still there unchanged *after* doing their job.

Chemical reactions also proceed to different extents. In some, the reactants are converted entirely to products; these reactions are said to go to *completion*. In others, the products react, re-forming the original starting materials. These reactions, outside the cell, come to equilibrium. In a living cell, equilibrium would be deadly. The cellular processes must go to completion—or very nearly so. Products can become reactants in the body, but not under equilibrium conditions.

In this chapter, we will examine energy changes, reaction rates, and equilibria. For the most part, we will deal with simple, nonliving systems. The principles developed, however, will be exceedingly important in later chapters, where we will deal with the more complex chemistry of living cells.

5.1 BALANCING CHEMICAL EQUATIONS

Chemistry is a study of matter and the changes it undergoes. More than that, it is a study of the energy that brings about those changes—or the energy that is released when those changes occur. In Chapter 4, we discussed the symbols and formulas that have been invented to represent elements and compounds. Now let's look at a shorthand way of describing chemical changes—the **chemical equation.**

Carbon reacts with oxygen to form carbon dioxide. In the chemical shorthand this reaction is written

$$C + O_2 \longrightarrow CO_2$$

The plus sign ($+$) indicates addition of carbon to oxygen (or vice versa) or a mixing of the two in some manner. The arrow ($\rightarrow$) is often read "yields." Substances on the left of the arrow are **reactants** or **starting materials.** Those on the right are the **products** of the reaction. The conventions here are like those we used in writing nuclear equations (Chapter 3). Now, however, the nucleus will remain untouched. The chemical reactions we are going to look at will involve only electron structures.

Chemical equations have meaning on the atomic and molecular level. The equation

$$C + O_2 \longrightarrow CO_2$$

means that one atom of carbon (C) reacts with one molecule of oxygen (O_2) to produce one molecule of carbon dioxide (CO_2).

Not all chemical reactions are so simply represented. Hydrogen reacts with oxygen to form water. We can write this reaction as

$$H_2 + O_2 \longrightarrow H_2O \qquad \text{(Not balanced)}$$

This representation, however, is *not consistent with the law of conservation of matter.* There are two oxygen atoms shown among the reactants (as O_2), and only one among the products (in H_2O). For the equation to represent correctly the chemical happening, it must be **balanced.** To balance the oxygen atoms, we need only place the coefficient 2 in front of the formula for water.

$$H_2 + O_2 \longrightarrow 2\,H_2O \qquad \text{(Not balanced)}$$

This coefficient means that there are *two* molecules of water involved. As is the case with subscripts (Chapter 4), a coefficient of 1 is understood where no other number appears. A coefficient preceding a formula multiplies everything in the formula. In the second equation, the coefficient 2 not only increases the number of oxygen atoms to two, but also increases the number of hydrogen atoms to four.

But that equation is still not balanced. We took care of oxygen at the expense of messing up hydrogen. To balance hydrogen, we place a coefficient 2 in front of the H_2.

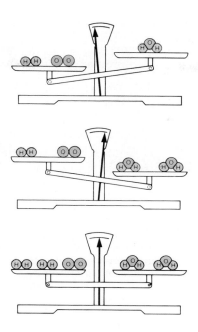

FIGURE 5.1
To balance the equation for the reaction of hydrogen with oxygen that forms water, the same number of each kind of atom must appear on each side (atoms are conserved). When the equation is balanced, there are four hydrogen atoms and two oxygen atoms on each side.

$$2\,H_2 + O_2 \longrightarrow 2\,H_2O \quad \text{(Balanced)}$$

Now there are enough hydrogen atoms on the left. In fact, there are four hydrogens and two oxygens on each side of the equation. Atoms are conserved; the equation is balanced. (See Figure 5.1.)

Note that we could not solve our problem by changing the subscript for oxygen in water.

$$H_2 + O_2 \longrightarrow H_2O_2 \quad \text{(Incorrect)}$$

The equation would be balanced, but it would not mean "hydrogen reacts with oxygen to form *water*." The formula H_2O_2 represents *hydrogen peroxide*, not water. The law of definite proportions tells us that water is always H_2O. We cannot change that merely for the convenience of balancing chemical equations.

Balance the following equation.

$$N_2 + H_2 \longrightarrow NH_3 \quad \text{(Not balanced)}$$

**example
5.1**

For this sort of problem we will again find the concept of the least common multiple useful. We'll balance the hydrogen first. There are two hydrogens on the left and three on the right. The least common multiple of 3 and 2 is 6. Six will be the smallest number of hydrogens that can be evenly converted from reactants to products. Thus, we need three molecules of H_2 and two of NH_3.

$$N_2 + 3H_2 \longrightarrow 2NH_3 \quad \text{(Balanced)}$$

We've balanced the hydrogens and, in the process, the nitrogens also. There are two on the left and two on the right. The entire equation is balanced! (See Figure 5.2.)

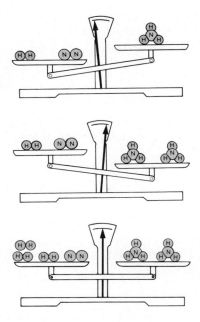

FIGURE 5.2
To balance the equation for the reaction of nitrogen with hydrogen that forms ammonia, the same number of each kind of atom must appear on each side of the equation. When the equation is balanced, there are two nitrogen atoms and six hydrogen atoms on each side.

example 5.2

Balance the following equation.

$$Fe + O_2 \longrightarrow Fe_2O_3 \quad \text{(Not balanced)}$$

Let's balance oxygen first. The least common multiple is 6. We need three molecules of O_2 and two of Fe_2O_3.

$$Fe + 3O_2 \longrightarrow 2Fe_2O_3 \quad \text{(Not balanced)}$$

We now have four atoms of iron on the right side. We can get four on the left by placing the coefficient 4 in front of Fe.

$$4\,Fe + 3\,O_2 \longrightarrow 2\,Fe_2O_3 \quad \text{(Balanced)}$$

The equation is now balanced.

Balance the following equation.

$$CH_4 + O_2 \longrightarrow CO_2 + H_2O \quad \text{(Not balanced)}$$

example
5.3

   In this equation, oxygen appears in two different products. In this case, we'll leave the oxygen for last and balance the other two elements first. Carbon is already balanced, with one atom on each side of the equation. The least common multiple of 2 and 4 is 4, so to balance hydrogen we place the coefficient 2 in front of H_2O and we have four hydrogens on each side.

$$CH_4 + O_2 \longrightarrow CO_2 + 2\,H_2O \quad \text{(Not balanced)}$$

Now for the oxygen. There are four oxygens on the right. If we place a 2 in front of O_2, the oxygen will balance.

$$CH_4 + 2\,O_2 \longrightarrow CO_2 + 2\,H_2O \quad \text{(Balanced)}$$

The whole equation is balanced.

Balance the following equation.

$$H_2SO_4 + NaCN \longrightarrow HCN + Na_2SO_4 \quad \text{(Not balanced)}$$

example
5.4

   Here we have an equation that involves compounds incorporating poly-atomic ions. The SO_4 group should be treated as a unit and balanced as a whole. The same is true of the CN group. As the equation is presently written, the SO_4 groups and the CN groups are balanced, but the hydrogens and the sodiums are not. The least common multiple for sodium is 2, so to get two sodiums on the left we place a 2 before NaCN. The least common multiple for hydrogen is 2, so to get two hydrogens on the right we place a 2 before HCN.

$$H_2SO_4 + 2\,NaCN \longrightarrow 2\,HCN + Na_2SO_4 \quad \text{(Balanced)}$$

It turns out that these same coefficients balance the CN groups and the SO_4 groups, and the equation as a whole is balanced.

We have made the task of balancing equations deceptively easy. Among the problems facing chemists is that of balancing equations for reactions such as the following:

$$H_2SO_4 + K_2Cr_2O_7 + C_9H_{10} \longrightarrow$$
$$Cr_2(SO_4)_3 + H_2O + C_7H_6O_2 + CO_2 + K_2SO_4 \qquad \text{(Not balanced)}$$

The answer is

$$32\,H_2SO_4 + 8\,K_2Cr_2O_7 + 3\,C_9H_{10} \longrightarrow$$
$$8\,Cr_2(SO_4)_3 + 38\,H_2O + 3\,C_7H_6O_2 + 6\,CO_2 + 8\,K_2SO_4 \qquad \text{(Balanced)}$$

Elaborate schemes have been developed for balancing such equations. Most of you reading this, however, are not going into careers in chemistry, and you will not be faced with balancing so complicated an equation. But you should understand what is meant by a balanced equation and know how to handle simple systems.

5.2 GASES: GAY-LUSSAC'S LAW OF COMBINING VOLUMES

Chemists generally cannot work with individual atoms and molecules. Even the tiniest speck of matter that we can see contains billions of billions of atoms. John Dalton postulated that atoms of different elements had different weights. Therefore, equal weights of different elements would contain different numbers of atoms. Consider the similar situation of golf balls and Ping-Pong balls. A kilogram of golf balls contains a smaller number of balls than a kilogram of Ping-Pong balls. One could determine the number of balls in each case simply by counting them. For atoms, however, such a straightforward method is not available. It was in the experiments of a French chemist and the mind of an Italian scientist that approaches to the problem of numbering atoms were found.

In 1809, Joseph Louis Gay-Lussac announced the results of some chemical reactions that he had carried out with gases. He found that, when all measurements were made at the same temperature and pressure, the volumes of gaseous reactants and products were in a small whole-number ratio. For example, when he allowed hydrogen to react with oxygen to form steam at $100\,^{\circ}C$, two volumes of hydrogen would unite with one volume of oxygen to give two volumes of steam (Figure 5.3). The small whole-number ratio was $2:1:2$.

In another experiment (Figure 5.4), Gay-Lussac found that two volumes of carbon monoxide combined with one volume of oxygen to give two volumes of carbon dioxide ($2:1:2$).

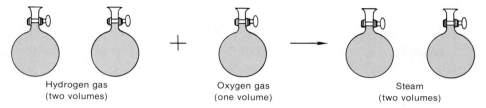

FIGURE 5.3 *Two volumes of hydrogen gas react with one volume of oxygen gas to give two volumes of steam.*

The ratio is not always 2:1:2. If hydrogen is permitted to react with nitrogen (Figure 5.5), the product formed is ammonia, and the combining volumes are three of hydrogen with one of nitrogen to give two of ammonia (3:1:2).

Gay-Lussac thought there must be some relationship between the *numbers* of molecules and the *volumes* of gaseous reactants and products. But it was the Italian chemist Amadeo Avogadro who first explained the **law of combining volumes.** His hypothesis, based on shrewd interpretation of experimental facts, was that equal volumes of all gases (at the same temperature and pressure) contain the same number of molecules (Figure 5.6). (It was also Avogadro who first suggested that certain elements such as hydrogen, oxygen, and nitrogen were made up of diatomic molecules, i.e., molecules containing two atoms each.)

The equation for the combination of hydrogen and oxygen to form water (steam) is

$$2\,H_2 + O_2 \longrightarrow 2\,H_2O$$

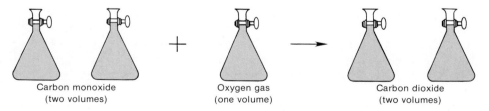

FIGURE 5.4 *Two volumes of carbon monoxide gas react with one volume of oxygen gas to give two volumes of carbon dioxide gas.*

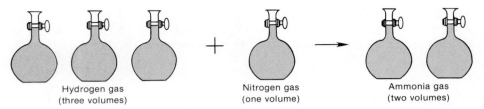

FIGURE 5.5   *Three volumes of hydrogen gas react with one volume of nitrogen gas to give two volumes of ammonia gas.*

The coefficients of the molecules are the same as the combining ratio of the gas volumes, $2:1:2$ (see Figure 5.3). For the reaction of carbon monoxide and oxygen (see Figure 5.4), the equation is

$$2\,CO + O_2 \longrightarrow 2\,CO_2$$

The coefficients again reflect the combining volumes. Finally, the formation of ammonia is described in the following equation.

$$3\,H_2 + N_2 \longrightarrow 2\,NH_3$$

The coefficients are identical to the factors of the combining ratio (see Figure 5.5). The equation says that a nitrogen molecule reacts with three hydrogen molecules to produce two ammonia molecules. It also indicates that if you had 1 million nitrogen molecules, you would need 3 million hydrogen molecules to produce 2 million ammonia molecules. The equation provides the combining

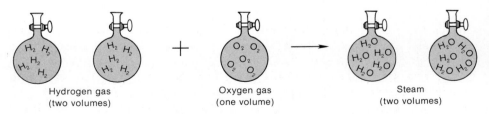

FIGURE 5.6   *Avogadro's explanation of Gay-Lussac's law of combining volumes. Equal volumes of each of the gases contain the same number of molecules.*

ratios. If identical volumes of gases contain identical numbers of molecules, then, according to the equation, one volume of nitrogen reacts with three volumes of hydrogen to produce two volumes of ammonia.

According to the equation

$$CH_4 \longrightarrow C + 2H_2$$

what volume of hydrogen would be obtained from 1 L of methane (CH_4)?
  The equation indicates that two volumes of hydrogen are obtained from the reaction of one volume of methane. Therefore, 1 L of methane would yield 2 L of hydrogen.

example 5.5

Using the same equation, calculate what volume of hydrogen would be obtained from the reaction of 25 L of methane.
  The ratio is one volume of methane to two volumes of hydrogen.

$$25 \text{ L of methane} \times \frac{2 \text{ L of hydrogen}}{1 \text{ L of methane}} = 50 \text{ L of hydrogen}$$

example 5.6

Using the same equation, calculate how much methane must have reacted to produce 10 L of H_2.

$$10 \text{ L of } H_2 \times \frac{1 \text{ L of } CH_4}{2 \text{ L of } H_2} = 5 \text{ L of } CH_4$$

example 5.7

5.3 AVOGADRO'S NUMBER

Recall (Chapter 2) that Dalton set up a table of *relative* atomic weights. This table has been modified through the years. The currently accepted relative atomic weights are listed on the inside front cover.
  Chemists can't weigh individual atoms or molecules. What they can do to obtain relative weights is weigh *equal numbers* of atoms or molecules of different substances. Avogadro's hypothesis (verified many times through the years) gave chemists a way to do that by measuring the weights of equal volumes of gases.

Once the relative weights of atoms are known, it is possible to plan reactions so no materials are wasted. The relative atomic weight of carbon is 12 (carbon weighs 12 times as much as hydrogen, which is assigned the weight of 1). On this basis, the relative weight of oxygen is 16. If we could (we can't) weigh out 12 amu of carbon and 16 amu of oxygen, we would have one carbon atom and one oxygen atom. From these we could make one molecule of carbon monoxide (CO). If we weigh out 12 g of carbon and 16 g of oxygen, we still have the proper *ratio* of atoms—and these quantities are easily weighed. We would obtain from these amounts of reactants 28 g of CO, with none of the reactants left over. Similarly, if we wished to make carbon dioxide (CO_2), we would weigh out 12 g of carbon and 32 g of oxygen in order to get exactly 44 g of CO_2. Again, all of the reactants would be consumed in the reaction.

How many carbon atoms are there in 12 g of carbon? Certain techniques have enabled scientists to determine (indirectly) the number of atoms or molecules in weighed samples of some materials. The number of atoms in 12 g of carbon is unimaginably large. In exponential form it is written 6.02×10^{23}. Scientists love to find ways of making very large numbers meaningful. Here's one: if you had a fortune worth 6.02×10^{23} dollars, you could spend a billion dollars each second of your entire life and have used only about 0.001% of your money. Here's another: if carbon atoms were the size of peas, 6.02×10^{23} of them would cover the surface of the earth to a depth of about 15 m, easily burying the Jolly Green Giant. That's a lot of peas!

The number 6.02×10^{23}—called **Avogadro's number**—is a fundamental one. There are 6.02×10^{23} atoms of oxygen in 16 g of oxygen and 6.02×10^{23} atoms of hydrogen in 1 g of hydrogen. The atomic weight of any element, expressed in grams, contains 6.02×10^{23} atoms.

5.4 MOLES OF MOLECULES, ATOMS, OR IONS

If we allow Avogadro's number of carbon atoms to react with Avogadro's number of oxygen *molecules* (each molecule containing two oxygen atoms), we get Avogadro's number of carbon dioxide molecules (Figure 5.7).

The International Bureau of Weights and Measures has defined the mole as one of seven basic SI units. A **mole** (mol) is the amount of a substance containing as many elementary units as there are atoms in exactly 12 g of the carbon-12 isotope ($^{12}_{6}C$). The elementary units must be specified by a chemical formula. They may be molecules (such as O_2 or CO_2 or even $C_{24}H_{50}O_4N_5Cl$), they may be atoms (such as C or O), they may be ions (such as SO_4^{2-} or K^+), or they may be pairs of ions (such as NaCl).

Since there are 6.02×10^{23} atoms of carbon-12 in 12 g of the isotope, we may define a mole of carbon dioxide as 6.02×10^{23} molecules of carbon dioxide.

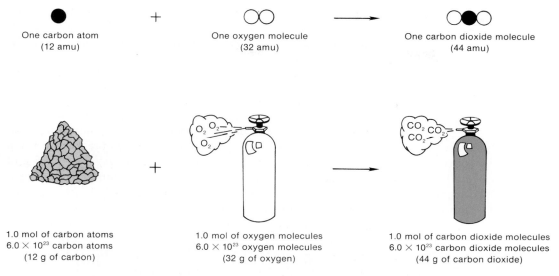

One carbon atom
(12 amu)

+

One oxygen molecule
(32 amu)

One carbon dioxide molecule
(44 amu)

1.0 mol of carbon atoms
6.0×10^{23} carbon atoms
(12 g of carbon)

+

1.0 mol of oxygen molecules
6.0×10^{23} oxygen molecules
(32 g of oxygen)

1.0 mol of carbon dioxide molecules
6.0×10^{23} carbon dioxide molecules
(44 g of carbon dioxide)

FIGURE 5.7 *We cannot weigh single atoms or molecules, but we can weigh equal numbers of these fundamental particles.*

A mole of sodium chloride is 6.02×10^{23} ion pairs of sodium chloride, that is, 6.02×10^{23} sodium ions and the same number of chloride ions.

The mole used in chemistry is something like the dozen we use every day. You can have a dozen white eggs, a dozen brown eggs, or a dozen chickens. In each case, you mean you have 12 of whatever you specified. A mole simply means 6.02×10^{23} of whatever you're talking about.

Now 6.02×10^{23} is, from one standpoint, not a pretty number. A million is a nice number; so, also, is a million million. However, the real beauty of 6.02×10^{23} lies in the fact that it is easy to calculate the mass of this number of particles, that is, the mass of a mole. All one has to do is measure out in grams one formula weight of a substance. The **formula weight** of a substance is merely the sum of the atomic weights of the atoms in the formula.

Calculate the formula weight of carbon dioxide (CO_2).

example 5.8

$$1 \times \text{the atomic weight of C} = 1 \times 12 = 12$$
$$2 \times \text{the atomic weight of O} = 2 \times 16 = 32$$

So the formula weight of CO_2 is $12 + 32 = 44$. Thus, 1 mol of CO_2 weighs 44 g.

5.5 CHEMICAL ARITHMETIC

Chemical equations, therefore, represent not only atom ratios but mass ratios as well. The equation

$$C + O_2 \longrightarrow CO_2$$

tells us that 12 g of carbon reacts with 32 g of oxygen to give 44 g of CO_2.

We need not use exactly 1 mol of each reactant. The important thing is to keep the ratio constant. For example, in the reaction above the mass ratio of oxygen to carbon is 32:12, or 8:3. We could use 8 g of oxygen and 3 g of carbon to produce 11 g of CO_2. In fact, to calculate the amount of oxygen needed to react with a given amount of carbon, we need only to multiply the amount of carbon by the factor $\frac{32}{12}$ (Figure 5.8).

example 5.9

Calculate the mass of oxygen needed to react with 10 g of carbon.

$$10 \text{ g of C} \times \frac{32 \text{ g of O}_2}{12 \text{ g of C}} = 27 \text{ g of O}_2$$

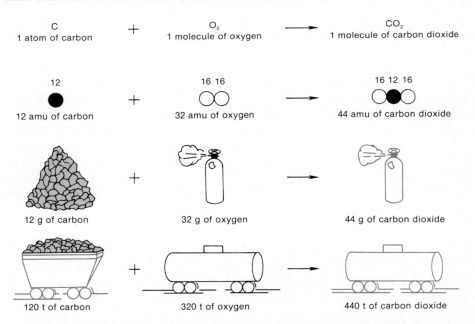

FIGURE 5.8   *Chemical equations express mass ratios as well as numbers of atoms and molecules.*

We can set up conversion factors for any two compounds involved in this reaction. (Conversion factors are reviewed in Special Topic A.) Typical problems will always ask you to calculate how much of one compound is equivalent to a given amount of one of the other compounds. All you need to do to solve such a problem is put together a conversion factor relating the two compounds. The equation supplies you with the data necessary for constructing the conversion factors. From the formulas in the equation

$$C + O_2 \longrightarrow CO_2$$

you know that the formula weights of carbon, oxygen gas, and carbon dioxide are, respectively, 12, 32, and 44. The following examples show the derivation of the conversion factors from the formula weights.

Calculate the mass of carbon that can be burned to carbon dioxide in the presence of 400 g of oxygen. (Hint: You are given the amount of oxygen and are asked for the amount of carbon. You need a conversion factor relating carbon to oxygen.)

$$400 \text{ g of } O_2 \times \frac{12 \text{ g of C}}{32 \text{ g of } O_2} = 150 \text{ g of C}$$

example 5.10

How many grams of carbon dioxide can be obtained from 48 g of carbon, assuming sufficient oxygen? (Hint: You are given the amount of carbon and are asked for the amount of carbon dioxide. You need a conversion factor relating carbon dioxide to carbon.)

$$48 \text{ g of C} \times \frac{44 \text{ g of } CO_2}{12 \text{ g of C}} = 176 \text{ g of } CO_2$$

example 5.11

How many grams of carbon are needed to produce 500 g of carbon dioxide? (Hint: You are given the amount of carbon dioxide and are asked for the amount of carbon. You need a conversion factor relating carbon to carbon dioxide.)

$$500 \text{ g of } CO_2 \times \frac{12 \text{ g of C}}{44 \text{ g of } CO_2} = 136 \text{ g of C}$$

example 5.12

5.6 THE MOLE METHOD

The examples of the previous section were based on a very simple reaction. In the balanced equation, there were no coefficients other than 1. That is, one molecule of oxygen reacted with one carbon atom to produce one carbon dioxide molecule. When a balanced equation does involve coefficients other than 1, that fact introduces a slight complication into the calculation. There are a number of ways of dealing with this complication. We shall illustrate the **mole method,** which is probably the most versatile. In principle we are doing the same thing we did in Section 5.5, only now we are going to get a mole ratio instead of a gram ratio from the chemical equation.

Before we work with an equation, let's try a conversion between moles and grams for a single compound.

example 5.13

Your friendly neighbor, an alchemist, is working on a secret solution. In response to your offer to help, he asks you to weigh out 0.5 mol of ammonium sulfate [$(NH_4)_2SO_4$]. How many grams would you weigh out?

Calculate the formula weight first, remembering that "$(NH_4)_2$" means that everything within the parentheses should be multiplied by 2.

$$2 \times \text{the atomic weight of N} = 2 \times 14 = 28$$
$$8 \times \text{the atomic weight of H} = 8 \times \ 1 = \ 8$$
$$1 \times \text{the atomic weight of S} = 1 \times 32 = 32$$
$$4 \times \text{the atomic weight of O} = 4 \times 16 = 64$$

So the formula weight of $(NH_4)_2SO_4$ is $28 + 8 + 32 + 64 = 132$, and

$$1 \text{ mol of } (NH_4)_2SO_4 = 132 \text{ g of } (NH_4)_2SO_4$$

Now you have the conversion factor to use in your calculation.

$$0.5 \text{ mol of } (NH_4)_2SO_4 \times \frac{132 \text{ g of } (NH_4)_2SO_4}{1 \text{ mol of } (NH_4)_2SO_4} = 66 \text{ g of } (NH_4)_2SO_4$$

Now we're ready to tackle an equation. We shall work out the sort of problem industrial chemists face every day. Whether chemists are making drugs, obtaining metals from their ores, synthesizing plastics, or whatever, they must use materials economically. The mole is the method.

Ammonia (for fertilizer and other uses) is made by causing hydrogen and nitrogen to react at high temperature and pressure. The equation is

example 5.14

$$N_2 + H_2 \longrightarrow NH_3 \quad \text{(Not balanced)}$$

How many grams of ammonia can be made from 60 g of hydrogen?
  First, we must balance the equation (see Example 5.1).

$$N_2 + 3 H_2 \longrightarrow 2 NH_3 \quad \text{(Balanced)}$$

Next, we calculate the formula weights of the compounds involved in the problem, that is, of hydrogen and ammonia.

$$H_2 \text{ has a mass of 2}$$
$$NH_3 \text{ has a mass of 17}$$

We are given an amount of hydrogen and are asked for a mass of ammonia. We must convert the mass given to moles.

$$60 \text{ g of } H_2 \times \frac{1 \text{ mol of } H_2}{2 \text{ g of } H_2} = 30 \text{ mol of } H_2$$

Next, we must use the coefficients given in the equation to determine the number of moles of ammonia that can be obtained from the calculated moles of hydrogen gas. The equation tells us that 3 mol of H_2 produce 2 mol of NH_3 Thus,

$$30 \text{ mol of } H_2 \times \frac{2 \text{ mol of } NH_3}{3 \text{ mol of } H_2} = 20 \text{ mol of } NH_3$$

  Finally, we must convert moles of ammonia to grams of ammonia.

$$20 \text{ mol of } NH_3 \times \frac{17 \text{ g of } NH_3}{1 \text{ mol of } NH_3} = 340 \text{ g of } NH_3$$

  Or we could do it all in one step.

$$60 \text{ g of } H_2 \times \frac{1 \text{ mol of } H_2}{2 \text{ g of } H_2} \times \frac{2 \text{ mol of } NH_3}{3 \text{ mol of } H_2} \times \frac{17 \text{ g of } NH_3}{1 \text{ mol of } NH_3} = 340 \text{ g of } NH_3$$

  The sequence of steps to be followed in the mole method is

1. Balance the equation.
2. Determine the formula weights of materials of interest.

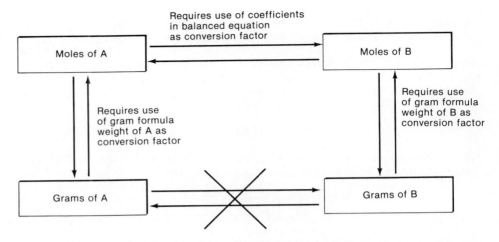

FIGURE 5.9 Outline of the procedure for using chemical equations to relate amounts of reactants and products. Direct gram-to-gram conversion is not possible.

3. Calculate number of moles from the number of grams *given* and the formula weight.
4. Calculate (using the ratio specified by the equation) the number of moles of the compound you were asked about from the number of moles calculated in Step 3.
5. Finally, convert the number of moles determined in Step 4 to the mass in grams, using the appropriate formula weight. (See Figure 5.9.)

example 5.15

How many grams of oxygen must be consumed if the following reaction is to produce 9 g of water?

$$H_2 + O_2 \longrightarrow H_2O \quad \text{(Not balanced)}$$

The balanced equation is

$$2\,H_2 + O_2 \longrightarrow 2\,H_2O$$

The formula weight for oxygen gas is 32, and the formula weight for water is 18.

$$9 \text{ g of } H_2O \times \frac{1 \text{ mol of } H_2O}{18 \text{ g of } H_2O} \times \frac{1 \text{ mol of } O_2}{2 \text{ mol of } H_2O} \times \frac{32 \text{ g of } O_2}{1 \text{ mol of } O_2} = 8 \text{ g of } O_2$$

5.7 STRUCTURE, STABILITY, AND SPONTANEITY

We saw at the beginning of Chapter 4 that some electron configurations are more stable than others. Sodium *ions* and chloride *ions,* arranged in a crystal lattice, are less reactive than sodium *atoms* and chlorine *molecules.* We do know that if sodium metal and chlorine gas are mixed, a vigorous reaction ensues.

$$2\,Na + Cl_2 \longrightarrow 2\,NaCl$$

A great deal of energy as heat and light is produced during the reaction. Sometimes this energy is listed as one of the products in the chemical equation.

$$2\,Na + Cl_2 \longrightarrow 2\,NaCl + energy$$

Energy changes associated with a particular reaction are quantitatively related to the *amount* of chemicals changed. For example, burning 16 g (1 mol) of methane to form carbon dioxide and water releases 192 kcal of energy as heat.

$$CH_4 + 2\,O_2 \longrightarrow CO_2 + 2\,H_2O + 192\ kcal$$

When this information is included in the equation, it is possible to calculate the amount of heat produced from a certain amount of reactant in much the same way as masses of material products are calculated.

The reaction between hydrogen and chlorine to form hydrogen chloride gas results in the release of 44 kcal of heat per mole of hydrogen consumed.

example 5.16

$$H_2 + Cl_2 \longrightarrow 2\,HCl + 44\ kcal$$

How much heat would be released by the reaction of 40 g of hydrogen with chlorine?

First, calculate the number of moles of hydrogen, using the formula weight.

$$40\ g\ of\ H_2 \times \frac{1\ mol\ of\ H_2}{2\ g\ of\ H_2} = 20\ mol\ of\ H_2$$

Then, using the information given by the equation, calculate the heat produced.

$$20\ mol\ of\ H_2 \times \frac{44\ kcal}{1\ mol\ of\ H_2} = 880\ kcal$$

Or

$$40\ g\ of\ H_2 \times \frac{1\ mol\ of\ H_2}{2\ g\ of\ H_2} \times \frac{44\ kcal}{1\ mol\ of\ H_2} = 880\ kcal$$

Chemical reactions that result in the release of heat are said to be **exothermic.** The burning of methane is an exothermic reaction, as is the burning of gasoline or coal. In each case, chemical energy is converted into heat energy. There are other reactions, such as the decomposition of water, for which energy must be supplied.

$$2\,H_2O + 137\text{ kcal} \longrightarrow 2\,H_2 + O_2$$

If the energy is supplied as heat, such reactions are said to be **endothermic.** It takes 137 kcal of energy to decompose 36 g (2 mol) of water into hydrogen and oxygen. It should be noted that the same amount of energy is released when enough hydrogen is burned to form 36 g (2 mol) of water.

$$2\,H_2 + O_2 \longrightarrow 2\,H_2O + 137\text{ kcal}$$

The energy released in an exothermic reaction is related to the *chemical energy* present in the reactants. This energy is a form of *potential energy.* Potential energy is released (as heat or light, for example) when reactants are brought together in a way that allows them to achieve more stable electron arrangements.

To say that a reaction is exothermic does not necessarily mean that it is instantaneous. For example, coal (carbon) has a lot of chemical energy. Carbon reacts with oxygen to form carbon dioxide with the release of considerable heat. But coal doesn't react very rapidly with oxygen at ordinary temperatures. In fact, one can store a pile of coal in air indefinitely without perceptible change. Before coal will react with oxygen to release its stored energy, it must be heated to a temperature of several hundred degrees. The coal must be supplied with a certain amount of energy (called the energy of activation; Section 5.8) before it will begin to burn steadily. Once this energy of activation is supplied, the heat evolved in the reaction will keep the coal burning brightly, and the energy eventually produced will exceed the amount of energy required to start the reaction. Overall, you get more energy out than you put in. The reaction is exothermic. In an endothermic reaction, more energy goes in than comes out.

We will return to the burning of coal in a moment, but first let's consider an analogy involving a more familiar situation. If you were in Browning, Montana (elevation 1300 m), and wished to travel to Kalispell (elevation 900 m), you could choose to drive the scenic Going-to-the-Sun Highway through Glacier National Park. To do this you would have to cross the continental divide at Logan Pass (elevation 2000 m). First, you would have to climb 700 m, but then it would be downhill the rest of the way (Figure 5.10).

The reaction of coal (carbon) with oxygen can be explained in much the same way. The reactants, carbon and oxygen, lie in a rather high potential energy valley (Figure 5.11). A certain amount of energy has to be put into the

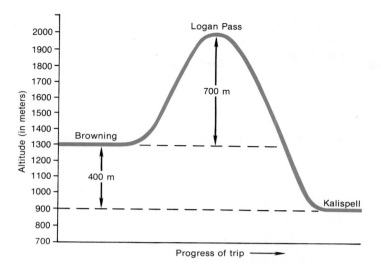

FIGURE 5.10   *To get from Browning, Montana, to Kalispell via Going-to-the-Sun Highway, we would first have to climb to Logan Pass, even though Kalispell is 400 m lower in elevation than Browning.*

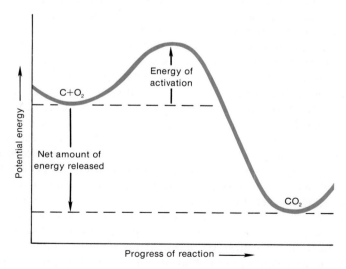

FIGURE 5.11   *To get from reactants (carbon and oxygen) to product (carbon dioxide), we must first put some energy into the system.*

system for it to get over the top of the "mountain." Once the reaction is under way, the heat released more than compensates for the energy needed to get the reaction going in the first place.

5.8 REVERSIBLE REACTIONS

Living cells take in oxygen and "burn" glucose to obtain energy.

$$C_6H_{12}O_6 + 6\,O_2 \longrightarrow 6\,CO_2 + 6\,H_2O + energy$$

Glucose

Green plants carry out the reverse reaction (an endothermic reaction called **photosynthesis**), using energy from sunlight to convert carbon dioxide and water to glucose and oxygen.

$$6\,CO_2 + 6\,H_2O + energy \longrightarrow C_6H_{12}O_6 + 6\,O_2$$

All life on our planet is based on this **reversible** reaction. But what makes it go one way in one case and the opposite way in another? One obvious answer is energy. Since energy is released when glucose is oxidized, the reactants (glucose and oxygen) must be at a higher energy level than the products (carbon dioxide and water). As we have indicated previously, this does not mean that the path is straight downhill from reactants to products. Indeed, if glucose and oxygen are mixed at ordinary temperatures outside a cell, no perceptible reaction occurs. As with the reaction of coal and oxygen, a certain amount of energy, the energy of activation, must be supplied before the reaction takes place. The potential energy diagram for this reaction (Figure 5.12) strongly resembles that for the coal and oxygen reaction (see Figure 5.11).

For reversible reactions, the energy hill, or the barrier to reaction, may be approached from either side. That is, we can read the diagram from left to right or right to left. With the burning of glucose, we are considering an exothermic reaction in which higher-energy reactants are converted to lower-energy products. When read in the reverse direction, the diagram presents the energy changes that occur in the endothermic reaction called photosynthesis. The climb to the top of the barrier (the energy of activation) is longer from one side than from the other. The **energy of activation** is the difference in energy between the level the reactants are on and the top of the energy hill. When the reactants are in the valley at the left (as in the forward reaction), the climb to the top is not so high. When the reactants are in the valley to the right (as in the reverse reaction), the climb to the top is much longer. The potential energy diagram shown here is much simplified. Just as one seldom crosses a mountain by going straight up one side and straight down the other, so reactions seldom proceed by a smooth, one-hump potential energy change.

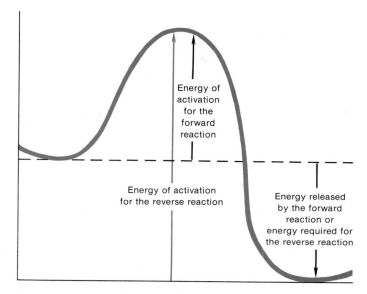

FIGURE 5.12 *Potential energy diagram for a reversible reaction.*

5.9 REACTION RATES: COLLISIONS, ORIENTATION, AND ENERGY

Before atoms, molecules, or ions can react, they must first get together: they must collide. Second, for all except the simplest particles, they must come together in the proper orientation. Third, the collision must provide a certain minimum energy, the energy of activation.

First, let's consider the frequency of collision, which is influenced by two factors, concentration and temperature. The more concentrated the reactants, the more frequently the particles will collide, simply because there are more of them in a given volume. Molecules also move faster at higher temperatures; thus, an increase in temperature also increases the frequency of collision. The effects of concentration and temperature on reaction rates will be discussed in more detail in following sections.

Now, though, let's take a closer look at the effect of orientation. It might not be obvious that orientation is important. Sometimes, though rarely, it's not. Consider a situation in which you want to knock someone down. You can tackle from the front, back, or side and still accomplish your objective. If, however, a kiss is the objective, then an orientation in which the participants are face-to-face works best. For chemical reactions, there are also a few instances in which orientation is not important. If two hydrogen atoms are to react to form a hydrogen molecule,

$$2\,H \longrightarrow H_2$$

then orientation is unimportant. The quantum mechanical view of hydrogen atoms depicts them as symmetrical (spherical) electron clouds. Front, back, top, and bottom all look the same. Therefore, all orientations of two hydrogen atoms are identical. If such atoms come into contact with one another, they can react.

For most particles, however, proper orientation is a prerequisite to reaction. Consider the molecule

```
      H   H   H   H
      |   |   |   |
  H—C—C—C—C—C=C—H
      |   |   |   |
      H   H   H   H   H   H
```

The two carbon atoms sharing the double bond can react with bromine to form

```
      H   H   H   H   Br  Br
      |   |   |   |   |   |
  H—C—C—C—C—C—C—H
      |   |   |   |   |   |
      H   H   H   H   H   H
```

Now, it is possible for bromine to strike the original compound anywhere along its length. However, only if the collision occurs in such a way as to bring bromine into contact with the electrons in the double bond will the reaction occur. It is even true that the bromine must approach the electrons of the double bond from a certain direction. Here is a case in which orientation is very important. We shall point out another example of proper orientation shortly.

The third factor, that collisions must provide a certain minimum energy, is a good deal more subtle. There is a temperature effect, which will be discussed in some detail in the next section. Certain substances, called catalysts, may substantially lower this activation energy, thus increasing the reaction rate. The effect of catalysts will be discussed more thoroughly in Section 5.11.

5.10 REACTION RATES: THE EFFECT OF TEMPERATURE

Reactions generally take place at a faster rate when temperatures are high. We have seen that carbon reacts very slowly with oxygen at room temperature. When heated, the coal reacts much faster. The effect of temperature on the rates of chemical reactions is explained by the kinetic-molecular theory. We shall consider this theory in more detail in the next chapter. For the moment, we will cite one of its postulates, which states that, at high temperatures, molecules move more rapidly. Thus, they collide more frequently, providing increased chance for reaction.

At high temperatures, the rapidly moving molecules also strike one another harder. The harder these collisions are, the more energy is involved in them. These harder collisions are more likely to supply the activation energy needed to break the chemical bonds and get the reaction going. Consider the reaction that takes place between hydrogen gas and chlorine gas. In order for hydrogen and chlorine to react, bonds between hydrogen atoms and between chlorine atoms must be broken.*

$$H\text{---}H + Cl\text{---}Cl \longrightarrow 2\,H\text{---}Cl$$

This is an exothermic reaction. Once it has started, the energy released by the formation of hydrogen-to-chlorine bonds more than compensates for that required to break H—H and Cl—Cl bonds. There is a net conversion of chemical energy to heat energy.

In endothermic reactions, the energy released by bond formation is less than that required to break the necessary bonds. For these reactions, energy must be supplied continuously from an external source, or the process will stop. A typical endothermic reaction is the decomposition of a salt called potassium chlorate ($KClO_3$) to give oxygen and another salt, potassium chloride (KCl).

$$2\,KClO_3 + heat \longrightarrow 2\,KCl + 3\,O_2$$

The potassium chlorate must be heated continuously. If the source of heat is removed the reaction quickly subsides.

We make use in our daily lives of our knowledge of the effect of temperature on chemical reactions. For example, we freeze foods to retard those chemical reactions that lead to spoilage. On the other hand, if we want to cook our food more rapidly, we turn up the heat. There is a general rule of thumb that reaction rates double for every temperature increase of 10 °C. (There are many exceptions to this rule, and it should be applied only if there are no other data available.) The chemical reactions that occur in our bodies generally do so at a constant temperature of 37 °C (98.6 °F). A few degrees' rise in temperature (fever) leads to an increase in respiration, pulse rate, and other physiological reactions. A drop in body temperature of a few degrees slows these same processes considerably, as is exemplified by the slowed metabolism of hibernating animals. Use is made of this phenomenon in some surgical procedures. In some cases of heart surgery, the body temperature of the patient is lowered to about 15 °C (60 °F). Ordinarily, the brain is permanently damaged when its oxygen

---

* This statement says nothing about the *order* in which the bonds are broken. This sequence, called the mechanism of a reaction, can be quite complicated (see p. 174).

supply is interrupted for more than 5 min. But at the lower temperature, metabolic processes slow considerably, and the brain may survive much longer periods of oxygen deprivation. The surgeon can stop the heartbeat, perform an hour-long surgical procedure on the heart, and then restart the heart and bring the patient's temperature back to normal (Figure 5.13).

It must be said that, despite these examples, the manipulation of reaction rates in living systems through changes in temperature is severely restricted.

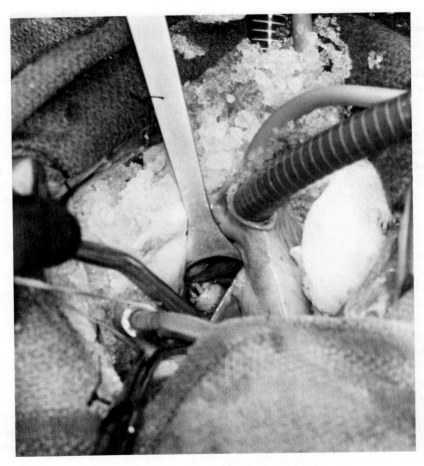

FIGURE 5.13 Heart surgery can be performed with the patient's temperature lowered to slow metabolic processes and thus minimize the possibility of brain damage. After cooling the heart with fluids, the area around the heart is often packed with ice to keep it cold. (Photo courtesy of J. Ernesto Molino, University of Minnesota Medical School.)

Increasing temperature, for example, may "kill" enzymes (Section 5.11) that mediate cellular chemistry and are essential to life. We kill germs by heat sterilization in autoclaves. For living cells, there is often a rather narrow range of optimum temperatures. Both higher and lower temperatures can be disabling, if not deadly.

5.11 REACTION RATES: CATALYSIS

Recall the endothermic reaction involving the decomposition of potassium chlorate to oxygen and potassium chloride (p. 169). For the oxygen to be produced at a useful rate, the potassium chlorate must be heated to over 400 °C. However, if we add a small amount of manganese dioxide (MnO_2), we can get the same rate of oxygen evolution by heating the reactant to just 250 °C. Further, after the reaction is complete, the manganese dioxide can be completely recovered, unchanged. A substance that, like manganese dioxide, changes the rate of a chemical reaction without itself being changed is called a **catalyst.** In general, catalysts act by lowering the activation energy required for the reaction to occur (Figure 5.14). If activation energy is lower, then the collisions of more slowly moving molecules will be sufficient to supply that energy. The lower temperature required for a catalyzed reaction reflects this. This energy of activation is lowered because the catalyst changes the *path* of the reaction.

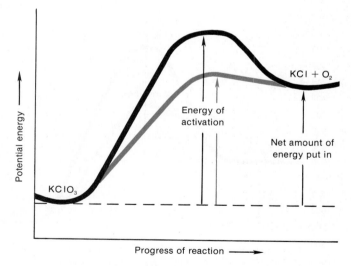

FIGURE 5.14 *In the decomposition of potassium chlorate, a catalyst acts to lower the energy of activation (colored arrow) compared with that required for the uncatalyzed reaction.*

To return to our analogy of the trip from Browning to Kalispell, it is possible to take an alternate route. U.S. Highway 2 crosses the continental divide through Marias Pass. This route involves a climb of only 300 m, compared with 700 m via Logan Pass (Figure 5.15). This alternate route is analogous to that provided by a catalyst in a chemical reaction.

Catalysts are of great importance in the chemical industry. A reaction that would otherwise be so slow as to be impractical can be made to proceed at a reasonable rate with the proper catalyst.

Catalysts are even more important in living organisms, where raising the temperature by 100 °C is not a feasible way of increasing the rate of critical reactions. If we raised our body temperature by 100 °C, we'd boil our blood, among other things, and our fatty tissue would melt (though that might not be undesirable). Biological catalysts, called **enzymes,** mediate nearly all the chemical reactions that take place in living systems. But these catalysts themselves may suffer irreversible damage when living cells are subjected to too high a temperature. Once a catalyst is deactivated (rendered inactive), the reactions that required the catalyst no longer proceed at a rate that maintains life. In addition to being examples of catalysts, enzymes offer a striking illustration of the importance of orientation in chemical reactions. These compounds are huge molecules, each possessing what is called an active site. The reactants must come into contact with this active site if the enzyme is to catalyze the reaction. If the reactants do not collide at the active site, no reaction occurs. So important are enzymes to life that we discuss them more fully in Chapter 24.

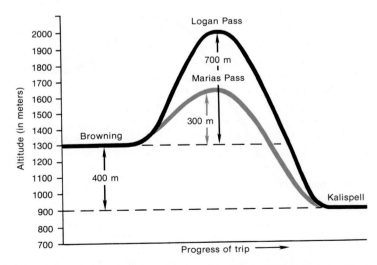

FIGURE 5.15 *To get from Browning to Kalispell via Highway 2 involves a lower energy barrier than the route via Logan Pass.*

5.12 REACTION RATES: THE EFFECT OF CONCENTRATION

Another factor affecting the rate of a chemical reaction is the concentration of reactants. The more reactant molecules there are in a volume of space, the more collisions will occur. The more collisions there are, the more reactions will occur. For example, you can light a wood splint and then blow out the flame. The splint will continue to glow as the wood reacts slowly with the oxygen of the air. If the glowing splint were placed in pure oxygen, the splint would burst into flame, indicating a much more rapid reaction. This more rapid reaction can be interpreted in terms of the concentration of oxygen. Air is about one-fifth oxygen. The concentration of O_2 molecules in pure oxygen is, therefore, about five times as great as in air. The caution against smoking in a hospital room where a patient is in an oxygen tent is not merely a concession to the sensitivity of nonsmokers. It is meant to prevent disaster.

For reactions in solution, the concentration of a reactant can be increased if more of it is dissolved. One of the first studies of reaction rate was done by Ludwig Wilhelmy in 1850. He studied the rate of reaction of sucrose (cane or beet sugar) with water. The products are two simpler sugars, glucose and fructose.

$$\text{Sucrose} + H_2O \xrightarrow{\text{HCl}} \text{glucose} + \text{fructose}$$

(We will not be concerned with the chemical formulas for these sugars at this time. The hydrochloric acid serves as a catalyst for the reaction. Chemists generally write formulas or names for catalysts above the arrow, since the catalyst is recovered unchanged.) Wilhelmy found that the rate of the reaction was proportional to the concentration of sucrose. If he dissolved 0.001 mol of sucrose in 1 L of water, the sucrose reacted at a certain rate. If he doubled the concentration of sucrose (i.e., if he dissolved 0.002 mol of sucrose in 1 L of water), the reaction rate was doubled.

In general, when the temperature is constant and the catalyst (if any) is present in fixed amount, the rate of the reaction can be related quantitatively to the amounts of reacting substances. The relationship is not necessarily a simple one, however. In order to know *which* reacting substances are involved in determining the rate, one must know the step-by-step detail of how the molecules collide, come apart, and recombine.

Consider again the photosynthesis reaction.

$$6\,CO_2 + 6\,H_2O + \text{energy} \longrightarrow C_6H_{12}O_6 + 6\,O_2$$

For this reaction to occur in one step, six carbon dioxide molecules and six water molecules would all have to come together at the same instant in an effective collision. Such an event is highly unlikely. In fact, even three-body collisions are rare. Most reactions proceed through a series of steps involving collisions of only two molecules each. Some steps may involve only the breaking of a bond of a single molecule. The photosynthesis reaction involves an extraordinarily complex web of intermediate steps, but each individual step is fairly simple. Virtually all steps require enzymes as catalysts.

The step-by-step process, or **mechanism,** by which a reaction takes place is determined by a study of the rate of the reaction as various factors, such as temperature, concentrations, and catalysts, are changed. Such studies are important in chemistry, for chemists want to get as much product as possible with a minimum expenditure of materials and energy. Investigations of mechanisms are also important in the chemistry of living cells. Many types of cancer are thought to be induced by chemicals. A lot of research is under way to work out the mechanism by which the chemicals act—or are acted upon—in the induction of cancer. A knowledge of the mechanism of various metabolic reactions has enabled scientists to determine how certain poisons work, and this knowledge has led to effective treatments in many instances. Certain diseases of genetic origin involve the disruption of a single step in the mechanism of a reaction essential to health. Again, this information has permitted physicians to deal successfully with what might otherwise have been a fatal defect.

5.13 EQUILIBRIUM IN CHEMICAL REACTIONS

For reversible reactions, those that proceed in both a forward and a backward direction, the concept of rate requires further consideration. In such reactions, the reverse reaction can't occur until some product molecules (that is, product with respect to the forward reaction) have been formed. Even then, because of the effect of concentration on rate, the reverse reaction will be slow until sufficient product has been formed to increase the chance of these molecules getting together to form reactants once more. Similarly, as the reactants change to products, there will be fewer collisions between reactant molecules, because their number becomes depleted. In isolated systems (ones that are cut off from outside influences), the rates of the forward and reverse reactions eventually become equal, and a condition called **equilibrium** is established (Figure 5.16). At equilibrium there *appears* to be no further reaction. In this case appearances are deceiving. Reactants are still changing to products, and products are still changing back to reactants. It is just that these changes are occurring at precisely the same rate. For every reactant molecule lost through the forward reaction, one is gained through the reverse reaction. Once equilibrium is established, we can measure no change in the concentration of reactants or products. Because molecules are still reacting, even though their concentrations don't change, we say that equilibrium is a dynamic situation.

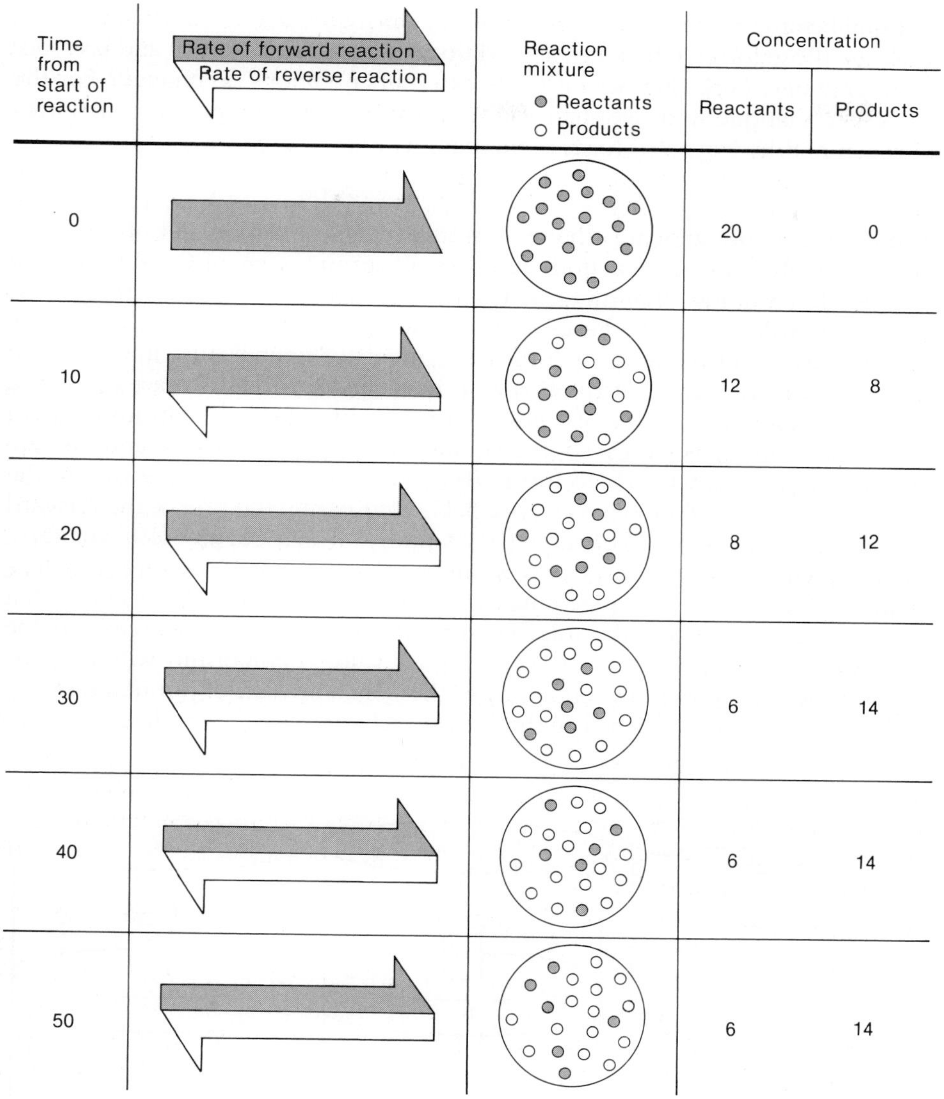

FIGURE 5.16 *Progress of reaction toward achieving equilibrium.*

Equilibrium, we mentioned, is established for reversible reactions in isolated systems, those for which external conditions such as temperature and pressure do not change. What happens if those external conditions do change? For the answer to that question, we shall once again use as an example the reaction of nitrogen and hydrogen to form ammonia.

$$N_2 + 3H_2 \rightleftharpoons 2NH_3 + energy$$

The double arrow indicates that this is a reversible reaction. All of the compounds involved in this reaction are gases. From the work of Gay-Lussac and Avogadro, we know that the coefficients given in the equation reflect the combining volumes of the gases. The reactants occupy four volumes and the products two. In Chapter 6, we study the effects of temperature and pressure on the volume of a gas. For the moment, let us just say that increased pressure tends to reduce the volume of a gas. If the pressure of the equilibrium system we are considering is increased, the equilibrium will be disrupted, and the rate at which N_2 reacts with H_2 to form NH_3 will increase; that is, the rate of the forward reaction will increase (Figure 5.17). Increasing the rate of the forward reaction has the effect of changing H_2 and N_2, which occupy four volumes, to NH_3, which occupies only two volumes. The increased pressure therefore has the effect of changing reactants that occupy four volumes to products that occupy only two. Thus, the total volume of the system will decrease. If the pressure is held constant at the new, higher value, equilibrium will again be established. But in the new equilibrium the concentration of ammonia (NH_3) will be higher than it was before the pressure changed. We sometimes describe

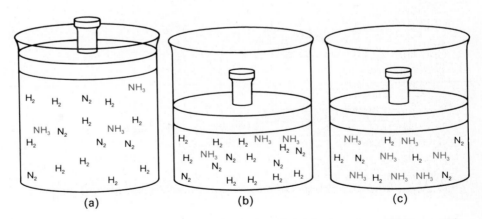

FIGURE 5.17 *Le Chatelier's principle illustrated. (a) System at equilibrium with 10 H_2, 5 N_2, and 3 NH_3, for a total of 18 molecules. (b) Same molecules forced into a smaller volume, creating a stress on the system. (c) Six H_2 and two N_2 have been converted to four NH_3. A new equilibrium has been established with 4 H_2, 3 N_2, and 7 NH_3, a total of 14 molecules. The stress is partially relieved by the reduction in the total number of molecules.*

this change by saying that the equilibrium has been shifted to the right. Thus, for the reversible reaction of nitrogen and hydrogen, an increase in pressure shifts the equilibrium to the right.

Observations of changes that occur in an equilibrium system when factors such as concentration, pressure, or temperature are changed were made in the nineteenth century. These effects were summarized in 1884 by a French chemist, Henri Louis Le Chatelier. His rule, still called **Le Chatelier's principle,** may be stated as follows: If a stress is applied to a system in equilibrium, the system rearranges in such a direction as to minimize the stress. If heat is added to the N_2—H_2—NH_3 system, the reaction will proceed to the left, using up heat (energy). If more nitrogen gas is added to the system, the reaction will go to the right, using up nitrogen. If additional hydrogen gas is introduced, the system will shift to the right, to use up hydrogen. Additional ammonia will cause the reaction to proceed to the left, using up ammonia.

The equilibrium can also be shifted by removal of one of the substances. Removal of ammonia will cause hydrogen and nitrogen to react, forming more ammonia. Removal of hydrogen will cause ammonia to break down and form more hydrogen (and, incidentally, more nitrogen). It should be noted, however, that a catalyst will not shift an equilibrium system. Catalysts change the rate of both the forward and reverse reactions. They do not change the position of the equilibrium, that is, the equilibrium concentrations of reactants and products.

What effect, if any, will each of the following changes have on the equilibrium in the reaction

$$2\,CO + O_2 \longrightarrow 2\,CO_2 + heat$$

a. addition of CO
b. removal of O_2
c. cooling the reaction mixture

d. increasing the pressure
e. adding a catalyst

_____

a. The reaction shifts to the right to use up the added CO.
b. The reaction shifts to the left to replace the O_2 that is removed.
c. The reaction shifts to the right to replace the lost heat.
d. The reaction shifts to the right to relieve the pressure by converting three molecules ($2\,CO + 1\,O_2$) into two molecules ($2\,CO_2$).
e. A catalyst has no effect on the position of equilibrium.

example
5.17

In Chapter 7, we discuss equilibria between liquid and solid phases and between liquid and vapor phases. In Chapter 9, we encounter equilibria involving solutions. The concept is quite important to our study of the chemistry of life.

PROBLEMS

1. Define or illustrate each of the following:
   a. formula weight
   b. mole
   c. Avogadro's number
   d. exothermic
   e. endothermic
   f. energy of activation
   g. catalyst
   h. reversible reaction
   i. equilibrium
   j. Le Chatelier's principle

2. Relate the law of conservation of matter to the need for working with balanced chemical equations.

3. Indicate how many hydrogen atoms per formula unit are in each of the following:
   a. NH_4NO_3
   b. $(NH_4)_2HPO_4$

4. How many atoms of each kind (Al, C, H, and O) are included in the notation 2 $Al(C_2H_3O_2)_3$?

5. Indicate whether the equations are balanced. You need not balance the equations. Just determine whether they are balanced as written.
   a. $Mg + H_2O \longrightarrow MgO + H_2$
   b. $FeCl_2 + Cl_2 \longrightarrow FeCl_3$
   c. $F_2 + H_2O \longrightarrow 2\,HF + O_2$
   d. $Ca + 2\,H_2O \longrightarrow Ca(OH)_2 + H_2$
   e. $2\,LiOH + CO_2 \longrightarrow Li_2CO_3 + H_2O$

6. Indicate whether the equations are balanced as written.
   a. $2\,KNO_3 + 10\,K \longrightarrow 6\,K_2O + N_2$
   b. $2\,NH_3 + O_2 \longrightarrow N_2 + 3\,H_2O$
   c. $4\,LiH + AlCl_3 \longrightarrow 2\,LiAlH_4 + 2\,LiCl$
   d. $SF_4 + 3\,H_2O \longrightarrow H_2SO_3 + 4\,HF$
   e. $4\,BF_3 + 3\,H_2O \longrightarrow H_3BO_3 + 3\,HBF_4$

7. Indicate whether the equations are balanced as written.
   a. $2\,Sn + 2\,H_2SO_4 \longrightarrow 2\,SnSO_4 + SO_2 + 2\,H_2O$
   b. $3\,Cl_2 + 6\,NaOH \longrightarrow 5\,NaCl + NaClO_3 + 3\,H_2O$

8. Balance the following chemical equations:
   a. $Al + O_2 \longrightarrow Al_2O_3$
   b. $C + O_2 \longrightarrow CO$
   c. $N_2 + O_2 \longrightarrow NO$
   d. $SO_2 + O_2 \longrightarrow SO_3$
   e. $NO + O_2 \longrightarrow NO_2$

9. Balance the following chemical equations:
   a. $Zn + HCl \longrightarrow ZnCl_2 + H_2$

   b. $H_2S + O_2 \longrightarrow H_2O + S$
   c. $Al_2(SO_4)_3 + NaOH \longrightarrow Al(OH)_3 + Na_2SO_4$
   d. $Zn(OH)_2 + HNO_3 \longrightarrow Zn(NO_3)_2 + H_2O$
   e. $NH_4OH + H_3PO_4 \longrightarrow (NH_4)_3PO_4 + H_2O$

10. Balance the following chemical equations:
    a. $Cu + H_2SO_4 \longrightarrow SO_2 + CuSO_4 + H_2O$
    b. $NH_4Cl + CaO \longrightarrow CaCl_2 + H_2O + NH_3$

11. Calculate the formula weight (to the nearest whole number) for each of these compounds:
    a. CH_4
    b. AlF_3
    c. UF_6

12. Calculate the formula weight (to the nearest whole number) for the following:
    a. NH_4NO_3
    b. $BaSO_4$
    c. H_3PO_4
    d. $KClO_3$

13. Calculate the formula weight (to the nearest whole number) for the following:
    a. $Ca(NO_3)_2$
    b. $Mg(OH)_2$
    c. $(NH_4)_2SO_4$

14. Calculate the number of moles of compound for each of the following:
    a. 32 g of CH_4
    b. 336 g of AlF_3
    c. 17.6 g of UF_6

15. Calculate the number of moles of compound for each of the following:
    a. 8.0 g of NH_4NO_3
    b. 0.233 g of $BaSO_4$
    c. 980 g of H_3PO_4
    d. 3.69 g of $KClO_3$

16. Calculate the number of moles of compound for each of the following:
    a. 1.64 g of $Ca(NO_3)_2$
    b. 5.8 g of $Mg(OH)_2$
    c. 0.0132 g of $(NH_4)_2SO_4$

17. Calculate the number of grams in each of these samples:
    a. 0.0010 mol of CH_4
    b. 6.00 mol of AlF_3
    c. 40.0 mol of UF_6

18. Calculate the number of grams in each of these samples:
    a. 0.05 mol of NH_4NO_3
    b. 3.00 mol of $BaSO_4$
    c. 10 mol of H_3PO_4
    d. 0.0200 mol of $KClO_3$

19. Calculate the number of grams in each of these samples:
    a. 1.50 mol of $Ca(NO_3)_2$

b. 5.8 mol of $Mg(OH)_2$
c. 0.25 mol of $(NH_4)_2SO_4$

20. Consider the reaction

$$S + O_2 \longrightarrow SO_2$$

a. How many moles of SO_2 would be formed by the burning of 4 mol of sulfur?
b. How many moles of SO_2 would be formed in the reaction of 0.6 mol of O_2?
c. How many moles of sulfur would be required to react with 0.0684 mol O_2?

21. Answer the following questions based on the same equation:

$$S + O_2 \longrightarrow SO_2$$

a. How many moles of SO_2 would form in the reaction of 32 g of S?
b. How many moles of S would react with 32 g of O_2?

22. Still using the same equation, answer the following questions:

$$S + O_2 \longrightarrow SO_2$$

a. How many grams of SO_2 would form in the reaction of 16 g of O_2?
b. How many grams of SO_2 would be formed by the burning of 8.0 g of sulfur?
c. How many grams of sulfur would be needed to produce 3.2 g of SO_2?

23. Consider the reaction

$$CH_4 + 2O_2 \longrightarrow CO_2 + 2H_2O$$

a. How many moles of water are produced in the reaction of 0.4 mol of CH_4?
b. How many moles of O_2 are required to produce 25 mol of CO_2?
c. How many moles of CH_4 are required to react with 10 mol of O_2?
d. If 0.86 mol of O_2 react, how many moles of H_2O form?

24. Answer the following questions based on the reaction used in Problem 23:

$$CH_4 + 2O_2 \longrightarrow CO_2 + 2H_2O$$

a. If 9 g of H_2O form, how many moles of CH_4 react?
b. If 2 mol of O_2 react, how many grams of CO_2 form?

25. Again consider the reaction

$$CH_4 + 2O_2 \longrightarrow CO_2 + 2H_2O$$

a. How many grams of O_2 are required to react completely with 8.0 g of CH_4?
b. How many grams of CO_2 are produced if 180 g of H_2O are formed?

26. Consider the equation

$$C_3H_8 + 5O_2 \longrightarrow 3CO_2 + 4H_2O$$

a. How many moles of CO_2 are produced in the reaction of 2 mol of C_3H_8?
b. How many moles of H_2O are produced in the reaction of 20 mol of O_2?

27. Again, use the equation

$$C_3H_8 + 5O_2 \longrightarrow 3CO_2 + 4H_2O$$

a. How many moles of O_2 react with 22 g of C_3H_8?
b. If 1.8 g H_2O is produced, how many moles of CO_2 are formed?

28. Answer the following questions based on the same reaction:

$$C_3H_8 + 5O_2 \longrightarrow 3CO_2 + 4H_2O$$

a. If 16 g of O_2 reacts, how many g of C_3H_8 react?
b. If 320 g of O_2 reacts, how many g of CO_2 are formed?

29. What work is accomplished by the activation energy required in both exothermic and endothermic reactions?

30. Refer to the following reaction diagram in answering the questions.

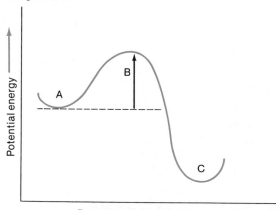

a. Which letter in the diagram refers to the products?

b. Which letter refers to the activation energy?

c. Which letter refers to the reactants?

31. Is the reaction diagrammed in Problem 30 endothermic or exothermic?

32. If the reaction diagrammed in Problem 30 were reversible, would the reverse reaction be endothermic or exothermic?

33. What effect does changing the temperature have on the rate of a reaction?

34. What is the effect of changing the concentration on the rate of a reaction?

35. If all other conditions are kept constant, what effect would decreasing the temperature of a reaction have on the rate of a reaction?

36. If all other conditions are kept constant, what effect would decreasing the concentration of reactants have on the rate of the reaction?

37. If all other conditions are kept constant, what effect would adding a catalyst have on the rate of the reaction?

38. Why does a catalyst increase the rate of a chemical reaction?

39. According to Le Chatelier's principle, what effect will increasing the temperature have on the following equilibria?

a. $H_2 + Cl_2 \rightleftarrows 2 HCl + energy$

b. $2 CO_2 + energy \rightleftarrows 2 CO + O_2$

c. $3 O_2 + heat \rightleftarrows 2 O_3$

40. What effect would decreasing the concentration of O_2 have on the equilibrium in Problem 39.b?

41. What would increasing the concentration of O_2 do to the equilibrium in Problem 39.c?

42. The Haber process for making ammonia is given in the equation

$$N_2(g) + 3 H_2(g) \rightleftarrows 2 NH_3(g) + heat$$

What would be the effect of each of the following on the equilibrium of that reaction?

a. the removal of ammonia

b. the addition of nitrogen

c. the removal of heat

d. the addition of a catalyst

43. Explain how the orientation of reactants can influence the rate of a reaction.

44. What is the "mechanism" of a reaction?

45. What weight of quicklime (calcium oxide) is formed when 100 g of limestone (calcium carbonate) is decomposed by heating?

$$CaCO_3 \longrightarrow CaO + CO_2$$

46. How many grams of hydrogen would be required to produce 68 g of ammonia (NH_3)?

$$N_2 + 3 H_2 \longrightarrow 2 NH_3$$

47. Some of the ammonia produced by the above process (Problem 46) is converted to nitric acid.

$$NH_3 + 2 O_2 \longrightarrow HNO_3 + H_2O$$

How much nitric acid can be made from 68 g of ammonia?

48. Nitric acid is used in the production of trinitrotoluene (TNT), an explosive.

$$C_7H_8 + 3 HNO_3 \longrightarrow C_7H_5N_3O_6 + 3 H_2O$$

Toluene   TNT

How much TNT can be made from 46 kg of toluene?

49. The burning of acetylene (C_2H_2) in pure oxygen produces a very hot flame. (This is the reaction in the oxyacetylene torch.) How much oxygen would be required to burn 52 kg of acetylene?

$$2 C_2H_2 + 5 O_2 \longrightarrow 4 CO_2 + 2 H_2O$$

50. Joseph Priestley discovered oxygen in 1774 by using heat to decompose "red calx of mercury," known today as mercury(II) oxide.

$$2 HgO \longrightarrow 2 Hg + O_2$$

How much oxygen is produced by the decomposition of 108 g of HgO?

51. How much iron can be converted to the magnetic oxide of iron, Fe_3O_4, by 800 g of pure oxygen?

$$3 Fe + 2 O_2 \longrightarrow Fe_3O_4$$

52. Dinitrogen monoxide (also called nitrous oxide) can be made by cautiously heating ammonium nitrate.

$$NH_4NO_3 \longrightarrow N_2O + 2 H_2O$$

How much nitrous oxide can be made from 40 g of NH_4NO_3?

53. Sometimes a small amount of hydrogen gas is made for laboratory use by allowing calcium metal to react with water.

$$Ca + 2 H_2O \longrightarrow Ca(OH)_2 + H_2$$

How many grams of hydrogen would be formed by the action of water on 10 g of calcium?

54. For many years it was thought that "inert" gases such as xenon (Xe) would not form chemical compounds. Following Neil Bartlett's breakthrough in 1962, a number of compounds involving "inert" gases have been synthesized. Xenon hexafluoride is made according to the equation

$$Xe + 3 F_2 \longrightarrow XeF_6$$

How many grams of fluorine are required to form 49 g of XeF_6?

55. If formed into a single piece, all the gold (Au) ever mined would form a cube 16 m on a side. The density of gold is 19.3 g/cm^3. What mass of gold has been mined? How many moles of Au is that?

56. Aspartame is an artificial sweetener about 160 times as sweet as sucrose. Aspartame breaks down in acidic solution to produce (among other products) methanol, which is fairly toxic. The equation is

$$C_{14}H_{18}N_2O_5 + 2 H_2O \xrightarrow{H^+}$$

Aspartame

$$C_4H_7NO_4 + C_9H_{11}NO_2 + CH_4O$$

Methanol

a. A typical can of pop contains about 40 g of sucrose. How much aspartame is required to obtain the same level of sweetness?

b. How much methanol is formed by the complete breakdown of that much aspartame?

c. How many cans of pop with decomposed aspartame would you have to drink to get the approximate lethal dose of 25 g of methanol?

57. When copper metal is placed in a solution of gold(III) chloride, metallic gold is plated out on the surface of the copper, and some of the copper goes into solution as copper(II) ions. How much gold will be plated out when 0.635 g of copper is displaced? The equation is

$$2 AuCl_3 + 3 Cu \longrightarrow 2 Au + 3 Cu^{2+} + 6 Cl^-$$

58. The gas phosphine (PH_3) is used as a fumigant to protect stored grain and other durable produce from pests. Phosphine is generated *in situ* by the action of water on aluminum phosphide or magnesium phosphide.

a. Write formulas for the two phosphides.

b. Write balanced chemical equations for the two reactions. (By-products are aluminum hydroxide and magnesium hydroxide.)

6

Gases

O ur astronauts have seen firsthand Earth's barren, airless moon. Our spaceships have photographed the desolation of Mercury from a few kilometers up and have measured the inhospitably high temperatures of Venus. They have given us close-up portraits of the crushing, turbulent atmospheres of Jupiter and Saturn. Our experiments on the harsh Martian surface failed to detect the presence of life there. As we enter the final decades of the twentieth century it is becoming increasingly clear that the Earth, a small island of green and blue in the vastness of space, is uniquely equipped to serve the needs of the life that inhabits it (Figure 6.1).

The life-support system of Spaceship Earth consists in part of a thin blanket of gases called the atmosphere. It is difficult to measure just how deep the atmosphere is. It does not end abruptly, but gradually fades as the distance from the surface of the Earth increases. We do know, though, that 99% of the atmosphere lies within 30 km of the surface of the Earth. That thin layer of air—which, if the Earth and its atmosphere were an apple, would be thinner than the peel—is all that stands between us and the emptiness of space. Our supply of air, once thought inexhaustible, now appears limited.

Air is so familiar, and yet so nebulous, that it is difficult to think of it as matter. But it is matter—matter in the gaseous state. All gases, air included, have mass and occupy space. Like other forms of matter, gases obey certain physical laws. In this chapter, we will examine some of those laws and see how they are related to certain vital processes—such as breathing!

FIGURE 6.1
Earth most likely is the only place in the solar system with an atmosphere that can support life.

183

TABLE 6.1 *Composition of the Earth's Dry Atmosphere (in Molecules of Each Substance per 10 000 Molecules)*

Substance	Formula	Number of Molecules
Nitrogen	N_2	7800
Oxygen	O_2	2100
Argon	Ar	93
Carbon dioxide	CO_2	3
All others	—	4
Total		10 000

6.1 AIR: A MIXTURE OF GASES

Air is a mixture of gases. Dry air is (by volume) about 78% nitrogen (N_2), 21% oxygen (O_2), and 1% argon (Ar). The amount of water vapor varies up to about 4%. There are a number of minor constituents, the most important of which is carbon dioxide (CO_2). The concentration of carbon dioxide in the atmosphere is believed to have increased from 296 parts per million (ppm) in 1900 to its present value of 338 ppm. It will most likely continue to rise as we burn more and more fossil fuels (coal, oil, and gas). The composition of the atmosphere is summarized in Table 6.1.

Air is a mixture of gases, but what are gases? Perhaps they are best understood in terms of the kinetic-molecular theory.

6.2 THE KINETIC-MOLECULAR THEORY

We have mentioned the kinetic-molecular theory previously. Now we are going to go into more detail in order to give you some idea of how this model enables scientists to visualize gases (and liquids and solids) and to understand the changes that occur in these materials.

The theory treats gases as collections of individual particles in rapid motion (hence the term **kinetic**). The particles of nitrogen gas, for example, are molecules (N_2); those of argon gas (Ar) are atoms. The distances between the particles are quite large when compared with the dimensions of the particles themselves. Therefore, gases, by comparison with solids and liquids, can be readily compressed. According to the theory, the individual particles of a solid or liquid are already in contact with one another, so they can't be compressed much.

The particles of a gas are in such rapid motion and are so light that gravity has little effect on them. They move up and down and sideways with ease and will not fall to the bottom of a container (as a liquid, for example, will). Any container of gas is completely filled. By *filled* we do not mean that the gas is packed tightly, but rather that it is distributed throughout the container's entire volume. A particle moves along a straight path unless it strikes something (another particle or the walls of the container). Then it may bounce off at an angle and travel from the point of collision along a straight path until it hits something else. These collisions are perfectly elastic; there is no tendency for the collection of particles to slow down and eventually stop. Two particles that are about to collide have a certain combined kinetic energy. After the collision the sum of their kinetic energies is the same. One of the particles may have been slowed down by the collision, but the other will have been speeded up just enough to compensate (Figure 6.2).

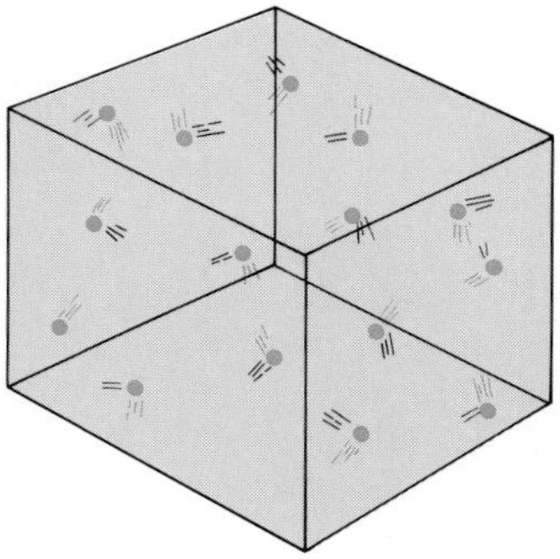

The kinetic-molecular theory explains what it is we are measuring when we measure temperature. According to the theory, temperature is just a reflection of the kinetic energy of the gas particles. The higher the kinetic energy, that is, the faster the particles are moving, the higher the temperature of the sample. On the average, the particles of a cold sample are moving more slowly than the particles of a hot sample. In any single sample, some particles are moving faster than others. Temperature reflects the *average* kinetic energy of the particles.

As a particle strikes a wall of its container, it gives the wall a little push. (If you were hit with a baseball or a brick, you would feel a push.) The sum of all these tiny pushes over a given area of the wall is what we call pressure.

FIGURE 6.2
According to the kinetic-molecular theory, particles of a gas are in constant motion, occasionally bouncing off one another and off the walls of their container.

6.3 ATMOSPHERIC PRESSURE

Molecules of air are constantly bouncing off each of us. Molecules are so tiny, though, that we don't feel the impact of individual ones. In fact, at ordinary altitudes we don't feel the molecules pushing on our skin, because there are molecules on the inside pushing out just as hard. When we increase our altitude rapidly by driving up a mountain or riding an elevator up to the top of a tall building, however, our ears pop because there are fewer molecules (because of the thinner air) on the outside pushing in than there are on the inside pushing out. Once we are at the top of the mountain or building, the pressures are soon equalized, and the popping stops.

The pressure of the atmosphere is measured by a device called a **barometer.** The simplest type of barometer is a long glass tube, closed at one end, filled with mercury, and inverted in a shallow dish containing mercury (Figure 6.3). Suppose the tube is 1 m long. Some of the mercury in the tube will drain into the dish, but *not all* of it. The mercury will drain out only until the pressure exerted by the mercury remaining in the tube exactly balances the pressure exerted by the atmosphere on the surface of the mercury in the dish. The mercury in the tube is trying to push its way out under the influence of gravity,

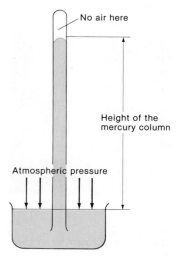

FIGURE 6.3
A mercury barometer.

and the air pressure is pushing it back in. At some point they reach a stalemate. Mercury is a very heavy (dense) liquid. On the average, at sea level, a column of mercury 760 mm high will balance the push of a column of air many kilometers high. The average atmospheric pressure at sea level is thus said to be 760 mm Hg. This pressure is called **1 atmosphere** (atm), or standard pressure. The unit **millimeters of mercury** is often called a **torr** (after the Italian physicist Evangelista Torricelli, the inventor of the mercury barometer).

$$1 \text{ atm} = 760 \text{ mm Hg} = 760 \text{ torr}$$

Several other units of pressure are widely used. Weather reports in the United States often include atmospheric pressure in inches of mercury.

$$1.00 \text{ atm} = 29.9 \text{ in. Hg} = 760 \text{ torr}$$

Engineers generally use pounds (of air) per square inch (lb/in.^{2}).

$$1.00 \text{ atm} = 14.7 \text{ lb/in.}^2$$

Another unit of pressure, used by respiratory therapists, is centimeters of water. Such measurements assume water to be the liquid in the barometer.

$$1.36 \text{ cm H}_2\text{O} = 1.00 \text{ torr}$$

The SI unit for pressure is the pascal (Pa), a unit that is interpreted in terms of newtons (N). (A newton is the force that when applied to a mass of 1 kg, will give that mass an acceleration of 1 m/s during each second.)

$$1 \text{ Pa} = 1 \text{ N/m}^2$$
$$1.00000 \text{ atm} = 101 \ 325 \text{ Pa} = 101.325 \text{ kP}$$

FIGURE 6.4
Robert Boyle.
(Duplication courtesy of the Smithsonian Institution, Washington, D.C.)

6.4 BOYLE'S LAW: PRESSURE-VOLUME RELATIONSHIP

The relationship between the volume and the pressure of a gas was first determined by Robert Boyle (Figure 6.4) in 1662. This relationship, called **Boyle's law,** states that, for a given mass of gas at constant temperature, the volume varies inversely with the pressure. When the pressure goes up, the volume goes down. When the pressure goes down, the volume goes up (Figure 6.5). Think

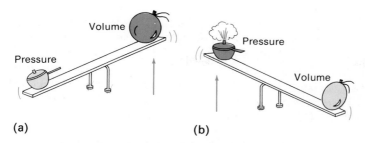

FIGURE 6.5   *The pressure-volume relationship is like a seesaw. (a) When the pressure goes down, the volume goes up. (b) When the pressure goes up, the volume goes down. (From an idea by Cindy Hill.)*

of gases as pictured in the kinetic-molecular theory. A sample of gas in a container exerts a certain pressure because the particles are bouncing against the walls at a certain rate with a certain force. If the volume of the container is expanded, the particles will have to travel longer distances before they strike the walls. Also, the surface area of the walls increases as the volume increases, so each unit of area is struck by fewer particles. The rate of strikes goes down, and the pressure goes down.

Mathematically, Boyle's law is written

$$V \propto \frac{1}{P}$$

where the symbol $\propto$ means "is proportional to." This may be changed to an equation by insertion of a **proportionality constant**, k.

$$V = \frac{k}{P} \quad \text{or} \quad VP = k$$

This is an elegant and precise, if somewhat abstract, way of summarizing a lot of experimental data. If the product of V times P is to be a constant, then if V goes up, P must come down. A typical Boyle's law experiment is diagrammed in Figure 6.6. Gas is enclosed in a cylinder fitted with a movable piston. In the example, the gas occupies a volume of 1 L under 1 atm of pressure. When the pressure is increased to 2 atm, the volume is reduced to 0.5 L. When the pressure is 4 atm, the volume becomes 0.25 L.

Boyle's law has a number of practical applications. These are perhaps best illustrated by some examples. Note that in applications of Boyle's law, any units for pressure or volume can be used, as long as the use is consistent.

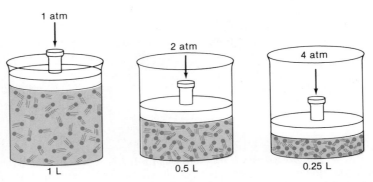

FIGURE 6.6 A diagram illustrating the effect of different pressures on the volume of a gas.

example 6.1

A cylinder of oxygen has a volume of 2.0 L. The pressure of gas is 1470 lb/in.^{2} at 20 °C. What volume will the oxygen occupy at standard atmospheric pressure (14.7 lb/in.^{2}), assuming no temperature change?

The simplest way to work problems of this type is to evaluate them as follows. You are asked for the volume that results when the pressure is changed from 1470 lb/in.^{2} to 14.7 lb/in.^{2}. The pressure has been decreased; therefore, the volume must increase. To find out how much, multiply by a fraction made up of the two pressures. To make the fraction greater than 1 (so the volume will increase), arrange the pressures with the larger one on top.

$$2.0 \text{ L} \times \frac{1470 \text{ lb/in.}^2}{14.7 \text{ lb/in.}^2} = 200 \text{ L}$$

Gases are stored for use under high pressure, even though they will be used at atmospheric pressure. This arrangement allows much oxygen to be stored in a small volume.

example 6.2

A space capsule is equipped with a tank of air that has a volume of 0.100 m^3. The air is under a pressure of 100 atm. After a space walk, during which the cabin pressure is reduced to 0, the cabin is closed and filled with the air from the tank. What will be the final pressure if the volume of the capsule is 12.5 m^3?

Since the volume in which the air is confined increases, the pressure must decrease; therefore, the multiplier must be a fraction less than 1.

$$100 \text{ atm} \times \frac{0.100 \text{ m}^3}{12.5 \text{ m}^3} = 0.800 \text{ atm}$$

example 6.3

A weather balloon is partially filled with helium gas. On the ground, where the atmospheric pressure is 740 torr, the volume of the balloon is 10.0 m³. What will the volume be when the balloon reaches an altitude of 5300 m, where the pressure is 370 torr, assuming the temperture is constant?

Since the pressure decreases, the volume will increase.

$$10.0 \text{ m}^3 \times \frac{740 \text{ torr}}{370 \text{ torr}} = 20.0 \text{ m}^3$$

The pressure-volume relationship can be used to explain the mechanics of breathing. When we breathe in (inspire), the diaphragm is lowered, and the chest wall is expanded (Figure 6.7). This increases the volume of the chest

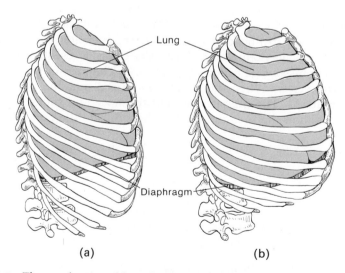

(a)   (b)

FIGURE 6.7   The mechanics of breathing. (a) Expiration. The diaphragm is relaxed and the rib cage is down. (b) Inspiration. The diaphragm is pulled down and the rib cage is lifted up and out, increasing the volume of the chest cavity.

cavity. According to Boyle's law, the pressure inside the cavity must decrease. Outside air enters the lungs because it is at a higher pressure than that in the chest cavity. When we breathe out (expire), the diaphragm rises and the chest wall contracts. This decreases the volume of the chest cavity. The pressure is increased, and some of the air is forced out (Figure 6.8).

During normal inspiration the pressure inside the lungs drops about 3 torr below atmospheric pressure. During expiration the internal pressure is about 3 torr above atmospheric pressure. About half a liter of air is moved in and out of the lungs in this process, and this normal breathing volume is referred to as the **tidal volume.** The **vital capacity** is the maximum amount of air that can be forced from the lungs and ranges in volume from 3 to 7 L, depending on the individual. A pressure inside the lungs 100 torr greater than the external pressure is not unusual during such a maximum expiration.

The lungs are never emptied completely, however. The space around the lungs is maintained at a slightly lower pressure than are the lungs themselves, causing the lungs to be kept partially inflated by the higher pressure within them. If a lung, the diaphragm, or the chest wall is punctured, allowing the two pressures to equalize, the lung will collapse. Sometimes a lung will be collapsed intentionally to give it time to heal. Closing the opening reinflates the lung.

People were breathing long before Boyle formulated his law, but it is satisfying to understand how the process works. We get more than just satisfaction out of science, though. An understanding of the pressure-volume relationship has enabled us to keep people alive. When paralysis prevents people from being able to breathe, they can be kept alive by artificial respirators. The iron lung, which kept many polio victims alive during the 1950s, is a sealed chamber connected to a compressor and bellows (Figure 6.9). The pressure in the chamber is varied rhythmically. When the bellows is moved out, the pressure in the chamber is reduced. The pressure around the nose and mouth (outside the tank) is greater than the pressure on the chest (inside the tank), so air flows

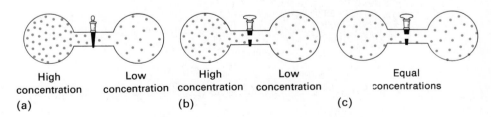

| High concentration | Low concentration | High concentration | Low concentration | Equal concentrations |
| (a) | | (b) | | (c) |

FIGURE 6.8 *A gas (air in the lungs, for example) flows from an area of high concentration to an area of low concentration. (a) With the stopcock closed, no flow is possible. (b) With the stopcock open, there is a net flow of gases from the area of higher concentration on the left to the area of lower concentration on the right. (c) After some time, there is no net flow of gas, because concentrations are equal.*

FIGURE 6.9
An iron lung uses changes in pressure to force air into and out of the lungs. (a) The older version enclosed the entire body except for the head. (b) A modern iron lung encloses the chest only. (Photos courtesy of J. H. Emerson Company, Cambridge, Mass.)

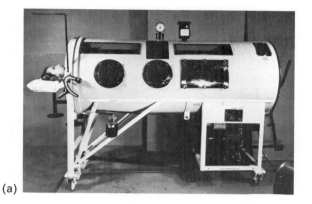

(a)

(b)

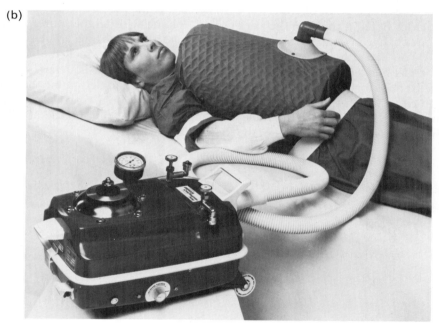

in and fills the lungs. When the bellows is moved in, pressure in the tank increases, and air is expelled from the lungs.

The iron lung, designed to enclose the patient completely (except for the head), is cumbersome and uncomfortable. It has been replaced by respirators that enclose the chest only. In fact, the whole area of respiratory therapy has become far more complicated in recent years. Specialists in this area are an indispensable part of the medical team.

6.5 CHARLES'S LAW: VOLUME-TEMPERATURE RELATIONSHIP

In 1787, the French physicist J. A. C. Charles studied the relationship between volume and temperature of gases. When a gas is cooled at constant pressure, its volume decreases. When the gas is heated, its volume increases. Temperature and volume vary directly, that is, they rise or fall together. But this law requires a bit more thought. If 1 L of gas is heated from 100 °C to 200 °C at constant pressure, the volume does not double but only increases to about 1.3 L. The law appears not to be as tidy as we had hoped.

Remember how temperature scales were defined? While zero pressure or zero volume really means zero—that is, there is zero (or no) pressure or volume to be measured, no matter what units are used—zero degrees Celsius means only the freezing point of water (0 °C). The zero point was arbitrarily set.

Charles noted a trend in the variation of volume with temperature. If you plot the change in temperature against the change in volume on a graph (Figure 6.10), you can extend the line, in theory, to the point at which the volume of the gas hits zero. Before a gas ever reaches this point, it liquefies, so this is an exercise for the imagination. From the graph you can determine the temperature at which the volume of the imaginary gas would reach zero. That temperature is −273 °C. This value was made the zero point on the absolute temperature scale, whose unit is the kelvin (K) (Section 1.6).

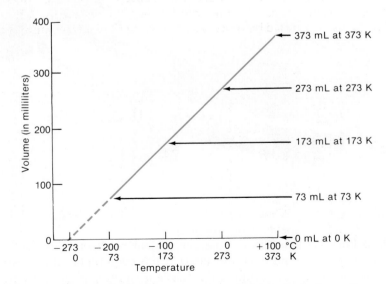

FIGURE 6.10 *The volume-temperature relationship for gases (with pressure constant).*

Charles compared changes in the volume of a gas with changes in temperature on the absolute scale. When he did this, he found that a doubling of the temperature (from 100 K to 200 K, for example) resulted in a doubling in the volume (from 1 L to 2 L, for example). Here is the simple relationship we are looking for.

Charles's law states that the volume of a gas is directly proportional to its temperature (on the absolute scale) if the pressure is held constant (Figure 6.11). Mathematically this relationship is expressed as

$$V \propto T$$

In the form of an equation, this becomes

$$V = kT \quad \text{or} \quad \frac{V}{T} = k$$

Again let us emphasize that the temperature increase (or decrease) must be based on the absolute temperature scale. In the case of temperature (unlike that of volume or pressure), the units of measurement for gas law calculations are restricted.

The kinetic-molecular model accounts for this relationship. If we heat a gas, we supply it with energy, and the particles of the gas begin moving faster. These speedier particles strike the walls of their container harder and more often. If the pressure is to remain constant, the volume of the container must increase. Increased volume means the particles take longer to travel from one wall to another, and the increased wall area means each unit of area will be hit less often. The pressure exerted by more slowly moving (low-temperature) particles

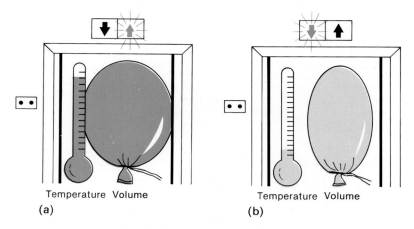

Temperature Volume
(a)

Temperature Volume
(b)

FIGURE 6.11   Temperature and volume are like passengers on an elevator. (a) When one goes up, the other must go up as well. (b) When one goes down, the other goes down, too. (From an idea by Cindy Hill.)

confined within the smaller volume will be the same as the pressure of the faster moving (high-temperature) particles contained within the larger volume.

example 6.4

A balloon, indoors where the temperature is 27 °C, has a volume of 2.0 L. What will its volume be outside where the temperature is −23 °C? (Assume no change in pressure.)

First, convert all temperatures to the absolute scale. The initial temperature is

$$273 + 27 = 300 \text{ K}$$

The final temperature is

$$273 - 23 = 250 \text{ K}$$

As the temperature decreases, the volume must also decrease.

$$2.0 \text{ L} \times \frac{250 \text{ K}}{300 \text{ K}} = 1.7 \text{ L}$$

example 6.5

What would be the final volume of the balloon in Example 6.4 if it were measured where the temperature was 47 °C? (Assume no change in pressure.)

The initial temperature is

$$273 + 27 = 300 \text{ K}$$

The final temperature is

$$273 + 47 = 320 \text{ K}$$

In this case, since the temperature increases, the volume must also increase.

$$2.0 \text{ L} \times \frac{320 \text{ K}}{300 \text{ K}} = 2.1 \text{ L}$$

6.6 GAY-LUSSAC'S LAW: PRESSURE-TEMPERATURE RELATIONSHIP

At about the same time that Charles was doing the experiments on which his law is based, Joseph Gay-Lussac, a French chemist, was also working with gases. His experiments showed that at constant volume, pressure is directly proportional to absolute temperature. Mathematically, this is expressed as

$$P \propto T$$

In the form of an equation, this becomes

$$P = kT \qquad \text{or} \qquad \frac{P}{T} = k$$

As the temperature increases at constant volume, so does the pressure. At the higher temperature, the particles move faster and hit the walls of their container harder and more often, and, thus, the pressure increases (Figure 6.12).

Automobile tires are filled to a pressure of 30 lb/in.^{2} at 20 °C. The tires become hot from high-speed driving and reach a temperature of 50 °C. What will the pressure be at that temperature? (Assume that the volume remains constant.)

example
6.6

The initial temperature is

$$273 + 20 = 293 \text{ K}$$

FIGURE 6.12
A hospital autoclave uses the principle of Gay-Lussac's law to achieve the high temperatures used to sterilize surgical instruments. The trend today is toward disposable equipment that does not require resterilization. (Photo by Raymond C. Carballada, Department of Medical Photography, Geisinger Medical Center.)

The final temperature is

$$273 + 50 = 323 \text{ K}$$

The temperature increases, so the pressure must also increase.

$$30 \text{ lb/in.}^2 \times \frac{323 \text{ K}}{293 \text{ K}} = 33 \text{ lb/in.}^2$$

example
6.7

At 127 °C the pressure in an autoclave is 2.0 atm. What will the pressure be at 27 °C?
The initial temperature is

$$273 + 127 = 400 \text{ K}$$

The final temperature is

$$273 + 27 = 300 \text{ K}$$

The temperature decreases; therefore, the pressure must also decrease.

$$2.0 \text{ atm} \times \frac{300 \text{ K}}{400 \text{ K}} = 1.5 \text{ atm}$$

6.7 THE COMBINED GAS LAWS

We have seen that the volume of a gas varies with both temperature and pressure. If we want to compare two samples, we have to do so at identical temperatures and pressures for the comparison to be meaningful. We can use Boyle's law and Charles's law, combined, to calculate the volume of a gas at any given temperature and pressure provided we know its volume at any other temperature and pressure.

example
6.8

A balloon is partially filled with helium on the ground at 27 °C and 740 torr pressure. Its volume is 10 m³. What would the volume be at an altitude of 5300 m, where the pressure is 370 torr and the temperature is −23 °C?
   First, since volume is inversely related to pressure, the pressure decrease would lead to a volume increase.

$$10 \text{ m}^3 \times \frac{740 \text{ torr}}{370 \text{ torr}} = 20 \text{ m}^3$$

Second, since the volume is directly proportional to the absolute temperature, the temperature decrease would lead to a volume decrease. The initial temperature is

$$273 + 27 = 300 \text{ K}$$

The final temperature is

$$273 - 23 = 250 \text{ K}$$

$$20 \text{ m}^3 \times \frac{250 \text{ K}}{300 \text{ K}} = 17 \text{ m}^3$$

It does not matter in which order you do these two calculations. The final answer will be the same. It is usual to combine both changes in one equation.

$$10 \text{ m}^3 \times \frac{740 \text{ torr}}{370 \text{ torr}} \times \frac{250 \text{ K}}{300 \text{ K}} = 17 \text{ m}^3$$

Because gases are so sensitive to changes in temperature and pressure, chemists have found it convenient to define standard conditions of temperature and pressure (referred to as STP) as $0\,°C$ (273 K) and 1 atm (760 torr).

example
6.9

What is the volume at STP of a sample of carbon dioxide whose volume at 25°C and 4.0 atm is 10 L?

The pressure decreases from 4.0 atm to the standard pressure of 1.0 atm; this should cause a volume increase.

$$10 \text{ L} \times \frac{4.0 \text{ atm}}{1.0 \text{ atm}} = 40 \text{ L}$$

The temperature decrease from 298 K (25 °C) to 273 K (0 °C) should be accompanied by a volume decrease.

$$40 \text{ L} \times \frac{273 \text{ K}}{298 \text{ K}} = 37 \text{ L}$$

Or

$$10 \text{ L} \times \frac{4.0 \text{ atm}}{1.0 \text{ atm}} \times \frac{273 \text{ K}}{298 \text{ K}} = 37 \text{ L}$$

Densities of gases usually are reported in the literature in grams per liter at STP. Recall (Section 5.4) that there are 6.02×10^{23} molecules in 1 mol of any gas. Recall also that the weight of 1 mol of gas is merely the formula weight expressed in grams. Now, if we divide the molecular weight (in grams) by the density at STP (in grams per liter), we will get the volume occupied by a mole of gas at STP. For example, the density of nitrogen gas (N_2) at STP is 1.25 g/L. The molecular weight of nitrogen gas is 28.0 g. Dividing, we get

$$\frac{28.0 \text{ g/mol}}{1.25 \text{ g/L}} = 22.4 \text{ L/mol}$$

For oxygen (O_2) the density at STP is 1.43 g/L. The molecular weight is 32.0. Dividing, we get

$$\frac{32.0 \text{ g/mol}}{1.43 \text{ g/L}} = 22.4 \text{ L/mol}$$

In fact, the volume occupied by a mole of most gases at STP is quite close to 22.4 L. This quantity is known as the **molar volume** of a gas. A box for a basketball has a volume of about 22.4 L. Such a box would hold 28 g of N_2 or 32 g of O_2. Thus, Gay-Lussac's law of combining volumes (Section 5.2) is explained; no matter what the gas, a mole of it occupies about the same volume as a mole of any other gas under the same conditions.

example
6.10

What is the molar volume of carbon dioxide, which has a density of 1.98 g/L at STP?
   The molecular weight is 44.0. Dividing, we get

$$\frac{44.0 \text{ g/mol}}{1.98 \text{ g/L}} = 22.2 \text{ L/mol}$$

6.8 THE IDEAL GAS LAW

So far we have done calculations in which the quantity of a gas does not change. Avogadro's hypothesis that equal volumes of gases at the same temperature and pressure contain equal numbers of molecules (Section 5.2) enables us to write a gas law that takes into account varying quantities of gas. This relationship is called the **ideal gas equation.**

$$PV = nRT$$

In this equation, the pressure is in atmospheres, the volume in liters, and the temperature in kelvins. The number of moles of the gas is given by n. The constant R, which has a value of

$$0.082 \frac{L \cdot atm}{mol \cdot K}$$

is called the **universal gas constant**.

The ideal gas law can be used to calculate any of the four quantities—P, V, n, or T—if the other three are known.

Use the ideal gas law to calculate the volume occupied by 1.0 mol of nitrogen gas at 244 K and 1.0 atm pressure.

example **6.11**

$$V = \frac{nRT}{P} = \frac{1.0 \text{ mol}}{1.0 \text{ atm}} \times \frac{0.082 \text{ L atm}}{\text{mol K}} \times 244 \text{ K} = 20 \text{ L}$$

Use the ideal gas law to calculate the pressure exerted by 0.50 mol of oxygen in a 15-L container at 303 K.

example **6.12**

$$P = \frac{nRT}{V} = \frac{0.50 \text{ mol}}{15 \text{ L}} \times \frac{0.082 \text{ L atm}}{\text{mol K}} \times 303 \text{ K} = 0.83 \text{ atm}$$

6.9 DALTON'S LAW OF PARTIAL PRESSURES

John Dalton is most renowned for his atomic theory (Section 2.3). But Dalton had wide-ranging interests, including meteorology. In trying to understand the weather, he did a number of experiments on water vapor in the air. In one experiment, he found that if he added water vapor at a certain pressure to dry air, the pressure exerted by the air would increase by an amount equal to the pressure of the water vapor. Based on this and other experiments, Dalton concluded that each of the gases in a mixture behaves independently of the other gases. Each gas exerts its own pressure. The total pressure of the mixture is equal to the sum of the *partial pressures* exerted by the separate gases (Figure 6.13).

Mathematically, **Dalton's law of partial pressures** is stated as

$$P_{total} = P_1 + P_2 + P_3 + \cdots$$

where the terms on the right side refer to the partial pressures of gases 1, 2, 3, and so on.

Gases such as oxygen, nitrogen, and hydrogen are nonpolar. They are only slightly soluble in water and are usually collected over water by the technique

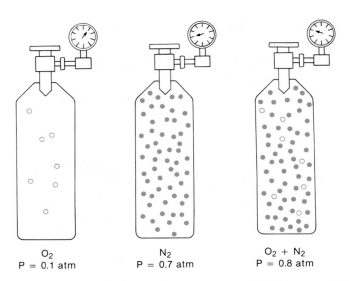

O_2
P = 0.1 atm

N_2
P = 0.7 atm

$O_2 + N_2$
P = 0.8 atm

FIGURE 6.13   Dalton's law of partial pressures states that the pressure of a mixture of gases is equal to the sum of the pressure that each gas would exert by itself.

of displacement. Such gases always contain water vapor, and the total pressure in the collection vessel is that of the gas plus that of water vapor.

The vapor pressure of water depends on the temperature of the water. (The **vapor pressure** of a substance is the partial pressure exerted by the molecules of the substance that are in the gas phase above the liquid phase of the substance.) The hotter the water, the higher the vapor pressure. If a gas is collected over water, we can make use of vapor pressure tables (Table 6.2) to calculate the pressure due to the gas alone. One need only subtract the vapor pressure of the water, as determined from the table, from the value for the total pressure within the collection vessel.

example
6.13

Oxygen is collected over water at 20 °C. The pressure inside the jar is 740 torr. What is the pressure due to oxygen alone?

$$P_{total} = P_{O_2} + P_{H_2O}$$

From Table 6.2 we find that the vapor pressure of water at 20 °C is 18 torr. Since the total pressure is equal to 740 torr, we have

$$740 \text{ torr} = P_{O_2} + 18 \text{ torr}$$
$$P_{O_2} = 740 - 18 = 722 \text{ torr}$$

TABLE 6.2   *Water Vapor Pressure at Various Temperatures*

Temperature (in degrees Celsius)	Water Vapor Pressure (in torr)
0	5
10	9
20	18
30	32
40	55
50	93
60	149
70	234
80	355
90	526
100	760

Humidity is a measure of the amount of water vapor in the air. Relative humidity compares the actual amount of water vapor in the air with the maximum amount the air could hold at the same temperature. If the temperature is 20 °C and the vapor pressure of water in the atmosphere is 12 torr, the relative humidity is

$$\frac{12 \text{ torr}}{18 \text{ torr}} \times 100 = 67\%$$

The 18 torr in the denominator was obtained from Table 6.2. When the humidity is 100%, the air is saturated with water vapor. (Note that 100% humidity does not mean the air is 100% water vapor, just that it is holding as much water as it can. At 20 °C and 100% relative humidity, only about 2 or 3 molecules in every 100 molecules of air are water.)

Cool air can hold less water vapor than warm air. As the temperature falls during the night, the atmosphere may become saturated. Water vapor condenses from the air as dew.

Respiratory therapists must concern themselves with the humidity of the gases they administer to patients. Normally, as one breathes, the inspired air is saturated with moisture as it passes through the nose and respiratory passages. Oxygen as it comes from a tank is quite dry. If oxygen is being administered over a long period of time, it must be humidified to prevent it from irritating the mucous linings of the nasal passages and the lungs. If the oxygen or mixture of gases is conducted through the nose, the therapist may merely assist the normal body processes by imparting about 30% humidity to the inspired gases. If the breathing mixture is conducted directly to the trachea (bypassing the nose), the therapist saturates the gas mixture with water vapor.

6.10 HENRY'S LAW: PRESSURE-SOLUBILITY RELATIONSHIP

In the 1760s, Joseph Priestley invented soda water by dissolving carbon dioxide gas in water. No doubt you have noticed the hissing sound and the formation of bubbles when you opened a bottle of pop. Carbon dioxide is dissolved in the liquid, and the bottle is capped under pressure. William Henry, a close friend of John Dalton, spent a great deal of time studying the solubility of gases in liquids. In 1801 he summarized his findings in the law we know as **Henry's law.** The solubility of a gas in a liquid at a given temperature, he found, is directly proportional to the pressure of the gas at the surface of the liquid. To get back to the bottle of pop: when the bottle was capped under pressure, a certain amount of carbon dioxide was dissolved in the pop. When you opened the bottle, the pressure was *reduced* (the hissing sound you heard was pressure being released), and the solubility of the carbon dioxide was *reduced* (the bubbles of gas you noticed were carbon dioxide escaping from solution).

The carbonated beverage industry is not the only group interested in Henry's law. Deep-sea divers get the *bends* from nitrogen dissolved in the blood. Very little nitrogen dissolves in our blood at normal pressures. In a diver, who breathes air under greater pressure, appreciable amounts dissolve. If the diver comes up too rapidly, the nitrogen comes out of solution as the pressure diminishes. Tiny bubbles of nitrogen form. These cause severe pain in the arms, legs, and joints, perhaps by disrupting nerve pathways. To prevent bends, divers sometimes use a mixture of helium and oxygen rather than air. Very little of the helium dissolves, even under increased pressure, and the problem of bubble formation as the diver ascends is minimized.

The pressure-solubility relationship is also used in therapy. In cases of carbon monoxide poisoning (see Section 6.12), the victim is placed in a hyperbaric (high-pressure) chamber. This chamber is a device that supplies oxygen at pressures of 3 or 4 atm. More oxygen is forced into the tissues at these pressures to compensate for the lack of oxygen that accompanies carbon monoxide poisoning.

Hyperbaric chambers were also used to treat infections by anaerobic bacteria (bacteria that live without air). Gangrene is one such disease. The organisms that cause gangrene cannot survive in an oxygenated atmosphere. If sufficient oxygen can be forced into the diseased tissues, the infection can be arrested.

6.11 PARTIAL PRESSURES AND RESPIRATION

When we breathe in, the inspired air becomes moistened and warmed to our body temperature of 37 °C. The air is drawn into our lungs where it enters a highly branched system of tubes that end in tiny air sacs called alveoli (Figure

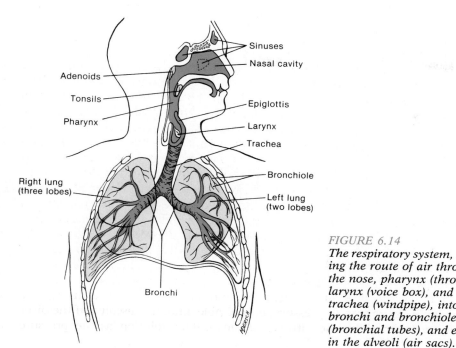

Right lung
(three lobes)

Bronchi

Sinuses
Nasal cavity
Adenoids
Tonsils
Pharynx
Epiglottis
Larynx
Trachea
Bronchiole
Left lung
(two lobes)

FIGURE 6.14
The respiratory system, show-
ing the route of air through
the nose, pharynx (throat),
larynx (voice box), and
trachea (windpipe), into the
bronchi and bronchioles
(bronchial tubes), and ending
in the alveoli (air sacs).

6.14). These thin-walled pouches are surrounded by blood vessels that are part of a circulatory system serving every cell in the body.

Inspired air is rich in oxygen (P_{O_2} = 150 torr) and poor in carbon dioxide (P_{CO_2} = 0.2 torr). The fluid in our cells is poor in oxygen (P_{O_2} = 6 torr) and rich in carbon dioxide (P_{CO_2} = 50 torr). Our cells use up oxygen in metabolic reactions designed to produce energy. Carbon dioxide accumulates in the cells as a waste product of these reactions. To maintain life, we must transfer the oxygen in the inspired air to our cells. At the same time, the carbon dioxide waste in our cells must be transferred to our lungs and then exhaled to the atmosphere. The transfer of both gases occurs through the process of **diffusion.** In diffusion, gases flow from regions of higher concentration to regions of lower concentration. In our bodies, oxygen makes its way to the cells through a pressure gradient, that is, in a series of steps in which oxygen diffuses from areas where its concentration is higher into areas where its concentration is lower. By the same method, carbon dioxide moves in the opposite direction. It makes its way from the cells, where its partial pressure is high, to the atmosphere, where its partial pressure is low. Figures 6.15 through 6.18 show the steps in the gradient for both gases. Thus, given the mechanical action of the chest and diaphragm (Section 6.4) to get air into and out of the lungs, respiration is all downhill as far as the gases are concerned.

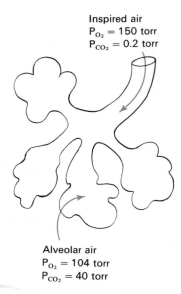

Inspired air
$P_{O_2} = 150$ torr
$P_{CO_2} = 0.2$ torr

Alveolar air
$P_{O_2} = 104$ torr
$P_{CO_2} = 40$ torr

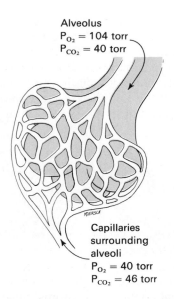

Alveolus
$P_{O_2} = 104$ torr
$P_{CO_2} = 40$ torr

Capillaries
surrounding
alveoli
$P_{O_2} = 40$ torr
$P_{CO_2} = 46$ torr

FIGURE 6.15 Oxygen flows from the inspired air into the alveolar air. Carbon dioxide flows in the opposite direction. In each case, the flow is from a region of high partial pressure to a region of low partial pressure.

FIGURE 6.16 Oxygen diffuses from the alveolus to the capillary. Carbon dioxide moves in the opposite direction.

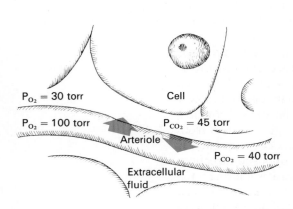

$P_{O_2} = 30$ torr   Cell

$P_{O_2} = 100$ torr $P_{CO_2} = 45$ torr

Arteriole

$P_{CO_2} = 40$ torr

Extracellular
fluid

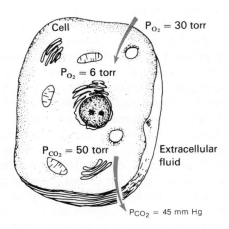

Cell $P_{O_2} = 30$ torr

$P_{O_2} = 6$ torr

$P_{CO_2} = 50$ torr   Extracellular
fluid

$P_{CO_2} = 45$ mm Hg

FIGURE 6.17 Oxygen diffuses from an arteriole into the extracellular fluid. Carbon dioxide flows in the opposite direction.

FIGURE 6.18 Oxygen diffuses into a cell from the extracellular fluid. Carbon dioxide moves in the opposite direction.

Normally, the carbon dioxide level in the blood (not the oxygen level) acts as a trigger for the breathing process. To oversimplify, when carbon dioxide levels build up, we take a breath; if they get too low, we don't. Thus, one of the concerns of a therapist is the partial pressure of carbon dioxide in the blood. It is possible, under certain unusual conditions, for the level of carbon dioxide to fall so low that it fails to trigger the breathing mechanism. A person with access to a plentiful supply of oxygen suffocates because he or she simply stops breathing.

6.12 HEMOGLOBIN: OXYGEN AND CARBON DIOXIDE TRANSPORT

So far we have explained the movement of blood gases on the basis of pressure gradients. We also discussed the solubility of gases in terms of Henry's law. Our bodies, however, need a lot more oxygen than can be dissolved in blood plasma alone. Most of the oxygen in the bloodstream is taken up by the red blood cells. In these cells it is combined with hemoglobin, a huge, complex molecule that contains iron atoms. Schematic diagrams of hemoglobin and its oxygen complex (called oxyhemoglobin) are shown in Figure 6.19.

The formation of oxyhemoglobin is favored by a high P_{O_2}, a relatively low temperature, and other conditions that exist in the capillaries that surround the alveoli of the lungs at inspiration. When the blood reaches the tissues, the oxyhemoglobin encounters opposite conditions, under which the oxyhemoglobin dissociates, allowing the oxygen to diffuse into the cells.

Hemoglobin also participates in the transport of carbon dioxide, but to a lesser extent than in oxygen transport. Most of the carbon dioxide is transported as bicarbonate ion (HCO_3^-) dissolved in the blood. Some is carried as the simple dissolved gas. In the lungs, the bound carbon dioxide is released. It diffuses into the alveoli and is exhaled.

Other chemical substances also react with hemoglobin. Most notorious, perhaps, is carbon monoxide (CO). This odorless, tasteless gas is formed in large quantities when carbonaceous fuels are burned in insufficient oxygen. Millions of tons of carbon monoxide (about 80% of it from automobile exhausts) are poured into our atmosphere each year. Carbon monoxide binds so tightly to hemoglobin that the latter becomes unavailable for oxygen transport. The symptoms of carbon monoxide poisoning are, therefore, those of oxygen deprivation. Drowsiness is an early symptom, followed by impairment of physical and mental processes. Excessive amounts, as would be found in a closed garage with an automobile running, can cause death. All except the most severe cases of carbon monoxide poisoning are reversible. Treatment with pure oxygen, in a hyperbaric chamber, if possible (Section 6.10), is the best antidote. Artificial respiration may help if pure oxygen is not available.

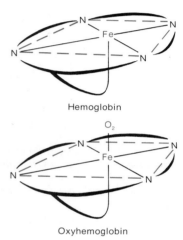

Hemoglobin

Oxyhemoglobin

FIGURE 6.19
Schematic representations of hemoglobin and oxyhemoglobin.

A number of commercially supplied gases are used in medicine, the majority in respiratory therapy. Table 6.3 lists these gases and some of their uses.

TABLE 6.3 *Ten Compressed Gases Used in Medicine*

Gas	Chemical Formula	Use
Air	N_2 and O_2 (mixture)	Life support
Carbon dioxide	CO_2	Laboratory tests, lung function tests
Carbon dioxide/oxygen mixtures	CO_2 and O_2	Diagnosis, inhalation therapy
Cyclopropane	C_3H_6	Anesthetic
Helium	He	Laboratory analyses
Helium/oxygen mixtures	He and O_2	Inhalation therapy, diagnostic tests
Nitrogen	N_2	Diagnostic testing, inhalation therapy
Nitrous oxide	N_2O	Anesthetic
Oxygen	O_2	Life support, medical emergencies, adjunct to anesthetics
Oxygen/nitrogen	O_2 and N_2 (mixture)	Treatment of obstructive lung diseases

PROBLEMS

1. Define or explain the following terms:
   a. kinetic-molecular theory
   b. mm Hg
   c. cm H_2O
   d. torr
   e. atmosphere
   f. pascal
   g. Boyle's law
   h. Charles's law
   i. Gay-Lussac's law
   j. Dalton's law
   k. Henry's law
   l. STP
   m. partial pressure
   n. diffusion
   o. relative humidity
   p. tidal volume
   q. vital capacity

2. List the four major gases found in dry air. Which of these are important in respiration?

3. Why is atmospheric pressure greater at sea level than on the top of a high mountain?

4. Explain how a mercury barometer works.

5. What effect will the following changes have on the volume of a gas?
   a. an increase in pressure
   b. a decrease in temperture

   c. a decrease in pressure coupled with an increase in temperature

6. What effect will the following changes have on the pressure of a gas?
   a. an increase in temperature
   b. a decrease in volume
   c. an increase in temperature coupled with a decrease in volume

7. According to the kinetic-molecular theory, what change in temperature is occurring if the particles of a gas begin to move more slowly, on the average?

8. According to the kinetic-molecular theory, what change in pressure occurs when the walls of the container are struck less often by particles of a gas?

9. Container A has twice the volume but holds twice as many gas particles as container B. Using the kinetic-molecular theory as the basis for your judgment, compare the pressures in the two containers.

10. Density is defined as mass per unit volume. For each of the following, indicate which sample you would expect to exhibit the higher density. (Hint: How many particles are there in equivalent volumes of the sample?)
    a. Containers A and B have the same volume and are at the same temperature, but the gas in container A is at a higher pressure.
    b. Containers A and B are at the same pressure and temperature, but the volume of container A is greater than that of container B.
    c. Containers A and B are at the same pressure and volume, but the temperature of the gas in container A is higher. (Hint: One can float in the air in a *hot* air balloon.)
11. Carry out the following conversions:
    a. 2.0 atm to torr
    b. 0.50 atm to torr
    c. 0.0030 atm to Pa
12. Carry out the following conversions:
    a. 76 torr to atm   c. 0.076 torr to Pa
    b. 320 torr to mm Hg
13. Carry out the following conversions:
    a. 10 atm to lb/in.^{2} c. 3.0 in. Hg to atm
    b. 10 torr to cm H$_2$O
14. Carry out the following conversions:
    a. 37 °C to K   c. −82 °C to K
    b. 420 K to °C
15. A tank contains 500 mL of helium at 1500 torr. What volume will the helium occupy at 750 torr, assuming no temperature change?
16. A hyperbaric chamber with a volume of 10 m^3 operates at an internal pressure of 4.0 atm. What volume would the air inside the chamber occupy at normal atmospheric pressure?
17. Oxygen used in respiratory therapy is stored under a pressure of 2200 lb/in.^{2} in gas cylinders with a volume of 60 L.
    a. What volume would the gas occupy at normal atmospheric pressure?
    b. If oxygen flow to the patient is adjusted to 8.0 L per minute, how long will the tank of gas last?
18. The pressure within a balloon with a volume of 2.5 L is 1.0 atm. If the volume of the balloon increases to 7.5 L, what will be the final pressure within the balloon?
19. During inhalation, does the chest cavity expand or

contract? Is the pressure inside the lungs decreased or increased? What happens during exhalation?
20. The cough reflex is designed to keep air passages clear. Typically, when a person coughs, he or she first inhales about 2.0 L of air. The epiglottis and the vocal cords then shut, trapping the air in the lungs. The air in the lungs is then compressed to a volume of about 1.7 L by the action of the diaphragm and chest muscles. The sudden opening of the epiglottis and vocal cords releases this air explosively. Just prior to the release, what is the pressure of the gas inside the lungs?
21. A gas at a temperature of 100 K occupies a volume of 100 mL. What will the volume be at a temperature of 10 K, assuming no change in pressure?
22. An automobile tire is inflated to 30 lb/in.^{2} at 27 °C. What will be the pressure at 127 °C, assuming no volume change?
23. A balloon is filled with helium. Its volume is 5.0 L at 27 °C. What will be its volume at −73 °C, assuming no pressure change?
24. A gas at a temperature of 300 K and a pressure of 1.0 atm is cooled to 250 K. Assuming no change in volume, calculate the change in pressure.
25. A sealed can with an internal pressure of 720 torr at 25 °C is thrown into an incinerator operating at 750 °C. What will the pressure inside the heated can be, assuming the container remains intact during incineration?
26. If a gas occupies 4.0 L at a temperature of 25 °C and a pressure of 2.0 atm, what volume will it occupy at a temperature of 200 °C and a pressure of 1.0 atm?
27. If a gas occupies 2.5 m^3 at a temperature of −15 °C and a pressure of 190 torr, what volume will it occupy at a temperature of 25 °C and 1140 torr pressure?
28. What volume will 500 mL of a gas, measured at 27 °C and 720 torr, occupy at STP?
29. If a gas has a volume of 55 cc at STP, what volume will it occupy at 100 °C and 76 torr?
30. If a gas has a volume of 732 mL at 760 torr and 25 °C, what volume will it occupy at 1.0 atm and 298 K?
31. A container holds oxygen at a partial pressure of 0.25 atm, nitrogen at a partial pressure of 0.50 atm, and helium at a partial pressure of 0.20 atm. What is the pressure inside the container?

32. A container is filled with equal numbers of nitrogen, oxygen, and carbon dioxide molecules. The total pressure in the container is 750 torr. What is the partial pressure of nitrogen in the container?

33. Oxygen is collected over water. If the temperature is 30 °C and the collected sample has a pressure of 740 torr, what is the partial pressure of the oxygen in the container? (You may refer to Table 6.2.)

34. The pressure of the atmosphere on the surface of Venus is about 100 atm. Carbon dioxide makes up about 97% of the atmospheric gases. What is the partial pressure of carbon dioxide in the atmosphere of Venus?

35. Atmospheric pressure on the surface of Mars is about 6.0 torr. The partial pressure of carbon dioxide is 5.7 torr. What percent of the Martian atmosphere is carbon dioxide?

36. A sample of intestinal gas was collected and found on analysis to consist of 44% CO_2, 38% H_2, 17% N_2, 1.3% O_2, and 0.003% CH_4. (The percentages do not add to exactly 100% because of rounding.) What is the partial pressure of each gas if the total pressure in the intestine is 820 torr?

37. What is the molar volume of methane (CH_4) gas, which has a density of 0.72 g/L at STP?

38. Calculate the approximate density of sulfur dioxide (SO_2) gas at STP.

39. Use the ideal gas law to calculate the volume occupied by 0.60 mol of a gas at a temperature of 310 K and a pressure of 0.80 atm.

40. What volume is occupied by 12 mol of hydrogen sulfide gas at a temperature of 620 K and a pressure of 18 atm?

41. What pressure is exerted by 0.010 mol of methane gas if its volume is 0.26 L at 373 K?

42. What pressure is exerted by 44 mol of carbon monoxide gas in a 36 L tank at 290 K?

43. Calculate the temperature of 0.78 mol of oxygen if its volume is 68 L and its pressure is 0.37 atm.

44. A hyperbaric chamber has a volume of 4200 L. How many moles of oxygen are needed to fill the chamber to a pressure of 3.0 atm at 290 K?

45. The interior volume of the Hubert H. Humphrey Metrodome in Minneapolis is 1.70×10^{10} L. The Teflon-coated fiberglass roof is supported by air pressure provided by 20 huge electric fans. How many moles of air are in the dome if the pressure is 1.02 atm and the temperature is 18 °C?

46. When air is inspired it becomes fully saturated with water vapor as it passes through the trachea on its way to the lungs. What is the approximate partial pressure of water vapor in the air in the alveoli? (Hint: At what temperature will the air be?)

47. If the P_{H_2O} in air is 12 torr on a day when the temperature is 20 °C, what is the relative humidity?

48. Two flasks are connected. Flask A contains only oxygen at a pressure of 460 torr. Flask B has oxygen at a partial pressure of 320 torr and nitrogen at a partial pressure of 240 torr.
    a. Which direction will the net flow of oxygen take?
    b. Which direction will the net flow of nitrogen take?

49. In which net direction, cells to lungs or lungs to cells, does oxygen move? What about carbon dioxide?

50. Why does oxygen flow from the alveoli to the pulmonary capillaries? Why does carbon dioxide flow in the reverse direction?

51. A person at rest breathes about 80 mL of air per second. How long does it take to breathe 22.4 L of air?

52. What effect does carbon monoxide have on oxygen transport?

CHAPTER
7

Liquids and Solids

In Chapter 4, we considered the subject of bonding. Our primary concern then was how atoms combined to form molecules and why elements reacted to form compounds. Now we're going to expand our consideration of bonding, and this time we will be looking for an answer to this question: What makes a substance a solid rather than a liquid or a liquid rather than a gas? Some force of attraction holds the particles of solids and liquids in contact with one another. The particles of a gas fly about at random; those of liquids or solids cling together. Gas particles interact so little with one another that a collection of them retains neither a specific volume nor a specific shape. But particles of a liquid are held together with sufficient force that a collection of them has a specific volume. And particles of a solid are so rigidly held together that not only the volume but the shape of a given sample is fixed. Are these mysterious forces of attraction important? They are if appearances are important to you. You are built of solids and liquids. If you think you've got it all together, you should thank the special forces of attraction that give shape and volume to the condensed forms of matter.

In this chapter, we're also going to consider changes in the state of matter—what happens when a solid is converted to a liquid or a liquid to a gas. Is that important? Well, consider perspiration. From the amount of advertising directed against this lowly liquid, you would think it was an unnecessary annoyance. However, were it not for the conversion of this liquid to a vapor on skin surfaces, we would find it difficult just to survive in the temperate and tropical zones of our planet, let alone to carry out vigorous physical activity.

7.1 STICKY MOLECULES

Gases are easily manipulated. Changes in pressure or temperature result in volume changes that are readily measured. Not so with liquids and solids. Whereas gas particles are rather far apart, allowing gases to be considerably compressed, the particles of liquids and solids are close together. The forces between rather distant gas particles are negligible, but those between particles in the liquid or solid state may be considerable.

To illustrate the vast difference in the spacing between molecules in the gaseous and the condensed states, let's consider a given volume of water vapor (a gas) and compare it with the same volume of liquid water. At STP, 22.4 L of water vapor contains 6.02×10^{23} molecules. Under the same conditions, 22.4 L of liquid water would contain many times that number of molecules. We can calculate just how many. Since 1 L of water has a mass of 1 kg, 22.4 L of water has a mass of 22.4 kg, or 22 400 g. We know that 1 mol of water has a mass of 18 g and contains 6.02×10^{23} molecules (Section 5.4). Therefore, we can calculate the number of molecules in 22.4 L of liquid water.

$$22\ 400 \text{ g} \times \frac{1 \text{ mol}}{18 \text{ g}} \times \frac{6.02 \times 10^{23} \text{ molecules}}{1 \text{ mol}} = 7490 \times 10^{23} \text{ molecules}$$

Does this mean that there are 1240 (7490/6.02) times as many molecules in the liquid as there are in the same volume of gas? Yes, and it makes for a rather crowded situation (Figure 7.1).

Even in the gaseous state, there is some attraction between molecules. This attraction may be sufficient to cause some deviation from the "ideal" behavior described in the gas laws. Generally, however, intermolecular forces are much weaker in the gaseous state than in the liquid and solid states.

Before studying the types of forces in detail, let's make some generalizations. First, all ionic compounds are solids at room temperature. It is possible to obtain them as liquids (by melting them), but generally only at high temperatures. Second, nearly all metals (mercury is a notable exception) are solids at room temperature. They, too, can be melted—some at fairly low temperatures, but others only at high temperatures.

It would be nice to have a third generalization to cover molecular substances. However, molecular substances exist in all three physical states at room temperature. Nitrogen (N_2) and carbon dioxide (CO_2) are gases, water (H_2O) and bromine (Br_2) are liquids, and sulfur (S_8) and glucose ($C_6H_{12}O_6$) are solids. The physical state of molecular substances is determined both by molecular weight (the size of the molecules) and by the type of force between molecules. If one of these two variables can be eliminated (or held constant), then simple generalizations are possible. For example, the Group VIIA elements all exist as nonpolar diatomic molecules. Intermolecular forces are similar; thus, any variation in properties can be attributed to variations in molecular weight. And

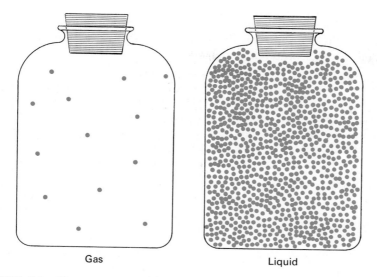

Gas   Liquid

FIGURE 7.1   *There are over a thousand times as many molecules in a given volume of a liquid as there are in the same volume of a gas.*

we do find such a trend. Fluorine (F_2), with a molecular weight of 38, and chlorine (Cl_2), with a molecular weight of 71, are gases at room temperature. Bromine (Br_2), with a molecular weight of 160, is a liquid, and iodine (I_2), with a molecular weight of 254, is a solid. A similar trend can be found for compounds that experience similar intermolecular forces, as is the case for the following compounds of carbon and the halogens.

Compound	CF_4	CCl_4	CBr_4	CI_4
Formula weight	88	154	332	520
Physical state (at room temperature)	Gas	Liquid	Solid	Solid

The types of intermolecular force are usually of overriding importance if we are comparing molecules that are subject to dissimilar forces. We will consider the various types of force in detail in the following sections.

7.2 INTERIONIC FORCES

One type of force, already mentioned in Section 4.4, is that between ions. Generally, interionic forces are the strongest of all the forces that hold solids and liquids together. In fact, nearly all ionic compounds are solids at room temperature. The energy available to most ionic compounds at room temperature is not sufficient to break the strong forces of attraction between them. Indeed, most ionic compounds melt only at extremely high temperatures.

Recall that ions are electrically charged. Those with opposite charges attract one another. The attraction is not usually limited to just a pair of ions. Rather, each ion is attracted by several oppositely charged ions that surround it (see Figure 4.2.) These **ionic bonds** are exceptionally strong. Forces are greater between ions of higher charge. The interionic force between mercury(II) ion (Hg^{2+}) and sulfide ion (S^{2-}) is greater than that between sodium ion (Na^+) and chloride ion (Cl^-). The force between aluminum ion (Al^{3+}) and nitride ion (N^{3-}) is greater still.

7.3 DIPOLE FORCES

In Section 4.8 we saw that certain unlike atoms shared electrons unequally, giving rise to polar covalent bonds. Unsymmetrical molecules containing polar bonds are themselves polar (Section 4.10). Such molecules contain a center of negative charge and a center of positive charge. They are called **dipoles.**

Two dipoles brought close enough attract one another. The positive end of one molecule attracts the negative end of another, much as the opposite poles of a magnet attract one another. Such forces may exist throughout the struc-

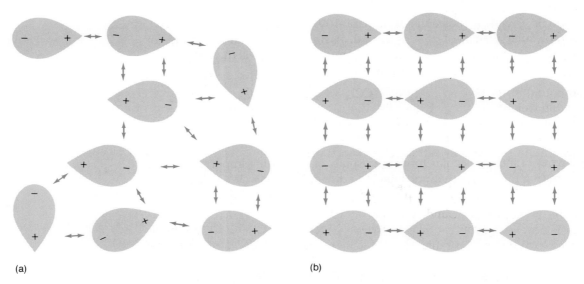

(a)   (b)

FIGURE 7.2   *An idealized representation of dipole forces in (a) a liquid and (b) a solid. In a real liquid or solid, interactions are more complex.*

ture of a liquid or a solid (Figure 7.2). In general, attractive forces between dipoles are fairly weak. They are, however, stronger than the forces between nonpolar molecules of comparable molecular weight. Compounds such as hydrogen chloride (HCl), hydrogen sulfide (H_2S), and sulfur dioxide (SO_2) consist of dipolar molecules.

Dipolar forces are weaker than ionic bonds. Whereas the attraction between dipolar molecules results from forces between centers of *partial* charge, in ions the attraction is between fully charged particles, those which have completely lost or gained control of electrons.

7.4 HYDROGEN BONDS

Certain molecules that have hydrogen atoms attached to fluorine, oxygen, or nitrogen have stronger attractive forces than are expected on the basis of dipolar attractive forces alone. These forces are strong enough to be given a special name, the **hydrogen bond.** Note that hydrogen bond is a somewhat misleading name since it emphasizes only the hydrogen component. Not all compounds containing hydrogen exhibit this strong attractive force. As we indicated in the first sentence of this paragraph, the hydrogen must be attached to oxygen, fluorine, or nitrogen if the molecules are to engage in hydrogen bonding. It is the presence of these atoms that permits us to offer an explanation for the extra strength of hydrogen bonds as compared with dipolar forces. Fluorine, oxygen, and nitrogen all have a high electron-attracting power (they are very electronegative). And these atoms are small, allowing the hydrogen atom on another molecule to approach closely for maximum attraction. If one pictures the bond between two atoms as an electron "cloud" (the orbital picture), then a hydrogen-fluorine bond (for example) is a much denser cloud at the fluorine end because of fluorine's power to attract electrons. This leaves the hydrogen somewhat exposed. The fluorine of another molecule can then approach the hydrogen (very closely because of the fluorine's small size) and "share" some of its wealth with the hydrogen.

Since fluorine forms only one covalent bond, there is only one pure fluorine-containing compound, hydrogen fluoride (HF), capable of intermolecular hydrogen bonding. Oxygen-containing compounds that can form hydrogen bonds include not only water but also the alcohols (Chapter 15), which are classed among the organic compounds discussed in later chapters of this book. Nitrogen-containing compounds that can form hydrogen bonds include ammonia (NH_3) and a number of organic compounds, among them the amines (Chapter 18).

Hydrogen bonds are generally represented by dotted lines (Figure 7.3). In gaseous hydrogen fluoride, the hydrogen bond is about 5% as strong as the hydrogen-to-fluorine covalent bond. In ice, the hydrogen bond is also about

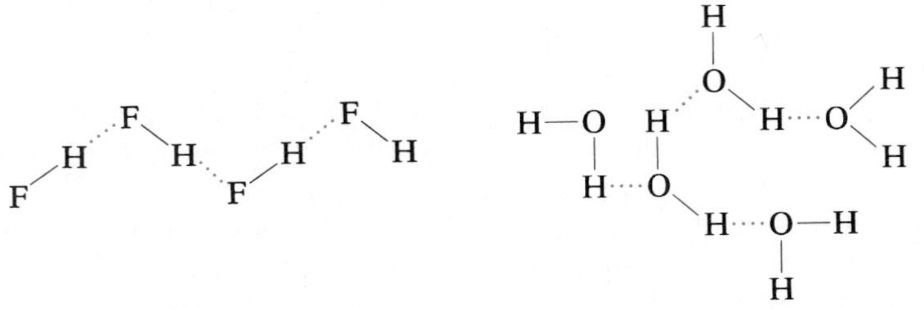

FIGURE 7.3   *Hydrogen bonding in hydrogen fluoride and in water.*

5% as strong as the covalent bond between hydrogen and oxygen. Hydrogen bonding appreciably affects the properties of compounds, as we shall see.

While the hydrogen bond may, at this point, seem merely an interesting piece of chemical theory, its importance to life and health is immense. The structure of proteins, chemicals essential to life, is determined, in part, by hydrogen bonding. And the heredity that one generation passes on to the next is dependent on an elegant application of hydrogen bonding.

7.5 DISPERSION FORCES

If one understands that positive attracts negative, then it is easy enough to understand how polar molecules are held together. But how can we explain the fact that nonpolar compounds such as bromine (Br_2) and iodine (I_2) exist in the liquid and solid states? Even hydrogen can exist as a liquid or a solid if the temperature is low enough. Something must hold these molecules together.

The answer arises from the fact that the electron cloud pictures that we talked about in Section 4.7 are only *average* positions. On the average, electrons spend more time near the fluorine atom in a fluorine-hydrogen bond, and this average, uneven sharing is what we have termed a dipole. On the average, the two electrons in the hydrogen molecule are between (and equidistant from) the two nuclei. The electrons are equally shared in this case. At any given instant, however, the electrons may be at one end of the molecule. At some other instant, the electrons may be at the other end of the molecule. Such electron motions give rise to momentary dipoles (Figure 7.4). One dipole, however momentary, can induce a similar momentary dipole in a neighboring molecule. At the instant that the electrons of one molecule are at one end, the electrons in the next molecule will move away from its adjacent end. Thus, at this instant there will be an attractive force between the electron-rich end of one molecule and

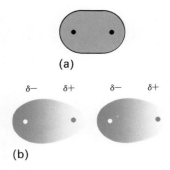

(a)

$\delta-$ $\delta+$ $\delta-$ $\delta+$

(b)

FIGURE 7.4
Electron cloud shapes on hydrogen molecules. (a) Average picture with no net dipole. (b) Instantaneous pictures with momentary dipoles. The transient dipoles are constantly changing, but the net result is attraction.

Hydrogen fluoride

Water

the electron-poor end of the next. These transient, attractive forces (usually small) between molecules are called **dispersion forces.***

Recall that fluorine (F_2) and chlorine (Cl_2) are gases, that bromine (Br_2) is a liquid, and that iodine (I_2) is a solid. Dispersion forces are greater for larger molecules than for smaller ones. Larger atoms have larger electron clouds. Their outermost electrons are farther from the nucleus than those of smaller atoms. These far out electrons are more loosely bound and can shift toward another atom more readily than the tightly bound electrons on a smaller atom. This makes molecules with larger atoms more polarizable than small ones. Iodine molecules are attracted to one another more strongly than bromine molecules are attracted to one another. Bromine molecules have greater dispersion forces than chlorine molecules, and chlorine molecules, in turn, have greater dispersion forces than fluorine molecules. If you look at the periodic table, you will see that this is the order in which these elements (called, as a group, the halogens) appear in Group VIIA, with iodine the largest and fluorine the smallest. Dispersion forces, to a large extent, determine the physical properties of nonpolar compounds (Section 7.7).

It should be noted that dispersion forces may be important even when other types of forces are present. Even though we might think of such forces as individually weaker than dipolar attractions or ionic bonds, in substances composed of large molecules the cumulative effect of dispersion forces may be considerable. For large ions, such as silver (Ag^+) and iodide (I^-), dispersion forces may play a significant role, even though interionic forces also exist. Dispersion forces play a major role in the presence of some dipolar forces. In hydrogen chloride (HCl), for example, dipolar forces may contribute as little as 15% to the intermolecular attraction; the rest is due to dispersion forces.

7.6 THE LIQUID STATE

Molecules of a liquid are much closer together than those of a gas. Consequently, liquids can be compressed only slightly. The molecules are in constant motion, but their movements are greatly restricted by neighboring molecules. One liquid can diffuse into another, but such diffusion is much slower than in gases because of the restricted molecular motion of liquids.

The shape of the molecules that make up a liquid has an effect on one of the properties of the liquid, its **viscosity,** or resistance to flow. Liquids with low

---

* Dispersion forces are sometimes called London forces, after Fritz London, professor of chemical physics at Duke University, who first treated them systematically in 1930. They are also called van der Waals forces, in honor of the Dutch physicist, Johannes D. van der Waals. Actually, van der Waals forces include all types of attractive forces between molecules, not just dispersion forces.

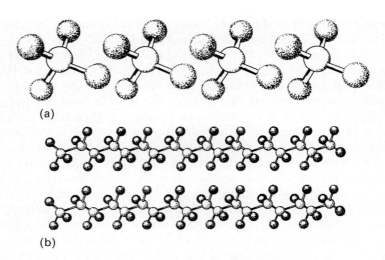

(a)

(b)

FIGURE 7.5   (a) Carbon tetrachloride (CCl_4) consists of small, symmetrical molecules with fairly weak intermolecular forces. It has a low viscosity. (b) Octadecane ($C_{18}H_{38}$) consists of long molecules with fairly strong intermolecular forces. It has a relatively high viscosity.

viscosity—"thin" liquids—generally consist of small, symmetrical molecules with weak intermolecular forces. Viscous liquids, on the other hand, are generally made up of large or unsymmetrical molecules with fairly strong intermolecular forces (such as hydrogen bonding) (Figure 7.5). Viscosity generally decreases with increasing temperature. Increased kinetic energy partially overcomes the intermolecular forces. Cooking oil, for example, as it pours from the bottle is thick and, well, oily. After it's been heated in a frying pan, it becomes thinner and more watery, that is, more like water in its consistency.

Another property of liquids is **surface tension.** A clean glass can be slightly overfilled with water before it spills over. A small needle, carefully placed, can be made to float horizontally on the surface of water—even though steel is several times denser than water is. A variety of insects can walk or skate across the surface of a pond with ease. These phenomena indicate something quite unusual about the surface of a liquid. There is a special force or tension there that resists being disrupted by the penetration of a needle or a water bug (Figure 7.6).

These surface forces can be explained by intermolecular forces. A molecule in the center of a liquid is pulled equally in all directions by the molecules surrounding it. A molecule on the surface, however, is attracted by molecules at its sides and below it only (Figure 7.7). There is no corresponding upward attraction. These unequal forces tend to pull inward at the surface of the liquid

FIGURE 7.6 *The surface tension of water enables this water strider to walk across the surface of ponds. Notice how the water is indented, but not penetrated, by the insect's legs. (© Laurence Pringle, The National Audubon Society Collection/PR.)*

and cause it to contract, much as a stretched sheet of rubber would tend to contract. A small amount of liquid will ''bead'' to minimize its surface area, and a drop will be spherical for the same reason. The smaller the surface area, the smaller the number of molecules experiencing the unequal pull. Soaps and other detergents act in part by lowering surface tension, enabling water to spread out and wet a solid surface.

7.7 FROM LIQUID TO GAS: VAPORIZATION

The molecules of a liquid are in constant motion, some moving fast, some more slowly. Occasionally one of these molecules has enough kinetic energy to escape from the liquid's surface and become a molecule of vapor. If a liquid, such as water, is placed in an open container, it will soon disappear through evaporation. The vapor molecules disperse throughout the atmosphere, and eventually

FIGURE 7.7
Molecules in the body of a liquid are attracted equally in all directions. Those at the surface, however, are pulled downward and sideways, but not upward.

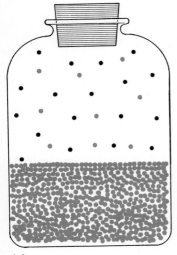

(a)

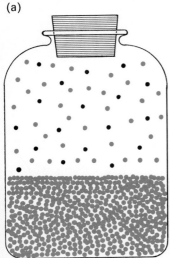

(b)

● = air molecule
● = water molecule
(liquid or vapor)

FIGURE 7.8
(a) Diagram of a liquid with its vapor at equilibrium in a closed container at a given temperature. Black circles represent molecules of air. (b) The same system at a higher temperature.

all the liquid molecules escape as they enter the vapor state. On the other hand, if the liquid is placed in a closed container, it does not go away. Some of the liquid is converted to vapor, but the vapor molecules are trapped within the container. Eventually the air above the liquid becomes saturated, and evaporation seems to stop. This vapor exerts a partial pressure (Section 6.8) that is constant at a given temperature.

It may well appear that nothing further is happening, but molecular motion has not ceased. Some molecules of liquid are still escaping into the vapor state. Vapor molecules in the space above the liquid occasionally strike the liquid's surface, are captured, and thus return to the liquid state. To begin with, there are lots of liquid molecules and no vapor molecules. So, at first, conversion of liquid to vapor (**vaporization**) is taking place, but conversion of vapor to liquid (**condensation**) is not. As more molecules pass into the vapor state, the rate of condensation increases. Eventually, the rate of condensation equals the rate of vaporization. This condition, called **equilibrium,** appears static but is, in fact, dynamic, as we have mentioned previously. Two opposing processes are taking place at exactly the same rate. This situation is analogous to that encountered in the case of reversible chemical reactions (Section 5.13).

At higher temperatures, more molecules of the liquid would have enough energy to escape from the liquid state. The vapor pressure would increase, but equilibrium would soon be reestablished at the higher temperature. The rate of vaporization would be greater, but so would the rate of condensation. At equilibrium, the rates would once again be equal (Figure 7.8), but the equilibrium vapor pressure would be higher at the higher temperature. Schematically, these processes may be illustrated as

$$\text{Liquid} \rightleftharpoons \text{vapor}$$

If a liquid is placed in an open container, the escape of molecules of the liquid is opposed by atmospheric pressure. If the liquid is heated, the vapor pressure will increase. Continued heating will eventually result in a vapor pressure equal to atmospheric pressure. At that temperature, the liquid will begin to boil. Vaporization will take place not only at the surface of the liquid but also in the body of the liquid, with vapor bubbles forming and rising to the surface. The **boiling point** of a liquid is the temperature at which its vapor pressure becomes equal to atmospheric pressure. Since the latter varies with altitude and weather conditions, boiling points of liquids also vary (Figure 7.9). The cooking of foods requires that they be supplied with a certain amount of energy. When water boils at 100 °C, an egg can be placed in the water and soft-boiled to perfection in 3 min. If, at reduced atmospheric pressure, water boils at a lower temperature, then it contains less heat energy with which to cook the egg. It would take a bit longer to boil an egg on top of Mount Everest. Ah, well, there's no wood for a fire up there anyway.

The boiling point is increased when external pressure is increased. Autoclaves and pressure cookers are based on this principle. We can achieve a higher

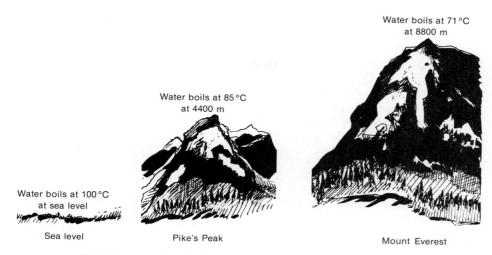

Water boils at 71 °C at 8800 m

Water boils at 85 °C at 4400 m

Water boils at 100 °C at sea level

Sea level   Pike's Peak   Mount Everest

FIGURE 7.9 *The boiling point of water at different altitudes.*

temperature at the higher pressures attained in these closed vessels. (Heat added to a liquid at atmospheric pressure would merely convert liquid to vapor. No increase in temperature would occur until all the liquid had vaporized.) The chemical reactions involved in the cooking of a tough piece of meat proceed more rapidly at the temperature that can be attained in a pressure cooker. Bacteria (even resistant spores) are killed more rapidly in an autoclave, not directly by the increased pressure but, rather, by the higher temperatures attained. Table 7.1 gives the temperatures attainable with pure water at various pressures.

The boiling point of a liquid is a useful physical property, often used as an aid in identifying compounds. Since boiling point varies with pressure, it is necessary to define the **normal boiling point,** that temperature at which a

TABLE 7.1 *Boiling Points of Pure Water at Various Pressures*

Pressure (in torr)	Temperature (in degrees Celsius)
760	100
816	102
875	104
938	106
1004	108
1075	110
1283	115
1535	121

TABLE 7.2   Boiling Points of Various Compounds
at 1 Atm

Compound	Boiling Point (in degrees Celsius)
Diethyl ether (anesthetic)	34.6
Methyl alcohol (wood alcohol)	64.5
Ethyl alcohol (grain alcohol)	78.3
Water	100.0
Mercury	356.6

liquid boils under standard pressure (1 atm, or 760 torr). Alternatively, one can specify the pressure at which the boiling point was determined. For example, the *Handbook of Chemistry and Physics* lists the boiling point of antipyrine (a pain reliever and fever reducer) as 319^{741}. This means that the compound boils at 319 °C under a pressure of 741 torr. Table 7.2 gives the normal boiling points of some familiar liquids.

Liquids can be purified by a process called **distillation.** Imagine a mixture of water and some nonvolatile material, that is, some material that will not vaporize readily. If the mixture is heated until it boils, the water will vaporize, but the nonvolatile material will not. The water vapor can then be condensed back to the liquid state and collected in a separate container. In such a distillation, the water is separated from the other component of the mixture and thereby purified.

Even if a mixture contains two or more volatile components, purification by distillation is possible. Let's consider a mixture of two components, one of which is somewhat more volatile than the other. At the boiling point of such a mixture, both components will contribute some molecules to the vapor. The more volatile component, because it is more easily vaporized, will have more of its molecules in the vapor state than will the less volatile component. When the vapor is condensed into another container, the resulting liquid will be richer in the more volatile component than was the original mixture. Thus, a purer sample of the more volatile component would have been produced.

Figure 7.10 shows a typical distillation apparatus.

Heat is required for the conversion of a liquid to a vapor. A liquid evaporating at room temperature absorbs heat from its surroundings. Most of us are familiar with this cooling effect of evaporation. Even on a warm day, we feel cool after a swim, because the water evaporating from our skin removes heat. Volatile liquids, such as ethyl chloride (C_2H_5Cl), which boils at 12.5 °C, may be used as local anesthetics. Rapid evaporation from the skin removes enough heat to freeze a small area, rendering it insensitive to pain. Alcohol rubs also act to cool the skin by their evaporation.

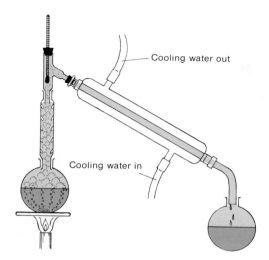

<figcaption>FIGURE 7.10

A distillation apparatus. A mixture is heated in the flask at the left. The vapors formed travel up the vertical column, are then condensed in the cooled tube angled downward toward the right, and are finally collected in the flask at the right.</figcaption>

— Cooling water out

Cooling water in

The amount of heat required to vaporize a given amount of liquid can be measured. The quantity of heat required to vaporize 1 mol of a liquid at a constant pressure is called the **molar heat of vaporization.** The heat of vaporization is characteristic of a given liquid. It depends to a large extent on the type of intermolecular force in the liquid. Water, with molecules strongly associated through hydrogen bonding, has a heat of vaporization of 9.7 kcal/mol. Methane, with molecules held together by weak dispersion forces only, has a heat of vaporization of only 0.232 kcal/mol. Heats of vaporization of several liquids are given in Table 7.3.

Given the molar heat of vaporization, one can calculate the heat of vaporization in calories per gram.

TABLE 7.3   *Heats of Vaporization (at the Normal Boiling Point) of Several Liquids*

Compound	Heat of Vaporization (in calories per mole)	Heat of Vaporization (in calories per gram)
Diethyl ether ($C_2H_5OC_2H_5$)	6200	84
Benzene (C_6H_6)	7300	94
Methyl alcohol (CH_3OH)	8400	260
Water	9700	540
Mercury	14 200	71

example
7.1

The molar heat of vaporization of ammonia (NH_3) is 556 cal. What is the heat of vaporization in calories per gram?

The molecular weight of ammonia is 17.0 g. Therefore,

$$\frac{556 \text{ cal/mol}}{17.0 \text{ g/mol}} = 33.0 \text{ cal/g}$$

example
7.2

How much heat would be required to vaporize 400 g of water at its boiling point?

The heat of vaporization of water is 540 cal/g.

$$\frac{540 \text{ cal}}{1.00 \text{ g}} \times 400 \text{ g} = 216\ 000 \text{ cal or } 216 \text{ kcal}$$

When a vapor condenses to a liquid, it gives up exactly the same amount of heat energy as was used up in converting the liquid to a vapor. A refrigerator operates by alternately vaporizing and condensing a fluid. The heat required to vaporize the fluid is drawn from the refrigerated compartment. The heat is released to the outside atmosphere when the fluid is condensed back to the liquid state.

7.8 THE SOLID STATE

Solids resemble liquids in that the particles (atoms, molecules, or ions) in them are close together, making them virtually incompressible. But these two physical states differ significantly in the motion of their particles. In the liquid state, particles are in constant (if somewhat restricted) motion. In solids, there is little motion other than gentle vibration about a fixed point. Consequently, diffusion in solids is generally extremely slow. An increase in temperature will increase the vigor of the vibrations in a solid. If the vibrations become violent enough, the solid will melt (Section 7.9).

In many solids, the particles are arranged in regular, systematic patterns. Such solids are said to be **crystalline,** and the structure is called a **crystal lattice.** The detail of a crystal lattice can be described in terms of a small, repeating segment called a **unit cell.** Extension of these unit cells into three-dimensional space results in the plane faces and definite angles of macroscopic crystals (such as those of quartz or uncut diamonds or rock salt).

In spite of the complex appearance of the many different crystalline solids, there are relatively few fundamental types of crystal lattices. The easiest to

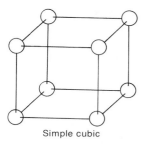 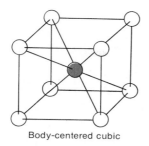 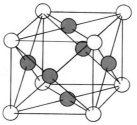

Simple cubic   Body-centered cubic   Face-centered cubic

FIGURE 7.11   *Three types of crystal lattices based on the cube.*

visualize is the simple cubic arrangement. This cell and two others readily derived from it are shown in Figure 7.11. In one of the derived structures, there is an additional particle (atom, ion, or molecule) at the center of the cube. This arrangement is called **body-centered cubic.** The other derived structure has particles at the center of each of the six faces of the cube. This structure is described as **face-centered cubic.** Other crystal systems are based on different geometric shapes. Two of these, the orthorhombic and the hexagonal, are shown in Figure 7.12.

Crystalline solids may also be classified on the basis of the types of forces holding the particles together. The four classes are ionic, molecular, covalent network, and metallic.

Ionic solids have ions occupying the lattice points in the crystal. A typical ionic solid is sodium chloride (NaCl), which we discussed in Section 4.4. In the lattice, each chloride ion is surrounded by six sodium ions, and each sodium ion is surrounded by six chloride ions. Interionic forces are very strong. Ionic solids, consequently, have high melting points and low vapor pressures and are quite hard.

Molecular crystals have discrete covalent molecules at the lattice points. These are held together by rather weak dispersion forces, as in crystalline iodine, by dispersion forces plus dipolar forces, as in iodine chloride (ICl), or by hydrogen bonds, as in ice. Molecular solids generally have low melting points and relatively high vapor pressures. Ice is somewhat exceptional; the water molecules are strongly associated by hydrogen bonding (Section 7.10). Even huge molecules, such as viruses, often have an ordered array, giving them a crystallike structure (Figure 7.13).

Covalent network crystals have atoms at the lattice points. These are joined into extensive networks by covalent bonds. Thus, each crystal is in essence one large molecule. Covalent network solids are generally extremely hard and nonvolatile and melt (with decomposition) at very high temperatures. Diamond is a familiar example. Carbon atoms occupy the lattice points. Each is covalently bonded to four other carbon atoms (Figure 7.14). Silicon carbide

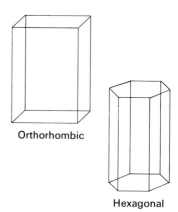

Orthorhombic

Hexagonal

FIGURE 7.12
Other types of crystal lattices are based on the orthorhombic and hexagonal systems.

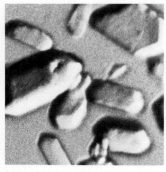

(a)

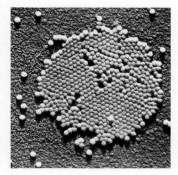

(b)

FIGURE 7.13
(a) Polio virus crystals, magnified 1000 times. (Courtesy of F. L. Schaffer.) (b) Spherical particles of polio virus, magnified 51 460 times. (Courtesy of the Virus Laboratory, University of California, Berkeley.)

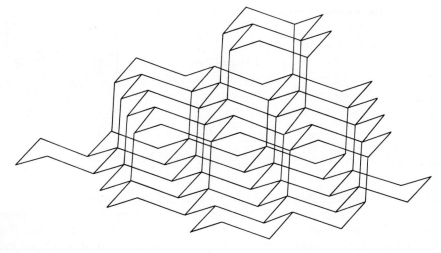

FIGURE 7.14   *The crystal structure of diamond, a covalent network solid. Each intersection of lines represents a carbon atom.*

(SiC), also called Carborundum, is another familiar compound with an extensive network of covalent bonds.

At the lattice sites in **metallic solids** are positive ions. These ions are formed when metal atoms, such as silver (Ag), lose their outer electrons. The electrons thus released are distributed throughout the lattice, almost like a fluid. These electrons, which can move freely about the lattice, make metals good conductors of heat and electricity. Some metals, such as sodium or potassium, are fairly soft and have low melting points. Others, such as manganese and iron, are hard and have high melting points. The extra electrons in iron and manganese atoms seem to lead to stronger forces between atoms. As a class, metals are malleable; that is, they can be shaped under the influence of pressure or heat. They can be rolled into bars, pressed into sheets, or extruded into wire.

Table 7.4 lists some characteristics of crystalline solids.

7.9 FROM SOLID TO LIQUID: FUSION

Most solids, when heated enough, will melt. The temperature at which this change of state occurs is called the **melting point.** The amount of heat required to convert 1 mol of a solid to a liquid at the melting point is called the **molar heat of fusion.** Generally, the heat of fusion is only a fraction of the heat of vaporization. The forces holding liquids together are not as strong as those holding solids together, but the difference between the forces holding

TABLE 7.4   *Some Characteristics of Crystalline Solids*

Crystal Type	Particles in Crystal	Principal Attractive Force Between Particles	Melting Point	Electrical Conductivity of Liquid	Characteristics of the Crystal	Examples
Ionic	Positive and negative ions	Electrostatic attraction between ions (very strong)	High	High	Hard, brittle, most dissolve in polar solvents	$NaCl$, CaF_2, K_2S, MgO
Covalent network	Atoms	Covalent bonds (very strong)	Generally do not melt	—	Very hard, insoluble	Diamond (C), SiC, AlN
Metal	Positive ions plus mobile electrons	Metallic bonds (strong)	Most are high	Very high	Most are hard, malleable, ductile, good electrical and thermal conductors, insoluble unless a reaction occurs	Cu, Ca, Al, Pb, Zn, Fe, Na, Ag
Molecular Hydrogen-bonded	Molecules with H on N, O, or F	Hydrogen bonds (intermediate)	Intermediate	Very low	Fragile, soluble in other H-bonding liquids	H_2O, HF, NH_3, CH_3OH
Polar	Polar molecules (no H-bonds)	Electrostatic attraction between dipoles (rather weak)	Low	Very low	Fragile, soluble in other polar and many nonpolar solvents	HCl, H_2S, $CHCl_3$, ICl
Nonpolar	Atoms or nonpolar molecules	Dispersion forces only (weak)	Very low	Extremely low	Soft, soluble in nonpolar or slightly polar solvents	Ar, H_2, I_2, CH_4, CO_2, CCl_4

solids together and the forces holding liquids together is much less than the difference between the forces holding liquids together and the forces holding gases together (Figure 7.15). The heat of fusion is the amount of energy that will disrupt the crystal lattice but still leave the particles in contact with one another and under the influence of their mutual attraction. A much larger amount of energy is required to vaporize the liquid because, in vaporization, the attraction between particles must be completely overcome (or very nearly so). Table 7.5 gives some representative heats of fusion.

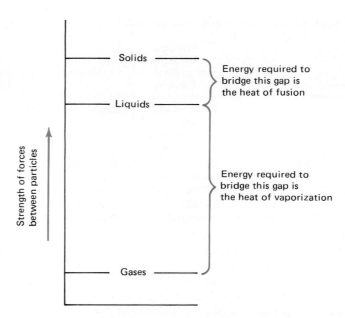

FIGURE 7.15   The heat of fusion for a substance is only a fraction of the heat of vaporization.

**example
7.3**

The heat of fusion of water is 80 cal/g. How much heat would be required to melt 400 g of ice?

$$400 \text{ g} \times \frac{80 \text{ cal}}{1.0 \text{ g}} = 32\ 000 \text{ cal or } 32 \text{ kcal}$$

TABLE 7.5   Heat of Fusion (at the Melting Point) for Several Solids

Compound	Melting Points (°C)	Heat of Fusion (in calories per mole)	Heat of Fusion (in calories per gram)
Ammonia (NH_3)	−78	1620	95
Water (H_2O)	0	1440	80
Benzene (C_6H_6)	−6	2550	33
Copper (Cu)	1083	3110	49
Sodium chloride (NaCl)	804	7000	120
Tungsten (W)	3410	8050	43

The molar heat of fusion of naphthalene ($C_{10}H_8$) is 4610 cal. What is the heat of fusion in calories per gram?

example
7.4

The molecular weight of naphthalene is 128 g.

$$\frac{4610 \text{ cal/mol}}{128 \text{ g/mol}} = 36.0 \text{ cal/g}$$

7.10 WATER: A MOST UNUSUAL LIQUID

Now that we have laid something of a theoretical foundation, let's look more closely at a very special liquid, water. Next to air, water is the most familiar substance on Earth. Even so, it is a most unusual compound. At room temperature, it is the only liquid compound with a molecular weight as low as 18 amu. The solid form of water (ice) is less dense than the liquid, a relatively rare situation. The consequences for life on this planet of this peculiar characteristic are immense. Ice forms on the surface of lakes when the temperature drops below freezing, and this ice insulates the lower layers of water, enabling fish and other aquatic organisms to survive the winters of the temperate zones. If ice were denser than liquid water, it would sink to the bottom as it formed. This would permit the new surface water to freeze and, in its turn, sink to the bottom. The repetition of this process would eventually result in a lake frozen from top to bottom. Even the deeper lakes of the northern latitudes would freeze solid in winter. Life in the northern lakes and rivers would be very different from what it is now, if indeed there were life in those waters.

The same property—the relative density of ice and liquid water—has dangerous consequences for living cells. Since ice has a lower density than liquid water, 1 g of ice occupies a larger volume than 1 g of water. As ice crystals form in living cells, the expansion ruptures and kills the cells. The slower the cooling, the larger the crystals of ice and the more damage to the cell. The frozen food industry takes into account this property of water. Food is "flash frozen," that is, frozen so rapidly that the ice crystals are kept very small and do minimum damage to the cellular structure of the food.

Another unusual property of water is its high specific heat. It takes 1 cal of heat to raise the temperature of 1 g of water 1 °C. That's 10 times as much energy as is required to raise the temperature of the same amount of iron 1 °C. The specific heats of a number of common materials are given in Table 7.6.

The reason cooking utensils are made of iron, copper, aluminum, or glass is that these materials have low heat capacities, or specific heats. Thus, they heat up very quickly. The reason the handle of a frying pan is made of wood or plastic is that these materials have high specific heats. When they are exposed to heat, their temperatures increase more slowly.

TABLE 7.6 *Specific Heats of Some Common Substances*

Substance	Specific Heat (in calories per gram per degree Celsius)
Water (liquid)	1.00
Water (solid)	0.50
Water (gas)	0.47
Ethyl alcohol (grain alcohol)	0.54
Wood	0.42
Glass	0.12
Iron	0.11
Aluminum	0.21
Copper	0.09
Silver	0.06
Gold	0.03

example 7.5

How much heat is required to increase the temperature of 10 g of water from 10 °C to 30 °C?

The specific heat of liquid water is 1 cal/g/°C. The temperature change is 20 °C (30 °C − 10 °C = 20 °C).

$$10 \text{ g} \times 20 \text{ °C} \times \frac{1 \text{ cal}}{(1 \text{ g})(1 \text{ °C})} = 200 \text{ cal}$$

example 7.6

How much energy is required to change 10 g of ice at −10 °C to steam at 100 °C?

This problem should be worked in parts. First calculate the energy required to raise the temperature of ice from −10 °C to 0 °C, a change of 10 °C. The specific heat of ice is 0.5 cal/g/°C.

$$10 \text{ g} \times 10 \text{ °C} \times \frac{0.5 \text{ cal}}{(1 \text{ g})(1 \text{ °C})} = 50 \text{ cal}$$

Next, calculate, using the heat of fusion of water (80 cal/g), the energy required to melt the ice at 0 °C.

$$10 \text{ g} \times \frac{80 \text{ cal}}{1 \text{ g}} = 800 \text{ cal}$$

Then calculate the heat required to change the temperature of the water, which is at 0 °C, to 100 °C. The specific heat of water is 1 cal/g/°C and the temperature change is 100 °C.

$$10 \text{ g} \times 100 \text{ °C} \times \frac{1 \text{ cal}}{(1 \text{ g})(1 \text{ °C})} = 1000 \text{ cal}$$

Now calculate the amount of heat required to change the water (at 100 °C) to steam (at 100 °C). The heat of vaporization for water is 540 cal/g.

$$10 \text{ g} \times \frac{540 \text{ cal}}{1 \text{ g}} = 5400 \text{ cal}$$

Finally, total all the calculated values.

To raise the temperature of ice from -10 °C to 0 °C	50 cal
To change ice to liquid water	800 cal
To raise the temperature of water from 0 °C to 100 °C	1000 cal
To change water to steam	5400 cal
Total	7250 cal

Note that almost 75% of the total energy is used in vaporizing the water.

The high specific heat of water means not only that much energy is required to raise the temperature of water but also that much heat is given off by water for even a small drop in temperature. The vast amounts of water on the surface of the Earth thus act as a giant thermostat to moderate daily temperature variations. We need only consider the extreme temperature changes on the surface of the waterless moon to appreciate this important property of water. The temperature of the moon varies from just above the boiling point of water (100 °C) to about -175 °C, a range of 275 °C. In contrast, temperatures on the Earth rarely fall below -50 °C (-58 °F) or rise above 50 °C (122 °F), a range of only 100 °C.

Water also has a higher density than many other familiar liquids, including petroleum products. As a consequence, a number of liquids that are insoluble in water float on the surface of it. This situation has caused problems in recent years. Gigantic oil spills, which occur when a tanker ruptures or an offshore well gets out of control, result in a slick on the water's surface. This oil covers the feathers of waterfowl and the coats of sea animals, such as the otter and the seal. The oil is often washed onto beaches, where it does considerable ecological and aesthetic damage. If oil were denser than water, it would sink to the bottom. The problem would be of a different nature, though not necessarily less acute.

Still another way in which water is unusual is that it has a very high heat of vaporization; that is, a large amount of heat is required to evaporate a small amount of water. This is of enormous importance to animals. Large amounts of body heat, produced as a by-product of metabolic processes, can be dissipated by the evaporation of small amounts of water (perspiration) from the skin. The heat of vaporization of this water is obtained from the body, and the body is cooled. Conversely, when steam condenses, considerable heat is released. For

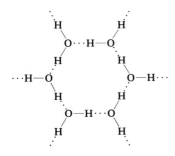

FIGURE 7.16
A two-dimensional representation of the structure of ice, showing a large hexagonal hole formed by six water molecules.

this reason, steam causes serious burns when it contacts the skin. We previously mentioned that water's high specific heat modifies the climate. Water's high heat of vaporization also contributes to the climate-modifying effect of lakes and oceans. A large portion of the heat that would otherwise warm up the land is used instead to vaporize water from the surface of a lake or the sea. Thus, in summer it is cooler near a large body of water than in interior land areas.

All of these fascinating properties of water result from the unique structure of the water molecule (Section 4.10). Recall that the water molecule is polar. In the liquid state, water molecules are strongly associated by hydrogen bonding (see Figure 7.4). These strong attractive forces account for the high heat of vaporization of water. They must be overcome if vaporization is to take place, and this is why a large amount of energy must be supplied to water for it to convert from liquid to vapor.

In the liquid state, water molecules are quite close together but randomly arranged. In the solid state, water molecules are in a much more ordered arrangement. But this orderly arrangement, as illustrated in Figure 7.16, is less compact than that achieved in the liquid state. Large hexagonal holes are incorporated into the ice lattice. It is this empty space that makes ice less dense than liquid water.

Because of its polar nature, water tends to dissolve ionic substances. Now, ionic solids are held together by strong ionic bonds. We have already indicated that to melt ionic solids and break these bonds, very high temperatures are required. Yet, simply by placing sodium chloride, an ionic solid, in water at room temperature, we can dissolve the salt (or, rather, the water can). And when such a solid dissolves, its bonds are broken. The difference between the two processes is the difference between brute force and persuasion. In the melting process, we are simply pouring in enough energy (as heat) to pull the crystal lattice apart. In the dissolving process, we offer the ions an attractive alternative to the crystal lattice. It works this way. Water molecules surround the lattice. As they approach a negative ion, they align themselves so that their positive ends point toward the ion. With a positive ion the process is reversed,

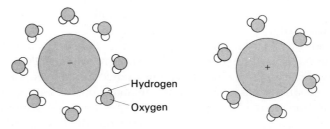

FIGURE 7.17 *The interaction of polar water molecules with ions.*

and the negative end of the water dipole points toward the ion. Still, the attraction between a dipole and an ion is not as strong as that between two ions, so, to compensate for their weaker attractive power, the water molecules gang up on the ions. Several molecules surround each ion, and in this way the many ion-dipole interactions overcome the single ion-ion interaction (Figure 7.17).

Chemical reactions in living cells take place in water solutions. Our bodies are about 65% water. The importance of solutions is such that we devote an entire chapter to the subject.

PROBLEMS

1. How do liquids and solids differ from gases in their compressibility, spacing of molecules, and intermolecular forces?

2. In what ways are liquids and solids similar? In what ways are they different?

3. Define or illustrate the following terms:
   a. dipolar molecule
   b. hydrogen bond
   c. dispersion forces
   d. viscosity
   e. surface tension
   f. vaporization
   g. condensation
   h. equilibrium
   i. boiling point
   j. normal boiling point
   k. molar heat of vaporization
   l. ionic crystal
   m. molecular crystal
   n. covalent network crystal
   o. metallic solid
   p. melting point
   q. molar heat of fusion
   r. specific heat

4. List four types of interactions between particles in the liquid and in the solid states. Give an example of each type.

5. Rank N_2, H_2O, and HCl in order of increasing strength of intermolecular interaction (weakest interaction first). What type of interaction is involved in each case?

6. Explain how oxygen can be liquefied at a low enough temperature, even though O_2 molecules are nonpolar.

7. What type of interaction exists between molecules of
   a. H_2 b. NO   c. HF

8. List three distinctive properties of water.

9. Explain why a salt such as sodium chloride dissolves in water.

10. Describe the effect of temperature on the viscosity of a liquid.

11. Describe, in terms of intermolecular forces, the cause of the phenomenon of surface tension.

12. In which of the following compounds would hydrogen bonding be an important intermolecular force?

   a. H—S
          \
           H

   b. H—C—N—H   with H's on C and N

   c. H—C—F   with H's

   d. H—C—O
          \
           H

   e. H—C—C—H   with H's

13. The normal boiling point of a substance depends on the molecular mass *and* on the type of intermolecular attractions. Rank the following sets in order of increasing boiling points (lowest boiling first):
   a. H_2S, H_2Se, H_2Te
   b. H_2O, CO, O_2
   c. CCl_4, CBr_4, CI_4

14. How does a pressure cooker work?
15. The heat of vaporization of bromine (Br_2) is 45 cal/g. What is the molar heat of vaporization of bromine?
16. The heat of vaporization of ammonia (NH_3) is 327 cal/g. What is the molar heat of vaporization of ammonia?
17. The molar heat of vaporization of acetic acid ($C_2H_4O_2$) is 5.81 kcal/mol. How much heat is required to vaporize 1.00 g of acetic acid?
18. The molar heat of vaporization of acetone (C_3H_6O) is 7.23 kcal/mol. How much heat is required to vaporize 5.80 g of acetone?
19. How is the heat of vaporization of a liquid related to intermolecular forces in the liquid?
20. The heat of fusion of water is 80 cal/g. How much heat would be required to melt a 15-kg block of ice?
21. The molar heat of fusion of acetone (C_3H_6O) is 2.58 kcal/mol. How much heat is required to melt 1.00 g of acetone?
22. The molar heat of fusion of water is 1.44 kcal/mol. How much heat is required to melt 40.0 g of ice?
23. There is a rule of thumb that says that for many liquids the molar heat of vaporization is approximately 21 times the normal boiling point in kelvin. Use this rule to calculate the molar heat of vaporization for benzene, which has a boiling point of 353 K. How does your calculated value compare with the experimental value given in Table 7.3?
24. The specific heat of silver is 0.06 cal/g/°C. That of gold is 0.03 cal/g/°C. Which metal will be hotter (i.e., will be at a higher temperature) if both are exposed to the same amount of heat?
25. The molar heats of vaporization of ethanol (C_2H_6O) and ethyl acetate ($C_4H_8O_2$) are 9.39 kcal/mol and 7.77 kcal/mol, respectively. Which liquid has stronger intermolecular forces?
26. Calculate the amount of heat required to raise the temperature of 25 g of water from 20 °C to 60 °C. The specific heat of water is 1 cal/g/°C.
27. How much energy will be expended in changing 100 g of ice at −5 °C to steam at 100 °C?
28. How much heat is required to convert 10.0 g of ice at −12.0 °C to steam at 130 °C?

29. How much heat is absorbed when 50 g of ammonia (NH_3) is vaporized? The heat of vaporization of ammonia is 327 cal/g.
30. Why does it take longer to boil an egg at high altitude than at sea level? Does it take longer to fry an egg at high altitude? Explain.
31. Why does ice float on liquid water?
32. Why does steam at 100 °C cause more severe burns than liquid water at the same temperature?
33. Which has a higher boiling point, methane (CH_4) or ethane (C_2H_6)? Why?
34. Which has a higher boiling point, ethane or methanol? Why?

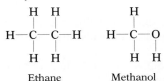

Ethane   Methanol

35. Which noble gas, neon (Ne) or xenon (Xe), has the higher boiling point? Why? (Both exist as monatomic gases.)
36. To obtain water each day, a bird in winter eats 5 g of snow at 0 °C. How many kcal (food Calories) of energy does it take to melt this snow and warm the liquid to the bird's body temperature of 40 °C?
37. To obtain water on a winter hike, a woman decides to eat snow. How many extra kcal (food Calories) of food would she have to take in each day to raise the 1500 g of snow that she needs from −10 °C to 0 °C, melt it, then raise the liquid water from 0 °C to her body temperature of 37 °C?
38. In which process is energy absorbed by the material undergoing the change of state?
    a. melting or freezing
    b. condensation or vaporization
39. Label the arrows with the term that correctly identifies the process described.

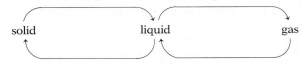

solid   liquid   gas

Oxidation and Reduction

hemical reactions can be classified in several ways. In this chapter, we will consider an important group of reactions called **reduction-oxidation** (or **redox**) **reactions.** The two processes—oxidation and reduction—always occur together. You can't have one without the other. One substance is oxidized; another is reduced. For convenience, however, we may choose to talk about only a part of the process—the oxidation part or the reduction part.

Our cells obtain energy to maintain themselves by oxidizing foods. Green plants, using energy from sunlight, produce food by the reduction of carbon dioxide. We win metals from their ores by reduction, then lose them again to corrosion as they are oxidized. We maintain our technological civilization by oxidizing fossil fuels (coal, natural gas, and petroleum) to obtain the chemical energy that was stored in these materials eons ago by green plants.

Reduced forms of matter—sugars, coal, gasoline—are high in energy (Figure 8.1). Oxidized forms—carbon dioxide and water—are low in energy. Let's examine the processes of oxidation and reduction in some detail, in order that we might better understand the chemical reactions that keep us alive and enable us to maintain our civilization.

8.1 OXYGEN: ABUNDANT AND ESSENTIAL

It would be inaccurate to say that any one element is the most important, for there are 20 or so elements essential to life. Nevertheless, in any list of important elements, oxygen would be at or near the top. It is the most abundant of all

FIGURE 8.1
Reduced forms of matter, such as foods and fossil fuels, are high in energy. The energy content is released through oxidation-reduction reactions.

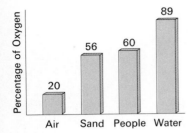

FIGURE 8.2
Oxygen is the most abundant element on Earth.

the elements on this planet, making up about one-half the Earth's crust. Oxygen occurs in each of the three subdivisions of the crust—the atmosphere, the hydrosphere, and the lithosphere. It occurs in the atmosphere (the gaseous mass surrounding the earth) as molecular oxygen (O_2). In the hydrosphere (oceans, seas, rivers, and lakes of the Earth), oxygen occurs combined with hydrogen in that remarkable compound, water (H_2O). In the lithosphere (the solid portion of the Earth), oxygen occurs combined with silicon (pure sand is largely SiO_2) and with a variety of metals.

Oxygen is found in most of the compounds that are important to living organisms. Foodstuffs—starches, fats, and proteins—all contain oxygen. The human body is approximately 65% (by weight) water. Since 89% of the weight of water is due to oxygen—and many other compounds in our bodies also contain oxygen—about 60% of the weight of each of us is oxygen.

Oxygen accounts for about 20% (by volume) of the atmosphere. (The other 80% is nitrogen, which enters into relatively few chemical reactions.) This oxygen, as O_2, is said to be free or uncombined, that is, not part of a compound containing another element. We have seen (Chapter 6) that oxygen (in air) is taken into the lungs. From there it passes into the bloodstream. The blood carries oxygen to the body tissues, and the food we eat combines with this oxygen. This chemical process provides us with energy in the form of heat to maintain our body temperature. It also gives us the energy we need for mental and physical activity.

Oxygen performs many other functions. Fuels such as natural gas, gasoline, and coal need oxygen in order to burn and release their stored energy. Not all that oxygen does is immediately desirable, however. Oxygen causes iron to rust and copper to corrode, and it aids in the decay of wood. All these—and many other—chemical processes are called **oxidative** reactions.

8.2 CHEMICAL PROPERTIES OF OXYGEN: OXIDATION

When iron rusts, it combines with oxygen from the atmosphere to form a reddish brown powder.

$$4\,Fe + 3\,O_2 \longrightarrow 2\,Fe_2O_3$$

The chemical name for iron rust (Fe_2O_3) is iron(III) oxide. Many other metals react with oxygen to form **oxides.**

Most nonmetals also react with oxygen to form oxides. For example, carbon reacts to form carbon dioxide.

$$C + O_2 \longrightarrow CO_2$$

Sulfur combines with oxygen to form sulfur dioxide.

$$S + O_2 \longrightarrow SO_2$$

FIGURE 8.3   *Cooking, breathing, and burning fuel all involve oxidation.*

At high temperatures, such as those that occur in automobile engines, even nitrogen, an element that is ordinarily quite unreactive, combines with oxygen.

$$N_2 + O_2 \longrightarrow 2\,NO$$

The product, which should be called nitrogen monoxide, is perhaps better known as nitric oxide.

Oxygen also reacts with many compounds. Methane, the principal ingredient of natural gas, burns in air to produce carbon dioxide and water.

$$CH_4 + 2\,O_2 \longrightarrow CO_2 + 2\,H_2O$$

Hydrogen sulfide, a gaseous compound with a rotten egg odor, burns, producing water and sulfur dioxide.

$$2\,H_2S + 3\,O_2 \longrightarrow 2\,H_2O + 2\,SO_2$$

In each example, oxygen combines with each of the elements in the compound.

The combination of elements and compounds with oxygen is called **oxidation** (Figure 8.3). The substances that combine with oxygen are said to be **oxidized.** Originally, the term oxidation was restricted to reactions involving combination with oxygen. Chemists came to recognize, though, that combination with chlorine (or bromine, or other elements in the upper right portion of the periodic table) was not all that different from reaction with oxygen. So they broadened the definition of oxidation. Oxidation is now defined in terms of **oxidation numbers.**

8.3 OXIDATION NUMBERS

Oxidation numbers are related to the electron density about an atom, but the relationship is not always a simple one. We will concern ourselves only with the rather arbitrary way in which these numbers are derived and with their use in studying chemical reactions involving oxidation.

Following are a set of rules for assigning oxidation numbers:

1. Free (uncombined) elements have an oxidation number of 0. Thus, oxygen, whether it occurs in the atomic form (O), in the form of a diatomic molecule (O_2), or as ozone (O_3), has an oxidation number of 0.
2. In compounds, elements in Group IA have oxidation numbers of $+1$, and elements in Group IIA have oxidation numbers of $+2$. For example, each sodium in sodium chromate (Na_2CrO_4) has an oxidation number of $+1$. By the same rule, calcium in calcium sulfate ($CaSO_4$) has an oxidation number of $+2$. In other words, the oxidation number is the same as the charge on the ion: Na^+ has $+1$, and Ca^{2+} has $+2$.
3. Hydrogen in compounds has an oxidation number of $+1$ except when combined with metals, in which case it is -1. In sulfuric acid (H_2SO_4), the hydrogen has an oxidation number of $+1$. In lithium aluminum hydride ($LiAlH_4$), since it is combined with the metals lithium and aluminum, hydrogen has an oxidation number of -1.
4. In most compounds, oxygen has an oxidation number of -2. Thus, oxygen in sulfuric acid or sodium chromate or water has an oxidation number of -2. In special compounds called **peroxides**, oxygen's oxidation number is -1. Examples are hydrogen peroxide (H_2O_2) and barium peroxide (BaO_2). Using the formula alone, it is not always easy to distinguish a peroxide from other oxygen-containing compounds. For example, BaO_2 is a peroxide, but CO_2 is not. And CH_2O_3 may or may not be a peroxide (there are two different compounds with this formula). How will you know which is which? If we are dealing with a peroxide, we will make note of the fact, although hydrogen peroxide (H_2O_2) should be recognized by its formula. In structural formulas, peroxides are identified by the presence of an oxygen-to-oxygen bond ($O-O$).
5. For compounds, the algebraic sum of all the oxidation numbers must be 0. For example, in the compound sodium hydroxide (NaOH), the sodium is $+1$, the oxygen is -2, and the hydrogen is $+1$. The sum is 0.

$$(+1) + (-2) + (+1) = 0$$

6. For polyatomic ions, the algebraic sum of all the oxidation numbers must be equal to the charge on the ion. For example, in the hydroxide ion (OH^-), the oxygen is -2, and the hydrogen is $+1$. The sum is -1.

$$(-2) + (+1) = -1$$

These rules are used to calculate the oxidation states of other elements in compounds. Rules 2, 3, and 4 are used first. Then Rule 5 or 6 is used to calculate the oxidation number of any remaining element. You can best learn the process by working examples.

What is the oxidation number of sulfur in SO_2?

    Oxygen is -2 (Rule 4). There are two oxygens, for a total of -4. Since the sum must be 0 (Rule 5), the sulfur must be $+4$.

example 8.1

What is the oxidation number of sulfur in H_2SO_4?

    Hydrogen is $+1$ (Rule 3) and oxygen is -2 (Rule 4). There are two hydrogens, for a total of $+2$, and four oxygens, for a total of -8. Since the sum must be 0 (Rule 5), we can write

$$2(+1) + 4(-2) + x = 0$$

where x is the unknown oxidation state of sulfur. Solving the equation we get

$$(+2) + (-8) + x = 0$$
$$-6 + x = 0$$
$$x = +6$$

You need not use algebra if you can see (without it) that sulfur must be $+6$.

example 8.2

What is the oxidation number of nitrogen in the nitrate ion (NO_3^-)?

    Oxygen is -2 (Rule 4). There are three oxygen atoms, for a total of -6. Since the sum of oxidation numbers must be -1 (Rule 6), we can write

$$(-6) + x = -1$$
$$x = +5$$

The nitrogen must be $+5$.

example 8.3

What is the oxidation number of nitrogen in the ammonium ion (NH_4^+)?

    Hydrogen is $+1$ (Rule 3). There are four hydrogen atoms, for a total of $+4$. Since the sum of oxidation numbers must be $+1$ (Rule 6), we can write

$$(+4) + x = +1$$
$$x = -3$$

The nitrogen is -3.

example 8.4

**example
8.5**

What is the oxidation number of carbon in C_2H_4O?
Hydrogen is $+1$. Four hydrogen atoms total $+4$. One oxygen is -2.
Since the sum must be 0,

$$(+4) + (-2) + x = 0$$
$$x = -2$$

To balance the oxidation numbers, we need -2, so the carbons total -2.
But there are two carbons to share the burden, so each carbon has an
oxidation number of -1.

**example
8.6**

What is the oxidation number of each element in Na_2O_2?
Rule 2 tells us that Na, a group I element, has an oxidation state of $+1$.
Rule 4 tells us that oxygen is *usually* -2, but mentions that oxygen is -1
in peroxides. No exceptions are mentioned in Rule 2, so let's assume that
Na is $+1$ and calculate the oxidation state for O.

$$2(+1) + 2x = 0$$
$$2x = -2$$
$$x = -1$$

We calculate that the oxidation state is -1. (The compound is sodium
peroxide.)

We now have a way to define oxidation: it is an increase in oxidation
number. We can use this definition to determine whether or not oxidation has
occurred in a chemical reaction. Consider the reaction

$$CaO + CO_2 \longrightarrow CaCO_3$$

Is carbon oxidized? In carbon dioxide there are two oxygen atoms (each with
an oxidation number of -2), for a total of -4. The carbon in carbon dioxide
must be $+4$. Now we calculate the oxidation number of carbon in calcium
carbonate ($CaCO_3$). Calcium is $+2$ (Rule 2). Each oxygen is -2, and there are
three oxygens, for a total of -6. The carbon, therefore, must be $+4$.

$$(+2) + (-6) + x = 0$$
$$x = +4$$

No, carbon is not oxidized in this reaction. Its oxidation number has not
changed. Now consider the reaction

$$2\,Na_2CO_3 + Cr_2O_3 + 3\,KNO_3 \longrightarrow 2\,CO_2 + 2\,Na_2CrO_4 + 3\,KNO_2$$

Is chromium (Cr) oxidized? The oxidation number of chromium in chromium(III) oxide (Cr_2O_3) is $+3$; in sodium chromate (Na_2CrO_4), it is $+6$. The oxidation number of chromium has increased. Therefore, it is oxidized in the reaction. Check this answer by calculating the oxidation numbers of chromium in chromium(III) oxide and in sodium chromate for yourself.

There are alternative ways (actually special cases of the oxidation number method) of defining oxidation. Three working definitions of possible utility are the following:

1. An element or compound is oxidized if it gains oxygen atoms. In the reactions encountered so far in this section, iron, carbon, sulfur, nitrogen, methane, and hydrogen sulfide all gain oxygen atoms. We say, then, that each of these substances is oxidized.
2. A compound is oxidized if it losses hydrogen atoms. Methyl alcohol, when passed over hot copper gauze, forms formaldehyde and hydrogen gas.

$$CH_4O \longrightarrow CH_2O + H_2$$
Methyl alcohol   Formaldehyde

Since methyl alcohol loses hydrogen atoms, it is said to be oxidized.
3. An element is oxidized if it loses electrons. When magnesium metal reacts with chlorine, magnesium ions and chloride ions are formed.

$$Mg + Cl_2 \longrightarrow Mg^{2+} + 2\,Cl^-$$

Since the magnesium atom obviously loses electrons, it is oxidized.

We will encounter many reactions that fit these definitions when we get to the section of this book that deals with biochemistry, for it is by oxidation of foodstuffs that organisms derive their life-sustaining energy.

Why so many different definitions of oxidation? Simply for convenience. Which do we use? Whichever is most convenient. For the reaction

$$C + O_2 \longrightarrow CO_2$$

we could assign oxidation numbers and determine that carbon is oxidized (it goes from 0 to $+4$), but it is more convenient to see that carbon is oxidized because it gains oxygen atoms. Similarly, for the reaction

$$CH_4O \longrightarrow CH_2O + H_2$$

we could assign oxidation numbers and determine that carbon is oxidized (it goes from -2 to 0, an *increase* of 2), but it is easier to see that carbon is oxidized because it loses hydrogen atoms. For the reaction involving chromium (p. 238), though, it would perhaps be best to calculate oxidation numbers.

8.4 HYDROGEN: OCCURRENCE, PREPARATION, AND PHYSICAL PROPERTIES

For every oxidation process, there is a complementary process called reduction. Before discussing reduction in detail, let's look at some of the chemistry of another important element—hydrogen.

By weight, hydrogen makes up only about 0.9% of the Earth's crust, placing it far down on the list of abundant elements.* Because hydrogen is the lightest of all elements, however, hydrogen atoms are quite abundant. If we consider a random sample of 10 000 atoms from the Earth's crust, 5330 are oxygen atoms, 1590 are silicon atoms, and 1510 are hydrogen atoms. Unlike oxygen, hydrogen is seldom found free in nature on Earth. Most of it is combined with oxygen in water. Nearly all compounds derived from living organisms contain hydrogen. Fats, starches, sugars, and proteins all contain combined hydrogen. Petroleum and natural gas are mixtures composed mainly of hydrocarbons, compounds of hydrogen and carbon.

Since elemental hydrogen does not occur in nature (again, we limit our view to this planet), it is necessary to make it. Small amounts are made for laboratory use by the reaction of zinc with hydrochloric acid,

$$Zn + \underset{\substack{\text{Hydrochloric} \\ \text{acid}}}{2\ HCl} \longrightarrow \underset{\substack{\text{Zinc} \\ \text{chloride}}}{ZnCl_2} + H_2$$

or by the reaction of calcium metal with water,

$$Ca + 2\ H_2O \longrightarrow \underset{\substack{\text{Calcium} \\ \text{hydroxide}}}{Ca(OH)_2} + H_2$$

The gaseous hydrogen is easily collected by displacement of water (Figure 8.4).

Commercially, hydrogen is usually obtained as a by-product of other processes. Much of it comes from petroleum refineries. We consider those reactions in Chapter 14.

Physically, hydrogen is a colorless and odorless gas. It is essentially insoluble in water. Hydrogen is the least dense of all substances: its density is only $\frac{1}{14}$ that of air under comparable conditions. For this reason it was at one time used in dirigibles to give them the necessary buoyancy in air. However, since a spark or flame can trigger the explosion of hydrogen (Section 8.5), the use of this gas represented a considerable danger. When a disastrous explosion occurred aboard a luxury airship called the Hindenburg in 1937, the use of

---

* Hydrogen ranks low in abundance on Earth. If we look beyond our home planet, however, hydrogen becomes much more significant. The sun, for example, is largely hydrogen. In fact, hydrogen is by far the most abundant element in the universe.

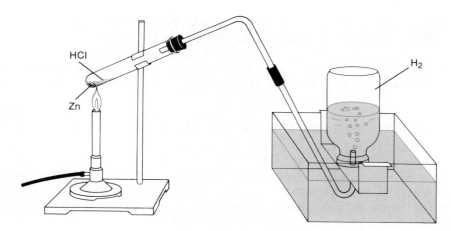

FIGURE 8.4 *Diagram of an apparatus for the laboratory synthesis of hydrogen. Because it is collected over water, the hydrogen gas is saturated with water vapor.*

hydrogen was discontinued, and so indeed was the commercial use of lighter-than-air ships (Figure 8.5). Those blimps (small dirigibles) still in use employ the nonreactive gas helium, while hot-air balloons use, as their name indicates, hot air, which is less dense than the atmosphere (recall Charles's law, from Section 6.5).

FIGURE 8.5 *Hydrogen is the most effective buoyant gas, but it is highly flammable. Disastrous fires in hydrogen-filled dirigibles led to the replacement of hydrogen by nonflammable helium. (Courtesy of National Air and Space Museum, Smithsonian Institution, Washington, D.C.)*

Certain metals, such as platinum (Pt), palladium (Pd), and nickel (Ni), are capable of collecting large volumes of hydrogen, in a condensed form, on their surface. This *adsorbed* hydrogen appears to be a great deal more reactive than ordinary molecular hydrogen, thus, these metals are often used as catalysts for reactions involving hydrogen gas as one of the reactants.

8.5 THE CHEMICAL PROPERTIES OF HYDROGEN: REDUCTION

A jet of hydrogen, when ignited, burns in air with an almost colorless flame. Hydrogen and oxygen may be mixed at room temperature with no perceptible reaction. However, if the mixture is ignited by a spark or flame, a tremendous explosion results. The product in both cases is water.

$$2\,H_2 + O_2 \xrightarrow{\text{spark}} 2\,H_2O$$

It is interesting to note that if a piece of platinum gauze is inserted into a container of hydrogen and oxygen, the two gases react at room temperature. The platinum acts as a catalyst: it lowers the activation energy (Section 5.11) for the reaction. The platinum will glow from the heat evolved in the initial reaction and then ignite the mixture, causing an explosion.

Hydrogen reacts with a variety of metal oxides to remove oxygen and give the free metal. For example, when hydrogen is passed over heated copper(II) oxide, metallic copper and water are formed.

$$CuO + H_2 \longrightarrow Cu + H_2O$$

With lead(II) oxide, the products are metallic lead and water.

$$PbO + H_2 \longrightarrow Pb + H_2O$$

This process, in which a compound is changed, or **reduced,** to an element, is called **reduction.**

Like oxidation, which once simply meant combination with oxygen, the term reduction has been broadened in meaning. Most generally, reduction is defined as a *decrease* in oxidation number. In the examples above, copper goes from $+2$ in copper(II) oxide to 0 in metallic copper, and lead goes from $+2$ in lead(II) oxide to 0 in metallic lead. We say that copper and lead are reduced.

We may also use working definitions complementary to those given for oxidation (p. 239).

1. A compound is reduced if it loses oxygen atoms. In the examples above, copper(II) oxide and lead(II) oxide obviously are reduced, since they lose oxygen. When potassium chlorate ($KClO_3$) is heated, it forms potassium chloride and oxygen gas.

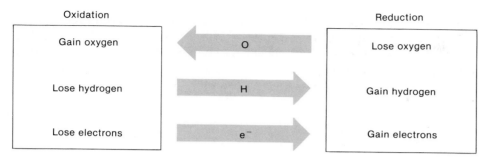

Oxidation

Gain oxygen

Lose hydrogen

Lose electrons

O

H

e$^-$

Reduction

Lose oxygen

Gain hydrogen

Gain electrons

FIGURE 8.6 *Three different definitions of oxidation and reduction.*

$$2 \; KClO_3 \xrightarrow{\text{heat}} 2 \; KCl + 3 \; O_2$$

The potassium chlorate loses oxygen; therefore, it is reduced.

2. A compound is reduced if it gains hydrogen atoms. For example, methyl alcohol is produced when carbon monoxide is allowed to react with hydrogen.

$$CO + 2 \; H_2 \xrightarrow{\text{catalyst}} CH_3OH$$

The carbon monoxide gains hydrogen; therefore, it is reduced.

3. An atom or ion is reduced if it gains electrons. When an electric current is passed through a solution containing copper(II) ions (Cu^{2+}), copper metal is plated out at the cathode.

$$Cu^{2+} + 2e^- \longrightarrow Cu$$

The copper obviously gains electrons; therefore, it is reduced.

(Figure 8.6 summarizes these definitions of oxidation and reduction.)

Oxidation and reduction go hand in hand. You can't have one without the other. When one substance is oxidized, another is reduced (Figure 8.7). For example, in the reaction

$$CuO + H_2 \longrightarrow Cu + H_2O$$

copper is reduced (from $+2$ to 0), and hydrogen is oxidized (from 0 to $+1$). Further, if one substance is being oxidized, the other must be causing it to be oxidized. In the example, copper(II) oxide is causing hydrogen gas to be oxidized. Therefore, copper(II) oxide is called the **oxidizing agent.** Conversely, hydrogen gas is causing copper(II) oxide to be reduced, so hydrogen gas is the **reducing agent.** Each oxidation-reduction (redox) reaction will have an oxidizing agent and a reducing agent among the reactants. The reducing agent is

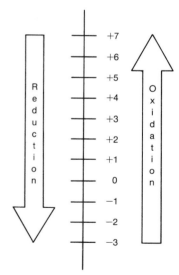

Reduction

Oxidation

+7
+6
+5
+4
+3
+2
+1
0
−1
−2
−3

FIGURE 8.7
Oxidation is an increase in oxidation number. Reduction is a decrease in oxidation number.

the substance being oxidized, and the oxidizing agent is the substance being reduced.

Reduction:
copper is being reduced;
CuO is the oxidizing agent

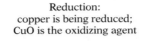

Oxidation:
hydrogen is being oxidized;
H_2 is the reducing agent

example 8.7

Circle the oxidizing agent and underline the reducing agent in the following reactions.

$$C + O_2 \longrightarrow CO_2$$
$$N_2 + 3\,H_2 \longrightarrow 2\,NH_3$$
$$SnO + H_2 \longrightarrow Sn + H_2O$$
$$NO + O_3 \longrightarrow NO_2 + O_2$$
$$Mg + Cl_2 \longrightarrow Mg^{2+} + 2\,Cl^-$$

The answers can be determined by means of any of the definitions previously listed.

$$\underline{C} + \boxed{O_2}$$

C gains oxygen and is oxidized (Definition 1, Section 8.3), so it must be the reducing agent. Therefore, O_2 is the oxidizing agent.

$$\boxed{N_2} + 3\,\underline{H_2}$$

N_2 gains hydrogen and is reduced (Definition 2, Section 8.5), so it must be the oxidizing agent. Therefore, H_2 is the reducing agent.

$$\boxed{SnO} + \underline{H_2}$$

SnO loses oxygen and is reduced (Definition 1, Section 8.5), so it is the oxidizing agent. H_2 is the reducing agent.

$$\underline{NO} + (O_3)$$

NO gains oxygen and is oxidized (Definition 1, Section 8.3), so it is the reducing agent. O_3 is the oxidizing agent.

$$\underline{Mg} + (Cl_2)$$

Mg loses electrons and is oxidized (Definition 3, Section 8.3.), so it is the reducing agent. Cl_2 is the oxidizing agent.

8.6 SOME FAMILIAR OXIDIZING AGENTS

Oxygen is undoubtedly the most common oxidizing agent. It oxidizes the wood in our campfires and the gasoline in our automobile engines. It rusts and corrodes the metals that we get by reducing ores. It even "burns" the foods we eat to give us energy to move about. It is involved in the rotting of wood and the weathering of rocks. Indeed, we live in an oxidizing atmosphere. Fortunately, though, oxygen is a mild oxidizing agent. It takes us about 70 years to "burn out."

Oxygen is sometimes used as a laboratory and industrial oxidizing agent. For example, acetylene (for cutting and welding torches) is made by the partial oxidation of methane.

$$4\,CH_4 + 3\,O_2 \longrightarrow 2\,C_2H_2 + 6\,H_2O$$
Methane   Acetylene

(Complete oxidation of methane would give carbon dioxide and water.) Often, though, it is more convenient to use other oxidizing agents. Potassium permanganate ($KMnO_4$) and sodium dichromate ($Na_2Cr_2O_7$) are frequently used.

Potassium permanganate is a black, shiny, crystalline solid. It dissolves in water to give deep purple solutions. This purple color disappears as the permanganate is reduced (remember, if permanganate is an oxidizing agent, it must be reduced). So potassium permanganate is often used as a test for oxidizable substances. For example, potassium permanganate is used to oxidize iron from Fe^{2+} to Fe^{3+}. The amount of iron(II) ion in a sample can be determined by its reaction with permanganate. One can add permanganate solution, which is deep purple, to a sample of iron(II) ion. As the permanganate is reduced, it is changed to manganese(II) ion, and the purple color disappears (Figure 8.8). When all the iron(II) ion has been oxidized, the addition of more permanganate will not be accompanied by the loss of the purple color because there will be no iron(II) ion left to reduce the permanganate. Thus, one can measure just

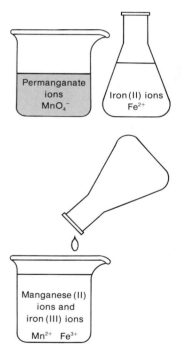

FIGURE 8.8
Purple permanganate ions (MnO_4^-) are reduced to Mn^{2+} by iron(II) ions (Fe^{2+}). The purple color disappears as the reaction proceeds.

how much iron(II) ion there was in the sample by keeping track of how much permanganate was reduced, that is, by measuring how much permanganate was added until the purple color remained. The equation for this reaction is quite complex.

$$MnO_4^- + 5\ Fe^{2+} + 8\ H^+ \longrightarrow Mn^{2+} + 5\ Fe^{3+} + 4\ H_2O$$

| Permanganate | Manganese (II) |
| ion (purple) | ion (colorless) |

Permanganate solutions can also be used to oxidize oxalic acid (a poisonous compound found in rhubarb leaves), sulfur dioxide (SO_2), and many other compounds.

Sodium dichromate is used to oxidize alcohols (Chapter 15) to compounds called aldehydes and ketones. For the oxidation of ethyl alcohol (found in alcoholic beverages) to acetaldehyde, the reaction is

$$8\ H^+ + Cr_2O_7^{2-} + 3\ C_2H_5OH \longrightarrow 2\ Cr^{3+} + 3\ C_2H_4O + 7\ H_2O$$

| Dichromate ion | Chromium(III) ion |
| (orange-red) | (green) |

Some of the tests for intoxication (the "breathalyzer," used to check for drunken drivers, for instance) depend on the oxidation of alcohol by an oxidizing agent that changes color as it is reduced (such as a chromium compound).

example 8.8

Calculate the oxidation number of manganese in MnO_4^- and of chromium in $Cr_2O_7^{2-}$. What are the oxidizing agents in the two preceding reactions? What are the reducing agents?

Mn is reduced;
MnO_4^- is the oxidizing agent

$$\overset{+7}{MnO_4^-} + 5\ \underset{+2}{Fe^{2+}} + 8\ H^+ \longrightarrow \overset{+2}{Mn^{2+}} + 5\ \underset{+3}{Fe^{3+}} + 4\ H_2O$$

Fe is oxidized;
Fe^{2+} is the reducing agent

Cr is reduced;
$Cr_2O_7^{2-}$ is the oxidizing agent

$$8\ H^+ + \overset{+6}{Cr_2}\underset{-2}{O_7^{2-}} + 3\ C_2H_5OH \longrightarrow 2\ \overset{+3}{Cr^{3+}} + 3\ C_2\underset{-1}{H_4}O + 7\ H_2O$$

C is oxidized;
C_2H_5OH is the reducing agent

Another common oxidizing agent is hydrogen peroxide (H_2O_2). Pure hydrogen peroxide is a syrupy, colorless liquid. It is usually used as an aqueous solution, though, with concentrations of 30% and 3% generally available. Some interesting uses of hydrogen peroxide will be discussed in Sections 8.8 and 8.9. For now, let's look at some of its simpler chemistry.

Hydrogen peroxide will oxidize sulfides (S^{2-}) to sulfates (SO_4^{2-}). When lead-based paints are exposed to polluted air containing hydrogen sulfide (H_2S), they turn black due to the formation of lead sulfide (PbS). Hydrogen peroxide will oxidize the black sulfide to the white sulfate.

$$PbS + 4\,H_2O_2 \longrightarrow PbSO_4 + 4\,H_2O$$
$$\text{(Black)} \qquad\qquad\qquad \text{(White)}$$

A nice advantage of hydrogen peroxide as an oxidizing agent is that, in most reactions, it is converted to water, an innocuous by-product.

The oxidation number of oxygen in hydrogen peroxide is -1. (Hydrogen is $+1$ and there are two hydrogens, so each oxygen must be -1.) When hydrogen peroxide acts as an oxidizing agent, it is reduced. The oxygen goes from -1 to -2.

Another oxidizing agent is nitric acid (HNO_3). For example, metallic copper will dissolve in nitric acid solutions because it is oxidized to copper(II) ions (Cu^{2+}).

$$Cu + 4\,HNO_3 \longrightarrow Cu(NO_3)_2 + 2\,NO_2 + 2\,H_2O$$

Is HNO_3 oxidized, reduced, or neither in the reaction above?

 There are four molecules of HNO_3 involved in the reaction. Two end up in $Cu(NO_3)_2$, and two end up as NO_2.

example 8.9

$$Cu(NO_3)_2 \longrightarrow Cu^{2+} + 2\,NO_3^-$$

No change
in oxidation state

$$\overset{+5 \qquad\qquad +5}{HNO_3 \longrightarrow NO_3^-}$$

HNO_3 is reduced

$$\overset{+5 \qquad\qquad +4}{HNO_3 \longrightarrow NO_2}$$

Two of the HNO_3 molecules are reduced, but two are neither oxidized nor reduced.

Other common oxidizing agents are the halogens—fluorine (F_2), chlorine (Cl_2), bromine (Br_2), and iodine (I_2). Chlorine, for example, oxidizes magnesium metal to magnesium ions.

$$Mg + Cl_2 \longrightarrow Mg^{2+} + 2\,Cl^-$$

In the process, the oxidation number of chlorine is reduced from 0 (in Cl_2) to -1 (in Cl^-).

We will encounter some other interesting oxidizing agents in other sections.

8.7 SOME REDUCING AGENTS OF INTEREST

There is no single reducing agent that stands out as oxygen does among oxidizing agents. Hydrogen will reduce many compounds, but it is relatively expensive for large-scale processes. Elemental carbon, or coke (obtained by driving off the volatile matter from coal), is frequently used to release metals from their ores. For example, tin can be obtained from tin(IV) oxide by reduction, with carbon used as the reducing agent.

$$SnO_2 + C \longrightarrow Sn + CO_2$$

Hydrogen may be used for the production of expensive metals such as tungsten (W). The ore is first converted to an oxide (WO_3) and then is reduced in a stream of hydrogen gas at 1200 °C.

$$WO_3 + 3\,H_2 \longrightarrow W + 3\,H_2O$$

Perhaps a more familiar reducing agent, by use if not by name, is the developer used in photography. Photographic film is coated with a silver salt (Ag^+Br^-), in which silver occurs in the $+1$ oxidation state. Silver ions that have been exposed to light react with the developer, a reducing agent such as the organic compound hydroquinone, to form metallic silver.

$$C_6H_4(OH)_2 + 2\,Ag^+ \longrightarrow C_6H_4O_2 + 2\,Ag + 2\,H^+$$

Hydroquinone   Silver metal

Those silver ions that were not exposed to light are not reduced by the developer. The film is then treated with "hypo," a solution of sodium thiosulfate ($Na_2S_2O_3$), which washes out unexposed silver bromide. That leaves the negative dark where the metallic silver has been deposited (where it was originally exposed to light) and transparent where light did not strike it. Light shone through the negative onto light-sensitive paper then makes the positive print.

Hydrogen peroxide, mentioned in the last section as a common oxidizing agent, is also a reducing agent. In the presence of a stronger oxidizing agent, such as permanganate, hydrogen peroxide accepts the role of reducing agent and is oxidized.

$$6 H^+ + 2 MnO_4^- + 5 H_2O_2 \longrightarrow 5 O_2 + 2 Mn^{2+} + 8 H_2O$$

Note that in this reaction the oxidation number of oxygen goes from -1 (in H_2O_2) to 0 (in O_2). When acting as an oxidizing agent, oxygen in hydrogen peroxide goes from -1 to -2 (in H_2O). Such behavior is characteristic of compounds with elements in intermediate oxidation states. They may go up or down the scale (be oxidized or reduced). Putting it another way, they may act as either reducing or oxidizing agents.

8.8 OXIDATION: ANTISEPTICS AND DISINFECTANTS

Many common antiseptics (compounds that are applied to living tissue to kill or prevent the growth of microorganisms) are mild oxidizing agents. Hydrogen peroxide, in the form of a 3% aqueous solution, finds use in medicine as a topical antiseptic. It can be used to treat minor cuts and abrasions on certain parts of the body. Potassium permanganate, in concentrations ranging from 0.01% to 0.2%, can be used as a topical antiseptic and an astringent.

Benzoyl peroxide $[(C_6H_5COO)_2]$, a powerful oxidizing agent, has long been used at 5% and 10% concentrations in ointments for treating acne. In addition to its antibacterial action, benzoyl peroxide acts as a skin irritant, causing old skin to slough off and be replaced with newer, fresher-looking skin. When used on areas exposed to sunlight, benzoyl peroxide is thought to promote skin cancer; such use is therefore discouraged.

Sodium hypochlorite (NaOCl), available in aqueous solution as a laundry bleach (Purex, Clorox, and the like), finds some use in the irrigation of wounds and for bladder infections. It is also used as a disinfectant, deodorizer, and bleach (Section 8.9). Solutions of iodine or iodine-releasing compounds are frequently used as antiseptics. In surgery, the area around the incision is usually disinfected with an iodine-containing solution. Because iodine solutions are colored, it is easy to see if any of the critical area has been missed.

The exact method of operation of these oxidizing agents as antiseptics is unknown. Their action is rather indiscriminate; they attack human cells as well as microorganisms. For many purposes, these simple inorganic compounds have been replaced by organic compounds such as the phenols (Chapter 15). The simplest of the phenols (carbolic acid) has itself been replaced by less poisonous modifications of its basic structure.

Other oxidizing agents are used as disinfectants. Calcium hypochlorite $[Ca(OCl)_2]$, called bleaching powder, is used to disinfect clothing and bedding. Chlorine (Cl_2) is used to kill pathogenic (disease-causing) microorganisms in drinking water. Wastewater is usually treated with chlorine also, before it is returned to a stream or lake. Such treatment has been quite effective in preventing the spread of infectious diseases such as typhoid fever. Use of chlorine in this manner came under criticism in 1974, when it was shown that the chlorine reacted with organic compounds (presumably from industrial wastes) to form toxic chlorinated compounds.

Ozone has also been used to disinfect drinking water. Many European cities, including Paris and Moscow, use ozone to treat their drinking water. Ozone is more expensive than chlorine, but less of it is needed. An added advantage is that ozone kills viruses on which chlorine has little, if any, effect. Tests in Russia have shown ozone to be a hundred times as effective as chlorine for killing polio virus.

Ozone (O_3) acts by transferring its "extra" oxygen to the contaminant. The oxidized contaminants are thought to be less toxic than the chlorinated ones. In addition, ozone imparts no "chemical taste" to the water. We may well see a shift from chlorine to ozone in the treatment of our drinking water and wastewater.

8.9 OXIDATION: BLEACHING AND STAIN REMOVAL

Bleaches are compounds that are used to remove unwanted color from white fabrics. Nearly any oxidizing agent will do the job. However, some also harm the fabric, some are unsafe, some produce undesirable by-products, and some are simply too expensive.

The most familiar laundry bleaches are aqueous solutions of sodium hypochlorite (NaOCl). These formulations generally have 5.25% NaOCl, yet they vary widely in price. Such price variations in standardized products usually result from costly advertisement of the more familiar brand name products.

Bleaching powder [$Ca(OCl)_2$] is generally preferred for large-scale bleaching operations. The paper industry uses it to make white paper, and the textile industry uses it to make whiter fabrics.

In both aqueous bleaches and bleaching powder, the active agent is the hypochlorite ion (ClO^-). Materials appear colored because loosely bound electrons are boosted to higher energy levels by absorption of visible light (Section 2.9). Bleaching agents do their work by removing or tying down these mobile electrons. For example, hypochlorite ion (in water) takes up electrons to form chloride ions and hydroxide ions.

$$ClO^- + H_2O + 2e^- \longrightarrow Cl^- + 2OH^-$$

Hypochlorite bleaches are safe and effective for cotton and linen fabrics. They should not be used for wool, silk, or nylon.

Other bleaching agents include hydrogen peroxide, sodium perborate (often formulated $NaBO_2 \cdot H_2O_2$ to indicate a loose association of $NaBO_2$ and H_2O_2), and a variety of chlorine-containing organic compounds that release chlorine molecules in water. When hydrogen peroxide is used to bleach hair, it acts as an oxidizing agent. In the process, the black or brown pigment in the hair, called melanin, is oxidized to colorless products.

Stain removal is not nearly so simple a process as bleaching. A few stain removers are oxidizing agents or reducing agents; others have quite different chemical natures. Nearly all stains require rather specific stain removers. Hydrogen peroxide in cold water removes blood stains from cotton and linen fabrics. Potassium permanganate will remove most stains from white fabrics (except for rayon). The permanganate stain can then be removed by treatment with oxalic acid.

$$5\ H_2C_2O_4 + 2\ MnO_4^- + 6\ H^+ \longrightarrow 10\ CO_2 + 2\ Mn^{2+} + 8\ H_2O$$

Oxalic   (Purple)   (Colorless)
acid

(Oxalic acid also removes rust spots, but its action in this case is as a complexing agent, not a reducing agent. The oxalic acid picks up the otherwise insoluble rust and carries it in solution to be washed away.)

Sodium thiosulfate ($Na_2S_2O_3$) readily removes iodine stains by reducing iodine to a colorless ion (I^-).

$$I_2\ \ + 2\ Na_2S_2O_3 \longrightarrow 2\ NaI + Na_2S_4O_6$$

(Brown)   (Colorless)

Many stain removers are simply adsorbents (e.g., cornstarch, which removes grease spots), solvents (e.g., amyl acetate or acetone, which removes ballpoint ink), or detergents (which remove mustard stains).

8.10 OXIDATION, REDUCTION, AND LIVING THINGS

We obtain energy for physical and mental activities by the slow, multistep oxidation of food. These processes will be discussed in detail in later chapters, but for now we will represent them schematically as follows.

$$\text{Carbohydrates (sugars and starches)} + O_2 \longrightarrow CO_2 + H_2O + \text{energy}$$
$$\text{Fats} + O_2 \longrightarrow CO_2 + H_2O + \text{energy}$$
$$\text{Proteins} + O_2 \longrightarrow CO_2 + H_2O + \text{urea} + \text{energy}$$

Green plants make these foods by reduction of carbon dioxide. Energy for the reaction, called photosynthesis, comes from the sun. The process, shown here for the formation of glucose, a simple sugar, may be written

$$6\ CO_2 + 6\ H_2O + \text{energy} \longrightarrow C_6H_{12}O_6 + 6\ O_2$$

(a)   (b)

FIGURE 8.9   *The food we eat is oxidized to provide energy for our activities. That energy originates in the sun and is trapped by plants through photosynthetic reactions that reduce carbon dioxide to carbohydrates. (Photo a courtesy of* The Cokato Enterprise, *Cokato, Minn. Photo b courtesy of Northrup King Co.)*

We eat either the plants or the animals that feed on the plants, and we obtain energy from the carbohydrates, fats, and proteins of those organisms. In fact, the photosynthesis reaction is precisely the reverse of that for the oxidation of carbohydrates. If we take the carbohydrate to be glucose, that reaction is

$$C_6H_{12}O_6 + 6O_2 \longrightarrow 6CO_2 + 6H_2O + \text{energy}$$

We can see that green plants carry out the redox reaction that makes possible all life on Earth. Animals can only oxidize the foods that plants provide. We might therefore consider crop farming a process of reduction. Energy captured in cultivated plants, whether the plants are used directly or are fed to animals, is the basis for human life (Figure 8.9).

example 8.10

Calculate the oxidation state of carbon in carbon dioxide (CO_2) and in glucose ($C_6H_{12}O_6$).

The oxidation state of carbon in CO_2 is $+4$. In $C_6H_{12}O_6$ it may be calculated as follows.

$$6x + 12(+1) + 6(-2) = 0$$
$$6x + 12 - 12 = 0$$
$$6x = 0$$
$$x = 0$$

Now you can see how green plants reduce carbon dioxide to provide the fuel for all life processes.

PROBLEMS

1. Write the formula for the products formed when each of the following burns (i.e., reacts with oxygen):
   a. $CH_4 + O_2 \longrightarrow$
   b. $C + O_2 \longrightarrow$
   c. $S + O_2 \longrightarrow$
   d. $CS_2 + O_2 \longrightarrow$
   e. $N_2 + O_2 \longrightarrow$
   f. $C_6H_{12}O_6 + O_2 \longrightarrow$

2. Define oxidation and reduction in terms of the following:
   a. oxygen atoms gained or lost
   b. hydrogen atoms gained or lost
   c. electrons gained or lost
   d. change in oxidation number

3. In which of the following is the reactant undergoing oxidation? (These are not complete chemical equations.)
   a. $Cl_2 \longrightarrow 2\,Cl^-$
   b. $WO_3 \longrightarrow W$
   c. $2\,H^+ \longrightarrow H_2$
   d. $CO \longrightarrow CO_2$

4. The following "equations" show only part of a chemical reaction. Indicate whether the reactant shown is being oxidized or reduced.
   a. $C_2H_4O \longrightarrow C_2H_4O_2$
   c. $Fe^{3+} \longrightarrow Fe^{2+}$
   b. $C_2H_4O \longrightarrow C_2H_6O$

5. Green grapes are exceptionally sour due to a high concentration of tartaric acid. As the grapes ripen, this compound is converted to glucose.
$$C_4H_6O_2 \longrightarrow C_6H_{12}O_6$$
   Tartaric   Glucose
acid

   Is the tartaric acid being oxidized or reduced?

6. Circle the oxidizing agent and underline the reducing agent in these reactions.
   a. $4\,Al + 3\,O_2 \longrightarrow 2\,Al_2O_3$
   b. $C_2H_2 + H_2 \longrightarrow C_2H_4$
   c. $2\,SO_2 + O_2 \longrightarrow 2\,SO_3$
   d. $2\,AgNO_3 + Cu \longrightarrow Cu(NO_3)_2 + 2\,Ag$

7. Assign oxidation numbers to the underlined elements.
   a. $\underline{S}_8$ **b.** $\underline{S}O_3$ **c.** $H_2\underline{S}$ **d.** $K\underline{I}$

8. Assign oxidation numbers to the underlined elements.
   a. $\underline{Al}_2O_3$ **b.** $H\underline{N}O_2$ **c.** $Na_2\underline{S}O_3$

9. Assign oxidation numbers to the underlined elements.
   a. $H\underline{C}O_3^-$ **b.** $H\underline{P}O_2^-$ **c.** $\underline{U}O_2^{2+}$

10. In the following reactions, which element is oxidized and which is reduced?
    a. $2\,HNO_3 + SO_2 \longrightarrow H_2SO_4 + 2\,NO_2$
    b. $2\,CrO_3 + 6\,HI \longrightarrow Cr_2O_3 + 3\,I_2 + 3\,H_2O$
    c. $5\,C_2H_6O + 4\,MnO_4^- + 12\,H^+ \longrightarrow$
    $$5\,C_2H_4O_2 + 4\,Mn^{2+} + 11\,H_2O$$
    d. $Cl_2 + 2\,KBr \longrightarrow 2\,KCl + Br_2$

11. List four common oxidizing agents.

12. List three common reducing agents.

13. Describe how a bleaching agent such as hypochlorite (ClO^-) works.

14. How can hydrogen peroxide act as both an oxidizing agent and a reducing agent?

15. Name some oxidizing agents used as antiseptics and disinfectants.

16. To test for an iodide ion (for example, in iodized salt), a solution is treated with chlorine to liberate iodine. The reaction is
$$2\,I^- + Cl_2 \longrightarrow I_2 + 2\,Cl^-$$
    Which substance is oxidized? Which is reduced?

17. What is the oxidation state of P in each of the following?
    a. PH_3
    b. P_4
    c. P_4O_6
    d. P_4O_{10}
    e. H_3PO_4
    f. H_3PO_2
    g. PH_4^+
    h. PO_4^{3-}
    i. PO_3^{3-}
    j. HPO_2^{2-}
    k. $H_2PO_4^-$
    l. Na_3PO_4
    m. KH_2PO_4
    n. Na_2HPO_3
    o. $NaPO_3$
    p. $K_2H_2P_2O_7$
    q. $Na_5P_3O_{10}$
    r. H_3PO_3
    s. Na_3P
    t. Ca_3P_2
    u. KH_2PO_3
    v. $P_2O_7^{4-}$

18. Molybdenum metal, used in special kinds of steel, can be manufactured by the reaction of its oxide with hydrogen. The reaction is
$$MoO_3 + 3\,H_2 \longrightarrow Mo + 3\,H_2O$$

Which substance is reduced? Which is the reducing agent?

19. Consider the equation

$$C_2H_4 + H_2O \longrightarrow C_2H_6O$$

Is carbon oxidized or reduced?

20. To oxidize 1 kg of fat, our bodies require about 2000 L of oxygen. A good diet contains about 80 g of fat per day. What volume of oxygen is required to oxidize that fat?

21. Unsaturated vegetable oils react with hydrogen to form saturated fats. A typical reaction is

$$C_{57}H_{104}O_6 + 3\,H_2 \longrightarrow C_{57}H_{110}O_6$$

Is the unsaturated oil oxidized or reduced? Explain.

22. The dye indigo (used to color blue jeans) is formed from indoxyl by exposure of the latter to air.

$$2\,C_8H_7ON + O_2 \longrightarrow C_{16}H_{10}N_2O_2 + 2\,H_2O$$

      Indoxyl   Indigo

What substance is oxidized? What is the oxidizing agent?

23. Calculate the oxidation state of Cl in $HClO_4$ (a powerful oxidizing agent).

24. Relate the chemistry of photosynthesis to the chemistry that provides energy for your heartbeat.

25. In the reaction

$$H_2CO + H_2O_2 \longrightarrow H_2CO_2 + H_2O$$

which substance is oxidized? Which is the oxidizing agent?

26. Ethylene, C_2H_4, reacts with hydrogen to form ethane, C_2H_6. Is the ethylene oxidized or reduced? Explain your answer.

27. The oxidizing agent we use to obtain energy from food is oxygen (from the air). If you breathe 15 times a minute (at rest), taking in and exhaling 0.5 L of air with each breath, what volume of air do you breathe each day? Air is 21% oxygen by volume. What volume of oxygen do you breathe each day?

28. When a water pump failed in the nuclear reactor at Three Mile Island in 1979, zirconium metal reacted with the very hot water to produce hydrogen gas.

$$Zr + 2\,H_2O \longrightarrow ZrO_2 + 2\,H_2$$

What substance was oxidized in the reaction? What was the oxidizing agent?

29. Which of the following are redox reactions?

a. $CaCl_2 + 2\,KF \longrightarrow CaF_2 + 2\,KCl$

b. $CaI_2 + Cl_2 \longrightarrow CaCl_2 + I_2$

c. $PbO_2 + 4\,HCl \longrightarrow PbCl_2 + Cl_2 + 2\,H_2O$

d. $CaCO_3 + 2\,HCl \longrightarrow CaCl_2 + CO_2 + H_2O$

Solutions

Y ou are a solution of sodium ions, potassium ions, calcium ions, bicarbonate ions, chloride ions, glucose, amino acids, fatty acids, glycerol, fats, proteins, acetylcholine, and lots of other goodies. Well, you aren't all in solution, or else you would wash away in the shower. But, except for a few semisolid parts, such as skin and muscle and bone, you are mostly water. The rest of you is floating around in that water.

Almost all living systems are made up of thin chemical "soups" in contact with membranes and small cellular parts called organelles. The membranes and organelles are made up of complex chemicals called lipids (fats and fatlike substances), carbohydrates (sugars, starches, and cellulose), proteins, and nucleic acids (DNA and RNA). Life processes occur in solutions and at the interfaces between solutions and semisolid organelles and membranes. Although the chemistry of these processes is now being rapidly unraveled, it is still poorly understood. Therefore, in this chapter we will deal mainly with the simpler solutions.

9.1 SOLUTIONS: DEFINITIONS AND TYPES

Put a teaspoonful of sugar in a cup of water. Stir until all the sugar is dissolved. Taste the sweetened water from one side of the cup and then from the other. Use a straw to taste it from the center and then from the bottom. If you did a proper job of mixing, the water has the same degree of sweetness throughout; that is, it is **homogeneous.** You could have added more sugar to make the

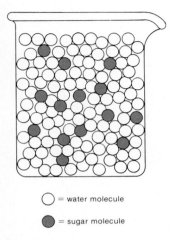

◯ = water molecule

⬤ = sugar molecule

FIGURE 9.1
In a solution of sugar in water, the sugar molecules are randomly distributed among the water molecules.

water sweeter, or you could have used less sugar to make it less sweet. Further, you could boil the water away and recover the sugar. The sugar and water have not reacted chemically. This mixture of sugar in water is called a **solution,** and we say that the sugar is **dissolved** in the water. We can define a solution as a homogeneous mixture of two or more kinds of atoms, molecules, or ions. In a true solution, the mixture is intimate right down to the molecular level. There are *not* clumps of sugar molecules, but single molecules randomly distributed among the water molecules in much the same way that black marbles can be distributed among white ones (Figure 9.1).

The components of a solution are given special names. The substance being dissolved (usually present in a lesser amount) is called the **solute.** The substance doing the dissolving (usually present in a greater amount) is called the **solvent.** Water is undoubtedly the most familiar solvent, and it dissolves many familiar substances such as sugar, salt, and alcohol. There are many other solvents. Gasoline dissolves grease. Many drugs are dissolved in alcohol. Banana oil is a solvent for the glue used in making model airplanes. In this chapter, though, we will deal for the most part with **aqueous** solutions, those in which the solvent is water.

A solvent need not be a liquid. Air is a solution of oxygen, argon, water vapor, and other gases in nitrogen gas. Steel is a solution of carbon (the solute) in iron (the solvent)—a solid in a solid. The most common types of solutions are summarized in Table 9.1.

TABLE 9.1   Types of Solutions

Solute	Solvent	Solution	Example
Gas	Gas	Gas	Air (O_2 in N_2)
Gas	Liquid	Liquid	Soda pop (CO_2 in H_2O)
Liquid	Liquid	Liquid	Wine (alcohol in H_2O)
Solid	Liquid	Liquid	Saline (NaCl in H_2O)
Solid	Solid	Solid	14-karat gold (Ag in Au)

9.2 QUALITATIVE ASPECTS OF SOLUBILITY

We say that sugar is **soluble** in water. Just what does that mean? Can we dissolve a teaspoonful of sugar in a cup of water? Can we dissolve 10 teaspoonfuls, or 100 teaspoonfuls? We know from everyday experience that there is a limit to the amount of sugar we can dissolve in a given volume of water. Nevertheless, we still find it convenient to say that sugar is soluble in water, for we can dissolve an appreciable amount.

There are a few substances that can be mixed in all proportions. Water and alcohol are familiar examples. We say that such substances are completely **miscible.** For most "soluble" substances, though, there is a limit to the amount that will dissolve in a given solvent. For others, which we call **insoluble,** that limit is near zero. Put an iron nail in a beaker of water. There is no apparent change. We say that iron is insoluble in water. Even insolubility is relative, however. If we had a method sensitive enough, we would find that some iron had dissolved. The amount might well be regarded as insignificant, and that is the sense in which the term insoluble is used. We will find terms such as soluble and insoluble useful, but they are imprecise and must be used with care.

Two other imprecise but sometimes useful terms are **dilute** and **concentrated.** A dilute solution is one that contains a little bit of solute in lots of solvent. A concentrated solution is one in which lots of solute is dissolved in a relatively small amount of solvent. If we dissolved a few milliliters of ethylene glycol, or antifreeze (Chapter 15), in several liters of water, the dilute solution would be quite "thin"—little changed in appearance from that of pure water. However, if we dissolved 1 L of ethylene glycol in 1 L of water, the concentrated solution would be rather syrupy—similar to pure ethylene glycol in consistency.

The terms dilute and concentrated are used in a quantitative way for solutions of acids and bases. We specify their meanings in this context in the next chapter.

9.3 THE SOLUBILITY OF IONIC COMPOUNDS

In Chapter 7, we saw that the unique structure of water not only results in relatively strong forces between water molecules but also enables water to dissolve ionic compounds. Further, we examined the different types of forces that exist between identical molecules in pure liquids. Now let's look at the types of forces that exist between the solute and solvent molecules in solutions. The solubility of a given solute depends on the relative attraction between particles in the pure substances and in the solution.

Water is a good solvent for compounds of the Group IA elements. Most lithium, sodium, and potassium salts are quite soluble in water. Examples are sodium chloride ($NaCl$), sodium sulfate (Na_2SO_4), potassium phosphate (K_3PO_4), and lithium bromide ($LiBr$). Further, nearly all nitrate salts are soluble, as are salts that incorporate the ammonium ion. Silver nitrate ($AgNO_3$), mercury(II) nitrate [$Hg(NO_3)_2$], aluminum nitrate [$Al(NO_3)_3$], ammonium chloride (NH_4Cl), ammonium sulfate [$(NH_4)_2SO_4$], and ammonium phosphate [$(NH_4)_3PO_4$] are examples. Why do these compounds dissolve in water? At the end of Chapter 7 (p. 230), we attempted to picture how water dissolves a salt such as sodium chloride. In essence, three things must happen. The attractive forces holding the salt ions together must be overcome. Similarly, the attractive forces holding at least some of the water molecules together must be overcome. Finally, the solute and solvent molecules must interact; that is, they must attract one another. Hydration (the process in which water molecules surround the solute ions) occurs when the energy released by the interaction of solute with solvent is greater than that needed to overcome the forces holding the ions together in the crystal lattice and the forces between the solvent molecules.

$$\text{Hydration of ions} > \frac{\substack{\text{separation of} \\ \text{solvent molecules}}}{\substack{+ \\ \text{breakdown of lattice}}} = \text{soluble}$$

Energy released   Energy absorbed

In some solids, the forces holding the ions together are so strong that they cannot be overcome by the hydration of the ions. Many solids in which both ions are doubly or triply charged are essentially insoluble in water. Examples are calcium carbonate (Ca^{2+} and CO_3^{2-}), aluminum phosphate (Al^{3+} and PO_4^{3-}) and barium sulfate (Ba^{2+} and SO_4^{2-}). The large electrostatic forces between the ions hold the particles together despite the attraction of solvent molecules.

$$\text{Hydration of ions} < \frac{\substack{\text{separation of} \\ \text{solvent molecules}}}{\substack{+ \\ \text{breakdown of lattice}}} = \text{insoluble}$$

Energy released   Energy absorbed

Table 9.2 summarizes the solubilities of a variety of ionic compounds.

TABLE 9.2   *Solubilities of Solid Ionic Compounds in Pure Water*

	NO_3^-	CH_3COO^-	Cl^-	SO_4^{2-}	OH^-	S^{2-}	CO_3^{2-}	PO_4^{3-}
NH_4^+	S*	S	S	S	N	S	S	S
Na^+	S	S	S	S	S	S	S	S
K^+	S	S	S	S	S	S	S	S
Ba^{2+}	S	S	S	I	S	D	I	I
Ca^{2+}	S	S	S	P	P	P, D	I	I
Mg^{2+}	S	S	S	S	I	D	I	I
Cu^{2+}	S	S	S	S	I	I	I	I
Fe^{2+}	S	S	S	S	I	I	I	I
Fe^{3+}	S	N	S	P	I	D	N	I
Zn^{2+}	S	S	S	S	I	I	I	I
Pb^{2+}	S	S	P	I	I	I	I	I
Ag^+	S	P	I	I	N	I	I	I
Hg_2^{2+}	S, D	P	I	I	N	I	I	I
Hg^{2+}	S	S	S	D	N	I	N	I

* The letters are defined as follows:

S = is soluble in water
P = is partially soluble in water
I = is insoluble in water
D = decomposes
N = does not exist as an ionic solid

9.4 THE SOLUBILITY OF COVALENT COMPOUNDS

An old but helpful rule states that like dissolves like. This means that nonpolar (or only slightly polar) solutes dissolve best in nonpolar solvents and that polar solutes dissolve best in polar solvents. The rule works well for nonpolar substances. Fats, oils, and greases (nonpolar or only slightly polar) dissolve well in nonpolar solvents such as benzene, C_6H_6. The forces that hold nonpolar molecules together are generally weak. Thus, the amount of energy needed to pull apart molecules of pure solute and to disrupt the attractive forces between molecules of pure solvent is small. And this energy can be balanced by that released through the interaction of solute and solvent molecules, although this too is slight.

$$\underset{\text{Energy released}}{\underbrace{\begin{array}{c}\text{Interaction of nonpolar solute} \\ \text{with nonpolar solvent}\end{array}}} > \underset{\text{Energy absorbed}}{\underbrace{\begin{array}{c}\text{separation of} \\ \text{solvent molecules} \\ + \\ \text{separation of} \\ \text{solute molecules}\end{array}}} = \text{solution}$$

The rule that like dissolves like is not as helpful for polar substances and, in particular, for aqueous solutions. The important thing for water solubility is the ability of water to form hydrogen bonds to the solute molecules. Thus, molecules containing a high proportion of nitrogen or oxygen atoms will dissolve in water because these are the elements that can form hydrogen bonds. One example is methyl alcohol (CH_3OH), which is completely miscible with water. Methylamine (CH_3NH_2) is also quite soluble in water. Figure 9.2 gives the structures of these molecules (and others mentioned in this section) and shows how they interact with water by forming hydrogen bonds.

There need not be a hydrogen atom on the oxygen (or nitrogen) atom of the solute molecules. Acetaldehyde (CH_3CHO) is completely miscible with water, even though none of the hydrogen atoms in the molecules is attached to the oxygen (see Figure 9.2). The hydrogen bonds depicted in the drawing incorporate hydrogen atoms covalently bonded to the oxygen of the water molecules.

Each nitrogen or oxygen atom in a solute molecule can carry along into solution about four attached carbon atoms. For alcohols (carbon-containing compounds featuring the —OH group), those with three or fewer carbons are completely miscible with water. Butyl alcohol ($CH_3CH_2CH_2CH_2OH$) is only partially soluble, and a compound with 12 carbon atoms and only 1 oxygen is essentially insoluble (see Figure 9.2). In its solubility, this molecule reflects the character of the nonpolar chain of carbons and hydrogens, rather than that of the single oxygen atom.

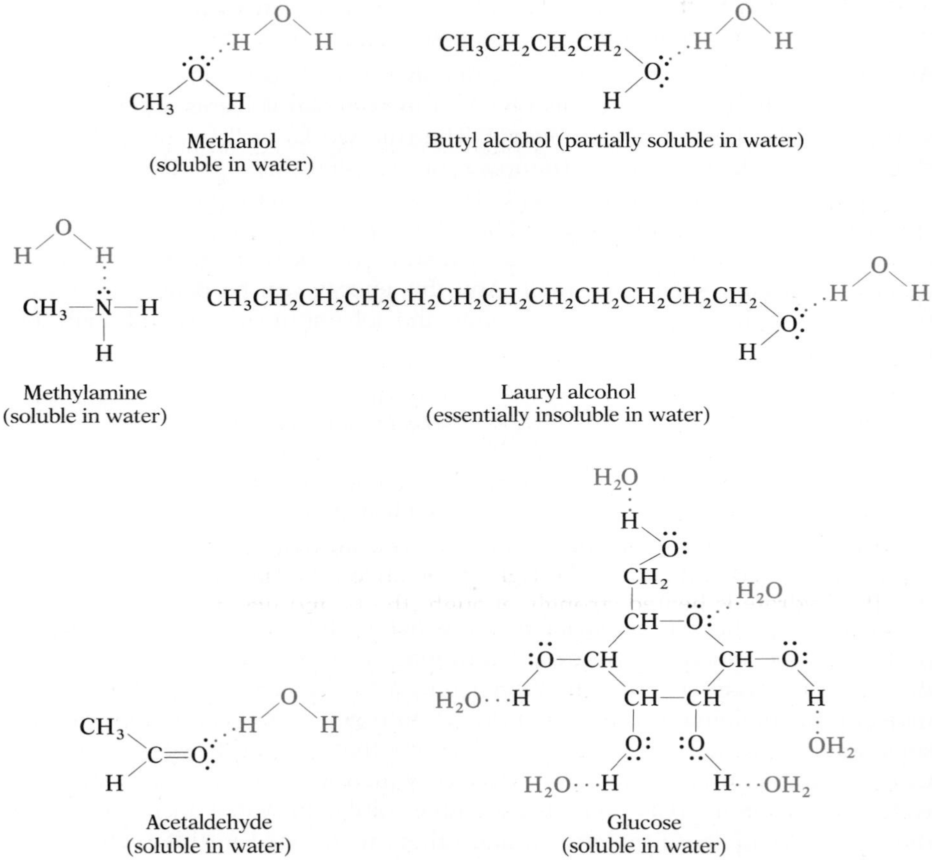

Methanol
(soluble in water)

Butyl alcohol (partially soluble in water)

Methylamine
(soluble in water)

Lauryl alcohol
(essentially insoluble in water)

Acetaldehyde
(soluble in water)

Glucose
(soluble in water)

FIGURE 9.2 *Water can form hydrogen bonds to molecules that contain nitrogen or oxygen atoms. Those molecules with four or fewer carbon atoms per nitrogen or oxygen atom are usually soluble in water.*

Some fairly complex molecules, such as those of the sugars, are quite soluble in water. Glucose, or blood sugar ($C_6H_{12}O_6$), contains six carbon atoms, but its six oxygen atoms permit it to hydrogen bond to many water molecules and, thus, confer on it water solubility.

Hydrogen bonding, then, is the important factor in water solubility. Polarity alone is not enough. Methyl chloride (CH_3Cl) and methyl alcohol (CH_3OH) have about the same polarity, yet methyl chloride is essentially insoluble in water while methyl alcohol is completely miscible with water. Methyl chloride does not engage in hydrogen bonding, and methyl alcohol does. A few polar substances, such as hydrogen chloride (HCl), dissolve in water because they react to form ions; these are discussed in subsequent chapters.

The importance of water as a solvent is reflected in the number of terms that have been coined to describe systems involving water. For example, the general term for the interaction of solvent with solute is **solvation,** but there is a special term for the interaction of water with a solute—**hydration.** Certain compounds, such as calcium sulfate, tend to hold on to some water molecules even when they crystallize from solution. These compounds with their bound water molecules are called **hydrates.** The formulas for these crystals are written in such a way as to indicate the number of attached water molecules. Plaster of Paris is $(CaSO_4)_2 \cdot H_2O$ (one water molecule for every two calcium sulfate units). If more water is available, $CaSO_4 \cdot 2\,H_2O$ is formed (now there are two water molecules for every $CaSO_4$ molecule). When a plaster cast is formed to immobilize a broken bone, the first hydrate is converted to the second. The powdery plaster of Paris changes to the rigid, protective material of the cast.

If a hydrate is heated strongly enough, the bound water can be driven off to produce the **anhydrous** compound, that is, the compound without water. Some hydrates lose their bound water simply on standing in dry air. Such compounds are said to be **efflorescent.** Other compounds form hydrates by picking up water from the atmosphere. These are described as **hygroscopic.** And finally, some compounds are so good at pulling water molecules from the air, that they eventually dissolve in the water thus accumulated. These compounds are said to be **deliquescent.**

9.5 DYNAMIC EQUILIBRIA

For most substances, there is a limit to how much can be dissolved in a given volume of solvent. This limit varies with the nature of the solute and of the solvent. Solubility also varies with temperature, generally (but not always) increasing with increasing temperature. We usually wash our clothes in hot water because most forms of "dirt" are more soluble at higher temperatures.*

---

* For some stains, such as that produced by blood, cold water is recommended. In this case, hot water causes a change in the structure of proteins in the blood that makes these more insoluble. Cold water does not change the blood in this way, and it is for this reason that cold water is recommended.

Solubilities are often expressed in terms of grams of solute per 100 g of solvent. Since solubility varies with temperature, it is necessary to indicate the temperature at which the solubility is measured. For example, 100 g of water will dissolve up to 109 g of sodium hydroxide (NaOH) at 20 °C. At 50 °C, 145 g of NaOH will dissolve in 100 g of water. In a shorthand method, the solubility of sodium hydroxide is expressed as 109^{20} and 145^{50} (the 100 g of water is understood).

The solubility of sodium chloride, or common table salt (NaCl), is 36 g per 100 g of water at 20 °C. Suppose that we place 40 g of NaCl in 100 g of water. What happens? Initially, many of the sodium (Na^+) ions and chloride (Cl^-) ions leave the surface of the crystals and wander about at random through the solvent. Some of the ions in their wanderings return to a crystal surface. These ions can even be trapped there, becoming once more a part of the crystal lattice. As more and more salt dissolves, the number of "wanderers" that return to be trapped once again in the solid state increases. Eventually (when 36 g of NaCl has dissolved), the number of ions leaving the surface of the undissolved crystals will just equal the number returning. A condition of **dynamic equilibrium** will have been established. The *net* amount of sodium chloride in solution will remain the same despite the fact that there is still a lot of activity as ions come and go from the surface of the crystals. The net amount of undissolved crystals also remains constant (in this example, 4 g), although individual crystals may change in shape and size as ions leave one part of the crystal to enter solution while others are deposited at another part of the lattice. Some small crystals may even disappear as others grow larger, yet the net amount of undissolved salt will not change. The rate of dissolution will just equal the rate of regrowth.

A solution that contains all the solute that it can at equilibrium and at a given temperature is said to be **saturated.** One that contains less than this amount is **unsaturated.** A solution with 24 g of NaCl in 100 g of water at 20 °C would be unsaturated because it could dissolve 12 g more at that temperature.

Equilibrium is established at a given temperature. If the temperature changes, more solute will dissolve or separate out until equilibrium is established at the new temperature. Consider once more a sodium hydroxide solution. If we add 145 g of NaOH to 100 g of water, 109 g of the NaOH will dissolve at 20 °C, leaving 36 g as undissolved solute. If the solution is then warmed, more solute will dissolve. Finally, at 50 °C all 145 g of NaOH will be in solution. The solubility of sodium hydroxide increases with increasing temperature.

Most solid compounds are increasingly soluble as the temperature is raised (Figure 9.3). This should not be surprising. As the temperature goes up, the motion of all the particles is increased. More ions are knocked loose from the lattice and go into solution. Further, it is more difficult for the crystal to recapture the ions that return to its surface, because they are moving at higher speeds. There are a few exceptions to this general rule of increased solubility

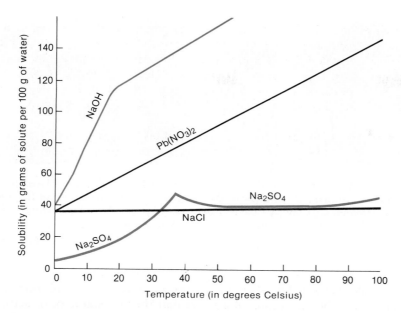

FIGURE 9.3   *The effect of temperature on the solubility of several solids in water.*

at higher temperatures (see the graph of Na_2SO_4 in Figure 9.3). These exceptions probably involve changes in the hydration of the ions as the temperature increases.

Notice that for some substances, such as sodium hydroxide (NaOH) and lead nitrate $[Pb(NO_3)_2]$, solubility increases rapidly as the temperature increases. The solubility of sodium chloride (NaCl), in contrast, changes very little over the indicated range of temperatures. The solubility of sodium sulfate (Na_2SO_4) first increases rather rapidly and then decreases, indicating a change in the hydration of the sodium (Na^+) ions and sulfate (SO_4^{2-}) ions.

If a saturated solution of lead nitrate (with excess solid lead nitrate present) is cooled, more solute precipitates until the equilibrium is once again established at the lower temperature. For example, consider a saturated solution of lead nitrate at 90 °C. For each 100 g of water, there will be 120 g of $Pb(NO_3)_2$ dissolved. When the solution is cooled to 20 °C, the solution at equilibrium can contain only 54 g of $Pb(NO_3)_2$. The excess, 66 g, will precipitate out, increasing the amount of undissolved solute.

Now consider what would happen if one started to cool a saturated solution of lead nitrate with *no* excess solute present. Would lead nitrate precipitate? It might. Then again, it might not. There is no equilibrium—no crystals to capture the wandering ions. One might well be able to cool the solution to 20 °C without precipitation. Such a solution, containing solute in excess of that

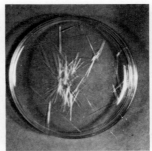

FIGURE 9.4 *Addition of a seed crystal induces rapid crystallization of excess solute from a supersaturated solution. (From Keenan, Charles W., Wood, Jesse H., and Kleinfelter, Donald C., General College Chemistry, 5th ed., New York: Harper & Row, 1976. Copyright © 1976 by Keenan, Wood, and Kleinfelter.*

which it could contain *if* it were at equilibrium, is said to be **supersaturated.** This system is not stable because it is not at equilibrium. Solute may precipitate when the solution is stirred or if the inside of the container is scratched with a glass rod. Addition of a "seed" crystal of solute will nearly always result in the sudden precipitation of all the excess solute. Equilibrium is established rather rapidly when there is a crystal to which the ions can attach themselves (Figure 9.4).

Supersaturated solutions are not unknown in nature. Honey is one example; the solute is sugar. If honey is left to stand, the sugar crystallizes. We say, not very scientifically, that the honey has "turned to sugar." Supersaturated sugar solutions are fairly common in cooking. Jellies are one example. The sugar often crystallizes from jelly that has been standing for a long time.

Some wines have high concentrations of potassium hydrogen tartrate ($KHC_4H_4O_6$). When chilled, the solution becomes supersaturated. Crystals eventually form and settle out, often when the wine is stored in the consumer's refrigerator. Modern wineries solve this problem—and render the wine less acidic—by chilling the wine to $-3\,°C$ and adding tiny seed crystals of $KHC_4H_4O_6$. Precipitation is complete in 2 or 3 hr, and the crystals are filtered off.

9.6 SOLUTIONS OF GASES IN WATER

You are no doubt familiar with a number of solutions in which gases are dissolved in water. Soda pop is a solution of carbon dioxide (and flavoring and sweetening agents) in water. Blood contains dissolved oxygen and carbon dioxide. Familiar household cleansers include those with ammonia (NH_3) dissolved in water. Formalin, used as a biological preservative, is an aqueous solution of formaldehyde (HCHO). Natural waters contain dissolved oxygen. Although only slightly soluble in water (0.0043 g of O_2 will dissolve in 100 g of water at 20 °C), this oxygen is vital to the survival of fish and other aquatic species.

Unlike most solid solutes, gases become less soluble as the temperature increases. Heat increases the molecular motion of both solute and solvent particles. In contrast to solid solutes, gaseous ones can escape from the solution when they reach the surface of a liquid in an open container.

We all heat water from time to time and know that, long before the water begins to boil, bubbles of dissolved gases appear (Figure 9.5). These soon rise to the surface and escape into the atmosphere. Figure 9.6 shows the effect of temperature on the solubility of oxygen in water.

The solubility of gases in water also varies with the pressure of the gas (Section 6.9). The higher the pressure, the more gas one can dissolve in a given amount of water. Figure 9.7 shows how the solubility of oxygen in water at 25 °C varies with pressure.

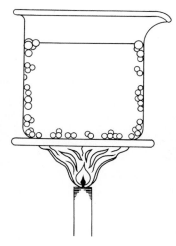

FIGURE 9.5
Bubbles of air form when a breaker of water is heated.

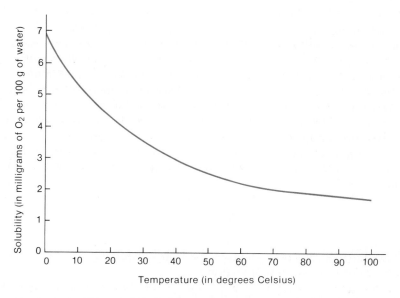

FIGURE 9.6 *The solubility of oxygen at various temperatures at 1 atm of pressure.*

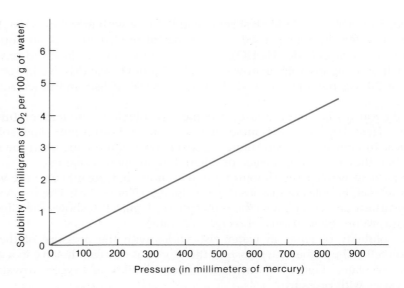

FIGURE 9.7 *The effect of pressure on the solubility of oxygen in water at constant temperature.*

9.7 MOLARITY

Most reactions of interest to us, including those in our bodies, take place in solution. A good cook may well get by with concentrations expressed as "a pinch of salt in a pint of water," but scientific work generally requires more precise measurement of amounts. We have already discussed the measurement of solubility in grams of solute per 100 g of solvent. Since substances react in small whole-number ratios of molecules or ions (Chapter 5), it is often convenient to measure the amount of solute in moles. The amount of solution is usually measured in liters or milliliters. A solution that contains 1 mol of solute per L of solution is called a 1 **molar** (or 1 *M*) solution. **Molarity,** then, is defined as the number of moles of solute divided by the number of liters of solution.

$$\text{Molarity } (M) = \frac{\text{moles of solute}}{\text{liters of solution}}$$

example 9.1

What is the molarity of a solution made by dissolving 3 mol of NaCl in enough water to make 6 L of solution?

$$\frac{3 \text{ mol of solute}}{6 \text{ L of solution}} = 0.5 \ M$$

Recall that the number of moles of a substance is calculated by division of mass in grams by formula weight in grams per mole.

$$\text{Moles} = \frac{\text{mass (g)}}{\text{formula weight (g/mol)}}$$

The formula weight of NaOH is $23 + 16 + 1 = 40$ g/mol. In 120 g of NaOH there would be $120 \div 40 = 3$ mol of NaOH.

What is the molarity of a solution of 300 g of $KHCO_3$ in enough water to make 10 L of solution?

The formula weight of $KHCO_3$ is $39 + 1 + 12 + (3 \times 16) = 100$ g/mol. The number of moles of $KHCO_3$ is $300 \div 100 = 3$. The molarity of the solution is

$$\frac{3 \text{ mol}}{10 \text{ L}} = 0.3 \; M$$

example 9.2

Frequently we need to know how much solute is required for the preparation of a specified amount of solution of a given molarity. Recall that

$$\text{Molarity} = \frac{\text{moles of solute}}{\text{liters of solution}}$$

We can rearrange this to get

$$\text{Moles of solute} = \text{molarity} \times \text{liters of solution}$$

How many grams of NaOH (which has a formula weight of 40) are required for the preparation of 500 mL (0.50 L) of 6.0 M solution?

We can use the rearranged equation to determine the number of *moles* of NaOH required.

$$\text{Moles of NaOH} = 6.0 \; M \times 0.50 \text{ L} = \frac{6.0 \text{ mol}}{1.0 \, \cancel{L}} \times 0.50 \, \cancel{L} = 3.0 \text{ mol}$$

But we were asked for the number of grams and not moles, so we must now calculate the number of grams in 3 mol.

$$\text{Moles of NaOH} = \frac{\text{grams of NaOH}}{\text{formula weight of NaOH}}$$

example 9.3

Rearranging, we get

$$\text{Grams of NaOH} = \text{moles of NaOH} \times \text{formula weight of NaOH}$$

$$= 3.0 \; \text{mol} \times \frac{40 \; \text{g}}{1 \; \text{mol}} = 120 \; \text{g}$$

Quite often solutions of known molarity are available. How would you calculate the volume you would need in order to get a certain number of moles of solute? We can again rearrange the definition of molarity to obtain

$$\text{Liters of solution} = \frac{\text{moles of solute}}{\text{molarity}}$$

example 9.4

How many liters of 12 *M* HCl solution would one have to take to get 0.48 mol of HCl?

$$\text{Liters of HCl solution} = \frac{0.48 \; \text{mol HCl}}{12 \; M} = \frac{0.48 \; \text{mol HCl}}{12 \; \text{mol HCl/L}} = 0.040 \; \text{L}$$

We would need 0.040 L (40 mL) of the solution to have 0.48 mol.

Remember that molarity is moles per liter of *solution*, not per liter of solvent. To make a liter of a 1 *M* solution, one weighs out 1 mol of solute and places it in a volumetric flask, which is a standard piece of laboratory glassware designed to contain a precisely specified volume of liquid (Figure 9.8). Enough water is added to dissolve the solute, and then more water is added to bring the volume up to the mark that indicates 1 L of *solution*. (Simply adding 1 mol of solute to 1 L of water would, in most cases, give more than 1 L of solution.)

9.8 PERCENT CONCENTRATIONS

For many purposes, it is not necessary to know the number of moles of solute but only the relative amounts of solute and solvent. One way of expressing this is in terms of the percent of solute. If both the solute and solvent are liquids, **percent by volume** is used.

$$\text{Percent by volume} = \frac{\text{volume of solute}}{\text{volume of solute} + \text{solvent}} \times 100$$

1000 mL
c 20 °C

FIGURE 9.8
A volumetric flask. When filled to the mark on the neck, the flask contains 1000 mL (1 L) of solution.

What is the percent by volume of a solution of 100 mL of alcohol dissolved in 300 mL of water?

example
9.5

$$\text{Percent by volume} = \frac{100 \text{ mL of alcohol}}{100 \text{ mL of alcohol} + 300 \text{ mL of water}} \times 100$$

$$= \frac{100 \text{ mL}}{400 \text{ mL}} \times 100 = 25\%$$

What volume of alcohol and water must one mix to obtain about 200 mL of a 40%-by-volume solution of alcohol in water?

example
9.6

$$40\% = \frac{\text{milliliters of alcohol}}{200 \text{ mL total volume}} \times 100$$

Rearranging, we get

$$\text{Milliliters of alcohol} = \frac{40 \times 200}{100} = 80 \text{ mL of alcohol}$$

To find the amount of water required, subtract the volume of alcohol from the total volume.

$$200 \text{ mL} - 80 \text{ mL} = 120 \text{ mL of water}$$

When the 80 mL of alcohol is mixed with 120 mL of water, the total volume may not be exactly 200 mL, but it will be close.

For solutions involving solutes that are solids, a more commonly used relationship is **percent by mass.**

What is the percent by mass of a solution of 5 g of NaCl dissolved in 495 g (495 mL) of water?

example
9.7

$$\text{Percent by mass} = \frac{5 \text{ g}}{495 \text{ g} + 5 \text{ g}} \times 100 = \frac{5 \text{ g}}{500 \text{ g}} \times 100 = 1\%$$

Note that, since 1 mL of water has a mass of 1 g, one can obtain 495 g of water simply by measuring 495 mL of water.

example
9.8

How would you prepare about 100 mL of a 5% NaOH solution?

For dilute aqueous solutions, the density of the solution is approximately equal to the density of water. Therefore, let's assume that 100 mL of the solution will have a mass of about 100 g.

$$5\% = \frac{\text{grams of NaOH}}{100 \text{ g total}} \times 100$$

Rearranging, we get

$$\text{Grams of NaOH} = \frac{5 \times 100}{100} = 5 \text{ g}$$

Percent means "parts per hundred." To make a hundred "parts" of solution, weigh out 5 g of NaOH. Add 95 g (95 mL) of water. (The final volume might not be exactly 100 mL, but it will be close to that.)

example
9.9

How would you prepare about 500 mL of a 10% NaHCO$_3$ solution?

$$10\% = \frac{\text{grams of NaHCO}_3}{500 \text{ g total}} \times 100$$

Rearranging, we get

$$\text{Grams of NaHCO}_3 = \frac{10 \times 500}{100} = 50 \text{ g}$$

Weigh out 50 g of NaHCO$_3$ and dissolve it in $500 - 50 = 450$ g (450 mL) of water. Again, the total volume will be about, rather than exactly, 500 mL.

Let us call attention to a special aspect of percent concentrations. If you are trying to calculate how much solute to weigh out for a particular percent concentration, it makes no difference what the solute is. A 10% solution of NaOH contains 10 g of NaOH per 100 g of total solution. Similarly, 10% HCl and 10% $(NH_4)_2SO_4$ and 10% $C_{110}H_{190}N_3O_2Br$ each contain 10 g of the specified solute per 100 g of solution. For molar solutions, however, the mass of solute in a solution of specified molarity is different for different solutes. A liter of a 0.1 M solution requires 4 g (0.1 mol) of NaOH, 3.7 g (0.1 mol) of HCl, 13.2 g (0.1 mol) of $(NH_4)_2SO_4$, and 166 g (0.1 mol) of $C_{110}H_{190}N_3O_2Br$.

Low concentrations of solutes, such as those in blood and urine, are often expressed in **milligram percent**, which is defined as milligrams per 100 mL of solution.

$$\text{Milligram percent} = \frac{\text{milligrams of solute}}{100 \text{ mL of solution}}$$

For example, normally there are about 320 to 350 mg of sodium ion per 100 mL of blood plasma. Thus, the normal concentration of sodium ion in blood plasma ranges from 320 to 350 mg%. The use of milligram percent avoids the sometimes cumbersome use of decimal numbers. For example, the above concentrations of sodium ion would be expressed as 0.320% to 0.350% by mass.

For some dilute solutions, concentrations are expressed in parts per million (ppm) or even parts per billion (ppb). For aqueous solutions, ppm is the same as milligrams per liter (mg/L), and ppb is identical to micrograms per liter (μg/L). These units are frequently used to measure extremely low levels of toxic materials. Current concern over the purity of our drinking water centers on minute amounts of potentially dangerous compounds. For example, benzene is a compound that has been shown to produce leukemialike symptoms in laboratory animals and humans. A 1984 Supreme Court decision dealt with the question of whether the concentration of benzene in the air breathed by workers should be limited to 10 ppm or 1 ppm. The court decided that industries could not be required to lower the concentration from 10 to 1 ppm unless the higher concentration was *proven* to be dangerous.

As our technology becomes more sophisticated, our ability to detect minute quantities of materials increases. This increase in the sensitivity of analytical techniques raises questions. When a substance is first detected in the ppm or ppb range, is it a new contaminant in our environment or has it been there all along at levels that were previously undetectable?

9.9 RATIO CONCENTRATIONS

The strength of a solution is sometimes presented as a ratio, for example, 1:1000 or 1:2000. Unfortunately, there is no universal agreement as to the meaning of the ratio. The ratio 1:1000 may be used to indicate that 1 g of pure solute (perhaps a drug) is to be dissolved in 1000 mL of solution. In the case of a liquid solute, the ratio may mean 1 mL of solute in 1000 mL of solution. Sometimes the ratio is presented as a dilution factor. Thus, 1:1000 may specify that 1 part of a sample is to be diluted with solvent until the volume of the solution is 1000 times the original volume. In specific instances, the meaning of the ratio may be clear, but because of the possibility for confusion, the use of ratio concentrations is being discouraged. Nonetheless, let us look at one example of the use of a ratio as a dilution factor.

example
9.10

The directions for use of a liquid drug indicate that it should be diluted 1:2000. How would 500 mL of the solution be prepared?

The ratio should be read as

$$\frac{1 \text{ part solute}}{2000 \text{ parts solution}}$$

The final volume of solution required is 500 mL (500 "parts").

$$\text{Volume of solute} = 500 \text{ mL (parts) of solution} \times \frac{1 \text{ part solute}}{2000 \text{ parts solution}}$$

$$= 0.25 \text{ part (mL) solute}$$

To prepare the solution, 0.25 mL of the drug (as supplied) should be diluted to a total volume of 500 mL by addition of the appropriate solvent. This gives exactly the same final concentration as the dilution of 1 mL of drug to a total volume of 2000 mL.

9.10 STRENGTHS OF DRUGS

Many drug manufacturers market their products for use in strengths that are not specified by concentration units discussed above. Instead, they provide explicit instructions on how to prepare the drug for administration (referred to as **reconstitution** of the drug).

example
9.11

A drug label reads, "Add 5.0 mL of sterile water; each mL of solution will contain 0.50 g." If the dose prescribed for a patient is 750 mg, how many mL of the reconstituted drug should be administered?

First, the drug must be reconstituted by addition of 5.0 mL of water to the container. After thorough mixing, the drug is ready for use. To calculate the dose in mL, the conversion factor given on the label (1.0 mL = 0.50 g) is used. Before that can be done, the specified dose (750 mg) must be converted to grams.

$$\text{Dose in mL} = 750 \text{ mg} \times \frac{1.0 \text{ g}}{1000 \text{ mg}} \times \frac{1.0 \text{ mL}}{0.50 \text{ g}} = 1.5 \text{ mL}$$

For some medications, the strength of the medication is reported literally in **units** (u.). A unit of the substance is defined as an amount that elicits a particular effect. U.S.P. units, for example, meet standards set by the United States Pharmacopeia. The strength of some vitamins and antibiotics, among other substances, are reported in this way.

example
9.12

The label on a container of crystalline penicillin G states, "RECONSTITU-TION: add 9.6 mL diluent to provide 100 000 u. per mL." If the prescribed dosage is 500 000 u., how much of the reconstituted drug should be administered?

Let us assume that you have correctly reconstituted the drug. The rest is easy.

$$\text{Dose in mL} = 500\ 000\ \text{u.} \times \frac{1\ \text{mL}}{100\ 000\ \text{u.}} = 5\ \text{mL}$$

9.11 COLLIGATIVE PROPERTIES OF SOLUTIONS

Solutions have higher boiling points and lower freezing points than the corresponding pure solvent. The antifreeze in automobile cooling systems is there precisely because of these effects. If water alone were used as the engine coolant, it would boil away in the heat of summer and freeze in the depths of northern winters. Addition of antifreeze to the water raises the boiling point of the coolant and also prevents the coolant from freezing solid when the temperature drops below $0\ °C$. Salt is thrown on icy sidewalks and streets because the salt dissolves in the ice and lowers its freezing point. Therefore, the ice melts because the outdoor temperature is no longer low enough to maintain it as a solid.

The extent to which freezing points and boiling points are affected by solutes is related to the number of solute particles present in solution. The higher the concentration of solute particles, the more pronounced the effect. **Colligative properties** of solutions are those properties, like boiling point elevation and freezing point depression, that depend on the number of solute particles present in solution. For living systems, perhaps the most important colligative property is osmotic pressure. Osmosis is a phenomenon we shall discuss in detail in the next section.

Before we do that, we must first consider a rather subtle aspect of solute concentration. See if you can answer the following questions. How many solute particles are there in 1 L of a 1 M glucose, $C_6H_{12}O_6$, solution? How many in 1 L of 1 M sodium chloride, NaCl, solution? In 1 M calcium chloride, $CaCl_2$, solution? The answer "should" be 6×10^{23}, right? All of the solutions contain 1 mol of their respective solute compounds. However, the question did *not* ask for the number of *formula units;* it asked for the number of *solute particles.* Glucose is a covalent compound; its atoms are all firmly tied together in molecules. In the glucose solution, each solute particle is a glucose molecule, and there *are* 6×10^{23} of these. But in the sodium chloride solution, each formula unit of NaCl consists of a separate sodium ion (Na^+) and chloride ion (Cl^-) in solution. When sodium chloride dissolves in water, individual ions are carried off into solution by solvent molecules. So 6×10^{23} formula units of NaCl produce 12×10^{23} particles in solution— 6×10^{23} sodium ions plus 6×10^{23} chloride ions. Each calcium chloride unit produces three particles in solution—one calcium ion plus two chloride ions. Thus, the effect of a 1 *M* NaCl solution on colligative properties is twice that of a 1 *M* glucose solution. A calcium chloride solution has about three times the effect of a glucose solution of the same molarity.

When colligative properties (specifically osmotic pressure, Section 9.12) are being discussed, concentration may be reported in terms of *osmol per liter* or *osmolarity* (osmol/L). An **osmol** is a mole of solute particles. A 1 *M* NaCl solution contains 2 osmol of solute per liter of solution; a 1 *M* $CaCl_2$ solution contains 3 osmol/L. The osmolarity of a 1 *M* glucose solution is 1 osmol/L. The concentration of body fluids is typically reported in milliosmoles per liter (mosmol/L).

9.12 SOLUTIONS AND CELL MEMBRANES: OSMOSIS

Solutions have a variety of characteristic properties. The particles are molecules, atoms, or ions. Once the solute and solvent are thoroughly mixed, the *solute does not settle out*. Molecular motion keeps the particles randomly distributed. The sugar in a bottle of pop, for instance, does not settle to the bottom on standing. The last drop is just as sweet as, but no sweeter than, the first. You cannot get the sugar out by passing the pop through a piece of filter paper. The sugar molecules go through the pores of the paper as readily as the water molecules. All true solutions can be filtered without removal of the solute.

Solutions may be colored, but they are *clear* and *transparent*. Copper sulfate ($CuSO_4$), dissolved in water, gives a beautiful clear blue solution. A beam of light will shine right through the solution but will not be visible in the solution. When the path of light through a solution is visible, then particles larger than those of a true solution are present (Section 9.13).

Solutions—both solute and solvent—will go through filter paper. Another way to describe the phenomenon is to say that the paper is **permeable** to water and other solvents and to solutions. Everyday experience tells us that other materials are **impermeable.** Water and solutions will not pass through the metal walls of cans nor through the glass walls of jars and bottles. Are there, perhaps, materials with intermediate properties? Materials that will pass solvent molecules but not solute molecules? Materials that are permeable to some solutes but not others? The answer is an emphatic yes! Many natural membranes are **semipermeable.** Cell membranes, the lining of the digestive tract, and the walls of blood vessels are all semipermeable; they allow certain substances to go through while holding others back.

FIGURE 9.9
The sieve model of osmosis holds that the semipermeable membrane has pores large enough to permit the passage of water molecules (small particles) but too small to permit the passage of larger molecules.

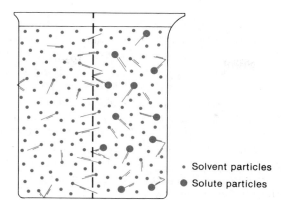

• Solvent particles

● Solute particles

The sieve model of osmosis pictures semipermeable membranes as having extremely small pores. The size of these pores is such that tiny water molecules can pass through, but larger particles, like sugar molecules, cannot. If a membrane with these characteristics separates a compartment containing pure water from one containing a sugar solution, an interesting thing happens. The volume of liquid in the compartment containing sugar increases while the volume in the pure water compartment decreases.

Use Figure 9.9 as a reference. On both sides of the membrane all molecules are moving about at random, occasionally bumping against the membrane. If a water molecule happens to hit one of the pores, it passes through the membrane into the other compartment. When the much larger sugar molecule strikes a pore, it bounces back instead of through. The more sugar molecules there are in solution (i.e., the more concentrated the solution), the smaller the chance that a water molecule will strike a pore. Thus, in our example, there will be a *net* flow of water from the left compartment into the right compartment. This net diffusion of water through a semipermeable membrane is called **osmosis.** During osmosis, there is always a *net* flow of solvent from the more dilute (or pure solvent) area to the compartment in which the solution is more concentrated (Figure 9.10). The net diffusion of water would be *from* a 5% sugar solution into a 10% sugar solution.

As the liquid level in the right compartment builds up and that in the left compartment drops, the increased weight of fluid in the right compartment exerts pressure that makes it more difficult for water molecules to enter that compartment. (See Section 6.3 on how a barometer works.) Eventually the buildup of pressure is sufficient to prevent further net flow of water into that compartment. Things have not really come to a standstill; the rates at which water molecules move back and forth are just equal.

Instead of waiting for the liquid level to build up and stop the net flow of water, we can apply an external pressure to the compartment containing the

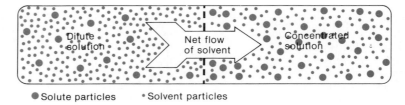

● Solute particles   • Solvent particles

FIGURE 9.10   Net solvent flow through a semipermeable membrane occurs spontaneously in only one direction, from the compartment containing dilute solution (or pure solvent) into the compartment of concentrated solution. Remember—ordinarily the terms dilute *and* concentrated *are used to describe the concentration of* solute. *The net flow of solvent is from where the solvent is more concentrated to where the solvent is less concentrated.*

more concentrated solution and accomplish the same thing. The precise amount of pressure needed to prevent the net flow of solvent from the dilute solution to the concentrated one is called the **osmotic pressure.** The magnitude of the osmotic pressure depends only on the concentration of solute particles, that is, on the osmolarity of the solution. The more particles (the higher the osmolarity), the greater the osmotic pressure. You can think of osmotic pressure or osmolarity as a measure of the tendency of a solution to draw solvent into itself.

Living cells can be regarded as selectively permeable bags filled with solutions of ions, small and large molecules, and still larger cell components. Normally, the fluid surrounding a cell has the same osmolarity as the fluid within the cell. Flow in and flow out are about equal. If a cell is surrounded by a solution of lower osmolarity (a **hypotonic solution**), there is a net flow of water into the cell. The cell swells. It may even burst. The rupture of a cell by a hypotonic solution is called **plasmolysis.** If the cell is a red blood cell, the more specific term **hemolysis** is used. What happens if a cell is placed in a solution of higher osmolarity (a **hypertonic solution**)? The net flow of water is then out of the cell. The cell wrinkles and shrinks (Figure 9.11). This shriveling is called **crenation,** a process that can also lead to the death of the cell.

Solutions that exhibit the same osmotic pressure as that of the fluid inside the cell are said to be **isotonic.** In replacing body fluids intravenously, it is important that the fluid be isotonic. Otherwise, hemolysis or crenation results, and the patient's well-being is seriously jeopardized. A 0.92% sodium chloride solution, called physiological saline, and a 5.5% glucose (also called dextrose or blood sugar) solution are isotonic with the fluid inside red blood cells. The "D5W" so often referred to by television's doctors and paramedics is this approximately 5% solution of dextrose (the *D*) in water (the *W*). A 0.92% sodium chloride solution is about 0.16 *M*. A 5.5% glucose solution is approximately 0.31 *M*. The osmolarity of the solutions is about the same.

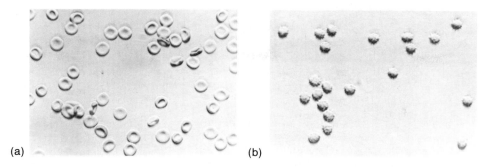

FIGURE 9.11   *(a) Normal human red blood cells. (b) After exposure to a hypertonic solution, the cells are wrinkled and shriveled. (Photos courtesy of A. M. Winchester, Ph.D., Greeley, Colo.)*

What are the osmolarities of these two isotonic solutions?

    The concentration of the glucose (dextrose) solution is 0.31 *M*. Each glucose molecule represents one particle in solution, therefore the osmolarity of the solution is 0.31 osmol/L or 310 mosmol/L. Each formula unit of NaCl provides two particles in solution. Therefore, the osmolarity of the 0.16 *M* solution is 0.32 osmol/L or 320 mosmol/L.

example 9.13

    Isotonic solutions have their limitations. Consider the case of a patient who is to be fed intravenously. There is a limit to the amount of water such a patient can handle in a day—about 3 L. If an isotonic solution of 5.5% glucose were to be used, 3.0 L of the solution would supply about 160 g of glucose. This would yield about 4.0 kcal/g to the patient, or a total of about 640 kcal, an amount woefully inadequate. Even a resting patient requires about 1400 kcal per day. And for a person suffering from serious burns, for example, requirements as high as 10 000 kcal per day have been recorded. We have oversimplified the situation. Other vital nutrients are required by such patients, and, in fact, a person can normally be given up to 1200 kcal through carefully formulated solutions. This still falls short of the requirements of many seriously ill people. One answer to the problem is to use very concentrated solutions (about six times as concentrated as isotonic solutions). Instead of being administered through the vein of an arm or a leg, this solution is infused directly through a tube into the superior vena cava, a large blood vessel leading to the heart (Figure 9.12). The large volume of blood flowing through this vein quickly dilutes the solution to levels that will not damage the blood. With this technique, patients have been given 5000 kcal a day and have even gained weight.

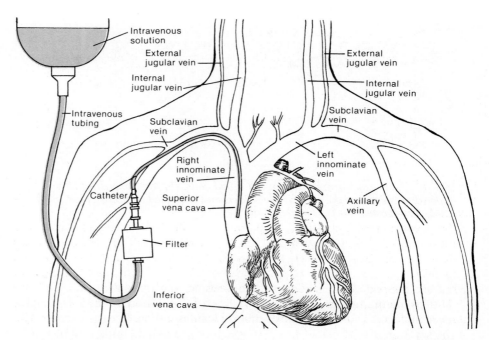

FIGURE 9.12 *Concentrated nutrient solution may be infused directly into the superior vena cava. The solution is quickly diluted to near-isotonic strength by the large volume of blood flowing through the vein.*

9.13 COLLOIDS

If sugar is dissolved in water, the molecules become intimately mixed. The solution is homogeneous; that is, it has the same properties throughout. The sugar cannot be filtered out by ordinary filter paper, nor does it settle out on standing. On the other hand, if one tries to dissolve sand in water, the two substances may momentarily appear to be mixed, but the sand rapidly settles to the bottom. The temporary dispersion of sand in water is called a **suspension.** By allowing the water to pass through a filter paper, one can trap the sand. The mixture is obviously heterogeneous, for part of it is clearly sand with one set of properties and part of it is water with another set of properties.

Is there nothing in between the true solution, with particles the size of ordinary molecules and ions, and suspensions, with gross chunks of insoluble matter? Yes, there is something else. And it is a highly important arrangement of matter at that—the **colloidal state.**

Colloids are defined not by the kind of matter they contain but by the *size of the particles* involved. True solutions have particles of about 0.05 to about

TABLE 9.3   *Properties of Solutions, Colloids, and Suspensions*

Property	Solution	Colloid	Suspension
Particle size	0.1–1.0 nm	1–100 nm	>100 nm
Settles on standing?	No	No	Yes
Filter with paper?	No	No	Yes
Separate by dialysis?	No	Yes	Yes
Homogeneous?	Yes	Borderline	No

0.25 nanometers (nm) in diameter (1 nm = 10^{-9} m). Suspensions have particles with diameters of 100 nm or more. Particles intermediate between these are said to be colloidal.

The properties of a **colloidal dispersion** are different from those of a true solution and also different from those of a suspension (Table 9.3). Colloidal particles cannot be filtered by filter paper, nor do they settle on standing. Colloidal dispersions usually appear milky or cloudy. Even those that appear clear will reveal a beam of light that passes through the dispersion (Figure 9.13). This phenomenon, called the **Tyndall effect,** is not observed in true solutions. Colloidal particles, unlike tiny molecules, are large enough to scatter

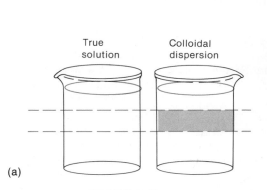

(a)   (b)

FIGURE 9.13   *(a) A beam of light passes through a true solution without noticeable effect. The same beam, passing through a colloidal solution, is clearly visible because the light is scattered by the colloidal particles. This phenomenon is called the Tyndall effect. (b) Searchlight beams through fog, which is a dispersion of water in air, show the Tyndall effect. (Photo courtesy of Black Star.)*

and reflect light off to the side. You have probably observed the Tyndall effect in a movie theater. The shaft of light that originates in the projection booth and ends at the movie screen is brought to you through the courtesy of colloidal dust particles in the air.

There are eight different kinds of colloids, based on the physical state of the particles themselves (the dispersed phase) and that of the "solvent" (the dispersing phase). These are listed, with examples of each, in Table 9.4.

The most important colloids in biological systems are emulsions—either liquids or solids dispersed in water. Living matter consists of colloidal particles as well as simple ions and molecules. Substances with high molecular weights, such as starches and proteins, often form colloidal dispersions rather than true solutions in water.

In most colloidal dispersions, the particles are charged. This charge is often due to the adsorption of ions on the surface of a particle. A given colloid will preferentially adsorb only one kind of ion (either positive or negative) on its surfaces; thus, all the particles of a given colloidal dispersion bear like charges. Since like charges repel, the particles tend to stay away from one another. They cannot come together and form particles large enough to settle out. These colloids can be made to coalesce and separate out by addition of ions of opposite charge, particularly those doubly or triply charged. Aluminum chloride ($AlCl_3$), which contains Al^{3+} ions, is great for breaking up colloids in which the particles are negatively charged.

Some colloids are stabilized by addition of a material that provides a protective coating. Oil is ordinarily insoluble in water, but it can be emulsified by soap. The soap molecules form a negatively charged layer about the surface of each tiny oil droplet. These negative charges keep the oil particles from coming together and separating out. In a similar manner, bile salts emulsify the fats we eat, keeping them dispersed as tiny particles that can be more efficiently

TABLE 9.4 *Types of Colloidal Dispersions*

Type	Particle Phase	Medium Phase	Example
Foam	Gas	Liquid[†]	Whipped cream
Solid foam	Gas	Solid	Floating soap
Aerosol	Liquid	Gas	Fog, hair sprays
Liquid emulsion	Liquid	Liquid	Milk, mayonnaise
Solid emulsion	Liquid	Solid	Butter
Smoke	Solid	Gas	Fine dust or soot in air
Sol*	Solid	Liquid	Starch solutions, jellies
Solid sol	Solid	Solid	Pearl

* Sols that set up in semisolid, jellylike form are called gels.
[†] By their very nature, gas mixtures always qualify as solutions. The size of particles in gas mixtures and their homogeneity fulfill the requirements of solutions.

digested. Milk is an emulsion in which fat droplets are stabilized by a coating of casein, a protein. Casein, soap, and bile salts are examples of **emulsifying agents,** substances that stabilize emulsions.

9.14 DIALYSIS

To live, an organism must take in food and get rid of toxic wastes. The nutrients necessary to life must be gotten into cells, and the wastes must be gotten out. Generally, then, cell membranes must permit the passage not only of water molecules but also of other small molecules and ions. At the same time, it is important that large molecules and colloidal particles not be lost from the cell. Membranes that pass small molecules and ions while holding back large molecules and colloidal particles are called **dialyzing membranes.** The process is called **dialysis.** It differs from osmosis in that osmotic membranes pass only solvent molecules. In dialysis, molecules and ions always diffuse from areas of higher concentration to areas of lower concentration.

Dialyzing membranes are used in laboratories to purify colloidal dispersions of smaller molecules and ions. The mixture to be purified is placed in a bag made of a dialyzing membrane. The bag is placed in a container of pure water (Figure 9.14). The ions and small molecules pass out through the membrane, leaving the colloidal particles behind. Pure water is continuously pumped past the bag, and the unwanted small particles are carried away. The dialyzing membrane may be an animal bladder or it may be an artificial one made from cellophane or from collodion, a semisynthetic plastic made by treatment of cellulose with nitric acid, alcohol, and camphor.

The kidneys are a complex dialyzing system responsible for the removal of certain potentially toxic waste products from the blood. By first gaining an

FIGURE 9.14
A simple apparatus
for dialysis.

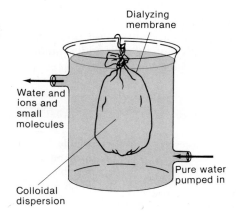

Dialyzing
membrane

Water and
ions and
small
molecules

Pure water
pumped in

Colloidal
dispersion

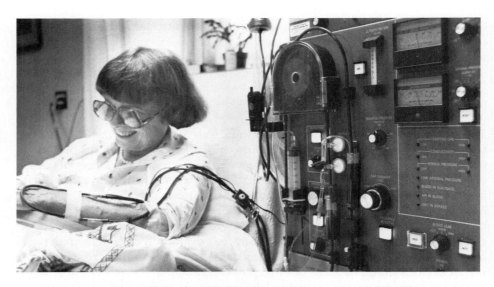

FIGURE 9.15   *In renal dialysis, elaborate machinery substitutes for human kidneys that are no longer capable of cleansing the blood. (Photo by Raymond C. Carballada, Department of Medical Photography, Geisinger Medical Center.)*

understanding of the function of living kidneys, scientists have been able to construct artificial ones (Figure 9.15). Artificial kidneys are more elaborate in structure than the simple apparatus shown in Figure 9.14, but their principle of operation is the same. We discuss the operation of kidneys in Chapter 26.

Many substances—medications, poisons, anesthetics—are thought to act by changing the permeability of membranes. Methyl mercury, a powerful poison that acts on the nervous system, is thought to act by making the membranes of nerve cells leakier than normal. A person going into shock has leaky capillaries that allow proteins and other colloidal particles as well as fluids to escape into the spaces between cells. If untreated, the cells may die from lack of oxygen and nutrients. As research increases our understanding of cellular membranes, we will undoubtedly gain a better understanding of drug action and of poisoning, of health, and of disease.

PROBLEMS

1. Define or explain and, where possible, illustrate these terms:
   a. solution
   b. solvent
   c. solute
   d. aqueous
   e. soluble
   f. insoluble
   g. miscible
   h. dilute
   i. concentrated
   j. saturated
   k. unsaturated
   l. supersaturated

m. percent by volume
n. percent by mass
o. milligram percent
p. ppm
q. ppb
r. molarity
s. U.S.P. unit
t. colligative property
u. permeable
v. semipermeable
w. impermeable
x. osmosis
y. osmolarity
z. osmotic pressure
aa. isotonic solution
bb. hypotonic solution
cc. hypertonic solution

dd. crenation
ee. plasmolysis
ff. hemolysis
gg. suspension
hh. colloidal dispersion
ii. Tyndall effect
jj. emulsifying agent
kk. dialysis
ll. hydrate
mm. anhydrous compound
nn. hygroscopic compound
oo. efflorescent compound
pp. deliquescent compound

2. Describe the properties of a true solution with respect to the following:
   a. size of solute particles
   b. distribution of solute and solvent particles
   c. filtration
   d. Tyndall effect
   e. color and clarity

3. Use the kinetic-molecular theory to explain why most solid solutes become more soluble with increasing temperature but gases become less soluble.

4. Fish live on oxygen dissolved in water. Would it be a good idea to thoroughly boil the water you place in a fishbowl? Explain.

5. Without referring to Table 9.2, indicate which compounds you would expect to be soluble in water. Explain each answer. You may use the periodic table.
   a. CH_3CH_2OH f. CS_2 k. $RbCl$
   b. CH_3CH_2Cl g. $C_{12}H_{22}O_{11}$ l. $BaCO_3$
   c. $CH_3CH_2NH_2$ h. $NaBr$ m. $SrSO_4$
   d. CH_2O i. $(NH_4)_2CO_3$ n. $C_{10}H_{20}O$
   e. CCl_4 j. $Ca(NO_3)_2$

6. In a dynamic equilibrium, two processes are occurring at the same rate. In the equilibrium involving a saturated solution, for which processes are the rates equal?

7. A supersaturated solution is maintained at constant temperature, and precipitation is induced by the addition of a seed crystal. When no more solid appears to precipitate, is the solution saturated, unsaturated, or supersaturated?

8. What is the percent-by-volume concentration of a mixture containing 60 mL of water and enough alcohol to give 600 mL of solution? Which component is the solvent and which is the solute?

9. How would you prepare exactly 2.0 L of a 2.0%-by-volume aqueous solution of acetic acid?

10. How much acetone and water should be mixed to give about 50 mL of an approximately 20%-by-volume aqueous solution of acetone?

11. If 4.0 g of NaOH is dissolved in 96 g of water, what is the percent-by-mass concentration of the solution?

12. If 100 g of glucose is mixed with 400 g of water, what is the percent-by-mass concentration of the solution?

13. If 2.0 g of $NaHCO_3$ is mixed with 38 mL of water, what is the percent-by-mass concentration of the solution?

14. How would you prepare 100 g of a 3.0%-by-mass aqueous solution of $C_{12}H_{22}O_{11}$?

15. Explain how you would prepare 5.0 kg of a 10%-by-mass solution of NaCl in water?

16. How would you prepare about 50 mL of a 2.0%-by-mass aqueous solution of Na_2CO_3?

17. Arrange the following in order of increasing concentration: 1%, 1 ppb, 1 ppm, 1 mg%.

18. On the average, glucose makes up about 0.10% by mass of human blood. What is the concentration in mg%?

19. A liquid drug is to be diluted for use. How would you prepare the following amounts of the drug?
   a. 100 mL of a 1:1000 dilution
   b. 1 L of a 1:2000 dilution

20. If 6.0 mol HCl is dissolved in 3.0 L of solution, what is the molarity of the solution?

21. If 0.50 mol NaCl is dissolved in water to give a total of 250 ml of solution, what is the molarity of the solution?

22. If 1.0 L of solution contains 9.8 g of H_2SO_4, what is the molarity of the solution?

23. If 500 mL of solution contains 60 g NaOH, what is the molarity of the solution?

24. How would you prepare 2.0 L of 0.20 *M* aqueous NaOH solution?

25. Describe the preparation of 100 mL of 1.00 *M* aqueous NH_4Cl solution.

26. Explain how to prepare 0.40 L of 0.40 *M* aqueous NaCl solution.

27. Describe how you would prepare each of these aqueous solutions. In each case, state how much solute you would measure and how much solvent you would use or how much solution you would make.
    a. exactly 100 mL of 30%-by-volume alcohol solution
    b. about 100 mL of 5.0%-by-mass NaOH solution
    c. exactly 10 g of 10%-by-mass $NaHCO_3$ solution
    d. exactly 1.00 L of 2.00 M NaCl solution
    e. about 500 mL of a 0.92%-by-mass NaCl solution
    f. exactly 500 mL of 0.50 M $C_6H_{12}O_6$ solution
28. Reconstitution of an enzyme preparation used in the treatment of infection gives a concentration of 25 000 u./2.5 mL. How much of this solution should be administered to provide a single dose of 5000 u.?
29. How many solute particles does each formula unit of the following give in solution? A quick review of covalent and ionic compounds (Chapter 4) might prove useful.
    a. KCl   c. $CaBr_2$ e. $(NH_4)_2SO_4$
    b. CH_3OH d. NaOH
30. How many osmol are there in 1 mol of each of the compounds shown in Problem 29?
31. To provide 1.0 osmol, how many moles of each of the compounds listed in Problem 29 are required?
32. What is the osmolarity of a 0.50 M $(NH_4)_3PO_4$ solution?
33. If two containers of gas at different pressures are connected, the net diffusion of gas is from the *higher* to the *lower* pressure. The net diffusion of water in osmosis is from the *lower* to the *higher* concentration solution. How would you explain this apparent difference?
34. Which has the higher osmotic pressure: a 1% NaCl solution or a 5% NaCl solution?
35. For each pair of solutions, indicate which has the higher osmotic pressure.
    a. 0.1 M $NaHCO_3$, 0.05 M $NaHCO_3$
    b. 1 M NaCl, 1 M glucose
    c. 1 M NaCl, 1 M $CaCl_2$
    d. 1 M NaCl, 3 M glucose
36. For each pair of solutions, indicate which has the higher osmotic pressure.
    a. 1 osmol/L NaCl, 1 osmol/L $CaCl_2$

    b. 1 osmol/L NaCl, 2 osmol/L glucose
    c. 1 osmol/L $NaHCO_3$, 0.5 osmol/L $NaHCO_3$
37. If red blood cells were placed in 5.5% NaCl solution, what would be the effect on the cells? What if they were placed in 0.92% glucose solution?
38. Compare and contrast true solutions, colloidal dispersions, and suspensions.
39. Compare and contrast osmosis and dialysis with respect to the kind of particles that pass through the membrane.
40. Complete the following table:

Concentration of Solute (in grams per liter)	Molarity	Formula Weight of Solute
98	1.0	—
32	0.5	—
0.74	0.01	—
—	0.1	26
—	0.025	80
120	—	40
17	—	68

41. Benzene (C_6H_6) is a nonpolar solvent. Would you expect NaCl to dissolve in benzene? Explain.
42. Motor oil is nonpolar. Would you expect it to dissolve in water? In benzene? Explain.
43. Bubbles of carbon dioxide escape when the cap is removed from a bottle of pop. Explain.
44. You have a stock solution of 6 M HCl. How many moles of HCl are there in 1.0 L? In 100 mL? In 1 mL?
45. On the average, glucose ($C_6H_{12}O_6$) makes up about 0.10% by weight of human blood. How much glucose is there in 1 kg of blood?
46. A single dose of ethanol, C_2H_5OH, raises the intracellular level of calcium ions. The Ca^{2+} released by 0.1 M ethanol kills rat liver cells [*Science* 224, 1361 (1984)]. What is this concentration of ethanol in g/L? In mass–volume %? The density of ethanol is 0.785 g/mL. What is the volume–volume %?
47. A cyanide solution, made by adding 1 lb of sodium cyanide (NaCN) to 1 ton of water, is used to leach gold from its ore. What is the concentration of the cyanide solution in (a) percent by mass, (b) parts per million, (c) g per kg, and (d) mol per L? (1 ton = 2000 lb)

10

Acids and Bases

O urs is a strange, sometimes perplexing, yet wondrous world. All of us practice a complicated chemistry, from the baking of bread to the treatment of heartburn and headache and beyond. Central to much of the chemistry that we practice are two special kinds of compounds called **acids** and **bases.** We use lots of acids and bases around the home (Figure 10.1). We use them for cleaning. We eat them and drink them. Our bodies produce them and consume them.

The quantitative aspects of acid-base chemistry are discussed in Chapter 11. In this chapter, we will look at some of the more qualitative properties of acids and bases—what they are, how they are produced, how they are named, and how they react chemically.

10.1 ACIDS: THE ARRHENIUS CONCEPT

Acidic substances have been known for millennia. However, it was only in 1887 that Svante Arrhenius, a young Swedish chemist still in graduate school, proposed definitions of acids and bases that are still in common use today. Acids were generally recognized as substances that, in aqueous solution, would

1. Turn the indicator dye litmus from blue to red.
2. React with active metals such as zinc, iron, and tin, dissolving the metal and producing hydrogen gas.

FIGURE 10.1
(a) Acids and (b) bases found in the home. (Photos © The Terry Wild Studio.)

(a)

(b)

3. Taste sour, if diluted enough to be tasted safely.
4. React with certain compounds called alkalis or bases to form water and compounds called salts.

Arrhenius proposed that these characteristic properties of acids are actually properties of the hydrogen ion (H^+) and that acids are compounds that yield hydrogen ions in aqueous solutions.

Arrhenius's definition is still useful—as long as we deal only with aqueous solutions. If you want to know whether a given compound is an Arrhenius acid, just dissolve some of it in water. Stick in a piece of blue litmus paper. If the litmus turns red, the compound is an acid.

Many foods are acidic. Vinegar contains 4% to 10% acetic acid. Citrus fruits—and many fruit-flavored drinks—contain citric acid. If a food tastes sour, it most likely contains one or more acids.

Arrhenius's theory has been modified somewhat through the years. We know, for example, that simple hydrogen ions do not exist in water solutions. When a hydrogen atom is stripped of its only electron, all that is left is the bare nucleus containing a single proton. Other positive ions, even the relatively small lithium ion (Li^+), still have their nuclei shielded by electrons remaining in the inner levels. Thus, the hydrogen ion (also referred to as a proton) is too reactive to exist as a stable ion in solution. We know today that the acidic properties of a solution are due to H_3O^+, in which each hydrogen atom shares a pair of electrons with oxygen. The electron configuration is

$$\left[\begin{matrix} H \colon \ddot{O} \colon H \\ \ddot{H} \end{matrix} \right]^+$$

This species is called the **hydronium ion.** Today, we might well say that an acid is a substance that produces hydronium ions when dissolved in water. Even the hydronium ion is an oversimplification, for it is hydrated by other water molecules. For most purposes, though, we can use the simple hydronium ion (H_3O^+) and ignore any further hydration. The hydronium ion is quite reactive; it can readily transfer a proton to other molecules and ions. Thus, to talk about protons when we mean their source (hydronium ions) is permissible as long as we understand that we are using a simplification of the real situation.

Some acids give up one hydrogen ion (proton) per molecule. These are called **monoprotic acids.** An example is hydrogen chloride. When this compound, a gas, is dissolved in water, we get one hydronium ion per molecule of hydrogen chloride.

$$HCl(g) + H_2O \longrightarrow H_3O^+(aq) + Cl^-(aq)$$

(The symbol *g* in parentheses indicates the gaseous state; *aq* in parentheses indicates an aqueous solution.)

Other acids may give up more than one hydrogen ion per molecule. Sulfuric acid (H_2SO_4) is a **diprotic acid,** for it is capable of giving up two protons. Similarly, phosphoric acid (H_3PO_4) is a **triprotic acid.** A general term for acids that give up more than one proton is **polyprotic.** You should not assume that all the hydrogen atoms in a compound are acidic, meaning that they are released in water solutions. For example, none of the hydrogens of methane (CH_4) is given up in aqueous solution. Only one of the hydrogens in acetic acid ($C_2H_4O_2$) is acidic. For this reason, the formula for acetic acid is frequently written CH_3COOH or $HC_2H_3O_2$ to emphasize that only one proton is released. You will gain experience in determining which hydrogen atoms in a molecule are acidic as you learn more about molecular structure.

10.2 STRONG AND WEAK ACIDS

Strong acids are those that ionize completely (each molecule in dilute solution donates its acidic protons to water). Acids that ionize nearly completely are also called strong acids. In an equation, the ionization of a strong acid is sometimes shown as an equilibrium between un-ionized and ionized forms.

$$HCl + H_2O \rightleftharpoons H_3O^+ + Cl^-$$

The unequal double arrow indicates that the equilibrium strongly favors the ionized form. Because strong acids are essentially completely ionized in solution, the equation is most often written with a single arrow.

Strong acids, if fairly concentrated, can cause serious damage to skin and flesh. They eat holes in clothes made of natural fibers such as cotton, silk, or wool. Strong acids also destroy most synthetics such as nylon, polyester, and acrylic fibers. Care should always be taken to prevent spills on either skin or clothing. In very dilute solutions, acids may well prove harmless. Hydrochloric acid, for example, is produced in dilute solution in our stomachs, where it aids in the digestion of certain foodstuffs. (Even here, under certain conditions, it can cause problems—as anyone with an ulcer can attest.)

Acids that ionize only slightly in dilute solution are called **weak acids.** Acetic acid (CH_3COOH) is an example. The equation for the ionization of this acid is drawn in such a way as to indicate a preference for the un-ionized form.

$$CH_3COOH + H_2O \longleftarrow CH_3COO^- + H_3O^+$$

Solutions of 4% to 10% acetic acid are called vinegar, a familiar condiment. We should be aware, though, that while it is safe to eat (oil and) vinegar on our salad, more concentrated acetic acid solutions can be corrosive to the skin and especially so to the digestive tract.

A sampling of strong and weak acids is given in Table 10.1. Many of them will be encountered again and again as we continue our study of chemistry.

TABLE 10.1 *Some Representative Acids*

Name	Formula	Classification	Number of Hydronium Ions From 1 000 000 Molecules in 0.01 M Solution
Sulfuric acid	H_2SO_4	Strong	1 220 000
Nitric acid	HNO_3	Strong	920 000
Hydrochloric acid	HCl	Strong	920 000
Phosphoric acid	H_3PO_4	Moderate	270 000
Lactic acid	$CH_3CHOHCOOH$	Weak	87 000
Acetic acid	CH_3COOH	Weak	13 000
Boric acid	H_3BO_3	Weak	Less than 1
Hydrocyanic acid	HCN	Weak	Less than 1

10.3 NAMES OF SOME COMMON ACIDS

In Chapter 4 you learned the names of some common anions (and if you didn't—now is the time). It is easy to learn the names of acids derived from these anions. Just follow these simple rules.

If the anion name ends in *-ide, -ate,* or *-ite,* then the acid name is, respectively, hydro____ic acid, ____ic acid, or____ous acid. Let's apply these rules to a few examples.

Ion	*Name*	*Acid*	*Name*
Cl^-	Chloride	HCl	Hydrochloric acid
$NO_3{}^-$	Nitrate	HNO_3	Nitric acid
$SO_4{}^{2-}$	Sulfate	H_2SO_4	Sulfuric acid
$PO_4{}^{3-}$	Phosphate	H_3PO_4	Phosphoric acid
$SO_3{}^{2-}$	Sulfite	H_2SO_3	Sulfurous acid
$CO_3{}^{2-}$	Carbonate	H_2CO_3	Carbonic acid
CN^-	Cyanide	HCN	Hydrocyanic acid
S^{2-}	Sulfide	H_2S	Hydrosulfuric acid
CH_3COO^-	Acetate	CH_3COOH	Acetic acid
$BO_3{}^{3-}$	Borate	H_3BO_3	Boric acid

Note that most of the names have the same stem in the acid as in the anion. A few have extra letters: for example, the *ur* in sulfuric and sulfurous, and the *or* in phosphoric. These you will have to learn as special cases if you haven't done so already.

Any acid with three or more replaceable hydrogens can exist in two forms, called **ortho** and **meta.** The meta form can be derived by subtracting two hydrogens and one oxygen (i.e., H_2O) from the ortho formula. For example, H_3PO_4 is really orthophosphoric acid, although the prefix is often dropped in everyday usage. Metaphosphoric acid is therefore HPO_3 (i.e., H_3PO_4 minus H_2O). Metaboric acid is derived in a similar manner from orthoboric acid (H_3BO_3). Subtraction of H_2O gives HBO_2. The corresponding anions are metaphosphate $(PO_3{}^-)$ and metaborate $(BO_2{}^-)$. Sodium metaphosphate and sodium metasilicate are used as water softeners.

The formula for phosphite ion is $PO_3{}^{3-}$. What is the formula for (ortho) phosphorous acid? For metaphosphorous acid? For metaphosphite ion?

Orthophosphorous acid is H_3PO_3 (add one H^+ for each negative charge on $PO_3{}^{3-}$). Metaphosphorous acid is HPO_2 (H_3PO_3 minus H_2O). The metaphosphite ion is $PO_2{}^-$.

example
10.1

We will encounter other acids as we go along. The names of most will be obvious from the preceding discussion. A few will have special, nonsystematic names. The names of important organic acids (of which acetic and lactic acid are but two examples) are given in Chapter 17.

10.4 SOME COMMON BASES

Like acids, basic compounds have been known for thousands of years. According to the biblical account, the Israelites, in their journey from Egypt to Palestine, came upon bitter waters at Marah (Exod. 15:23). Although the writers may not have meant to, they thus recorded for posterity the existence of an aqueous solution of base. By the time of Arrhenius, bases, or alkaline compounds, were generally recognized as those substances that, when dissolved in water, would

1. Turn the indicator dye litmus from red to blue.
2. Feel slippery or soapy on the skin.
3. Taste bitter (as the Israelites found at Marah).
4. React with acids to form water and salts.

Arrhenius's theory was that these were actually properties of the hydroxide ion (OH^-). Bases, he said, are compounds that yield **hydroxide ions** in aqueous solutions.

Perhaps the most familiar strong base is sodium hydroxide ($NaOH$), sometimes called lye. Even in the solid state, sodium hydroxide is completely ionic; that is, it exists as sodium ions and hydroxide ions. In solution, the hydroxide ions enter into the characteristic reactions that we refer to when we say a solution is basic.

Other common bases are potassium hydroxide (KOH), calcium hydroxide [$Ca(OH)_2$], and magnesium hydroxide [$Mg(OH)_2$]. All of these are completely ionic. Potassium hydroxide, like sodium hydroxide, is quite soluble in water. Therefore, aqueous solutions of the compound are rich in hydroxide ions. Calcium hydroxide is only sparingly soluble in water; the solution thus contains relatively less dissolved hydroxide ion. Magnesium hydroxide is so nearly insoluble that it can be safely taken internally as an antacid (Section 10.8).

Ammonia (NH_3) is a familiar weak base. It is a gas at room temperature but readily dissolves in water to give a basic solution. It reacts with water to a slight extent to produce ammonium ions (NH_4^+) and hydroxide ions.

$$NH_3 + H_2O \rightleftharpoons NH_4^+ + OH^-$$

Ammonia is called a weak base. The solution of ammonia contains a relatively low concentration of hydroxide ions; once ammonia is in solution it tends *not* to ionize. The solution so formed is sometimes called **ammonium hydroxide,** but since most of the nitrogen atoms are still in the form of NH_3, **aqueous ammonia** is perhaps a better name. Many familiar household cleansers contain ammonia, a compound easily detected by its characteristic odor.

10.5 ACIDS, METALS, AND THE ACTIVITY SERIES

Acids and bases have characteristic reactions. One such reaction for acids is that with certain metals. For example, if zinc metal is treated with hydrochloric acid, hydrogen gas is produced along with a salt, zinc chloride (Figure 10.2). The equation may be written

$$Zn(s) + 2\,HCl(aq) \longrightarrow ZnCl_2(aq) + H_2(g)$$

(The symbol s in parentheses indicates the solid state.) Since zinc chloride is ionic and hydrogen chloride ionizes in aqueous solution, the equation may also be written

$$Zn(s) + 2\,H_3O^+ + 2\,Cl^- \longrightarrow Zn^{2+} + 2\,Cl^- + H_2(g) + 2\,H_2O$$

In this form, the equation clearly indicates that the chloride ions are not really involved in the reaction. They appear as reactants and, unchanged, as products. We can eliminate them from the equation and still accurately record the changes that have occurred. The *net* ionic equation becomes

$$Zn + 2\,H_3O^+ \longrightarrow Zn^{2+} + H_2(g) + 2\,H_2O$$

This equation tells us that zinc will react with an acid (i.e., a source of hydronium ions) to produce zinc ions and hydrogen gas. The following equation also can be reduced to a net ionic equation identical to the one shown above.

$$Zn(s) + H_2SO_4(aq) \longrightarrow ZnSO_4(aq) + H_2(g)$$

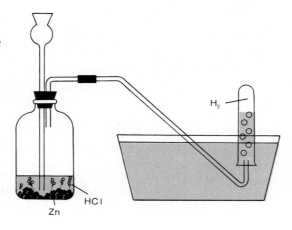

FIGURE 10.2
Zinc reacts with hydrochloric acid to produce hydrogen gas. The hydrogen can be collected by displacement of water from a test tube inverted in a pan of water.

Zinc is one of a number of **active metals** that react with acids in this way. Not all these metals react at the same rate, however. Some, such as sodium and potassium, react violently with plain water—without any added acid. The reaction may produce enough heat to ignite the hydrogen gas, causing an explosion. With sodium, the by-product is a solution of sodium hydroxide.

$$2\,Na + 2\,H_2O \longrightarrow 2\,NaOH(aq) + H_2$$

A similar reaction occurs with potassium.

$$2\,K + 2\,H_2O \longrightarrow 2\,KOH(aq) + H_2$$

All the alkali metals (those in Group IA) react in this manner. Knowing the reaction for sodium, we can write the equation for the reaction of cesium with water.

$$2\,Cs + 2\,H_2O \longrightarrow 2\,CsOH(aq) + H_2$$

We could have substituted lithium or rubidium for cesium and still have been correct. Similar reactions involving members of the same chemical family are a part of the framework that helps make chemistry more understandable. We discuss the chemistry of some of the families of elements in Chapter 13. A word of caution, though. Even members of a family differ. They do react at different rates. Sometimes even the *kind* of reaction is different. But these exceptions will not concern us very much here.

In contrast to the alkali metals, metals like zinc or iron do not react appreciably when placed in water. However, if the temperature is raised and zinc is brought into contact with steam, then hydrogen gas is produced. Tin won't react with steam, but it will release hydrogen on contact with acids. Gold won't even react with acids.

It is possible to arrange metals in an **activity series** with the most reactive metals (such as potassium) at the top and the least reactive (such as gold) at the bottom (Table 10.2). The position of a metal in the table reflects its tendency to give up electrons to form ions. All of those listed above hydrogen in the series (the **active metals**) will react with acids to produce hydrogen gas. These metals will give their electrons to the hydrogen ions produced by the acids, thus becoming ions. Those metals below hydrogen will not give up electrons to hydrogen ions and hence will not liberate hydrogen gas from acids.

A lot of chemical information is summarized in Table 10.2. Remember that the arrangement of the table indicates how readily these metallic elements give up their electrons. If one of the metals is brought into contact with the ions of another metal, one of two things can happen. Either the metal can transfer electrons to the ions (a reaction occurs) or the ions do not accept the electrons (no reaction occurs). We can use the table to predict what will happen. A metal can transfer electrons to the ions of any metal appearing lower in the table. For

TABLE 10.2 *An Activity Series of the Metals*

Potassium Calcium Sodium	React with cold water, releasing hydrogen
Magnesium Aluminum Zinc Chromium Iron	React with steam, releasing hydrogen
Nickel Tin Lead	React with acids, releasing hydrogen
Hydrogen	
Copper Silver Mercury Platinum Gold	Do not react with acids to form hydrogen

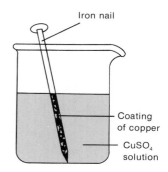

FIGURE 10.3
An iron nail placed in a solution of copper sulfate becomes coated with copper.

example, an iron nail, placed in a dilute solution of copper(II) sulfate ($CuSO_4$), gradually becomes coated with copper metal (Figure 10.3). Since iron atoms have a greater tendency to give up electrons than do copper atoms, the iron atoms transfer electrons to the copper ions.

$$Fe + Cu^{2+} \longrightarrow Fe^{2+} + Cu$$

What we observe is the copper metal plating out of solution as it is formed. Some of the iron dissolves as iron(II) ions, but this is not as obvious to the observer.

If we use another system, it is possible to see both processes as they occur. A strip of copper metal placed in a solution of silver nitrate ($AgNO_3$) becomes coated with crystals of metallic silver (Figure 10.4). The silver ions grab electrons from the more reactive copper atoms.

$$2\,Ag^+ + Cu \longrightarrow 2\,Ag + Cu^{2+}$$

There is also visual evidence that the copper is dissolving. A solution of copper ions is blue, whereas a solution of silver ions is colorless. Thus, as the silver metal crystallizes out of solution, the copper ions form and slowly turn the solution blue.

Note that in one of the examples copper loses electrons and in the other it gains them. That is because there are elements both above and below copper in the activity series, and what happens depends only on the *relative* reactivity of the elements involved.

FIGURE 10.4
A chemical tree made from copper wire and a silver nitrate solution. Copper ions replace silver ions in solution. (Reprinted with permission from Toler, Richard J., "A Chemical Christmas Tree," Chemistry, December 1974. Copyright © 1974 by the American Chemical Society.)

10.6 NEUTRALIZATION REACTIONS

As a characteristic property of both acids and bases, we listed the reaction of these compounds with one another to form water and a salt. Sodium hydroxide reacts with hydrochloric acid to produce water and sodium chloride, a salt (in this case, ordinary table salt).

$$NaOH(aq) + HCl(aq) \longrightarrow H_2O + NaCl(aq)$$

If equivalent amounts of acid and base (e.g., 100 mL each of 0.1 M solution) are used, the resulting solution no longer affects litmus paper. Nor does the new solution taste bitter or sour; it tastes salty. The solution is neither acidic nor basic. It is neutral, and the reaction is called **neutralization**.

Both aqueous sodium hydroxide and aqueous hydrogen chloride are ionic. The solution formed by mixing of the two contains sodium chloride, which is also ionic. Rewriting the equation in ionic form, we get

$$Na^+ + OH^- + H_3O^+ + Cl^- \longrightarrow 2\,H_2O + Na^+ + Cl^-$$

In a similar manner we can write the equation for the reaction of potassium hydroxide with nitric acid.

$$KOH(aq) + HNO_3(aq) \longrightarrow H_2O + KNO_3(aq)$$

Or

$$K^+ + OH^- + H_3O^+ + NO_3^- \longrightarrow 2\,H_2O + K^+ + NO_3^-$$

Note that in each reaction, hydrogen ions come together with hydroxide ions to form water. The other ions are left unchanged in the solution. Thus, for any neutralization, we can write the *net* ionic reaction as

$$H_3O^+ + OH^- \longrightarrow 2\,H_2O$$

The other ions don't just disappear. If the solution after neutralization is evaporated, a crystalline ionic solid—a **salt**—is obtained. In general, then, we can say that an acid reacts with a base to form a salt and water.

10.7 REACTIONS OF ACIDS WITH CARBONATES AND BICARBONATES

Add vinegar to baking soda, and you get a vigorous fizzing action. Some antacid preparations are designed to effervesce. What causes the fizz? Generally, it is the reaction of a carbonate or bicarbonate salt with an acid.

Carbonates and bicarbonates are salts of carbonic acid (H_2CO_3). This diprotic acid is a very weak one; it holds on to its protons quite tightly. (Notice the apparent contradiction: it's the weak acids that hold tightly to their protons and the strong acids that don't.) Conversely, carbonate ions and bicarbonate ions pick up protons readily to form carbonic acid. Further, carbonic acid is quite unstable. It decomposes to carbon dioxide and water.

$$H_2CO_3(aq) \longrightarrow H_2O + CO_2(g)$$

This reaction is not related to the acidity of carbonic acid. It simply indicates that the compound, which happens to be an acid, is not stable. One can't purchase a bottle of carbonic acid; it is too unstable to be isolated and bottled. It can, however, be formed in solution. But as soon as it is formed, it tends to decompose.

If sodium bicarbonate is dissolved in water, a solution of sodium ions and bicarbonate ions is formed. If hydrochloric acid is added, the bicarbonate ions come into contact with the hydronium ions from the acid and immediately grab a proton to form carbonic acid.

$$Na^+ + HCO_3^- + H_3O^+ + Cl^- \longrightarrow H_2CO_3 + Na^+ + Cl^- + H_2O$$

But carbonic acid is unstable and immediately falls apart, forming carbon dioxide and water. The gaseous carbon dioxide bubbles out of solution (the fizz). If the remaining solution is evaporated, sodium chloride is left. Even if acid is poured on solid sodium bicarbonate, carbon dioxide is released. The reaction may be written

$$NaHCO_3(s \text{ or } aq) + HCl(aq) \longrightarrow NaCl(aq) + CO_2(g) + H_2O$$

The net ionic equation is simpler.

$$HCO_3^- + H_3O^+ \longrightarrow H_2CO_3 \longrightarrow H_2O + CO_2$$
$$+$$
$$H_2O$$

Similarly, if hydrochloric acid is added to sodium carbonate, bubbles of carbon dioxide and a solution of sodium chloride are formed.

$$Na_2CO_3(s \text{ or } aq) + 2\,HCl(aq) \longrightarrow 2\,NaCl(aq) + CO_2(g) + H_2O$$

The net ionic equation shows that the doubly negative carbonate ion picks up two protons to form carbonic acid. The latter breaks down to form carbon dioxide and water.

$$CO_3^{2-} + 2\,H_3O^+ \longrightarrow H_2CO_3 \longrightarrow H_2O + CO_2$$
$$+$$
$$2\,H_2O$$

Limestone and marble are mainly calcium carbonate ($CaCO_3$). Both are important building stones. Marble is also used in statues, monuments, and sculptures. The calcium carbonate in them is readily attacked by acids in the atmosphere or in rain. And the atmosphere has been made increasingly acidic

(a)

(b)

FIGURE 10.5
This stone sculpture on the exterior of Herten Castle, near Recklinghausen, in Westphalia, Germany, shows the effects of acidic smog. The castle was constructed in 1702. (a) The sculpture as it looked in 1908. (b) The sculpture as it looked in 1969. (Reprinted with permission from Winkler, E. M., Stone: Properties, Durability in Man's Environment, Berlin: Springer-Verlag, 1973. Copyright © 1973 by E. M. Winkler.)

in recent years, mainly by the burning of sulfur-containing coal. The sulfur combines with oxygen to form sulfur dioxide.

$$S + O_2 \longrightarrow SO_2$$

Some of the sulfur dioxide reacts further with oxygen to form sulfur trioxide.

$$2\,SO_2 + O_2 \longrightarrow 2\,SO_3$$

The sulfur trioxide then reacts with water to form sulfuric acid.

$$SO_3 + H_2O \longrightarrow H_2SO_4$$

The sulfuric acid, in the form of an aerosol mist or diluted by rainwater, furnishes the hydronium ions that dissolve the marble and limestone.

$$CaCO_3(s) + 2\,H_3O^+ \longrightarrow Ca^{2+}(aq) + CO_2(g) + 3\,H_2O$$

The acid mists and acidic rainwater also attack and dissolve many metals (Section 10.5). Damage to our buildings, automobiles, and other structures and machines from air pollution amounts to billions of dollars per year (Figure 10.5). And even then the story is not complete. These acid pollutants are also damaging to crops and to human health, as we shall see in Section 10.10.

It should be noted that natural rainwater is slightly acidic. Water falling through the air dissolves carbon dioxide and forms carbonic acid.

$$H_2O + CO_2 \longrightarrow H_2CO_3$$

This mild acid has little effect compared to the strong acids formed in polluted air.

10.8 ANTACIDS

The stomach secretes hydrochloric acid (HCl) as an aid in the digestion of food. Sometimes overindulgence or emotional stress leads to a condition called hyperacidity (too much acid). A number of basic compounds called **antacids** are available to treat this condition, many of them aggressively advertised. Indeed, sales of antacids in the United States are estimated to be about $650 million each year. Let's look at some popular antacids from the standpoint of acid-base chemistry.

Sodium bicarbonate, or baking soda ($NaHCO_3$), is an old standby antacid. It is probably safe and effective for occasional use by most people. Overuse will make the blood too alkaline, a condition called **alkalosis.** Sodium bicarbonate is not recommended for those with hypertension (high blood pressure), because high concentrations of sodium ion tend to aggravate the condition. The antacid in Alka-Seltzer is sodium bicarbonate. This popular remedy also contains citric acid and aspirin. When Alka-Seltzer is placed in water, the familiar fizz occurs, due to the reaction of bicarbonate ions with hydronium ions from the acid.

$$HCO_3^- + H_3O^+ \longrightarrow CO_2(g) + 2\,H_2O$$

Alka-Seltzer may be dangerous because the aspirin in it might be harmful to people with ulcers and other stomach disorders.

Calcium carbonate, commonly called precipitated chalk ($CaCO_3$), is another common antacid ingredient. It is fast acting and safe in small amounts, but regular use can cause constipation. It also appears that calcium carbonate can cause *increased* acid secretion after a few hours. Temporary relief may be achieved at the expense of a worse problem later. Tums is simply flavored calcium carbonate. Alka 2 and Di-Gel liquid suspension also contain calcium carbonate as the antacid ingredient. The carbonate ion neutralizes acid by the following reaction:

$$CO_3^{2-} + 2\,H_3O^+ \longrightarrow CO_2 + 3\,H_2O$$

Aluminum hydroxide [$Al(OH)_3$] is another popular antacid ingredient. Like calcium carbonate, it can cause constipation when taken in large doses. The hydroxide ions neutralize acid.

$$OH^- + H_3O^+ \longrightarrow 2\,H_2O$$

There is concern that antacids containing aluminum ions deplete the body of essential phosphate ions. The aluminum phosphate formed is insoluble and is excreted from the body.

$$Al^{3+} + PO_4^{3-} \longrightarrow AlPO_4(s)$$

Aluminum hydroxide is the only antacid in Amphojel. It occurs in combination in many popular products.

Magnesium compounds constitute the fourth category of antacids. These include magnesium carbonate ($MgCO_3$) and magnesium hydroxide [$Mg(OH)_2$]. Milk of magnesia is a suspension of magnesium hydroxide in water. It is sold under a variety of brand names, the best known of which is probably Phillips. In small doses, magnesium compounds act as antacids. In large doses, they act as laxatives. Magnesium ions are poorly absorbed in the digestive tract. Rather, these small, dipositive ions draw water into the colon (large intestine), causing the laxative effect.

A variety of popular antacids combine aluminum hydroxide, with its tendency to cause constipation, and a magnesium compound, which acts as a laxative. These tend to counteract one another. Maalox and Mylanta are familiar brands.

Rolaids, a highly advertised antacid, contains the complex substance aluminum sodium dihydroxy carbonate [$AlNa(OH)_2CO_3$]. Both the hydroxide ion and the carbonate ion consume acid.

Antacids interact with other medications. Anyone taking any type of medicine should consult a physician before taking antacids. Generally, antacids are safe and effective for occasional use in small amounts. All antacids are *basic* compounds. If you are otherwise in good health, you can choose a base on the basis of price.

Anyone with severe or repeated attacks of indigestion should consult a physician. Self-medication in such cases might be dangerous.

10.9 THE BRØNSTED-LOWRY ACID-BASE THEORY

The Arrhenius theory of acids and bases was useful for many years. It is still useful today for many aqueous solutions. But by the 1920s chemists were working with solvents other than water. Compounds were found that acted like bases yet did not have OH in their formulas. A new theory was needed. One was suggested in 1923, at almost the same time, by J. N. Brønsted, a colleague of Neils Bohr (Chapter 2) at Copenhagen, and by Thomas M. Lowry, of England.

The **Brønsted-Lowry theory** defines an **acid** as a **proton donor** (H^+ donor) and a **base** as a **proton acceptor** (H^+ acceptor). All the substances that Arrhenius considered acids are also acids by this definition. For instance, the reaction of hydrogen chloride with water has hydrogen chloride acting as a proton donor. All the Arrhenius bases are also bases in the new theory, but there are also new bases. In the hydrogen chloride–water reaction, the proton acceptor is water.

$$H-Cl + H_2O \longrightarrow H_3O^+ + Cl^-$$

Proton   Proton
donor   acceptor

Thus, water becomes a base in the Brønsted-Lowry system. The driving force for the reaction is the formation of a weaker acid (one that has less tendency to give up a proton) and a weaker base (one that has less tendency to accept a proton). A stronger acid reacts with a stronger base to produce a weaker base and a weaker acid.

$$HCl + H_2O \longrightarrow H_3O^+ + Cl^-$$

Stronger   Stronger   Weaker   Weaker
acid   base   acid   base

It may not be obvious that hydrogen chloride is a stronger acid than hydronium ion, but keep in mind that a stronger acid is one that succeeds in giving away its proton. It is sort of the opposite of the "keep-away" game that children play; it's a "give-away" game. It may also seem odd to think of water as a base. Remember the definition, though. A base is a proton acceptor. Not only is water a base in that sense, but it is a stronger base than chloride ion. In the "give-away" game, water gets the proton every time.

Hydrogen chloride and chloride ion make up a **conjugate** acid-base pair. Hydrogen chloride is an acid; chloride ion is its conjugate base (the conjugate base is the acid minus its proton). Water and the hydronium ion make up another conjugate pair. Water is a Brønsted-Lowry base; hydronium ion is its conjugate acid.

Conjugate
acid-base pair

$$HCl + H_2O \longrightarrow H_3O^+ + Cl^-$$

Conjugate acid-base pair

Let's take another look at the aqueous ammonia system in terms of the Brønsted-Lowry theory. Ammonia reacts with water (to a slight extent) to form ammonium ions and hydroxide ions. In this system, ammonia acts as a proton acceptor (base) and water acts as a proton donor (acid!).

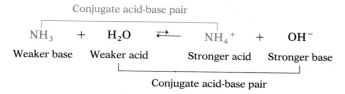

$$NH_3 \quad + \quad H\,)\,OH \quad \rightleftarrows \quad NH_4^+ + OH^-$$

Base   Acid
(proton acceptor)   (proton donor)

Ammonia and hydroxide ions compete for protons. Which is the stronger base? Hydroxide ion gets the proton most of the time. A dynamic equilibrium is established among all four species present. However, at any instant there are a lot more ammonia and water molecules than there are ammonium and hydroxide ions.

Ammonia and ammonium ion make up one conjugate pair, and water and hydroxide ion make up another.

Conjugate acid-base pair

$$NH_3 \quad + \quad H_2O \quad \rightleftarrows \quad NH_4^+ \quad + \quad OH^-$$

Weaker base   Weaker acid   Stronger acid   Stronger base

Conjugate acid-base pair

Since the driving force is toward the weaker base and the weaker acid, it is easy to see that there would be more ammonia than ammonium ion and more water than hydroxide ion. Note that in such systems conjugate pairs always consist of a stronger and a weaker partner. This is certainly not a coincidence.

TABLE 10.3   *The Relative Strengths of Some Brønsted-Lowry Acids and Their Conjugate Bases*

		Acid		Base	
		Name	Formula	Name	Formula
↑ Stronger acids →		Perchloric acid	$HClO_4$	Perchlorate ion	ClO_4^-	↑ Weaker bases
		Sulfuric acid	H_2SO_4	Hydrogen sulfate ion	HSO_4^-	
		Hydrogen chloride	HCl	Chloride ion	Cl^-	
		Nitric acid	HNO_3	Nitrate ion	NO_3^-	
		Hydronium ion	H_3O^+	Water	H_2O	
		Sulfurous acid	H_2SO_3	Hydrogen sulfite ion	HSO_3^-	
		Phosphoric acid	H_3PO_4	Dihydrogen phosphate ion	$H_2PO_4^-$	
Weaker acids →		Acetic acid	CH_3COOH	Acetate ion	CH_3COO^-	Stronger bases →
		Carbonic acid	H_2CO_3	Hydrogen carbonate ion	HCO_3^-	
		Ammonium ion	NH_4^+	Ammonia	NH_3	
		Water	H_2O	Hydroxide ion	OH^-	
↓		Ammonia	NH_3	Amide ion	NH_2^-	↓

What makes an acid stronger is its greater tendency not to hold its proton. It is necessarily true, therefore, that its conjugate base must not eagerly seek a proton and so is classified weaker.

To summarize, a strong Brønsted-Lowry acid is one that readily gives up protons, and a weak acid is one that holds on to its protons tightly. A strong Brønsted-Lowry base is one that readily accepts protons and then binds them tightly. A weak base accepts a proton less readily and then holds it rather loosely. Table 10.3 lists a number of acids and their conjugate bases in order of their relative strengths. Notice that the strongest acid has the weakest conjugate base and the strongest base has the weakest conjugate acid.

10.10 ACIDS, BASES, AND HUMAN HEALTH

Strong acids and bases cause damage on contact with living cells. Their action is nonspecific. All cells, regardless of type, are damaged more or less equally. These **corrosive poisons** produce what are known as chemical burns. The injuries, once the offending agent is neutralized or removed, are similar to burns from heat. They are often treated the same way.

Sulfuric acid (H_2SO_4) is by far the leading chemical product of U.S. industry. About 40 billion kg are produced annually. Most of this acid is used by industry. A major use is the conversion of phosphate rock to soluble compounds for use as fertilizer. Only a small portion is used in or around the home.

Automobile batteries contain sulfuric acid. It is also the major ingredient in one type of drain cleaner, presumably because it dissolves drain-clogging hair. Sulfuric acid is a powerful dehydrating agent. It takes up water from cellular fluid, rapidly killing the cell. The sulfuric acid molecules dehydrate by actually reacting with water to form hydronium ions and hydrogen sulfate ions.

$$H_2SO_4 + H_2O \longrightarrow H_3O^+ + HSO_4^-$$

Hydration of these ions and other secondary reactions may also be involved in the dehydration process.

In Section 10.7 we saw how the burning of sulfur-containing coal produces aerosol mists of sulfuric acid. When this airborne pollutant comes into contact with the alveoli of the lungs, the cells are broken down. The alveoli lose their resilience, making it difficult for them to remove carbon dioxide. Such lung damage may contribute to pulmonary emphysema, a condition characterized by increasing shortness of breath. Emphysema is the fastest-growing cause of death in the United States. Most likely the principal factor in the rise of emphysema is cigarette smoking. Air pollution is known to be a factor, however. For instance, the incidence of the disease among smokers is three times as great in St. Louis, where air pollution is rather heavy, as in Winnipeg, Manitoba, where air pollution is rather mild.

Hydrochloric acid (also called muriatic acid) is used in homes to clean calcium carbonate deposits from toilet bowls and in industry to clean excess mortar from bricks and to clean metal products. Concentrated solutions (about 38% HCl) cause severe burns, but more dilute solutions are considered safe enough for use around the home. Even dilute solutions, however, can cause skin irritation and inflammation.

Ingestion of sulfuric, hydrochloric, or any other strong acid causes corrosive damage to the digestive tract. As little as 10 mL of concentrated (98%) H_2SO_4, taken internally, may be fatal.

Phosgene, used in the gas warfare of World War I, acts by reacting with moisture in the lungs to produce hydrochloric acid.

$$\underset{\displaystyle Cl-\overset{\textstyle O}{\overset{\|}{C}}-Cl}{} + H_2O \longrightarrow CO_2 + 2\,HCl$$

The irritated lung tissue gives up water to the dehydrating acid, and the lungs full with fluid. The victim dies of suffocation.

Sodium hydroxide (lye) is by far the most common of the strong bases. It is used to open clogged drains and as an oven cleaner. It destroys tissue rapidly, causing severe chemical burns. Several detergent additives, such as carbonates, silicates, and borates, form strongly basic solutions when dissolved in water. These, too, can cause corrosive damage to tissues, particularly those of the digestive tract and the eyes.

Both acids and bases, even in dilute solutions, break down the protein molecules in living cells. Generally, the fragments are not able to carry out the functions of the original protein. In cases of severe exposure, the fragmentation continues until the tissue has been completely destroyed.

Acids and bases, misused, can be damaging to human health. But acids and bases affect human health in more subtle—and ultimately more important—ways, as we shall see.

PROBLEMS

1. List four general properties of acidic solutions.
2. To what ion did Arrhenius attribute the properties of acidic solutions?
3. What is a hydronium ion? How does it differ from a hydrogen ion (proton)?
4. Identify the first compound in each equation as an acid or a base.

   a. $C_5H_5N + H_2O \longrightarrow C_5H_5NH^+ + OH^-$
   b. $C_6H_5OH + H_2O \longrightarrow C_6H_5O^- + H_3O^+$
   c. $CH_3COCOOH + H_2O \longrightarrow$
   $\qquad\qquad\qquad CH_3COCOO^- + H_3O^+$
   d. $C_6H_5SH + H_2O \longrightarrow C_6H_5S^- + H_3O^+$
   e. $CH_3NH_2 + H_2O \longrightarrow CH_3NH_3^+ + OH^-$
   f. $C_6H_5SO_2NH_2 + H_2O \longrightarrow$
   $\qquad\qquad\qquad C_6H_5SO_2NH^- + H_3O^+$

5. Give examples of a monoprotic, a diprotic, and a triprotic acid.
6. Indicate whether the acid is monoprotic, diprotic, or triprotic.

   a. $H_2SO_3 + 2 H_2O \longrightarrow SO_3^{2-} + 2 H_3O^+$
   b. $CH_3CHOHCOOH + H_2O \longrightarrow$
   $\qquad\qquad\qquad CH_3CHOHCOO^- + H_3O^+$
   c. $CH_2(COOH)_2 + 2 H_2O \longrightarrow$
   $\qquad\qquad\qquad CH_2(COO^-)_2 + 2 H_3O^+$
   d. $H_3AsO_4 + 3 H_2O \longrightarrow AsO_4^{3-} + 3 H_3O^+$
   e. $H_4P_2O_7 + 4 H_2O \longrightarrow P_2O_7^{4-} + 4 H_3O^+$

7. Give formulas and names of three strong acids and three weak acids.
8. List four general properties of basic solutions.
9. To what ion did Arrhenius attribute the properties of basic solutions?
10. Give formulas and names of two strong bases and two weak bases.

11. Magnesium hydroxide is completely ionic, even in the solid state, yet it can be taken internally as an antacid. Explain why it does not cause injury as sodium hydroxide would.
12. If Br^- is bromide ion, what is the name of the acid HBr?
13. If Se^{2-} is selenide ion, what is the name of the acid H_2Se?
14. If NO_2^- is nitrite ion, what is the name of the acid HNO_2?
15. If PO_3^{3-} is phosphite ion, what is the name of the acid H_3PO_3?
16. If $C_2O_4^{2-}$ is oxalate ion, what is the name of the acid $H_2C_2O_4$?
17. If $C_6H_5COO^-$ is benzoate ion, what is the name of the acid C_6H_5COOH?
18. If H_4SiO_4 is orthosilicic acid, what is the formula for metasilicic acid? For metasilicate ion? For sodium metasilicate?
19. If H_3AsO_4 is orthoarsenic acid, what is the formula for metaarsenic acid? For metaarsenate ion? For potassium metaarsenate?
20. Each of the following equations presents an equilibrium between two acids. In each case, identify the acids and indicate which is stronger.

    a. $HBr + F^- \rightleftharpoons HF + Br^-$
    b. $HCN + F^- \rightleftharpoons HF + CN^-$
    c. $HIO_3 + H_2PO_4^- \rightleftharpoons H_3PO_4 + IO_3^-$
    d. $HCOOH + CCl_3COO^- \rightleftharpoons$
    $\qquad\qquad\qquad HCOO^- + CCl_3COOH$
    e. $HClO_3 + NO_3^- \rightleftharpoons HNO_3 + ClO_3^-$

21. Bases are proton acceptors. Each equilibrium in Problem 20 also includes a pair of bases. Identify the bases in each equation and indicate which is stronger.

22. Complete the following by writing formulas for the expected products and balancing the equations:
    a. $HCl + LiOH \longrightarrow$
    b. $Al(OH)_3 + HCl \longrightarrow$
    c. $H_2SO_4 + Mg(OH)_2 \longrightarrow$
23. Complete the following by writing formulas for the expected products and balancing the equations:
    a. $NaHCO_3 + HNO_3 \longrightarrow$
    b. $CaCO_3 + HBr \longrightarrow$
    c. $H_2SO_4 + K_2CO_3 \longrightarrow$
24. Complete the equation: $NH_3 + HI \longrightarrow$
25. Write the *net ionic equation* for a neutralization reaction involving a strong acid and a strong base in aqueous solution.
26. Write the equation for the *decomposition* of carbonic acid.
27. Write the *net ionic equation* for the reaction of a bicarbonate salt with an acid.
28. Write the *net ionic equation* for the reaction of a carbonate salt with an acid.
29. Complete the following by writing formulas for the expected products, and then balance the equations:
    a. $Mg + HCl \longrightarrow$ d. $Zn + H_2SO_4 \longrightarrow$
    b. $Al + HCl \longrightarrow$ e. $Al + H_2SO_4 \longrightarrow$
    c. $Mg + H_2SO_4 \longrightarrow$
30. Complete the following by writing formulas for the expected products, and then balance the equations:
    a. $Na + H_2O \longrightarrow$ c. $Li + H_2O \longrightarrow$
    b. $K + H_2O \longrightarrow$ d. $Ca + H_2O \longrightarrow$
31. In each reaction below, label each reactant and product as acid or base (in the Brønsted-Lowry sense). Which acid is stronger? Which base is stronger? Indicate the pairs of conjugate acid and base.
    a. $HBr + H_2O \longrightarrow H_3O^+ + Br^-$
    b. $HF + H_2O \rightleftarrows H_3O^+ + F^-$
    c. $N_2H_4 + H_2O \rightleftarrows N_2H_5^+ + OH^-$

32. According to the equation, is phenol, C_6H_5OH, an acid or base? Should it be classified as strong or weak?
$$C_6H_5OH + H_2O \rightleftarrows C_6H_5O^- + H_3O^+$$
33. According to the equation, is aniline, $C_6H_5NH_2$, an acid or base? Should it be classified as strong or weak?
$$C_6H_5NH_2 + H_2O \rightleftarrows C_6H_5NH_3^+ + OH^-$$
34. Write an equation that describes the effect of acid rain on marble.
35. Name some of the active ingredients in antacids.
36. Should a person suffering from hypertension be advised to use baking soda or milk of magnesia as an antacid? Why?
37. How do corrosive acids and alkalis destroy cells? (Give two ways.)
38. How does magnesium hydroxide, in large doses, act as a laxative? Would other magnesium compounds have a similar effect?
39. Strong acids and weak acids both have properties characteristic of hydronium ions. How do strong and weak acids differ?
40. Lime (CaO) has been used to neutralize water in lakes that have become too acidic from acid precipitation. Assume that the acid is sulfuric acid and write the equation for the reaction. How many metric tons of lime does it take to neutralize 10 t of sulfuric acid?
41. What is aqueous ammonia? Why is it sometimes called ammonium hydroxide?
42. How does phosgene exert its damaging effect?
43. Examine the labels of at least five antacid preparations. Make a list of the ingredients in each. What kind of chemical compound is each of the active ingredients?
44. Define and illustrate the following terms:
    a. neutralization   b. salt

More Acids and Bases

If our bodies are largely solutions—and they surely are—they are very special solutions. Delicate balances must be maintained among the many solutes in our blood and other body fluids. And no balance is more important than that between acids and bases. If the acidity of the blood changes very much, the blood loses its capacity to carry oxygen. Since many bodily processes produce acids, the control of acidity is—quite literally—a matter of life or death.

Our bodies have developed a marvelously complex yet efficient mechanism for maintaining the proper acid-base balance. Before we can talk about this mechanism in a meaningful way, however, we need to develop a few concepts—concepts that are more quantitative in nature than those developed in Chapter 10. It is important to know how exact concentrations of acids and bases are determined and expressed. We must understand the concept of pH, particularly as it relates to the chemistry of the blood and other body fluids. Most important, we must see just how, through substances called buffers, the level of acidity is controlled.

11.1 CONCENTRATIONS OF ACIDS AND BASES

While 1 mol of HCl just neutralizes 1 mol of NaOH, 1 mol of H_2SO_4 will neutralize 2 mol of NaOH.

$$HCl(aq) + NaOH(aq) \longrightarrow NaCl(aq) + H_2O$$
$$\text{1 mol} \qquad \text{1 mol}$$

$$H_2SO_4(aq) + 2\,NaOH(aq) \longrightarrow Na_2SO_4(aq) + 2\,H_2O$$

$$\underset{1\ mol}{} \qquad\qquad \underset{2\ mol}{}$$

In neutralization reactions, then, 1 mol of H_2SO_4 is equivalent to 2 mol of HCl, that is, will do the work of 2 mol of HCl. Alternatively, we may say that 0.5 mol of H_2SO_4 is equivalent to 1 mol of HCl. For acids and bases, some people find it more convenient to use **equivalent weights** than formula weights.

$$\text{Equivalent weight of an acid} = \frac{\text{formula weight of the acid}}{\text{number of replaceable hydrogens per molecule}}$$

$$\text{Equivalent weight of a base} = \frac{\text{formula weight of the base}}{\text{number of available hydroxides per formula unit}}$$

These definitions are perhaps best understood through examples.

example 11.1

What is the equivalent weight of H_2SO_4?
    The formula weight of H_2SO_4 is 98. There are two replaceable hydrogens; thus, the equivalent weight of H_2SO_4 is

$$\frac{98}{2} = 49 \text{ g/eq}$$

example 11.2

What is the equivalent weight of aluminum hydroxide [$Al(OH)_3$]?
    The formula weight of $Al(OH)_3$ is 78. There are three available hydroxides per formula; therefore, the equivalent weight of $Al(OH)_3$ is

$$\frac{78}{3} = 26 \text{ g/eq}$$

For monoprotic acids and for bases with one hydroxide per formula unit, the equivalent weight and the formula weight are the same.
    To express concentrations of acids and bases, scientists sometimes use the term **normality** (N) rather than **molarity** (M). A 1 N solution of an acid or a base is one that contains 1 equivalent weight (expressed in grams) of acid or base in 1 L of solution.

$$\text{Normality} = \frac{\text{equivalent of solute}}{\text{liters of solution}}$$

What is the normality of an H_2SO_4 solution that contains 3 eq of H_2SO_4 in 5 L of solution?

example 11.3

$$\frac{3 \text{ eq}}{5 \text{ L}} = 0.6 \ N$$

What is the normality of a calcium hydroxide $[Ca(OH)_2]$ solution that contains 7.4 g of $Ca(OH)_2$ in 2 L of solution?

example 11.4

The formula weight of $Ca(OH)_2$ is 74 g/mol. The equivalent weight is $74 \div 2 = 37$ g/eq. Thus, the number of equivalents of $Ca(OH)_2$ is

$$\frac{7.4 \text{ g}}{37 \text{ g/eq}} = 0.20 \text{ eq}$$

The concentration of the solution is

$$\frac{0.2 \text{ eq}}{2 \text{ L}} = 0.1 \ N$$

In Chapter 9, the terms **dilute** and **concentrated** were introduced to indicate relative amounts of solute for a given amount of solvent. For solutions of acids and bases, these same terms are often used in a more precise manner. For example, in a laboratory, a stock solution of "concentrated" hydrochloric acid is one that is 12 N. A stock bottle of hydrochloric acid labeled "dilute" contains a 6 N HCl solution. A concentrated sulfuric acid solution is 36 N H_2SO_4; dilute sulfuric acid is 6 N. Table 11.1 gives a summary of concentrations of common

TABLE 11.1 *Concentrations of Stock Acids and Bases*

Name	Formula	Normality Dilute	Normality Concentrated	Molarity Dilute	Molarity Concentrated
Hydrochloric acid	HCl	6	12	6	12
Nitric acid	HNO_3	6	16	6	16
Sulfuric acid	H_2SO_4	6	36	3	18
Phosphoric acid	H_3PO_4	—	45	—	15
Acetic acid	CH_3COOH	6	17	6	17
Ammonium hydroxide*	NH_4OH*	6	15	6	15
Sodium hydroxide	NaOH	6	—	6	—

* A better name, perhaps, is aqueous ammonia (NH_3); however, stock solutions are still labeled "ammonium hydroxide (NH_4OH)."

stock acids and bases. Notice that in the table all stock solutions labeled "dilute" are 6 N. In contrast, the normality of "concentrated" solutions is different for different solutes. In general, the concentrated solutions are the strongest solutions commercially available.

Table 11.1 also indicates that the normality of an acid equals the molarity times the number of replaceable hydrogens. If we know either the normality or the molarity of an acid, it is a simple matter to calculate the unknown concentration.

example 11.5

What is the normality of a 0.15 M H_2SO_4 solution?
The normality is $0.15 \times 2 = 0.30\ N$.

Sometimes it is necessary for a laboratory worker to make a more dilute solution from a more concentrated one. To do so, one can use the following relationship.

$$\begin{matrix}\text{Volume of the} \\ \text{concentrated} \\ \text{solution}\end{matrix} \times \begin{matrix}\text{concentration of the} \\ \text{concentrated} \\ \text{solution}\end{matrix} = \begin{matrix}\text{volume of the} \\ \text{dilute} \\ \text{solution}\end{matrix} \times \begin{matrix}\text{concentration} \\ \text{of the dilute} \\ \text{solution}\end{matrix}$$

Once again, it is easier to show the use of this equation through examples.

example 11.6

How much concentrated H_2SO_4 (36 N) would you use to make 500 mL of dilute H_2SO_4 (6.0 N)?

$$V_{conc} \times 36\ N = 500\text{ mL} \times 6.0\ N$$

$$V_{conc} = \frac{500\text{ mL} \times 6.0\ \cancel{N}}{36\ \cancel{N}} = 83\text{ mL}$$

example 11.7

How much 6.0 N HCl would you use to make 1.0 L of 0.50 N HCl?

$$V_{conc} \times 6.0\ N = 1.0\text{ L} \times 0.50\ N$$

$$V_{conc} = \frac{1.0\text{ L} \times 0.50\ \cancel{N}}{6.0\ \cancel{N}} = 0.083\text{ L, or 83 mL}$$

The relationship works just as well for solutions whose concentrations are given in molarity (or percent, for that matter).

How much 18 M H_2SO_4 would you need to make 100 mL of 3.0 M H_2SO_4?

example
11.8

$$V_{conc} \times 18\ M = 100\ \text{mL} \times 3.0\ M$$

$$V_{conc} = \frac{100\ \text{mL} \times 3.0 M}{18\ M} = 17\ \text{mL}$$

11.2 ACID-BASE TITRATIONS

Sometimes it is necessary to determine just how much acid (or base) there is in a solution of unknown concentration. This is often done by a process called **titration.** If the unknown is acidic, a certain volume of the acid is carefully measured from a buret into a flask (Figure 11.1). The buret is a piece of laboratory glassware designed to deliver known amounts of liquid into another container. An indicator dye (Section 11.3) is added to the measured volume of acid. Then a basic solution of known concentration (a **standard base**) is added from another buret, slowly and carefully, until finally just one additional drop of base changes the color of the indicator dye. This is the **end point** of the titration.

For the determination of the amount of hydroxide ion in a base of unknown concentration, the procedure is reversed. A quantity of the basic solution is measured from one buret into a flask. Then standard acid is added from the second buret until the indicator changes color. If everything is done

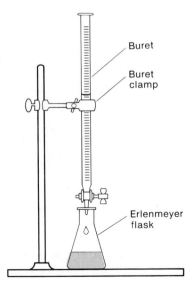

FIGURE 11.1
An apparatus for titration.
A sample of unknown acid is
measured into the flask. Base
is added from a buret until
an indicator changes color.

Buret

Buret
clamp

Erlenmeyer
flask

properly, the color change (end point) occurs when the number of equivalents of base are just equal to the number of equivalents of acid (the **equivalence point**).

$$\text{Number of equivalents of acid} = \text{number of equivalents of base}$$

Recall that normality (N) is equal to the number of equivalents per liter. Rearranging the equation, we get

$$\text{Number of equivalents} = \text{normality } (N) \times \text{volume (in liters)}$$

By substituting this definition of number of equivalents into the preceding equation, we obtain a new expression for the equivalence point.

$$\text{Normality of acid} \times \text{volume of acid} = \text{normality of base} \times \text{volume of base}$$

(Notice the similarity between this equation and that used in Section 11.1 for dilutions of concentrated solutions.) This equation is most useful for calculations involving titrations. We developed the equation for volumes expressed in liters, but we may use other units as long as they are the same for both acid and base. Burets are usually calibrated in milliliters, and most calculations involve this unit.

example 11.9

Calculate the normality of 20 mL of HCl, which is just neutralized by 28 mL of 0.10 N NaOH solution

$$N_{\text{HCl}} \times 20 \text{ mL} = 0.10 \ N \times 28 \text{ mL}$$

$$N_{\text{HCl}} = \frac{0.10 \ N \times 28 \text{ mL}}{20 \text{ mL}} = 0.14 \ N$$

example 11.10

What is the concentration of an H_2SO_4 solution if 30 mL of it requires 20 mL of 0.15 N NaOH for neutralization?

$$N_{\text{H}_2\text{SO}_4} \times 30 \text{ mL} = 0.15 \ N \times 20 \text{ mL}$$

$$N_{\text{H}_2\text{SO}_4} = \frac{0.15 \ N \times 20 \text{ mL}}{30 \text{ mL}} = 0.10 \ N$$

While the volume units may vary in problems such as these, it is important that concentrations be expressed as normality. Now, it is true that it will sometimes make no difference whether one uses normality or molarity. In Example 11.9, one would obtain the same (and correct) answer if molarity were used. But that's because molarity equals normality for both hydrochloric acid and sodium hydroxide. The molarity of sulfuric acid solutions, however, is half their normality. This difference would introduce an error if molarity were used in Example 11.10. In a titration, *equivalent* amounts of acid and base are present at the *equivalence* point. It takes only 0.5 mol of an acid (such as H_2SO_4) to supply a number of protons equivalent to the number of hydroxide ions released by 1 mol of a base (such as NaOH). By using normality, you are assured that this factor has been taken into account.

Titration is a powerful technique for the determination of the concentration of many kinds of chemicals, not just acids and bases. All that is required is a detectable change at the equivalence point. Titration can help you determine the concentration of oxidizing agents and reducing agents, the amount of silver in an ore sample, or even the activity of an enzyme.

11.3 THE pH SCALE

When we think of water, we think of H_2O molecules. But even the purest water isn't all H_2O. About 1 molecule in 500 million transfers a proton to another, giving a hydronium ion and a hydroxide ion.

$$H_2O + H_2O \rightleftharpoons H_3O^+ + OH^-$$

In other words, water is in equilibrium with hydronium ion and hydroxide ion, although the equilibrium lies far to the left. The concentration of hydronium ion in pure water is 0.000 000 1, or 1×10^{-7}, mol/L (*M*). The concentration of hydroxide ion is also 1×10^{-7} *M*. Pure water, then, is a neutral solution, having an *excess* of neither hydronium ions nor hydroxide ions.

The product of the hydronium ion concentration and the hydroxide ion concentration is

$$(1 \times 10^{-7})(1 \times 10^{-7}) = 1 \times 10^{-14}$$

This product, called the **ion product** of water, is sometimes represented by the symbol K_w. K_w is a constant. If you add acid to pure water, thereby increasing the concentration of hydronium ion, the concentration of hydroxide ion will fall until the product of the concentrations of the two ions equals 1×10^{-14}. Adding base to pure water will result in a similar adjustment in ion concentrations. These facts can be summarized in a single equation.

$$K_w = [H_3O^+][OH^-] = 1 \times 10^{-14}$$

The bracketed symbols are a shorthand for the concentration of the enclosed ions in moles per liter. Thus $[H_3O^+]$ should be read as the molarity of hydronium ion. The equation is valid for all aqueous solutions. And since it relates the hydronium and hydroxide ion concentrations, if the concentration of one ion is known, the concentration of the other can be calculated.

example
11.11

Lemon juice has a $[H_3O^+]$ of 0.01 M. What is the $[OH^-]$?

In exponential form, 0.01 M is written 1×10^{-2} M. Since $[H_3O^+]$ $[OH^-] = 1 \times 10^{-14}$,

$$[OH^-] = \frac{1 \times 10^{-14}}{1 \times 10^{-2}}$$

To divide exponential numbers, one has only to subtract the exponent in the denominator from that in the numerator, so $-14 - (-2) = -14 + 2 = -12$.

$$[OH^-] = 1 \times 10^{-12} \ M$$

example
11.12

A sample of bile has a $[OH^-]$ of 1×10^{-6} M. What is the $[H_3O^+]$?

$$[H_3O^+] = \frac{1 \times 10^{-14}}{1 \times 10^{-6}} = 1 \times 10^{-8} \ M$$

It is inconvenient to use exponential numbers to express the often minute concentrations of hydronium and hydroxide ion. In 1909, S. P. L. Sørensen, working in the Carlsberg Laboratory in Denmark, proposed the convenient pH scale, where

$$pH = \log \frac{1}{[H_3O^+]} = -\log [H_3O^+]$$

The reaction of most people on first encountering this definition of pH is "This is more convenient?" Well, it really is. To determine the pH of a solution, you need only take the exponent of the hydronium ion concentration and reverse its sign. The pH of a solution whose hydronium ion concentration is 1×10^{-4} M is 4. The point is this: it is easier to say, "The pH is 4," than to say, "The hydronium ion concentration is 1×10^{-4} M." They mean exactly the same thing.

example
11.13

What is the pH of the bile sample in Example 11.12?

$$pH = -\log [H_3O^+]$$
$$= -\log (1 \times 10^{-8})$$

To take the logarithm of 1×10^{-8}, we need to know that

$$\log (a \times b) = \log a + \log b$$

So we can write

$$\log (1 \times 10^{-8}) = \log 1 + \log 10^{-8}$$

The logarithm of 1 is 0 ($10^0 = 1$) and the logarithm of 10^{-8} is -8. So the problem can be written

$$\begin{aligned} pH &= -\log 1 - \log 10^{-8} \\ &= 0 - (-8) \\ &= 8 \end{aligned}$$

example 11.14

Calculate the pH of a solution that has a hydronium ion concentration of 1×10^{-3}.

$$\begin{aligned} pH &= -\log [H_3O^+] \\ &= -\log (1 \times 10^{-3}) \\ &= -\log 1 - \log 10^{-3} \\ &= 0 - (-3) \\ &= 3 \end{aligned}$$

example 11.15

What is the pH of a solution that has a hydronium ion concentration of 4.5×10^{-3}?

We need to take the logarithm of 4.5×10^{-3}.

$$\log (4.5 \times 10^{-3}) = \log 4.5 + \log 10^{-3}$$

Using a scientific calculator with a base 10 log key, we find that the logarithm of 4.5, to two significant figures, is 0.65. Thus we have

$$\begin{aligned} pH &= -\log [H_3O^+] \\ &= -\log (4.5 \times 10^{-3}) \\ &= -\log 4.5 - \log 10^{-3} \\ &= -0.65 - (-3) \\ &= 2.35 \end{aligned}$$

example 11.16

Calculate the pH of a solution that has a hydronium ion concentration of 2.0×10^{-8}.

$$\begin{aligned} pH &= -\log [H_3O^+] \\ &= -\log (2.0 \times 10^{-8}) \\ &= -\log 2.0 - \log 10^{-8} \\ &= -0.30 - (-8) \\ &= 7.70 \end{aligned}$$

TABLE 11.2 *The Relationship Between pH and $[H_3O^+]$ and Between pOH and $[OH^-]$ (at 20°C)*

$[H_3O^+]$	pH	$[OH^-]$	pOH	
1×10^0	0	1×10^{-14}	14	↑
1×10^{-1}	1	1×10^{-13}	13	
1×10^{-2}	2	1×10^{-12}	12	
1×10^{-3}	3	1×10^{-11}	11	Acidic solutions
1×10^{-4}	4	1×10^{-10}	10	
1×10^{-5}	5	1×10^{-9}	9	
1×10^{-6}	6	1×10^{-8}	8	↓
1×10^{-7}	7	1×10^{-7}	7	Neutral solution
1×10^{-8}	8	1×10^{-6}	6	↑
1×10^{-9}	9	1×10^{-5}	5	
1×10^{-10}	10	1×10^{-4}	4	
1×10^{-11}	11	1×10^{-3}	3	Basic solutions
1×10^{-12}	12	1×10^{-2}	2	
1×10^{-13}	13	1×10^{-1}	1	
1×10^{-14}	14	1×10^0	0	↓

The pH scale has been universally adopted. The relationship between pH and $[H_3O^+]$ is given in Table 11.2. Note from the table that

$$pH + pOH = 14$$

where

$$pOH = -\log [OH^-]$$

A pH of 7 represents a neutral solution. A pH of less than 7 represents an acidic solution, and one of more than 7 represents a basic solution. The lower the pH, the more acidic the solution; the higher the pH, the more basic the solution (Figure 11.2).

The cells in our bodies are bathed in solutions of fairly constant pH. Indeed, even small changes in pH may be fatal. Many chemical reactions, especially those in living cells, are quite sensitive to pH. The control of pH, by compounds called buffers, is described in Sections 11.5 and 11.6.

A variety of dyes, some synthetic and some from plant and animal sources, have the property of changing color over a range of pH values. We noted earlier that litmus, a vegetable dye, turns red in acidic solutions and blue in basic solutions. Actually the change occurs over a range of pH from 4.5 to 8.3. Certain combinations of indicator dyes will exhibit a whole range of colors as the pH changes from strongly basic to strongly acidic (Figure 11.3). By selecting the proper indicator, we can determine the pH of almost any clear, colorless aqueous solution.

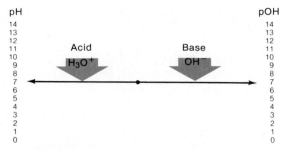

(a) When acid and base are in balance, the solution is neutral.

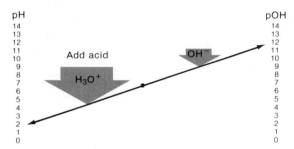

(b) When excess acid is added, pH decreases, and pOH increases.

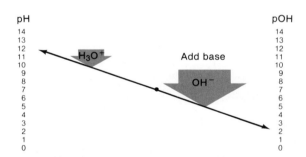

(c) When excess base is added, pH increases, and pOH decreases.

FIGURE 11.2   Effect of adding acid or base on the pH and pOH of a solution.

(a)

(b)

FIGURE 11.3
Two common pH papers.
(a) Wide range. (b) Narrow range. (Courtesy of Micro Essential Laboratory, Inc., Brooklyn, N.Y.)

More accurate measurement of pH can be made electrically with pH meters. Generally, these instruments can measure pH to a precision of about 0.01 pH unit. Also, colors and turbidity generally do not interfere with electrical measurement of pH as they do with indicator color changes. Thus, pH meters can be used on blood, urine, and other complex mixtures without prior treatment of the specimen. A typical pH meter is illustrated in Figure 11.4.

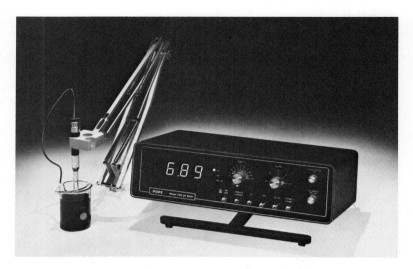

FIGURE 11.4   *A pH meter. (Courtesy of Pope Scientific, Inc., Menomonee Falls, Wis.)*

Keep in mind that the pH scale is logarithmic; for each unit change in pH, there is a 10-fold change in $[H_3O^+]$. Acid rain with a pH of 4 is ten times as acidic as rain with a pH of 5.

Table 11.3 gives the approximate pH of some solutions.

TABLE 11.3   *The Approximate pH of Some Solutions*

Solution	pH	
0.1 M HCl	1.0	↑
Gastric juices	1.6–1.8	
Lemon juice	2.3	
Vinegar	2.4–3.4	
Soft drinks	2.0–4.0	Acidic
Milk	6.3–6.6	
Urine	5.5–7.5	
Rainwater (unpolluted)	5.6	
Saliva	6.2–7.4	
Pure water	7.0	Neutral
Blood	7.35–7.45	↑
Egg white (fresh)	7.6–8.0	
Bile	7.8–8.6	Basic
Milk of magnesia	10.5	
Washing soda	12	
0.1 M NaOH (lye)	13	↓

11.4 SALTS IN WATER: ACIDIC, BASIC, OR NEUTRAL?

When an acid reacts with a base, the products are a salt and water. The process is called neutralization, but is the solution neutral? What if we simply take a salt and dissolve it in water? Would the solution be acidic, basic, or neutral? Although we might well expect a salt solution to be neutral, not all of them are. There are many neutral salts, including the familiar sodium chloride and others such as potassium sulfate and sodium nitrate. Some, such as ammonium nitrate and aluminum chloride, are acidic. Others, such as sodium acetate and potassium cyanide, are basic.

How do we tell if a salt is acidic, basic, or neutral? Simple. Just dissolve some of the salt in water and test the solution with indicator paper or with a pH meter. That is the experimental way, and who can argue with it? There is another way, though. We can *predict* whether a solution of a salt will be acidic, basic, or neutral by considering the relative strengths of the acid and base from which the salt was made. Just think how much more convenient it is to apply a rule than to go into a laboratory and do an experiment. The rules are as follows:

1. The salt of a strong acid and a strong base forms a neutral solution.
2. The salt of a strong acid and a weak base forms an acidic solution.
3. The salt of a weak acid and a strong base forms a basic solution.
4. The salt of a weak acid and a weak base may form an acidic, a basic, or (by chance) a neutral solution.

To apply these rules you must be able to recognize strong acids, strong bases, weak acids, and weak bases. After reviewing Sections 10.2 and 10.4, look at the formula of the salt. Pair the positive ion with the number of hydroxide ions equivalent to the positive charge to get the base from which the salt could be formed. Then put one or more protons (H^+) with the negative ion of the salt to get the acid.

From what acid and what base is the salt NaCl formed? Does NaCl form an acidic, basic, or neutral solution?

The positive ion in NaCl is Na^+. Na^+ requires one OH^-; the base is NaOH. The negative ion in NaCl is Cl^-. Cl^- requires one H^+; the acid is HCl. Since NaOH is a strong base and HCl is a strong acid, then, according to Rule 1, NaCl is a neutral salt.

example 11.17

Is a solution of $(NH_4)_2SO_4$ acidic, basic, or neutral?

The base is NH_4OH (really aqueous NH_3), and the acid is H_2SO_4. Since aqueous NH_3 is a weak base and H_2SO_4 is a strong acid, the solution is acidic (Rule 2).

example 11.18

example
11.19

Is a solution of CH_3COOK acidic, basic, or neutral?
    The base, KOH, is a strong one. Acetic acid, CH_3COOH, is a weak acid. The solution is basic (Rule 3).

example
11.20

Is a solution of CH_3COONH_4 acidic, basic, or neutral?
    The base, NH_4OH (aqueous NH_3), is a weak one. The acid, CH_3COOH, is also weak. There is no way to tell from the rules whether the solution would be acidic, basic, or neutral (Rule 4).

Why would a salt be acidic or basic? Let's consider the case of a particular salt, sodium acetate. Solutions of this salt contain sodium ions and acetate ions, each type surrounded by water molecules. Now, the acetate (CH_3COO^-) ion is the conjugate base of acetic acid (Section 10.9). And acetic acid is a weak acid, one that does not like to give up a proton. The acetate ion looks around, sees all those lovely water molecules with their hydrogens, and grabs a proton from one.

$$CH_3COO^- + H_2O \longrightarrow CH_3COOH + OH^-$$

This makes the acetic acid molecule fairly content because it has gained back the proton it hates to lose. Unfortunately, the water molecule that has just given up the proton is not at all happy. Water is also a weak acid, even weaker than acetic acid, and it will fight effectively to keep the proton and usually will win.

$$CH_3COO^- + H_2O \rightleftharpoons CH_3COOH + OH^-$$

All this means that there is an equilibrium established in which the two acids, water and acetic acid, compete for the proton, and the equilibrium is shifted strongly to the left. Nonetheless, an equilibrium exists, and some undissociated acetic acid and some excess hydroxide ions are present in the solution. Those hydroxide ions are what makes the solution basic.

It is important to note that no analogous equilibrium exists that results in extra hydronium ions to balance the hydroxide ions. The sodium ions, the other half of the salt in our example, have no tendency to react with water to produce extra hydronium ions. We could write an equation to suggest the possibility.

$$Na^+ + 2\,H_2O \xrightarrow{\quad\times\quad} NaOH + H_3O^+$$

But sodium hydroxide is a strong base and is completely ionized, so to be correct we would have to write

$$Na^+ + 2\,H_2O \longrightarrow Na^+ + OH^- + H_3O^+$$

Since sodium ion appears on both sides of the equation, this reaction would boil down to

$$2\,H_2O \longrightarrow H_3O^+ + OH^-$$

which is simply a restatement of the fact that water itself ionizes slightly. The slight ionization of water gives equal amounts of hydronium ion and hydroxide ion, whereas the equilibrium involving the acetate ion (p. 318) gives a small amount of excess hydroxide ion. The salt of a strong base and a weak acid gives a slightly basic solution.

For a solution of the salt of a weak base and a strong acid, precisely the opposite happens. Thus, for ammonium chloride (NH_4Cl), the ammonium ion exists in equilibrium with its conjugate base.

$$NH_4^+ + H_2O \rightleftharpoons H_3O^+ + NH_3$$

Ammonia, although a weak base, is a stronger base than water, so the equilibrium is shifted to the left. However, some excess hydronium ion is present in the solution because of this equilibrium. The chloride ion from ammonium chloride has no tendency to pick up a proton from water to form undissociated hydrochloric acid and hydroxide ion, because hydrochloric acid is a strong acid and chloride ion simply won't accept the proton. The solution of a salt of a strong acid and a weak base contains excess hydronium ion and is slightly acidic.

In a solution of sodium chloride, neither ion has a tendency to enter equilibrium with water molecules, and neither excess hydroxide ion nor hydronium ion is present. The solution is neutral.

Finally, both ions from a salt of a weak acid and a weak base enter equilibria. The solution of such a salt can be acidic, basic, or neutral, depending on which equilibrium predominates, that is, how much to the left or right each equilibrium is shifted.

11.5 BUFFERS: CONTROL OF pH

Who cares about the pH of a salt solution anyway? You do, that's who, for some of these salts play a vital role in the control of the pH of body fluids. If they should fail to function, so will you. Our bodies are acid factories. Our stomachs produce hydrochloric acid. Our muscles produce lactic acid. Starches and sugars produce pyruvic acid when metabolized. Carbon dioxide from respiration produces carbonic acid in the blood. Our bodies must eliminate or neutralize these acids, because excess acidity in the wrong place would kill us rather quickly.

A **buffer solution** is one in which the pH remains relatively constant even if acid or base is added. Chemically, a buffer solution is one that contains a weak acid and one of its salts (or a weak base and one of its salts), usually in approximately equal concentrations. For example, 1 L of a solution that contains 0.1 mol of acetic acid (CH_3COOH) and 0.1 mol of sodium acetate (CH_3COONa) acts as a buffer at pH 4.74, an acidic value.* This buffer can absorb significant amounts of additional acid or base without appreciable change in pH. For instance, the addition of 10 mL of 0.1 M NaOH to this buffer causes a pH change of only 0.01 pH unit. If that amount of base were added to 1 L of pure water, the pH would change by 4 full pH units. The buffer system is what prevents such a large change in pH.

How does a buffer work? It may seem strange that a solution can absorb acid or base without the pH changing appreciably. The explanation, however, is fairly simple. It follows Le Chatelier's principle (Section 5.13). The buffer solution has a large reservoir of both acid molecules and the conjugate base of the acid (anions from the salt). If a strong acid is added, the hydronium ions from the added acid will donate protons to the anions of the buffer to form the weak acid and water.

$$H_3O^+ + CH_3COO^- \; \rightleftharpoons \; CH_3COOH + H_2O$$

Although the reaction is reversible to a slight extent, most of the protons are removed from the solution as they are added, and the pH changes hardly at all.

When a strong base is added, the hydroxide ions will react with the hydronium ions formed in the solution by the acetic acid of the buffer.

$$OH^- + H_3O^+ \; \longrightarrow \; 2\,H_2O$$

The added hydroxide ions are tied up, and the hydronium ions removed from the solution are immediately replaced by further ionization of the acetic acid in the buffer.

$$CH_3COOH + H_2O \; \rightleftharpoons \; CH_3COO^- + H_3O^+$$

The concentration of hydronium ions returns to approximately the original value, and the pH is only slightly changed.

There are many important buffer solutions. Most biochemical reactions, whether they occur in a laboratory or in our bodies, are carried out in buffered solutions. The buffers that control the pH of our blood will be discussed in the next section. Table 11.4 lists some buffers of interest and the pH range in which they operate.

---

* This may be a bit confusing since we have just spent some time considering the fact that a solution of sodium acetate is slightly basic (i.e., has a pH greater than 7). However, it is a solution containing *only* the salt that is basic. The buffer solution consists not only of a salt of acetic acid but also of some acetic acid itself. It is the presence of this acid that makes this buffer solution acidic.

TABLE 11.4 *Some Important Buffers*

Buffer Components	Buffer System Names	pH*
$CH_3CHOHCOOH/CH_3CHOHCOO^-$	Lactic acid/lactate ion	3.86
CH_3COOH/CH_3COO^-	Acetic acid/acetate ion	4.76
$H_2PO_4^-/HPO_4^{2-}$	Dihydrogen phosphate ion/ monohydrogen phosphate ion	7.20
H_2CO_3/HCO_3^-	(Carbon dioxide) carbonic acid/ bicarbonate ion	6.46†
NH_4^+/NH_3	Ammonium ion/ammonia	9.25

* The values listed are for solutions that are 0.1 M in each compound at 25 °C.
† This value includes dissolved CO_2 molecules as undissociated H_2CO_3. The value for H_2CO_3 alone is about 3.8.

11.6 BUFFERS IN BLOOD

The pH of the blood of higher animals is held remarkably constant. In humans, blood plasma normally varies from 7.35 to 7.45 in pH. Should the pH rise above 7.8 or fall below 6.8, due to starvation or disease, the person may suffer irreversible damage to the brain or even die. Fortunately, human blood has not one, but at least three, buffering systems. Of these, the bicarbonate/carbonic acid (HCO_3^-/H_2CO_3) buffering system is the most important.

If acids are put into the blood, hydronium ions are taken up by the bicarbonate ions to form undissociated carbonic acid and water.

$$HCO_3^- + H_3O^+ \longrightarrow H_2CO_3 + H_2O$$

As long as there is sufficient bicarbonate to take up the added acid, the pH will change little.

The carbonic acid is a weak acid and exists in equilibrium with hydronium and bicarbonate ions.

$$H_2CO_3 + H_2O \rightleftharpoons H_3O^+ + HCO_3^-$$

If bases come into the bloodstream, reacting with hydronium ions to form water, more carbonic acid molecules will ionize to replace the removed hydronium ions. Further, as the carbonic acid molecules are used up, more carbonic acid can be formed from the large reservoir of dissolved carbon dioxide in the blood.

$$CO_2 + H_2O \rightleftharpoons H_2CO_3$$

Thus, bicarbonate/carbonic acid buffers the blood against either added base or added acid.

Another blood buffer is the dihydrogen phosphate/monohydrogen phosphate ($H_2PO_4^-$/HPO_4^{2-}) system. Any acid reacts with monohydrogen phosphate to form dihydrogen phosphate.

$$HPO_4^{2-} + H_3O^+ \longrightarrow H_2PO_4^- + H_2O$$

The dihydrogen phosphate is a weak acid and exists in equilibrium with hydronium ions and monohydrogen phosphate.

$$H_2PO_4^- + H_2O \rightleftharpoons H_3O^+ + HPO_4^{2-}$$

Any base that comes into the blood would react with hydronium ions to form water. However, more dihydrogen phosphate would ionize to replace these hydronium ions, leaving the pH essentially unchanged.

Proteins act as a third type of blood buffer. These complex molecules (Chapter 22) contain —COO^- groups, which, like acetate ions (CH_3COO^-), can act as proton acceptors. Proteins also contain —NH_3^+ groups, which, like ammonium ions (NH_4^+), can donate protons. If acid comes into the blood, hydronium ions can be neutralized by the —COO^- groups.

$$\overset{|}{C}OO^- + H_3O^+ \longrightarrow \overset{|}{C}OOH + H_2O$$

If base is added, it can be neutralized by the —NH_3^+ groups.

$$\overset{|}{N}H_3^+ + OH^- \longrightarrow \overset{|}{N}H_2 + H_2O$$

These three buffers (and perhaps others) act to keep the pH of the blood constant. Buffers can be overridden by large amounts of acid or base; their capacity is not infinite. The blood buffers can be overwhelmed if the body's metabolism goes badly awry.

11.7 ACIDOSIS AND ALKALOSIS

Have you ever had your muscles hurt after prolonged physical activity? If so, you have had your blood buffers somewhat overloaded. Muscle contraction produces lactic acid. This acid ionizes somewhat more strongly than carbonic acid and thus tends to lower the pH of the blood (it tends to release more hydronium ions into the blood). Moderate amounts of lactic acid can be handled by the blood buffers. For bicarbonate, the reactions would be

$$\overset{\overset{\textstyle OH}{|}}{CH_3CHCOOH} + HCO_3^- \longrightarrow \overset{\overset{\textstyle OH}{|}}{CH_3CHCOO^-} + H_2CO_3$$

Lactic acid   Lactate ion

Excessive amounts of lactic acid overload the buffers, however, and the pH is lowered. Nerve cells respond to the increased acidity by sending a message of pain to the brain.

If the pH of the blood falls below 7.35, the condition is called **acidosis.** If the pH of the blood rises above 7.45, **alkalosis** sets in. These pathological conditions can be caused by faulty respiration or by metabolic problems. In severe cases of starvation, the body gets its energy by the oxidation of stored fats. The products of fat metabolism are acidic, and prolonged starvation leads to acidosis. Fad diets, such as those that severely limit the intake of carbohydrates, can also lead to acidosis.

The body's excretory system tries to compensate for acidosis or alkalosis by selectively excreting certain compounds. Conversely, the conditions can be brought on by kidney failure or other excretory problems. We discuss these pathological problems further in later chapters.

PROBLEMS

1. Define or illustrate the following terms:
   a. indicator
   b. end point
   c. equivalence point
   d. K_w
   e. pH
   f. pOH
   g. equivalent
   h. titration
   i. buffer
   j. acidosis
   k. alkalosis

2. How many equivalents are there in
   a. 1 mol H_2SO_4?   b. 1 mol $Al(OH)_3$?

3. How many equivalents are there in
   a. 2 mol H_3PO_4?
   b. 0.5 mol NH_3?
   c. 10 mol CH_3COOH?

4. How many moles are required to provide 1 eq of each of the following?
   a. HNO_3 b. $Mg(OH)_2$

5. How many moles are required to provide the indicated number of equivalents of each of the following?
   a. 6 eq H_3PO_4 c. 1.5 eq KOH
   b. 0.2 eq H_2SO_4

6. How many equivalents are there in each of the following?
   a. 162 g HBr   b. 6.0 g LiOH   c. 3.4 g H_2S

7. How many equivalents are there in each of the following?
   a. 1.71 g $Ba(OH)_2$ c. 15 g CH_3COOH
   b. 49 g H_3PO_4

8. Would 1 mol of NaOH neutralize
   a. 1 mol HNO_3?   c. 3 mol H_3PO_4?
   b. 1 mol CH_3COOH?   d. 0.5 mol H_2SO_4?

9. Would 0.50 eq of $Ca(OH)_2$ neutralize all of the acid in
   a. 0.50 eq of H_3PO_4?   c. 0.25 eq H_3BO_3?
   b. 1.0 eq of HCl?

10. Calculate the formula weight and equivalent weight of each of the following compounds:
    a. HBr
    b. LiOH
    c. H_2S
    d. $Ba(OH)_2$
    e. H_3PO_3
    f. H_2SiO_3
    g. CH_3COOH

11. Calculate the normality of each of these solutions.
    a. 3 equivalents of H_3AsO_4 in 6 L of solution
    b. 20 g of NaOH in 0.5 L of solution
    c. 9.8 g of H_2SO_4 in 0.2 L of solution
    d. 342 g of $Ba(OH)_2$ in 10 L of solution
    e. 12 g of CH_3COOH in 100 mL of solution

12. What is the normality of a 3.0 *M* solution of $Ba(OH)_2$?

13. What volume of concentrated (36 *N*) sulfuric acid would be required for the following to be made?
    a. 12 L of 6.0 *N* solution?
    b. 500 mL of 0.10 *N* solution?

14. You have a stock solution of 12 *M* HCl. How would you prepare the following solutions?
    a. 100 mL of 1.2 *M* HCl solution
    b. 5.0 L of 6.0 *M* HCl solution
    c. 1.50 L of 1.00 *M* HCl solution

15. In an ideal titration, the end point occurs at the equivalence point. What is the difference between an end point and an equivalence point?

16. Write an equation for the equilibrium established when CH_3COO^- is placed in water.

17. Write an equation for the equilibrium established when NH_4^+ is placed in water.

18. Classify the aqueous solution of each of these salts as acidic, basic, or neutral.
    a. KCl   c. NH_4CN e. $(NH_4)_2SO_4$
    b. NaCN   d. CH_3COOK f. Na_2SO_4

19. A weak acid is titrated with a strong base. Would the solution at the equivalence point be acidic, basic, or neutral? Explain.

20. Calculate the normality of an HCl solution if 20 mL of it requires the following for neutralization.
    a. 40 mL of 0.25 N NaOH
    b. 10 mL of 0.50 N KOH

21. A 20-mL sample of gastric fluid is neutralized by 25 mL of 0.10 N NaOH. What is the normality of HCl in the fluid? Assume that all the acidity of the gastric fluid is due to HCl.

22. If 25.0 mL of a sulfuric acid solution requires 15.0 mL of 0.500 N sodium hydroxide to neutralize it, what is the normality of the H_2SO_4 solution?

23. A 40.0 mL sample of a 0.150 N solution of KOH is titrated with 0.600 N HCl solution. What volume of the HCl solution is required to reach the end point?

24. A 50.0 mL sample of an acid is titrated with 0.250 N NaOH. It takes 12.5 mL of the base to reach the end point. What is the normality of the acid?

25. A 35.0 mL sample of a base is titrated with 0.100 N HCl. It takes 42.5 mL of the acid just to neutralize the base. What is the normality of the base?

26. When the stomach isn't being stimulated by food to make more, it produces 0.0023 eq of HCl and 30 mL to 60 mL of total juices per hour. What range of concentrations of HCl in the stomach does this represent?

27. What is the pH of each of the following solutions?
    a. 1×10^{-2} M HCl   b. 1×10^{-3} M HNO_3

28. What is the pH of each of the following solutions?
    a. 0.0001 M HCl   c. 0.1 M HBr
    b. 0.00001 M HNO_3

29. What is the pOH of each of the following solutions?
    a. 1×10^{-1} M NaOH   b. 1×10^{-5} M KOH

30. What is the pOH of each of the following solutions?
    a. 0.001 M NaOH   b. 0.01 M KOH

31. What is the pOH of each of the solutions in Problems 27 and 28?

32. What is the pH of each of the solutions in Problems 29 and 30?

33. Indicate whether each of the following pH values represents an acidic, basic, or a neutral solution:
    a. 11   c. 7
    b. 4   d. 3.4

34. Answer Problem 33 assuming that the values given are for pOH.

35. Use acetic acid and acetate ion to explain how a buffer controls pH.

36. Use ammonia and ammonium chloride to explain how a buffer controls pH.

37. Name three buffer systems operating in the blood.

38. What groups on protein buffers react with added acid and base?

39. If acid is added to an unbuffered solution, will the pH increase or decrease?

40. If someone is suffering from alkalosis, is the blood pH too high or too low?

Note: The following problems require the use of a table of logarithms or a scientific calculator with a base 10 logarithm function.

41. Calculate the pH of solutions with the following molar hydronium ion concentrations:
    a. 3×10^{-3} b. 5×10^{-7} c. 8×10^{-10}

42. Calculate the pH of a 3.6×10^{-4} M HCl solution.

43. Calculate the pH of a 8.8×10^{-2} M HNO_3 solution.

44. Calculate the pH of a 3.0×10^{-3} M NaOH solution.

45. Calculate the pH of a blood solution that has a hydronium ion concentration of 4.6×10^{-8} M.

46. Calculate the pH of a urine sample that has a hydronium ion concentration of 2.3×10^{-6} M.

47. Calculate the pH of an ammonia solution that has a hydronium ion concentration of 2.0×10^{-12} M.

48. Calculate the pH of a sample of gastric juice that has a hydronium ion concentration of 0.12 M.

D

Equilibrium Calculations

I n Chapters 10 and 11, we described the equilibria established when a weak
acid, a weak base, a salt of either, or some combination of these is dissolved in water. Our discussion was entirely qualitative. It is possible to
treat these equilibria in a more quantitative fashion. That we will now do, in
this special topic, for those who are interested or whose work might require it.

D.1 EQUILIBRIUM CONSTANT EXPRESSIONS

The proportions of reactants and products at equilibrium are determined by a
simple relationship. Let's write a generalized reaction,

$$a\,A + b\,B \rightleftharpoons c\,C + d\,D$$

where A and B are reactants and C and D are products. The small letters are
the coefficients for the substances. The relationship at a given temperature is
given by the expression

$$K = \frac{[C]^c \times [D]^d}{[A]^a \times [B]^b}$$

The quantities within brackets stand for molar concentration at equilibrium.
The quantity **K** is called the **equilibrium constant.** The entire expression is
called the **equilibrium constant expression.** The coefficients in the generalized
reaction become exponents in the equilibrium constant expression.

Let's consider a specific reaction.

$$H_2 + Cl_2 \rightleftharpoons 2\,HCl$$

The equilibrium constant expression for this reaction is

$$K = \frac{[HCl]^2}{[H_2] \times [Cl_2]}$$

For the reaction

$$N_2 + 3\,H_2 \rightleftharpoons 2\,NH_3$$

The equilibrium expression is

$$K = \frac{[NH_3]^2}{[N_2] \times [H_2]^3}$$

**example
D.1**

Write the equilibrium constant expression for each of the following reactions:

$$\text{a. } H_2 + F_2 \rightleftharpoons 2\,HF$$
$$\text{b. } 2\,NO + O_2 \rightleftharpoons 2\,NO_2$$
$$\text{c. } CO + H_2O \rightleftharpoons CO_2 + H_2$$

The expressions are:

$$\text{a. } K = \frac{[HF]^2}{[H_2] \times [F_2]} \qquad \text{b. } K = \frac{[NO_2]^2}{[NO]^2 \times [O_2]} \qquad \text{c. } K = \frac{[CO_2] \times [H_2]}{[CO] \times [H_2O]}$$

All of the above examples involve gaseous reactants and products. The same treatment is applied to substances in aqueous solution in the next section. Special cases involving a solid reactant or product are considered in Chapter 12.

D.2 THE IONIZATION OF WEAK ACIDS

Now let's take another look at the ionization of acetic acid. The reaction is

$$CH_3COOH + H_2O \rightleftharpoons H_3O^+ + CH_3COO^-$$

At equilibrium, the equilibrium constant expression is

$$K = \frac{[H_3O^+] \times [CH_3COO^-]}{[CH_3COOH] \times [H_2O]}$$

For dilute solutions, the molar concentration of water, H_2O, stays nearly constant at 55 M*. We can substitute this value into the equation:

$$K = \frac{[H_3O^+] \times [CH_3COO^-]}{[CH_3COOH] \times (55)}$$

Rearranging, we get

$$K \times 55 = \frac{[H_3O^+] \times [CH_3COO^-]}{[CH_3COOH]} = K_a$$

The product of the two constants, $K \times 55$, is a constant. This new constant, K_a, is called the **ionization constant** for the weak acid.

Sometimes the involvement of water is omitted, and the ionization of acetic acid is written

$$CH_3COOH \xleftrightarrow{\hspace{1cm}} H^+ + CH_3COO^-$$

The ionization constant becomes

$$K_a = \frac{[H^+] \times [CH_3COO^-]}{[CH_3COOH]}$$

If you recognize that the hydronium ion is simply a hydrated proton, you will see that the two K_a expressions are the same.

The strength of an acid is related to the degree of ionization. The greater the degree of ionization, the stronger the acid. (Strong acids, such as HCl, are 100% ionized in dilute solution.) Also, the greater the degree of ionization, the larger the K_a. Table D.1 lists the K_a values for several common acids. Don't forget: The larger the K_a (less negative exponent), the stronger the acid.

Ionization constants can be calculated from measurements of the hydrogen ion concentration as shown in the following example.

Find the K_a for an organic acid, HOrg, if the hydrogen ion concentration of a 0.0200 M solution is 3.0×10^{-4} M. The simplified ionization reaction is

$$HOrg \xleftrightarrow{\hspace{1cm}} H^+ + Org^-$$

example D.2

---

* In 1 L of dilute aqueous solution there is approximately 1 L of water, which weighs 1000 g and contains

$$\frac{1000 \text{ g}}{18 \text{ g/mol}} = 55 \text{ mol of water}$$

Thus, the concentration of water in 1 L of dilute solution is approximately 55 M.

TABLE D.1 K_a Values for Several Acids in Water at 25°C

Acid	Simplified Reaction	K_a
Hydrochloric acid	$HCl \rightleftharpoons H^+ + Cl^-$	Very large
Sulfuric acid	$H_2SO_4 \rightleftharpoons H^+ + HSO_4^-$	Very large
	$HSO_4^- \rightleftharpoons H^+ + SO_4^{2-}$	1.2×10^{-2}
Hydrofluoric acid	$HF \rightleftharpoons H^+ + F^-$	6.6×10^{-4}
Hydrocyanic acid	$HCN \rightleftharpoons H^+ + CN^-$	6.2×10^{-10}
Formic acid	$HCOOH \rightleftharpoons H^+ + HCOO^-$	1.8×10^{-4}
Acetic acid	$CH_3COOH \rightleftharpoons H^+ + CH_3COO^-$	1.8×10^{-5}
Benzoic acid	$C_6H_5COOH \rightleftharpoons H^+ + C_6H_5COO^-$	6.6×10^{-5}
Carbonic acid	$H_2CO_3 \rightleftharpoons H^+ + HCO_3^-$	4.2×10^{-7}
	$HCO_3^- \rightleftharpoons H^+ + CO_3^{2-}$	4.8×10^{-11}
Phosphoric acid	$H_3PO_4 \rightleftharpoons H^+ + H_2PO_4^-$	7.5×10^{-3}
	$H_2PO_4^- \rightleftharpoons H^+ + HPO_4^{2-}$	6.2×10^{-8}
	$HPO_4^{2-} \rightleftharpoons H^+ + PO_4^{3-}$	4.4×10^{-13}

and the equilibrium expression is

$$K_a = \frac{[H^+][Org^-]}{[HOrg]}$$

Ionization of the acid produces equal concentrations of H^+ and Org^-.

$$[H^+] = [Org^-] = 3.0 \times 10^{-4}\ M = 0.00030\ M$$

At equilibrium, the concentration of nonionized acid equals its original concentration minus the amount that has ionized to form H^+ and Org^-.

$$[HOrg] = 0.0200 - 0.00030 = 0.0197\ M = 1.97 \times 10^{-2}\ M$$

$$K_a = \frac{(3.0 \times 10^{-4})(3.0 \times 10^{-4})}{1.97 \times 10^{-2}} = 4.6 \times 10^{-6}$$

Usually, though, we calculate hydrogen ion concentrations from K_a values found in tables. This approach is illustrated in the following examples.

example D.3

Calculate the $[H^+]$ in a 0.10 M solution of acetic acid. The equation is $CH_3COOH \rightleftharpoons H^+ + CH_3COO^-$.

First, write the equilibrium expression. (The K_a of acetic acid is given in Table D.1.)

$$K_a = \frac{[H^+][CH_3COO^-]}{[CH_3COOH]} = 1.8 \times 10^{-5}$$

For every CH_3COOH that ionizes, one H^+ and one CH_3COO^- are formed. We can therefore let $x = [H^+] = [CH_3COO^-]$.

Since each H^+ formed represents one CH_3COOH ionized, the equilibrium concentration of CH_3COOH will be $0.10 - x$. Substitution gives

$$\frac{(x)(x)}{(0.10 - x)} = 1.8 \times 10^{-5}$$

The amount of CH_3COOH ionized is very small compared with the total amount of CH_3COOH that we started with, so let's assume that

$$0.10 - x \cong 0.10 = 1.0 \times 10^{-1}$$

We then have

$$\frac{x^2}{1.0 \times 10^{-1}} = 1.8 \times 10^{-5}$$

$$x^2 = (1.8 \times 10^{-5})(1.0 \times 10^{-1}) = 1.8 \times 10^{-6}$$

$$x = 1.3 \times 10^{-3} \, M = [H^+]$$

Now let's check our assumption. Is 0.10 changed significantly by subtracting 0.0013?

$$0.10 - 0.0013 \overset{?}{=} 0.10$$

Certainly no significant error was introduced.

Calculate the $[H^+]$ in a 0.0010 M solution of HCN.

example D.4

$$HCN \rightleftharpoons H^+ + CN^-$$

$$K_a = \frac{[H^+][CN^-]}{[HCN]} = 6.2 \times 10^{-10}$$

Let $x = [H^+] = [CN^-]$

$$[HCN] = 0.0010 - x \cong 0.0010$$

Substituting:

$$\frac{(x)(x)}{1.0 \times 10^{-3}} = 6.2 \times 10^{-10}$$

$$x^2 = 6.2 \times 10^{-13} = 62 \times 10^{-14}$$

$$x = 7.9 \times 10^{-7} = [H^+]$$

Is x small compared with 0.0010? *Yes!*

Approximations such as those made in the preceding examples are good only if the degree of ionization is small. Such an assumption should always be checked.

D.3 EQUILIBRIA INVOLVING WEAK BASES

Equilibrium constant expressions can also be written for weak bases. For ammonia, the reaction is

$$NH_3 + H_2O \rightleftharpoons NH_4^+ + OH^-$$

and the equilibrium expression is

$$K = \frac{[NH_4^+][OH^-]}{[NH_3][H_2O]}$$

Since the concentration of water is quite constant for dilute solutions, we may include $[H_2O]$ with K to give a new constant, K_b.

$$K_b = \frac{[NH_4^+][OH^-]}{[NH_3]}$$

Ionization constants for several weak bases are given in Table D.2.

example D.5

Calculate the hydroxide ion concentration in a 0.010 *M* solution of aniline.

$$K_b = \frac{[C_6H_5NH_3^+][OH^-]}{[C_6H_5NH_2]} = 4.2 \times 10^{-10}$$

Let $x = [OH^-] = [C_6H_5NH_3^+]$

$$C_6H_5NH_2 = 0.010 - x \cong 0.010 = 1.0 \times 10^{-2}$$

TABLE D.2   *Ionization Constants for Some Weak Bases in Water at 25°C*

Base	Reaction	K_b
Ammonia	$NH_3 + H_2O \rightleftharpoons NH_4^+ + OH^-$	1.8×10^{-5}
Methylamine	$CH_3NH_2 + H_2O \rightleftharpoons CH_3NH_3^+ + OH^-$	4.2×10^{-4}
Aniline	$C_6H_5NH_2 + H_2O \rightleftharpoons C_6H_5NH_3^+ + OH^-$	4.2×10^{-10}
Phosphate ion	$PO_4^{3-} + H_2O \rightleftharpoons HPO_4^{2-} + OH^-$	4.5×10^{-2}
Carbonate ion	$CO_3^{2-} + H_2O \rightleftharpoons HCO_3^- + OH^-$	1.8×10^{-4}
Bicarbonate ion	$HCO_3^- + H_2O \rightleftharpoons H_2CO_3 + OH^-$	2.3×10^{-8}
Zinc hydroxide	$Zn(OH)_2 \rightleftharpoons ZnOH^+ + OH^-$	9.6×10^{-4}
Silver hydroxide	$AgOH \rightleftharpoons Ag^+ + OH^-$	1.1×10^{-4}

Substituting, we have

$$\frac{(x)(x)}{1.0 \times 10^{-2}} = 4.2 \times 10^{-10}$$

$$x^2 = 4.2 \times 10^{-12}$$

$$x = 2.0 \times 10^{-6} \, M = [OH^-]$$

D.4 CALCULATIONS INVOLVING BUFFERS

Let's take another look at solutions that contain both a weak acid and a salt of that acid. We saw in Section 11.5 that such a solution acts as a buffer to resist a change in pH. Now we can calculate the hydrogen ion concentration in such a system.

If sodium acetate is added to a solution of acetic acid, ionization of the acid is considerably decreased. In effect, we are increasing the concentration of acetate ion in the system.

$$CH_3COOH \rightleftharpoons H^+ + CH_3COO^-$$

The acetate ions from sodium acetate react with the H^+, forming more un-ionized acetic acid molecules. A greater proportion of the total acetic acid is in the molecular form; ionization is decreased. This is an example of the **common ion effect**. If an ion common to those in the equilibrium is added, the degree of ionization is decreased. A new equilibrium is established with more acetate ions but fewer H^+. The value of K_a remains unchanged.

What is the $[H^+]$ in a solution which is 0.10 M in acetic acid and 0.10 M in sodium acetate?

example D.6

Use the K_a expression for acetic acid.

$$K_a = \frac{[H^+][CH_3COO^-]}{[CH_3COOH]} = 1.8 \times 10^{-5}$$

Let $x = [H^+]$

Sodium acetate is completely ionized; thus the concentration of acetate ion equals the sum of that from the salt and that from the ionization of acetic acid.

$$[CH_3COO^-] = 0.10 + x \cong 0.10 = 1.0 \times 10^{-1}$$

The concentration of acetic acid at equilibrium is the original concentration minus the amount ionized.

$$[CH_3COOH] = 0.10 - x \cong 0.10 = 1.0 \times 10^{-1}$$

$$\frac{(x)(1.0 \times 10^{-1})}{(1.0 \times 10^{-1})} = 1.8 \times 10^{-5}$$

$$x = 1.8 \times 10^{-5} \, M = [H^+]$$

Was our assumption valid that x was small enough to be ignored when added to or subtracted from 0.10? Yes, 1.8×10^{-5} is negligibly small compared with 0.10, that is, with 1.0×10^{-1}. (This buffer solution tends to keep the hydrogen ion concentration constant at a value of 1.8×10^{-5} mol/L.)

The ionization of a weak base is also decreased by the addition of a common ion. Adding ammonium chloride to a solution of ammonia decreases the amount of hydroxide ion in the system.

$$NH_3 + H_2O \rightleftharpoons NH_4^+ + OH^-$$

example D.7

Calculate the $[OH^-]$ and $[H^+]$ of a solution that is 0.50 M in NH_3 and 0.10 M in NH_4^+.

Use the K_b expression for NH_3.

$$K_b = \frac{[NH_4^+][OH^-]}{[NH_3]} = 1.8 \times 10^{-5}$$

Let $y = [OH^-]$

$$[NH_4^+] \cong 0.10 = 1.0 \times 10^{-1}$$

$$[NH_3] \cong 0.50 = 5.0 \times 10^{-1}$$

$$\frac{(1.0 \times 10^{-1})(y)}{(5.0 \times 10^{-1})} = 1.8 \times 10^{-5}$$

$$y = 9.0 \times 10^{-5} = [OH^-]$$

Now use the relationship

$$K_w = [H^+][OH^-] = 1.0 \times 10^{-14}$$

to calculate $[H^+]$.

$$[H^+] = \frac{1.0 \times 10^{-14}}{9.0 \times 10^{-5}} = 1.1 \times 10^{-10}$$

Buffered solutions are widely used in analytical chemistry, medicine, and biochemistry and in many industrial applications such as leatherworking and dyeing. A rearrangement of the equilibrium constant expression, called the Henderson-Hasselbach equation, is available for calculating the pH of a buffer.

For a weak acid, HA, the equation is

$$pH = pK_a + \log \frac{[A^-]}{[HA]}$$

In the equation, pK_a is a logarithmic term similar to pH.

$$pK_a = -\log K_a$$

If large amounts of strong acid or base are added, the capacity of a buffer is exceeded, and the pH changes rapidly. A buffer works most effectively when [HA] and [A⁻] are equal. Under these conditions, the buffer is said to operate at its optimum pH. If [A⁻] = [HA], then

$$\frac{[A^-]}{[HA]} = 1$$

The logarithm of 1 is 0, so

$$\log \frac{[A^-]}{[HA]} = 0$$

Therefore, according to the Henderson-Hasselbach equation, the optimum pH of a buffer is equal to the pK_a of the acid in the buffer.

$$pH = pK_a + \log \frac{[A^-]}{[HA]} = pK_a + 0 = pK_a$$

What is the pH of a solution that is 0.2 M in H_2S and 0.2 M in HS^-? The K_a for H_2S is 1×10^{-7}. (Ignore the second ionization of H_2S.)

example D.8

Since $[H_2S] = [HS^-]$, the Henderson-Hasselbach equation reduces to

$$pH = pK_a$$
$$pK_a = -\log K_a = -\log (1 \times 10^{-7}) = 7$$

Therefore,

$$pH = 7$$

What is the pH of a solution that is 0.50 M in HF and 0.50 M in F^-? The K_a for HF is 6.6×10^{-4}.

example D.9

$$pH = pK_a + \log \frac{[F^-]}{[HF]}$$
$$= -\log (6.6 \times 10^{-4}) + \log \frac{0.50}{0.50}$$
$$= -\log 6.6 - \log (10^{-4}) + \log 1$$
$$= -0.82 - (-4.00) + 0$$
$$= 3.18$$

example
D.10

What is the pH of a solution that is 0.10 M in HCN and 0.50 M in CN^-? The K_a for HCN is 6.2×10^{-10}.

$$pH = pK_a + \log \frac{[CN^-]}{[HCN]}$$

$$= -\log (6.2 \times 10^{-10}) + \log \frac{0.50}{0.10}$$

$$= -\log 6.2 - \log 10^{-10} + \log 5.0$$

$$= -0.79 - (-10.00) + 0.70$$

$$= 9.91$$

PROBLEMS

1. Write the equation for the ionization of each of the following acids:
   a. HBO_2
   b. $HClO_2$
   c. $HC_9H_7O_4$
   d. H_2Se (first ionization only)
2. Write equations for the ionization of each of the following bases (i.e., for the reaction of the base with water to form ions):
   a. $C_4H_9NH_2$ b. $C_{11}H_{21}O_4N$ c. C_3H_5N
3. Find the K_a for the acid HZ if the $[H^+]$ of a 0.100 M solution of HZ is 0.0002 M.
4. Find the K_b for the base QOH if the $[OH^-]$ of a 0.0200 M solution of QOH is 0.0004 M. The ionization reaction is

$$QOH \rightleftharpoons Q^+ + OH^-$$

5. Write an equilibrium constant expression for each of the following reactions:
   a. $HOCl \rightleftharpoons H^+ + OCl^-$
   b. $HC_6H_7O_6 \rightleftharpoons H^+ + C_6H_7O_6^-$
   c. $HCO_2H \rightleftharpoons H^+ + HCO_2^-$
6. Write an equilibrium constant expression for each of the following reactions:
   a. $C_5H_5N + H_2O \rightleftharpoons C_5H_5NH^+ + OH^-$
   b. $C_2H_5NH_2 + H_2O \rightleftharpoons C_2H_5NH_3^+ + OH^-$
   c. $HPO_4^{2-} + H_2O \rightleftharpoons H_2PO_4^- + OH^-$
7. Calculate the $[H^+]$ of each of these solutions. You may refer to Table D.1 for K_a values.
   a. 0.001 M CH_3COOH

   b. 0.02 M C_6H_5COOH
   c. 0.50 M HCN
8. Calculate the $[H^+]$ for each of these solutions. You may refer to Table D.1 for K_a values.
   a. 0.01 M HCOOH   c. 0.005 M HCN
   b. 0.1 M HF
9. What is the $[OH^-]$ in each of the solutions in Problem 7?
10. What is the $[OH^-]$ in each of the solutions in Problem 8?
11. Calculate the $[OH^-]$ for each of these solutions. You may refer to Table D.2 for K_b values.
    a. 0.025 M NH_3 c. 0.10 M $C_6H_5NH_2$
    b. 0.10 M CH_3NH_2
12. Calculate the $[OH^-]$ for each of these solutions. You may refer to Table D.2 for K_b values.
    a. 0.010 M $Zn(OH)_2$ c. 0.030 M CO_3^{2-}
    b. 0.15 M HCO_3^-
13. What is the $[H^+]$ in each of the solutions in Problem 11?
14. What is the $[H^+]$ in each of the solutions in Problem 12?
15. What is the $[H^+]$ in each of the following buffer solutions? You may refer to Table D.1 for K_a values.
    a. 0.25 M HCN and 0.25 M KCN
    b. 0.50 M HF and 0.50 M NaF
    c. 0.033 M C_6H_5COOH and 0.033 M $C_6H_5COO^-$
16. What is the $[H^+]$ in each of the following buffer solutions? You may refer to Table D.1 for K_a values.
    a. 0.1 M HCN and 0.2 M KCN

b. 0.5 M HF and 0.2 M NaF
c. 0.4 M C_6H_5COOH and 0.2 M $C_6H_5COO^-$
17. Calculate the $[OH^-]$ in a solution that is 0.4 M in NH_3 and 0.4 M in NH_4^+. What is the $[H^+]$ in the solution?
18. Calculate the $[OH^-]$ in a solution that is 0.04 M in NH_3 and 0.02 M in NH_4^+. What is the $[H^+]$ in the solution?
19. What is the pH of a buffer solution that is 0.1 M in HOC_6H_5 and 0.1 M in $NaOC_6H_5$? The K_a for HOC_6H_5 is 1×10^{-10}.
20. What is the pH of a buffer solution that is 0.05 M in $C_5H_{11}COOH$ and 0.05 M in $C_5H_{11}COO^-$? The K_a for $C_5H_{11}COOH$ is 1×10^{-8}.

Note: The following problems require the use of a table of logarithms or a scientific calculator with a base 10 logarithm function. You may refer to Table D.1 for K_a values.
21. Calculate the pH of a solution that is 0.040 M in HCN and 0.040 M in CN^-.
22. Calculate the pH of a solution that is 0.20 M in HCOOH and 0.20 M in $HCOO^-$.
23. Calculate the pH of a solution that is 0.15 M in benzoic acid and 0.15 M in benzoate ion.
24. Calculate the pH of a solution that is 0.20 M in HF and 0.50 M in F^-.
25. Calculate the pH of a solution that is 0.15 M in HCOOH and 0.60 M in $HCOO^-$.
26. Calculate the pH of a solution that is 0.11 M in benzoic acid and 0.96 M in benzoate ion.

12

Electrolytes

The term **ion** was introduced in Chapter 4 to describe charged particles. We have come a long way since then, and along the way we've discussed ionic bonds, ionic solids, and solutions of ions. The nineteenth-century scientists whose experiments laid the foundation for the ionic theory were working mainly for the joy of discovery and recognition gained with success. We can only speculate whether they imagined the importance of ions in living systems.

The fluids in our bodies are like the saltwater of the oceans. In our fetal development, we have gill-like organs, hands that look like fins, and a shape much like that of a fish. Until we are born and breathe the air, we float in the watery darkness of the womb. Even after birth, fluids bathe our tissues and transport the materials that keep them alive. It is now well established that messages are sent to and from the brain in the form of electrical signals. These messages are often carried by ions through cellular and intercellular fluids. Certain ions are essential to the proper functioning of all living organisms. They must be present in proper concentrations, however. Too few or too many can be dangerous. We will discuss some of these ions and their properties in this chapter.

12.1 EARLY ELECTROCHEMISTRY

In 1800, an Italian physicist, Alessandro Volta, invented a battery that produced an electric current. Electrical phenomena had already been subjected to considerable study before this time. (In 1752, Benjamin Franklin determined that lightning was a form of electricity by flying a kite in a thunderstorm.) The importance of Volta's invention lay in the fact that the battery provided a convenient source of electricity, one that could be used by other scientists who wished to study the interaction of matter and electricity.

Within 6 weeks of its invention, a battery was used by two English chemists, William Nicholson and Anthony Carlisle, to decompose water into hydrogen gas and oxygen gas. They accomplished the decomposition by passing an electric current through the water sample. Sir Humphry Davy, another English chemist, used an electric current to liberate potassium metal from potassium hydroxide (KOH), sodium metal from sodium hydroxide (NaOH), and metallic magnesium, strontium, barium, and calcium from their respective compounds.

Sir Humphry's protégé, Michael Faraday, termed the process of splitting compounds by means of electricity **electrolysis,** a word of Greek origin that literally means "loosing by electricity." It was Faraday who was responsible for many of the terms used in electrochemistry today. Some of these terms were introduced in Chapter 2, but we offer a brief review here. The carbon rods or metal strips that are connected to a source of electricity (the battery) and inserted into solutions under study are called **electrodes** (Figure 12.1).

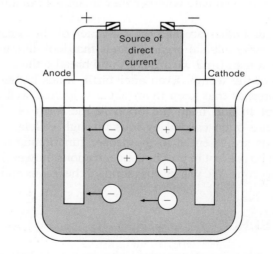

FIGURE 12.1   *When electrical current is passed through an electrolyte, positive ions move to the cathode, and negative ions move to the anode. Reduction occurs at the cathode, and oxidation takes place at the anode.*

Electricity, we know now, is a flow of electrons. In systems used by electrochemists, electrons flow from one electrode to the other. The electrode that has lost electrons and is therefore positively charged is called the **anode.** The electrode that has gained electrons and is negatively charged is called the **cathode**. Electrolysis involves oxidation and reduction reactions. *Oxidation* occurs at the *anode*, where substances give up electrons to the positively charged electrode. *Reduction* occurs at the *cathode*, where substances can pick up electrons from the electrode, which has an excess of them.

In Faraday's studies, electrodes were placed in certain solutions and melts (Section 12.2). When this was done, the electric "circuit" was completed. The battery provided the driving force (the electrical potential or voltage), and electric current flowed from one electrode to the other, through the solution, and back to the first electrode, that is, around the circuit. Faraday hypothesized that the electric current was carried through the solution (or melt) by atoms that had electrical charges and that he called **ions,** a term we have used extensively in preceding chapters. You will recall that positively charged ions, which are attracted to the negatively charged electrode (the cathode), are called **cations** and that negatively charged ions, which are attracted to the positively charged anode, are called **anions.**

12.2 ELECTRICAL CONDUCTIVITY

The electricity with which we are most familiar usually flows in metallic wires. The outer electrons of the metal atoms flow through the wire, while the nucleus and inner level electrons of the atom remain nearly fixed in place. Most nonmetals are nonconductors. The outer level electrons of these elements are tightly bound or shared with neighboring atoms. They are not free to roam around. Likewise, in most covalent compounds, the electrons are firmly held and cannot conduct electricity. A few polar covalent compounds, such as hydrochloric acid, react with water to form ions.

$$HCl + H_2O \longrightarrow H_3O^+ + Cl^-$$

Solutions in which these ions are formed are then able to conduct electricity.

Solid ionic compounds such as salts, however, do not conduct electricity. The ions occupy fixed positions in the crystal lattice and are unable to move very much in an electric field. When the lattice is broken down by heat (melting) or by the process of dissolution, the ions are freed. These charged particles can then move in an electric field. Substances that conduct electricity as melts or in solutions are called **electrolytes.**

How do we know an electrolyte when we see one? Simple. Just dissolve some of the compound in water and test it with a conductivity apparatus

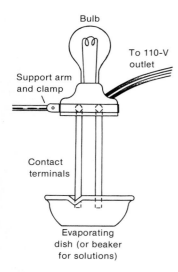

Bulb

To 110-V outlet

Support arm and clamp

Contact terminals

Evaporating dish (or beaker for solutions)

FIGURE 12.2
An apparatus for testing the electrical conductivity of solutions.

(Figure 12.2). If a solution that conducts electricity is placed in the dish, thus completing the circuit, electric current will flow in the apparatus, and the light bulb will glow. Those compounds that, in aqueous solution, are good conductors (i.e., those that cause the bulb to glow brightly) are called **strong electrolytes.** Those compounds whose aqueous solutions produced only a dim light are called **weak electrolytes.** Other compounds, whose solutions don't light the bulb at all, are called **nonelectrolytes.** These compounds produce few if any ions in solution and conduct no appreciable current.

Isn't there an easier way to determine whether a compound is an electrolyte? Can we predict electrical conductivity? Indeed we can, at least in many cases. All strong acids and strong bases (Chapter 10) and all salts that are appreciably soluble in water are strong electrolytes. All weak acids and weak bases and some slightly soluble salts are weak electrolytes. Molecular substances, such as sugar and alcohol, which are neither acids nor bases, are nonelectrolytes. The deciding factor in all cases is the ability of the compound being considered to provide ions in aqueous solution. Table 12.1 lists some familiar compounds and the behavior of their solutions toward electrical current.

The work of Faraday and other pioneers in electrochemistry generated a great deal of interest and excitement among scientists whose concern was the fundamental structure of matter. It is worth noting, however, that the interest of the general public was excited by a purely fictional experiment involving electricity and described by a 21-year-old woman. It is Mary Wollstonecraft Shelley's novel *Frankenstein*, published in 1818, which has captured the imagination of nineteenth- and twentieth-century readers by relating electricity to life (Figure 12.3).

TABLE 12.1   *A Selection of Strong Electrolytes, Weak Electrolytes, and Nonelectrolytes*

Compound Name	Formula	Kind of Compound	Electrical Conductivity
Hydrochloric acid	HCl	Strong acid	Strong
Sulfuric acid	H_2SO_4	Strong acid	Strong
Sodium hydroxide	NaOH	Strong base	Strong
Sodium chloride	NaCl	Salt	Strong
Calcium nitrate	$Ca(NO_3)_2$	Salt	Strong
Acetic acid	CH_3COOH	Weak acid	Weak
Ammonia	NH_3	Weak base	Weak
Calcium sulfate	$CaSO_4$	Slightly soluble salt	Weak
Copper(II) sulfide	CuS	Insoluble salt	None
Sugar (sucrose)	$C_{12}H_{22}O_{11}$	Molecular solid	None
Ethyl alcohol	C_2H_5OH	Molecular liquid	None

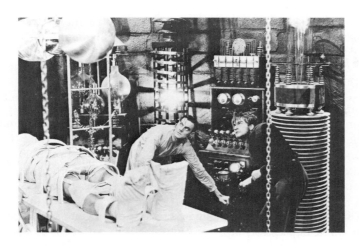

FIGURE 12.3 *The laboratory of Frankenstein, with its apparatus for the production of electric current. (The Bettmann Archive.)*

12.3 THE THEORY OF ELECTROLYTES: IONIZATION AND DISSOCIATION

In 1887, Svante Arrhenius proposed a general theory to explain the properties of electrolytes. We encountered a part of that theory in Chapter 10 in our study of acids and bases. Further, Arrhenius's theory has been modified somewhat through the years to account for new data. The modernized theory is summarized here.

1. An electrolyte, when dissolved in water, dissociates into ions, each of which carries an electric charge. For salts, this process may be summarized by an equation. For example,

$$Na_2SO_4(s) \longrightarrow 2\,Na^+(aq) + SO_4{}^{2-}(aq)$$
$$CaCl_2(s) \longrightarrow Ca^{2+}(aq) + 2\,Cl^-(aq)$$

For acids and some bases, the process is actually one of *ion formation* or **ionization.**

$$HCl(g) + H_2O \longrightarrow H_3O^+(aq) + Cl^-(aq)$$
$$NH_3(g) + H_2O \longrightarrow NH_4{}^+(aq) + OH^-(aq)$$

For strong bases, the process is usually one of **dissociation,** that is, separation of already existing ions.

$$NaOH(s) \longrightarrow Na^+(aq) + OH^-(aq)$$

2. When an ionic solid dissociates or when a polar molecule ionizes, the algebraic sum of all positive charges and all negative charges is 0.

$$K_3PO_4(s) \longrightarrow 3\,K^+(aq) + PO_4^{3-}(aq)$$
$$3(+1) + (-3) = 0$$

Solutions as a whole are therefore electrically neutral.

3. Each ion, regardless of size, charge, or shape, has the same effect on boiling-point elevation, freezing-point depression, and osmotic pressure (Chapter 9) as an undissociated molecule would have. This assumption holds quite well for dilute solutions but is not strictly true for concentrated ones. In concentrated solutions, each ion is rather closely surrounded by others of opposite charge. This hinders the movement of the ions; they are not completely free. This decreases the expected activity somewhat. For example, the freezing-point depression of a solution of 1 mol of NaCl in 1 kg of water is 1.8 times (not quite twice) that of a nonelectrolyte at the same concentration.

4. Weak electrolytes react with water to a limited extent and hence provide only a limited number of ions in solution.

5. Nonelectrolytes exist in molecular form in solution or are insoluble salts; they produce no appreciable concentration of ions.

12.4 ELECTROLYSIS: CHEMICAL CHANGE CAUSED BY ELECTRICITY

If sodium chloride is heated until it melts, the molten salt conducts electricity, as one can see by testing the melt in the conductivity apparatus in Figure 12.2. In addition to causing the light bulb to glow, the electric current, as it passes through the melt, causes observable chemical changes. Yellow-green chlorine gas forms at the anode. At the cathode, silvery metallic sodium is formed and is rapidly vaporized by the hot melt. The molten, ionic sodium chloride is decomposed by electrical energy into elemental sodium and chlorine.

$$2\,NaCl + energy \longrightarrow 2\,Na + Cl_2$$

In crystalline form, sodium chloride does not conduct electricity. The ions occupy relatively fixed positions in the lattice and do not move very much, even under the influence of an electrical potential. When sodium chloride melts, the ions are freed to move around. When a battery is connected to the melt through a pair of electrodes, the sodium ions are attracted to the electron-rich cathode, where they pick up electrons (Figure 12.4).

$$Na^+ + e^- \longrightarrow Na$$

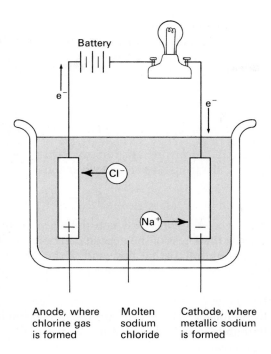

FIGURE 12.4
The electrolysis of molten sodium chloride.

Anode, where chlorine gas is formed

Molten sodium chloride

Cathode, where metallic sodium is formed

The chloride ions migrate to the electron-poor anode, where they give up electrons.

$$2\, Cl^- \longrightarrow Cl_2 + 2e^-$$

The battery in the circuit is responsible for seeing to it that this exchange of electrons does not eventually lead to neutralized electrodes. Electrons picked up by the anode from chloride ions are immediately shunted, under the influence of the battery, to the cathode, which has been losing electrons to sodium ions. As long as sodium ions and chloride ions are present in the melt, current will flow. When all of the salt has been converted to elemental sodium and chlorine, the circuit will be broken and current will cease.

This electrolytic reaction is an oxidation-reduction process. Oxidation occurs at the anode, where the oxidation number of chlorine goes from -1 to 0. At the cathode, the oxidation number of sodium is reduced from $+1$ to 0.

Electrolysis, then, is a process of using electricity to bring about chemical change. The process is useful for the preparation and purification (refining) of many metals. Electrolysis is also used for coating one metal with another, an operation called **electroplating.** Usually the object to be electroplated, such as a spoon, is cast of a cheaper metal. It is then coated with a thin layer of a more attractive and more corrosion-resistant metal, such as gold or silver. The cost of the finished product is far less than that of a corresponding item made entirely of silver or gold. A cell for the electroplating of silver is shown in Figure 12.5. The silver is made the anode, and the spoon is made the cathode. A solution of silver nitrate is used as the electrolyte. Under the influence of the battery in the system (or any voltage source), the silver ions (Ag^+) are attracted to the cathode (spoon), where they pick up electrons and are deposited as silver atoms.

$$Ag^+ + e^- \longrightarrow Ag$$

At the anode, electrons are removed from the silver bar. Some of the silver atoms lose electrons to become silver ions.

$$Ag \longrightarrow Ag^+ + e^-$$

The net process is one in which the silver from the bar is transferred to the spoon. The thickness of the deposit can be controlled by accurate measurement of the amount of current flow and of the duration of the process.

Electrolysis finds some biological applications, including the removal of unwanted hair. In this process, a tiny wire needle is used to supply a mild electrical current to the hair root. The chemical changes engendered there kill the living follicle. This is perhaps the only permanent method of hair removal. In the hands of a skilled technician, the method can be clean and safe. It is tedious, however, for each hair root must be treated individually. Similar procedures are sometimes used for the removal of warts or other growths.

FIGURE 12.5
An electrolytic cell for the plating of silver.

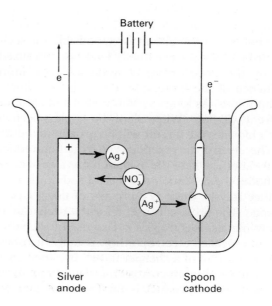

12.5 ELECTROCHEMICAL CELLS: BATTERIES

Electricity can cause chemical change. Conversely, chemical change can produce electricity. That's what batteries are all about.

If a strip of zinc metal is placed in a solution of zinc sulfate, zinc atoms exhibit a tendency to go into solution as zinc ions.

$$Zn \longrightarrow Zn^{2+} + 2e^-$$

The two electrons are left behind on the zinc strip. On the other hand, the zinc ions in solution have a tendency to migrate to the zinc strip, acquire two electrons, and be deposited as zinc atoms.

$$Zn^{2+} + 2e^- \longrightarrow Zn$$

The system rapidly comes to equilibrium.

$$Zn \rightleftharpoons Zn^{2+} + 2e^-$$

This equilibrium is affected by temperature, concentrations, and other factors described in Chapter 5. For example, adding more zinc ions would shift the reaction to the left, using up zinc ions.

If a copper strip is added to a solution of copper sulfate, a similar equilibrium is established.

$$Cu \rightleftharpoons Cu^{2+} + 2e^-$$

Each metal differs in its tendency to give up electrons (remember the activity series of Chapter 10). Copper is less likely to do so than is zinc. Now, if we connect the copper strip to the zinc strip, a current (electrons) will flow from the zinc to the copper (Figure 12.6). In order to ionize, zinc metal gives its electrons to the copper ions. When we discussed the activity series in Chapter 10, we described a similar reaction. The difference between the present reaction and that previously considered is that, in the reaction of Chapter 10, the zinc metal was placed in direct contact with the copper ions in solution, so it could simply hand its electrons to the copper ions. Here, the copper ions are in one compartment, and the zinc is in another. The electrons have to pass through the wire connecting the electrodes (the metal strips) to get to the copper ions. This movement of electrons is an electric current. The chemical reaction produces electricity.

This cell is slightly more complicated than indicated by the diagram in Figure 12.6. The compartment containing zinc ions also contains sulfate ions. The salt, zinc sulfate, is what is dissolved in the solution. Similarly the solution of copper ions also contains sulfate ions (from copper sulfate). The sulfate ions don't enter into any chemical reaction, but they do cause complications. Consider the zinc sulfate solution. During the reaction, zinc atoms are converted to zinc ions, so the solution becomes more concentrated in zinc ions. However,

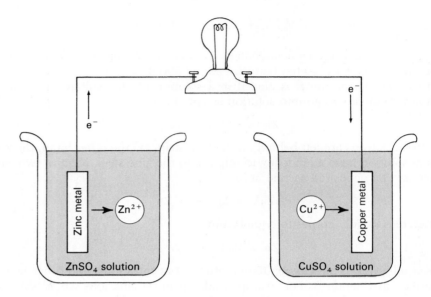

FIGURE 12.6 *An attempt to convert chemical energy to electricity. In the left compartment, zinc atoms would give up electrons, and, in the right, copper(II) ions would pick up electrons. This electrochemical cell will not work, however. For an explanation, see the text.*

nothing happens to the sulfate ions. We are faced with the possibility of the solution containing more positively charged ions (Zn^{2+} ions) than negatively charged ions (SO_4^{2-} ions). This electrical imbalance is an unstable situation, and if something isn't done about it, the reaction will stop. In the other compartment of Figure 12.6, the concentration of copper ions is being reduced while the concentration of sulfate ions remains constant. Thus, there is a tendency for this compartment to develop an excess of negatively charged ions. One way to deal with the problem is to connect the two solutions through a porous plate, made of a material that will retard the mixing of the metal ions in solution and still permit passage of sulfate ions (Figure 12.7). When a zinc atom is converted to a zinc ion in the left compartment, a copper ion changes to a copper atom in the right, and a sulfate ion moves through the porous plate from the solution in the right compartment to the solution in the left. The charges in the two solutions stay balanced. Why bother with the porous plate? Why not just put both solutions and both electrodes in a single compartment so the sulfate ions don't have to move through pores? If we were to do that, the electrons wouldn't have to make the trip through the wire; they could just be passed from zinc atoms to copper ions, which would now be in contact with one another. And unless the electrons go through the wire, the light bulb won't glow (nor

FIGURE 12.7
A simple electrochemical cell.
A battery may consist of one
or more such cells. Variants
of this **Zn/Cu** *cell, called*
Daniell cells, were used to
power telegraph relays about
100 years ago.

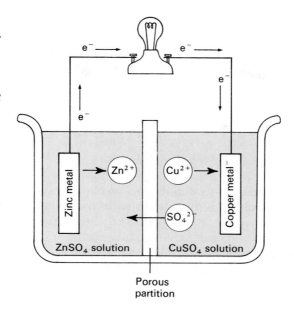

will battery-operated portable radios, hand calculators, or cardiac pacemakers work).

The net reaction for this electrochemical cell is

$$Zn + Cu^{2+} \longrightarrow Zn^{2+} + Cu$$

If 1 M solutions of $ZnSO_4$ and $CuSO_4$ are used, the system will produce about 1.1 volts (V) at 25 °C. The **volt** is a measure of electrical potential, or of the tendency of the electrons in a system to flow.

A battery is technically a series of such cells, the word **battery** meaning a group of similar things (as in a battery of tests). The familiar 12-V automobile storage battery consists of six cells wired in series (i.e., one after another), each cell consisting of two electrodes. The cells do not incorporate the zinc-copper couple just described but a somewhat more complicated system involving lead, lead oxide, and sulfuric acid. A distinctive feature of the lead storage battery is its capacity for being recharged. It will supply electricity (i.e., it will discharge) when you turn on the ignition to start the car or when the motor is off and the lights are on. But it can be recharged when the car is moving and an electric current is supplied *to* the battery by the mechanical action of the car (a process we won't detail here).

Ionic processes in and around cells give rise to electrical potentials. These processes are especially important in muscle and nerve cells. Instruments are available that measure small potential differences and small changes in current

FIGURE 12.8
An electroencephalograph measures changes in the flow of current associated with the activity of the brain. (Courtesy of Biomedical Graphics, University of Minnesota Hospitals, Minneapolis.)

flow in the human body. The **electrocardiograph** (ECG) monitors the electrical changes associated with the beating of the heart. The **electroencephalograph** (EEG) measures the electrical activity of the brain. The EEG can be connected to the scalp by means of electrodes (Figure 12.8). The ECG is connected to the body through a suspension of electrolyte on the skin. The changes in current flow associated with the heart's activity are conducted to the ECG through the electrolyte.

12.6 CORROSION

An oxidation-reduction reaction of particular economic importance is corrosion of metals. It is estimated that in the United States alone, corrosion costs $70 billion a year. Perhaps 20% of all the iron and steel production in the United States each year goes to replace corroded items. Let's look first at the corrosion of iron.

In moist air, iron is oxidized, particularly at a nick or scratch.

$$Fe \longrightarrow Fe^{2+} + 2e^-$$

In order for iron to be oxidized, oxygen must be reduced.

$$O_2 + 2H_2O + 4e^- \longrightarrow 4OH^-$$

The net result, initially, is the formation of insoluble iron(II) hydroxide.

$$2Fe + O_2 + 2H_2O \longrightarrow 2Fe(OH)_2$$

This product is usually further oxidized to iron(III) hydroxide.

$$4Fe(OH)_2 + O_2 + 2H_2O \longrightarrow 4Fe(OH)_3$$

The latter, sometimes written as $Fe_2O_3 \cdot 3H_2O$, is the familiar iron rust.

The corrosion of iron appears to be an electrochemical reaction. Oxidation and reduction often occur at separate points on the metal surface. Electrons are transferred through the iron metal. The circuit is completed by an electrolyte in aqueous solution. In the snowbelt, this solution is often the slush from road salt and melting snow. The metal is pitted in an **anodic** area, where iron is oxidized to Fe^{2+}. These ions migrate to the **cathodic** area, where they react with the hydroxide ions formed by reduction of oxygen.

$$Fe^{2+} + 2OH^- \longrightarrow Fe(OH)_2$$

As indicated above, this iron(II) hydroxide is then oxidized to $Fe(OH)_3$, or rust. This process is diagrammed in Figure 12.9. Notice that the anodic area is protected from oxygen by the water film, while the cathodic area is exposed to air.

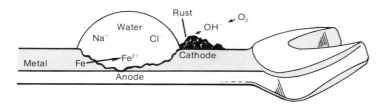

FIGURE 12.9   *The corrosion of iron requires water, oxygen, and an electrolyte.*

Aluminum is more reactive than iron. We might expect it to corrode more rapidly, but billions of beer cans testify to the fact that it doesn't. How can that be? It so happens that freshly prepared aluminum quickly forms a thin, hard film of aluminum oxide on its surface. This film is impervious to air and protects the underlying metal from further oxidation.

The tarnish on silver is another example of corrosion. Silver is oxidized to silver ions.

$$Ag \longrightarrow Ag^+ + e^-$$

These silver ions react with sulfur compounds such as hydrogen sulfide to form black silver sulfide.

$$2\,Ag^+ + H_2S \longrightarrow Ag_2S + 2\,H^+$$

You can use a silver polish to remove the tarnish, but in doing so, you also lose part of the silver. An alternative method involves the use of aluminum metal to reduce the silver ions back to silver metal.

$$3\,Ag^+ + Al \longrightarrow 3\,Ag + Al^{3+}$$

This reaction also requires an electrolyte. Sodium bicarbonate ($NaHCO_3$) is usually used. The tarnished silver is placed in contact with aluminum foil, covered with a solution of sodium bicarbonate, and heated. A precious metal is conserved at the expense of a cheaper one.

12.7 PRECIPITATION: THE SOLUBILITY PRODUCT RELATIONSHIP

We have thus far focused our attention on strong electrolytes, those compounds that supply high concentrations of ions in solution. There are ionic compounds whose solubility in water is quite low. For example, the solubility of barium sulfate ($BaSO_4$) is 0.0002 g per 100 g of water at 25 °C. This makes barium sulfate a very weak electrolyte in aqueous solution. The fact that a salt is a weak electrolyte or a nonelectrolyte does not necessarily mean that it is unimportant in the chemistry of living systems. On the contrary, for some physiological processes, the relative insolubility of certain salts is of critical importance. To understand these processes, it will be necessary to consider first a new way of describing solubility, or, more correctly, of determining when a salt will start precipitating from solution because its solubility has been exceeded.

When solid barium sulfate is added to water, an equilibrium is established between the undissolved solute and the ions.

$$BaSO_4(s) \rightleftharpoons Ba^{2+}(aq) + SO_4^{2-}(aq)$$

An extremely small amount goes into solution. Even though barium salts are quite toxic, large amounts of barium sulfate can be swallowed or given by enema because very little will dissolve in the solutions of the body. Since barium sulfate is opaque to X rays, technicians can use it to outline the stomach or intestines for X-ray photographs (Chapter 3). The undissolved barium salt scattered throughout the intestine blocks the X rays. The X-ray film is unexposed in these areas, which, therefore, appear white in the developed negative. The barium sulfate is later voided from the body unchanged.

Even though we say that barium sulfate is insoluble, the small amount that does dissolve can be measured (as we indicated above). The concentration of barium ions (Ba^{2+}) and sulfate ions (SO_4^{2-}) in a saturated solution of barium sulfate is 0.000 01 M (or 1×10^{-5} M) each at 18 °C.

The product of the concentrations of the ions in a saturated solution,

$$[Ba^{2+}][SO_4^{2-}]$$

is a constant. At 18 °C, the product is

$$(1 \times 10^{-5})(1 \times 10^{-5}) = 1 \times 10^{-10}$$

This value is called the **solubility product constant** for barium sulfate. It is this constant that we will find useful in predicting whether precipitation will occur if a solution containing barium ions is mixed with one containing sulfate ions. Precipitation will occur if the product $[Ba^{2+}][SO_4^{2-}]$ is greater than 1×10^{-10}. Barium sulfate will continue to separate until the product is just equal to 1×10^{-10}. At that value, the solution will be saturated. If the $[Ba^{2+}][SO_4^{2-}]$ is less than 1×10^{-10}, no barium sulfate precipitate will be formed. The solution will be unsaturated in barium sulfate.

As in the case of the ion product of water, K_w (Section 11.3), it is the *product* of the ion concentrations that is important. If barium sulfate is simply placed in water, the concentration of barium ion equals the concentration of sulfate ion. It is, however, possible to prepare a solution in which the two concentrations are not equal (Example 12.1). The barium ion concentration could be higher than the sulfate ion concentration, or vice versa. Precipitation will always occur when the *product* of the two concentrations exceeds 1×10^{-10}, but precipitation will not necessarily coincide with concentrations of 1×10^{-5} for each ion.

If 0.001 mol of $BaCl_2$ and 0.0001 mol of Na_2SO_4 are added to 1 L of water, will $BaSO_4$ precipitate? (Assume no volume change.)

$BaCl_2$ and Na_2SO_4 are soluble salts; each one will completely dissolve. The 0.001 mol of $BaCl_2$ will yield 0.001 M Ba^{2+} (as well as 0.002 M Cl^-). The 0.0001 mol of Na_2SO_4 will yield 0.0001 M SO_4^{2-} (as well as 0.0002 M Na^+). The product of the barium and sulfate ion concentrations is

$$(0.001)(0.0001) = (1 \times 10^{-3})(1 \times 10^{-4}) = 1 \times 10^{-7}$$

Since 1×10^{-7} is greater than 1×10^{-10},* $BaSO_4$ will precipitate until the product of the $[Ba^{2+}]$ and $[SO_4^{2-}]$ is just equal to 1×10^{-10}. The sodium and chloride ions in solution have no tendency to precipitate as sodium chloride salt, because sodium chloride is quite soluble in water.

example 12.1

The solubility product principle can be used in the preparation of certain compounds. If silver nitrate solution is mixed with sodium chloride solution, slightly soluble silver chloride will precipitate.

$$\underbrace{Ag^+ + NO_3^-}_{\text{Soluble}} + \underbrace{Na^+ + Cl^-}_{\text{Soluble}} \longrightarrow \underbrace{AgCl(s)}_{\text{Insoluble}} + \underbrace{Na^+ + NO_3^-}_{\text{Soluble}}$$

The silver chloride can be removed by filtration, that is, by pouring the mixture through a porous paper that will trap the solid silver chloride but permit passage of the solution. Further, if equimolar quantities of silver nitrate and sodium chloride are used, the filtrate (the solution going through the filter paper) can be evaporated, and the other ions can be obtained as sodium nitrate in cystalline form. The solubility product of silver chloride, $[Ag^+][Cl^-]$, is equal to 1.6×10^{-10} at 25 °C. That means that a very small number of silver ions and chloride ions in solution will pass through the filter paper. But the vast majority of these ions will be trapped on the paper as the solid salt.

The solubility products for many salts are known. A few values are given in Table 12.2.

---

* Remember that the smaller negative exponent corresponds to the larger number. For example, $10^{-1} = 0.1$ and $10^{-2} = 0.01$.

TABLE 12.2 *Selected Solubility Product Constants at 25°C*

Compound	Formula	Solubility Product Constant
Barium carbonate	$BaCO_3$	2.0×10^{-9}
Barium sulfate	$BaSO_4$	1.1×10^{-10}
Calcium carbonate	$CaCO_3$	4.8×10^{-9}
Copper(II) sulfide	CuS	8.7×10^{-36}
Lead chromate	$PbCrO_4$	2.0×10^{-14}
Magnesium carbonate	$MgCO_3$	2.0×10^{-8}
Mercury(II) sulfide	HgS	3.0×10^{-53}
Silver acetate	$AgC_2H_3O_2$	2.0×10^{-3}
Silver bromide	$AgBr$	5.0×10^{-13}
Silver chloride	$AgCl$	1.2×10^{-10}

Precipitation is important in the formation of many minerals in nature. Geologists are not the only ones concerned with the formation of mineral deposits, however. Our teeth and bones are largely calcium phosphate salts. One such salt is $Ca_3(PO_4)_2$, sometimes called tricalcium phosphate. Teeth and bones are formed by the precipitation of calcium phosphate salts from solution. In order for this precipitation to occur, the concentrations of ions must exceed the solubility product in the immediate area of deposition. If we assume that the precipitation reaction is

$$3\,Ca^{2+} + 2\,PO_4{}^{3-} \rightleftharpoons Ca_3(PO_4)_2(s)$$

the solubility product is

$$[Ca^{2+}][Ca^{2+}][Ca^{2+}][PO_4{}^{3-}][PO_4{}^{3-}]$$

Each calcium phosphate unit provides three calcium ions (Ca^{2+}) and two phosphate ions ($PO_4{}^{3-}$). More simply, we may write the expression as

$$[Ca^{2+}]^3[PO_4{}^{3-}]^2$$

At a temperature of 37 °C (normal body temperature), the solubility product constant for calcium phosphate is about 4×10^{-27}. In the blood, the concentration of free calcium ions is about $0.0012\ M$ ($1.2 \times 10^{-3}\ M$), and the concentration of phosphate ions is about $1.6 \times 10^{-8}\ M$. Plugging these values into the solubility product relationship, we get

$$(1.2 \times 10^{-3})^3 \times (1.6 \times 10^{-8})^2 = (1.7 \times 10^{-9})(2.6 \times 10^{-16}) = 4.4 \times 10^{-25}$$

(A brief review on the manipulation of exponential numbers is offered in Appendix II.) Since this value is *larger* than the solubility product (4×10^{-27}), we expect precipitation of calcium phosphate and the subsequent growth of bone and tooth to occur.

In growing children, the foregoing description may closely represent the actual mechanism. In adults, however, teeth and bones are no longer growing. What keeps them from getting ever larger? The pH right at the area of growth is a little below 7.4 due to metabolic processes in the cells. Hydronium ions thus produced tie up phosphate ions as hydrogen phosphate ions.

$$PO_4{}^{3-} + H_3O^+ \rightleftharpoons HPO_4{}^{2-} + H_2O$$

The phosphate ion concentration is reduced to a value just satisfying the solubility product relationship at the constant value of 4.0×10^{-27}. Normal bone and tooth maintenance takes place, with neither growth nor diminution.

Two pathological conditions of the mouth and teeth are worthy of mention. If the salivary glands are removed or destroyed, the teeth are no longer bathed by saliva. At the usual pH values, saliva provides just the right concentration of calcium ions and phosphate ions to prevent dissolution of the teeth. With the salivary glands gone, the concentration of the ions rapidly falls below the solubility product constant. When this happens, more calcium phosphate dissolves, and the teeth, if not removed, erode away.

A similar situation occurs when children suffer from chronic acidosis. The blood pH may be as low as 7.1, and the pH in the immediate areas of bone formation is probably even lower. Hydronium ions tie up phosphate ions, lowering the concentration of the latter. Bone growth is greatly hindered, and the child's skeleton is badly formed.

Tooth decay (caries) can also be related to solubility. A combination carbohydrate-protein called **mucin** forms a film, called plaque, on teeth. If it is not removed by brushing and flossing, buildup of plaque continues. Food and bacteria trapped in the plaque ferment carbohydrates, producing acid. Saliva does not penetrate plaque and hence cannot buffer against the buildup of acid. The pH at the surface of the tooth may go as low as 4.5, and the concentration of phosphate ions in solution is rapidly depleted as hydrogen phosphate ions are formed. The calcium phosphate of the tooth dissolves to replenish the phosphate ion, leaving a cavity in the tooth.

Teeth are also eroded in patients suffering from **bulimia,** a condition characterized by binge eating followed by vomiting. Hydrochloric acid vomited from the stomach acts in the mouth to tie up phosphate ions. The pH may drop as low as 1.5, and erosion can be much more rapid than in caries.

12.8 THE CASE OF THE DISAPPEARING PRECIPITATE

In the preceding section, we saw how certain combinations of ions cause precipitates to form. Knowledge of solubility product relationships can also be used to bring slightly soluble salts into solution.

If a relatively concentrated solution of calcium chloride ($CaCl_2$) is added to another with a fair concentration of sodium oxalate ($Na_2C_2O_4$), a precipitate of calcium oxalate (CaC_2O_4) is formed (some kidney stones are primarily calcium oxalate).

$$\underbrace{Ca^{2+} + 2\,Cl^-}_{\text{Soluble}} + \underbrace{2\,Na^+ + C_2O_4^{2-}}_{\text{Soluble}} \longrightarrow \underbrace{CaC_2O_4(s)}_{\text{Insoluble}} + \underbrace{2\,Na^+ + 2\,Cl^-}_{\text{Soluble}}$$

If hydrochloric acid is now added to the precipitate, the precipitate disappears. The solid dissolves because the oxalate ions ($C_2O_4^{2-}$) are tied up by the hydronium ions (from the HCl) to form soluble, slightly ionized oxalic acid.

$$C_2O_4^{2-} + 2\,H_3O^+ \rightleftharpoons H_2C_2O_4 + 2\,H_2O$$

This decreases the concentration of oxalate ions so that the solubility product constant for calcium oxalate is no longer exceeded. The precipitate dissolves.

Other precipitates can be dissolved by reactions that form slightly soluble gases. Zinc sulfide is ever so slightly soluble in water.

$$ZnS(s) \rightleftharpoons Zn^{2+} + S^{2-}$$

The product of ion concentrations at equilibrium is

$$[Zn^{2+}][S^{2-}] = 1.1 \times 10^{-21}$$

Zinc sulfide dissolves in hydrochloric acid, however, because the hydronium ions react with sulfide ions (S^{2-}) to form hydrogen sulfide, which escapes as a gas.

$$S^{2-} + 2\,H_3O^+ \longrightarrow H_2S(g) + 2\,H_2O$$

Removal of sulfide ion soon leads to a situation in which the solubility product constant is no longer exceeded, and the precipitate dissolves.

Unfortunately, strong hydrochloric acid solutions can't be used to dissolve kidney stones from kidneys. The acid would be much too corrosive to our cells. There are chemicals, though, that have shown modest success in dissolving kidney stones. If the stones are calcium oxalate, removing foods containing oxalates from the diet might be of some help.

12.9 THE SALTS OF LIFE: MINERALS

A variety of inorganic compounds are necessary for the proper growth and repair of our tissues. It is estimated that such minerals represent about 4% of human body weight. Some of these, such as the chlorides (Cl^-), phosphates (PO_4^{3-}), bicarbonates (HCO_3^-), and sulfates (SO_4^{2-}), occur in the blood and other body fluids. Others, such as iron (as Fe^{2+}) in hemoglobin and phosphorus in the nucleic acids (DNA and RNA), are constituents of complex organic compounds.

Minerals essential to one or more living organisms contain the elements sodium (Na), magnesium (Mg), potassium (K), phosphorus (P), sulfur (S), chlorine (Cl), calcium (Ca), manganese (Mn), iron (Fe), copper (Cu), cobalt (Co), zinc (Zn), iodine (I), fluorine (F), silicon (Si), tin (Sn), vanadium (V), chromium (Cr), selenium (Se), molybdenum (Mo), and arsenic (As). These, with the structural elements carbon, hydrogen, nitrogen, and oxygen, make up the 25 chemical elements of life. Other elements are sometimes found in body fluids and tissues but are not known to be essential. These include aluminum (A1), lithium (Li), nickel (Ni), and boron (B). It might yet be discovered that one or more of these is essential.

These minerals serve a variety of functions. Perhaps the most dramatic is that of iodine. A small amount of iodine is necessary to the proper function of the thyroid gland. A deficiency of iodine produces dire effects, of which **goiter** is perhaps the best known (Figure 12.10). Iodine is available in seafood. To guard against iodine deficiency, a small amount of sodium iodide (NaI) is added to table salt (NaCl). Iodized salt has greatly reduced the incidence of goiter.

Iron(II) ions (Fe^{2+}) are necessary for the proper function of the oxygen-transporting compound hemoglobin. Without sufficient iron, there will be a shortage of oxygen supplied to the body tissues. The resulting weakened condition is called **anemia.** Foods especially rich in iron compounds include red meats and liver.

As we have seen, calcium and phosphorus are necessary for the proper development of bones and teeth. Growing children need about 1.5 g of each per day. These elements are available in plentiful quantities in milk. The need of adults for these elements is less widely known but very real. For example, calcium ions are necessary for the coagulation of blood (to stop bleeding) and for maintaining the rhythm of the heartbeat. Phosphorus is necessary for the body to metabolize carbohydrates (Chapter 27). Without phosphorus compounds, we couldn't get *any* energy from those "quick energy" foods. Compounds containing phosphorus play many other essential roles, also. We will encounter a number of these compounds in subsequent chapters.

Sodium chloride in moderate amounts is essential to life. It is important in the exchange of fluids between cells and plasma, for example. The presence of salt increases water retention. A high volume of retained fluids can cause swelling and high blood pressure **(hypertension)**. Over 120 million prescriptions are written each year in the United States for diuretics, drugs that induce urination in an attempt to reduce the volume of retained fluids. Another 30 million prescriptions are written for potassium (K^+) supplements to replace that washed out in the excess urine. There are an estimated 36 million people in the United States suffering from hypertension. And most physicians agree that our diets generally contain too much salt.

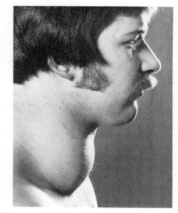

FIGURE 12.10
A man affected by goiter. The swollen thyroid gland in the neck results from a lack of the trace element iodine in the diet. (Photo by Joseph J. Mentrikoski, Department of Medical Photography, Geisinger Medical Center.)

Iron, copper, zinc, cobalt, manganese, molybdenum, calcium, and magnesium are essential to the proper functioning of certain enzymes. These metalloenzymes are necessary to life. The functions of some other minerals are quite complex. Some things are known about how they operate, but a great deal remains to be learned about the role of inorganic chemicals in our bodies. Bioinorganic chemistry is a flourishing area of research.

PROBLEMS

1. Define or illustrate these terms:
   a. cathode
   b. anode
   c. cation
   d. anion
   e. electrolysis
   f. strong electrolyte
   g. weak electrolyte
   h. nonelectrolyte
   i. ionization
   j. ion dissociation
   k. solubility product constant

2. Classify the following as strong electrolytes, weak electrolytes, or nonelectrolytes in solution:
   a. KCl
   b. HNO_3
   c. H_2CO_3
   d. Na_2SO_4
   e. KOH
   f. CH_3OH
   g. AgCl
   h. CCl_4
   i. $CaCl_2$

3. Why does a solution of 1 mol of NaCl in 1 kg of water freeze at a lower temperature than a solution of 1 mol of sugar in 1 kg of water? Why is the freezing-point depression for the salt solution not quite twice that for the sugar solution?

4. Hydrogen chloride is a covalent molecule. Why does a solution of the gas in water conduct electricity?

5. What new substances would be formed if electricity were passed through the following?
   a. molten KBr   b. molten LiCl   c. molten Al_2O_3

6. Both $Ba(OH)_2$ and H_2SO_4 are strong electrolytes. When equimolar solutions of the two are mixed, the resulting solution is essentially nonconducting. Write the equation for the reaction and explain the observation.

7. Write an equation that shows why hydrogen iodide (HI), a gas, forms an aqueous solution that conducts electricity.

8. Lead chromate ($PbCrO_4$) is an ionic compound, yet a saturated aqueous solution of lead chromate does not conduct electricity. Explain.

9. State the main ideas of the modernized version of Arrhenius's theory of electrolytes.

10. How does an electrolytic cell differ from an electrochemical cell?

11. Solid sodium chloride does not conduct electricity, but the molten salt does. Explain.

12. Formaldehyde (CH_2O) is a nonelectrolyte. What type of bonding exists in formaldehyde?

13. Classify each of the following electrolytes as an acid, a base, or a salt. Write equations to show what happens when each is dissolved in water.
    a. $Ca(OH)_2$ b. HBr   c. $Sr(NO_3)_2$

14. Why are water, an electrolyte, and oxygen all required for the corrosion of iron?

15. Why does aluminum corrode more slowly than iron, even though aluminum is more reactive than iron?

16. How does silver tarnish? How can the tarnish be removed without the loss of silver?

17. Will a precipitate form if 0.001 mol of $MgCl_2$ and 0.001 mol of Na_2CO_3 are added to 1.0 L of water? The net ionic reaction is

$$Mg^{2+} + CO_3{}^{2-} \rightleftharpoons MgCO_3(s)$$

You may refer to Table 12.2 for the K_{sp} value.

18. Will a precipitate form if 0.001 mol of $AgNO_3$ and 0.0001 mol of NaCl are added to 1.0 L of water? The net ionic reaction is

$$Ag^+ + Cl^- \rightleftharpoons AgCl(s)$$

19. Will a precipitate occur if 1×10^{-6} mol of $Pb(NO_3)_2$ and 1×10^{-5} mol of Na_2CrO_4 are added to 1.0 L of water? The net ionic reaction is

$$Pb^{2+} + CrO_4{}^{2-} \rightleftharpoons PbCrO_4(s)$$

20. Would an object to be electroplated be made the anode or the cathode in an electrolytic cell? Why?

21. If a concentrated solution of sodium acetate is mixed with a solution of silver nitrate, a precipitate of silver acetate is formed. The precipitate dissolves readily when nitric acid is added. Write equations to explain what happens.
22. Boiler scale ($CaCO_3$) is insoluble in water, yet it readily dissolves in hydrochloric acid. Write the equation that explains what happens.
23. What is goiter? How is it prevented?
24. What is anemia? How is it prevented?
25. Describe the role of calcium and phosphorus in the proper development of bones and teeth.
26. What is hypertension?
27. What is a diuretic?
28. What causes tooth decay?
29. What is bulimia? How is it related to erosion of the teeth?
30. What happens to bone formation in children who suffer from chronic acidosis? Why?

31. What happens to the teeth when the salivary glands are removed or destroyed? Why?
32. Nineteen centuries ago, the Romans added calcium sulfate to wine. It clarifies the wine, but also removes any dissolved lead. If 1 L of wine is saturated with calcium sulfate, the concentration of SO_4^{2-} is 0.014 M. What concentration of Pb^{2+} would remain in solution? The K_{sp} for lead sulfate is 1.1×10^{-8}, and the net ionic equation is

$$Pb^{2+} + SO_4^{2-} \rightleftharpoons PbSO_4(s)$$

33. Phosphates can be removed from water by precipitation using iron(III) sulfate.

$$2\,PO_4^{3-}(aq) + Fe_2(2O_4)_3 \longrightarrow 2\,FePO_4 + 3\,SO_4^{2-}(aq)$$

An estimated 7 t per day of phosphates enters the East Anglian (United Kingdom) water system. How much iron(III) sulfate would be required to remove this phosphate?

13

Inorganic Chemistry

I t almost seems a contradiction of terms. **Inorganic** means nonliving and not derived from life. **Organic** means living, having the characteristics of life, or derived from life. At least, those are the everyday definitions of the terms. So how could inorganic chemicals be important to living organisms?

In the old days, chemists used the terms *organic* and *inorganic* in much the same way as everyone else. They believed that, while they could make many different inorganic chemicals, they could not hope to make organic compounds in the laboratory. They believed, with everyone else, that only living organisms, within their tissues, could make organic compounds. Some **vital force,** they thought, was necessary for the synthesis of organic substances.

A series of experiments in the early 1800s led to the overthrow of the vital force theory. Perhaps the most important single step was made by the young German chemist Friedrich Wöhler in 1828. He made some urea (NH_2CONH_2), long recognized as a typical organic compound, simply by heating a solution of ammonium cyanate ($NH_4^+CNO^-$), a compound generally regarded to be inorganic. Although a few die-hard vitalists held out for several decades, the vital force theory was practically dead by the middle of the nineteenth century. **Organic chemistry** is now defined as the chemistry of the compounds of carbon. **Inorganic chemistry** is that of all the other elements. Several later chapters are devoted to organic chemistry. In this chapter, we will study some important inorganic chemistry. The elements and their compounds will be organized by family or group, following the periodic table.

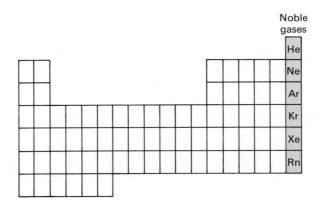

FIGURE 13.1 *The noble gases.*

13.1 THE NOBLE GASES

When Dmitri Mendeléev presented the periodic table in 1869 (Chapter 2), he confidently left spaces in it for elements that he felt should exist but that had yet to be discovered. In succeeding years, his predictions were substantiated by the isolation and identification of the missing elements. In the last decade of the nineteenth century, however, a series of elements were discovered that made up an entirely new family, one completely unexpected by Mendeléev and his contemporaries. Nonetheless, this new group, called the noble gases, fit neatly between the very reactive alkali metals (Group IA) and the highly active nonmetals of Group VIIA. In the usual flat form of the periodic table, the noble gases are placed to the far right, in the column immediately following Group VIIA (Figure 13.1).

The six noble gases are helium, neon, argon, krypton, xenon, and radon. All, with the exception of radioactive radon, are components of the atmosphere (Table 13.1). Argon is rather abundant, making up nearly 1% by volume of the atmosphere. Xenon, on the other hand, is quite rare, making up only 1 part in 11 000 000 of the atmosphere.

TABLE 13.1 *Noble Gases in the Atmosphere*

Gas	Number of Atoms in 11 000 000 Molecules of Air
Helium	55
Neon	15
Argon	100 000
Krypton	11
Xenon	1

Helium is found in natural gas deposits, particularly those under the Great Plains of the United States. Presumably, this helium was formed by the *alpha* decay of radioactive elements within the Earth. Alpha particles (helium nuclei) become helium atoms by acquiring two electrons. Although some helium is continually being made in the Earth's interior, the process is very slow. The helium we tap has been accumulated over eons and, therefore, must be regarded as an essentially nonrenewable resource. Our future supplies are threatened, because it is not economical at present to collect and store all that is produced in commercial operations. Much helium is simply released into the atmosphere.

Helium is used to fill balloons and dirigibles (Figure 13.2). Its lifting power is more than 90% that of hydrogen, the lightest of all the gases, and helium has the advantage of being nonflammable. Helium is also used to provide an inert atmosphere for the welding of metals that otherwise might be attacked by oxygen in the air. Liquid helium is used to achieve extremely low temperatures. It boils at $-268.9\,°C$, only $4.2\,°C$ above absolute zero.

Another interesting use of helium is in breathing mixtures for deep-sea divers (see Section 6.9). Mixtures of helium and oxygen have also been used in the treatment of asthma, emphysema, and other conditions involving respiratory obstruction. The same low formula weight that gives helium its lifting power also permits it to diffuse more rapidly than nitrogen. The oxygen-helium mixture puts less strain on the muscles involved in breathing than does air.

Neon is used in lighted signs for advertising (Figure 13.3). A tube with electrodes is shaped into letters or symbols and is filled with neon at low pressure. An electric current is passed through the tube, causing the atoms to emit their characteristic spectral lines (Section 2.9). With a spectroscope, we would see the individual lines. With the unaided eye, however, we merely see the familiar orange-red glow. Other colors are obtained with mixtures of argon and neon or mercury vapor and neon or by use of colored glass for the tube.

Argon, as Table 13.1 shows, is the most plentiful of the noble gases. It can be separated from air rather inexpensively. Argon is used to fill incandescent light bulbs. Unlike nitrogen and oxygen, it does not react with the tungsten filament. It also decreases the tendency of the filament of vaporize, thus extending the filament's life. Fluorescent lights are filled with a mixture of argon and mercury vapor. Argon also provides an inert atmosphere for carrying out chemical reactions in which one or more of the reactants is sensitive to air oxidation.

Krypton and xenon are too expensive to have many important commercial applications, although krypton has found some use in light bulbs. Radon, although exceedingly rare in the atmosphere, can be collected from the radioactive decay of radium.

$$^{226}_{88}\text{Ra} \longrightarrow\ ^{4}_{2}\text{He} + ^{222}_{86}\text{Rn}$$

FIGURE 13.2
The helium-filled Goodyear blimp is a familiar sight to those who watch sports events on television. (Courtesy of The Goodyear Tire and Rubber Co., Akron, Ohio.)

FIGURE 13.3
The color of neon signs is due to changes in electronic energy levels in neon atoms. (Photo courtesy Stock, Boston; © Barbara Alper.)

FIGURE 13.4
Crystals of xenon
tetrafluoride, the first binary
compound of a noble gas.
(Courtesy of Argonne
National Laboratory,
Argonne, Ill.)

This radon may be sealed in small vials and used for radiation therapy of certain malignancies.

The noble gases are exceptionally resistant to chemical reaction. Other elements undergo reactions in order to achieve the more stable electronic configurations of the noble gases (Chapter 4). All the noble gases exist in elemental form as monatomic species. The other gases we have encountered, such as hydrogen (H_2), nitrogen (N_2), and chlorine (Cl_2), exist as diatomic molecules so that each atom can achieve a stable outer electron configuration, but the individual atoms of the noble gases already have this arrangement. Because of their stability, the noble gases were often called "inert gases." But, in 1962, it was discovered that a few compounds of krypton and xenon could be formed, necessitating the change from "inert" to "noble." (Presumably, radon also would form compounds, but it is relatively rare and highly radioactive. The dangers involved in working with this compound have restricted research involving it.) The first binary (two-element) compound of a noble gas was prepared at Argonne National Laboratory by the simple procedure of heating a mixture of xenon and fluorine (the most reactive element) at 400 °C. Colorless crystals of xenon tetrafluoride (XeF_4) were obtained (Figure 13.4).

$$Xe(g) + 2\,F_2(g) \longrightarrow XeF_4(s)$$

The xenon tetrafluoride crystals are stable at room temperature. In subsequent years, more than 20 compounds of xenon and krypton have been made. As yet, despite many attempts, no compounds have been made of the lighter elements, helium, neon, and argon. While this family of elements is no longer called "inert," its nobility is still unquestioned. More than any other family, the noble gases disdain interactions with the masses of other elements.

13.2 GROUP IA: THE ALKALI FAMILY

The Group IA elements (Figure 13.5) are called the alkali metals (from Arabic words describing a source of these elements). Because the hydroxides of these elements are strongly basic, basic solutions are said to be alkaline. There are six elements in the group, lithium, sodium, potassium, rubidium, cesium, and francium. The last is highly radioactive, and little is known of its properties. It will not be discussed further here.

In the elemental form, the alkali metals are soft solids with low melting points. Indeed, on a hot day, cesium would be a liquid, for it melts at 29 °C. These metals, when freshly cut, are bright and shiny, but they tarnish readily as they become oxidized by the atmosphere. All are highly reactive, showing a great tendency to give up electrons and form 1+ ions. For example,

$$Na \longrightarrow Na^+ + 1e^-$$
$$K \longrightarrow K^+ + 1e^-$$

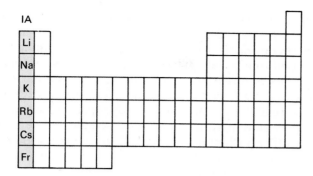

FIGURE 13.5 Group IA, the alkali metals.

In compounds, the alkali metals occur as the $1+$ ions. A variety of compounds are of considerable importance.

Lithium salts are found in certain naturally occurring brines. Lithium carbonate (Li_2CO_3) is used in medicine to level out the dangerous "manic" highs that occur in manic-depressive psychoses. Some practitioners also recommend lithium carbonate for the depression stage of the cycle. It appears to act by affecting the transport of chemical substances across cell membranes in the brain.

Sodium salts are quite common. Ordinary table salt (NaCl) supplies the body with chloride ion, necessary for the production of hydrochloric acid by our stomachs. Living tissues require a balance of sodium ion and potassium ion. In animals, the former predominates; in plants, the latter. By salting our vegetables, we maintain the proper balance (although it is safe to say that few people have precisely this objective in mind when they pick up the saltshaker).

Potassium ion is an essential nutrient for plants. It is generally abundant and is readily available to plants, except in soil depleted by high-yield agriculture. The precise function of potassium ion in plant cells has been difficult to pinpoint. It seems to be involved in the formation and transport of carbohydrates. It may also be necessary for the buildup of proteins from smaller compounds called amino acids. The usual form of potassium in commercial fertilizers is potassium chloride (KCl). Vast deposits of this salt occur at Strassfurt, in Germany. For years this source supplied nearly all the world's potassium fertilizer. With the coming of World War I, the United States sought supplies within its own borders. Deposits at Searles Lake, California, and Carlsbad, New Mexico, now supply most of the needs of the United States. Canada has vast deposits in Saskatchewan and Alberta. Beds of potassium chloride up to 200 m thick lie about 1 km below the prairies.

In animals, potassium ions are the principal positive ions inside cells. They serve to maintain the osmotic pressure and electrical potential of cells. Sodium ions are the principal positive ions in the extracellular fluid. They are involved in fluid retention and blood pressure. Thus, individuals who are hypertensive or who are suffering from edema usually are advised to reduce salt (NaCl) intake. Sodium ions also serve to regulate the excitability of nerves and muscles.

Many acidic drugs are used in the form of their sodium or potassium salts. These ionic forms are more soluble in water than the "free" acids. For example, benzylpenicillinic acid (free penicillin G) is sparingly soluble in water. The potassium salt (potassium penicillin G) is quite soluble.

We encounter other sodium and potassium salts in other chapters. Salts of rubidium and cesium are expensive and rare.

13.3 GROUP IIA: THE ALKALINE EARTH FAMILY

The oxides and hydroxides of the Group IIA (Figure 13.6) elements are basic. Only those of the heavier elements are appreciably soluble in water. In the early days of chemistry, the name **earth** referred to substances unaffected by the heat and insoluble in water. All of these facts together account for the name assigned to the Group IIA elements, the alkaline earth metals.

The six alkaline earth metals are beryllium, magnesium, calcium, strontium, barium, and radium. In the elemental form, these metals are fairly soft and reactive. They show a tendency to give up electrons and form 2+ ions. For example,

$$Mg \longrightarrow Mg^{2+} + 2e^-$$
$$Ca \longrightarrow Ca^{2+} + 2e^-$$

In compounds, these metals occur almost exclusively as the 2+ ions. Many of the compounds are of great importance.

Beryllium is something of an oddball member of the family. Unlike the others, it does not react with water. Indeed, beryllium tends to form covalent,

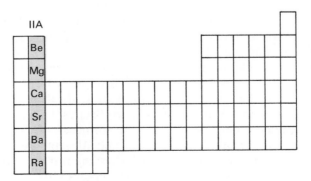

FIGURE 13.6 Group IIA, the alkaline earth family.

rather than ionic, bonds. The metal itself is rather hard, rigid, and strong. Its lightness makes it quite valuable in structural alloys. The element is poisonous in all its forms. Inhalation of the metal or of beryllium oxide may cause *berylliosis,* a serious and sometimes fatal lung disease that resembles the infamous black lung disease of coal miners.

Magnesium ions are essential to both plants and animals. In plants, magnesium ions are incorporated into chlorophyll molecules. Thus, this ion is essential to photosynthesis. Both calcium and magnesium ions are found primarily within the cells. They are essential for proper functioning of the nerves that control muscles.

Calcium ions are necessary for the proper development of bones and teeth. For this reason, growing children are usually encouraged to drink milk, a rich source of calcium. Adults also require calcium, which is necessary for clotting of blood and maintenance of a regular heartbeat.

Calcium carbonate (limestone) and other rocks containing calcium ions (Ca^{2+}), magnesium ions (Mg^{2+}), or iron ions (Fe^{2+} or Fe^{3+}) are widely distributed in nature. Most compounds containing these ions are only slightly soluble, yet enough of the ions dissolve in the natural water in some areas to cause problems. Water containing calcium, magnesium, or iron ions is what is known as **hard water.** The ions react with the negative ions in soaps to form precipitates. For example,

$$Ca^{2+}(aq) + 2\,C_{17}H_{35}COO^-(aq) \longrightarrow Ca(C_{17}H_{35}COO)_2(s)$$

$$\text{Soap} \qquad\qquad\qquad \text{Bathtub ring}$$

This curdy precipitate clings to clothes, hair, and skin, leaving them dingy and dirty in appearance and feel even when they've been freshly washed.

13.4 GROUP IIIA: ALUMINUM

Group IIIA (Figure 13.7) consists of the nonmetal boron and the metals aluminum, gallium, indium, and thallium. Boron tends to form covalent compounds. Boric acid, H_3BO_3, is a familiar ingredient of mild, antiseptic eye rinses. The other elements in Group IIIA are typical metals and tend to form $3+$ ions.

$$Al \longrightarrow Al^{3+} + 3e^-$$
$$Ga \longrightarrow Ga^{3+} + 3e^-$$

Aluminum is the most abundant metal in the Earth's crust, but it is tightly bound in its compounds in nature. Much energy is required to extract aluminum metal from its ores. The principal ore of aluminum is bauxite, a mineral in which aluminum oxide is associated with one, two, or three water molecules per Al_2O_3 unit. The formula for bauxite often is written $Al_2O_3 \cdot xH_2O$. The bauxite also contains iron oxides, silicates, and other impurities. It is extracted with

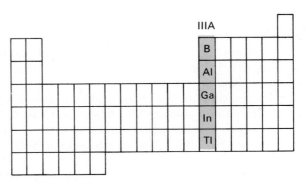

FIGURE 13.7   *Group IIIA.*

a strong base and then heated to form the oxide Al_2O_3. The oxide is melted and separated by passing electricity through it.

$$2 Al_2O_3 \xrightarrow{\text{electricity}} 4 Al + 3 O_2$$

It takes 2 t of aluminum oxide and 17 000 kWh of electricity to produce 1 t of aluminum.

Aluminum is light and strong. It is widely used in airplanes. As we try to decrease the weight of automobiles in order to increase gas mileage, aluminum replaces steel in many automobile parts. Aluminum corrodes much more slowly than iron, and, for many purposes, that is a considerable advantage.

13.5 GROUP IVA: CARBON AND SOME OF ITS COMPOUNDS

Group IVA (Figure 13.8) is made up of carbon, silicon, germanium, tin, and lead. Of these, carbon is easily the most important. Carbon forms thousands—perhaps millions—of compounds with hydrogen. These compounds, called hy-

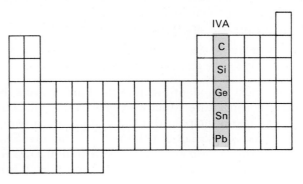

FIGURE 13.8   *Group IVA.*

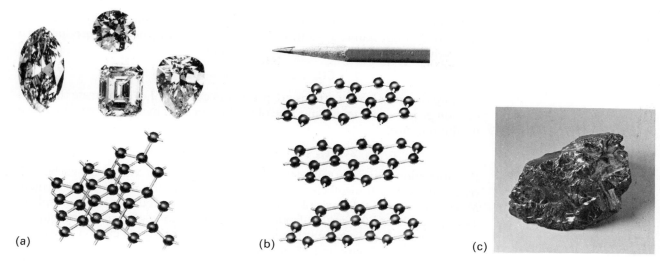

FIGURE 13.9   *The allotropic forms of carbon. (a) Diamond. (b) Graphite.
(c) Amorphous. Coal has no regular molecular structure. (a and b printed with
permission from Cotton, F. A., Darlington, C. L., and Lynch, L. D., Chemistry: An
Investigative Approach, rev. ed., Boston: Houghton Mifflin, 1976. Photo of coal by
Dennis Tasa.)*

drocarbons, are discussed in detail in Chapter 14. Hydrocarbons and their derivatives are called organic chemicals, and their study is called organic chemistry.
There are a few simple compounds of carbon, though, that are more like inorganic than organic chemicals. Among these are carbon monoxide (CO), carbon
dioxide (CO_2), and such minerals as limestone and marble (calcium carbonate,
$CaCO_3$).

Carbon exists in three allotropic forms (Figure 13.9). **Allotropes** are modifications of an element that can exist in more than one form in the same physical
state. The solid element can be found in the form of graphite (the "lead" of
pencils), carbon black (the soot that forms on the bottom of a casserole warmed
over a candle flame), and diamond (the precious jewel). Coal is composed of
varying amounts of elemental carbon, from about 6% in peat up to 88% or
more in anthracite. When burned, the carbon in coal is oxidized to carbon
dioxide.

$$C + O_2 \qquad CO_2$$

Heat is given off in this process. Not all the carbon is completely oxidized. Some
of it winds up as carbon monoxide.

$$2\,C + O_2 \longrightarrow 2\,CO$$

Still other carbon, essentially unburned, ends up as soot.

Hydrocarbons are also burned as fuels. Methane, the simplest hydrocarbon, burns with a hot flame. If sufficient oxygen is present, the main products are relatively innocuous carbon dioxide and water.

$$CH_4 + 2O_2 \longrightarrow CO_2 + 2H_2O$$

More complex hydrocarbons are present in gasoline. Let's illustrate the combustion process with an octane, one of the hundreds of hydrocarbons that make up the mixture.

$$2C_8H_{18} + 25O_2 \longrightarrow 18H_2O + 16CO_2$$

If combustion occurs in sufficient oxygen, the major products are carbon dioxide and water. If insufficient oxygen is present, carbon monoxide is formed. Millions of tons of this invisible but deadly gas are poured into the atmosphere each year; about 80% of that from human activity comes from automobile exhausts. The U.S. government has set danger levels of 9 ppm carbon monoxide (average) over 8 hours and 35 ppm (average) over 1 hour. Even in off-street urban areas, levels often average 7 to 8 ppm. On streets, danger levels are exceeded much of the time. Such levels do not cause immediate death, but, over a long period, exposure can cause physical and mental impairment.

Carbon monoxide is an invisible, odorless, tasteless gas. There is no way for a person to tell that it is around (without using test reagents or instruments). Drowsiness is usually the only symptom, and that is not an unpleasant one. How many auto accidents are caused by drowsiness or sleep induced by carbon monoxide? No one knows for sure. Cigarette smoke also contains a fairly high concentration of carbon monoxide.

You can't necessarily escape carbon monoxide by going inside. Office buildings and apartments have been found to have essentially the same concentration of carbon monoxide as the air outside. In some areas, this level exceeds federal standards. It has been recommended that buildings in high-traffic areas be tightly sealed at lower levels. They should also be spaced in such a way that the wind can disperse the pollutants.

Carbon monoxide exerts its insidious effect by tying up the hemoglobin in the blood. The normal function of hemoglobin is to transport oxygen (Figure 13.10). Carbon monoxide binds tenaciously to hemoglobin—once on, it refuses to get off. The hemoglobin is thus prevented from transporting oxygen. The symptoms of carbon monoxide poisoning are therefore those of oxygen deprivation. All except the most severe cases of carbon monoxide poisoning are reversible. The best antidote is the administration of pure oxygen. Greatly increasing the concentration of oxygen favors the oxygen-hemoglobin reaction at the expense of the carbon monoxide–hemoglobin combination. Artificial respiration may help if a tank of oxygen is not available.

Carbon monoxide is very much a local pollution problem. It is a severe threat to urban areas with heavy traffic, but it does not appear to be a global threat.

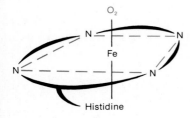

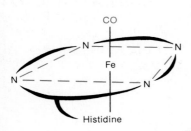

FIGURE 13.10
Schematic representations of a portion of the hemoglobin molecule. Histidine is an amino acid. Carbon monoxide bonds much more tightly than oxygen.

In laboratory tests, carbon monoxide will survive about 3 years in contact with air. But nature is somehow able to prevent an overall buildup, despite the large amounts being poured into the environment. In fact, it is estimated that up to 80% of the total carbon monoxide in the atmosphere comes from natural sources. Except for those highly localized situations (they're bad enough!), nature seems to have the carbon monoxide situation under control.

The Minipolluters: Cigarettes, Smokers, and Nonsmokers. The smoke-filled room is well known in political mythology as the place where decisions are made. The health effects of smoking on the smoker are also widely known. These have been publicized following a series of reports by the Surgeon General. The first report was published in 1964 and led to a ban on television advertisements of cigarettes. Later reports have included studies of the effect of tobacco smoke as an air pollutant. They may yet end the era of the smoke-filled room.

A series of studies have shown that the air quality in such a room is very poor. The level of carbon monoxide, even in a well-ventilated room, is often equal to or greater than the legal limits permitted for ambient air quality. These levels have been shown to impair time-interval discrimination. In severe cases, performance on psychomotor tests was impaired. Health effects on people already suffering from heart or lung disease might be quite severe. Nonsmokers are also exposed to significant levels of "tar" and nicotine. These may be harmful, but little research has been done in this area. Some states have already acted to ban smoking in meeting rooms, waiting rooms, and other public places.

Not only is carbon dioxide a product of combustion, it is also produced in respiration. Generally, it is regarded as innocuous. Certainly any immediate effect on us is slight. But what about long-term changes? No matter how "clean" an engine or a factory is, as long as it burns coal or petroleum products, it will produce carbon dioxide and water. The concentration of carbon dioxide in the atmosphere has increased 14% in this century. It continues to increase at an expanding rate through the increased burning of carbon fuels.

Both water vapor and carbon dioxide produce a **greenhouse effect.** These chemicals let the sun's rays (visible light) in to warm the surface of the Earth, but when the Earth tries to radiate this energy back out into space as heat (infrared energy), the energy is trapped by the molecules of carbon dioxide and water (Figure 13.11). Water vapor spewed into the atmosphere soon falls back to Earth as rain. It therefore affects the climate mostly at the local level. Carbon dioxide poured into the air is not so readily dissipated. It hangs around to affect the climate of the entire world. The concentration of carbon dioxide

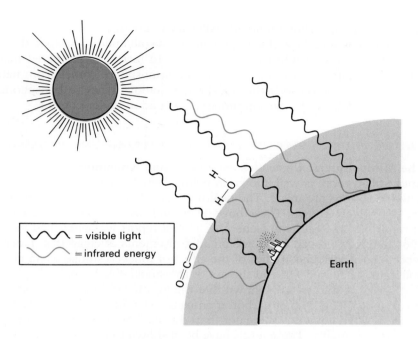

FIGURE 13.11 The greenhouse effect.

is now increasing at a rate of 1 ppm per year. Some scientists predict a warming trend that will melt the polar ice caps, thus flooding coastal cities. The fears have increased in recent years. It has been found that methane and other trace gases contribute to the greenhouse effect. The concentration of methane in the atmosphere has been increasing since 1977.

At present the Earth's atmosphere seems to be cooling, not warming. We also pour soot, smoke, and dust into the atmosphere. These, perhaps aided by the vapor trails of jet aircraft, screen out the sun's light and tend to lower the Earth's temperature. We may not always be so lucky as to balance one effect against another.

13.6 GROUP VA: NITROGEN COMPOUNDS

Group VA (Figure 13.12) includes nitrogen, phosphorus, arsenic, antimony, and bismuth. We will discuss the chemistry of nitrogen and some of its compounds in this section. In Section 13.7, we will pay particular attention to the role of nitrogen oxides in air pollution. The element nitrogen occurs as N_2 in the atmosphere. It also occurs in the combined form in nature. It is an essential element, occurring in all proteins and in many other molecules important to living organisms. We discuss some organic compounds of nitrogen in Chapter

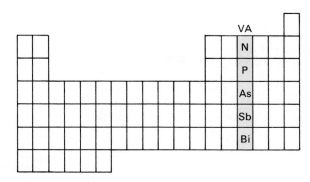

FIGURE 13.12 *Group VA.*

18. Biological compounds containing nitrogen atoms are a major part of our study in other chapters.

Although nitrogen makes up 78% of the atmosphere, the molecules of nitrogen gas cannot be used directly by higher plants or by animals. They first have to be "fixed," that is, converted to a more readily used form. Certain types of bacteria convert atmospheric nitrogen to nitrates. Other bacteria convert the nitrogen in compounds back to nitrogen gas. A nitrogen cycle (Figure 13.13) is thus established. Lightning also serves to "fix" nitrogen by causing

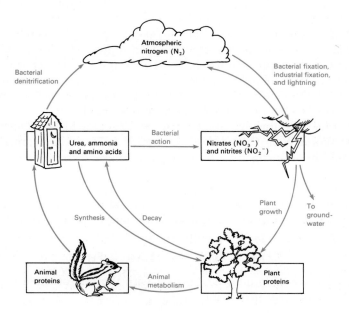

FIGURE 13.13 *The nitrogen cycle.*

it to combine with oxygen. Nitric oxide (NO) and nitrogen dioxide (NO_2) are formed. The equations are

$$N_2 + O_2 \xrightarrow{\text{lightning}} 2\,NO$$

$$2\,NO + O_2 \longrightarrow 2\,NO_2$$

The latter reacts with water to form nitric acid (HNO_3).

$$3\,NO_2 + H_2O \longrightarrow 2\,HNO_3 + NO$$

The nitric acid falls in rainwater, adding to the supply of available nitrates in sea and soil. Scientists have undertaken substantial intervention in the nitrogen cycle by industrial fixation—the manufacture of nitrogen fertilizers. This intervention has greatly increased our food supply, for the availability of fixed nitrogen is often the limiting factor in the production of food. Not all the consequences of this intervention have been favorable. Excessive runoff of nitrogen fertilizer has led to serious water pollution problems in some areas, but modern methods of high-yield farming demand "chemical" fertilizers. Plants usually take up nitrogen in the form of nitrates (NO_3^-) or ammonium ion (NH_4^+). These, combined with carbon compounds from photosynthesis, form amino acids, the building blocks of proteins. For years farmers were dependent on manure as a source of nitrates. Discovery of deposits of sodium nitrate, called Chile saltpeter, in the deserts of northern Chile led to exploitation of this substance as a supplemental source of nitrogen.

A rapid rise in population in the late nineteenth and early twentieth centuries led to increasing pressure on the available food supply. This in turn led to an increasing demand for nitrogen fertilizers. The atmosphere offered a seemingly inexhaustible supply—if only it could be converted to a form useful to humans. Every flash of lightning forms some nitric acid in the air. This process probably contributes about 9 kg of nitrogen per hectare of land.

The first real breakthrough in nitrogen fixation came in Germany on the eve of the First World War. The process, developed by Fritz Haber, made possible the combination of nitrogen with hydrogen to make ammonia (Section 1.1).

$$3\,H_2 + N_2 \longrightarrow 2\,NH_3$$

By 1913, one plant was in production and several more were under construction. The Germans were able to make ammonium nitrate (NH_4NO_3), an explosive, by oxidizing part of the ammonia to nitric acid.

$$2\,NH_3 + 4\,O_2 \longrightarrow 2\,HNO_3 + 2\,H_2O$$

The nitric acid was then allowed to react with the rest of the ammonia, and ammonium nitrate was produced.

$$HNO_3 + NH_3 \longrightarrow NH_4NO_3$$

The Germans were mainly interested in ammonium nitrate as an explosive, but it turned out to be a valuable nitrogen fertilizer as well. The same discovery that enabled the Germans to prolong World War I probably helped postpone predicted famine for several decades.

Increasingly, ammonia is applied directly as fertilizer. A gas at room temperature, ammonia is soluble in water and is also easily compressed into a liquid that can be stored and transported in tanks. Much of it, as we mentioned, is converted to ammonium nitrate, a crystalline solid. Some is converted to solid ammonium sulfate $[(NH_4)_2SO_4]$ by reaction with sulfuric acid.

$$2\,NH_3 + H_2SO_4 \longrightarrow (NH_4)_2SO_4$$

These may be applied separately or combined with other plant nutrients to make a more complete fertilizer.

It is largely through nitrogen fertilizers that we are able to feed the great human population that inhabits the Earth. Unfortunately, our ways of fixing nitrogen require great quantities of energy. The petroleum shortages and higher prices of recent years have lead to scarcity and to inflation of fertilizer prices. If only we could find a way to fix nitrogen gas at ambient temperatures, the way microorganisms do . . .

Fritz Haber was awarded the Nobel Prize for chemistry in 1918. There are a number of ironies in this. Alfred Nobel, a Swedish inventor and chemist who died in 1896, endowed the Nobel Prize (including the peace prize) with a fortune derived from his own work with explosives. During his lifetime, he was bitterly disappointed that the explosives he developed for excavation and mining were put to such disastrous use in wars. And Haber, a man who helped his country during World War I, was exiled from his native land in 1933. He was a Jew. Nazi racial laws forced him out of his position as director of the Kaiser Wilhelm Institute of Physical Chemistry. He accepted a post at Cambridge University in England, but his life was ended by a stroke less than a year later.

13.7 NITROGEN OXIDES AND AIR POLLUTION

Automobiles also fix nitrogen. Indeed, any time air is subjected to high temperatures, some of the nitrogen and oxygen combine. Unfortunately, most automobiles operate in urban areas, not on farms. And the nitrogen oxides undergo reactions with other pollutants to form a variety of noxious chemicals, not just fertilizers. This type of air pollution is called Los Angeles smog, or, more

FIGURE 13.14
Downtown Los Angeles on a smoggy day in 1956. Smog is trapped by a temperature inversion with a layer of warm air only 100 m above the ground. The upper portion of Los Angeles City Hall can be seen in the clear air above the base of the inversion. (Courtesy of the Southern California Air Pollution Control District, Los Angeles.)

properly, **photochemical smog.** In contrast to the London type (Section 13.9), which requires cold, damp air, photochemical smog usually occurs in dry, sunny climates. The principal culprits are unburned hydrocarbons and oxides of nitrogen from automobiles. The warm, sunny climate that has drawn so many people to the Los Angeles area is also the perfect setting for photochemical smog (Figure 13.14).

The combustion process in an automobile engine leads to the formation of nitric oxide (NO).

$$N_2 + O_2 \longrightarrow 2\,NO$$

Nitric oxide is oxidized slowly by oxygen to nitrogen dioxide (NO_2).

$$2\,NO + O_2 \longrightarrow 2\,NO_2$$

Nitrogen dioxide is an amber-colored gas. Smarting eyes and a brownish haze are excellent indicators of Los Angeles smog. It is this nitrogen dioxide that plays a crucial role in photochemical smog. It absorbs a photon of sunlight and breaks down into nitric oxide and very reactive oxygen atoms.

$$NO_2 + \text{sunlight} \longrightarrow NO + O$$

This oxygen atom reacts with other components of the automobile exhaust and the atmosphere to produce a variety of irritating and toxic chemicals (Figure 13.15).

At concentrations currently found in air, the oxides of nitrogen are not particularly dangerous in themselves. Nitric oxide at high concentrations will react with hemoglobin. As with carbon monoxide, this leads to oxygen deprivation.

The initiator:

$$NO_2 + \text{sunlight} \rightarrow NO + O$$

Secondary reactions:

$$O + O_2 \rightarrow O_3$$

$$O + \text{hydrocarbons} \rightarrow \text{aldehydes}$$

Tertiary reactions:

$$O_3 + \text{hydrocarbons} \rightarrow \text{aldehydes} \left(R-C \begin{matrix} \nearrow O \\ \searrow H \end{matrix} \right)$$

$$\text{Hydrocarbons} + \text{oxygen} + NO_2 \rightarrow \text{PAN} \left(R-C \begin{matrix} \nearrow O \\ \searrow O-O-NO_2 \end{matrix} \right)$$

FIGURE 13.15 A summary of some of the principal reactions in the formation of photochemical smog. Most reactive intermediates have been omitted from this simplified scheme.

Such levels are seldom, if ever, reached from ordinary air pollution but might be reached near a specific industrial source. Nitrogen dioxide serves as an irritant to the eyes and respiratory system. Tests with laboratory animals indicate that chronic exposure to levels in the range of 10 to 25 ppm might lead to emphysema or other degenerative diseases of the lungs.

The most serious environmental effect of the nitrogen oxides is their role in smog formation. However, these gases do contribute to the fading and discoloration of fabrics. By forming nitric acid, nitrogen oxides contribute to the acidity of rainwater, accelerating the corrosion of metals. They also contribute to crop damage, although specific effects are difficult to separate from those of other pollutants.

13.8 GROUP VIA: OXYGEN AND OZONE

Group VIA (Figure 13.16) includes oxygen, sulfur, selenium, tellurium, and polonium. The most important of these is oxygen, which we will discuss in this section. Second in importance is sulfur. We will discuss that element and some of its compounds in Section 13.9.

The element oxygen occurs in the atmosphere mainly as the diatomic molecule O_2. In Chapters 8 and 12, we saw how oxygen is involved in oxidation processes such as combustion (rapid burning) and rusting and other forms of corrosion, and in respiration. Oxygen reacts rapidly with more active metals. Magnesium, for example, burns with a brilliant white flame when ignited in air.

$$2\,Mg + O_2 \longrightarrow 2\,MgO + heat + light$$

At room temperature, a reaction occurs on the surface of freshly prepared magnesium metal. This magnesium oxide forms a thin, transparent coating that is impervious to air, preventing further oxidation. Such oxide coatings are common in metals, making it possible for us to use those otherwise quite reactive

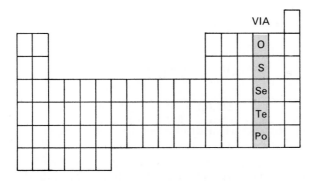

FIGURE 13.16   *Group VIA.*

metals in utensils and machines. Magnesium, aluminum, and titanium are familiar examples of highly reactive metals that can be used as structural materials because of their ability to form protective oxide coatings.

Iron also forms a passive oxide coating. This coating adequately protects iron from further reaction as long as the surface is dry or protected from air. Iron cookware won't rust as long as it is kept dry and covered with a protective grease coating. If the iron is left wet or the grease film is destroyed, and particularly if electrolytes such as salt are present, the iron will be converted quickly to iron(III) oxide, or rust. This oxide is not tightly bound to the surface of the metal. It flakes off, and corrosion continues.

Metallic oxides generally form basic solutions when dissolved in water. Barium oxide, for example, reacts with water to form barium hydroxide.

$$BaO(s) + H_2O \longrightarrow Ba(OH)_2(aq)$$

Lithium oxide reacts with water to form lithium hydroxide.

$$Li_2O(s) + H_2O \longrightarrow 2\,LiOH(aq)$$

Generally, oxygen reacts with metals by acquiring electrons and forming oxide ions.

$$O_2(g) + 4e^- \longrightarrow 2\,O^{2-}$$

The oxide ions react with water to form hydroxide ions.

$$O^{2-} + H_2O \longrightarrow OH^-$$

Oxides of metals, then, are generally **basic oxides.**

Oxygen also reacts with many nonmetals. Sulfur, for example, burns in air to form sulfur dioxide.

$$S(s) + O_2(g) \longrightarrow SO_2(g)$$

When dissolved in water, sulfur dioxide reacts to form sulfurous acid.

$$SO_2(g) + H_2O \longrightarrow H_2SO_3(aq)$$

Generally, the oxides of nonmetals form acidic solutions; thus, we call them **acidic oxides.**

Small amounts of oxygen occur as ozone (O_3), an allotropic form of oxygen. Ozone is quite unstable. At room temperature, it breaks down to ordinary oxygen.

$$2\,O_3(g) \longrightarrow 3\,O_2(g) + heat$$

Ozone is formed by electrical discharges through oxygen and by ultraviolet lamps. The pungent odor around electrical equipment is due to ozone.

In the upper atmosphere, ozone exists in equilibrium with diatomic oxygen. There it serves to shield us from harmful ultraviolet radiation by absorbing much of the ultraviolet light produced by the sun. The small percentage of ultraviolet light that escapes interaction with ozone is still sufficiently strong to give a bad case of sunburn. You can imagine the effect on life if the ozone were not there to protect us. Some recent laboratory studies have suggested that many of the compounds once used as propellants in aerosol spray cans (spray deodorants, spray waxes, hair sprays, and so on) and still used as refrigerants may react with ozone. As these materials reach the upper atmosphere, the ozone layer could be seriously depleted. The U.S. National Research Council predicts a 2% to 5% increase in skin cancer for each 1% depletion of the ozone layer.

When it occurs near the surface of the Earth, ozone is a harmful pollutant, for it is quite toxic. Ozone is formed as a component of photochemical smog (Section 13.7). At low concentrations, ozone causes eye irritation. At high concentrations, it may cause pulmonary edema, hemorrhage, and even death. The long-term effects of exposure to low levels of ozone are more difficult to evaluate. Inhalation of ozone is particularly dangerous during vigorous physical activity. Members of a New Jersey high school football team had to be hospitalized after collapsing during a severe pollution episode. School children in Los Angeles are not allowed to play outside when oxidants reach dangerous levels—as they often do. At concentrations as low as 0.15 ppm, ozone causes damage to vegetation within an hour. Exposure of animals to 1 ppm of ozone for 8 hours a day for a year has produced in them bronchial inflammation and irritation of fibrous tissues. It is not known whether the same thing occurs in humans. It is known that ozone levels have occasionally reached 0.5 ppm in southern California. Levels of 0.15 ppm are frequently exceeded.

In addition to adversely affecting health, ozone causes economic damage. It causes rubber to harden and crack. This shortens the life of automobile tires and other rubber items. Ozone causes extensive damage to crops. Tobacco and tomatoes are particularly susceptible.

13.9 GROUP VIA: SULFUR, SO_x, AND ACID RAIN

Sulfur occurs in nature in both the combined and elemental forms. Free sulfur occurs as S_8, a ring of eight atoms. For simplicity, however, sulfur is usually represented in equations only by the letter S, just as if it were monatomic.

Sulfur atoms can accept two electrons to form sulfide ions (S^{2-}).

$$S + 2e^- \longrightarrow S^{2-}$$

Sulfur gains electrons from the more active metals to form sulfides.

$$Ca(s) + S(s) \longrightarrow CaS(s)$$
$$Zn(s) + S(s) \longrightarrow ZnS(s)$$

FIGURE 13.17
London smog is often highly visible, with soot and fly ash forming a dark pall of smoke. (Courtesy of the Minnesota Environmental Control Citizens Association, St. Paul.)

Sulfur can also share electrons with other elements. With two hydrogen atoms, it forms hydrogen sulfide (H_2S). Hydrogen sulfide gas is toxic in fairly low concentrations. It is to our advantage, therefore, that it stinks too. The odor is most often described as that of rotten eggs. The odor will ordinarily drive people from an area before the gas builds up to dangerous levels. At high levels, however, H_2S deadens our sense of smell.

In the sulfides, sulfur has an oxidation number of -2. Sulfur can also have positive oxidation numbers, for example, $+4$ in sulfur dioxide (SO_2) and $+6$ in sulfur trioxide (SO_3). Sulfur burns in air to form sulfur dioxide.

$$S(s) + O_2(g) \longrightarrow SO_2(g)$$

The sulfur dioxide may react further with oxygen to form sulfur trioxide.

$$2\,SO_2(g) + O_2(g) \longrightarrow 2\,SO_3(g)$$

The sulfur trioxide can react in turn with water to form sulfuric acid.

$$SO_3(g) + H_2O \longrightarrow H_2SO_4$$

The two sulfur oxides, collectively called SO_x, and sulfuric acid are the major culprits of one type of smog. **London smog,** so called because of a dreadful pollution episode in that city in 1952, is a mixture of smoke, fog, soot, sulfur oxides, and often sulfuric acid (Figure 13.17).

The chemistry of London smog is fairly simple. Coal is mainly carbon, but it contains as much as 3% sulfur, with varying amounts of mineral matter. The sulfur in coal burns, producing acrid, choking sulfur dioxide. This gas is readily absorbed in the respiratory system, where it serves as a powerful irritant. Sulfur dioxide is known to aggravate the symptoms of asthma, bronchitis, emphysema, and other lung diseases. Sulfuric acid may react with ammonia to form a solid material, ammonium sulfate [$(NH_4)SO_4$].

$$2\,NH_3 + H_2SO_4 \longrightarrow (NH_4)_2SO_4$$

The sulfuric acid, in the form of minute liquid droplets, and the finely divided solid ammonium sulfate are easily trapped in the lungs. The harmful effect of these pollutants is considerably magnified by interaction. A particular level of sulfur dioxide, without the presence of particulate matter (finely divided solids), might be reasonably safe. A certain level of particulate matter, without sulfur dioxide around, might be fairly harmless. But take the same levels of the two together, and the effect might well be deadly. An interaction in which the total effect of two ingredients is greater than the sum of the effects of the two ingredients taken separately is **synergistic.** Synergistic effects are quite common whenever chemicals get together. We also encounter synergism in our study of the action of drugs.

The oxides of sulfur affect vegetation and other materials in addition to people. Leaves of plants show a bleached, splotchy effect, and both the yield and the quality of crops can be affected by exposure to oxides of sulfur. Rain in industrialized countries (and those downwind from them) has become more acidic in recent years. This increase has been attributed to increased amounts of sulfur oxides released into the atmosphere. Acidic rainwater can corrode metals, eat holes in nylon stockings, decompose stone buildings and statuary, and render lakes so acidic that all life is destroyed.

The element sulfur is essential to life. It is found in most proteins, including the enzymes that mediate cellular processes. We look at the chemistry of organic sulfur compounds in Special Topic H. The importance of sulfur in biochemistry is illustrated many times in other chapters.

13.10 GROUP VIIA: THE HALOGENS

Fluorine, chlorine, bromine, iodine, and astatine make up Group VIIA, a family of elements called the halogens (Figure 13.18). We saw in Chapter 4 that sodium reacts with chlorine to form sodium chloride, the familiar table salt. Indeed, it is characteristic of the entire group that they react with metals to form salts. The word **halogen** is derived from Greek words meaning "salt former." We will limit our discussion to the first four halogens, for the fifth, astatine, is rare and highly radioactive.

Each halogen atom has seven electrons in its outer energy level. Their electron dot structures differ only in the central symbol.

$$:\overset{..}{\underset{..}{F}}\cdot \quad :\overset{..}{\underset{..}{Cl}}\cdot \quad :\overset{..}{\underset{..}{Br}}\cdot \quad :\overset{..}{\underset{..}{I}}\cdot$$

All, in the elemental form, exist as diatomic molecules, formed by the sharing of a pair of electrons.

$$:\overset{..}{\underset{..}{F}}:\overset{..}{\underset{..}{F}}: \quad :\overset{..}{\underset{..}{Cl}}:\overset{..}{\underset{..}{Cl}}: \quad :\overset{..}{\underset{..}{Br}}:\overset{..}{\underset{..}{Br}}: \quad :\overset{..}{\underset{..}{I}}:\overset{..}{\underset{..}{I}}:$$

Note that each atom in the molecule has eight electrons in its valence level, giving it a noble gas configuration.

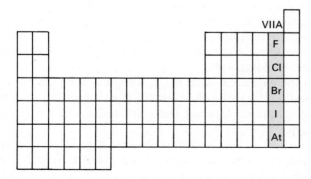

FIGURE 13.18 Group VIIA, the halogens.

Another way halogen atoms can achieve a noble gas configuration is by gaining an electron to form a negative ion.

$$:\!\overset{..}{\underset{..}{F}}\!\cdot\; +\; e^- \;\longrightarrow\; [:\!\overset{..}{\underset{..}{F}}\!:]^-$$

This tendency, combined with that of many metals to give up electrons readily, is responsible for the ability of the halogens to form so many salts.

Fluoride salts, in moderate to high concentrations, are acute poisons. Indeed, sodium fluoride (NaF) is used as a poison against such pests as roaches and rats. Small amounts of fluoride ion, however, are essential for our well-being. Up to a point, the hardness of our tooth enamel can be correlated with the amount of fluoride present. Tooth enamel is a complex calcium phosphate called hydroxyapatite. Fluoride ions replace some of the hydroxides, forming a harder mineral called fluorapatite.

$$Ca_{10}(PO_4)_6(OH)_2 + 2\,F^- \longrightarrow Ca_{10}(PO_4)_6F_2 + 2\,OH^-$$

Concentrations of 0.7 to 1.0 ppm, by weight, of fluoride (usually as H_2SiF_6 or Na_2SiF_6) have been added to the drinking water of many communities. Evidence indicates that such fluoridation results in a reduction in the incidence of dental caries (cavities) by as much as 50% in some areas. Interpretation of the statistics is complicated, however, by the varying occurrence of fluorides in the diet and by the fact that people retain fluorides at remarkably different rates. One thing that you don't have to worry about is acute poisoning from fluoridated drinking water. You would have to drink 4000 L of water containing 1.0 ppm of fluoride to approach the lethal dose of 4 g. And that 4000 L would have to be consumed in a fairly short period of time.

There is some concern about cumulative effects of consuming fluorides in drinking water, in the diet, in toothpaste, and from other sources. Excessive fluoride consumption during early childhood can cause mottling of the tooth enamel (Figure 13.19). The enamel becomes brittle in certain areas and gradually discolors. Fluorides in high doses also interfere with calcium metabolism, with kidney action, with thyroid function, and with the action of other glands and organs. Although there is little or no evidence that optimal fluoridation causes problems such as these, fluoridation of public water supplies will most likely remain a subject of controversy.

Chlorine, in the elemental form (Cl_2), is used to kill bacteria in water treatment plants. Sodium hypochlorite (NaOCl) is an ingredient of common household bleaching solutions. Most of these, regardless of price, contain 5.25% NaOCl by weight.

Chlorine is present in our bodies as chloride ion (Cl^-). This ion is essential to life. Many chlorine-containing organic compounds, though, are destructive of life. These include insecticides, herbicides, and even war gases such as phosgene. We discuss a variety of these compounds in Special Topic E.

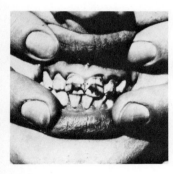

FIGURE 13.19
Excessive fluoride consumption in early childhood can cause mottling of tooth enamel. A severe case is shown here. (Courtesy of the National Institute of Dental Research, Washington, D.C.)

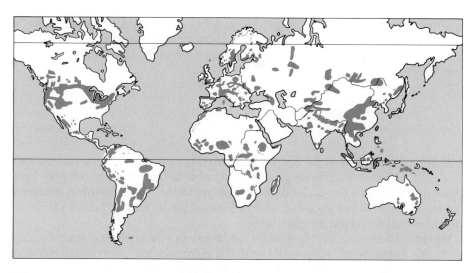

FIGURE 13.20 *This map shows the areas of the world where goiter was once endemic. The addition of sodium iodide or potassium iodide to table salt has largely eradicated goiter in the developed countries.*

Several compounds of bromine are of importance. Silver bromide (AgBr) is sensitive to light and is used in photographic film. Sodium bromide (NaBr) and potassium bromide (KBr) have been used medicinally as sedatives. Bromide ions do depress the central nervous system, but prolonged intake can cause mental deterioration, acneform skin eruptions, and, possibly, habituation. Bromides have been largely replaced by other, presumably safer, pain relievers.

Iodine is another essential nutrient. Compounds of iodine are necessary for the proper action of the thyroid gland. Because the major dietary source of iodine was seafood, goiter (an iodine-deficiency disease) was once quite prevalent in inland areas (Figure 13.20). It has been largely eradicated in the United States by the use of iodized salt. This product is made by addition of about 10 ppm of NaI to ordinary table salt. Iodine is also used as a topical antiseptic. Tincture of iodine is a solution of elemental iodine (I_2) in a mixture of alcohol and water. Iodine-releasing compounds, called iodophors, are often used as antiseptics in hospitals.

The halogens also react with hydrogen to form the hydrogen halides. In water solutions, these compounds are referred to as the hydrohalic acids. Hydrochloric acid is an example of a hydrohalic acid. These acids are classified as strong, and concentrated solutions are corrosive to tissue. The gaseous hydrogen halides can have disastrous effects on the respiratory system if they occur in significant concentration in the atmosphere. Hydrogen fluoride is extremely toxic. It will etch glass and is used to achieve the decorative frosted designs on some glassware. This same property, however, requires that solutions of hydrogen fluoride not be stored in glass containers. Hydrochloric acid, as we have previously indicated, is present in the stomach and is involved in the digestive process.

13.11 THE B GROUPS: SOME REPRESENTATIVE TRANSITION ELEMENTS

Most of our attention so far has been focused on the A groups of the periodic table. The chemistry of the B groups is also important, if perhaps more complicated. The B groups, collectively, often are called the **transition elements** (Figure 13.21). All are metals in the elemental form. They conduct electricity and have characteristic metallic luster.

The transition elements, in general, are those in which *inner* electron energy levels are being filled. (Group IB and IIB elements are exceptions, yet they share many properties with the other B groups and are often included as transition elements.) Recall from Chapter 2 that the third period (or row) of the periodic table ends with argon, which has 8 electrons in its outermost energy level. The fourth period begins with potassium, which has 1 electron in its fourth energy level. The next element, calcium, has 2 electrons in its fourth energy level. Recall, however, that the $2n^2$ rule (introduced in Section 2.10) predicts a maximum of 18, not 8 electrons for the third energy level ($2n^2 = 2 \times 3^2 = 2 \times 9 = 18$). The fourth energy level has begun to fill before the third one is full. Things return to "normal" with the next element, scandium. This elements skips back and continues to fill the third energy level. The first transition series (from scandium to zinc) corresponds to a completion of the filling of the third energy level with up to its maximum capacity of 18 electrons. Zinc has the structure represented in Table 13.2. With gallium (Ga), whose atomic

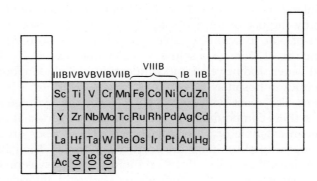

FIGURE 13.21   The B groups, or the transition elements.

TABLE 13.2 *Electron Configurations of Some Transition Metals (Color) and of Some Neighboring Elements*

Element	Electron Configuration
Argon (Z = 18)	$1s^2 2s^2 2p^6 3s^2 3p^6$
Potassium (Z = 19)	(Argon core) $4s^1$
Calcium (Z = 20)	(Argon core) $4s^2$
Scandium (Z = 21)	(Argon core) $3d^1 4s^2$
Vanadium (Z = 23)	(Argon core) $3d^3 4s^2$
Iron (Z = 26)	(Argon core) $3d^6 4s^2$
Zinc (Z = 30)	(Argon core) $3d^{10} 4s^2$
Gallium (Z = 31)	(Argon core) $3d^{10} 4s^2 4p^1$
Krypton (Z = 36)	(Argon core) $3d^{10} 4s^2 4p^6$
Zirconium (Z = 40)	(Krypton core) $4d^2 5s^2$

number is 31, we return to an A group element. All inner energy levels are filled, and the next electron enters the outer energy level.

Physical properties vary widely in the transition series. For example, mercury (Hg) is a liquid at room temperature (its melting point is $-38\,°C$). Tungsten (W), however, melts at $3410\,°C$. Chemically, the elements also exhibit a variety of properties. Most exist in more than one oxidation state in compounds. Iron is a familiar example. It readily forms both iron(II) and iron(III) ions. Some important transition elements are listed in Table 13.3.

TABLE 13.3 *Some Properties of Representative Transition Elements*

Element	Symbol	Electron Configuration	Common Oxidation Numbers	Melting Point (in degrees Celsius)	Density (in grams per milliliter)
Vanadium	V	(Argon core) $3d^3 4s^2$	2+, 3+, 4+, 5+	1900	6.1
Chromium	Cr	(Argon core) $3d^5 4s^1$	3+, 6+	1875	7.19
Manganese	Mn	(Argon core) $3d^5 4s^2$	2+, 3+, 4+, 7+	1245	7.43
Iron	Fe	(Argon core) $3d^6 4s^2$	2+, 3+	1536	7.86
Cobalt	Co	(Argon core) $3d^7 4s^2$	2+, 3+	1495	8.9
Nickel	Ni	(Argon core) $3d^8 4s^2$	2+	1453	8.9
Copper	Cu	(Argon core) $3d^{10} 4s^1$	1+, 2+	1083	8.96
Zinc	Zn	(Argon core) $3d^{10} 4s^2$	2+	420	7.14
Molybdenum	Mo	(Krypton core) $4d^5 5s^1$	2+, 3+, 4+, 5+, 6+	2610	10.2
Silver	Ag	(Krypton core) $4d^{10} 5s^1$	1+	961	10.5
Cadmium	Cd	(Krypton core) $4d^{10} 5s^2$	2+	321	8.65
Tungsten (wolfram)	W	(Xenon core) $5d^4 6s^2$	2+, 3+, 4+, 5+, 6+	3410	19.3
Gold	Au	(Xenon core) $5d^{10} 6s^1$	1+, 3+	1063	19.3
Mercury	Hg	(Xenon core) $5d^{10} 6s^2$	1+, 2+	−38	13.6

Many compounds of the transition metals are colored, often brilliantly so. The color varies with the oxidation state of the metal in the compound. An aqueous solution containing manganese ions (Mn^{2+}) is a faint pink. Solutions of MnO_4^{2-} (in which the oxidation number of manganese is $+6$) are an intense green, and those of MnO_4^- (in which the oxidation number of manganese is $+7$) are deep purple. Aqueous solutions of Fe^{2+} are often pale green; those of Fe^{3+} are often yellow.

Transition metals known to be essential to life are iron, copper, zinc, cobalt, manganese, vanadium, chromium, and molybdenum. Nickel is sometimes found in body tissues but has not been shown to be essential. Other transition metals, most notably cadmium and mercury, are dangerous posions.

Iron is essential for the proper functioning of hemoglobin, the red protein molecule involved in oxygen transport. This iron must be in the $+2$ oxidation state. If it is oxidized to the $+3$ state, the resulting compound (called **methemoglobin**) is incapable of carrying oxygen. The oxygen-deficiency disease that results is called methemoglobinemia. In infants, this condition is called the blue-baby syndrome.

Cobalt is a component of vitamin B_{12}. Deficiency leads to pernicious anemia. Vitamin B_{12} is found only in meat and meat products. One hazard of a strict vegetarian diet is vitamin B_{12} deficiency.

Iron, cobalt, copper, zinc, manganese, and molybdenum are essential to the proper functioning of certain enzymes. These metalloenzymes are necessary to life, but the exact function of the metal ion is not always known.

PROBLEMS

1. Define each term:
   a. inorganic chemistry
   b. organic chemistry
   c. noble gas
   d. halogen
   e. alkali metal
   f. alkaline earth metal
   g. transition element

2. Give the electron dot symbol for each of these elements:
   a. neon
   b. oxygen
   c. fluorine
   d. potassium
   e. barium
   f. nitrogen

3. What is the outstanding chemical property of the noble gases? What structural feature accounts for this property?

4. Why is helium preferred to hydrogen for filling dirigibles, even though the latter has greater lifting power? What advantage does helium have over nitrogen in breathing mixtures for deep-sea divers? For someone with emphysema?

5. What function does argon serve in an electric light bulb?

6. How does a neon sign work?

7. What similarities in properties are shared by the alkali metals? What structural feature do atoms of these elements share?

8. What similarities in properties are shared by the alkaline earth metals? What structural feature do atoms of these elements share?

9. Which alkaline earth metal is most different from the others? What are some of those differences?

10. What alkali metal compound is used to treat manic-depressive psychoses?

11. What is the most abundant metal in the Earth's crust?

12. Which Group IIIA element has the greatest tendency to form covalent compounds (rather than ionic compounds)? Why?

13. Draw an electron dot symbol for gallium (Ga). Draw an electron dot symbol for gallium ion.
14. Why is aluminum replacing steel wherever possible in automobiles?
15. Write a balanced equation for the preparation of aluminum metal from aluminum oxide.
16. Give the electron dot symbol for each of the following:
    a. fluoride ion
    b. iodide ion
    c. oxide ion
    d. sulfide ion
    e. potassium ion
    f. strontium ion
17. List one use for each element or compound:
    a. chlorine
    b. iodine
    c. NaF
    d. AgBr
    e. NaI
    f. NH_3
    g. NH_4NO_3
    h. NaOCl
    i. NaCl
18. What is the effect on tooth enamel of small amounts of fluoride? Of excess fluoride in the diet?
19. What are basic oxides? Use calcium oxide to illustrate your answer.
20. What are acidic oxides? Use sulfur trioxide to illustrate your answer.
21. Complete and balance the following equations:
    a. $Li + O_2 \longrightarrow$
    b. $Ca + O_2 \longrightarrow$
    c. $S + O_2 \longrightarrow$
    d. $CaO + H_2O \longrightarrow$
    e. $SO_2 + H_2O \longrightarrow$
    f. $Ca + S \longrightarrow$
22. What is photochemical smog? What chemical compound starts the formation of photochemical smog by absorbing sunlight?
23. What is synergism?
24. What are allotropes? Name two sets of allotropes.
25. How do oxygen atoms, oxygen molecules, and ozone differ in structure and properties?
26. Explain the difference, in terms of effects on health, of ozone in the upper atmosphere and at ground level.
27. How does the burning of sulfur-containing coal lead to acid rain?
28. What is London smog? How is it formed?
29. How do weather conditions differ in episodes of Los Angeles smog as compared to London smog?
30. What is the greenhouse effect?
31. What was the vital force theory? How was it overthrown?
32. What is the origin of helium in natural gas wells?
33. Why are the noble gases no longer called the "inert gases"?
34. What are the halogens? Why are they so called?
35. What is nitrogen fixation? Why is it important?
36. How can a petroleum shortage lead to a scarcity of fertilizer?
37. What condition(s) lead(s) to formation of carbon monoxide during combustion?
38. How does carbon monoxide exert its posionous effect?
39. What two alkali metal ions play a major role in maintaining fluid balance in the body?
40. What is berylliosis?
41. What important molecule in plants incorporates magnesium?
42. Name two functions of calcium ions in the body.
43. What is hard water? How does it affect the action of soaps?
44. List three distinguishing characteristics of transition metals.
45. List four transition metals essential to life. Indicate their function in the body.

CHAPTER

14

Hydrocarbons

Scientists of the eighteenth and nineteenth centuries studied compounds isolated from rocks and ores, from the atmosphere and oceans, and labeled them *inorganic* because they were obtained from nonliving systems. Compounds obtained from plants and animals were called *organic* because they were isolated from organized (living) systems. The early chemists believed that only living organisms could synthesize organic compounds. Although by the middle of the nineteenth century a number of "organic" compounds had been prepared using ordinary laboratory techniques, the labels *organic* and *inorganic* remained (Figure 14.1).

FIGURE 14.1 *The word* **organic** *has different meanings. Organic fertilizer is organic in the original sense that it is derived from a living organism. There is no legal definition of organic foods, but the term generally means foods grown without pesticides or synthetic fertilizers. Organic chemistry is the chemistry of compounds of carbon. (Photo © Zeva Oelbaum, New York.)*

Today, **organic chemistry** is defined simply as the chemistry of the compounds of carbon. It may seem strange that we divide chemistry into two branches, one of which considers the chemistry of one element while the other handles the chemistry of the more than 100 remaining elements. However, this division seems more reasonable when one discovers that, of the 7 million or more compounds that have been characterized, the overwhelming majority contain carbon. Carbon has a unique ability to form stable, covalent bonds with itself and with other elements in infinite variations. The molecules thus produced may contain only one or over a million carbon atoms. So complex is the chemistry of carbon that we shall approach its study by dividing its millions of compounds into families. We'll study one family at a time and begin by concentrating on the simpler members of each family. Eventually we will move to a consideration of those molecules that deserve to be called organic in the old sense—complex, carbon-containing molecules that determine the form and function of living systems.

We might pause to ponder how far science has come since Wöhler's synthesis of urea in 1828 (p. 359). Before that year, scientists believed they could not synthesize even the simplest organic molecule. In 1980, the United States Supreme Court ruled that new life forms, created in a laboratory, could be patented. The patent under consideration described a new organism, designed to consume spilled oil and made by changing the genetic heritage of an existing life form. Was this "creation of life in a test tube"? Not quite, but each new experiment in genetic engineering brings that achievement closer.

14.1 ORGANIC VERSUS INORGANIC

Organic compounds, like inorganic ones, obey all the natural laws. Often there is no clear distinction in chemical or physical properties between organic and inorganic molecules. Nevertheless, it may be useful to compare and contrast typical members of each class. Table 14.1 lists a variety of properties of the inorganic compound sodium chloride (NaCl, common table salt) and the organic compound benzene (C_6H_6, a common solvent once widely used to strip furniture for refinishing).

Most organic compounds have relatively low melting points. Many, like benzene, are liquids at room temperature. The typical inorganic substance, like sodium chloride, is a crystalline solid with a high melting point.

The typical organic compound is insoluble in water. Most are less dense than water and will float on top of water if we attempt to dissolve them. The typical inorganic compound is readily soluble in water.

The typical organic compound is highly flammable, a point important to remember when working in the laboratory. Examples are anesthetic ether, which forms explosive mixtures with air, and gasoline, a mixture of compounds

TABLE 14.1   *Comparison of an Inorganic and an Organic Compound*

	Benzene	Sodium Chloride
Formula	C_6H_6	NaCl
Solubility in H_2O	Insoluble	Soluble
Solubility in gasoline	Soluble	Insoluble
Flammable?	Yes	No
Melting Point	5.5 °C	801 °C
Boiling Point	80 °C	1413 °C
Density	0.88 g/cm³	2.7 g/cm³ (crystal)
Bonding	Covalent	Ionic

used as fuel in internal combustion engines. Typical inorganic compounds are nonflammable. Indeed, inorganics such as water, baking soda (sodium bicarbonate, $NaHCO_3$), and borax ($Na_2B_4O_7 \cdot 10\,H_2O$) are used in fighting fires. Sodium bicarbonate is one of the few carbon-containing compounds classified with the inorganics.

These and other typical properties reflect the fact that most organic compounds are composed of molecules with covalent bonds. The typical inorganic compound is ionic. Don't forget, though, that there are covalent inorganic compounds (water, for example). And we will encounter a few ionic organics.

14.2 ALKANES: THE SATURATED HYDROCARBONS

Before one can even begin to understand the large, complex molecules on which life is based, it is necessary to learn something about simpler molecules. We will start with organic compounds containing only two elements, carbon and hydrogen. These compounds are called **hydrocarbons.** In Sections 4.12 and 4.14, we encountered compounds called methane (CH_4) and ethane (C_2H_6). Models of molecules of these two compounds are shown in Figure 14.2. These are the first two members of a series of related compounds called **alkanes** or **saturated hydrocarbons.** Saturated, in this case, means that each carbon atom is bonded to four other atoms; there are no double or triple bonds in the molecules. Structurally, methane and ethane are generally represented as below.

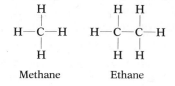

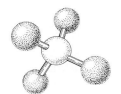

Methane

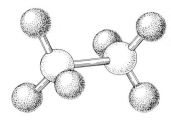

Ethane

FIGURE 14.2
Ball-and-stick models of methane and ethane.

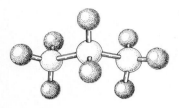

FIGURE 14.3
Ball-and-stick model
of propane.

These representations make no attempt to accurately portray bond angles in particular, or molecular geometry in general. You should keep the models in mind when you look at these flat pictures. (The bond angles and molecular geometry of methane are described in detail in Section 4.12.)

Natural gas consists largely of methane, with some ethane and smaller amounts of other hydrocarbons containing three to five carbon atoms.

The three-carbon hydrocarbon (C_3H_8) is called propane. A ball-and-stick model of this compound is shown in Figure 14.3. In two dimensions, the structure is generally written as shown below.

$$
\begin{array}{c}
\text{H} \quad \text{H} \quad \text{H} \\
| \quad\; | \quad\; | \\
\text{H--C--C--C--H} \\
| \quad\; | \quad\; | \\
\text{H} \quad \text{H} \quad \text{H}
\end{array}
$$

Propane

Notice that methane, ethane, and propane form a series in which the adjacent members differ by one carbon atom and two hydrogen atoms, that is, by a CH_2 unit (Figure 14.4). Compounds related in this way are said to make up a **homologous series.** Such a series has properties that vary in a regular and predictable manner. The principle called **homology** gives meaning to organic chemistry in much the same way that the periodic table gives organization and meaning to the chemistry of the elements. Instead of a bewildering array of individual carbon compounds, we can study a few members of a homologous series (called **homologs**), and from them we can deduce the properties of other compounds in the series.

Methane (CH₄), Ethane (C₂H₆), Propane (C₃H₈) structures:

$$
\begin{array}{c}
\text{H} \\
| \\
\text{H--C--H} \\
| \\
\text{H}
\end{array}
\qquad
\begin{array}{c}
\text{H} \quad \text{H} \\
| \quad\; | \\
\text{H--C--C--H} \\
| \quad\; | \\
\text{H} \quad \text{H}
\end{array}
$$

Methane (CH_4)   Ethane (C_2H_6)

$$
\begin{array}{c}
\text{H} \quad \text{H} \quad \text{H} \\
| \quad\; | \quad\; | \\
\text{H--C--C--C--H} \\
| \quad\; | \quad\; | \\
\text{H} \quad \text{H} \quad \text{H}
\end{array}
$$

Propane (C_3H_8)

FIGURE 14.4
Members of a homologous series. Each succeeding formula incorporates one carbon and two hydrogens more than the previous formula.

Continuing the homologous series, we add another carbon atom and a pair of hydrogens to get C_4H_{10}. It is rather easy to write a structure corresponding to this formula. We merely string 4 carbon atoms in a row,

$$\text{--C--C--C--C--}$$

and then we add enough hydrogen atoms to give each carbon atom four bonds.

$$
\begin{array}{c}
\text{H} \quad \text{H} \quad \text{H} \quad \text{H} \\
| \quad\; | \quad\; | \quad\; | \\
\text{H--C--C--C--C--H} \\
| \quad\; | \quad\; | \quad\; | \\
\text{H} \quad \text{H} \quad \text{H} \quad \text{H}
\end{array}
$$

There is a compound called butane that has this structure. But there is another way to put 4 carbons and 10 hydrogens together. String out 3 of the 4 carbon atoms, and then branch the other one off the middle carbon of the chain.

$$
\begin{array}{c}
\text{--C--C--C--} \\
| \\
\text{C}
\end{array}
$$

Now we add enough hydrogen atoms to give each carbon four bonds.

$$
\begin{array}{ccccc}
 & H & H & H & \\
 & | & | & | & \\
H-&C&-C&-C&-H \\
 & | & | & | & \\
 & H & | & H & \\
 & & H-C-H & & \\
 & & | & & \\
 & & H & &
\end{array}
$$

And, fortunately for the sake of our structural theory, there is another gaseous hydrocarbon that corresponds to this structure. There are two compounds, then, that have the same molecular formula, C_4H_{10}. One boils at 0 °C; the other, at −12 °C. Different compounds with the same molecular formula are called **isomers.** To give the two butanes unique names, we call the one with the continuous chain of carbon atoms butane. The one with the branched chain is called isobutane. Figure 14.5 shows ball-and-stick models of the two isomeric butanes.

The three smaller homologs of this series do not exist in isomeric forms. There is only one methane, one ethane, and one propane. We could draw what might appear to be different propanes.

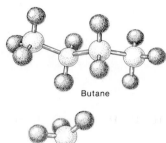
Butane

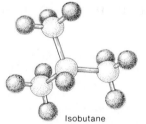
Isobutane

FIGURE 14.5
Ball-and-stick models of
butane and isobutane.

But these drawings just demonstrate the limitations of flat pictures. If we made a ball-and-stick model of propane, the one model could be made to resemble both of the flat structural formulas. We could also offer several alternate drawings of the butane formula.

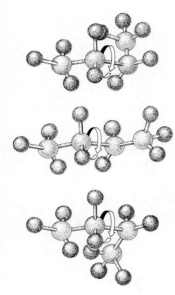

FIGURE 14.6
Rotation about single bonds. The right-hand portion of the molecule is spinning about the single bond between the second and third carbon atoms. The same molecule can thus assume different arrangements.

As in the case of propane, the variety of drawings of butane represent the same compound. All the structural formulas show a continuous four-carbon chain. This chain may be twisted or bent, but it is still one continuous chain. The different arrangements for the four-carbon chain result from the fact that the various parts of the molecule are free to spin about the single bonds that join them to the rest of the molecule. Look at Figure 14.6, and focus your attention on the bond between the second and third carbon atoms in the chain. We are spinning the right end of the molecule about this bond. You can see that the relative position of the carbon atom on the right end changes. Spinning can take place about any of the bonds in this compound, and that is what permits us to draw seemingly distinct formulas. The formulas do not represent isomers, but the same compound. Isobutane is a true isomer of butane. The formula of isobutane shows a continuous chain of three carbon atoms only, with the fourth carbon atom attached as a branch off the middle carbon of the continuous chain. If Figures 14.5 and 14.6 do not make this point clear, you should make models to convince yourself. (Borrow ball-and-stick models or create some of your own from marshmallows and toothpicks.)

Propane and the butanes are familiar fuels. They are usually supplied under pressure in tanks. Although they are gases at ordinary temperatures and under normal atmospheric pressure, they are liquefied under pressure and are sold as liquefied petroleum (LP) gas. Butane, liquefied under pressure, can be seen in disposable butane cigarette lighters. When the release lever is pressed on these lighters, the butane comes under atmospheric pressure, and some of it vaporizes and is ignited by a spark.

One more step up the homologous series gets us to C_5H_{12}. There are three compounds with this molecular formula. Collectively, these compounds are called pentanes. We can name compound I pentane, because it has all five carbons in a continuous chain. Compound II can be called isopentane, because, like isobutane, it has a single carbon atom branched off the second carbon of the continuous chain. But what shall we call compound III? Let's name it the way the chemists did when it was discovered in 1870. Since the other two pentanes were characterized first, this one was called neopentane (from the Greek *neos*, "new").

I

II

III

Hydrocarbon chemistry seems quite complex, but there is some system to it. For example, the first part of the name indicates the number of carbon atoms in the molecule. For five or more carbon atoms, the prefixes are derived from Greek or Latin names for the numbers (Table 14.2).

The ending of the name also has meaning. Note that all end in *-ane*. This indicates that all are alkanes. The name heptane, then, means an alkane with seven carbon atoms. To draw the structure of heptane, we merely write out a string of seven carbon atoms,

$$-C-C-C-C-C-C-C-$$

and then attach enough hydrogens to give each carbon a valence of four. This requires three hydrogens on each end carbon and two on each of the others.

$$
\begin{array}{ccccccc}
 & H & H & H & H & H & H & H \\
 & | & | & | & | & | & | & | \\
H- & C- & C- & C- & C- & C- & C- & C- & H \\
 & | & | & | & | & | & | & | \\
 & H & H & H & H & H & H & H
\end{array}
$$

In this manner we can readily extend our list of structures of continuous-chain hydrocarbons to decane. These alkanes are shown in Table 14.3. Notice that each molecular formula in that table differs from the one preceding it by a CH_2 unit, illustrating the principle of homology. Notice also that the number of isomers increases rapidly with increasing carbon number. There are 5 hexanes, 9 heptanes, and 18 octanes. There are over 4 billion isomers with the molecular formula $C_{30}H_{62}$. Fortunately, not all of those have been isolated or characterized. We've already run into a problem of giving unique names to the 5 hexanes, let alone the large numbers of isomers of higher alkanes.

14.3 CONDENSED STRUCTURAL FORMULAS

Formulas such as we have used so far, showing all the carbon and hydrogen atoms and how they are attached to one another, are called **structural formulas.** These structures convey much more information than simple chemical formulas. For example, the formula C_4H_{10} doesn't even tell us whether we are dealing with butane or with isobutane. The formulas

$$
\begin{array}{cccc}
H & H & H & H \\
| & | & | & | \\
H-C-C-C-C-H \\
| & | & | & | \\
H & H & H & H
\end{array}
\quad \text{and} \quad
\begin{array}{ccc}
H & H & H \\
| & | & | \\
H-C-C-C-H \\
| & | & | \\
H & | & H \\
 & H-C-H \\
 & | \\
 & H
\end{array}
$$

identify the specific isomers by the order of attachment of the various atoms.

TABLE 14.2 *Prefixes That Indicate the Number of Carbon Atoms in Organic Molecules*

Prefix	Number
Meth-	1
Eth-	2
Prop-	3
But-	4
Pent-	5
Hex-	6
Hept-	7
Oct-	8
Non-	9
Dec-	10

TABLE 14.3 *The First 10 Continuous-Chain Alkanes*

Name	Molecular Formula	Structural Formula	Number of Possible Isomers
Methane	CH_4	H—C—H with H above and below (C bonded to 4 H)	1
Ethane	C_2H_6	H—C—C—H with H above and below each C	1
Propane	C_3H_8	H—C—C—C—H with H above and below each C	1
Butane	C_4H_{10}	H—C—C—C—C—H with H above and below each C	2
Pentane	C_5H_{12}	H—C—C—C—C—C—H with H above and below each C	3
Hexane	C_6H_{14}	H—C—C—C—C—C—C—H with H above and below each C	5
Heptane	C_7H_{16}	H—C—C—C—C—C—C—C—H with H above and below each C	9
Octane	C_8H_{18}	H—C—C—C—C—C—C—C—C—H with H above and below each C	18
Nonane	C_9H_{20}	H—C—C—C—C—C—C—C—C—C—H with H above and below each C	35
Decane	$C_{10}H_{22}$	H—C—C—C—C—C—C—C—C—C—C—H with H above and below each C	75

Unfortunately, structural formulas are hard to type and take up a lot of space. Chemists often use **condensed structural formulas** to alleviate these problems. The condensed structures show the hydrogen atoms right next to the carbon atoms to which they are attached. The two isomeric butanes become

$$CH_3-CH_2-CH_2-CH_3 \quad and \quad CH_3-CH-CH_3$$
$$| $$
$$CH_3$$

Sometimes these are further simplified by omitting some (or all) of the bond lines.

$$CH_3CH_2CH_2CH_3 \quad and \quad CH_3CHCH_3$$
$$|$$
$$CH_3$$

We shall shortly abandon the full structural formulas in favor of the more efficient condensed versions. First, though, a few more examples in the first part of the next section.

14.4 THE UNIVERSAL LANGUAGE: IUPAC NOMENCLATURE

To bring order to the chaotic naming of newly discovered compounds, the *International Union of Pure and Applied Chemistry* (IUPAC—sometimes pronounced "you pack") held what was to be the first of several meetings on nomenclature (i.e., a system for naming) in 1892. This conference established formal rules for naming compounds. Some of the rules for naming alkanes are summarized here.

1. The names of individual members end in **-ane,** indicating that they are alkanes. The names of the continuous-chain members having up to 10 carbon atoms are given in Table 14.3.
2. The names of branched-chain alkanes are made up of two parts. The end of the name is taken from what is sometimes referred to as the parent compound. For example, the compound

$$\quad\quad H \quad H \quad H \quad H \quad H$$
$$\quad\quad | \quad\ | \quad\ | \quad\ | \quad\ |$$
$$H-C-C-C-C-C-H \quad (or\ CH_3CHCH_2CH_2CH_3)$$
$$\quad\quad | \quad\ | \quad\ | \quad\ | \quad\quad\quad\quad\quad |$$
$$\quad\quad H \quad | \quad H \quad H \quad H \quad\quad\quad\quad CH_3$$
$$\quad\quad\ \ H-C-H$$
$$\quad\quad\quad\quad |$$
$$\quad\quad\quad\quad H$$

would be a derivative of pentane because there are five carbon atoms in the longest continuous chain. The second part of the name would be **pentane.**

3. The first part of the name consists of prefixes that indicate the groups attached to the parent chain. If the group contains only carbon and hydrogen with no double or triple bonds, it is called an **alkyl** group. The **alk-** indicates that these groups are similar to alkanes; the **-yl** indicates that a group of atoms is attached to some parent chain. Specific alkyl groups are named after the alkane with the same number of carbons. For example, the group

$$H-\underset{\displaystyle \overset{\displaystyle H}{|}}{\underset{\displaystyle \underset{\displaystyle H}{|}}{C}}- \qquad (\text{or } CH_3-)$$

is derived from methane and is called **methyl.** The alkyl group derived from ethane is

$$H-\underset{\displaystyle \overset{H}{|}}{\underset{\displaystyle \underset{H}{|}}{C}}-\underset{\displaystyle \overset{H}{|}}{\underset{\displaystyle \underset{H}{|}}{C}}- \qquad (\text{or } CH_3CH_2-)$$

It is called an **ethyl** group. *Two* alkyl groups can be derived from propane.

$$H-\underset{\displaystyle \overset{H}{|}}{\underset{\displaystyle \underset{H}{|}}{C}}-\underset{\displaystyle \overset{H}{|}}{\underset{\displaystyle \underset{H}{|}}{C}}-\underset{\displaystyle \overset{H}{|}}{\underset{\displaystyle \underset{H}{|}}{C}}- \qquad (\text{or } CH_3CH_2CH_2-)$$

$$H-\underset{\displaystyle \overset{H}{|}}{\underset{\displaystyle \underset{H}{|}}{C}}-\underset{\displaystyle \overset{H}{|}}{\underset{\displaystyle \underset{|}{|}}{C}}-\underset{\displaystyle \overset{H}{|}}{\underset{\displaystyle \underset{H}{|}}{C}}-H \qquad (\text{or } CH_3CHCH_3)$$

These are called propyl and isopropyl, respectively. Remember that there is only one alkane named propane, but a chain of three carbon atoms can be attached to a longer chain in two different ways. One has the attachment through an end carbon of the three-carbon chain; the other, through the middle carbon. There are many other alkyl groups. The ones that we are most likely to encounter are listed in Table 14.4. Notice that the groups have one less hydrogen than the corresponding alkanes. This hydrogen has to be removed so there will be a free bond at which the group can connect to the parent chain.

4. Arabic numerals are used to indicate the position to which the substitutents (the alkyl groups) are attached on the longest chain. Thus, to name the

TABLE 14.4 *Common Alkyl Groups*

Name	Structural Formula	Condensed Structural Formula
Methyl	$$H-\overset{\displaystyle H}{\underset{\displaystyle H}{C}}-$$	CH_3-
Ethyl	$$H-\overset{\displaystyle H}{\underset{\displaystyle H}{C}}-\overset{\displaystyle H}{\underset{\displaystyle H}{C}}-$$	CH_3CH_2-
Derived from propane:		
Propyl	$$H-\overset{\displaystyle H}{\underset{\displaystyle H}{C}}-\overset{\displaystyle H}{\underset{\displaystyle H}{C}}-\overset{\displaystyle H}{\underset{\displaystyle H}{C}}-$$	$CH_3CH_2CH_2-$
Isopropyl	$$H-\overset{\displaystyle H}{\underset{\displaystyle H}{C}}-\overset{\displaystyle H}{C}-\overset{\displaystyle H}{\underset{\displaystyle H}{C}}-H$$	CH_3CHCH_3
Derived from butane:		
Butyl	$$H-\overset{H}{\underset{H}{C}}-\overset{H}{\underset{H}{C}}-\overset{H}{\underset{H}{C}}-\overset{H}{\underset{H}{C}}-$$	$CH_3CH_2CH_2CH_2-$
Secondary butyl (*sec*-butyl)	$$H-\overset{H}{\underset{H}{C}}-\overset{H}{C}-\overset{H}{\underset{H}{C}}-\overset{H}{\underset{H}{C}}-H$$	$CH_3CHCH_2CH_3$
Derived from isobutane:		
Isobutyl	$$\begin{array}{c}H-C-H \\ \\ H-C-C-C- \end{array}$$	$$CH_3CHCH_2-$$ with CH_3 above
Tertiary butyl (*tert*-butyl)	$$\begin{array}{c}H-C-H \\ \\ H-C-C-C-H \end{array}$$	$$CH_3-C-CH_3$$ with CH_3 above

compound

$$H-\underset{\underset{\displaystyle H-\underset{\displaystyle H}{C}-H}{\overset{\displaystyle H}{|}}}{\overset{\displaystyle H}{C}}-\underset{H}{\overset{H}{C}}-\underset{H}{\overset{H}{C}}-\underset{H}{\overset{H}{C}}-\underset{H}{\overset{H}{C}}-H \qquad \text{(or } CH_3CHCH_2CH_2CH_2CH_3\text{)}$$
$$CH_3$$

we first identify the longest continuous chain.

$$H-\underset{\underset{\underset{\displaystyle H}{|}}{\overset{\displaystyle H}{C}-H}}{\overset{\displaystyle H}{C}}-\underset{H}{\overset{H}{C}}-\underset{H}{\overset{H}{C}}-\underset{H}{\overset{H}{C}}-\underset{H}{\overset{H}{C}}-H \qquad \text{(or } CH_3CHCH_2CH_2CH_2CH_3\text{)}$$
$$CH_3$$

There are six carbon atoms in this chain. The compound is therefore a derivative of hexane. The group attached to the chain (CH_3—) is methyl. It is on the second carbon atom from the left end. Thus, the compound is 2-methylhexane.*

5. If two or more identical groups are attached to the main chain, a number is required to specify the location of each. Further, we must indicate whether there are two, three, or four identical groups hanging from the parent chain, and we do this by using the prefix **di-** for two, the prefix **tri-** for three, and the prefix **tetra-** for four. And even if two identical groups are located at the same position, the number must be repeated for each group.

$$CH_3-\underset{\underset{\displaystyle CH_3}{|}}{\overset{\overset{\displaystyle CH_3}{|}}{C}}-CH_2-CH_2-CH_2-CH_3 \qquad CH_3-\underset{\underset{\displaystyle CH_3}{|}}{\overset{\overset{\displaystyle CH_3}{|}}{C}}-CH_2-\underset{\underset{\displaystyle CH_3}{|}}{CH}-CH_3$$

2,2-Dimethylhexane   2,2,4-Trimethylpentane

(Notice from these examples that commas are used to separate numbers from each other and that hyphens are used to separate numbers from words.)

---

* The numbering of the main chain is always started at the end that will provide the lowest number for the position of the substituent. If we numbered the parent chain starting at the right, the substituent would be located at carbon number 5 of the parent chain. In order to obtain the lowest possible number for the substituent, then, we must count from left to right in this instance.

The official IUPAC rules state that the groups should be listed in alphabetical order, although many chemists still list them in order of increasing size and you may sometimes hear 4-ethyl-2-methylhexane called "2-methyl-4-ethylhexane."

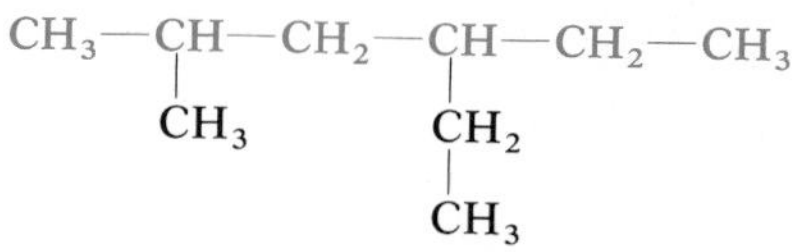

$$CH_3-\underset{\underset{CH_3}{|}}{CH}-CH_2-\underset{\underset{\underset{\underset{CH_3}{|}}{CH_2}}{|}}{CH}-CH_2-CH_3$$

4-Ethyl-2-methylhexane

The best way to learn how to name alkanes is by working out examples, not just by memorizing rules. It's easier than it sounds. Try the following.

Name the compound

$$CH_3CH_2-\underset{\underset{CH_3}{|}}{CH}-\underset{\underset{CH_3}{|}}{CH}-CH_3$$

The longest continuous chain has five carbon atoms. There are two methyl groups attached to the second and third carbon atoms (not the third and fourth); use the lowest combination of numbers, counting from one end. The correct name is 2,3-dimethylpentane.

example 14.1

Name the compound

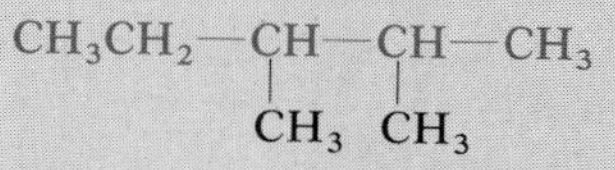

$$CH_3-\underset{\underset{\underset{\underset{CH_3}{|}}{CH_2}}{|}}{CH}-CH_2-\underset{\underset{CH_3}{|}}{CH}-CH_3$$

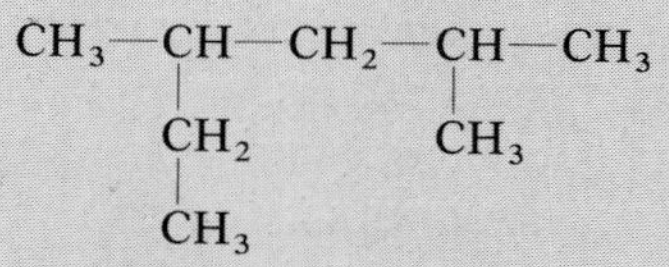

The correct name is 2,4-dimethylhexane, not 2-ethyl-4-methylpentane. This is a fooler. The parent compound is the longest continuous chain, not necessarily the chain drawn straight across the page. In this compound, the longest chain contains six, not five, carbon atoms.

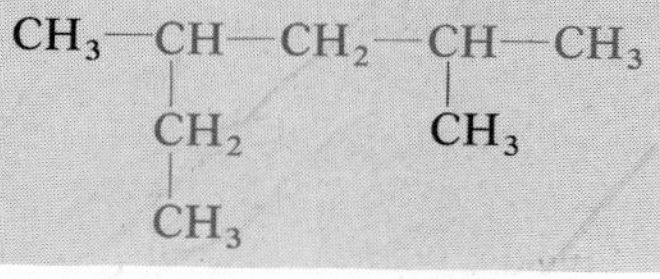

$$CH_3-\underset{\underset{\underset{\underset{CH_3}{|}}{CH_2}}{|}}{CH}-CH_2-\underset{\underset{CH_3}{|}}{CH}-CH_3$$

example 14.2

example 14.3

Name the compound

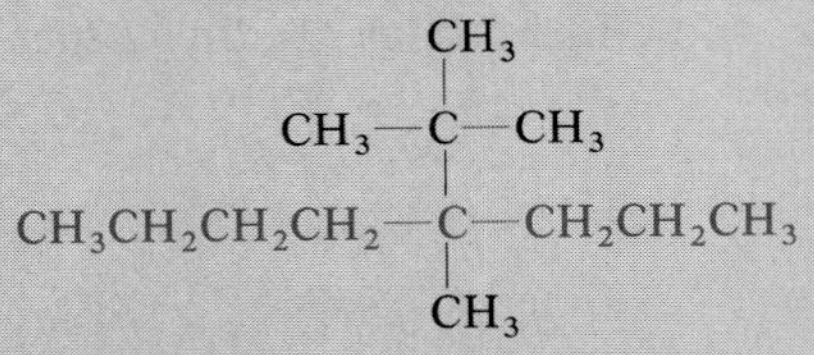

$$CH_3$$
$$CH_3-C-CH_3$$
$$CH_3CH_2CH_2CH_2-C-CH_2CH_2CH_3$$
$$CH_3$$

The correct name is 4-*tert*-butyl-4-methyloctane.

example 14.4

Draw 4-isopropyl-2-methylheptane.

In drawing compounds, always start with the parent chain, heptane in this case.

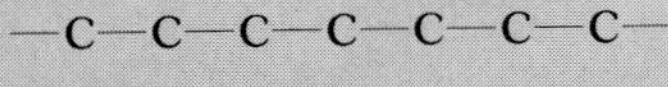

$$-C-C-C-C-C-C-C-$$

Then add the groups at their proper positions. You can number the parent chain from either direction as long as you are consistent (don't change directions in the middle of a problem).

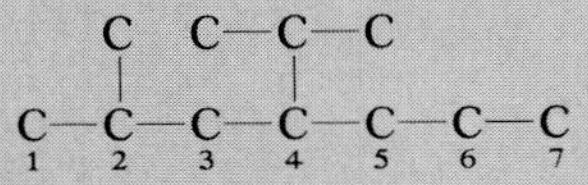

$$
\begin{array}{ccccc}
 & C & C-C-C & & \\
C-C-C-C-C-C-C \\
1 & 2 & 3 & 4 & 5 & 6 & 7
\end{array}
$$

Finally, fill in all the hydrogens (each carbon atom must have four bonds).

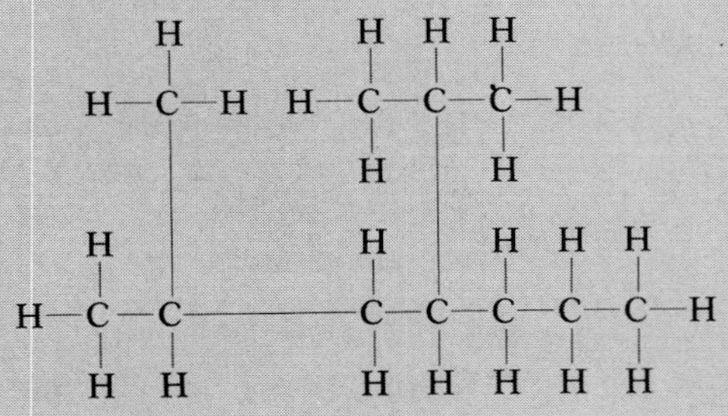

You can condense this formula by writing the hydrogens right next to the carbons to which they are attached.

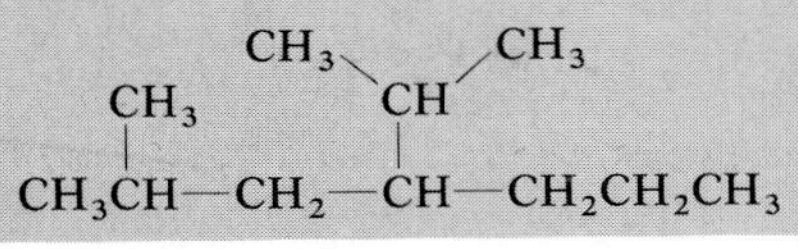

$$
\begin{array}{c}
CH_3 \quad\quad CH_3 \\
CH_3 \quad\quad CH \quad CH_3 \\
CH_3CH-CH_2-CH-CH_2CH_2CH_3
\end{array}
$$

TABLE 14.5 *Physical Properties of Selected Alkanes*

Name	Molecular Formula	Melting Point (in degrees Celsius)	Boiling Point (in degrees Celsius)	Density at 20 °C (in grams per milliliter)
Methane	CH_4	−183	−162	(Gas)
Ethane	C_2H_6	−172	−89	(Gas)
Propane	C_3H_8	−187	−42	(Gas)
Butane	C_4H_{10}	−138	0	(Gas)
Pentane	C_5H_{12}	−130	36	0.626
Hexane	C_6H_{14}	−95	69	0.659
Heptane	C_7H_{16}	−91	98	0.684
Octane	C_8H_{18}	−57	126	0.703
Nonane	C_9H_{20}	−54	151	0.718
Decane	$C_{10}H_{22}$	−30	174	0.730
Undecane	$C_{11}H_{24}$	−26	196	0.740
Dodecane	$C_{12}H_{26}$	−10	216	0.749
Tridecane	$C_{13}H_{28}$	−6	235	0.757
Tetradecane	$C_{14}H_{30}$	6	254	0.763
Pentadecane	$C_{15}H_{32}$	10	271	0.769
Hexadecane	$C_{16}H_{34}$	18	280	0.775
Heptadecane	$C_{17}H_{36}$	22	302	(Solid)
Octadecane	$C_{18}H_{38}$	28	316	(Solid)
Nonadecane	$C_{19}H_{40}$	32	330	(Solid)
Eicosane	$C_{20}H_{42}$	37	343	(Solid)

14.5 PHYSICAL PROPERTIES OF THE ALKANES

The alkanes form a homologous series. The properties of the members vary in a regular and predictable manner. For example, their boiling points show a fairly regular increase of from 20 to 30 °C as we go up the series (Table 14.5).

You should note from the table that at room temperature alkanes having from 1 to 4 carbon atoms per molecule are gases, alkanes having from 5 to about 16 carbon atoms per molecule are liquids, and alkanes having more than 16 carbon atoms per molecule are solids. Note also that the densities of the liquid alkanes are less than that of water (1.0 g/mL). The alkanes are essentially insoluble in water and hence will float on top of water. Alkanes dissolve many organic substances of low polarity, such as fats, oils, and waxes. Mixtures of alkanes are therefore frequently used as organic solvents.

14.6 PHYSIOLOGICAL PROPERTIES OF ALKANES

The physiological properties of alkanes vary in a regular way as we proceed through the homologous series. Methane appears to be totally physiologically inert. We could breathe a mixture of 80% methane and 20% oxygen without ill effect. Such a mixture would be flammable, however, and no fire or spark of any kind could be permitted in an atmosphere consisting of methane. Breathing an atmosphere of pure methane (the ''gas'' of a gas-operated stove) can lead to death not so much because of the presence of methane but because of the absence of oxygen (asphyxia). The other gaseous alkanes (and vapors of volatile liquid ones) act as anesthetics in high concentrations. They can also produce asphyxiation by excluding oxygen.

Liquid alkanes have varied effects depending on the part of the body exposed. On the skin, alkanes dissolve body oils. Repeated contact may cause dermatitis. Swallowed, alkanes do little harm while in the stomach. However, in the lungs, alkanes cause ''chemical pneumonia'' by dissolving fatlike molecules from the cell membranes in the alveoli. The cells become less flexible, and the alveoli are no longer able to expel fluids. The buildup of fluids is similar to that which occurs in bacterial or viral pneumonia. People who swallow gasoline, petroleum distillates, or other liquid alkane mixtures should not be made to vomit. That would increase the chance of their getting the alkane into the lungs.

Heavier liquid alkanes, when applied to the skin, act as emollients (skin softeners). Such alkane mixtures as mineral oil can be used to replace natural skin oils washed away by frequent bathing or swimming. Petroleum jelly (Vaseline is one brand) is a semisolid mixture of hydrocarbons that can be applied as an emollient or simply as a protective film. Water and water solutions (e.g., urine) will not dissolve such a film, which explains why petroleum jelly protects a baby's tender skin from diaper rash.

14.7 CHEMICAL PROPERTIES: LITTLE AFFINITY

The alkanes are the least reactive of all organic compounds. They are generally unreactive toward strong acids (such as sulfuric acid), strong bases (such as sodium hydroxide), most oxidizing agents (such as potassium dichromate), and most reducing agents (such as sodium metal). In fact, the alkanes undergo so few reactions that they are sometimes called **paraffins,** from the Latin words meaning ''little affinity.'' Paraffin wax is a mixture of solid alkanes frequently used as a seal for homemade preserves. The inertness of the wax makes it ideal for such use.

The alkanes do undergo a few important reactions. When mixed with oxygen at room temperature, alkanes give no apparent reaction. However, when a match flame or spark supplies sufficient energy to get things started (the energy of activation), an exothermic (heat-producing) reaction proceeds vigorously. The reaction, which is called **combustion,** is illustrated for methane.

$$CH_4 + 2\,O_2 \longrightarrow CO_2 + 2\,H_2O + \text{heat}$$

If the reactants are adequately mixed and there is sufficient oxygen, the products are carbon dioxide, water, and the all-important heat (for cooking foods, heating homes, and drying clothes). Such ideal conditions are rarely met, however, and other products than carbon dioxide, water, and heat are frequently formed. When the oxygen supply is limited, carbon monoxide is a by-product.

$$2\,CH_4 + 3\,O_2 \longrightarrow 2\,CO + 4\,H_2O$$

This reaction is responsible for dozens of deaths each year from unventilated or improperly adjusted gas heaters. (Similar reactions with similar results occur with kerosene heaters.) At the high temperatures achieved in some combustion reactions, some nitrogen (which, you will recall, makes up 80% of the atmosphere in which the fuel is burned) is converted to oxides. Further, petroleum usually contains some sulfur compounds, and combustion reactions involving sulfur will yield sulfur dioxide. Thus, the same combustion reactions that heat our homes, power our industries, and propel our automobiles also produce most of our air pollution.

In addition to undergoing combustion, alkanes react with fluorine, chlorine, and bromine. Such reactions produce halogenated hydrocarbons (Special Topic E).

14.8 ALKENES: STRUCTURE AND NOMENCLATURE

Not all hydrocarbons are as resistant to reaction as the alkanes. In fact, the next family that we will consider is quite reactive. This family, called **alkenes** (note the **-ene** ending), is characterized by the presence of a carbon-to-carbon double bond. Names, structures, and physical properties of a few representative alkenes are given in Table 14.6.

TABLE 14.6 *Physical Properties of Some Selected Alkenes*

IUPAC Name	Molecular Formula	Condensed Structure	Melting Point (in degrees Celsius)	Boiling Point (in degrees Celsius)
Ethene	C_2H_4	$CH_2{=}CH_2$	-169	-104
Propene	C_3H_6	$CH_3CH{=}CH_2$	-185	-47
1-Butene	C_4H_8	$CH_3CH_2CH{=}CH_2$	-185	-6
1-Pentene	C_5H_{10}	$CH_3CH_2CH_2CH{=}CH_2$	-138	30
1-Hexene	C_6H_{12}	$CH_3(CH_2)_3CH{=}CH_2$	-140	63
1-Heptene	C_7H_{14}	$CH_3(CH_2)_4CH{=}CH_2$	-119	94
1-Octene	C_8H_{16}	$CH_3(CH_2)_5CH{=}CH_2$	-102	121

We have used only condensed structural formulas in this table. Thus, $CH_2{=}CH_2$ stands for

$$H\diagdown \atop H\diagup C{=}C{\diagup H \atop \diagdown H}$$

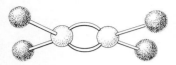

FIGURE 14.7
Ball-and-stick model of
ethylene.

The double bond is shared by the two carbon atoms and does not involve the hydrogens, although the condensed formula does not make this point obvious. Figure 14.7 shows a ball-and-stick model of this compound, ethylene.

Compare the molecular formulas of the alkenes in Table 14.6 with those for the alkanes in Table 14.3. The molecular formula for ethane is C_2H_6. The formula for the two-carbon alkene is C_2H_4. Each alkene has two fewer hydrogens than the corresponding alkane. Compared with the carbon atoms in an alkane, each of two carbon atoms in the alkene must bond to one less hydrogen atom in order to form the double bond. Since these carbons are sharing two of their bonds with one atom (and thus bonding to fewer different atoms), the compounds are called **unsaturated hydrocarbons.**

The first two alkenes of Table 14.6, ethene and propene, are most often called by their common names, ethylene and propylene, respectively. Ethylene ($CH_2{=}CH_2$) is one of the most important of all commercial chemicals. The United States chemical industry produces about 13 billion kg of ethylene annually, making it the most important of all synthetic organic chemicals. Over one-third of this ethylene went into the manufacture of polyethylene, one of the most familiar plastics (Chapter 19). Another one-sixth was converted to ethylene glycol, the major component of most brands of antifreeze for automobile radiators.

Propylene ($CH_3CH{=}CH_2$) is also an important industrial chemical. It is converted to plastics, isopropyl alcohol (Chapter 15), and a variety of other end products and intermediates for synthesis.

Before we consider the rest of the alkenes in Table 14.6, let's pause to discuss the nomenclature of alkenes. While there is only one alkene with the formula C_2H_4 (ethene) and only one alkene with the formula C_3H_6 (propene), there are four alkenes with the formula C_4H_8. Common names are hardly helpful in naming the many isomers of the higher alkenes. For the most part, the IUPAC system is used for these. Some of the IUPAC rules for alkenes are as follows:

1. All have names ending in **-ene.**
2. The longest chain of atoms *containing the double bond* is the parent compound. The name has the same stem as the corresponding alkane (i.e., the alkane with the same number of carbon atoms), but the ending is changed from **-ane** to **-ene.** Thus, the compound $CH_3CH{=}CH_2$, with three carbon atoms, is named **propene.**

3. When it is necessary to indicate the position of the double bond, the first carbon of the two that are doubly bonded is given the lowest possible number (i.e., the carbons are counted from the end of the chain nearer the first carbon of the double bond). The compound CH_3CH=$CHCH_2CH_3$, for example, has the double bond between the second and third carbon atoms. Its name is 2-pentene.
4. Substituent groups are named as usual. Their position is indicated by a number. Thus,

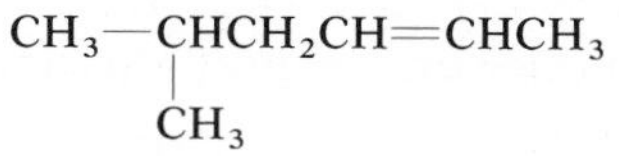

$$CH_3-\underset{\underset{\displaystyle CH_3}{|}}{CH}CH_2CH=CHCH_3$$

is 5-methyl-2-hexene. Note that the numbering of the parent chain is always done in such a way as to give the double bond the lowest number, even if that forces a substituent to have a higher number. We say the double bond has priority in numbering.

The rules are more easily learned through examples.

Name the compound

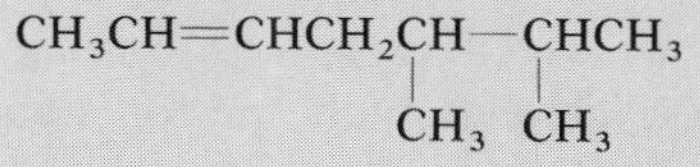

$$CH_3CH=CHCH_2\underset{\underset{\displaystyle CH_3}{|}}{CH}-\underset{\underset{\displaystyle CH_3}{|}}{CH}CH_3$$

example
14.5

The longest continuous chain has seven carbon atoms. To give the first carbon of the double bond the lowest number, we start numbering from the left.

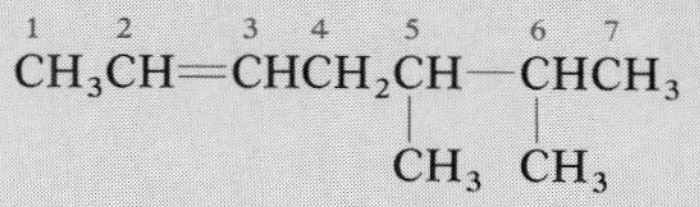

$$\overset{1}{C}H_3\overset{2}{C}H=\overset{3}{C}H\overset{4}{C}H_2\underset{\underset{\displaystyle CH_3}{|}}{\overset{5}{C}H}-\underset{\underset{\displaystyle CH_3}{|}}{\overset{6}{C}H}\overset{7}{C}H_3$$

The name of the compound is 5,6-dimethyl-2-heptene.

Name the compound

$$CH_2=\underset{\underset{\displaystyle CH_3-CH_2}{|}}{C}-CH_2CH_3$$

example
14.6

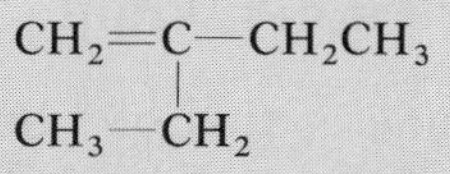

The name is 2-ethyl-l-butene. The longest continuous chain in the molecule contains five carbon atoms. However, the longest continuous chain *containing the double bond* incorporates only four carbon atoms, and this four-carbon chain serves as the parent compound.

example
14.7

Draw 3,4-dimethyl-2-pentene.
    To draw this compound, first write down the parent chain of five carbons.

$$C—C—C—C—C$$

Then add the double bond between the second and third carbons (this is 2-pentene).

$$\overset{1}{C}—\overset{2}{C}{=}\overset{3}{C}—\overset{4}{C}—\overset{5}{C}$$

Now add the groups at their proper positions and fill in all the hydrogens.

$$\left(\text{or } CH_3{-}CH{=}C\overset{\overset{CH_3}{|}}{}{-}CH\overset{\overset{CH_3}{|}}{}{-}CH_3\right)$$

There are four isomeric alkenes with the formula C_4H_8. Let's write their structures. The first, 1-butene, is easy.

$$CH_2{=}CHCH_2CH_3$$

This structure corresponds to a gaseous compound that melts at $-185\,°C$ and boils at $-6\,°C$. We can also draw a structure analogous to that of isobutane.

$$CH_2{=}\overset{\overset{CH_3}{|}}{C}{-}CH_3$$

This compound is frequently called isobutylene, but its IUPAC name is methylpropene. Next we can write a structure for 2-butene.

$$CH_3CH{=}CHCH_3$$

The carbon skeleton is the same as that of 1-butene, but the double bond is shared by the second and third carbon atoms instead of the first and second. The only problem is that there are *two* compounds that correspond to this formula for 2-butene. One melts at $-139\,°C$ and boils at $+4\,°C$; the other melts at $-106\,°C$ and boils at $+1\,°C$. The existence of two 2-butenes is explained by restricted rotation about the carbon-carbon double bond.

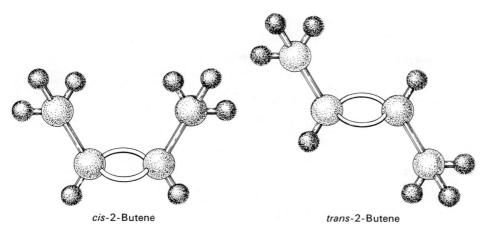

cis-2-Butene   *trans*-2-Butene

FIGURE 14.8   *Models of the 2-butenes. While the carbons joined by single bonds are free to spin about these bonds, the doubly bonded carbons are restricted in this regard.*

Remember that atoms and groups of atoms are free to spin about single bonds (Section 14.4). However, two atoms connected by a double bond are not free to spin relative to one another. While the models in Figure 14.8 don't really "look" like the actual molecules (we can't see individual molecules), they do offer a mechanical equivalent of restricted rotation. The two connectors between the doubly bonded carbon atoms prevent these carbons from spinning as the singly bonded carbons are free to do. Thus, the two arrangements of 2-butene shown in Figure 14.8 really represent two different compounds. One cannot be converted to the other unless the double bond is broken first. To distinguish the two different isomers, one is called *cis*-2-butene and the other is *trans*-2-butene. The **cis** isomer is the one with both methyl groups on the same side of the molecule. In the **trans** isomer, the methyl groups are on opposite sides.

Table 14.7 summarizes the structures and some properties of the four isomeric butylenes.

We can draw two *seemingly* different propylenes (structures IV and V).

$$
\underset{\text{IV}}{\overset{\displaystyle \begin{array}{c} H \\ \diagdown \\ H \diagup \end{array} C=C \begin{array}{c} CH_3 \\ \diagup \\ \diagdown H \end{array}}{}}
\qquad
\underset{\text{V}}{\overset{\displaystyle \begin{array}{c} H \\ \diagdown \\ H \diagup \end{array} C=C \begin{array}{c} H \\ \diagup \\ \diagdown CH_3 \end{array}}{}}
$$

TABLE 14.7 *Physical Properties of the Butylenes*

Name	Structure	Melting Point (in degrees Celsius)	Boiling Point (in degrees Celsius)
1-Butene	$CH_2{=}CHCH_2CH_3$	-185	-6
cis-2-Butene	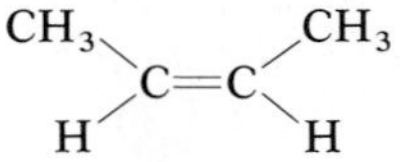	-139	$+4$
trans-2-Butene	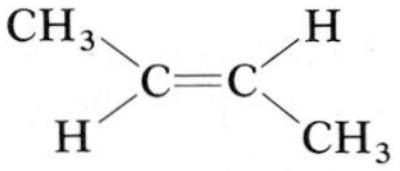	-106	$+1$
2-Methylpropene (isobutylene)	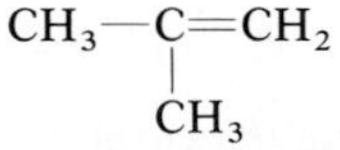 $CH_3{-}C{=}CH_2$ $\quad\quad\;$ CH_3	-140	-7

However, the second structure is not really different from the first. If you could pick it up from the page and flip it over, you would see that the two formulas were identical.

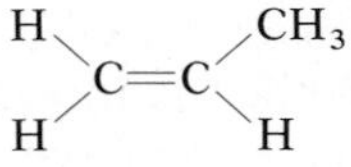

IV

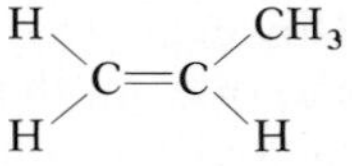

V
(flipped over)

The same thing *cannot* be done with *cis* and *trans* isomers. If we start with drawings of the two 2-butenes (VI and VII)

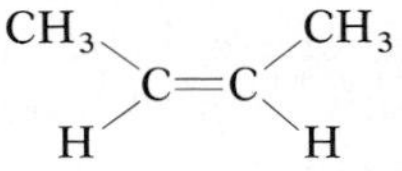

VI
cis-2-Butene

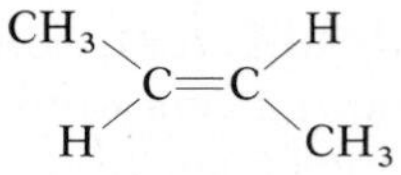

VII
trans-2-Butene

and then flip the *trans* isomer over, we see that the resulting structure is still clearly different from the *cis* isomer.

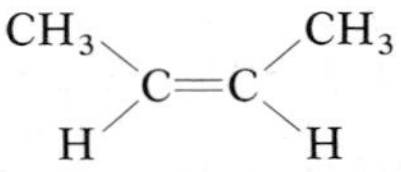

VI

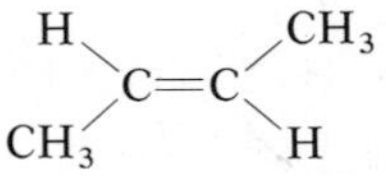

VII
(flipped over)

As the case of propylene proves, the mere presence of a double bond is not sufficient for **cis-trans,** or **geometric, isomerism.** The requirements for such isomerism are (1) that rotation be restricted in the molecule, *and* (2) that there be two nonidentical groups on *each* of the doubly bonded carbons. For 2-butene (CH_3CH=$CHCH_3$), the doubly bonded carbon on the left has a hydrogen group and a methyl group (two different groups), and the doubly bonded carbon on the right has a hydrogen group and a methyl group (two different groups). Thus, 2-butene exists as a *cis* and a *trans* isomer. Propene (CH_3CH=CH_2) has a doubly bonded carbon with two hydrogens (two identical groups) attached. The second requirement for geometric isomerism is not fulfilled, therefore, and this compound does *not* exist as *cis* and *trans* isomers. One of the doubly bonded carbons in propene does have two different groups attached, but the rules require that *both* carbons have two different groups.

example 14.8

Draw all alkenes with the formula C_5H_{10} and indicate which exist as *cis* and *trans* isomers. Give the IUPAC names for the isomers.

First we'll draw the various possible carbon skeletons incorporating a double bond.

$$CH_2\text{=}CHCH_2CH_2CH_3 \qquad CH_3CH\text{=}CHCH_2CH_3$$

1-Pentene
VIII

"2-Pentene"
IX

$$CH_2\text{=}\overset{\overset{\displaystyle CH_3}{|}}{C}\text{—}CH_2CH_3 \qquad CH_3\text{—}\overset{\overset{\displaystyle CH_3}{|}}{C}\text{=}CH\text{—}CH_3$$

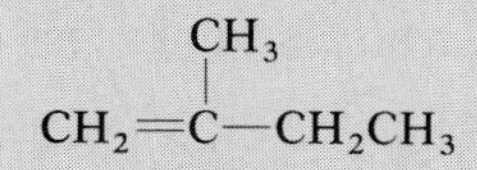

2-Methyl-1-butene
X

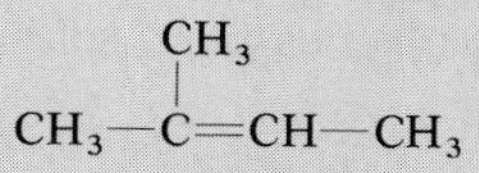

2-Methyl-2-butene
XI

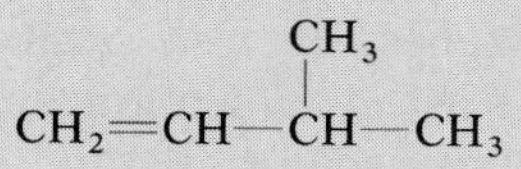

$$CH_2\text{=}CH\text{—}\overset{\overset{\displaystyle CH_3}{|}}{C}H\text{—}CH_3$$

3-Methyl-1-butene
XII

Of these, only IX exists as *cis* and *trans* isomers.

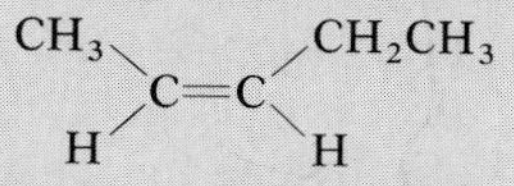

cis-2-Pentene 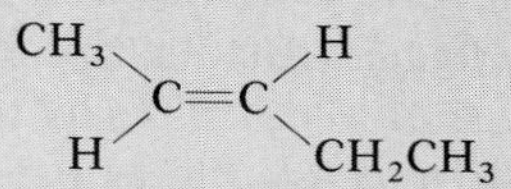   *trans*-2-Pentene

Structures VIII, X, and XII each have two hydrogens on one of their doubly bonded carbon atoms, and structure XI has two methyl groups on one of its doubly bonded carbons. Note that

$$
\begin{array}{c}
CH_3 \\
| \\
CH_3 - C = CH_2 \qquad \text{(Incorrect structure)} \\
| \\
CH_3
\end{array}
$$

is not a possible isomer. The central carbon atom has five bonds in this structure, and carbon can form only four covalent bonds.

While geometric isomerism may seem of little practical importance to us, it is extremely important to houseflies. The female housefly secretes *cis*-9-tricosene as an attractant for the male. The latter has little, if any, affinity for the *trans* isomer.

$$
\begin{array}{cc}
CH_3(CH_2)_7 \diagdown \qquad \diagup (CH_2)_{12}CH_3 & CH_3(CH_2)_7 \diagdown \qquad \diagup H \\
C = C & C = C \\
H \diagup \qquad \diagdown H & H \diagup \qquad \diagdown (CH_2)_{12}CH_3
\end{array}
$$

cis-9-Tricosene   *trans*-9-Tricosene

Indeed, in most biological systems, the geometry of a molecule is of utmost importance.

Physical properties of alkenes are quite similar to those of corresponding alkanes. Alkenes with 1 to 4 carbons are gases at room temperature. Those with 5 to 18 carbons are liquids, and those with more than 18 are solids. Like the alkanes, the alkenes are insoluble in water and are less dense than water.

14.9 ALKENES AND LIVING THINGS

The physiological properties of the alkenes are also similar to those of the alkanes. Ethylene has found some use as an inhalation anesthetic. Like the gaseous alkanes, ethylene can cause unconsciousness and even death by asphyxiation. Large amounts of liquid and solid (or mixtures of liquid and solid) alkenes are seldom encountered. They would probably act on or in our bodies much as the alkanes do.

Alkenes occur widely in nature. Ripening fruits and vegetables give off ethylene, which triggers further ripening. Fruit suppliers artificially introduce ethylene to hasten the normal ripening process; 1 kg of tomatoes can be ripened by exposure to as little as 0.1 mg of ethylene for 24 hours.

Other alkenes found in nature include 1-octene, found in lemon oil, and octadecene ($C_{18}H_{36}$), found in fish liver. Dienes (which have two double bonds) and polyenes (which have many double bonds) are also common. Butadiene ($CH_2 = CH - CH = CH_2$) is found in coffee. A hexadecadiene ($C_{16}H_{30}$) occurs in olive oil. Lycopene and the carotenes are isomeric polyenes ($C_{40}H_{56}$) that give the attractive red, orange, and yellow colors to watermelons, tomatoes, carrots, and other vegetables and fruits. Vitamin A, essential to good vision, is derived from a carotene. Vitamin A is converted in the body to *trans*-retinene. The latter absorbs visible light and is converted to *cis*-retinene. It is this process, which occurs on the retina of the eye, that is responsible in part for vision. The world would be a much darker place without the chemistry of the alkenes.

14.10 CHEMICAL PROPERTIES OF THE ALKENES

Like the alkanes—and all other hydrocarbons—the alkenes burn. While this is a hazard that you should remember when working with alkenes, these compounds are not commercially important as fuels. We can write an equation for the combustion of ethene that is similar to that for the combustion of an alkane.

$$C_2H_4 + 3\,O_2 \longrightarrow 2\,CO_2 + 2\,H_2O + heat$$

The typical reactions of the alkenes are **addition reactions.** One of the bonds in the double bond is broken, permitting each of the involved carbon atoms to bond to an additional atom or group. The originally doubly bonded carbons are still attached by the remaining single bond. Perhaps the simplest addition reaction is that which occurs with hydrogen in the presence of a nickel (Ni), platinum (Pt), or palladium (Pd) catalyst.

| Ethene | Hydrogen | Ethane |

The product of this reaction is an alkane with the same carbon skeleton as the original alkene. **Hydrogenation** was once widely used in industry in the conversion of unsaturated vegetable oils into saturated fats. Hydrogenation of the liquid oil yields a solid fat and makes vegetable shortening resemble the more familiar animal fat, lard. The difference in the liquid oil and the solid fat is due to a difference in the number of double bonds present; there are more in the unsaturated oils and fewer in the saturated fats. Vegetable oils are now more readily accepted and, in fact, are frequently preferred by the consumer, but some oils are still hydrogenated so that the vegetable product will resemble an animal product. Margarine, for example, resembles butter by virtue of hydrogenation.

Alkenes readily add halogen molecules. Indeed, the reaction with bromine is often used to test for alkenes. Solutions of bromine are brownish red. When an alkene is added to such a solution, the color disappears because the alkene reacts with the bromine.

$$\underset{\text{Ethene}}{\underset{H}{\overset{H}{>}}C=C\underset{H}{\overset{H}{<}}} + \underset{\substack{\text{Bromine} \\ \text{(brownish red)}}}{Br-Br} \longrightarrow \underset{\substack{\text{1,2-Dibromoethane} \\ \text{(colorless)}}}{H-\overset{\overset{\displaystyle H}{|}}{\underset{\underset{\displaystyle Br}{|}}{C}}-\overset{\overset{\displaystyle H}{|}}{\underset{\underset{\displaystyle Br}{|}}{C}}-H}$$

Another important addition reaction of the alkenes is that with water. This reaction, called **hydration,** requires the presence of a mineral acid, such as sulfuric acid (H_2SO_4), as a catalyst.

$$\underset{\text{Ethene}}{\underset{H}{\overset{H}{>}}C=C\underset{H}{\overset{H}{<}}} + \underset{\text{Water}}{H-OH} \xrightarrow{H_2SO_4} \underset{\text{Ethyl alcohol}}{H-\overset{\overset{\displaystyle H}{|}}{\underset{\underset{\displaystyle H}{|}}{C}}-\overset{\overset{\displaystyle H}{|}}{\underset{\underset{\displaystyle OH}{|}}{C}}-H}$$

Vast quantities of ethyl alcohol, for use as an industrial solvent, are made from ethylene. This alcohol is structurally identical to that used in alcoholic beverages. However, federal law requires that all drinking alcohol be produced by the natural process called fermentation (Chapter 15).

The most important alkene reaction of all, perhaps, is polymerization. This topic is discussed in Chapter 19.

example 14.9	Write equations for the reaction of $CH_3CH=CHCH_3$ with each of the following:

a. H_2 (Ni catalyst)
b. Br_2
c. H_2O (H_2SO_4 catalyst)

In each reaction, the reagent adds across the double bond. The answers are

a. $CH_3CH=CHCH_3 + H_2 \xrightarrow{\text{Ni}} CH_3\underset{\underset{\displaystyle H}{|}}{CH}-\underset{\underset{\displaystyle H}{|}}{CH}CH_3$ or $CH_3CH_2CH_2CH_3$

b. $CH_3CH=CHCH_3 + Br_2 \longrightarrow CH_3\underset{\underset{\displaystyle Br}{|}}{CH}-\underset{\underset{\displaystyle Br}{|}}{CH}CH_3$

c. $CH_3CH=CHCH_3 + H_2O \xrightarrow{H_2SO_4} CH_3CH_2\underset{\underset{\displaystyle OH}{|}}{CH}CH_3$

14.11 THE ALKYNE SERIES

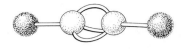

FIGURE 14.9
Ball-and-stick model of acetylene.

In alkenes, carbon atoms in the double bond share two pairs of electrons. Carbon atoms can also share three pairs of electrons, forming triple bonds. Compounds containing such bonds are called **alkynes.** The common name of the simplest alkyne is acetylene (C_2H_2) (Figure 14.9). Its structure is

$$H—C≡C—H$$

About 10% of all acetylene produced is used in oxyacetylene torches for cutting and welding metals. The flame from such a torch can attain a very high temperature. Most acetylene, however, is converted to chemical intermediates that are in turn used to make vinyl and acrylic plastics, fibers, and resins and a variety of other chemical products.

The alkynes are similar to the alkenes in both physical and chemical properties. For example, they undergo many of the typical addition reactions of alkenes. Like ethene, acetylene has been used as an anesthetic for surgery. At higher concentrations, it causes narcosis and asphyxia. The IUPAC nomenclature for alkynes parallels that of the alkenes, except that the family ending is **-yne** rather than **-ene.** The official name for acetylene is ethyne. Notice that the ending of the common name, acetylene, is deceptive. It sounds very much like ethylene or propylene. Remember that acetylene is an alkyne and that ethylene and propylene are alkenes.

Name the following alkynes:

a. $CH_3C≡CH$
b. $CH_3CH_2C≡CH$
c. $CH_3C≡CCH_3$

The names are

a. propyne
b. 1-butyne
c. 2-butyne

example
14.10

14.12 CYCLOALKANES

The hydrocarbons we have encountered so far have been composed of open-ended chains of carbon atoms. Carbon and hydrogen atoms can also hook up in other arrangements, some of them quite interesting. There is a synthetic hydrocarbon that has the formula C_3H_6. The three carbons are joined in a **ring,** or **cycle.** The compound is called cyclopropane (Figure 14.10).

In addition to having an interesting structure, cyclopropane has some intriguing properties. It is an excellent anesthetic, for when inhaled, it renders

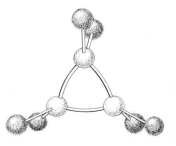

FIGURE 14.10
Ball-and-stick model of cyclopropane.

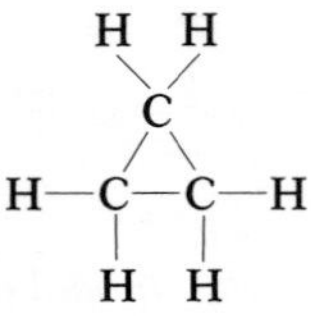

Cyclopropane

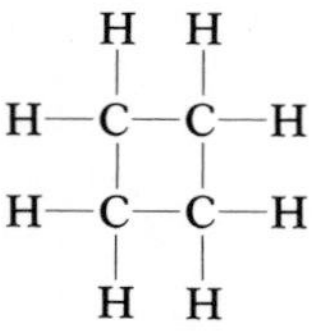

Cyclobutane

the patient unconscious and unaware of pain. It is also explosive when mixed with air; thus, care must be taken to exclude the possibility of a flame or spark from the operating room when this material is in use.

Names of cycloalkanes are formed by addition of the prefix **cyclo-** in front of the name of the open-chain compound with the same number of carbon atoms as are in the ring. Thus, the name for the cyclic compound C_4H_8 is cyclobutane. Names and structures of several cycloalkanes are given in Figure 14.11. Note that the carbon atoms of each compound form a regular geometric figure. For example, the three carbon atoms of cyclopropane form a triangle. Therefore, a triangle is frequently used to represent cyclopropane. Similarly, the five carbons of cyclopentane form a pentagon, and we use a pentagon to represent cyclopentane. It is understood that the carbon atoms occur at the angles of the particular figure and that each carbon atom is attached to sufficient hydrogens to give the carbon atom four bonds. Cyclic compounds may also contain attached substituent groups (see Figure 14.11).

example 14.11

Draw structures for the following cyclic hydrocarbons:

a. cyclooctane
b. methylcyclobutane

a.
$$CH_2-CH_2$$
$$CH_2 \qquad CH_2$$
$$CH_2 \qquad CH_2$$
$$CH_2-CH_2$$

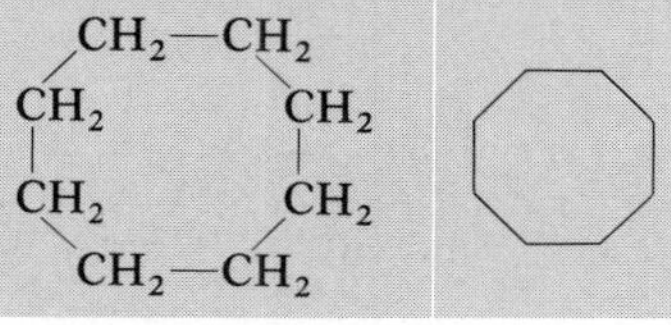

b.
$$CH_3$$
$$CH_2-CH$$
$$CH_2-CH_2$$

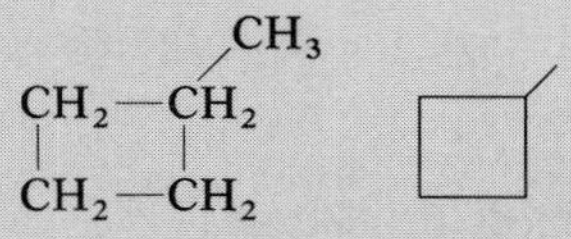

The physical, chemical, and physiological properties of these cyclic hydrocarbons are generally quite similar to those of the corresponding open-chain

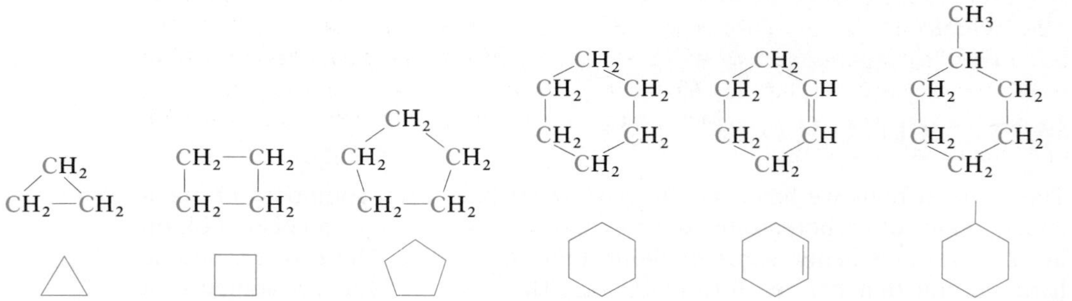

FIGURE 14.11 *Some cyclic hydrocarbons.*

Cyclopropane   Cyclobutane   Cyclopentane   Cyclohexane   Cyclohexene   Methylcyclohexane

compounds. Cycloalkanes (with the exception of cyclopropane) act very much like ordinary alkanes. Cycloalkenes, like other alkenes, undergo addition reactions. Cyclic hydrocarbons with five- and six-membered rings occur in petroleum from certain areas; California crude, for instance, is particularly rich in these compounds. Like all other hydrocarbons, cyclic hydrocarbons burn.

14.13 BENZENE

Still another type of hydrocarbon is represented by the compound benzene. Benzene was first isolated in 1825 from a by-product of whale oil by Michael Faraday. (This is the same Michael Faraday whose electrolysis experiments are described in Chapters 2 and 12). The molecular formula of benzene was soon (in 1834) determined to be C_6H_6. We can write many structures corresponding to the formula C_6H_6. Three such structures are

The real substance, benzene, does not have the properties that one would predict from these structures. For example, if benzene really contained double or triple bonds, it would be expected to undergo addition reactions readily. It does not.

In 1865, the third structure was proposed by Friedrich August Kekulé, a German chemist. It is accurate in some respects. Benzene does have a six-member ring structure. It is also an unsaturated molecule; that is, it contains something other than single bonds. To account for benzene's inertness toward the typical reactions of unsaturated compounds, Kekulé's structure has been modified. In order for benzene to behave as it does, the "extra" or double-bond electrons cannot be located between specific carbon atoms as they would be in ordinary double bonds. All six of these electrons are regarded as being equally shared by all six carbon atoms in the ring. Therefore, the structure of benzene is now drawn where the circle represents the six "extra" electrons. There are no real double bonds but a very stable ring of electrons, which resists being disrupted. Benzene does not enter addition reactions because that would destroy the ring of electrons.

It takes a good deal of time to draw out structures for benzene showing all the carbons and hydrogens. We note that, as with the cyclic hydrocarbons in the preceding section, the six carbon atoms of benzene are located at the corners of a hexagon. Therefore, a hexagon with a circle to represent the six unassigned electrons is used for benzene.

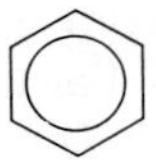

The Kékulé structure is still found in some texts.

Remember that the above structures are not really benzene (a gasolinelike liquid that boils at 80 °C). No one has ever seen a benzene molecule. The structure represents a *concept* of the benzene molecule.

14.14 AROMATIC HYDROCARBONS: STRUCTURE AND NOMENCLATURE

There are many compounds chemically similar to benzene. Several of those discovered in the early days had pleasant odors and were known as aromatic compounds. The label *aromatic* is also used today: an **aromatic compound** simply means a compound that has, like benzene, an unusually stable ring of electrons. (All the nonaromatic hydrocarbons that we have considered, the alkanes, alkenes, etc., are referred to collectively as **aliphatic compounds** to distinguish them from aromatic compounds. "Aliphatic" originally meant that the source of the compound was a fat. Today, however, it simply means "not aromatic.")

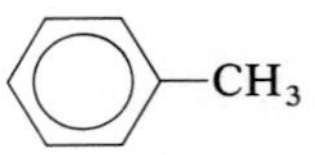

A number of aromatic compounds can be derived from benzene by substitution of a variety of groups for one or more of the hydrogen atoms. Substitution of a methyl group for one hydrogen gives methylbenzene, better known as **toluene** (pronounced to rhyme with "doll, you mean").

Remember that in this formula those points that are not shown attached to a substituent carry a hydrogen atom. Toluene is an important solvent and a starting material for the synthesis of other aromatic compounds.

Toluene

Substitution of an ethyl group for a hydrogen atom gives ethylbenzene. This compound is an important intermediate in the synthesis of styrene, from which the common plastic polystyrene is made (Chapter 19).

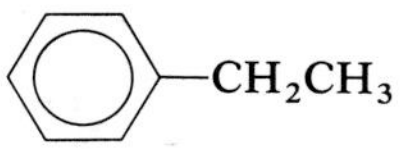

$-CH_2CH_3$

Ethylbenzene

Name the following compound:

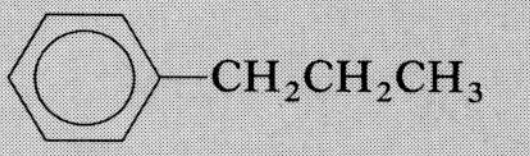

$-CH_2CH_2CH_3$

The alkyl group is propyl; the compound is propylbenzene.

example 14.12

Occasionally, when the group attached to the benzene ring does not have a simple name, it is the benzene ring that is named as a substituent group. In these cases, the ring is referred to as the **phenyl group.** Phenyl is C_6H_5- or

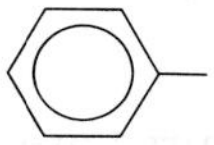

Thus,

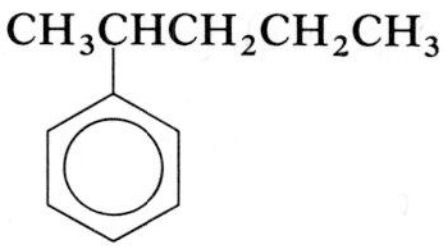

$CH_3CHCH_2CH_2CH_3$

is named 2-phenylpentane.

When two substituents are attached to the benzene ring, we must use some way of indicating their relative positions. There are two methods of doing this. One is the familiar method of using numbers. The other method uses the prefixes *ortho-*, *meta-*, and *para-* to indicate relative positions. The prefix *ortho-* is abbreviated *o-* and indicates substituents on adjacent carbon atoms. An ortho-substituted compound is a 1,2-disubstituted benzene. The prefix *meta-* (*m-*) is used for 1,3-disubstituted benzenes, and the prefix *para-* (*p-*) is used with 1,4-disubstituted benzenes. All four of the following structures represent the same compound, *m*-dinitrobenzene (NO_2 is called the *nitro* group).

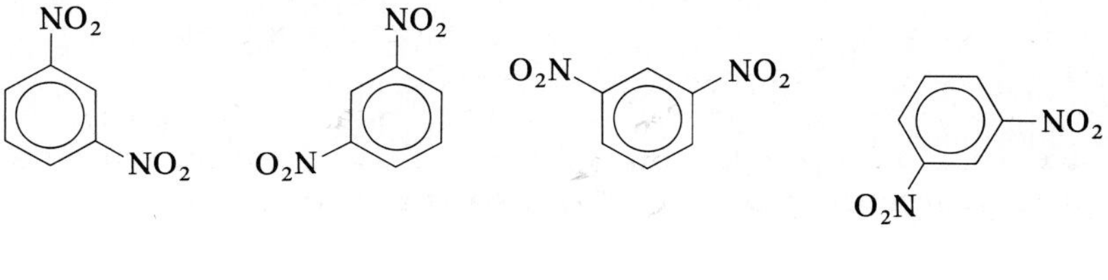

o-Xylene
(1,2-dimethylbenzene)

m-Xylene
(1,3-dimethylbenzene)

p-Xylene
(1,4-dimethylbenzene)

o-Nitrotoluene
(2-nitrotoluene)

1,3,5-Trimethylbenzene

2,4,6-Trinitrotoluene
(TNT)

FIGURE 14.12   *Some aromatic hydrocarbons and derivatives.*

Naphthalene

p-Dichlorobenzene

3,4-Benzpyrene

For three or more substituents, the numbering system must be used. Examples are shown in Figure 14.12.

In another group of aromatic hydrocarbons, two or more benzene rings are "fused" together; that is, the rings have two or more carbon atoms in common. The simplest of these is naphthalene ($C_{10}H_8$). Naphthalene has been used as a moth repellent, as has p-dichlorobenzene. One of the most distinctive physical properties of naphthalene is its "mothball" odor.

Aromatic compounds with several fused rings often are **carcinogenic,** that is, they can induce cancer. Typical of these compounds is 3,4-benzpyrene (benz[a]pyrene). This compound is known to induce cancer in laboratory animals. It is found in cigarette smoke and automobile exhaust fumes.

14.15 PROPERTIES OF AROMATIC HYDROCARBONS

Benzene, toluene, and the xylenes are liquids at room temperature. The compounds involving fused rings, the **polycyclic** hydrocarbons, are solids. All are insoluble in water. They are generally soluble in organic solvents such as hexane.

Like all hydrocarbons, aromatic hydrocarbons burn. In other chemical reactions, aromatic hydrocarbons behave unlike the unsaturated hydrocarbons we have previously encountered. Aromatic hydrocarbons resist addition reactions, but they do undergo a number of substitution reactions. Benzene, for example, can be chlorinated in the presence of a catalyst (iron, Fe, in this case).

$$C_6H_6 + Cl_2 \xrightarrow{\text{Fe}} C_6H_5Cl + HCl$$

We have written this equation using molecular formulas to emphasize that this is a substitution reaction. One of benzene's hydrogen atoms is replaced by a chlorine atom. Many organic chemists would write the equation as follows:

$$\text{benzene} \xrightarrow[\text{Fe}]{\text{Cl}_2} \text{chlorobenzene}$$

Because the hydrogen atoms are not explicitly shown in these structures, it is easy to forget that this is a substitution reaction. Remember—the chlorine atom *replaces* a hydrogen atom. Notice also that only the organic starting material and product are emphasized. The inorganic product is ignored completely. (It is still formed; it is simply not shown.) Not only the catalyst, but the inorganic reagent, too, is written by the arrow. You will encounter this form of a chemical equation again as we proceed through organic chemistry.

Most of the aromatic hydrocarbons present toxic hazards. Benzene, whether inhaled or ingested, may cause convulsions or even death by respiratory failure. Chronic exposure to lower levels of benzene can depress the formation of blood cells by bone marrow. Prolonged chronic exposure can cause death. Benzene is thought to cause leukemia in workers exposed to it over the years.

Benzene was once widely used as a laboratory solvent and in consumer products such as those used to strip the finish from old furniture. It is still an extremely important industrial chemical. United States production is about 5 billion kg annually. For school laboratories and in consumer products, benzene is now being replaced in many instances by toluene, which is somewhat less toxic. Toluene acts as a narcotic, more powerful than benzene, in high concentrations, but the body can metabolize and excrete small amounts of toluene. The methyl group on toluene offers the body a "handle" on which to work. Benzene does not have this "handle." The xylenes are similar to toluene in their physiological properties. As we have seen, many of the polycyclic aromatic hydrocarbons are carcinogenic. All aromatic compounds should be handled only after adequate evaluation of the hazards involved. Volatile solvents such as benzene and toluene should certainly be used only with adequate ventilation.

14.16 NATURAL GAS AND PETROLEUM

Compounds of carbon are the basis of all life on our planet. They are essential to all life processes and constitute by far our major energy source. They are the basis of many of our structural building materials and of nearly all our medicines. Some organic compounds are still obtained from plants and animals, but most come ultimately from the fossilized carbon materials coal and petroleum. The latter is largely a mixture of alkanes, with the molecules having from 1 to 40 (or more) carbon atoms each. Most of the **petrochemicals** so vital to our modern economy are derived from these alkanes. A portion of the remaining petrochemicals, the **aromatics** (Section 14.14), are derived from coal.

Petroleum, as it comes from the ground, is of limited use. So that it will better suit our needs, we separate it into fractions by boiling it in a distillation column (Figure 14.13). The lighter molecules, those with 1 to 4 carbon atoms each, come off the top of the column. The next fraction contains, for the most part, molecules having from 5 to 12 carbon atoms. These and other fractions are listed in Table 14.8.

Since gasoline is generally the fraction most in demand, the fractions with higher boiling points are often in excess supply. These can be converted to gasoline by heating in the absence of air. This process, called **cracking,** breaks

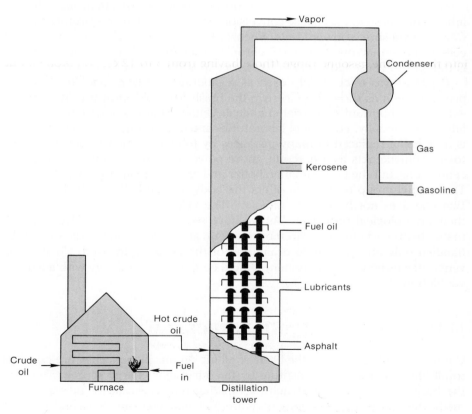

FIGURE 14.13   *The fractional distillation of petroleum.*

TABLE 14.8   *Typical Petroleum Fractions*

Fraction	Typical Range of Hydrocarbons	Approximate Range of Boiling Point (in degrees Celsius)	Typical Uses
Natural gas	CH_4 to C_4H_{10}	Less than 40	Fuel, starting materials for plastics
Gasoline	C_5H_{12} to $C_{12}H_{26}$	40–200	Fuel, solvents
Kerosene	$C_{12}H_{26}$ to $C_{16}H_{34}$	175–275	Diesel fuel, jet fuel, home heating; cracking to gasoline
Heating oil	$C_{15}H_{32}$ to $C_{18}H_{38}$	250–400	Industrial heating, cracking to gasoline
Lubricating oil	$C_{17}H_{36}$ and up	Above 300	Lubricants
Residue	$C_{20}H_{42}$ and up	Above 350 (some decomposition)	Paraffin, asphalt

the big molecules apart. The process is illustrated in Figure 14.14, where $C_{14}H_{30}$ is used as an example. Not only does cracking convert some of the molecules into those in the gasoline range (those having from 5 to 12 carbon atoms), but it results in a variety of useful by-products. The unsaturated hydrocarbons (Section 14.8) are starting materials for the manufacture of many plastics, detergents, and drugs—indeed a whole host of petrochemicals. Present and

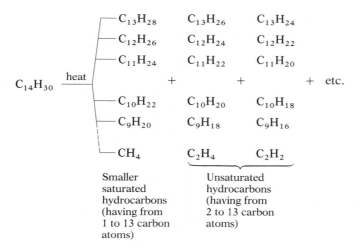

FIGURE 14.14   *Formulas of some of the products formed when $C_{14}H_{30}$ (a typical molecule in kerosene) is cracked. You need only note that a great variety of hydrocarbons with fewer carbon atoms are formed.*

future shortages of petroleum will mean a great deal more than just scarce, high-priced gasoline.

Benzene and other aromatic compounds are made by catalytic reforming. For example, hexane is converted to benzene by heating with a catalyst.

$$CH_3CH_2CH_2CH_2CH_2CH_3 \xrightarrow[\text{heat}]{\text{catalyst}} \bigcirc + 4H_2$$

$$(C_6H_{14}) \qquad\qquad\qquad (C_6H_6)$$

The cracking and reforming processes described here are crude, yet illustrative, examples of how chemists modify nature's materials to meet our needs and desires. Starting with coal tar or petroleum, the chemist can create a dazzling array of substances with a wide variety of properties. These include plastics, painkillers, antibiotics, stimulants, depressants, and detergents, to name just a few. Many of these materials are discussed in other chapters.

PROBLEMS

1. List three ways in which a "typical" organic compound differs from a "typical" inorganic one.
2. Define, illustrate, or give an example for each of these terms:
   - **a.** hydrocarbon
   - **b.** alkane
   - **c.** paraffin
   - **d.** saturated
   - **e.** unsaturated
   - **f.** substituent
   - **g.** alkene
   - **h.** alkyne
   - **i.** alkyl group
   - **j.** isomers
   - **k.** geometric isomers
   - **l.** aromatic compound
   - **m.** aliphatic compound
   - **n.** phenyl group
   - **o.** cracking
   - **p.** catalytic reforming
   - **q.** homologous series
3. Classify the following compounds as organic or inorganic:
   - **a.** C_6H_{10}
   - **b.** $CoCl_2$
   - **c.** $C_{12}H_{22}O_{11}$
   - **d.** CH_3NH_2
   - **e.** $NaNH_2$
   - **f.** $Cu(NH_3)_6Cl_2$
4. Which member of each pair has a higher melting point?
   - **a.** CH_3OH and $NaOH$
   - **b.** CH_3Cl and KCl
   - **c.** $C_{20}H_{42}$ and $C_{40}H_{82}$
5. You find a jar without a label containing a solid material. The substance melts at 48 °C. It ignites readily and burns cleanly. The substance is insoluble in water and floats on the surface of the water. Is the substance likely to be organic or inorganic?
6. How many carbon atoms are there in each of the following?
   - **a.** ethane
   - **b.** heptane
   - **c.** butane
   - **d.** nonane
7. Indicate whether the structures in each set represent the same compound or isomers.

   a. CH_3CH_3

   $\overset{\displaystyle CH_3}{\underset{\displaystyle CH_3}{|}}$

   b. $\overset{\displaystyle CH_3}{\underset{\displaystyle CH_3CH_2}{|}}$ $\qquad$ $CH_3CH_2CH_3$

   c. $\overset{\displaystyle CH_3}{\underset{\displaystyle CH_3CH_2CHCH_2CH_3}{|}}$ $\qquad$ $\overset{\displaystyle CH_3}{\underset{\displaystyle CH_3CHCH_2CH_2CH_3}{|}}$

   d. $\overset{\displaystyle CH_3}{\underset{\displaystyle CH_3CHCH_2CH_3}{|}}$ $\qquad$ $\overset{\displaystyle CH_3}{\underset{\displaystyle CH_3CH_2CH}{|}}$

   $\underset{\displaystyle CH_3}{|}$

   e. $\underset{\displaystyle CH_3CH_2CH—CH_2}{\overset{\displaystyle CH_3 \quad CH_3}{| \quad\; |}}$ $\qquad$ $\underset{\displaystyle CH_2CH_2CHCH_3}{\overset{\displaystyle CH_3 \quad\; CH_3}{| \qquad |}}$

8. Draw the structural formulas of the 4-carbon alkanes (C_4H_{10}). Identify butane and isobutane and give the IUPAC name for the latter.

9. Write structures for the five isomeric hexanes (C_6H_{14}). Name each by the IUPAC system.

10. Draw the following alkyl groups:
  a. ethyl
  b. isopropyl
  c. *tert*-butyl
  d. isobutyl

11. Classify each of the following compounds as saturated or unsaturated.

  a. $CH_3C{=}CH_2$
        $|$
        CH_3

  b.
        CH_3
        $|$
  $CH_3{-}C{-}CH_3$
        $|$
        CH_3

  c. $CH_3C{\equiv}CCH_3$

  d. ⬠

12. Indicate whether the structures in each set represent the same compound or isomers.

  a. $CH_3CH{=}CHCH_3$ $CH_3CH_2CH{=}CH_2$

  b.
  CH_3 CH_3
  $|$ $|$
  $CH_3C{=}CCH_3$ $CH_3C{=}CCH_3$
                                $|$ $|$
                                CH_3 CH_3

  c.
                        CH_3
                        $|$
  $CH_3CH_2CH_2CH{=}CCH_3$
                        CH_3
                        $|$
                $CH_3CHCH{=}CHCH_2CH_3$

  d.
        CH_3 CH_2CH_3
        $|$ $|$
  $CH_2{=}CCH_2CH_3$ $CH_2{=}CCH_3$

  e.
  CH_3\   /CH_2CH_3
      $C{=}C$
  CH_3CH_2/   \$CH_2CH_2CH_3$

  CH_3\   /$CH_2CH_2CH_3$
      $C{=}C$
  CH_3CH_2/   \CH_2CH_3

13. Including geometric isomers, there are five isomeric alkenes with this basic carbon skeleton and the molecular formula C_6H_{12}.

Draw all five isomers and name them by IUPAC rules.

14. Draw the three alkyne isomers with the formula C_5H_8. Give IUPAC names for the compounds.

15. Draw structural formulas for the following:
  a. 1,1-dimethylcyclobutane
  b. cyclobutene

16. Name the following compounds:

  a. ⬡
  b. ⬠—CH_3

17. What is the molecular formula for each of the following compounds?

  a. ⬡
  b. △—CH_3

18. Indicate whether the compound is aromatic or aliphatic.

  a. ⬡
  b. ⬡

  c. ⬡
  d. ⬡⬡

19. Identify the substitution pattern as *meta, ortho,* or *para*.

  a.   **b.**   **c.**

20. Give structural formulas for the following:
  a. heptane
  b. isopentane
  c. 2,2,5-trimethylhexane
  d. 4-ethyl-3-methyloctane

21. Give structural formulas for the following:
 a. propylene
 b. 3-ethyl-2-pentene
 c. 3-*tert*-butyl-1-hexene
 d. 2,3-dimethyl-2-butene

22. Give structural formulas for the following:
 a. *cis*-2-hexene   **b.** *trans*-3-hexene

23. Give structural formulas for the following:
 a. acetylene   **b.** 1-butyne

24. Give structural formulas for the following:
 a. toluene   **d.** *p*-dichlorobenzene
 b. *m*-diethylbenzene   **e.** 2,4-dinitrotoluene
 c. naphthalene   **f.** 1,2,4-trimethylbenzene

25. Give structural formulas for the following:
 a. cyclohexane   **b.** cyclopentene

26. Name the following compounds by the IUPAC system:

 a. $CH_3CH_2CHCH_2CH_2CH_3$
        $\underset{|}{\overset{|}{C}H_3}$

 b. $CH_3CH_2CH_2CHCH_2CH_2CH_3$
         $\underset{\underset{CH_3\ CH_3}{|}}{CH}$

27. Name the following compounds by the IUPAC system:

 a. $CH_2{=}CCH_2CH_2CH_3$
         $\underset{|}{\overset{|}{C}H_3}$

 b. $CH_3C{=}CHCH_2CHCH_3$
         $\underset{CH_3}{|}\quad\underset{CH_3}{|}$

 c. $CH_3CH_2\underset{H}{\overset{}{\diagdown}}C{=}C\underset{H}{\overset{CH_2CH_3}{\diagup}}$

28. Name the following compounds by the IUPAC system:
 a. $CH_3CH_2CH_2C{\equiv}CH$ **b.** $CH_3C{\equiv}CCH_3$

29. Name the following compounds by the IUPAC system:

 a. $\triangleright{-}CH_3$ **b.** $\square$

30. Name the following compounds by the IUPAC system:

 a.   CH_2CH_3

 b.   CH_3 ... NO_2

 c.   NO_2 ... O_2N ... NO_2

31. Give the reagents required for the following transformations:

 a. $HC{\equiv}CCH_3 \longrightarrow CH_3CH_2CH_3$

 b. $\longrightarrow$ OH

 c. $CH_3CH{=}CHCH_3 \longrightarrow CH_3\overset{OH}{\underset{|}{C}}HCH_2CH_3$

 d. $CH_2{=}CHCH_3 \longrightarrow \overset{Cl\ \ Cl}{\underset{|\ \ \ |}{CH_2CHCH_3}}$

32. What starting materials are required to complete the transformations shown?

 a. $\xrightarrow[\text{Ni}]{H_2}$

 b. $\xrightarrow{2\ Cl_2}$ $H{-}\overset{Cl}{\underset{Cl}{\overset{|}{\underset{|}{C}}}}{-}\overset{Cl}{\underset{Cl}{\overset{|}{\underset{|}{C}}}}{-}H$

 c. $\xrightarrow[H_2SO_4]{H_2O}$ $CH_3\overset{OH}{\underset{|}{C}}HCH_3$

 d. $\xrightarrow[\text{Fe}]{Cl_2}$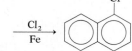

33. Which reactions in Problem 32 are addition reactions?
34. Which reactions in Problem 32 are substitution reactions?
35. Write an equation for the combustion of propane.
36. Write an equation for the hydrogenation of methylpropene (isobutylene).
37. Write an equation for the addition of bromine to ethene.
38. Write an equation for the addition of 2 mol of bromine to ethyne.
39. Write an equation for the bromination of benzene in the presence of iron.
40. Write an equation for the hydration of cyclohexene.
41. What is the danger in swallowing liquid alkanes?
42. Distinguish between lighter and heavier liquid alkanes in terms of their effect on the skin.
43. Name a physiological effect of some polycyclic aromatic hydrocarbons.
44. What are some of the hazards associated with the use of benzene? Why is toluene thought to be less toxic?

45. What physiological effect is shared by ethene and cyclopropane?
46. How is crude petroleum modified to better meet our needs and wants?
47. What are some possible sources of petrochemicals—for manufacturing drugs, plastics, detergents, and such—when our petroleum reserves are gone?
48. Write equations for the complete combustion of each of the following:
    a. natural gas (methane)
    b. a typical petroleum hydrocarbon (such as octane)
49. Write equations for the incomplete combustion that forms carbon monoxide from each of the substances in Problem 48.
50. The complete combustion of benzene forms carbon dioxide and water.

$$C_6H_6 + O_2 \longrightarrow CO_2 + H_2O$$

Balance the equation. What weight of carbon dioxide is formed by the complete combustion of 39 g of benzene?

E

Halogenated Hydrocarbons

Humans seem to have a greater liking for halogens in their organic compounds than does nature itself. Some halogen-containing organic compounds occur in nature, but chemists have created a dazzling variety of such compounds. And the general public has come to depend on a large number of these. Medicines, building materials, farm products, and a myriad of convenience products incorporate the halogen-carbon combination.

As each passing year demonstrates, nature can get upset over our attempts to correct its deficiencies. The environmental and health hazards posed by the sudden influx of synthetic materials have proven to be highly significant. Unfortunately, the answer to the problem is not simply a matter of stopping all production of synthetic compounds—not unless we are willing to put up with dramatic increases in pestilence, disease, and famine. What, then, is the answer? It is the same as for any difficult problem—to take a cold, hard look at what we gain and what we lose in each case, to weigh these against one another, and then to retain, eliminate, or modify the materials on the basis of the balance.

We consider some huge halogen-containing molecules in Chapter 19. For now, we'll concentrate on some smaller ones.

E.1 ALKYL HALIDES: STRUCTURE AND NOMENCLATURE

In general, halogenated hydrocarbons are compounds in which one or more hydrogens of a hydrocarbon have been replaced by halogen atoms. The halogen may be fluorine (F), chlorine (Cl), bromine (Br), or iodine (I). Each of the following is a halogenated hydrocarbon:

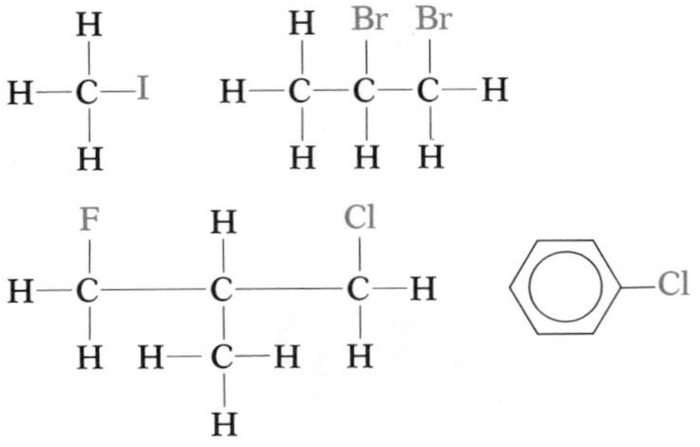

$$
\begin{array}{ccc}
& \text{H} & \text{H} \quad \text{Br} \quad \text{Br} \\
& | & | \quad | \quad | \\
\text{H}-\text{C}-\text{I} & \text{H}-\text{C}-\text{C}-\text{C}-\text{H} \\
& | & | \quad | \quad | \\
& \text{H} & \text{H} \quad \text{H} \quad \text{H}
\end{array}
$$

More specifically, replacement of one hydrogen atom of an alkane with a halogen gives an **alkyl halide.** These compounds are given common names that consist of two parts. The first is the name of the alkyl group; the second is the stem of the name of the halogen, with the ending **-ide.** Refer back to Table 14.4 for a list of the common alkyl groups.

---

example E.1

What is the common name for the compound with this formula?

$$CH_3CH_2Br$$

The alkyl group (CH_3CH_2—) is ethyl. The halogen is bromine. The compound is therefore ethyl bromide.

---

example E.2

What is the common name for this compound?

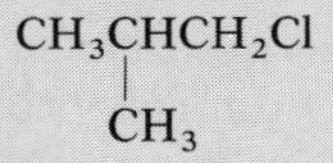

$$
\begin{array}{c}
CH_3CHCH_2Cl \\
| \\
CH_3
\end{array}
$$

The alkyl group is isobutyl. The halogen is chlorine. The compound is named isobutyl chloride.

Sometimes we may wish to speak of alkyl halides in general. Sometimes (later on in this book) we may wish to refer to other kinds of compounds that contain alkyl groups without bothering to indicate a specific alkyl group. In these instances, we represent the alkyl group by the letter R, as in R—Cl (read "alkyl chloride"). The **R** stands for any alkyl group—methyl, ethyl, isopropyl, whatever.

Names, structures, properties, and uses of some common alkyl halides are given in Table E.1.

TABLE E.1   *Some Alkyl Halides*

Compound	Structure	Boiling Point (in degrees Celsius)	Uses
Methyl chloride	CH_3Cl	−24	Refrigerant; chemical intermediate for manufacturing of silicones, methyl cellulose, and so on
Methyl bromide	CH_3Br	4	Poison gas for insect and rodent control
Ethyl chloride	CH_3CH_2Cl	13	Intermediate for synthesis of tetraethyl lead; local anesthetic; emergency general anesthetic
Butyl chloride	$CH_3CH_2CH_2CH_2Cl$	79	Intermediate for synthesis of butyl cellulose; killing of intestinal worms in dogs
Pentyl chlorides	$C_5H_{11}Cl$ (various isomers)	—	Chemical intermediates

The IUPAC system differs from the common nomenclature in that halogen substituents are indicated by the prefixes **fluoro-, chloro-, bromo-,** and **iodo-.** The prefix is used with the name of the parent alkane, with numbers to indicate the position of the halogen if necessary.

Give the IUPAC name for

example E.3

$$\overset{1}{C}H_3\overset{2}{C}H\overset{3}{C}H_2\overset{4}{C}H_2\overset{5}{C}H_3$$
$$\underset{Cl}{|}$$

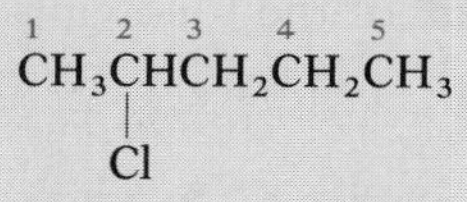

The parent alkane is pentane. The name of the compound is 2-chloro-pentane.

Give the IUPAC name for

example E.4

$$\overset{1}{C}H_3\overset{2}{C}H\overset{3}{C}H_2\overset{4}{C}H\overset{5}{C}H_2\overset{6}{C}H_3$$
$$\underset{CH_3}{|}\quad\underset{Br}{|}$$

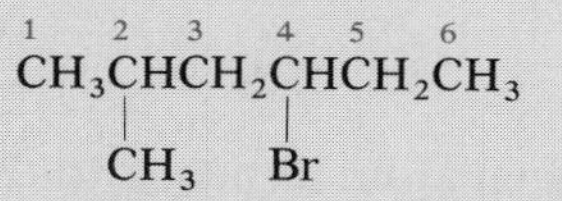

The parent alkane is hexane. The compound is named 4-bromo-2-methylhexane.

Vinyl chloride

(or $CH_2=CH-$)

The vinyl group

Chlorinated compounds can also contain doubly bonded carbons. Replacement of one hydrogen of ethylene with chlorine gives a compound called vinyl chloride. The IUPAC name for this compound is chloroethene. Its common name is patterned after that of the alkyl halides—the carbon portion is named as a substituent and the halogen as a halide. Vinyl is the name given to the carbon group (margin). A vinyl fluoride, a vinyl bromide, and a vinyl iodide also exist, but it is vinyl chloride that has had the most significant impact on our society. It is the starting material for the synthesis of the important plastic poly(vinyl chloride), the "vinyl" one finds on seat covers, in phonograph records, in house sidings, and in many other places (Chapter 19). In 1974, B. F. Goodrich Company announced that three workers who cleaned reactors in which the poly(vinyl chloride) was prepared had died of angiosarcoma, a rare form of liver cancer. During the following months, other cases were reported, and evidence was presented that linked the malignancy to long-term (20 to 30 years' duration), high-level exposure to vinyl chloride and not to the plastic itself. The plastics industry and the government undertook the establishment of safety guidelines for industrial workers engaged in operations involving the gaseous vinyl chloride. It was recommended that a safe breathing atmosphere be considered one for which tests revealed no detectable level of the gas.

E.2 ARYL HALIDES: STRUCTURE AND NOMENCLATURE

Just as alkyl halides are derived from alkanes, aryl halides are derived from aromatic hydrocarbons. As for alkyl halides, it is occasionally useful to have a general formula for compounds with halogen attached to an aromatic ring. For this purpose chemists use the symbol **Ar**, which stands for the **aryl** group. An aryl halide is any compound in which a halogen is attached directly to an aromatic ring. For example, all of the following are aryl chlorides:

Replacement of one of the hydrogen atoms of benzene with a bromine atom gives bromobenzene (C_6H_5Br).

Replacement of a hydrogen atom attached to the toluene ring with chlorine gives one of three isomeric chlorotoluenes (margin). Their names assume the methyl group of toluene to be on the first carbon atom of the ring.

Chlorobenzene (whose boiling point is 132 °C) and *o*-chlorotoluene (whose boiling point is 159 °C) both find some use as solvents, particularly where a liquid with a high boiling point is desired. All these compounds serve as chemical intermediates. For example, chlorobenzene is used in the production of aniline (Chapter 18), phenol (Chapter 15), and DDT (Section E.6).

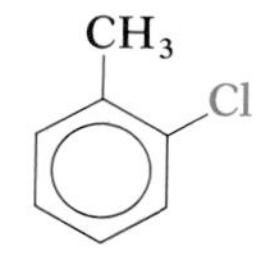

2-Chlorotoluene
(*o*-chlorotoluene)

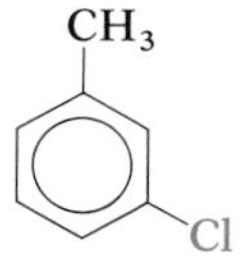

3-Chlorotoluene
(*m*-chlorotoluene)

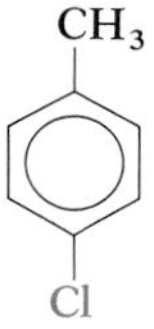

4-Chlorotoluene
(*p*-chlorotoluene)

E.3 POLYHALOGENATED HYDROCARBONS: STRUCTURE AND NOMENCLATURE

More than one hydrogen on a hydrocarbon molecule can be replaced by halogen atoms. This leads to a wide variety of interesting and often useful compounds.

Consider methane, the simplest hydrocarbon. One hydrogen can be replaced by chlorine, yielding methyl chloride (Section E.1). A second hydrogen can be substituted to yield dichloromethane or methylene chloride (CH_2Cl_2). This compound is a common solvent. It boils at 40 °C; hence, it can easily be removed by distillation if one wishes to recover the solute. Although the name methylene chloride makes it sound like the compound contains a double bond (remember ethylene and propylene?), there is no double bond in the molecule, which contains only one carbon atom (Figure E.1).

A third methane hydrogen can be replaced by chlorine, giving chloroform or trichloromethane ($CHCl_3$). Chloroform was one of the first anesthetics and was once widely used. It has been largely replaced by safer, less toxic chemicals. Chloroform is still an important commercial and industrial solvent.

Replacing all four of methane's hydrogens with chlorine gives carbon tetrachloride (CCl_4), also called tetrachloromethane. Carbon tetrachloride has been used as a dry-cleaning solvent and in fire extinguishers. It is no longer recommended for either use. Exposure to carbon tetrachloride (or most of the other chlorinated hydrocarbons, for that matter) can cause severe damage to the liver. Even the vapor, breathed in small amounts, can cause serious illness if the exposure is prolonged. Use of a carbon tetrachloride fire extinguisher in conjunction with water to put out a fire can be deadly. Carbon tetrachloride reacts with water at high temperatures to form phosgene ($COCl_2$), an extremely poisonous gas.

Chloroform and carbon tetrachloride have, through tests made on animals, been identified as possible carcinogens.

With ethane, from one to six hydrogen atoms can be replaced by chlorine. The various derivatives are shown in Figure E.2. Replacement of only one hydrogen gives ethyl chloride. Replacement of two hydrogens can give either of two isomers. 1,2-Dichloroethane is commonly used as a solvent, particularly for rubber.

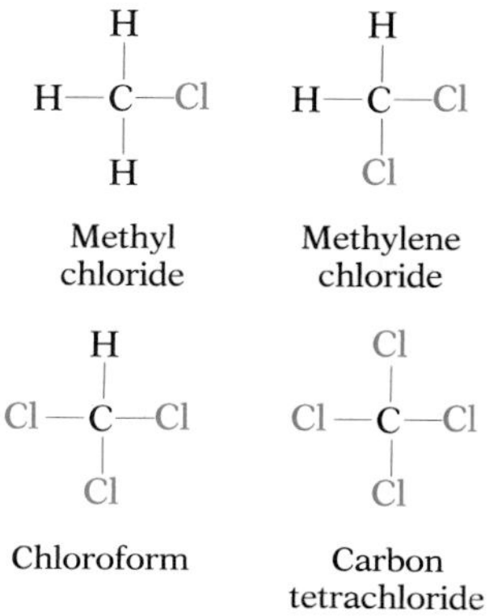

Methyl
chloride

Methylene
chloride

Chloroform

Carbon
tetrachloride

FIGURE E.1
The chlorine derivatives of methane.

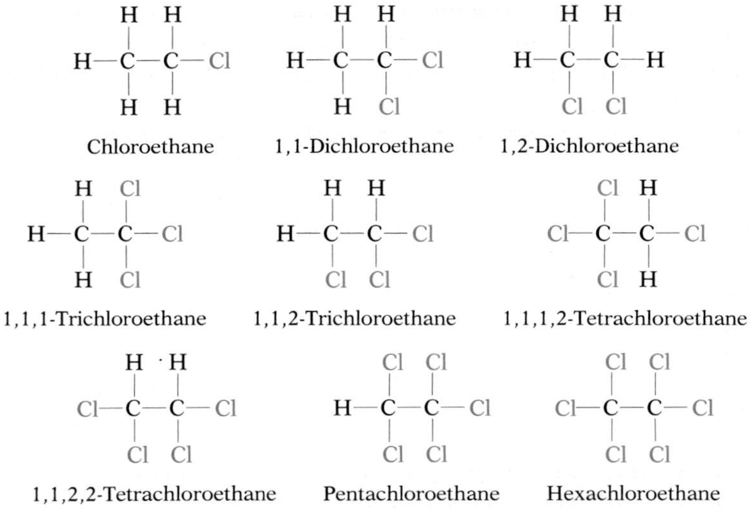

Chloroethane, 1,1-Dichloroethane, 1,2-Dichloroethane, 1,1,1-Trichloroethane, 1,1,2-Trichloroethane, 1,1,1,2-Tetrachloroethane, 1,1,2,2-Tetrachloroethane, Pentachloroethane, Hexachloroethane

FIGURE E.2 Nine compounds can be derived from ethane by replacing one or more hydrogen atoms with chlorine atoms.

1,2-Dibromoethane, commonly called ethylene dibromide (EDB), was once widely used as a fumigant for grain. It was banned in 1984 after residues were found in flour. Animal tests indicate that EDB is a carcinogen.

Replacement of three of the hydrogen atoms of ethane with chlorine atoms also gives two isomers. Both of the trichloroethanes are used as commercial and industrial solvents. 1,1,1-Trichloroethane is used for the cleaning of molds used in the fabrication of plastics.

Replacement of four hydrogens of ethane with chlorine also leads to isomers. Only one is of importance, however—1,1,2,2-tetrachloroethane. This compound is widely used but highly toxic. It dissolves rubber, cellulose acetate, and other complex organic materials. It is also used to sterilize soil and as a component of weed killers and insecticide preparations.

example E.5

How many isomeric tetrachloroethanes are there? How many pentachloroethanes are there? Write the structures and give IUPAC names for each isomer.

There are only two tetrachloroethanes, 1,1,2,2-tetrachloroethane and 1,1,1,2-tetrachloroethane (Figure E.2). There is only one possible pentachloroethane; hence, pentachloroethane is the IUPAC name, and it is not necessary to use numbers to indicate a form (Figure E.2).

Polyhalogenated alkenes also exist. Trichloroethene (also called trichloro-ethylene) is one of the most important of the chlorinated hydrocarbon solvents. It is used as a dry-cleaning solvent and (in industry) for degreasing metal parts. It has recently come under attack as a suspected carcinogen.

Tetrachloroethylene (also called perchloroethylene*) is also used as a dry-cleaning solvent, particularly for synthetic fabrics.

Polyhalogenated aromatic compounds are also commonly encountered. Replacement of two hydrogens of benzene with chlorine gives three isomeric dichlorobenzenes. *p*-Dichlorobenzene has been used to protect garments from the larvae of the clothes moth and to protect peach trees from borer larvae.

Trichloroethylene

Tetrachloroethylene

o-Dichlorobenzene
(1,2-dichlorobenzene)

m-Dichlorobenzene
(1,3-dichlorobenzene)

p-Dichlorobenzene
(1,4-dichlorobenzene)

E.4 PHYSICAL PROPERTIES OF THE HALOGENATED HYDROCARBONS

The physical properties of the halogenated hydrocarbons are much like those of the alkanes. Most are essentially insoluble in water. Generally, they are good solvents for other organic compounds such as fats, greases, oils, waxes, and resins and for some plastics and elastomers (Chapter 19).

Monofluoro and monochloro compounds are less dense than water, but monobromo and monoiodo compounds with only a few carbon atoms are denser than water (in contrast to the hydrocarbons). Polyhalogenated hydrocarbons are generally denser than water. Chloroform and carbon tetrachloride, for example, have densities of 1.5 g/mL and 1.6 g/mL, respectively.

The boiling points of many chlorinated hydrocarbons are close to those of alkanes of comparable molecular weight. For example, the boiling point of ethyl chloride (which has a molecular weight of 64.5) is 13 °C, between that of butane (whose molecular weight is 58.0 and whose boiling point is 0 °C) and pentane (whose molecular weight is 72 and whose boiling point is 36 °C). Chlorobenzene (with a molecular weight of 112.5) boils at 132 °C, while ethyl-benzene (with a molecular weight of 106.0) boils at 136 °C. As with most of our general rules, there are many exceptions to this one, but the correlation between the boiling points of these two classes of compounds is worthy of comment. Boiling points reflect the strength of the forces holding the molecules of a liquid together. For compounds of comparable molecular weight, the higher the boiling point, the stronger the forces of attraction between molecules. Since carbon-chlorine bonds are polar, it may seem surprising that the interaction of these dipoles doesn't result in higher boiling points for the chlorinated hydrocarbons. Recall, however, that dispersion forces for larger molecules can override forces between dipoles (Section 7.5). Recall also that most

* The prefix **per-** indicates the maximum substitution. In perhalo compounds, all hydrogens have been replaced by halogen atoms.

liquids that have high boiling points and that are composed of small molecules are associated by hydrogen bonding.

Physical properties, particularly solubility and volatility, determine to a large extent where halogenated hydrocarbons go in our bodies—and what they do when they get there.

E.5 PHYSIOLOGICAL PROPERTIES OF THE HALOGENATED HYDROCARBONS

As in their physical properties, the halogenated hydrocarbons are much like the alkanes in their physiological properties. Gaseous compounds—and vapors of volatile liquids—often act as anesthetics. Chloroform (Section E.3) is an effective anesthetic. It is powerful and fast acting. It has the advantage (over ether and cyclopropane) of being nonflammable. Unfortunately, however, the effective dose for anesthesia is quite near the lethal dose. It is seldom used these days on humans. It was largely replaced by 2-bromo-2-chloro-1,1,1-trifluoroethane (also called halothane), which shares the advantages of chloroform. The safety of this replacement with regard to toxicity, particularly for the workers in the operating room, has recently been questioned.

Some halogenated hydrocarbons are narcotic in high concentrations, perhaps accounting for the abuse of some of these compounds by glue sniffers. Such use is exceedingly dangerous. Aside from the hazard of immediate death, repeated exposure to these compounds often leads to extensive damage to the liver and kidneys.

On the skin, liquid halogenated hydrocarbons can cause dermatitis by washing away natural oils. Some are particularly irritating to the skin or to the eyes, the mucosa, or the respiratory tract. Many are readily absorbed through the skin, and spills on the skin can ultimately lead to liver damage. Several halogenated hydrocarbons have been shown to cause cancer in laboratory animals. Repeated tests, however, have failed to establish the carcinogenicity of others. Though many members of the halocarbon family are hazardous, it is difficult to generalize. Some individual compounds can be used safely if proper precautions are taken.

Fluorinated compounds have found some interesting uses. Some have been used as blood extenders. Oxygen is quite soluble in perfluorocarbons. These compounds can therefore serve as temporary substitutes for hemoglobin, the oxygen-carrying protein in blood. Fluosol-SA, a mixture of perfluorodecalin and perfluorotripropylamine, has been tested in hundreds of patients in Japan. In the United States, it has been used mainly for those who reject transfusions of normal blood for religious reasons. (Teflon, a perfluorinated polymer, is discussed in Chapter 19.)

Halothane

E.6 HALOGENATED HYDROCARBONS IN THE ENVIRONMENT

Chlorinated hydrocarbons make up an important class of insecticides. Perhaps the best known of these is dichlorodiphenyltrichloroethane, or DDT. Today, in the developed countries, DDT is known mostly for its harmful environmental effects. It interferes with calcium metabolism. Birds are threatened because eggshells are composed mainly of calcium compounds. Eggs of birds that have ingested DDT have thin shells that are poorly formed and easily broken. Even a few parts per billion (ppb) of DDT interfere with the growth of plankton and the reproduction of crustaceans such as shrimp. As bad as DDT sounds, though, it has probably saved the lives of more people than any other chemical substance. Not only has it successfully controlled crop-destroying insects, but it has been highly effective against insects that carry diseases such as malaria.

The use of DDT has been largely banned in developed countries, but the chemical is still widely used in less developed areas (Figure E.3). Unfortunately, it is no longer effective against many of the pests it once killed readily. Many insects have developed **resistance** to this and many other pesticides. The resistant insects can now detoxify the chemical compounds that were once deadly to their kind.

Chlorinated hydrocarbons generally are not very reactive. This lack of reactivity was a major advantage of DDT. Sprayed on a crop, it stayed there and killed insects for weeks. This persistence was also a major disadvantage. It did not break down readily in the environment. Rather, it built up in concentration over the years, threatening fish, birds, and other wildlife—perhaps even humans. DDT has been largely replaced by less persistent pesticides.

Another group of chlorinated hydrocarbons, the polychlorinated biphenyls (PCBs), has been used widely in industry and is now ubiquitous in the environment. Chemically, the PCBs are derived from a hydrocarbon called biphenyl $(C_{12}H_{10})$ by replacement of anywhere from 1 to 10 of the hydrogens with chlorine. The structures of the parent compound and some of the PCBs are shown in Figure E.4. Note the structural similarity of the PCBs to DDT.

PCBs were once widely used as plasticizers to soften hard, brittle plastics (Chapter 19) and in consumer products such as carbonless copy paper. Their major use today is in electrical transformers. All production of PCBs in the United States has been halted, but many transformers containing them remain in service.

As devices containing PCBs are discarded, the chemicals escape into the environment. PCB residues have been found in fish, birds, water, and sediments. Poultry and eggs have been found to have concentrations greater than the 5 ppm allowed by the Food and Drug Administration. Under the law these have to be destroyed. Fish, taken from major rivers from coast to coast

DDT

FIGURE E.3
Malaria is controlled in developing countries by spraying of the walls of the houses, where mosquitoes alight after their meal of blood. (Courtesy of the U.S. Agency for International Development, Washington, D.C.)

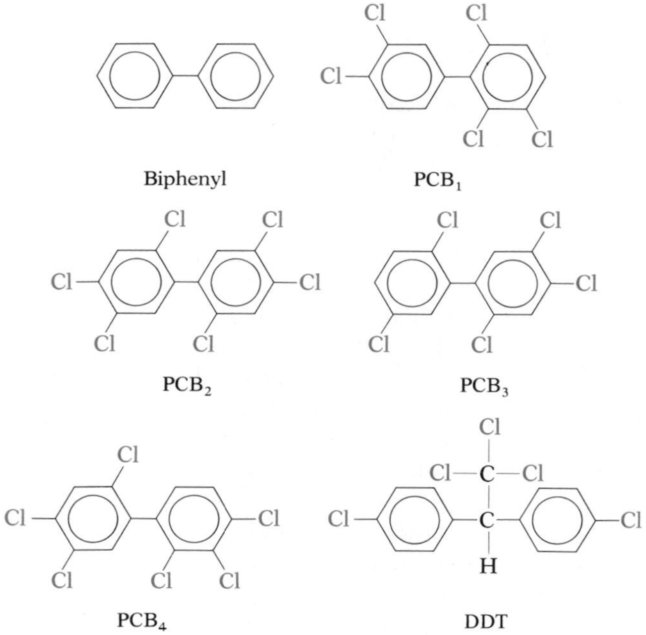

FIGURE E.4   *Biphenyl and some of the PCBs derived from it. Note that PCB$_1$ and PCB$_3$ are isomers with the formula $C_{12}H_5Cl_5$. These are but a few of the hundreds of possible PCBs. DDT is shown for comparison.*

in the United States, have been found to be contaminated with PCBs, often above the 5-ppm maximum.

Like DDT and other chlorinated hydrocarbons, PCBs undergo few reactions and are insoluble in water. They remain unchanged in the environment for years.

Recall that chlorinated hydrocarbons are good solvents for fats. The reverse is also true; fats are good solvents for chlorinated hydrocarbons such as DDT and PCBs. When these compounds are ingested as contaminants in food or water, they are extracted and concentrated in fatty tissues. Their fat-soluble nature causes chlorinated hydrocarbons to be concentrated up the food chain. This biological magnification was graphically demonstrated in California in 1957. Clear Lake, about a hundred miles north of San Francisco, was sprayed with DDT in an effort to control gnats. The water, after spraying, contained only 0.02 ppm of DDT. The microscopic plant and animal life contained 5 ppm— 250 times as much. Fish feeding on these microorganisms contained up to 2000 ppm. Grebes, the diving birds that ate the fish, died by the hundreds (Figure E.5). PCBs are concentrated in a similar manner.

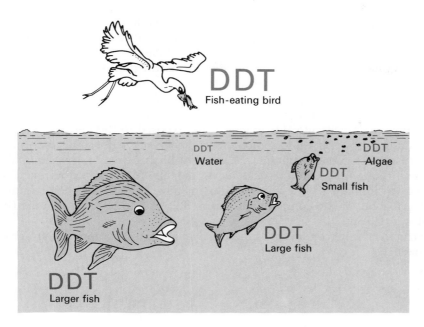

FIGURE E.5 *Concentration of DDT up the food chain. Animals at the top of the chain have the highest concentration of insecticide.*

Even human fat contains PCBs and other chlorinated hydrocarbons. Mother's milk often contains low levels of these compounds, leading some people to question the safety of breast-feeding babies. Unless there is evidence of high levels of contamination, though, breast-feeding is probably best for most babies.

DDT and PCBs are nerve poisons. They are concentrated in the fatlike compounds that make up nerve sheaths and somehow interfere with the transmission of electrical impulses along these sheaths. No substantial harm to humans has ever been demonstrated from DDT, but accidental ingestion of PCBs has caused skin, liver, and gastrointestinal problems. Experiments with monkeys show that levels of less than 5 ppm in food interfere with reproduction. The possible harm from long-term exposure to low levels of PCBs may not be known for years, but appears to be slight.

E.7 POLYCHLORINATED PRACTICALLY EVERYTHING

Pentachloronitrobenzene is used as an agricultural fungicide. Pentachlorophenol is a wood preservative, an algicide, and a fungicide. Hexachlorophene is a germicide, especially effective against staphylococcus infections. It has also caused neurological damage and even death in infants. 2,4-Dichloro-phenoxyacetic acid (2,4-D) and 2,4,5-trichlorophenoxyacetic acid (2,4,5-T) have been widely used as herbicides and defoliants. These last compounds, in a formulation called Agent Orange, were used extensively in Vietnam in an attempt to remove enemy cover and to destroy crops that maintained enemy armies. In addition to causing vast ecological damage, 2,4-D and 2,4,5-T were suspected of causing birth defects in Vietnamese children and in babies later born to American soldiers who were exposed to the herbicides. Laboratory studies show that these compounds, when pure, do not cause abnormalities in fetuses of laboratory animals. Extensive birth defects in test animals are caused, though, by contaminants called dioxins. Although no adverse health effects on humans of low levels of dioxins have been established, continuing concern about dioxin contamination caused 2,4,5-T to be banned in the United States in 1985.

Structural representatives of these polychlorinated compounds are given in Figure E.6.

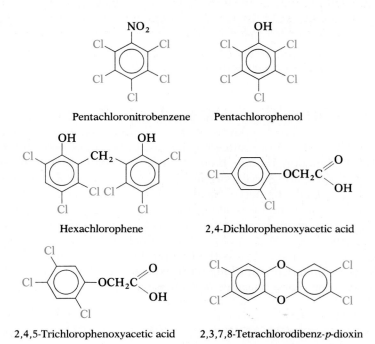

FIGURE E.6 *Polychlorinated potpourri. There is the potential for good and bad in most of these compounds, depending on how they are used.*

In the 1970s, still another problem with chlorinated compounds arose. Organic compounds were found in the drinking water of several major cities. Some 66 such compounds were found in New Orleans drinking water in 1974. Among them were several chlorinated compounds. Chloroform, for example, was present at 0.1 ppm. Chlorinated phenols were also found. Some of the compounds, including chloroform, are suspected carcinogens.

Most of the chlorinated compounds were probably formed by the reaction of organic wastes with the chlorine that is used to kill bacteria in the water. Even more frightening, people who drank New Orleans water (from the Mississippi River) had a 15% higher incidence of cancer than those who got their water from wells. The Environmental Protection Agency began a survey of water supplies in other cities. People became quite perturbed by the idea of getting cancer from drinking water. To put things somewhat in perspective, however, we might note that several popular cough medicines at that time contained chloroform at a thousand times the concentration found in New Orleans drinking water. Increasing concern over the physiological effects of chloroform, however, has led to the removal of chloroform from these formulations too.

Halogenated hydrocarbons were long regarded with lack of interest. Unlike the alcohols in Chapter 15, they had no fascinating history or clearly perceived relation to life. We have tried to indicate here just how wrong that early judgment was.

PROBLEMS

1. What is meant by R—?
2. What is meant by Ar—?
3. Write structures for the two isomers that have the molecular formula C_4H_9Cl. Give the common name and the IUPAC name of each.
4. Write structures for the four isomers that have the molecular formula C_4H_9Cl. Give the common name and the IUPAC name of each.
5. Draw structures for the following:
   a. perfluoropropane   b. perfluoropropene
6. Draw structures for the following:
   a. vinyl fluoride   b. vinyl bromide
7. Draw structures for the following:
   a. methylene chloride   c. carbon tetrachloride
   b. chloroform
8. Name each of the compounds in Problem 7 by the IUPAC system.

9. Draw structures for the following:
   a. 1,1-dichloropropene
   b. 2-chloro-4-methylheptane
10. Give the IUPAC name for each of the following:
    a. $CH_3CHCH_2CH_2CH_3$ b. $CHCl_2CH_2Cl$
       |
       Br
11. Give the IUPAC name for each of the following:
    a. $CH_3CHCH_2CHCH_2CHCH_3$
          | | |
          Cl   Cl   Cl
    b. $CH_3CH—CHCH_2CH_2CH_3$
          | |
          CH_3 Cl
12. Draw structures for the following:
    a. *m*-dibromobenzene   b. *p*-difluorobenzene

13. Draw structures for the following:
    a. *o*-chlorotoluene   b. 2,4-dibromotoluene
14. Give the IUPAC name for each of the following:

a.   b.

15. Describe a medical use of perfluorinated hydrocarbons.
16. Compare halogenated hydrocarbons with hydrocarbons with respect to boiling point, density, and solubility.
17. As an anesthetic, what advantage did chloroform possess over ether? Why is chloroform no longer used as an anesthetic?
18. What two properties of halogenated hydrocarbons, besides their toxicity, make these compounds useful as pesticides as well as dangerous in the environment?
19. What is meant by the sentence "DDT is concentrated up the food chain"?

20. Name two animal classes (excluding insects) particularly susceptible to the lingering effects of pesticides in the environment.
21. Why is DDT no longer useful against some insect pests?
22. In early attempts to determine contaminants in biological systems, PCBs were often mistaken for DDT or other chlorinated hydrocarbon pesticides. Why might this be expected?
23. How much DDT would it take to kill a person weighing 60 kg if the lethal dose were 0.5 g per kilogram of body weight?
24. Parathion, an organic phosphorus compound that is less persistent in the environment than chlorinated hydrocarbons, is often used as a replacement for DDT. How much parathion would it take to kill the person in Problem 23 if the lethal dose were 5 mg per kilogram of body weight?
25. The minimum lethal dose of dioxin is estimated to be 6 μg per kilogram of body weight. What is the smallest amount that might kill a 50-kg person?

CHAPTER
15

Alcohols, Phenols, and Ethers

The families of organic compounds discussed in this chapter occur widely in nature. The human race has been quick to adapt these materials to its own use. The earliest written histories record the isolation and use by primitive peoples of the compound known as alcohol. According to Genesis, Noah planted a vineyard after the flood, drank wine from its grapes, and became drunk.

Human ingenuity may have reached some sort of peak in finding sources of *aqua vitae*, the water of life. Alcohol has been obtained from the fermentation of fruits, grains, potatoes, rice, and even cacti. It was prescribed as medicine in the twelfth century but has been most frequently used without such justification. What we know as alcohol is actually only one member of a family known by the same name. The family includes among its members such familiar substances as cholesterol and the carbohydrates.

The name of another family of organic compounds considered in this chapter, the ethers, has become almost synonymous with anesthesia. And anyone who has ever been in a hospital would recognize the pungent, antiseptic odor of phenol, the simplest member of the third family to be introduced in this chapter (Figure 15.1).

Why are we taking up three different families in this one chapter? We do so because each can be considered an organic derivative of water. Water is by far the most important inorganic compound we have studied. It should not be surprising, therefore, that organic compounds derived from water are also of critical importance to life and health. Some, like the carbohydrates, deserve to be, and are, considered in a separate chapter (Chapter 20). For now, we shall, as usual, deal with the simpler members of each family.

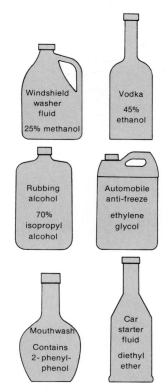

FIGURE 15.1
Alcohols, phenols, and ethers find many uses.

15.1 THE FUNCTIONAL GROUP

Alcohols, phenols, and ethers can be viewed as organic derivatives of water. Consider the water molecule.

$$\overset{\displaystyle O}{H \diagup \diagdown H}$$

It is a bent molecule with the central oxygen attached to two hydrogen atoms. If one of these hydrogens were removed and replaced with an alkyl group (R—), we would have

$$\overset{\displaystyle O}{R \diagup \diagdown H}$$

This is the general formula for the **alcohol** family. The alkyl group may be methyl, ethyl, *tert*-butyl, or an aliphatic group too complicated to have a simple name. As long as the carbon attached to the **hydroxyl group** (—OH) is aliphatic, the compound is an alcohol.

If the hydroxyl group is attached directly to an aromatic ring, a different family of compounds is produced. Compounds in which an aryl group (Ar—) is attached to a hydroxyl group are called **phenols.**

$$\overset{\displaystyle O}{Ar \diagup \diagdown H}$$

The chemistry of the phenols is sufficiently different from that of the alcohols to justify treatment of the two classes of compounds as separate, if closely related, families. Nonetheless, for both families, the chemistry is largely determined by the hydroxyl group. Nearly all the characteristic reactions of alcohols and many of those of the phenols take place at the hydroxyl group. Even the physical properties are determined to a large extent by the presence of the hydroxyl group. Such a group of atoms, which confers characteristic chemical and physical properties on a family of organic compounds, is called a **functional group.** The hydroxyl group is only one functional group. We have already encountered others. The carbon-carbon double bond in alkenes and the carbon-carbon triple bond in alkynes are functional groups. In both instances, these structural features confer on the members of the families a particular chemical reactivity. For example, alkenes and alkynes tend to undergo addition reactions. The halogens in halogenated hydrocarbons are functional groups, although we did not consider in detail the particular reactions associated with these groups. The alkanes are characterized by their *lack* of a distinct functional group. Other functional groups will serve as unifying concepts for the next three chapters. Some of the more important functional groups are listed in Table 15.1. For ready reference, this table is also reproduced on the inside back cover.

TABLE 15.1 Selected Organic Functional Groups

Name of Class	Functional Group	General Formula of Class
Alkane	None	$R-H$
Alkene	$-C=C-$	$R-\underset{}{\overset{R\quad R}{C=C}}-R$
Alkyne	$-C\equiv C-$	$R-C\equiv C-R$
Alcohol	$-C-O-H$	$R-O-H$
Ether	$-C-O-C-$	$R-O-R$
Aldehyde	$-\overset{O}{\overset{\|}{C}}-H$	$R-\overset{O}{\overset{\|}{C}}-H\;\cdot$
Ketone	$-\overset{O}{\overset{\|}{C}}-$	$R-\overset{O}{\overset{\|}{C}}-R$
Amine	$-C-N-$	$R-\overset{H}{\underset{}{N}}-H$
		$R-\overset{H}{\underset{}{N}}-R$
		$R-\overset{R}{\underset{}{N}}-R$
Carboxylic acid	$-\overset{O}{\overset{\|}{C}}-O-H$	$R-\overset{O}{\overset{\|}{C}}-O-H$
Ester	$-\overset{O}{\overset{\|}{C}}-O-C-$	$R-\overset{O}{\overset{\|}{C}}-O-R$
Amide	$-\overset{O}{\overset{\|}{C}}-N-$	$R-\overset{O}{\overset{\|}{C}}-\underset{H}{N}-H$
		$R-\overset{O}{\overset{\|}{C}}-\underset{H}{N}-R$
		$R-\overset{O}{\overset{\|}{C}}-\underset{R}{N}-R$

We have shown how both alcohols and phenols can be derived from a water molecule. We can also start with the water molecule and derive a general structure for a third family of organic compounds, the **ethers.** The ether family is the only one of the three that does not contain the hydroxyl functional group. To accomplish this, starting with a water molecule, we must substitute carbon groups for both of water's hydrogens. There are three possible ways of doing this.

$$\underset{R}{}\overset{O}{\diagdown}\underset{R}{} \qquad \underset{R}{}\overset{O}{\diagdown}\underset{Ar}{} \qquad \underset{Ar}{}\overset{O}{\diagdown}\underset{Ar}{}$$

All of the above formulas represent ethers. A compound is an ether as long as there are two carbon groups attached to the oxygen, whether the groups are aliphatic or aromatic. No distinction is made in this case because no distinctive difference in chemical or physical properties is observed.

15.2 THE ALCOHOLS: STRUCTURES AND NAMES

A compound containing a hydroxyl group attached to an aliphatic carbon atom is an alcohol. The simplest alcohol has the molecular formula CH_4O. Its structural formula is

$$H-\underset{\underset{H}{|}}{\overset{\overset{H}{|}}{C}}-O-H \qquad \text{or} \qquad CH_3OH$$

Notice that the compound can be thought of as a methyl group (CH_3-) joined to a hydroxyl group ($-OH$). For simple alcohols like this, common names are often used. We name the alkyl group and then add the word *alcohol* to indicate the presence of the hydroxyl group. The simplest alcohol is *methyl alcohol.*

Ethyl alcohol is

$$H-\underset{\underset{H}{|}}{\overset{\overset{H}{|}}{C}}-\underset{\underset{H}{|}}{\overset{\overset{H}{|}}{C}}-O-H \qquad \text{or} \qquad CH_3CH_2OH$$

Ethyl alcohol is the beverage alcohol. It is discussed in detail in subsequent sections.

As there are two propyl groups (Table 14.4), there are two propyl alcohols (margin). Isopropyl alcohol, in 70% and 91% solutions, is used as a topical antiseptic and in body rubs, after-shave lotions, and similar preparations.

The names of the butyl groups were first introduced in Table 14.4. The present discussion of alcohols offers us an opportunity to explain where two of those names originated. Alcohols are subdivided into three classes, called

$CH_3CH_2CH_2OH$

Propyl alcohol

$CH_3\underset{\underset{OH}{|}}{C}HCH_3$

Isopropyl alcohol

primary (1°), secondary (2°), and tertiary (3°). These classes are based on the number of carbon atoms attached to the carbon that bears the hydroxyl group. The structure

$$\begin{array}{c} -\overset{|}{\underset{|}{C}}- \\ H-\overset{|}{\underset{|}{C}}-H \\ OH \end{array}$$

is classified as a primary alcohol, because the hydroxyl group is attached to a primary carbon atom, that is, a carbon atom that is attached to only one other carbon (and to two hydrogens).

The alcohol

$$\begin{array}{c} -\overset{|}{\underset{|}{C}}- \\ -\overset{|}{\underset{|}{C}}-\overset{|}{\underset{|}{C}}-H \\ OH \end{array}$$

is secondary, because the hydroxyl group is attached to a secondary carbon, one that is bonding to two other carbons (and to one hydrogen). Finally, if the hydroxyl group is attached to a carbon bonded to three other carbon atoms, the alcohol is called tertiary.

$$\begin{array}{c} -\overset{|}{\underset{|}{C}}- \\ -\overset{|}{\underset{|}{C}}-\overset{|}{\underset{|}{C}}-\overset{|}{\underset{|}{C}}- \\ OH \end{array}$$

Some of the reactions of the alcohols, notably oxidation (Section 15.6), differ from one class to another.

But we were considering the four butyl alcohols. Two butyl groups are derived from butane. Removal of a hydrogen atom from an end carbon on butane gives the butyl group. Removal of a hydrogen from one of the interior carbon atoms gives the *secondary* butyl group (abbreviated *sec*-butyl). It is now easy to see why this group is called secondary butyl; the attachment of this butyl group to a longer chain or to a functional group, such as a hydroxyl group, is made through a carbon atom that is attached to two other carbon atoms, that is, through a secondary carbon atom.

Two other butyl groups are derived from isobutane. Removal of any of the nine hydrogens on an end carbon atom of isobutane gives an isobutyl group. Removal of the one hydrogen on the interior carbon atom gives the *tertiary* butyl group (*tert*-butyl). Here the name reflects the fact that the attachment is made through a tertiary carbon atom, that is, one that is attached to three other carbon atoms.

$CH_3CH_2CH_2CH_2-$

The butyl group

$CH_3CH_2\overset{|}{\underset{|}{C}}HCH_3$

The *sec*-butyl group

$CH_3\overset{|}{\underset{|}{C}}HCH_2-$
$\quad\quad CH_3$

The isobutyl group

$$CH_3-\overset{\overset{\displaystyle CH_3}{|}}{\underset{\underset{\displaystyle CH_3}{|}}{C}}-$$

The *tert*-butyl group

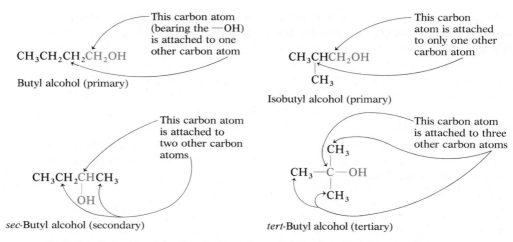

FIGURE 15.2   *The four butyl alcohols and their classification as primary, secondary, and tertiary.*

The four butyl alcohols are shown in Figure 15.2.

example
15.1

Classify each of the following alcohols as primary, secondary, or tertiary: ethyl alcohol, isopropyl alcohol, propyl alcohol.
   Ethyl alcohol,

$$CH_3CH_2OH$$

is a primary alcohol. Isopropyl alcohol,

$$CH_3CHCH_3$$
$$\quad\quad |$$
$$\quad\quad OH$$

is a secondary alcohol. Propyl alcohol,

$$CH_3CH_2CH_2OH$$

is a primary alcohol.

example
15.2

How would you classify methyl alcohol?
   Methyl alcohol can't be classified under this system. There is only one carbon atom in methyl alcohol; therefore, it is attached to no other carbon atoms and is neither primary, secondary, nor tertiary.

More complex alcohols are generally given IUPAC names. These systematic names are often used even for the simplest members of the family. By the IUPAC system, alcohols are named for the alkane corresponding to the longest continuous chain of carbon atoms. The final **-e** of the alkane name is dropped and replaced by the ending **-ol.** If necessary, the position of the hydroxyl group is indicated by a number placed immediately in front of the name of the longest, parent chain. The molecule is always numbered from the end nearer the functional group, giving the hydroxyl group the lowest possible number.

Give the IUPAC names for methyl alcohol and ethyl alcohol.

Methyl alcohol (CH_3OH) is named as a derivative of methane (CH_4). Merely drop the -e, add -ol, and you have it—methanol. To name CH_3CH_2OH, drop the final -e from the name of the corresponding alkane, ethane (CH_3CH_3), and add -ol. The IUPAC name is ethanol.

example 15.3

Give IUPAC names to the two propyl alcohols.

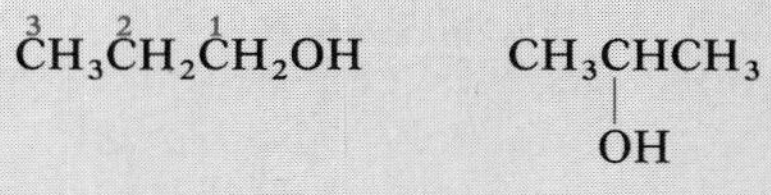

$$\overset{3}{C}H_3\overset{2}{C}H_2\overset{1}{C}H_2OH \qquad CH_3CHCH_3$$
$$\qquad\qquad\qquad\qquad\quad | $$
$$\qquad\qquad\qquad\qquad\ OH$$

The names are, respectively, 1-propanol (not 3-propanol) and 2-propanol.

example 15.4

Give the IUPAC name for *tert*-butyl alcohol.

tert-Butyl alcohol is

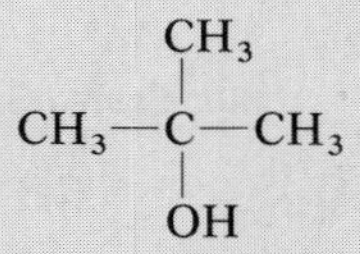

$$\begin{array}{c} CH_3 \\ | \\ CH_3-C-CH_3 \\ | \\ OH \end{array}$$

The longest continuous chain is three carbon atoms long, and the hydroxyl group is on the second carbon of this chain, yielding, for the moment, 2-propanol. There is also a methyl group hanging from the second carbon of the parent chain, giving 2-methyl-2-propanol.

example 15.5

example
15.6

Give the IUPAC name for

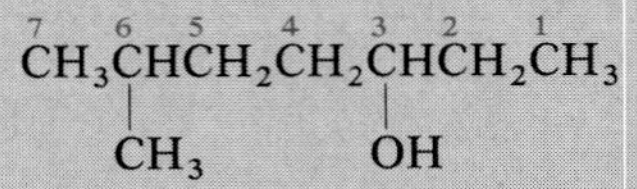

$$\overset{7}{C}H_3\overset{6}{C}H\overset{5}{C}H_2\overset{4}{C}H_2\overset{3}{C}H\overset{2}{C}H_2\overset{1}{C}H_3$$

$$\underset{CH_3}{|} \quad\quad \underset{OH}{|}$$

The name is 6-methyl-3-heptanol (*not* 2-methyl-5-heptanol)

15.3 PHYSICAL PROPERTIES OF THE ALCOHOLS

Most of the common alcohols are liquids at room temperature. The simplest one, methanol, boils at 65 °C. The hydrocarbon ethane, which has nearly the same molecular weight as methanol, boils at −89 °C and is a gas at room temperature. Here we see the pronounced effect of hydrogen bonding (Section 7.4) on boiling points. Figure 15.3 shows how alcohol molecules may be associated through hydrogen bonding. Such association is less extensive in alcohols than in water (compare Figure 7.3), because an alkyl group (R—) has replaced one of the hydrogen atoms. No hydrogen bond can occur through that alkyl group. Because of this, methanol has a lower boiling point than water even though the molecular weight of methanol is about twice that of water.

Alcohols of low molecular weight are soluble in water. Indeed, methyl, ethyl, and the two propyl alcohols can be mixed with water in all proportions; that is, they are completely miscible with water. As the size of the alkyl group increases, however, the alcohols become more like alkanes (they become more insoluble in water) and less like water itself. 1-Octanol ($CH_3CH_2CH_2CH_2CH_2CH_2CH_2CH_2OH$) is soluble only to the extent of 0.05 g in 100 g of water. The hydroxyl group's ability to form hydrogen bonds is almost totally overshadowed by the lack of attraction between water molecules

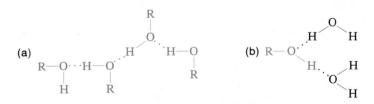

FIGURE 15.3   *(a) Alcohols have high boiling points relative to the hydrocarbons because of strong intermolecular hydrogen bonding. (b) Alcohols of low molecular weight are soluble in water because the molecules can hydrogen bond to water molecules.*

TABLE 15.2 *Physical Properties of Some Common Alcohols*

Name	Formula	Boiling Point (in degrees Celsius)	Solubility (in grams per 100 g of water)
Methyl alcohol	CH_3OH	64	∞
Ethyl alcohol	CH_3CH_2OH	78	∞
Propyl alcohol	$CH_3CH_2CH_2OH$	97	∞
Butyl alcohol	$CH_3CH_2CH_2CH_2OH$	118	7.9
Pentyl alcohol	$CH_3(CH_2)_3CH_2OH$	138	2.3
Hexyl alcohol	$CH_3(CH_2)_4CH_2OH$	156	0.6
Octyl alcohol	$CH_3(CH_2)_6CH_2OH$	195	0.05
Decyl alcohol	$CH_3(CH_2)_8CH_2OH$	228	(Insoluble)
Isopropyl alcohol	$CH_3CHOHCH_3$	82	∞
Isobutyl alcohol	$(CH_3)_2CHCH_2OH$	108	10.2
sec-Butyl alcohol	$CH_3CHOHCH_2CH_3$	99	12.5
tert-Butyl alcohol	$(CH_3)_3COH$	83	∞

and the long alkanelike hydrocarbon portion of the organic compound. With more than eight carbons, the simple alcohols are essentially insoluble in water.

To generalize, molecules that can effectively hydrogen bond to water will dissolve in water. Each functional group, such as the hydroxyl, that can form hydrogen bonds to water can carry along into solution an alkyl group of up to four or five carbon atoms. Thus, we will frequently find that the borderline of water solubility in a family of organic compounds occurs at four or five carbon atoms (Table 15.2). For isomeric alcohols, the more branched structures lead to greater solubility (consider the four butyl alcohols).

15.4 PREPARATION OF ALCOHOLS

Most simple alcohols are made by the hydration of alkenes (the addition of water to the double bond—Section 14.11). Since alkenes are made by the cracking of petroleum, our supply of most alcohols is dependent to a large extent on the availability of oil.

Ethanol is made by the hydration of ethylene in the presence of a mineral acid.

In a similar manner, isopropyl alcohol is produced by the addition of water to propylene.

$$H-\overset{\overset{\displaystyle H}{|}}{\underset{\underset{\displaystyle H}{|}}{C}}-\overset{\overset{\displaystyle H}{|}}{\underset{\underset{\displaystyle H}{|}}{C}}=C\overset{H}{\underset{H}{}} \; + \; H-OH \;\; \xrightarrow{H^+} \;\; H-\overset{\overset{\displaystyle H}{|}}{\underset{\underset{\displaystyle H}{|}}{C}}-\overset{\overset{\displaystyle H}{|}}{\underset{\underset{\displaystyle OH}{|}}{C}}-\overset{\overset{\displaystyle H}{|}}{\underset{\underset{\displaystyle H}{|}}{C}}-H$$

With isobutylene, the product is *tert*-butyl alcohol.

$$\overset{H}{\underset{H}{}}C=C\overset{CH_3}{\underset{CH_3}{}} \; + \; H-OH \;\; \xrightarrow{H^+} \;\; CH_3-\overset{\overset{\displaystyle CH_3}{|}}{\underset{\underset{\displaystyle OH}{|}}{C}}-CH_3$$

Note that in the last two reactions, the hydrogen goes on the carbon atom (of the two involved in the double bond) that has the most hydrogens already bonded to it. The hydroxyl group goes on the carbon with fewer hydrogens. Thus, addition of water to propylene always gives isopropyl alcohol, never propyl alcohol.

$$CH_3CH=CH_2 + H-OH \xrightarrow{H^+} \begin{array}{l} \xrightarrow{\text{always}} CH_3\underset{\underset{\displaystyle OH}{|}}{CH}CH_3 \\[2em] \xrightarrow{\text{never}} CH_3CH_2CH_2OH \end{array}$$

The above rule, in a more general form, was first formulated in 1870 by Vladimir V. Markovnikov, a Russian chemist. It is widely known as **Markovnikov's rule.** Sometimes the rule is stated (somewhat facetiously) as "the rich get richer."

Methanol is made commercially mainly by reaction of carbon monoxide with hydrogen.

$$CO + 2\,H_2 \longrightarrow CH_3OH$$

The reaction is carried out at a high temperature and pressure and in the presence of a catalyst. Unlike the hydration reaction, this industrial process does not represent a general approach to the synthesis of a variety of alcohols.

Although of considerably less (but still great) economic importance, the process of making alcohol by fermentation may be of greater interest to many people. Most alcoholic beverages are made by fermentation of starches or sugars (Figure 15.4). On an industrial scale, either molasses from sugarcane or starches from various types of grain are fermented by yeast to ethanol. It is ethanol that most people are referring to when they say "alcohol," meaning liquor or that which is imbibed.

In the preparation of ethanol from grain, the seeds are ground and cooked, producing **mash. Malt** (the dried sprouts of barley) or a mold such as *Aspergillus oryzae* is added as a source of the enzyme *diastase*. This enzyme catalyzes the conversion of starch to malt sugar, *maltose.*

FIGURE 15.4 *Alcohol can be made by the fermentation of nearly any type of starchy or sugary material. (Photo © The Terry Wild Studio.)*

$$(C_6H_{10}O_5)_{2x} + xH_2O \xrightarrow{\text{diastase}} xC_{12}H_{22}O_{11}$$

Starch   Maltose

The maltose solution is diluted to 10% concentration, and a pure yeast culture is added. The yeast cells produce two enzymes that catalyze the remainder of the fermentation process. First, the enzyme *maltase* converts the maltose to glucose.

$$C_{12}H_{22}O_{11} + H_2O \xrightarrow{\text{maltase}} 2C_6H_{12}O_6$$

Maltose   Glucose

Then several enzymatic reactions convert glucose to ethanol and carbon dioxide.

$$C_6H_{12}O_6 \longrightarrow 2CH_3CH_2OH + 2CO_2$$

In the production of beer or malt liquor, part of the carbon dioxide is retained, making a carbonated beverage. The fermentation process is capable of yielding aqueous solutions containing up to 18% alcohol. As the alcohol concentration builds to these levels, it inhibits the catalytic activity of the yeast cultures, and the system shuts itself down.

In the production of strong alcoholic beverages, the fermented mash is filtered (to remove solid materials) and then distilled (to increase the concentration of alcohol, usually to 40% or 50%). Such concentrated solutions are called **distilled spirits** and are prepared in distilleries (legally) or backwoods stills (illegally). Fermented alcohol can be concentrated to as much as 95% by distillation. Such grain alcohol is frequently used as a solvent for drugs meant for internal consumption.

15.5 PHYSIOLOGICAL PROPERTIES OF THE ALCOHOLS

The simple alcohols are poisonous to some degree. In an attempt to quantify the degree of toxicity, scientists use the term **LD_{50}** to indicate the lethal dose of a chemical to 50% of a population of test animals. Like humans, individual animals respond differently to various poisons. Some are killed by amounts much smaller than the LD_{50}; others survive considerably larger amounts. The LD_{50} term, then, is only approximate for animals. Extrapolation to human toxicities can introduce even larger errors. Table 15.3 lists LD_{50} values for alcohols administered orally to rats. Note that no LD_{50} is given for methanol (wood alcohol). While its acute (short-term) toxicity is not terribly high, it can cause permanent blindness or death, even in small concentrations. Each year many accidents are attributed to this alcohol, which is frequently mistaken for its less harmful relative ethyl alcohol.

The reason methanol is so dangerous* is that humans and other primates have liver enzymes that oxidize primary alcohols to compounds called aldehydes. Ethanol, for example, is oxidized to acetaldehyde.

$$CH_3CH_2OH \xrightarrow{\text{liver enzymes}} \underset{CH_3}{\overset{H}{>}}C=O$$

   Ethanol   Acetaldehyde

The acetaldehyde is in turn oxidized to acetic acid, a normal constituent of cells. The acetic acid can then be oxidized to carbon dioxide and water.

Similarly, methanol is oxidized to formaldehyde.

$$CH_3OH \xrightarrow{\text{liver enzymes}} \underset{H}{\overset{H}{>}}C=O$$

   Methanol   Formaldehyde

Formaldehyde reacts rapidly with the components of cells. It causes proteins to be coagulated, in much the same way that an egg is coagulated by cooking. It is this property of formaldehyde that accounts for the great toxicity of

---

* It should be noted that methanol is not particularly toxic to horses, rats, and some other animals. These animals are deficient in the enzymes that oxidize alcohols to aldehydes. These facts should make it abundantly clear that toxicity studies in other animals cannot always be extrapolated to humans.

TABLE 15.3 *Lethal Doses (Orally) of Some Alcohols in Rats*

Alcohol	Structure	LD_{50} (in grams per kilogram of body weight)
Methanol	CH_3OH	—
Ethanol	CH_3CH_2OH	10.3
1-Propanol	$CH_3CH_2CH_2OH$	1.9
2-Propanol	$CH_3CHOHCH_3$	5.8
1-Butanol	$CH_3CH_2CH_2CH_2OH$	4.4
1-Hexanol	$CH_3(CH_2)_4CH_2OH$	4.6
Ethylene glycol	$HOCH_2CH_2OH$	8.5
Glycerol	$HOCH_2CHOHCH_2OH$	>25

Data derived from Windholz, Martha, ed., *The Merck Index*, 10th ed., Rahway, N.J.: Merck, 1983.

methanol. The LD_{50} for formaldehyde administered orally to rats is 0.07 g per kilogram of body weight. That for acetaldehyde under the same conditions is 1.9 g per kilogram of body weight. (Keep in mind that the *smaller* the LD_{50}, the *more toxic* the substance.) Thus, formaldehyde is about 27 times as toxic to rats as acetaldehyde. Indeed, the antidote for methanol poisoning has long been ethanol, administered intravenously. The ethanol preferentially loads up the liver enzymes in humans and other primates. If the enzymes are tied up oxidizing ethanol to acetaldehyde, they cannot catalyze the oxidation of the methanol to the dangerously toxic formaldehyde. Thus, the unoxidized methanol is gradually excreted from the body.

Despite its toxicity, methanol is a valuable industrial solvent. It is also a raw material for the production of other chemicals.

But ethanol is toxic, too. A pint of pure ethanol, rapidly ingested, would kill most people. Even the strongest alcoholic beverages, however, are seldom more than 90 proof (45%).* Excessive ingestion over a long period of time leads to deterioration of the liver and loss of memory—and possibly to strong physiological addiction.

In addition to its use as a beverage, ethanol is used in great quantities as an industrial solvent and as a starting material for other chemical products. Such industrial alcohol is usually made by the addition of water to ethylene. The product is structurally identical to the ethanol produced through fermentation. Alcoholic beverages are generally highly taxed. To avoid the diversion of untaxed industrial alcohol to use in beverages, the industrial alcohol is **denatured** by addition of other chemicals that make the alcohol unfit to drink. Among the

---

* The "proof" is merely twice the percentage of alcohol by volume. The term has its origin in an old seventeenth-century method for testing whiskey. Dealers were perhaps too often tempted to increase profits by adding water to the booze. A qualitative method for testing the whiskey was to pour some of it on gunpowder and ignite it. If the gunpowder ignited after the alcohol had burned away, that was considered "proof" that the whiskey did not contain too much water.

TABLE 15.4 *Relationship Between Drinks Consumed, Blood Alcohol Level, and Behavior**

Number of Drinks[†]	Blood-Alcohol Level (percent by volume)	Behavior
2	0.05	Mild sedation; tranquility
4	0.10	Lack of coordination
6	0.15	Obvious intoxication
10	0.30	Unconsciousness
20	0.50	Possible death

* The data would be approximate for a 70-kg person who is a moderate drinker.
† Based on 30-mL (1-oz) "shots" of 90-proof whiskey or 360-mL (12-oz) bottles of beer, consumed rapidly.

common denaturants are gasoline and methanol. When necessary for laboratory work, pure ethanol is available for use. In the United States, this untaxed alcohol is strictly controlled by the federal government.

Ethanol acts as a mild hypnotic (sleep inducer). Perhaps this is fortunate, for a heavy drinker usually "passes out" before ingesting a lethal dose. Although it generally acts as a depressant (i.e., reduces the level of consciousness and the intensity of our reactions to environmental stimuli), ethanol in small amounts seems sometimes to act as a stimulant. Any such effect, however, is probably due to the alcohol's action in relaxing tensions and relieving inhibitions. Table 15.4 shows some effects of alcohol on a moderate drinker.

Isopropyl alcohol, an ingredient in rubbing alcohol and other products for external use, is sometimes mistaken for ethanol and is ingested. Though more toxic than ethanol, it seldom causes fatalities. Instead, it induces vomiting. It doesn't stay down long enough to kill you (though perhaps it stays down long enough to make you wish you were dead). Other higher alcohols behave in a similar manner.

The lower alcohols have a mild antiseptic action. Ethanol or isopropyl alcohol is often used to clean the skin before an injection, the drawing of blood, or minor surgery.

15.6 CHEMICAL PROPERTIES OF THE ALCOHOLS

The reactions of the alcohols occur mainly at the functional group. They may, however, involve hydrogen atoms attached to the carbon bearing the hydroxyl group or even those on an adjacent carbon. We will discuss three major kinds of reactions of the alcohols. Dehydration and oxidation are considered here. Esterification is covered in Chapter 17.

Dehydration (removal of water) is usually accomplished by addition of concentrated sulfuric acid to the alcohol and heating of the resulting mixture. The hydroxyl group is removed from the alcohol carbon and a hydrogen atom is removed from an adjacent carbon, giving an alkene.

Ethanol   Ethylene

Under the proper conditions, it is possible to perform a dehydration involving two molecules of alcohol. In this case, the hydroxyl group of one alcohol is removed, and only the hydrogen of the hydroxyl group of the second alcohol molecule is removed. The two organic groups remaining combine to form an ether molecule.

Two molecules of ethanol   Diethyl ether

Thus, depending on conditions, one can prepare either alkenes or ethers by dehydration of alcohols. At 180 °C, dehydration of ethanol gives ethylene as the main product. At 140 °C, the main product of dehydration of ethanol is diethyl ether.

Dehydration reactions also take place in biological systems. For example, in carbohydrate metabolism (Chapter 27), citric acid is dehydrated to *cis*-aconitic acid.

Citric acid   *cis*-Aconitic acid

Look carefully at this equation. The compounds involved are more complex than the ethanol and ethylene we used in our previous example, but the reaction is not. Ignore all that stuff hanging off to the right of the molecules. These parts of the starting material remain unchanged (in this reaction) in the product.

The only thing that has happened is that a hydrogen and a hydroxyl group have been eliminated from the starting material, and the product contains a double bond. The point to be made is that if you know the chemistry of a particular functional group, you know the chemistry of a thousand or a hundred thousand different individual compounds. Alcohols have a potential for undergoing dehydration, and you will find that big ones, little ones, and ones that incorporate other functional groups all dehydrate if conditions are right.

Dehydration to form simple ethers in biological systems is perhaps less common than the reaction to form an alkene. However, many important reactions, such as the formation of glycosides from sugars (Chapter 20), are at least technically dehydrations leading to etherlike compounds.

Primary and secondary alcohols are readily oxidized. In the preceding section we saw how methanol and ethanol are oxidized by liver enzymes to form aldehydes. Such reactions can also be performed in the laboratory with chemical oxidizing agents. For example, in acid solution, potassium dichromate oxidizes ethyl alcohol to acetaldehyde. The reaction is

$$8\,H^+ + Cr_2O_7{}^{2-} + 3\,C_2H_5OH \longrightarrow 2\,Cr^{3+} + 3\,C_2H_4O + 7\,H_2O$$

Dichromate ion   Chromium(III) ion
(orange-red)   (green)

(This is the breathalizer reaction, p. 246.) Similarly, propyl alcohol is oxidized to propionaldehyde. The balanced equation for this reaction is quite complicated, even if we write only the net ionic equation.

$$3\,CH_3CH_2\,CH_2OH + 8\,H^+ + Cr_2O_7{}^{2-} \longrightarrow 3\,CH_3CH_2\,CHO + 2\,Cr^{3+} + 7\,H_2O$$

In situations like this, organic chemists have a tendency to simplify everything until only the change involving the organic molecules is shown. Thus, the above reaction would be simplified to

$$CH_3CH_2\,\boxed{CH_2OH} \xrightarrow{\;K_2Cr_2O_7,\ H^+\;} CH_3CH_2\,\boxed{\overset{\textstyle H}{\underset{\textstyle}{|}}C{=}O}$$

Propyl alcohol   Propionaldehyde

The required inorganic reagents are written above the arrow (a place reserved for catalysts by inorganic chemists). The inorganic by-products are ignored (they're still formed, but we just ignore them in this form of the equation). In this way, all attention is focused on the organic starting material and product, and less time is spent balancing the frequently complicated equations.

The abbreviated form of this particular equation indicates that a primary alcohol is oxidized to an aldehyde. We shall see, in Chapter 16, that aldehydes are even more easily oxidized than alcohols and yield as products carboxylic acids. If one wishes to isolate the aldehyde initially formed in the oxidation of the alcohol, it is necessary to remove it from contact with the oxidizing agent. This can be done by distillation of the aldehyde from the reaction mixture as it forms.

Secondary alcohols are oxidized to compounds called ketones. Oxidation of isopropyl alcohol by dichromate gives acetone.

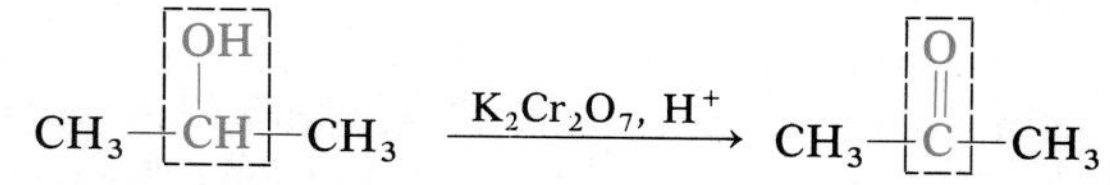

$$CH_3-\boxed{CH}-CH_3 \xrightarrow{K_2Cr_2O_7, \ H^+} CH_3-\boxed{\overset{O}{\overset{\|}{C}}}-CH_3$$

Isopropyl alcohol   Acetone
(a secondary alcohol)   (a ketone)

Unlike aldehydes, ketones are relatively resistant to further oxidation (Chapter 16), and special precautions to isolate the product of this reaction are not necessary.

As we saw in the preceding section, oxidation of alcohols is important in living organisms. Indeed, enzyme-controlled oxidation reactions provide the energy whereby cells can do useful work. One step in the metabolism of carbohydrates (Chapter 27) involves the oxidation of the secondary alcohol group in isocitric acid to a ketone group.

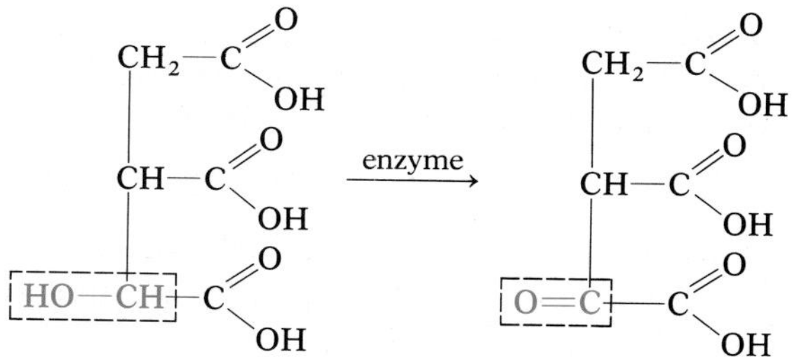

Isocitric acid   Oxalosuccinic acid

Again note that the overall reaction is identical to that of the conversion of isopropyl alcohol to acetone. The complications of structure that distinguish isocitric acid in no way interfere with the characteristic reaction of its secondary alcohol group.

Tertiary alcohols are resistant to oxidation. Ordinary oxidizing agents, such as dichromate, bring about no change in this class of alcohols, which lacks a hydrogen on the carbon atom bonded to the hydroxyl group.

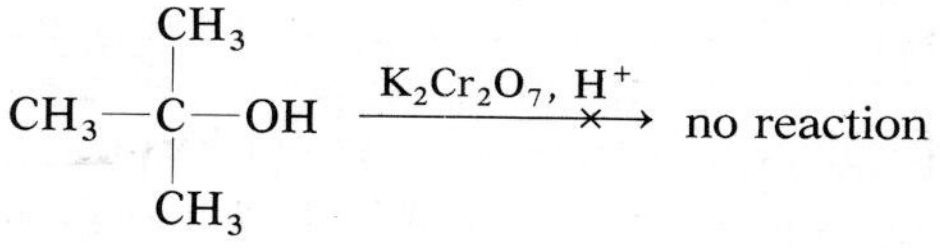

$$CH_3-\underset{\underset{CH_3}{|}}{\overset{\overset{CH_3}{|}}{C}}-OH \xrightarrow[\times]{K_2Cr_2O_7, \ H^+} \text{no reaction}$$

tert-Butyl alcohol

The oxidation reactions we have described involve the formation of a carbon-oxygen double bond. Thus, the carbon atom bonded to the oxygen in the alcohol must be able to release one of the atoms attached to it so it can form the double bond with oxygen. Hydrogen readily leaves the carbon under the conditions of oxidation. Both primary and secondary alcohols have a hydrogen on the carbon holding the hydroxyl group, and both these classes of alcohols are readily oxidized. The tertiary alcohol lacks such a hydrogen and is not easily oxidized. An analogous structural feature explains why aldehydes tend to oxidize further while ketones don't. The aldehydes still have a hydrogen left on the carbon holding the oxygen; the ketones do not. The following equations summarize the differences in the oxidation (indicated by the symbol [O]) of the various classes of alcohols.

$$\underset{\text{A primary alcohol}}{R-\overset{\overset{\displaystyle H}{|}}{\underset{\underset{\displaystyle H}{|}}{C}}-O_{\diagdown H}} \xrightarrow{[O]} \underset{\text{Aldehyde}}{R-\overset{\overset{\displaystyle H}{|}}{C}=O} \xrightarrow{[O]} \underset{\text{Carboxylic acid}}{R-\overset{\overset{\displaystyle OH}{|}}{C}=O}$$

$$\underset{\text{A secondary alcohol}}{R-\overset{\overset{\displaystyle R}{|}}{\underset{\underset{\displaystyle H}{|}}{C}}-O_{\diagdown H}} \xrightarrow{[O]} \underset{\text{Ketone}}{R-\overset{\overset{\displaystyle R}{|}}{C}=O}$$

$$\underset{\text{A tertiary alcohol}}{R-\overset{\overset{\displaystyle R}{|}}{\underset{\underset{\displaystyle R}{|}}{C}}-O_{\diagdown H}} \xrightarrow[\times]{[O]} \text{no reaction}$$

15.7 MULTIFUNCTIONAL ALCOHOLS: GLYCOLS AND GLYCEROL

The simple alcohols that we have met so far contain only one hydroxyl group each. They are called **monohydric** alcohols. Several important compounds that are frequently encountered contain more than one hydroxyl group per molecule. They are called **polyhydric** alcohols. Those with two such groups are said to be **dihydric.** Substances with three hydroxyl groups are called **trihydric** alcohols.

Dihydric alcohols are often called **glycols.** The most important of these is ethylene glycol. This compound is the main ingredient in permanent antifreeze mixtures for automobile radiators. Ethylene glycol is a sweet, somewhat vis-

$$\underset{\text{Ethylene glycol}}{\overset{\displaystyle CH_2-CH_2}{\underset{\displaystyle OH \quad\ OH}{|\qquad |}}}$$

TABLE 15.5   *Properties of Some Polyhydric Alcohols*

Name	Formula	Boiling Point (in degrees Celsius)	Solubility in Water
Ethylene glycol	CH_2OHCH_2OH	198	∞
Propylene glycol	$CH_3CHOHCH_2OH$	188	∞
Glycerol	$CH_2OHCHOHCH_2OH$	290 (Decomposes)	∞
Sorbitol	$CH_2OH(CHOH)_4CH_2OH$	(Solid)	∞

cous liquid. With two hydroxyl groups, extensive intermolecular hydrogen bonding exists. Thus, ethylene glycol has a high boiling point and does not boil away when used as antifreeze. It is also completely miscible with water.

Ethylene glycol is quite toxic. As with methanol, its toxicity is due to a metabolite. Liver enzymes oxidize the ethylene glycol to oxalic acid.

$$\underset{\text{Ethylene glycol}}{\overset{\displaystyle \overset{OH}{|} \quad \overset{OH}{|}}{CH_2-CH_2}} \xrightarrow{\text{liver enzymes}} \underset{\text{Oxalic acid}}{HO-\overset{\displaystyle \overset{O}{\|}}{C}-\overset{\displaystyle \overset{O}{\|}}{C}-OH}$$

This compound crystallizes as its calcium salt, calcium oxalate (CaC_2O_4), in the kidneys, leading to renal damage. Such injury can lead to kidney failure and death. As with methanol poisoning, the usual treatment for ethylene glycol poisoning is ethanol, administered to load up and thus block the liver enzymes from catalyzing the conversion of ethylene glycol to oxalic acid.

Another common dihydric alcohol is propylene glycol. The physical properties of this compound are quite similar to those of ethylene glycol (Table 15.5). Its physiological properties, however, are quite different.

$$\underset{\text{Propylene glycol}}{\overset{\displaystyle \quad\;\; \overset{}{CH_3CH-CH_2}}{\underset{OH \;\; OH}{}}}$$

Propylene glycol is essentially nontoxic, and it can be used as a solvent for drugs. It is also used as a moisturizing agent for foods. Like other alcohols, propylene glycol can be oxidized by liver enzymes.

$$\underset{\text{Propylene glycol}}{CH_3-\overset{\displaystyle \overset{OH}{|}}{CH}-\overset{\displaystyle \overset{OH}{|}}{CH_2}} \xrightarrow{\text{liver enzymes}} \underset{\text{Pyruvic acid}}{CH_3-\overset{\displaystyle \overset{O}{\|}}{C}-\overset{\displaystyle \overset{O}{\|}}{C}-OH}$$

In this case, however, the product is pyruvic acid, a normal intermediate in carbohydrate metabolism.

Glycerol (glycerin) is the most important trihydric alcohol. It is a sweet, syrupy liquid. Essentially nontoxic, it is a product of the hydrolysis of fats (Special Topic L).

$$\underset{\text{Glycerol}}{\overset{\displaystyle CH_2-CH-CH_2}{\underset{OH \;\; OH \;\; OH}{}}}$$

Certainly the carbohydrates are the most important polyhydric alcohols of all. We consider the structure of these compounds in Chapter 20 and their metabolism in Chapter 27.

15.8 THE PHENOLS

Phenol
(carbolic acid)

CH₂CH₂CH₂CH₂CH₂CH₃

4-Hexylresorcinol

o-Phenylphenol

o-Benzyl-p-chlorophenol

Hexachlorophene

Compounds with a hydroxyl group attached directly to an aromatic ring are called **phenols.** The parent compound, C_6H_5OH, is itself called phenol. Other compounds may be named as derivatives of phenol, but most of those of interest to us are best known by special, nonsystematic names.

The phenols generally are solids with low melting points or oily liquids at room temperature. Most are only sparingly soluble in water. By their very nature, phenols never contain fewer than six carbons. They have found wide use as germicides or antiseptics (substances that kill microorganisms on living tissue) and as disinfectants (substances intended to kill microorganisms on furniture, fixtures, floors, and around the house in general).

The first widely used antiseptic was phenol itself, which was also called carbolic acid. Joseph Lister used it for antiseptic surgery in 1867. Unfortunately, phenol doesn't kill only undesirable microorganisms. It kills all types of cells. Applied to the skin, it can cause severe burns. In the bloodstream, it is a systemic poison, that is, one that is carried to and affects all parts of the body. Its severe side effects led to searches for safer antiseptics, a number of which have been found.

One of the most active phenolic antiseptics is 4-hexylresorcinol. It is much more powerful than phenol as a germicide and has fewer undesirable side effects. Indeed, it is safe enough to be used as the active ingredient in some mouthwashes.

Prominent among disinfectants are the compounds *o*-phenylphenol and *o*-benzyl-*p*-chlorophenol. These compounds are the main active ingredients in preparations such as Lysol.

Hexachlorophene was once widely used in germicidal cleaning solutions (Phisohex) and as an ingredient in deodorant soaps and other cosmetics. Structurally, hexachlorophene is a phenol, but it also resembles DDT and the PCBs (Special Topic E). In the United States, products contained at most 3% hexachlorophene. The compound was generally considered a safe and effective antibacterial agent. In 1972, however, the picture changed rapidly. An outbreak of neurological disease among infants in northeastern France was traced to a baby powder called Bébé that contained over 20% hexachlorophene. Over 30 of the infants died. The U.S. Food and Drug Administration acted quickly. Hexachlorophene was banned from all products intended for over-the-counter sales. It is still available for prescription use and for use in hospitals—in concentrations not to exceed 3%.

Phenols are slightly acidic. They react with (and dissolve in) sodium hydroxide solutions.

They are not acidic enough, however, to react with sodium bicarbonate solutions.

$$\langle\bigcirc\rangle\!-\!OH + NaHCO_3(aq) \longrightarrow \text{No reaction}$$

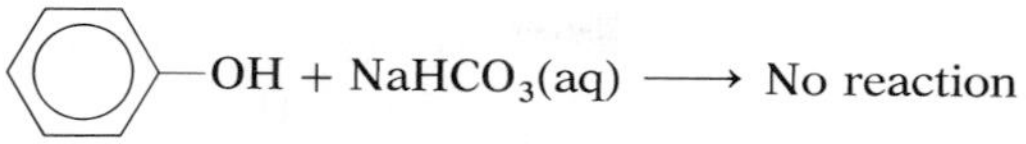

The latter reaction serves to distinguish the phenols from the carboxylic acids (Chapter 17), which do react with bicarbonate solutions.

15.9 THE ETHERS

Alcohols and phenols may be considered derivatives of water in which one of the hydrogen atoms has been replaced by an alkyl or an aryl group. **Ethers** may be thought of as compounds in which both hydrogens of water have been replaced by alkyl or aryl groups.

$$R\!-\!O\diagdown_R \qquad R\!-\!O\diagdown_{Ar} \qquad Ar\!-\!O\diagdown_{Ar}$$

Simple ethers are simply named. Just name the groups attached to oxygen and then add the generic name *ether* (Figure 15.5). For symmetrical ethers (those in which both organic groups are identical), the group name should be preceded by the prefix *di-*, although the prefix is sometimes dropped in common usage. The names *methyl ether* and *dimethyl ether* refer to the same compound, but the latter is preferred.

Ether molecules have no hydrogen atom on oxygen; hence, molecules in the pure liquid are incapable of intermolecular hydrogen bonding. Given their molecular weight, then, the ethers have quite low boiling points. Indeed, ethers have boiling points about the same as those of alkanes of comparable molecular weight. For example, diethyl ether (whose molecular weight is 74) boils at $35\,^{\circ}C$, and pentane (whose molecular weight is 72) boils at $36\,^{\circ}C$. Ether molecules do have an oxygen atom, however. They can participate in hydrogen

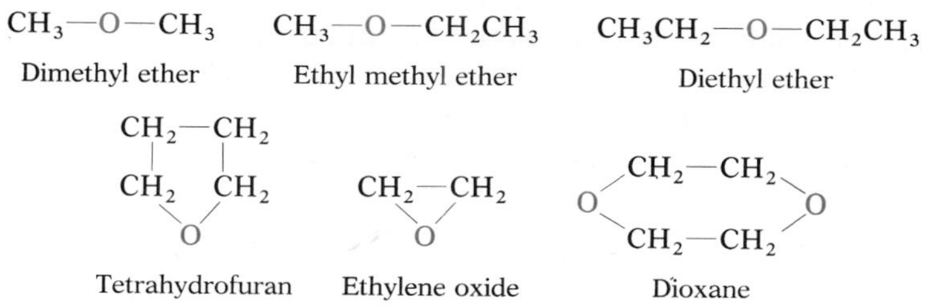

$$CH_3\!-\!O\!-\!CH_3 \qquad CH_3\!-\!O\!-\!CH_2CH_3 \qquad CH_3CH_2\!-\!O\!-\!CH_2CH_3$$

Dimethyl ether   Ethyl methyl ether   Diethyl ether

Tetrahydrofuran   Ethylene oxide   Dioxane

FIGURE 15.5 *Some representative ethers. Notice that in some ethers (ethylene oxide, dioxane, and tetrahydrofuran), the oxygen is incorporated as part of a ring.*

bonds if some other kind of molecule, such as water, will supply the appropriate kind of hydrogen.

$$R—O\cdots\cdots H—O\diagdown_H$$
$$\diagdown_R$$

Consequently, the ethers have about the same water solubilities as their isomeric alcohols. Both butyl alcohol and diethyl ether have the molecular formula $C_4H_{10}O$ (i.e., the ether and alcohol are isomers), and each is soluble to the extent of about 8 g in 100 g of water.

Chemically, the ethers are quite inert. Like the alkanes, they do not react with the usual oxidizing agents, reducing agents, or bases. The inertness of ethers makes them excellent solvents for organic materials. Diethyl ether is often used in the extraction of organic compounds from plant and animal materials or from mixtures of organic and inorganic substances. The volatile ether is then easily removed by evaporation, and the desired organic components are left behind. Dioxane and tetrahydrofuran (see Figure 15.5) are also important organic solvents.

Diethyl ether is perhaps best known to the public as a general anesthetic.* When vapors of this volatile liquid are inhaled, the compound acts as a central nervous system (CNS) depressant. As side effects it may produce nausea, vomiting, and respiratory arrest. Just as the word *alcohol* is sometimes used to designate ethyl alcohol, the word *ether* is often used to refer to diethyl ether.

Use of diethyl ether in the laboratory or hospital produces unusual hazards. The compound is quite volatile and extremely flammable. The vapors form an explosive mixture with air. They are also heavier than air and can travel long distances along a tabletop or the floor to reach a flame or spark and set off an explosion. Ether fires cannot be extinguished with water because the ether is less dense than water and will float on top of it. The use of carbon dioxide fire extinguishers is recommended.

Diethyl ether should not be stored in ordinary refrigerators. Even at low temperatures, it has sufficient vapor pressure to form explosive mixtures with air. A spark can ignite the vapors. Special explosion-proof refrigerators, with sealed electrical equipment to prevent contact of spark and flammable vapors, are available for safe storage of volatile, flammable liquids.

Still another hazard with ethers is that, upon standing, they react with oxygen from the air to form peroxides.

$$CH_3CH_2—O—CH_2CH_3 + O_2 \longrightarrow CH_3\underset{\underset{O—O—H}{|}}{CH}—O—CH_2CH_3$$

Diethyl ether   A peroxide

---

* A general anesthetic produces unconsciousness and renders one insensitive to pain. These substances are discussed in Special Topic G.

These peroxides are less violatile than the ether and are concentrated in the residue left behind during a distillation or evaporation. But these concentrated peroxides are highly explosive and sensitive to both shock and heat. Hospitals avoid these problems by buying only those amounts of ether sufficient for immediate use and by keeping containers tightly closed and away from strong light, which catalyzes peroxide formation. Ethers suspected of containing peroxides should be treated with a reducing agent (as well as a great deal of respect). The peroxides are destroyed by reagents such as alkaline ferrous sulfate solution.

PROBLEMS

1. What is a functional group?
2. Give the structure of and name the functional group in
   a. alkenes   **b.** alcohols
3. Give structural formulas for the eight isomeric pentyl alcohols ($C_5H_{12}O$). (Hint: Three are derived from pentane, four are derived from isopentane, and one is derived from neopentane.) Name the alcohols by the IUPAC system.
4. Classify the alcohols in Problem 3 as primary, secondary, or tertiary.
5. Draw and name the isomeric ethers with the formula $C_5H_{12}O$.
6. Name these compounds:
   a. $CH_3CH_2CH_2CH_2CHCH_3$
                               |
                              OH
   b. CH_3CHCH_2OH
             |
            CH_3
   c. CH_3CHCH_3
             |
            OH
   d. $CH_3CH_2CH_2CH_2CH_2CH_2OH$
7. Name these compounds:
   a. $CH_3CH_2CH_2OCH_2CH_2CH_3$
   b. $CH_3CH_2OCHCH_2CH_3$
                      |
                     CH_3

8. Name these compounds:

   a. OH
          (benzene ring with Cl)

   b. OH
          (benzene ring with NO_2)

9. Give structural formulas for the following:
   a. *tert*-butyl alcohol
   b. 3-hexanol
   c. 3,3-dimethyl-2-butanol
   d. pentyl alcohol
   e. 4-methyl-2-hexanol
10. Give structural formulas for the following:
    a. propylene glycol   **b.** glycerol
11. Give structural formulas for the following:
    a. ethyl methyl ether   **c.** isopropyl phenyl ether
    b. diisopropyl ether
12. Give structural formulas for the following:
    a. *m*-iodophenol
    b. *p*-methylphenol (*p*-cresol)
    c. 2,4,6-trinitrophenol (picric acid)
13. Draw the structural formula for carbolic acid and give another name for this compound.
14. Give the IUPAC names for the compounds referred to by these names:
    a. grain alcohol   **c.** rubbing alcohol
    b. wood alcohol
15. Which of the compounds in Problem 14 is most toxic? Which is least toxic?
16. What is denatured alcohol? Why is some alcohol denatured?

17. Is 90-proof liquor the product of simple fermentation, or should it be called distilled spirits?

18. Why is methanol so much more toxic to humans than ethanol?

19. Why is ethylene glycol so much more toxic to humans than propylene glycol?

20. What chemical compound is used in the treatment of acute methanol or ethylene glycol poisoning? How does it work?

21. Classify each of these conversions as oxidation, dehydration, or hydration. (Only the organic starting material and product are shown.)

a. $CH_3OH \longrightarrow$ H—$\overset{\displaystyle H}{\underset{}{C}}$=O

b. $CH_3\overset{\displaystyle OH}{\underset{}{C}}HCH_3 \longrightarrow CH_3CH$=$CH_2$

c. $CH_3\overset{\displaystyle OH}{\underset{}{C}}HCH_3 \longrightarrow CH_3\overset{\displaystyle O}{\underset{}{C}}CH_3$

d. $HOOCCH$=$CHCOOH \longrightarrow HOOCCH_2\overset{\displaystyle OH}{\underset{}{C}}HCOOH$

e. $2\ CH_3OH \longrightarrow CH_3OCH_3$

22. Each of the butyl alcohols is treated with potassium dichromate in acid. Draw the product (if any) expected from each of the four isomeric alcohols.

23. Write an equation for the dehydration of 2-propanol to yield an alkene.

24. Write an equation for the dehydration of 2-propanol to yield an ether.

25. Draw the ether that would form from the *intramolecular* dehydration of $HOCH_2CH_2CH_2CH_2CH_2OH$.

26. Draw the alkene that would form from the dehydration of cyclohexanol.

27. Without consulting tables, arrange the compounds in order of increasing boiling points: ethanol, 1-propanol, methanol.

28. Without consulting tables, arrange the compounds in order of increasing boiling points: butane, ethylene glycol, 1-propanol.

29. Without consulting tables, arrange the compounds in order of increasing boiling points: diethyl ether, propylene glycol, 1-butanol.

30. Without consulting tables, arrange the compounds in order of increasing solubility in water: methanol, 1-butanol, 1-octanol.

31. Without consulting tables, arrange the compounds in order of increasing solubility in water: pentane, propylene glycol, diethyl ether.

32. What is Markovnikov's rule?

33. What is the product of each of the following reactions?

a. CH_2=$CHCH_2CH_3 \xrightarrow{\ H^+,\ H_2O\ }$

b. $\xrightarrow{\ H^+,\ H_2O\ }$

34. Primary alcohols can be oxidized to aldehydes. The aldehydes will boil out of the oxidation reaction mixture at a temperature that will leave the unreacted alcohol behind. Explain how this difference in boiling points could be predicted from the structure of the following compounds.

$$R—CH_2OH \qquad R—\overset{\displaystyle O}{\underset{\displaystyle H}{C}}$$

35. What reagents are necessary to carry out the following conversions?

a. CH_3CH=$CH_2 \xrightarrow{\ ?\ } CH_3\overset{\displaystyle OH}{\underset{}{C}}HCH_3$

b. CH_2=$\overset{\displaystyle}{\underset{\displaystyle CH_3}{C}}$—$CH_3 \xrightarrow{\ ?\ } CH_3\overset{\displaystyle OH}{\underset{\displaystyle CH_3}{C}}$—$CH_3$

c. $CH_3CH_2OH \xrightarrow{\ ?\ } CH_2$=$CH_2$

d. $CH_3CH_2CH_2OH \xrightarrow{\ ?\ } CH_3CH_2\overset{\displaystyle O}{\underset{}{C}}$—H

e. $CH_3\overset{\displaystyle}{\underset{\displaystyle OH}{C}}HCH_3 \xrightarrow{\ ?\ } CH_3\overset{\displaystyle O}{\underset{}{C}}CH_3$

f. $2\,CH_3CH_2OH \xrightarrow{?} CH_3CH_2OCH_2CH_3$

36. Methanol is not particularly toxic to rats. If methanol were newly discovered and tested for toxicity in laboratory animals, what would you conclude about its safety for human consumption?

37. In addition to ethanol, the fermentation of grain produces other organic compounds collectively called fusel oils (FO). The four principal FO components are 1-propanol, isobutyl alcohol, 3-methyl-1-butanol, and 2-methyl-1-butanol. Draw a structural formula for each of these alcohols. (FO is quite toxic and accounts in part for hangovers.)

38. Tetrahydrocannabinol (THC) is the principal active ingredient in marijuana. What functional groups are present in the THC molecule?

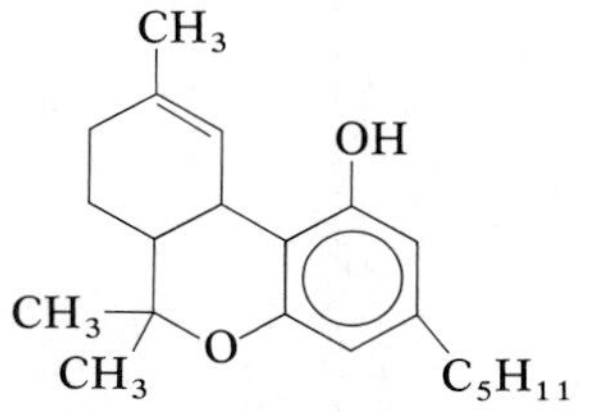

Tetrahydrocannabinol
(THC)

39. Give the structure of the alkene from which each of the following alcohols is made by reaction with water in acidic solution:

a. CH_3CHCH_3
　　　$|$
　　　OH

b. CH_3CH_2OH

c. $CH_3\overset{\displaystyle CH_3}{\underset{\displaystyle CH_3}{\overset{|}{\underset{|}{C}}}}\!-OH$

d. $CH_3CHCH_2CH_3$
　　　$|$
　　　OH

e. 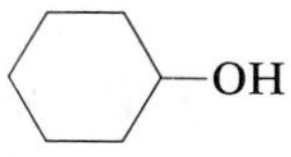 —OH

40. Write an equation for the reaction (if any) of phenol with aqueous
　a. NaOH　　　　**b.** $NaHCO_3$

41. What is a polyhydric alcohol?

42. What is a glycol?

43. What precautions must be taken when using diethyl ether as a solvent in a laboratory experiment?

44. Thymol is used in medicine as a topical antifungal agent and for preserving anatomical specimens and urine samples. What is the IUPAC name for thymol?

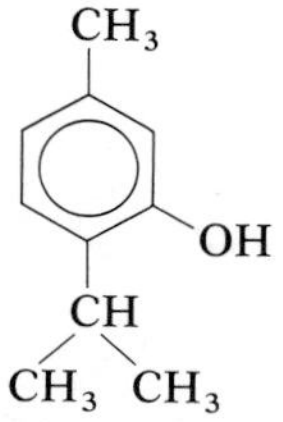

Thymol

45. Write the equation for the production of ethyl alcohol by the addition of water to ethylene. How much ethyl alcohol can be made from 14 t of ethylene?

46. The label on a bottle of light wine indicates that 100 mL of the wine furnishes 70 [food] Cal, 0.2 g of protein, 5.77 g of carbohydrates, and 0.0 g of fat. Assuming that carbohydrates and proteins furnish 4 Cal/g each, alcohol furnishes 7 Cal/g, and that no other caloric nutrients are present, make the following calculations:

　a. How many Calories are provided by the alcohol in a 100-mL serving of the wine? What percentage of the total Calories are provided by alcohol?

　b. How many grams of alcohol are there in each 100-mL serving?

　c. The density of alcohol is 0.789 g/mL. How many mL of alcohol are there in each 100-mL serving? What is the percentage alcohol by volume?

16

Aldehydes and Ketones

W hat do certain hormones, vanilla flavor, a biological tissue preservative, and fresh cucumbers have in common? A carbonyl functional
group, that's what. The carbonyl group is characteristic of aldehydes and ketones, the families we shall consider in this chapter. As the opening
list indicates, this functional group and these two families of compounds are
found in a most diverse company of products. Both the tempting aromas associated with cinnamon, vanilla, and fresh, buttered baked goods and the sicken-
ingly sweet smell of some rancid foods are associated with the carbonyl group.
The chemistry of the group is as interesting as its physical properties, and we
shall spend considerable time studying some aspects of that chemistry.

The aldehydes and ketones offer us an opportunity to study the carbonyl
group in its simplest surroundings. In Chapter 17, we consider somewhat more
complicated functional groups incorporating the carbonyl group. And we ultimately find ourselves running into this ubiquitous grouping of atoms in carbohydrates, fats, proteins, nucleic acids, hormones, vitamins, and the host of
organic compounds critical to the functioning of living systems. But first things
first. Let's begin by focusing on the carbonyl group in aldehydes and ketones.

16.1 THE CARBONYL GROUP: A CARBON-OXYGEN DOUBLE BOND

We first encountered a functional group containing a double bond in the
alkenes. In the members of this family, two carbon atoms share four electrons

(two pairs) to form a carbon-carbon double bond. In the alcohols, we saw a functional group in which an oxygen atom was attached to a carbon atom. The carbonyl group incorporates a feature of each of these other functional groups. It involves carbon bonded to oxygen and a double bond—a carbon-oxygen double bond.

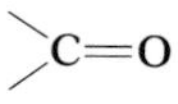

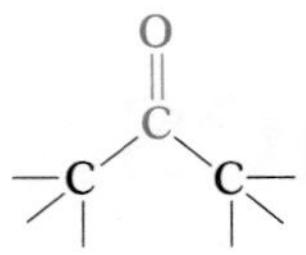

$$>C{=}O$$

The carbonyl double bond, like the alkene double bond, tends to undergo addition reactions. But, unlike the alkene double bond, it involves an oxygen atom and is highly polar. That polarity confers certain special properties on aldehydes and ketones.

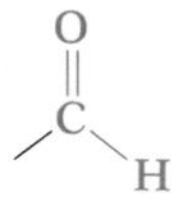

A ketone

An aldehyde

What is the difference between a ketone and an aldehyde? It appears to be rather trivial at first sight. In **ketones,** two carbon groups are attached to the carbonyl carbon. In **aldehydes,** at least one of the attached groups must be hydrogen. Thus, these general formulas all represent ketones.

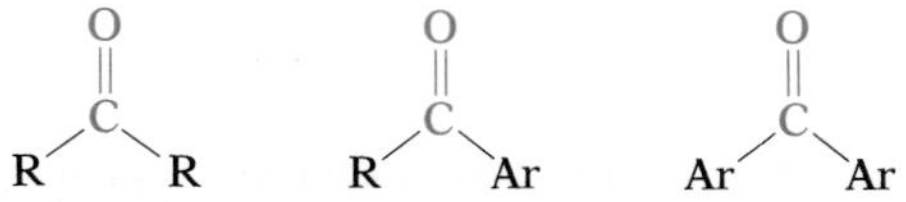

These compounds are all aldehydes.

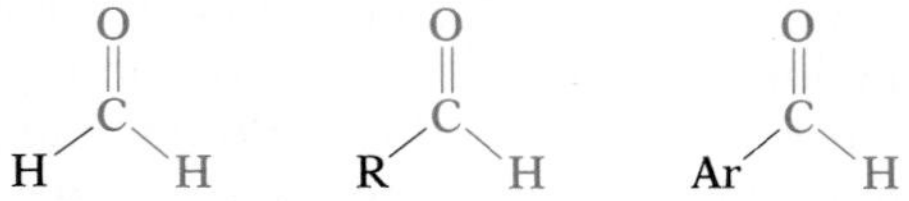

Aldehydes and ketones share many common properties, as one might expect for compounds with the same functional group. But they are different in other respects, different enough to warrant their classification into two families.

16.2 HOW TO NAME THE COMMON ALDEHYDES

As with most compounds isolated from natural sources, aldehydes came to be known by common names long before IUPAC rules were set up. And the common aldehyde names were adapted from common names for the carboxylic acids. This common naming system is, in many respects, as organized as the official IUPAC system. For the present, we shall simply give the common names of aldehydes of interest and cite one of the "rules" of the common system. In Chapter 17, the derivation of the common names is discussed.

The simplest aldehyde, and the only one with two hydrogens attached to the carbonyl group, contains only one carbon atom and is known as formaldehyde.

This compound is a gas at room temperature, but it is available as a 40% aqueous solution called formalin. This solution is a familiar biological preservative. You have most likely experienced the pungent odor of formaldehyde at some time or another during your studies in the biology laboratory.

The aldehyde with two carbon atoms is called acetaldehyde. Those with three, four, and five carbon atoms are named propionaldehyde, butyraldehyde, and valeraldehyde, respectively. These aldehydes and a few others of interest are shown in Figure 16.1.

In the common nomenclature, the position of substituents along the parent chain is indicated by Greek letters rather than numbers. Since an aldehyde must have a hydrogen attached to the carbonyl group, it is not possible to have a substituent located at the carbonyl carbon (all of its bonds are already in use). Therefore, the first Greek letter is assigned to the carbon atom *next to* the carbon in the carbonyl group.

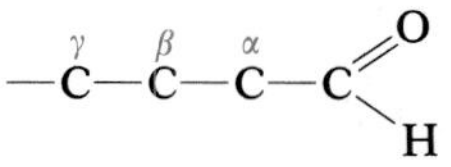

$$-\overset{\gamma}{C}-\overset{\beta}{C}-\overset{\alpha}{C}-C\overset{\displaystyle O}{\underset{\displaystyle H}{}}$$

There is rarely a need to specify a position beyond these. This system is illustrated by the three chlorobutyraldehydes in Figure 16.1.

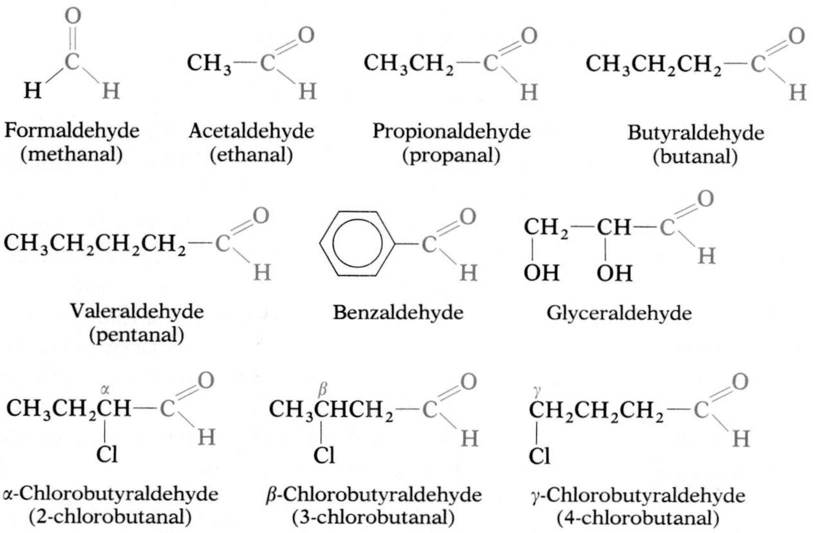

Formaldehyde (methanal)   Acetaldehyde (ethanal)   Propionaldehyde (propanal)   Butyraldehyde (butanal)

Valeraldehyde (pentanal)   Benzaldehyde   Glyceraldehyde

α-Chlorobutyraldehyde (2-chlorobutanal)   β-Chlorobutyraldehyde (3-chlorobutanal)   γ-Chlorobutyraldehyde (4-chlorobutanal)

FIGURE 16.1   *Some aldehydes of interest.*

example
16.1

Write the structural formula for γ-bromovaleraldehyde. Valeraldehyde has five carbon atoms.

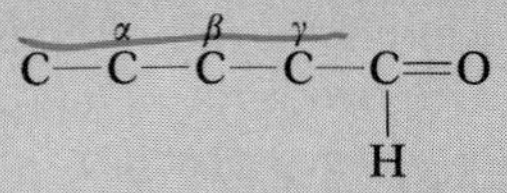

$$C—\overset{\alpha}{C}—\overset{\beta}{C}—\overset{\gamma}{C}—C=O$$
$$|$$
$$H$$

The gamma carbon is the third from the functional group (not counting the carbonyl carbon). Thus, the structure is

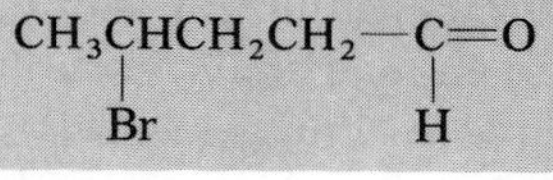

$$CH_3CHCH_2CH_2—C=O$$
$$\quad\quad |\quad\quad\quad\quad\quad |$$
$$\quad\quad Br\quad\quad\quad\quad\ H$$

example
16.2

Name the compound

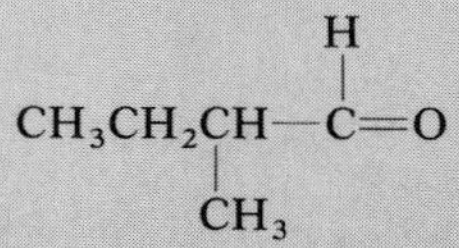

$$\quad\quad\quad\quad\quad\quad H$$
$$\quad\quad\quad\quad\quad\quad |$$
$$CH_3CH_2CH—C=O$$
$$\quad\quad\quad\quad |$$
$$\quad\quad\quad\quad CH_3$$

The compound may be named as a derivative of butyraldehyde (longest continuous chain including the carbonyl carbon contains four carbon atoms). There is a methyl group on the alpha carbon, the first one after the carbonyl carbon. A suitable name is α-methylbutyraldehyde.

The IUPAC names of aldehydes are derived from those of the corresponding alkanes. Select the longest continuous chain of carbon atoms that contains the functional group. Take the name of the alkane with that number of carbon atoms, drop the **-e,** and add the ending **-al.** Because the IUPAC ending for alcohols is **-ol,** there is occasionally some confusion unless great care is exercised in writing and pronouncing the IUPAC names of these two families. The one-carbon alcohol is methanol, with the ending pronounced like the *ol* in old. The one-carbon aldehyde is methanal, with the ending pronounced like the man's name *Al.* IUPAC names of the first five aldehydes, along with the names of other aldehydes of interest, are given in Figure 16.1.

In the IUPAC system the position of substituents is indicated by a number. The carbonyl carbon is *always* taken as the first carbon.

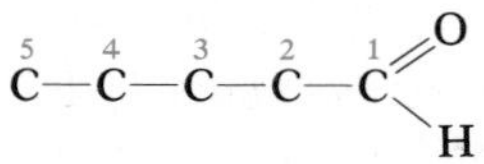

$$\overset{5}{C}—\overset{4}{C}—\overset{3}{C}—\overset{2}{C}—\overset{1}{C}\overset{\diagup O}{\diagdown H}$$

Note that the second carbon in the IUPAC system corresponds to the alpha carbon in the common system.

example
16.3

What is the IUPAC name for the compound in Example 16.2?
    There are four carbon atoms in the longest continuous chain. There is a methyl group on the second (alpha) carbon.

$$\overset{4}{CH_3}\overset{3}{CH_2}\overset{2}{CH}-\overset{1}{C}=O$$
$$\quad\quad\quad\; | \quad\; |$$
$$\quad\quad\quad CH_3 \;\; H$$

The name is 2-methylbutanal.

example
16.4

Write the structure for 7-chlorooctanal.
    There are eight carbon atoms in the longest continuous chain. There is a chlorine atom on the seventh carbon atom, numbering from the functional group and counting the carbonyl carbon as the first carbon atom.

$$CH_3CHCH_2CH_2CH_2CH_2CH_2-C=O$$
$$\quad\; | \quad\quad\quad\quad\quad\quad\quad\quad\quad |$$
$$\quad Cl \quad\quad\quad\quad\quad\quad\quad\quad\; H$$

16.3 NAMING THE COMMON KETONES

The carbonyl group in a ketone must be attached to two carbon groups. There-fore, the simplest ketone has three carbon atoms. It is known far and wide by the name **acetone.** The name is unique and does not correspond to the first in a series of similar common names. Generally, ketones are given common names consisting of the names of the groups attached to the carbonyl group, followed by the word ketone. (Note the similarity to the naming of ethers.) Another name for acetone, then, is dimethyl ketone. With four carbon atoms, we have ethyl methyl ketone.

$$\quad\quad\quad O$$
$$\quad\quad\quad \|$$
$$\quad\quad\quad C$$
$$CH_3 \quad\quad CH_3$$

Acetone

$$\quad\quad\quad O$$
$$\quad\quad\quad \|$$
$$\quad\quad\quad C$$
$$CH_3 \quad\quad CH_2CH_3$$

Ethyl methyl ketone

    If names for the groups attached to the carbonyl group are known, this com-mon naming system can be applied.
    In the IUPAC system, the longest continuous chain to which the oxygen is doubly bonded is selected as the parent chain. The **-e** ending of the correspond-ing alkane name is dropped and replaced with **-one.** Acetone thus becomes propanone, and ethyl methyl ketone is called butanone. In higher ketones, a number indicates the position of the doubly bonded oxygen. The locations of any substituents are also indicated by numbers. Figure 16.2 shows an assortment of ketones and lists both IUPAC and common names.

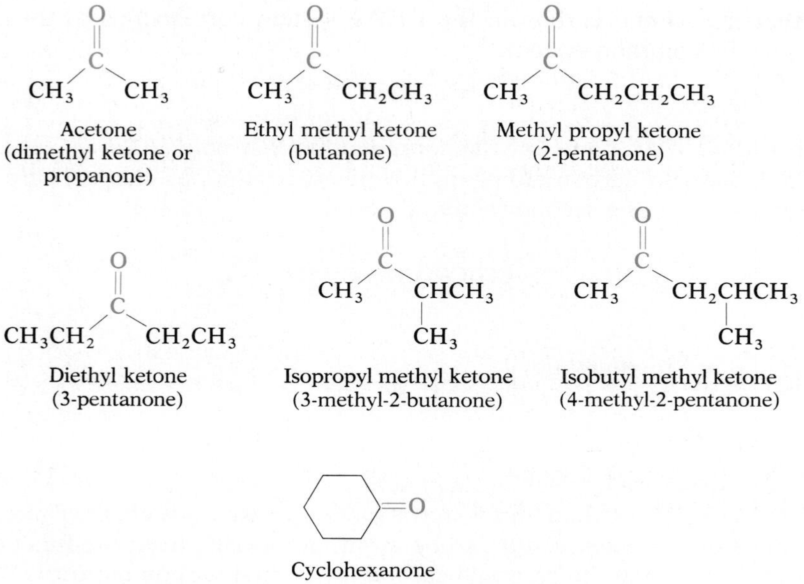

Acetone
(dimethyl ketone or propanone)

Ethyl methyl ketone
(butanone)

Methyl propyl ketone
(2-pentanone)

Diethyl ketone
(3-pentanone)

Isopropyl methyl ketone
(3-methyl-2-butanone)

Isobutyl methyl ketone
(4-methyl-2-pentanone)

Cyclohexanone

FIGURE 16.2   *Some common ketones.*

example 16.5

Write the structure for 4-methyl-3-hexanone.
The longest chain has six carbon atoms with a doubly bonded oxygen located at the third carbon and a methyl group at the fourth carbon.

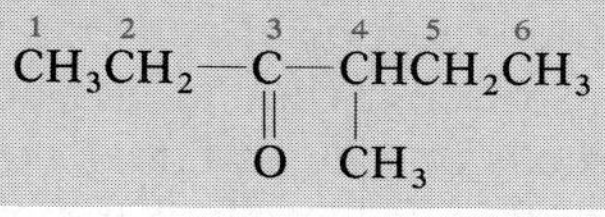

$$\overset{1}{C}H_3\overset{2}{C}H_2—\overset{3}{C}—\overset{4}{C}H\overset{5}{C}H_2\overset{6}{C}H_3$$
$$\quad\quad\quad\;\;\; \overset{\|}{O}\;\;\; \overset{|}{C}H_3$$

example 16.6

Give a common name for the ketone in Example 16.5.
The two alkyl groups attached to the carbonyl group are ethyl and *sec*-butyl. The common name is *sec*-butyl ethyl ketone.

example 16.7

Write the structure for diisopropyl ketone.

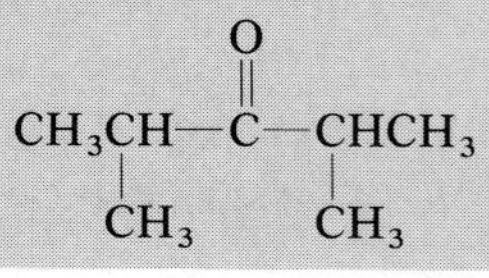

$$CH_3CH—\overset{\overset{\displaystyle O}{\|}}{C}—CHCH_3$$
$$\quad\;\; \overset{|}{C}H_3 \quad\quad \overset{|}{C}H_3$$

What is the IUPAC name for the ketone in Example 16.7?

example
16.8

$$\overset{1}{C}H_3\overset{2}{C}H-\overset{3}{C}-\overset{4}{C}H\overset{5}{C}H_3$$

with CH_3, O, CH_3 substituents

The name is 2,4-dimethyl-3-pentanone.

16.4 PHYSICAL PROPERTIES OF ALDEHYDES AND KETONES

The carbon and oxygen of the carbonyl group share two pairs of electrons, but they do not share them equally. The electronegative oxygen has a much greater attraction for the bonding pairs. Thus, the electron density is greater at the oxygen end of the bond and less at the carbon end. The carbon is left with a partial positive charge; the oxygen, with a partial negative charge. (Polar bonds were encountered earlier in Sections 4.8 and 4.9.)

$$\overset{\delta^+}{>}C=\overset{\delta^-}{O}$$

The polarity of the carbon-oxygen double bond is greater than that of the carbon-oxygen single bond. Indeed, double-bond polarity is great enough to affect the boiling points of aldehydes and ketones, whereas the polar single bonds in ethers (Chapter 15) have little effect on boiling points (Table 16.1). Such dipolar forces, however, are still not comparable to the hydrogen bonding that exists between molecules of an alcohol.

Although there can be no intermolecular hydrogen bonding in pure aldehydes or ketones,* there can be hydrogen bonds to water molecules. These

* Remember that the hydrogen that must be attached to the carbonyl group in an aldehyde is attached to the carbon atom, not the oxygen, of the functional group. Such a hydrogen is not capable of intermolecular hydrogen bonding.

TABLE 16.1 *Boiling Points of Compounds With Similar Molecular Weights and Different Types of Intermolecular Forces*

Compound	Molecular Weight	Type of Intermolecular Forces	Boiling Point (in degrees Celsius)
$CH_3CH_2CH_2CH_3$	58	Dispersion only	0
$CH_3OCH_2CH_3$	60	Weak dipole	6
$CH_3CH_2\overset{O}{\overset{\|}{C}}H$	58	Strong dipole	49
$CH_3CH_2CH_2OH$	60	Hydrogen bonding	97

TABLE 16.2 *Physical Properties of Selected Aldehydes and Ketones*

Compound	Formula	Boiling Point (in degrees Celsius)	Solubility in Water (in grams per 100 g of water)
Formaldehyde	HCHO	−21	Very soluble
Acetaldehye	CH_3CHO	20	∞
Propionaldehyde	CH_3CH_2CHO	49	16
Butyraldehyde	$CH_3CH_2CH_2CHO$	76	7
Valeraldehyde	$CH_3CH_2CH_2CH_2CHO$	103	Slightly soluble
Benzaldehyde	C_6H_5CHO	178	0.3
Acetone	CH_3COCH_3	56	∞
Ethyl methyl ketone	$CH_3COCH_2CH_3$	80	26
Methyl propyl ketone	$CH_3COCH_2CH_2CH_3$	102	6.3
Diethyl ketone	$CH_3CH_2COCH_2CH_3$	101	5

families are thus about as soluble in water as alcohols of comparable weight. The borderline of solubility occurs at about four carbon atoms per oxygen atom.

Physical properties of selected aldehydes and ketones are summarized in Table 16.2.

16.5 HOW TO MAKE ALDEHYDES AND KETONES

Careful oxidation of a primary alcohol gives an aldehyde (Chapter 15). For example, when propyl alcohol is warmed with an acidic solution of potassium dichromate, propionaldehyde is formed.

$$CH_3CH_2CH_2OH \xrightarrow{K_2Cr_2O_7,\ H^+} CH_3CH_2-C\underset{\displaystyle H}{\overset{\displaystyle O}{}}$$

Propyl alcohol
(boiling point, 97 °C)

Propionaldehyde
(boiling point, 49 °C)

If the solution is kept above 49 °C and below 97 °C, the propionaldehyde is distilled off as it is formed. The method is general for aldehydes and may be written

$$R-CH_2OH \xrightarrow{[O]} R-C\underset{\displaystyle H}{\overset{\displaystyle O}{}}$$

A primary alcohol

An aldehyde

where, once again, the symbol [O] represents oxidation.

Ketones are formed by the oxidation of secondary alcohols. For example, isopropyl alcohol is oxidized to acetone.

$$CH_3\overset{\overset{\displaystyle OH}{|}}{C}HCH_3 \xrightarrow{\text{K}_2\text{Cr}_2\text{O}_7,\ \text{H}^+} \underset{CH_3}{\overset{\overset{\displaystyle O}{\|}}{C}}CH_3$$

Isopropyl alcohol   Acetone

Unlike aldehydes, ketones resist further oxidation. It is not necessary to remove the ketone as it is formed. Oxidation of secondary alcohols is a general method for the preparation of ketones. The process may be represented as

$$R\overset{\overset{\displaystyle OH}{|}}{-}CH-R' \xrightarrow{[O]} R\overset{\overset{\displaystyle O}{\|}}{-}C-R'$$

A secondary   A ketone
alcohol

Chemists have many other ways of introducing carbonyl groups into molecules. Our discussion, though, is limited to oxidation processes because these are similar to the reactions occurring in our bodies.

16.6 OXIDATION OF ALDEHYDES

Aldehydes are themselves readily oxidized, yielding carboxylic acids. Ketones are resistant to oxidation.

$$R\overset{\overset{\displaystyle O}{\|}}{-}C-H \xrightarrow{[O]} R\overset{\overset{\displaystyle O}{\|}}{-}C-OH$$

An aldehyde   Carboxylic acid

$$R\overset{\overset{\displaystyle O}{\|}}{-}C-R' \xrightarrow[\times]{[O]} \text{no reaction}$$

A ketone

The aldehydes are, in fact, among the most easily oxidized of organic compounds, and this fact helps chemists identify them. Through the use of oxidizing agents, aldehydes can be distinguished not only from ketones but also from alcohols if the reagent is gentle enough. One such test reagent was invented by Professor Bernhard Tollens (1841–1918) at the University of Göttingen in Germany. Tollens's reagent employs silver ion as the mild oxidizing agent. In order for the silver ion to be kept in solution, it must be complexed by two ammonia molecules.

$$H_3N-Ag^+-NH_3$$

When Tollens's reagent oxidizes an aldehyde, the silver ion is reduced to free silver.

$$RCHO + 2\,Ag(NH_3)_2{}^+ + 2\,OH^- \longrightarrow RCOO^- + 2\,Ag(s) + NH_4{}^+ + 3\,NH_3 + H_2O$$

| An aldehyde | Tollens's reagent | | Salt of a carboxylic acid | Silver mirror |

The silver, when deposited on a clean glass surface, produces a beautiful mirror. Indeed, mirrors are often silvered by means of the Tollens reaction. The reducing agent of choice is often the sugar glucose (which contains an aldehyde functional group) rather than a simple aldehyde. Ordinary ketones do not react with Tollens's reagent.

Although ketones are resistant to oxidation by ordinary laboratory oxidizing agents, it is possible to force their oxidation. And, in particular, it should be recognized that both aldehydes and ketones will undergo combustion, that is, will burn. Acetone is a common organic solvent. Neither it nor any other volatile, flammable organic solvent should be used around open flames, heating elements, or other sources of possible ignition.

16.7 HYDRATED CARBONYL COMPOUNDS

Formaldehyde is a gas at room temperature, yet it dissolves readily in water. In fact, formaldehyde actually *reacts* with water.

The process is an addition reaction, analogous to the hydration of the carbon-carbon double bond of an alkene. The net result is that a hydrogen from water is added to the carbonyl oxygen and a hydroxyl group from water becomes attached to the carbonyl carbon. The product is called a **hydrate.** It readily breaks down to re-form formaldehyde and water. At equilibrium at 20 °C, the hydrate predominates. Indeed, only 1 molecule in 10 000 exists as free formaldehyde. The other 9999 are in the form of the hydrate.

Acetaldehyde is also hydrated in aqueous solution, but to a lesser extent than formaldehyde.

Out of 10 000 molecules, about 4200 are in the form of the free aldehyde at equilibrium. Still, that leaves 5800 in the hydrated form. Generally, higher aldehydes and ketones are even less hydrated, existing primarily in the free aldehyde form at equilibrium in water.

In most cases, it is impossible to isolate the hydrates from solution. Attempts to do so result in loss of water and regeneration of the carbonyl group. An exception is the hydrate of trichloroacetaldehyde.

$$\underset{\text{Trichloroacetaldehyde}}{\underset{CCl_3}{\overset{O}{\overset{\|}{C}}}\diagdown H} + H_2O \longrightarrow \underset{\text{Chloral hydrate}}{CCl_3-\overset{OH}{\underset{H}{\overset{|}{C}}}-OH}$$

The product, called chloral hydrate, is a stable crystalline solid. It is a powerful sedative and soporific (sleep-inducing drug). Chloral hydrate has had wide use in medicine. It is perhaps even better known in fictional mystery stories. Slipped into someone's drink, the mixture is called a "Mickey Finn" or "knockout drops." Such combinations of alcohol and chloral hydrate—two "downers"—are exceedingly dangerous. A little too much and the unfortunate victim may be put to sleep permanently.

16.8 THE ADDITION OF ALCOHOLS: HEMIACETALS AND ACETALS

Alcohols add to the carbonyl group of aldehydes and ketones in much the same way as water.

$$\underset{R}{\overset{O}{\overset{\|}{C}}}\diagdown H \quad H \quad O-R' \quad \rightleftarrows \quad R-\overset{OH}{\underset{H}{\overset{|}{C}}}-OR'$$

The product, called a **hemiacetal,** is generally rather unstable. It readily reverts to the aldehyde and alcohol.

Hemiacetals are important in carbohydrate chemistry (Chapter 20). Simple sugars, such as glucose, have aldehyde functional groups and alcohol groups on the same molecule. These interact to form intramolecular hemiacetals. Such cyclic hemiacetals are relatively stable. In aqueous solution, they are frequently the predominant form of the sugar.

Hemiacetals can be made to react further with alcohols. If dry hydrogen chloride gas is bubbled into a solution of aldehyde in excess alcohol, an **acetal** is formed. The reaction for acetaldehyde and methanol is

A hemiacetal   An acetal

Unlike hemiacetals and hydrates, acetals are stable. They can be isolated in pure form. Notice that the second part of the reaction, the formation of the acetal from the hemiacetal, is analogous to the dehydration reaction that yields an ether (Chapter 15). Thus, first an alcohol molecule adds to the double bond of the aldehyde to form the hemiacetal. Then the hydroxyl group of the hemiacetal and the hydrogen from the hydroxyl group of a second alcohol molecule are eliminated, and the two remaining pieces combine to form the acetal.

Acetal, or ketal (from a ketone), formation is often used to "protect" the functional group of aldehydes or ketones while other chemical operations are performed on the molecules, because acetals are resistant to oxidation whereas aldehydes themselves are not. An aldehyde could be converted to an acetal, an oxidation reaction could then be carried out on another part of the molecule, and finally the aldehyde could be regenerated from the acetal. The carbonyl group is easily regenerated by aqueous acid.

In an interesting application, the antibiotic chloramphenicol is treated with acetone, and a protective cyclic ketal is formed that masks the bitter taste of the drug. Both of the alcohol hydroxyl groups required to form the ketal are attached to a single molecule in this product.

Chloramphenicol (bitter)   Acetone   A cyclic ketal (not bitter)

The cyclic ketal is converted back to the free chloramphenicol by acids in the digestive system.

Chloramphenicol is a powerful, but hazardous, antibiotic. It is used only when other, less dangerous, drugs are ineffective. In about 1 person in 20 000 to 40 000 (depending on dosage), chloramphenicol causes fatal aplastic anemia.

16.9 THE ADDITION OF AMMONIA AND ITS DERIVATIVES

As with water and alcohols, ammonia readily adds to carbonyl compounds. The addition product is unstable. It eliminates a molecule of water to form a compound called an **imine.**

Unstable intermediate   An imine

Such imines are highly reactive. They usually undergo further reaction, often to complex products.

Certain ammonia derivatives, however, react to give stable, often crystalline products. An example is phenylhydrazine. It reacts with aldehydes and ketones to give yellow to orange crystalline solids called phenylhydrazones. Butyraldehyde, for example, gives butyraldehyde phenylhydrazone.

Butyraldehyde   Phenylhydrazine

Butyraldehyde phenylhydrazone

A quick way to determine what the product of these reactions will look like is to remove water from the two reactants. If the carbonyl oxygen is removed and the two hydrogens attached to nitrogen are removed, then the remaining pieces can be combined to give the correct product structure.

The phenylhydrazones are often used to identify specific aldehydes and ketones. The carbonyl compound, often a liquid, is converted to a solid derivative. The solid is carefully purified, and a melting point is determined. The value is compared with known values recorded in the chemical literature. This melting point is just one piece of data used in establishing the identity of the aldehyde or ketone.

Reactions similar to the addition of ammonia are important in the biosynthesis of certain amino acids. In the presence of a nitrogen source and certain enzymes, a carbohydrate metabolite, for example, may be converted into an amino acid (Chapter 29). In this way, the body can use carbohydrates as an ultimate source for some of the building blocks of proteins.

A common dipstick test for ketones in urine (Chapter 26) involves sodium nitroprusside, also known as sodium nitroferricyanide [$Na_2Fe(CN)_5NO$]. In alkaline media and in the presence of a nitrogen compound, this system gives a red to purple color if ketones are present.

16.10 THE HYDROGEN SHIFT: TAUTOMERISM

A carbonyl compound with a hydrogen atom on the alpha carbon exists in equilibrium with an isomeric form in which that hydrogen has shifted to the carbonyl oxygen atom.

Keto form   Enol form

The isomeric form is called an **enol,** for it contains a double bond (alk*ene*) and an alco*hol* functional group. We call this type of isomerism **tautomerism.**

For simple aldehydes and ketones, the equilibrium lies very far toward the keto form. Consider as an example the equilibrium between acetaldehyde and its tautomer, vinyl alcohol.

Acetaldehyde   Vinyl alcohol

In the pure liquid, only 1 vinyl alcohol molecule exists in over 10 million molecules. In liquid acetone, about 6 out of 1 billion molecules exist in the enol form.

$$CH_3-\underset{\underset{\displaystyle CH_3}{|}}{\overset{\overset{\displaystyle O}{\|}}{C}} \quad \rightleftarrows \quad CH_3-\underset{\underset{\displaystyle CH_2}{}}{\overset{\overset{\displaystyle OH}{|}}{C}}$$

Keto form   Enol form

There are compounds in which a substantial portion of the molecules exist in the enol form. For example, acetylacetone exists predominantly in the enol form.

$$CH_3-\overset{\overset{\displaystyle O}{\|}}{C}-CH_2-\overset{\overset{\displaystyle O}{\|}}{C}-CH_3 \quad \rightleftarrows \quad CH_3-\overset{\overset{\displaystyle OH}{|}}{C}=CH-\overset{\overset{\displaystyle O}{\|}}{C}-CH_3$$

Keto form   Enol form

Out of every 100 molecules, about 85 are in the enol form, and only about 15 are in the keto form.

Keto-enol shifts are important in carbohydrate chemistry, also. Fructose, which has a ketone functional group, can rearrange to glucose, which has an aldehyde functional group, through tautomeric shifts (Chapter 20).

16.11 THE ALDOL CONDENSATION

Generally, hydrogen attached to carbon has no appreciable acidity, that is, it tends to stay attached where it is. As the phenomenon of keto-enol tautomerism suggests, however, hydrogen atoms on an alpha carbon (i.e., on the carbon nearest the carbonyl group) are slightly acidic. They are sufficiently acidic to be pulled off by a strong base.

$$-\overset{\overset{\displaystyle H}{|}}{\underset{\underset{\displaystyle |}{}}{C}}-C\overset{\overset{\displaystyle O}{\diagup}}{\diagdown} \quad + \quad OH^- \quad \rightleftarrows \quad -\overset{..}{\underset{\underset{\displaystyle |}{}}{C}}-C\overset{\overset{\displaystyle O}{\diagup}}{\diagdown} \quad + H_2O$$

A carbonyl compound   A carbanion

The resulting ion, having a negative charge on the carbon, is called a **carbanion.** It should be emphasized that carbonyl compounds are not acidic enough in water to affect litmus. These compounds are much weaker acids than water itself, as the above equation indicates. Only in the presence of a strong base can a significant number of molecules be converted to anions. Even then the equilibrium lies far in the direction of the un-ionized form.

The carbanions formed by the reaction of base with a carbonyl compound can add to other molecules of the aldehyde or ketone. This type of reaction, in which small molecules combine to form larger ones, is called a **condensation.**

For acetaldehyde, the condensation may be written as

$$2 \ CH_3-C\overset{O}{\underset{H}{<}} \quad \xrightarrow{base} \quad CH_3-\overset{OH}{\underset{H}{\overset{|}{C}}}-CH_2-C\overset{O}{\underset{H}{<}}$$

Acetaldehyde   An aldol

The product, which is both an <u>aldehyde</u> and an <u>alcohol</u>, is called an **aldol.** The reaction is an **aldol condensation.** It can be viewed as the addition of one aldehyde molecule to the double bond of another.

$$-\overset{H}{\underset{|}{\overset{|}{C}}}-C\overset{O}{<} \ + \ \overset{H}{\underset{|}{\overset{|}{C}}}-C\overset{O}{\underset{H}{<}} \longrightarrow -\overset{H}{\underset{|}{\overset{|}{C}}}-\overset{OH}{\underset{|}{\overset{|}{C}}}-\overset{|}{\underset{H}{\overset{|}{C}}}-C\overset{O}{\underset{H}{<}}$$

Processes similar to the aldol condensation are important in the synthesis of many important biomolecules. That is the way, for example, that three-carbon sugars such as glyceraldehyde and dihydroxyacetone (in the form of phosphate esters) can be combined into six-carbon sugars such as fructose.

16.12 SOME COMMON CARBONYL COMPOUNDS

If you were accidentally to drink methanol, your liver would oxidize it to formaldehyde (Chapter 15). The formaldehyde would begin to "fix" your protein. It would, among other things, deactivate enzymes and render the lenses of your eyes opaque. In that way the methanol would cause blindness and—if you drank enough—death.

If you drank ethanol, your liver would oxidize it to acetaldehyde. This aldehyde is less toxic than formaldehyde. Further, acetaldehyde can be oxidized to acetic acid, which can be oxidized to carbon dioxide and water.

The liver oxidizes many compounds. Aldehydes and ketones are often intermediates—or even end products—of such oxidations. Nicotine, for example, is detoxified by oxidation. This occurs by way of an intermediate alcohol. The end product, cotinine, is less toxic than nicotine. With the additional polar group, it is more soluble in water, and thus more readily excreted in the urine.

Many familiar substances contain aldehydes or ketones as the active principles (Figure 16.3). Benzaldehyde is the major component of oil of bitter almond. Cinnamaldehyde is oil of cinnamon. Biacetyl contributes to the aroma and taste of fresh butter. Camphor is a bicyclic ketone. Irone is a ketone with the odor of violets; it is used in many perfumes. Vanillin is the active principle of vanilla flavoring; it is produced synthetically for use in imitation vanilla. Muscone, formed in special glands of the musk deer, is used in perfumes.

Even the odor of green leaves is due in part to carbonyl compounds. Most green leaves contain *cis*-3-hexenal (Figure 16.3). The compound *trans*-2-*cis*-6-nonadienal has a cucumber odor. These and other carbonyl compounds (with related acetals, ketals, and alcohols) impart a "green," herbal odor to shampoos and other cosmetics.

Several of the steroid hormones (Special Topic K) have the carbonyl functional group as an integral part of their structure. Progesterone is a hormone secreted by the ovaries. It stimulates the growth of cells in the wall of the uterus, preparing the uterine wall for attachment of the fertilized egg. Testosterone is the main male sex hormone. These (and other) sex hormones affect our development and our lives in most fundamental ways.

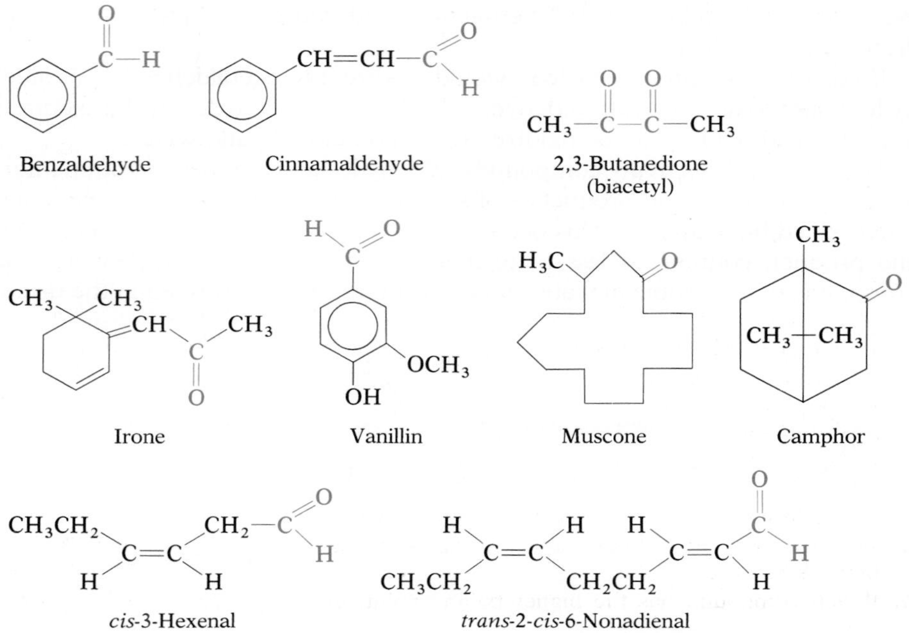

FIGURE 16.3   *Some interesting aldehydes and ketones. Benzaldehyde is an oil found in almonds. Cinnamaldehyde is oil of cinnamon. 2,3-Butanedione is responsible for the flavor of butter, and irone is responsible for the odor of violets. Vanillin, of course, gives vanilla its flavor. Muscone is musk oil, a common ingredient of perfumes. cis-3-Hexenal provides an herbal odor. trans-2-cis-6-Nonadienal gives a cucumber odor.*

PROBLEMS

1. Draw structures and give common and IUPAC names for the four isomeric aldehydes having the formula $C_5H_{10}O$.

2. Draw structures and give common and IUPAC names for the three isomeric ketones having the formula $C_5H_{10}O$.

3. Give suitable names to the following:

   a.

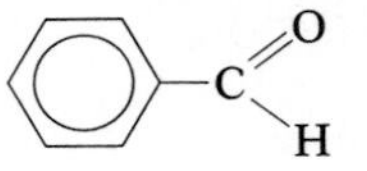

   b. 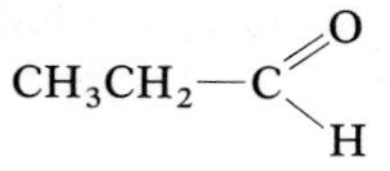 $CH_3CH_2-C\overset{O}{\underset{H}{\lessgtr}}$

   c. $CH_3\overset{CH_3}{\underset{CH_3}{C}}-CH_2C\overset{O}{\underset{H}{\lessgtr}}$

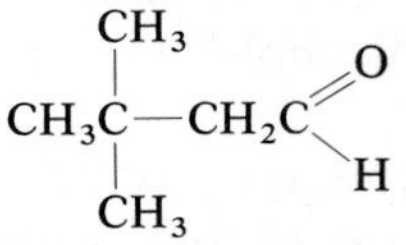

4. Give suitable names to the following:

   a. 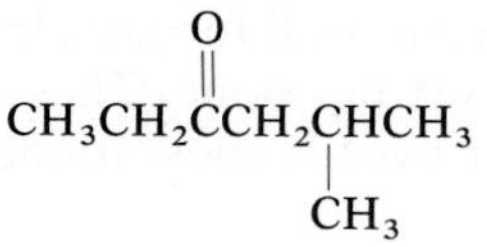 $CH_3CH_2\overset{O}{\overset{\|}{C}}CH_2\underset{CH_3}{CH}CH_3$

   b.

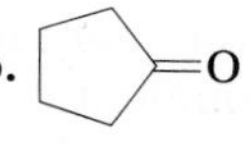

5. Give suitable names to the following:

a. 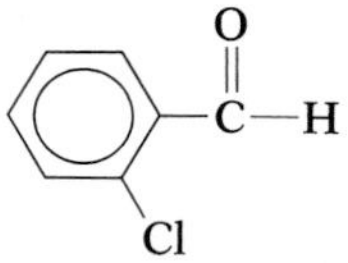 [structure: benzaldehyde with Cl substituent, —C(=O)—H]   **b.** 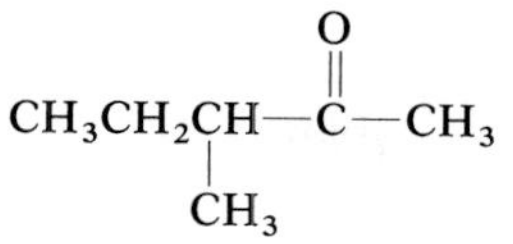 $CH_3CH_2CH-\overset{\displaystyle O}{\overset{\|}{C}}-CH_3$ with CH_3 substituent

6. Give structural formulas for the following:
   a. valeraldehyde   **c.** *p*-nitrobenzaldehyde
   b. 3-methylheptanal

7. Give structural formulas for the following:
   a. 4-methylcyclohexanone
   b. isobutyl *tert*-butyl ketone
   c. 2,4-dimethyl-3-pentanone

8. Give structural formulas for the following:
   a. β-chloropropionaldehyde
   b. α-methylvaleraldehyde
   c. γ-bromobutyraldehyde

9. Which compound has the higher boiling point: acetone or propanol?

10. Which compound has the higher boiling point: butanal or butanol?

11. Which compound has the higher boiling point: dimethyl ether or acetaldehyde?

12. Give the structure of the alcohol from which each of the following aldehydes and ketones can be made by oxidation:

a. 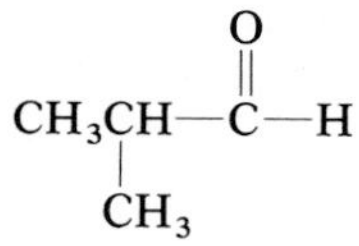 $CH_3CH-\overset{\displaystyle O}{\overset{\|}{C}}-H$ with CH_3 substituent   **d.** [cyclopentanone structure]

b. $CH_3CH_2CH_2\overset{\displaystyle O}{\overset{\|}{C}}CH_3$ **e.** $CH_3CH-\overset{\displaystyle O}{\overset{\|}{C}}-CHCH_3$ with CH_3 and CH_3 substituents

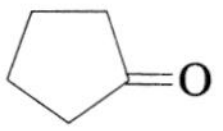

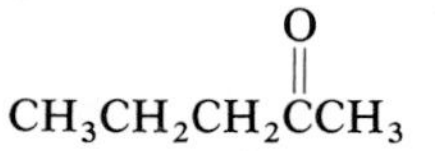

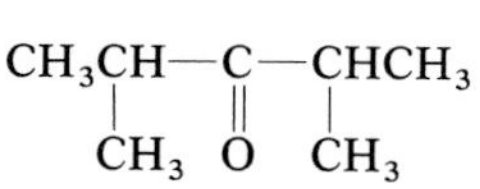

c. $H-\overset{\displaystyle O}{\overset{\|}{C}}-H$

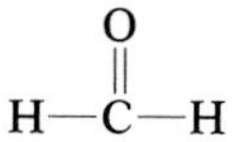

13. Write the equation for the reaction of acetaldehyde with each of the following:
   a. 1 mol of CH_3OH
   b. 2 mol of CH_3OH, with dry HCl present
   c. 1 mol of $HOCH_2CH_2OH$, with dry HCl present
   d. $Ag(NH_3)_2{}^+$

   e. 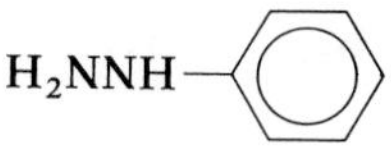 H_2NNH—[benzene ring]

f. $K_2Cr_2O_7$
g. dilute OH^- (aldol condensation)

14. Write the equation for the reaction, if any, of acetone with the reagents in Problem 13.

15. Indicate whether Tollens's reagent could be used to distinguish between the compounds in each set. Explain your reasoning.
   a. 1-pentanol and pentanal
   b. 2-pentanol and 2-pentanone
   c. pentanal and 2-pentanone
   d. pentanal and pentane
   e. 2-pentanone and pentane

16. Assume a *stronger* oxidizing agent, such as $K_2Cr_2O_7$, could be used as a test for distinguishing among compounds. For each set in Problem 15, indicate whether this reagent would distinguish between the two compounds. Explain your reasoning.

17. What reagent would you use to distinguish between 2-pentanone and 2-pentanol? What would be observed when the reagent was added?

18. Draw the carbanion that would be formed by the action of OH^- on acetone.

19. Write the structure for the enol form of propanal.

20. Which of the compounds in Figure 16.3 would give a positive Tollens test? Which would react with phenylhydrazine to give a phenylhydrazone?

21. Name three aldehydes or ketones that serve as active principles in flavors or aromas.

22. List the reagents necessary to carry out the following conversions:

a. 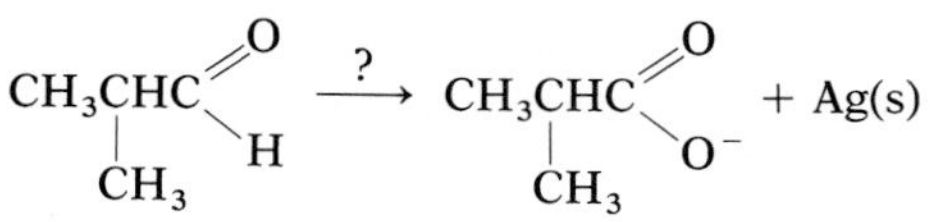 $CH_3CHC\overset{\displaystyle O}{\diagdown_H}$ with CH_3 substituent $\overset{?}{\longrightarrow}$ $CH_3CHC\overset{\displaystyle O}{\diagdown_{O^-}}$ with CH_3 substituent $+ Ag(s)$

b. 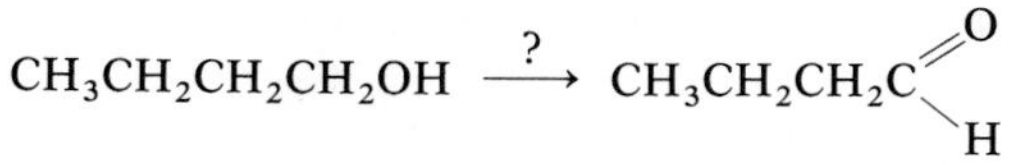 $CH_3CH_2CH_2CH_2OH \overset{?}{\longrightarrow} CH_3CH_2CH_2C\overset{\displaystyle O}{\diagdown_H}$

c. 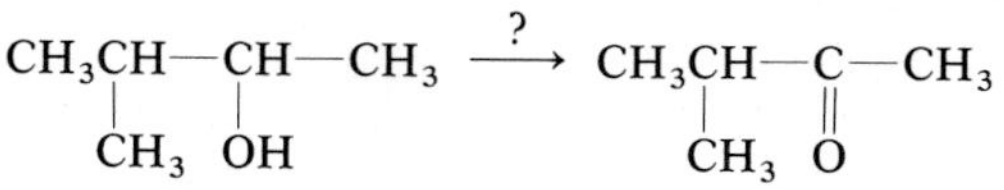 $CH_3CH-CH-CH_3$ with CH_3 and OH substituents $\overset{?}{\longrightarrow}$ $CH_3CH-\overset{\displaystyle O}{\overset{\|}{C}}-CH_3$ with CH_3 substituent

d. 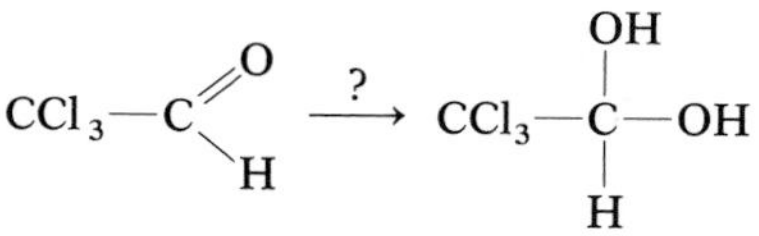 $CCl_3-C\overset{\displaystyle O}{\diagdown_H} \overset{?}{\longrightarrow} CCl_3-\overset{\displaystyle OH}{\underset{\displaystyle H}{C}}-OH$

e. 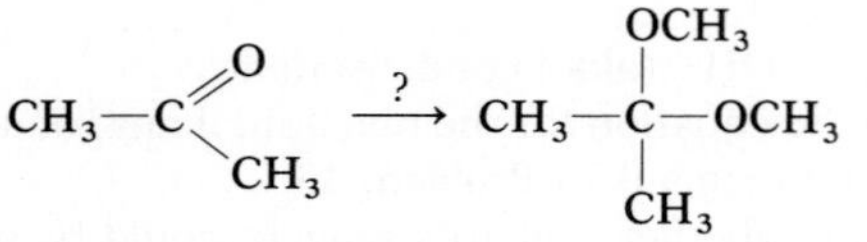 $CH_3-C\overset{O}{\underset{CH_3}{\big<}} \xrightarrow{?} CH_3-\overset{OCH_3}{\underset{CH_3}{\overset{|}{\underset{|}{C}}}}-OCH_3$

f. 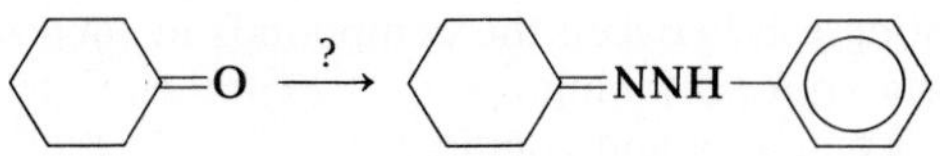 $\xrightarrow{?}$ =NNH—

g. 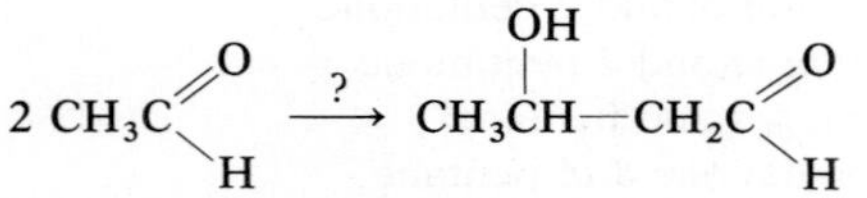 $2\ CH_3C\overset{O}{\underset{H}{\big<}} \xrightarrow{?} CH_3\overset{OH}{\underset{|}{CH}}-CH_2C\overset{O}{\underset{H}{\big<}}$

23. Draw the hemiacetal formed from the intramolecular reaction of

$$HOCH_2CH_2CH_2\overset{H}{\underset{|}{C}}=O$$

24. Name the three functional groups on the vanillin molecule (Figure 16.3).

25. Name the three functional groups on the testosterone molecule.

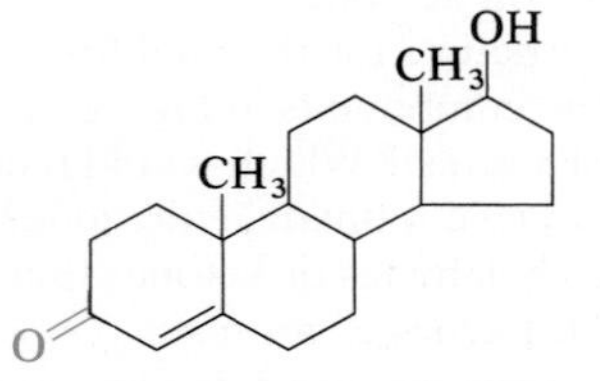

Testosterone

26. Chloral (CCl₃CHO), which forms a stable hydrate, also forms a stable hemiacetal. Give the structure of the hemiacetal formed by the reaction of chloral with methanol.

27. Which of the following are hemiacetals?

a. $CH_3CH_2\overset{}{\underset{OH}{\overset{|}{CH}}}OCH_3$ b. $CH_3CH_2\overset{}{\underset{OCH_3}{\overset{|}{CH}}}OCH_3$

c. $CH_3CH_2\overset{}{\underset{OH}{\overset{|}{CH}}}OH$

d. 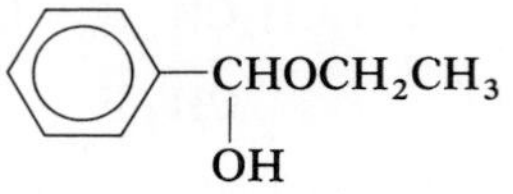 —$\overset{}{\underset{OH}{\overset{|}{CH}}}OCH_2CH_3$

e. 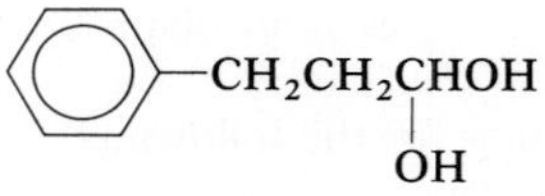 —$CH_2CH_2\overset{}{\underset{OH}{\overset{|}{CH}}}OH$

f. 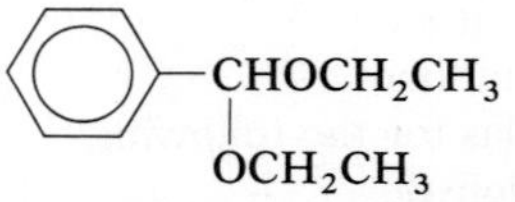 —$\overset{}{\underset{OCH_2CH_3}{\overset{|}{CH}}}OCH_2CH_3$

28. Which of the substances in Problem 27 are acetals?

29. Which of the substances in Problem 27 are hydrates?

30. The Maillard reaction, which occurs in the browning of toasted bread, involves the reaction of the —NH₂ group of an amino acid with the carbonyl group of a sugar. Write the structure of the product formed when the sugar glucose reacts with the amino acid alanine.

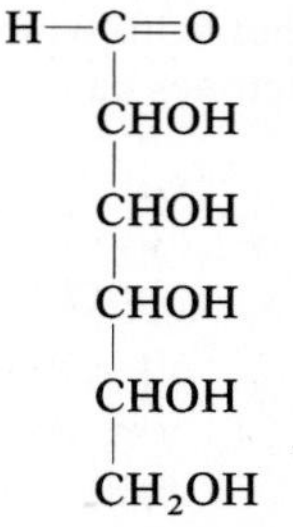

$$\begin{array}{ll} H-C=O & H_2N-CH-COOH \\ CHOH & \qquad\quad| \\ CHOH & \qquad\ CH_3 \\ CHOH & \\ CHOH & \text{Alanine} \\ CH_2OH & \\ \text{Glucose} & \end{array}$$

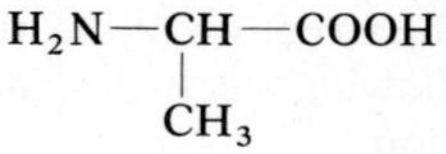

17

Carboxylic Acids and Derivatives

Organic acids were known long before the inorganic acids were isolated. We studied some inorganic acids (HCl and H_2SO_4) first; however, primitive tribes were more familiar with organic acids, such as that obtained when their fermentation reactions went awry and produced not alcohol but vinegar. Naturalists of the seventeenth century knew that the sting of a red ant's bite was due to an organic acid which that pest injected into the wound. And it was long recognized that the crisp, tart flavor of citrus fruits was produced by an organic compound appropriately called citric acid. The acetic acid of vinegar, the formic acid of red ants, and the citric acid of fruits all belong to the same family of compounds, the carboxylic acids.

A number of derivatives of carboxylic acids are also important. The amides, of which proteins (Chapter 22) are perhaps the most spectacular example, and the esters, which include fats (Chapter 21), are two classes of acid derivatives that we shall consider most carefully. Two synthetic fibers (Chapter 19) are also classed within these two families of derivatives. Nylon, like silk and wool, is a polyamide. Dacron is a polyester.

In this chapter, we shall look at simple carboxylic acids and at esters and amides, two kinds of acid derivatives. The more complex worlds of proteins and synthetic polymers we shall save for later chapters.

17.1 ACIDS AND THEIR DERIVATIVES: THE FUNCTIONAL GROUPS

We spoke of the carbonyl group in Chapter 16, and there we noted that it was this functional group that determined the chemistry of the aldehydes and

$$-C\overset{\displaystyle O}{\underset{\displaystyle OH}{\Big\langle}} \quad (or -COOH)$$

The carboxyl group

ketones. The carbonyl group is also incorporated in carboxylic acids and the derivatives of carboxylic acids. However, in these compounds, the carbonyl group is only one part of the functional group that characterizes these families.

The functional group of the **carboxylic acids** is the **carboxyl** group. This group can be considered a combination of the carbonyl group ($C=O$) and the hydroxyl group ($-OH$), but it has characteristic properties of its own.

The **amide** functional group has nitrogen attached to the carbonyl. The properties of the amide functional group are different from those of the simple carbonyl group and those of simple nitrogen-containing compounds, called amines (Chapter 18).

The functional group of the **esters** looks a little like that of an ether and a little like that of a carboxylic acid. As you should now suspect, compounds incorporating this group react neither like acids nor like ethers, but rather like a distinctive family of compounds.

We keep talking about the derivatives of carboxylic acids. All of the families we shall discuss in this chapter, excluding the carboxylic acids themselves, are regarded as derived from the acid. In each case, the hydroxyl group of the acid's functional group is replaced with some other group in the derivative. Table 17.1 gathers all of these functional groups in one location to permit you to compare and contrast the various groups more easily. The table also offers an example (with names) for each type of compound. We shall consider nomenclature in more detail as we take up each of these families separately.

TABLE 17.1 *Carboxylic Acid Derivatives*

Family	Functional Group	Example	Common Name	IUPAC Name
Carboxylic acid	$-\overset{O}{\overset{\|}{C}}-OH$	$CH_3C\overset{O}{\underset{OH}{\diagdown}}$	Acetic acid	Ethanoic acid
Amide	$-\overset{O}{\overset{\|}{C}}-\overset{\|}{N}-$	$CH_3C\overset{O}{\underset{NH_2}{\diagdown}}$	Acetamide	Ethanamide
Ester	$-\overset{O}{\overset{\|}{C}}-O-\overset{\|}{C}-$	$CH_3C\overset{O}{\underset{OCH_3}{\diagdown}}$	Methyl acetate	Methyl ethanoate

17.2 SOME COMMON ACIDS: STRUCTURES AND NAMES

Most of the organic acids that we will consider are derived from natural sources. These acids have common names, often associated with the natural source, that are widely used. These common names, as we mention in Chapter 16, have been adapted for use with other families of organic compounds, such as the aldehydes. The names of the acid derivatives are also based on the common names of the acids. In many ways, the common nomenclature of carboxylic acids will serve the same basic function as the IUPAC rules for naming alkanes. Once you've learned the common names for the acids, the naming of derivatives will involve only slight modifications of the original acid names.

The simplest organic acid is formic acid (from the Latin *formica*, "ant"). It was first obtained by the destructive distillation of ants. The structure of formic acid is

$$H-C{\overset{\displaystyle O}{\underset{\displaystyle OH}{}}} \qquad or \qquad HCOOH$$

Acetic acid is the two-carbon acid. It can be made by the **aerobic** ("with air") fermentation of a mixture of cider and honey. This reaction produces a solution of vinegar that contains about 4% to 10% acetic acid plus a number of other compounds that give the vinegar flavor. The structure of acetic acid is

$$H-\overset{\displaystyle H}{\underset{\displaystyle H}{C}}-C{\overset{\displaystyle O}{\underset{\displaystyle OH}{}}} \qquad or \qquad CH_3COOH$$

It is probably the most familiar *weak* acid used in educational and industrial chemistry laboratories.

Table 17.2 lists several members of the family of carboxylic acids and the derivation of their common names.

Although IUPAC names are seldom encountered in everyday use, they are readily derived from the names of alkanes with the same number of carbon atoms. Just drop the **-e** from the alkane name and add **-oic acid.** Thus, the IUPAC name for formic acid ($HCOOH$) is methanoic acid, and that for acetic acid (CH_3COOH) is ethanoic acid. The names of several common aliphatic acids are shown in Table 17.2. Nature prefers an even number of carbon atoms in its acids. Therefore, beyond six carbon atoms, we have listed only the commonly isolated, even-numbered acids.

If substituents are attached to the parent chain of the acid, the rules outlined in Section 16.2 are applied. Greek letters are used with common names; numbers are used with IUPAC names. The carbon of the carboxyl group is always considered the first carbon.

TABLE 17.2 *Some Common Aliphatic Acids*

Condensed Formula	IUPAC Name	Common Name	Derivation of Common Name
$HCOOH$	Methanoic acid	Formic acid	Latin *formica*, "ant"
CH_3COOH	Ethanoic acid	Acetic acid	Latin *acetum*, "vinegar"
CH_3CH_2COOH	Propanoic acid	Propionic acid	Greek *protos*, "first," and *pion*, "fat"
$CH_3CH_2CH\ COOH$	Butanoic acid	Butyric acid	Latin *butyrum*, "butter"
$CH_3(CH_2)_3COOH$	Pentanoic acid	Valeric acid	Latin *valere*, "powerful"
$CH_3(CH_2)_4COOH$	Hexanoic acid	Caproic acid ⎱	
$CH_3(CH_2)_6COOH$	Octanoic acid	Caprylic acid ⎬	Latin *caper*, "goat"
$CH_3(CH_2)_8COOH$	Decanoic acid	Capric acid ⎰	
$CH_3(CH_2)_{10}COOH$	Dodecanoic acid	Lauric acid	Laurel tree
$CH_3(CH_2)_{12}COOH$	Tetradecanoic acid	Myristic acid	*Myristica fragrans* (nutmeg)
$CH_3(CH_2)_{14}COOH$	Hexadecanoic acid	Palmitic acid	Palm tree
$CH_3(CH_2)_{16}COOH$	Octadecanoic acid	Stearic acid	Greek *stear*, "tallow"

example 17.1

Give the common and IUPAC names for

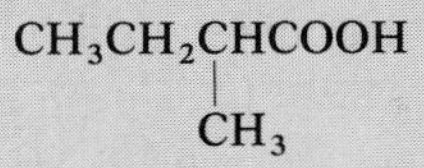

$$CH_3CH_2\underset{\underset{\displaystyle CH_3}{|}}{C}HCOOH$$

The longest continuous chain contains four carbon atoms; the compound is therefore named as a substituted butyric (or butanoic) acid. The methyl substitutent is at the alpha carbon (in the common system), or at the second carbon (in the IUPAC system). The compound is α-methylbutyric acid, or 2-methylbutanoic acid.

example 17.2

Draw α, β-dichloropropionic acid.
   Propionic acid contains three carbons

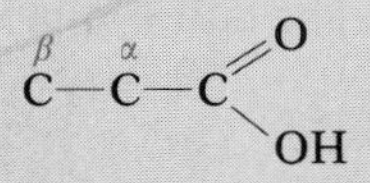

Two chlorine atoms must be attached to the parent chain, one at the alpha carbon and one at the beta carbon.

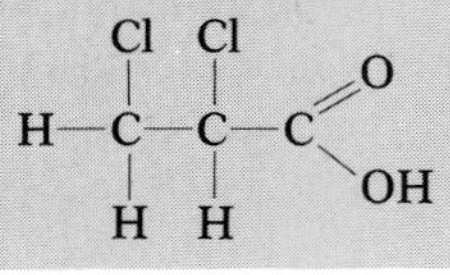

TABLE 17.3 *Some Common Dicarboxylic Acids*

Structure	IUPAC Name	Common Name	Derivation of Common Name
HOOC—COOH	Ethanedioic acid	Oxalic acid	Greek *oxys,* "sharp" or "acid"
HOOC—CH_2—COOH	Propanedioic acid	Malonic acid	From malic acid (Latin *malum,* "apple")
HOOC$(CH_2)_2$COOH	Butanedioic acid	Succinic acid	Latin *succinum,* "amber"
HOOC$(CH_2)_3$COOH	Pentanedioic acid	Glutaric acid	From glutamic acid (Latin *gluten,* "glue")
HOOC$(CH_2)_4$COOH	Hexanedioic acid	Adipic acid	Latin *adeps,* "fat"
	1,2-Benzenedi- carboxylic acid	Phthalic acid	From na*phthal*ene
	1,4-Benzenedi- carboxylic acid	Terephthalic acid	From na*phthal*ene

The simplest aromatic acid is called benzoic acid (margin). Others are called by common names or may be named as derivatives of benzoic acid.

A number of dicarboxylic acids, both aliphatic and aromatic, are worthy of mention. Seven of these are given, with names, in Table 17.3.

Benzoic acid

17.3 CARBOXYLIC ACIDS: PREPARATION AND PHYSICAL PROPERTIES

Few carboxylic acids occur free in nature. Many aliphatic acids, particularly those with an even number of carbon atoms, are found combined with glycerol in fats (Chapter 21). Some of the acids are thus available from the hydrolysis of fats.

A general method for the preparation of carboxylic acids is by the oxidation of primary alcohols (Chapter 15). For example, propyl alcohol can be oxidized, by an acidic solution of potassium dichromate, to propionic acid.

$$CH_3CH_2CH_2OH \xrightarrow{K_2Cr_2O_7,\ H^+} CH_3CH_2-C\overset{\displaystyle O}{\underset{\displaystyle OH}{\Big\langle}}$$

In a similar manner, 1,4-butanediol is oxidized to succinic acid.

$$HO-CH_2CH_2CH_2CH_2-OH \xrightarrow{\;K_2Cr_2O_7,\ H^+\;} HO-\overset{\displaystyle O}{\overset{\|}{C}}-CH_2CH_2-\overset{\displaystyle O}{\overset{\|}{C}}-OH$$

Carboxylic acids are highly polar and exhibit strong intermolecular hydrogen bonding. Consequently, these compounds have higher boiling points than even the alcohols of comparable molecular weights. Ethyl alcohol (with a molecular weight of 46) boils at 78 °C, while formic acid (with the same molecular weight) boils at 100 °C. Similarly, propyl alcohol (with a molecular weight of 60) boils at 97 °C, while acetic acid (with the same molecular weight) boils at 118 °C.

There is good evidence that, even in the vapor phase, some of the hydrogen bonds between acid molecules are not broken. The particular structure of the carboxyl group permits two molecules to hydrogen bond very strongly to one another.

In many situations, the interaction is so strong that the **dimer** (the two-molecule unit) acts as a single particle. Osmotic pressures, freezing-point depressions, and other properties are frequently less than one would expect when carboxylic acids are the solute. That is because we are counting as two separate molecules a combination that really is behaving as one piece.

The carboxyl groups readily hydrogen bond to water molecules. The acids having one to four carbon atoms are colorless liquids that are completely miscible with water. Solubility decreases with increasing number of carbon atoms. Hexanoic acid ($C_6H_{12}O_2$) is soluble only to the extent of 1.0 g per 100 g of water. Palmitic acid ($C_{16}H_{32}O_2$) is essentially insoluble.

Most of the carboxylic acids have irritating, obnoxious odors. Generally, the odors get more unpleasant as one goes up the homologous series, depending, of course, on how one defines unpleasant. Formic acid (CH_2O_2) has a very sharp and penetrating odor. Valeric acid ($C_5H_{10}O_2$) is not quite so aggressive, but it is nonetheless quite bad. It is most gently described as clinging or persistent and far more descriptively pinpointed as "essence of old gym sneakers." This trend in odors is alleviated somewhat by the decreasing tendency of the acids to vaporize as molecular weight increases. For acids with 12 or more carbon atoms, odors become progressively weaker.

Pure acetic acid freezes at 16.6 °C. Since this is only slightly below normal room temperature (about 20 °C), acetic acid solidifies when cooled only slightly. In the poorly heated laboratories of a century or so ago in northern North America and Europe, acetic acid often froze on the reagent shelf. For that reason, pure acetic acid (sometimes referred to as concentrated acetic acid) came to be known as **glacial acetic acid,** a name that survives to this day.

17.4 CHEMICAL PROPERTIES OF CARBOXYLIC ACIDS

In Chapter 10, we define an acid as a compound that (1) turns blue litmus red, (2) neutralizes bases, (3) reacts with active metals to give off hydrogen, and (4) tastes sour. Of the four properties of acids listed, you are probably most familiar with the last—sour taste. Vinegar is sour because it contains acetic acid. Grapefruits and lemons are sour because they contain citric acid. Sour milk contains lactic acid. The acids we eat are, for the most part, organic acids. The strong acids that we encounter in Chapter 10, such as hydrochloric acid (HCl), nitric acid (HNO_3), and sulfuric acid (H_2SO_4), are made from minerals and hence are called **mineral acids.** Historically, the first organic acids came from plant or animal matter, that is, from organisms. Now many organic acids are synthesized in the laboratory from petroleum products.

Those carboxylic acids that are water soluble form moderately acidic solutions. They will change litmus from blue to red. Carboxylic acids, whether water soluble or not, will react with aqueous solutions of sodium hydroxide, sodium carbonate, and sodium bicarbonate to form salts.

$$RCOOH + NaOH(aq) \longrightarrow RCOO^-Na^+(aq) + H_2O$$
$$2\,RCOOH + Na_2CO_3(aq) \longrightarrow 2\,RCOO^-Na^+(aq) + H_2O + CO_2$$
$$RCOOH + NaHCO_3(aq) \longrightarrow RCOO^-Na^+(aq) + H_2O + CO_2$$

In these reactions, the carboxylic acids act as typical acids; they neutralize basic compounds. With solutions of carbonate and bicarbonate ions they also form carbon dioxide gas.

The carboxylic acids are weak acids. They tend to ionize only slightly in aqueous solution.

$$RCOOH + H_2O \rightleftharpoons RCOO^- + H_3O^+$$

Acetic acid is one of the weak acids listed in Chapter 10. We need to distinguish between degrees of weakness. We can order some organic compounds (and some inorganic ones) according to their relative acidities.

Strongest acid						Weakest acid

$$H_2SO_4,\ HNO_3,\ HCl > RCOOH > H_2CO_3 > ArOH > H_2O > ROH > RH$$

Mineral acids	Carboxylic acids	Carbonic acid	Phenols	Water	Alcohols	Alkanes

Water is frequently used as a dividing line for acidity and basicity. Because we live in a water world, we quite naturally regard water as neutral. In aqueous solutions, if something is more acidic than water, it is treated as an acid (as are phenols and carboxylic acids). If a compound is less acidic than water (as are alcohols and alkanes), it is not regarded as an acid. Neither alcohols nor alkanes affect the pH of aqueous solutions.

Because of the difference in relative acidities among organic compounds, solubility behavior is often an identifying feature of carboxylic acids. Carboxylic acids that are insoluble in water dissolve in aqueous hydroxide, carbonate, or bicarbonate because the insoluble acids react to form ionic salts that are water soluble. The behavior of decanoic acid is illustrated in Figure 17.1. Solution in aqueous bicarbonate, with the formation of carbon dioxide bubbles, is characteristic of carboxylic acids.

The relationship between the names of acids and their corresponding anions is discussed in Section 10.3. The naming of salts in general was first encountered in Section 4.6. You have been working with the salt of at least one organic acid for quite a while—sodium acetate or, more generally, the acetate ion. To name

$$CH_3-C \underset{O^-Na^+}{\overset{O}{\diagup}}$$

Sodium acetate
(sodium ethanoate)

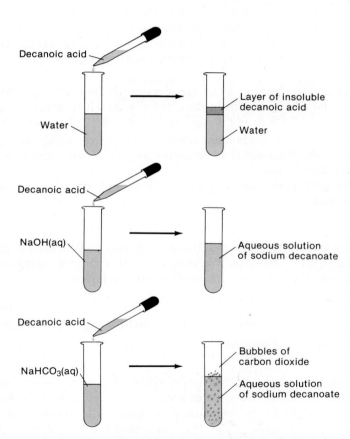

FIGURE 17.1   *Decanoic acid is insoluble in water but soluble in aqueous sodium hydroxide or bicarbonate.*

the salts of carboxylic acids, simply name the cation first and then name the anion by changing the **-ic** ending of the acid to **-ate**.

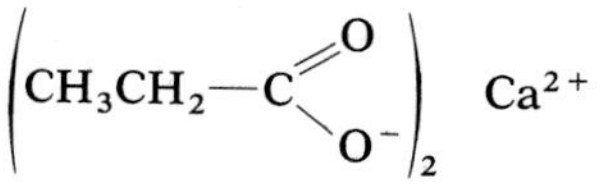

$$\left(CH_3CH_2-C\!\!\begin{array}{c}{}^{O}\\{}_{O^-}\end{array} \right)_2 Ca^{2+}$$

Calcium propionate
(Calcium propanoate)

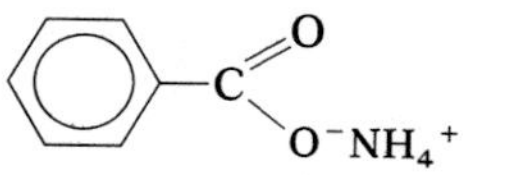

17.5 AN ESTER BY ANY OTHER NAME . . .

Esters generally have pleasant odors and are often responsible for the characteristic fragrances of fruits and flowers. Once a flower or fruit has been chemically analyzed, flavor chemists can attempt to duplicate the natural odor or taste. They are seldom completely successful, but they often get close enough for practical purposes. Artificial fruit flavors are often mixtures of esters. Several esters of interest are shown in Figure 17.2.

Ammonium benzoate

Esters are easy to name—if you follow the right procedure. First imagine the ester to have been made from an acid and an alcohol through the loss of water.

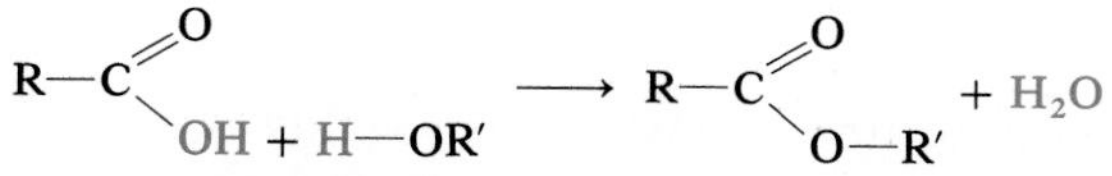

$$R-C\!\!\begin{array}{c}{}^{O}\\{}_{OH}\end{array} + H-OR' \longrightarrow R-C\!\!\begin{array}{c}{}^{O}\\{}_{O-R'}\end{array} + H_2O$$

Then name the alkyl group from the alcohol, designated R′ in the formula. (Notice that the alkyl group from the alcohol is attached directly to an oxygen by a single bond.) Finally, name the part derived from the acid,

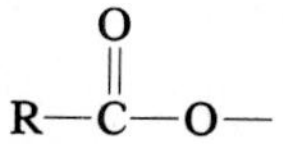

$$R-\overset{\overset{\textstyle O}{\|}}{C}-O-$$

as if it were an anion (Section 17.4).

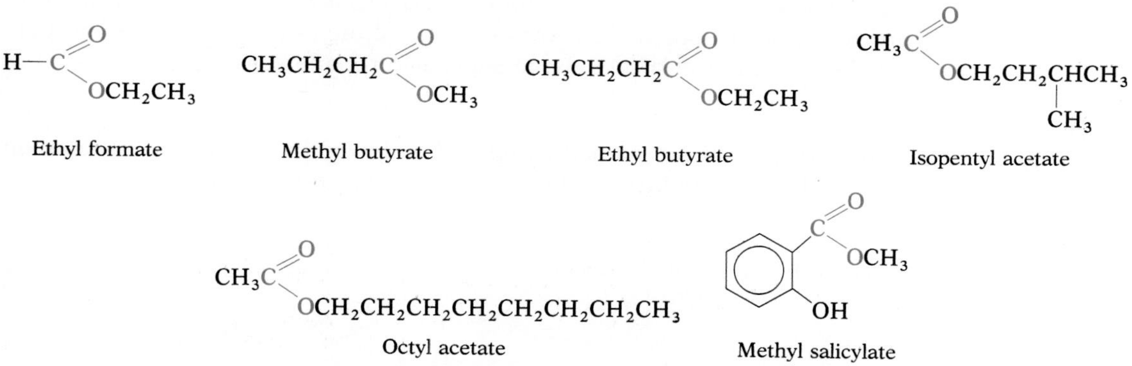

Ethyl formate   Methyl butyrate   Ethyl butyrate   Isopentyl acetate

Octyl acetate   Methyl salicylate

FIGURE 17.2 *Some esters of interest. Ethyl formate is an artificial rum flavor. Methyl butyrate occurs in apples, and ethyl butyrate occurs in pineapples. Isopentyl acetate is banana oil, used as a solvent and in flavoring. Octyl acetate occurs in oranges. Methyl salicylate is oil of wintergreen, used as a rub for sore muscles.*

example 17.3

What are the common and IUPAC names for this ester?

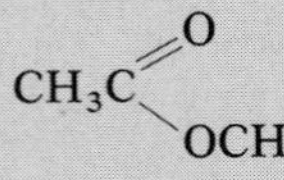

$$CH_3C\overset{O}{\underset{OCH_3}{\diagup}}$$

The alkyl group attached to oxygen is methyl. The

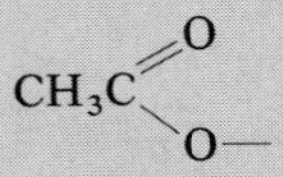

$$CH_3C\overset{O}{\underset{O-}{\diagup}}$$

is derived from acetic acid (which has two carbons). Its name is acetate. The compound is methyl acetate. The IUPAC name for the two-carbon acid unit is ethanoate. The IUPAC name for the ester is methyl ethanoate.

example 17.4

What are the common and IUPAC names for this ester?

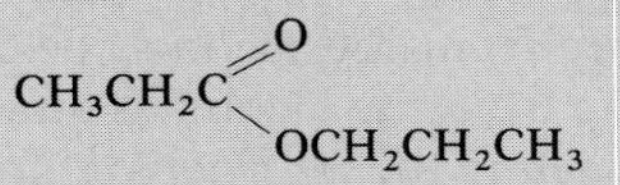

$$CH_3CH_2C\overset{O}{\underset{OCH_2CH_2CH_3}{\diagup}}$$

The alkyl group (attached directly to oxygen) is propyl. The part of the molecule derived from the acid,

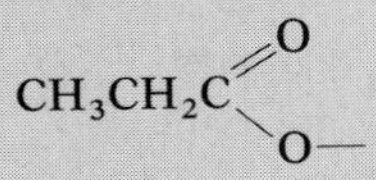

$$CH_3CH_2C\overset{O}{\underset{O-}{\diagup}}$$

has three carbon atoms. It is thus called propionate, or, by IUPAC terminology, propanoate. The common name of the ester is, therefore, propyl propionate, and the IUPAC name is propyl propanoate.

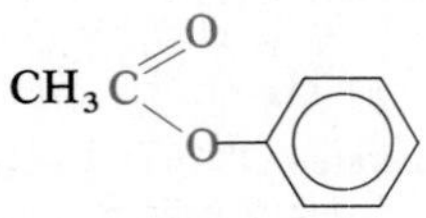

$$CH_3C\overset{O}{\underset{O}{\diagup}}\!-\!\!\bigcirc$$

Phenyl acetate

Esters of phenol are named as *phenyl*, followed by the name of the anion of the acid. The structure of phenyl acetate is given in the margin.

example 17.5

What is the name of this ester?

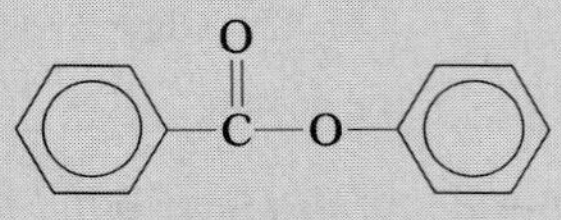

$$\bigcirc\!-\!\overset{O}{\underset{}{\overset{\|}{C}}}\!-\!O\!-\!\bigcirc$$

The group attached by a single bond to oxygen is phenyl. The acid portion corresponds to the benzoate group (from benzoic acid). Therefore, the compound is phenyl benzoate.

Draw the structure for ethyl valerate.

It is easier to start with the acid portion. Draw the valerate (five-carbon) group first.

example
17.6

$$CH_3CH_2CH_2CH_2C{\overset{\displaystyle O}{\underset{\displaystyle O-}{\diagdown}}}$$

Then, simply attach the ethyl group to the bond that ordinarily holds the hydrogen in the free acid.

$$CH_3CH_2CH_2CH_2C{\overset{\displaystyle O}{\underset{\displaystyle O-CH_2CH_3}{\diagdown}}}$$

17.6 PHYSICAL PROPERTIES OF THE ESTERS

The molecules of an ester are polar but are incapable of intermolecular hydrogen bonding with one another. Esters thus have considerably lower boiling points than the isomeric carboxylic acids. As one might expect, the boiling points of esters are about intermediate between those of ketones and ethers of comparable molecular weight.

Ester molecules are capable of hydrogen bonding to water molecules, rendering esters of low molecular weight somewhat water soluble. Borderline solubility occurs in those molecules having from three to five carbon atoms.

Esters dissolve many organic substances. They are often employed as solvents. Cellulose nitrate is dissolved in ethyl acetate and butyl acetate to form lacquers. The solvent evaporates as the lacquer "dries," leaving a thin protective film on the surface to which the lacquer was applied. Esters with high boiling points are used as softeners (plasticizers) for brittle plastics (Chapter 19).

$$CH_3C{\overset{\displaystyle O}{\underset{\displaystyle OCH_2CH_3}{\diagdown}}}$$

Ethyl acetate ($C_4H_8O_2$)
(boiling point, 77 °C)

$$CH_3CH_2CH_2C{\overset{\displaystyle O}{\underset{\displaystyle OH}{\diagdown}}}$$

Butyric acid ($C_4H_8O_2$)
(boiling point, 164 °C)

17.7 PREPARATION OF ESTERS

Some esters are prepared by direct esterification of the carboxylic acid. This is accomplished through the heating of a carboxylic acid with an alcohol in the presence of a mineral acid catalyst.

$$R-C{\overset{\displaystyle O}{\underset{\displaystyle OH}{\diagdown}}} + R'-OH \underset{}{\overset{H^+}{\rightleftharpoons}} R-C{\overset{\displaystyle O}{\underset{\displaystyle OR'}{\diagdown}}} + H_2O$$

An alcohol molecule condenses with an acid molecule, splitting out water to form an ester. The reaction is reversible, and it soon comes to equilibrium. To get a good yield of ester, it is necessary to apply a stress to the system, forcing the reaction to the right (a technique that was discussed in Chapter 5).

If the reaction involves an inexpensive alcohol, such as methanol (CH_3OH), excess alcohol can be used. This will drive the reaction toward completion. In the preparation of methyl benzoate, for example, 10 mol of CH_3OH may be used for each mole of benzoic acid.

$$\text{C}_6\text{H}_5-\underset{\text{OH}}{\overset{\text{O}}{\text{C}}} + CH_3OH \overset{H^+}{\rightleftharpoons} \text{C}_6\text{H}_5-\underset{\text{OCH}_3}{\overset{\text{O}}{\text{C}}} + H_2O$$

(Large excess)

Forcing of the reaction in this way can give a 75% yield of methyl benzoate.

Similarly, if the acid is cheap, it can be employed in excess. In the preparation of butyl acetate, acetic acid is used in a molar ratio of $2:1$ (or greater).

$$CH_3\underset{\text{OH}}{\overset{\text{O}}{\text{C}}} + CH_3CH_2CH_2CH_2OH \overset{H^+}{\rightleftharpoons} CH_3\underset{\text{OCH}_2\text{CH}_2\text{CH}_2\text{CH}_3}{\overset{\text{O}}{\text{C}}} + H_2O$$

(Excess)

A third method of driving the reaction toward completion involves removal of the product water as it is formed. This is easily accomplished if the acid, the alcohol, and the ester all boil at temperatures well above 100 °C.

$$CH_3CH_2CH_2\underset{\text{OH}}{\overset{\text{O}}{\text{C}}} + CH_3CH_2CH_2CH_2OH \overset{H^+}{\rightleftharpoons}$$

Butyric acid (boiling point, 164 °C)   Butyl alcohol (boiling point, 118 °C)

$$CH_3CH_2CH_2\underset{\text{OCH}_2\text{CH}_2\text{CH}_2\text{CH}_3}{\overset{\text{O}}{\text{C}}} + H_2O$$

Butyl butyrate (boiling point, 165 °C)   Distilled from mixture (boiling point, 100 °C)

Because of problems with equilibrium systems, esters are often prepared by way of derivatives called acyl chlorides. Thionyl chloride ($SOCl_2$) replaces the —OH group of a carboxylic acid with a —Cl.

$$R-\underset{\text{OH}}{\overset{\text{O}}{\text{C}}} + SOCl_2 \longrightarrow R-\underset{\text{Cl}}{\overset{\text{O}}{\text{C}}} + SO_2 + HCl$$

An acid   Thionyl chloride   An acyl chloride

The acyl chloride reacts rapidly and quantitatively with alcohols to form esters.

$$\underset{\substack{\\ }}{R-\overset{\displaystyle O}{\overset{\|}{C}}-Cl} + H-OR' \longrightarrow R-\overset{\displaystyle O}{\overset{\|}{C}}-OR' + HCl$$

As a specific example, butyryl chloride (from butyric acid) reacts with ethanol to form ethyl butyrate.

$$CH_3CH_2CH_2\overset{\displaystyle O}{\overset{\|}{C}}-Cl + HOCH_2CH_3 \longrightarrow CH_3CH_2CH_2\overset{\displaystyle O}{\overset{\|}{C}}OCH_2CH_3 + HCl$$

Similarly, benzoyl chloride (from benzoic acid) reacts with isobutyl alcohol to form isobutyl benzoate.

$$\underset{}{\bigcirc}-\overset{\displaystyle O}{\overset{\|}{C}}-Cl + HOCH_2\underset{\underset{CH_3}{|}}{CH}CH_3 \longrightarrow \underset{}{\bigcirc}-\overset{\displaystyle O}{\overset{\|}{C}}OCH_2\underset{\underset{CH_3}{|}}{CH}CH_3 + HCl$$

Esters of acetic acid usually are prepared by use of acetic anhydride.* For example, acetic anhydride reacts with isopentyl alcohol to give isopentyl acetate:

$$CH_3\overset{\displaystyle O}{\overset{\|}{C}}-O-\overset{\displaystyle O}{\overset{\|}{C}}CH_3 + CH_3\underset{\underset{CH_3}{|}}{CH}CH_2CH_2OH \longrightarrow$$

$$CH_3\overset{\displaystyle O}{\overset{\|}{C}}-OCH_2CH_2\underset{\underset{CH_3}{|}}{CH}CH_3 + CH_3C\overset{\displaystyle O}{\underset{OH}{\diagup}}$$

Isopentyl acetate

Similarly, octyl acetate can be made from octyl alcohol.

$$CH_3\overset{\displaystyle O}{\overset{\|}{C}}-O-\overset{\displaystyle O}{\overset{\|}{C}}CH_3 + CH_3CH_2CH_2CH_2CH_2CH_2CH_2CH_2OH \longrightarrow$$

$$CH_3C\overset{\displaystyle O}{\underset{OCH_2CH_2CH_2CH_2CH_2CH_2CH_2CH_3}{\diagup}} + CH_3C\overset{\displaystyle O}{\underset{OH}{\diagup}}$$

---

* Acetic anhydride can be considered as derived from acetic acid by the removal of one molecule of water from two molecules of acid.

$$CH_3\overset{\displaystyle O}{\overset{\|}{C}}\boxed{OH + H}O-\overset{\displaystyle O}{\overset{\|}{C}}CH_3 \longrightarrow CH_3\overset{\displaystyle O}{\overset{\|}{C}}-O-\overset{\displaystyle O}{\overset{\|}{C}}CH_3 + H_2O$$

17.8 CHEMICAL PROPERTIES OF ESTERS: HYDROLYSIS

Esters are neutral compounds, unlike the acids from which they are formed. As neutral compounds they exhibit neither acidic nor basic properties. Esters typically undergo chemical reactions in which the alkoxy (—OR′) group is replaced by another group. One such reaction is hydrolysis, or splitting with water. Hydrolysis is catalyzed by either acid or base. Acidic hydrolysis is simply the reverse of the esterification reaction (p. 497).

$$R-C\overset{O}{\underset{OR'}{}} + H_2O \overset{H^+}{\rightleftharpoons} R-C\overset{O}{\underset{OH}{}} + R'OH$$

As in esterification, the reaction comes to equilibrium, and the position of the equilibrium can be shifted by addition or removal of one or another of the species involved.

Basic or alkaline hydrolysis, on the other hand, goes to completion.

$$R-C\overset{O}{\underset{OR'}{}} + NaOH(aq) \longrightarrow R-C\overset{O}{\underset{O^-Na^+}{}} (aq) + R'OH$$

The free acid is not obtained in this reaction. In a basic solution, the salt of the acid is always isolated. Treatment of methyl benzoate with aqueous sodium hydroxide gives sodium benzoate and methyl alcohol.

$$\text{C}_6\text{H}_5-C\overset{O}{\underset{OCH_3}{}} + NaOH(aq) \longrightarrow \text{C}_6\text{H}_5-C\overset{O}{\underset{O^-Na^+}{}} (aq) + CH_3OH$$

Alkaline hydrolysis of fats and oils (esters of glycerol with long-chain carboxylic acids) is called saponification (from the Latin *sapon*, "soap"). The sodium salts of such fatty acids are soaps (Chapter 21).

The amide group

A simple amide

A substituted amide

17.9 AMIDES: STRUCTURES AND NAMES

In the amide functional group, a nitrogen is attached to a carbonyl group. If the two remaining bonds to nitrogen are attached to hydrogen atoms, the compound is called a simple amide. If one or both of the two remaining bonds to nitrogen are attached to alkyl or aryl groups, the compound is called a substituted amide.

Names for simple amides are formed by dropping the ending **-ic** (or **-oic**) from the name of the acid and adding the suffix **-amide.**

Name the compound

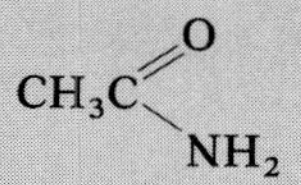

$$CH_3C \underset{NH_2}{\overset{O}{<}}$$

example
17.7

This amide is derived from acetic acid. Drop the -ic suffix, attach the ending -amide, and you have the name—acetamide (or ethanamide in the IUPAC system).

Name the compound

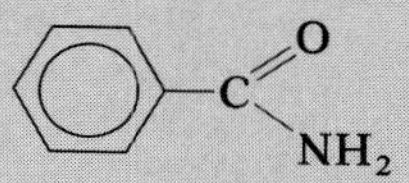

example
17.8

This amide is derived from benzoic acid. Drop the -oic, add -amide, and you have it—benzamide.

$$CH_3CH_2CH_2C \underset{NHCH_2CH_3}{\overset{O}{<}}$$

N-Ethylbutyramide

$$H-C \underset{\underset{CH_3}{|}}{\overset{O}{<}} N-CH_3$$

N,N-Dimethylformamide

In substituted amides, alkyl groups attached to the nitrogen atom are named as substituents. Instead of using a Greek letter or a number to specify location, chemists indicate the group's attachment to nitrogen by a capital letter *N*. If the substituent on nitrogen is phenyl, the compound is named as an anilide. The **-ic** or **-oic** ending of the acid name is replaced with **-anilide** instead of **-amide.**

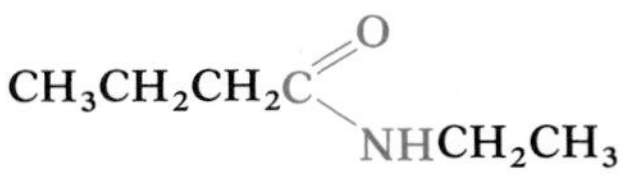

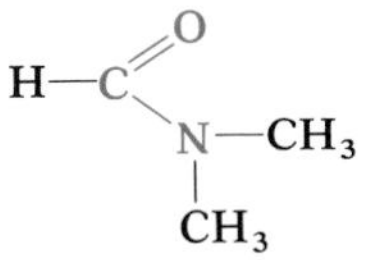

$$CH_3C \underset{NH-}{\overset{O}{<}}$$

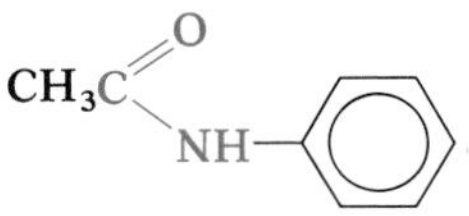

Acetanilide

Name the compound

$$CH_3C\underset{\underset{O}{||}}{-}NH\underset{\underset{CH_3}{|}}{C}HCH_3$$

example
17.9

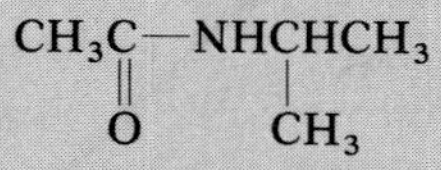

The acid portion of the molecule (the portion incorporating the carbonyl group and to one side of the nitrogen) contains two carbon atoms and is derived from acetic acid. The compound will, therefore, be named as a substituted acetamide. The substituent attached directly to the nitrogen is an isopropyl group. The name of the compound is N-isopropylacetamide. (The IUPAC name is N-isopropylethanamide.)

example
17.10

Draw butyranilide.

The compound is a substituted four-carbon amide (derived from butyric acid).

$$CH_3CH_2CH_2C \overset{O}{\underset{N-}{\diagdown}}$$

The suffix -anilide indicates that there is a phenyl group substituted for one of the two hydrogens on the nitrogen of the simple amide.

$$CH_3CH_2CH_2C \overset{O}{\underset{NH-\bigcirc}{\diagdown}}$$

17.10 PHYSICAL PROPERTIES OF THE AMIDES

With the exception of the simplest amide (formamide, whose melting point is 3 °C), the amides of the type

$$R-C \overset{O}{\underset{NH_2}{\diagdown}}$$

are solids at room temperature. These amides have both relatively high melting points and high boiling points because of strong intermolecular hydrogen bonding.

$$R-C \overset{O}{\underset{\underset{H}{N-H}}{\diagdown}} \cdots \cdots O{=}C \overset{R}{\underset{\underset{H}{N-H}}{\diagup}} \cdots \cdots$$

Similar hydrogen bonding plays a critical role in determining the structure and properties of proteins, DNA, RNA, and other giant molecules so important to life processes. Amides are quite soluble in water, with borderline solubility occurring in those having five or six carbon atoms.

17.11 SYNTHESIS OF AMIDES

Amides are usually prepared from acids by a two-step process. First the acid is treated with thionyl chloride, which converts it to an acyl chloride (Section 17.7). The acyl chloride is treated with ammonia, forming the amide.

$$R-C \overset{O}{\underset{Cl}{<}} + 2\ NH_3 \longrightarrow R-C \overset{O}{\underset{NH_2}{<}} + NH_4{}^+Cl^-$$

To make benzamide, for example, one allows benzoyl chloride to react with ammonia.

$$\bigcirc\!\!-C\overset{O}{\underset{Cl}{<}} + 2\ NH_3 \longrightarrow \bigcirc\!\!-C\overset{O}{\underset{NH_2}{<}} + NH_4{}^+Cl^-$$

The reaction of the acyl chloride with ammonia produces both an amide molecule and a molecule of hydrochloric acid, which immediately reacts with more ammonia to form the salt ammonium chloride (NH_4Cl).

Similarly, nicotinamide can be made from nicotinic acid.

Nicotinic acid		Nicotinamide
(niacin)		(niacinamide)

Nicotinic acid and nicotinamide are essential B vitamins having antipellagra activity (Special Topic J).

17.12 CHEMICAL PROPERTIES OF THE AMIDES: HYDROLYSIS

Generally, the amides are neutral compounds, showing neither appreciable acidity nor significant basicity in water. Further, the amides resist hydrolysis in plain water, even upon prolonged heating. In the presence of added acid or base, however, hydrolysis proceeds at a moderate rate. Acidic hydrolysis of a simple amide gives a carboxylic acid and an ammonium salt.

$$CH_3CH_2C\overset{O}{\underset{NH_2}{<}} + HCl(aq) + H_2O \longrightarrow CH_3CH_2C\overset{O}{\underset{OH}{<}} + NH_4Cl(aq)$$

Basic hydrolysis gives a salt of the carboxylic acid and ammonia.

$$CH_3CH_2C\overset{O}{\underset{NH_2}{<}} + NaOH(aq) \longrightarrow CH_3CH_2C\overset{O}{\underset{O^-Na^+}{<}} + NH_3$$

It may be easier to see why the products of the two reactions differ if we consider the hydrolysis products that would form if the reaction could be carried out in the absence of added acid or base.

$$CH_3CH_2C\underset{NH_2}{\overset{O}{\diagup}} + H_2O \longrightarrow CH_3CH_2C\underset{OH}{\overset{O}{\diagup}} + NH_3$$

The products of this reaction are an acid (the carboxylic acid) and a base (ammonia). If we actually carry out the hydrolysis in the presence of hydrochloric acid, some of the hydrochloric acid will react with the basic product ammonia to form ammonium chloride. If, instead, we use sodium hydroxide to speed the reaction, some of this base will react with the carboxylic acid product to form the sodium salt of the carboxylic acid. There are two points to be made here. One is that the several hydrolysis reactions discussed in this chapter are closely related and should be considered as variations on a theme rather than as separate reactions. The second point is that chemical principles are not confined within chapters. If acids react with bases to form salts in Chapter 10, they will also react in Chapter 17. If a reaction under consideration happens to produce an acidic product, that product will always exhibit those properties we have previously ascribed to acids.

The hydrolysis of amides is of more than theoretical interest. Digestion of proteins (Special Topic L) involves the hydrolysis of amide bonds. Clothes made of nylon, another polyamide, have been known to disintegrate in air polluted with sulfuric acid mist. The acid catalyzes the hydrolysis of the amide bonds that hold the long chains of the nylon molecules together. Perhaps it is worth a thought to consider what that same polluted air does to the proteins of our lungs.

17.13 ESTERS OF INORGANIC ACIDS

The esters we discussed in Section 17.5 are organic esters, made from organic acids and alcohols. There are also inorganic esters derived from mineral acids and alcohols. Several of these are of considerable importance in biochemistry.

Butyl nitrite, a vasodilator, is derived from 1-butanol and nitrous acid.

$$\underset{\text{1-Butanol}}{CH_3CH_2CH_2CH_2OH} + \underset{\text{Nitrous acid}}{HO-N=O} \longrightarrow \underset{\text{Butyl nitrite}}{CH_3CH_2CH_2CH_2O-N=O} + H_2O$$

Glyceryl trinitrate, better known as nitroglycerin, is derived from glycerol and nitric acid.

$$\begin{array}{ccc} CH_2OH & & CH_2ONO_2 \\ | & & | \\ CHOH & + 3\ HONO_2 \longrightarrow & CHONO_2 & + 3\ H_2O \\ | & & | \\ CH_2OH & & CH_2ONO_2 \\ \text{Glycerol} & \text{Nitric acid} & \text{Nitroglycerin} \end{array}$$

Nitroglycerin is a powerful explosive. It is extremely sensitive to shock. In 1866, Alfred Nobel (Figure 17.3) discovered that this sensitivity is reduced when nitroglycerin is absorbed in diatomaceous earth, a claylike material. Nobel called this stabilized material **dynamite.** He earned an enormous fortune from the manufacture of explosives. It was from this fortune that he endowed the Nobel Prizes (Section 13.6).

Strange as it may seem, nitroglycerin is also used in medicine. It is used to relieve the sharp chest pains, called **angina pectoris,** caused by insufficient blood supply to the heart muscle. Nitroglycerin relaxes the smooth muscles of the arterial walls, allowing the arteries to carry more blood and thus relieve the pain.

Perhaps the most important inorganic esters in biochemistry are those of phosphoric acid and two of its anhydrides, pyrophosphoric acid and triphosphoric acid (Figure 17.4a). Phosphoric acid is triprotic and may form monoalkyl, dialkyl, and trialkyl esters as one or more hydrogen atoms are replaced by alkyl groups (Figure 17.4b).

FIGURE 17.3
Alfred Nobel, inventor of dynamite and founder of the famed Nobel Prizes. (Duplication courtesy of the Smithsonian Institution, Washington, D.C.)

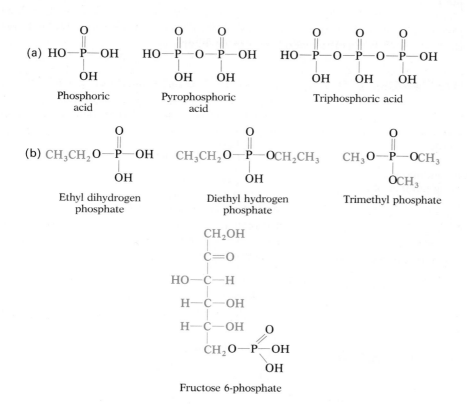

Fructose 6-phosphate

FIGURE 17.4   *Three phosphoric acids (a) and some phosphate esters (b).*

Before carbohydrates, such as the sugar fructose, can be metabolized, they must be converted to their phosphate esters (Chapter 27). Derivatives of the two anhydrides of phosphoric acid are involved in the chemistry of such transformations. Pyrophosphoric acid, as its name suggests (from the Greek *pyr*, "fire"), is formed by heating of phosphoric acid until water is driven off.

$$
\underset{\text{Pyrophosphoric acid}}{HO-\underset{\underset{OH}{|}}{\overset{\overset{O}{\|}}{P}}-\boxed{OH + H}-O-\underset{\underset{OH}{|}}{\overset{\overset{O}{\|}}{P}}-OH \xrightarrow{\text{heat}} HO-\underset{\underset{OH}{|}}{\overset{\overset{O}{\|}}{P}}-O-\underset{\underset{OH}{|}}{\overset{\overset{O}{\|}}{P}}-OH + H_2O}
$$

Pyrophosphoric acid, which is indeed an acid, is also an anhydride (of phosphoric acid). More complex anhydrides are possible; the other of interest at this point is triphosphoric acid (formed from three molecules of phosphoric acid).

Recall that acetic anhydride is an acetylating agent (Section 17.7); that is, it transfers the acetyl group to other molecules. The anhydrides of phosphoric acid, in the form of derivatives called adenosine triphosphate (ATP) and adenosine diphosphate (ADP), are phosphorylating agents; they transfer phosphate groups to molecules, such as the sugars, which can then be metabolized. We discuss these important reactions in other chapters.

PROBLEMS

1. Draw the functional groups in
   a. aldehydes
   b. ketones
   c. carboxylic acids
   d. esters
   e. ethers
   f. amides
2. Give common names for the continuous-chain carboxylic acids containing the given number of carbon atoms.
   a. 2   b. 3   c. 4   d. 5
3. Draw structural formulas for the following compounds:
   a. heptanoic acid
   b. 3-methylbutanoic acid
   c. 3-chloropentanoic acid
4. Draw structural formulas for the following compounds:
   a. *o*-nitrobenzoic acid
   b. *p*-chlorobenzoic acid
   c. 2,3-dibromobenzoic acid
5. Draw structural formulas for the following compounds:
   a. oxalic acid
   b. phthalic acid

6. Give the common and the IUPAC names for each of the following:
   a. HCOOH
   b. $CH_3CHCH_2CHCOOH$
             | |
            CH_3 CH_3
   c. $CH_3(CH_2)_8COOH$
   d. $HOOCCH_2CH_2CH_2COOH$
7. Draw structural formulas for the following:
   a. α-chloropropionic acid
   b. β-bromobutyric acid
   c. γ-fluorovaleric acid
8. Draw the following salts:
   a. potassium acetate   b. calcium oxalate
9. Name each of the following salts:

   a. $\underset{\text{(benzene ring)}}{\overset{\overset{O}{\|}}{C}}-O^-Na^+$

   b. $(CH_3CH_2\overset{\overset{O}{\|}}{C}-O^-)_2Ca^{2+}$

c. $CH_3\overset{O}{\underset{\|}{C}}-O^-NH_4^+$

d. $(CH_3CH_2CH_2\overset{O}{\underset{\|}{C}}-O^-)_2Zn^{2+}$

e. (benzene ring with two $\overset{O}{\underset{\|}{C}}-O^-$ groups) Ca^{2+}

10. Draw structural formulas for the following esters:
   a. methyl acetate   **c.** phenyl acetate
   b. heptyl acetate
11. Draw structural formulas for the following esters:
   a. ethyl pentanoate
   b. ethyl 3-methylhexanoate
12. Draw structural formulas for the following esters:
   a. ethyl benzoate   **c.** phenyl benzoate
   b. isopropyl benzoate
13. Draw structural formulas for the following esters:
   a. ethyl butyrate   **c.** diethyl oxalate
   b. isopropyl propionate
14. Name the following esters:

a. (benzene ring)$-\overset{O}{\underset{\|}{C}}-O-CH_3$

b. $CH_3-O-\overset{O}{\underset{\|}{C}}-H$

c. $CH_3CH_2\overset{O}{\underset{\|}{C}}-O-CH_2CH_3$

d. $CH_3CH_2CH_2O-\overset{O}{\underset{\|}{C}}-CH_3$

e. $CH_3CH_2\overset{O}{\underset{\|}{C}}-OCH_2CH_2CH_3$

f. $CH_3CH_2CH_2\overset{O}{\underset{\|}{C}}-O-$(benzene ring)

15. Draw structural formulas for the following amides:
   a. butanamide   **b.** hexanamide
16. Draw structural formulas for the following amides:
   a. valeramide   **b.** propionamide
17. From what alcohol might each acid be prepared via oxidiation with acidic dichromate (or liver enzymes!)?
   a. CH_3CH_2COOH **c.** $HCOOH$
   b. $HOOCCOOH$ **d.** $CH_3\underset{\underset{CH_3}{|}}{C}HCH_2COOH$
18. Arrange in order of increasing acidity, with the least acidic compound first.
   a. toluene
   b. benzoic acid
   c. benzyl alcohol ($C_6H_5CH_2OH$)
   d. phenol
19. Draw structural formulas for the following amides:
   a. N-methylacetamide
   b. N-ethylbenzamide
   c. N,N-dimethylbenzamide
20. Draw structural formulas for the following amides:
   a. acetanilide   **b.** benzanilide
21. Name the following amides:

a. (benzene ring)$-\overset{O}{\underset{\|}{C}}-NH_2$

b. $CH_3\underset{\underset{CH_3}{|}}{C}HNH-\overset{O}{\underset{\|}{C}}CH_3$

c. $CH_3\overset{O}{\underset{\|}{C}}-NH_2$

d. $Cl-$(benzene ring)$-\overset{O}{\underset{\|}{C}}-NH_2$

e. $H_2N-\overset{O}{\underset{\|}{C}}-CH_2-\overset{O}{\underset{\|}{C}}-NH_2$

f. $CH_3CH_2\overset{O}{\underset{\|}{C}}-\underset{\underset{CH_3}{|}}{N}-CH_3$

g. $CH_3CH_2CH_2\overset{\overset{\displaystyle O}{\|}}{C}-NH-\langle\bigcirc\rangle$

22. Which compound would have the higher boiling point?

$CH_3CH_2CH_2-O-CH_2CH_3 \qquad CH_3CH_2CH_2\overset{\overset{\displaystyle O}{\|}}{C}-OH$

23. Which compound would have the higher boiling point?

$CH_3CH_2CH_2CH_2CH_2OH \qquad CH_3CH_2CH_2\overset{\overset{\displaystyle O}{\|}}{C}-OH$

24. Which compound would have the higher boiling point?

$CH_3CH_2CH_2\overset{\overset{\displaystyle O}{\|}}{C}-NH_2 \qquad CH_3\overset{\overset{\displaystyle O}{\|}}{C}-O-CH_2CH_3$

25. Which compound would have the higher boiling point?

$CH_3CH_2CH_2\overset{\overset{\displaystyle O}{\|}}{C}-OH \qquad CH_3CH_2\overset{\overset{\displaystyle O}{\|}}{C}-O-CH_3$

26. Without consulting tables, arrange these in order of increasing boiling point:
  a. butyl alcohol   **c.** pentane
  b. propionic acid   **d.** methyl acetate

27. Which compound is more soluble in water?

$CH_3\overset{\overset{\displaystyle O}{\|}}{C}-OH \qquad CH_3CH_2CH_2CH_3$

28. Which compound is more soluble in water?

$CH_3C\equiv CCH_3 \qquad CH_3\overset{\overset{\displaystyle O}{\|}}{C}-NH_2$

29. Which compound is more soluble in water?

$CH_3\overset{\overset{\displaystyle O}{\|}}{C}-O-CH_3$

$CH_3CH_2CH_2CH_2\overset{\overset{\displaystyle O}{\|}}{C}-O-CH_2CH_3$

30. Which compound is more soluble in water?

$\langle\bigcirc\rangle-\overset{\overset{\displaystyle O}{\|}}{C}-OH \qquad \langle\bigcirc\rangle-\overset{\overset{\displaystyle O}{\|}}{C}-O^-Na^+$

31. Of the families of compounds discussed in this chapter, which has members with characteristically unpleasant odors? Which group has characteristically pleasant aromas?

32. Write an equation for the reaction of decanoic acid with aqueous NaOH and with aqueous $NaHCO_3$.

33. Write an equation for the reaction of benzoic acid with aqueous NaOH and with aqueous $NaHCO_3$.

34. Benzoic acid is insoluble in water. If the reactions described in Problem 33 were carried out in test tubes, what would you observe?

35. Write an equation for the acid-catalyzed hydrolysis of ethyl acetate.

36. Write an equation for the base-catalyzed hydrolysis of ethyl acetate.

37. Write an equation for the acid-catalyzed hydrolysis of benzamide.

38. Write an equation for the base-catalyzed hydrolysis of benzamide.

39. Complete these equations:

a. $CH_3CH_2\overset{\overset{\displaystyle O}{\|}}{C}-OH \xrightarrow{\text{NaOH}}$

b. $\langle\bigcirc\rangle-COOH \xrightarrow{\text{KOH}}$

c. $HOOC-COOH \xrightarrow{\text{Excess NaOH}}$

d. $\langle\bigcirc\rangle\overset{\textstyle -COOH}{\underset{\textstyle -COOH}{}} \xrightarrow{\text{Excess NaHCO}_3}$

40. Complete these equations:

a. $CH_3\overset{\overset{\displaystyle O}{\|}}{C}-OH + CH_3CH_2CH_2OH \xrightarrow{H^+}$

b. $HO-\overset{\overset{\displaystyle O}{\|}}{C}-CH_2\overset{\overset{\displaystyle O}{\|}}{C}-OH + 2\ CH_3OH \xrightarrow{H^+}$

41. Complete these equations:

a. 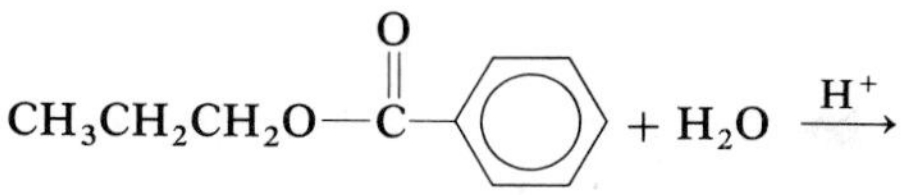 $CH_3CH_2CH_2O-\overset{O}{\overset{\|}{C}}-\bigcirc + H_2O \xrightarrow{H^+}$

b. $CH_3\underset{\underset{CH_3}{|}}{CH}-\overset{O}{\overset{\|}{C}}-O-CH_2CH_3 + H_2O \xrightarrow{H^+}$

42. Complete these equations:

a. $CH_3CH_2\overset{O}{\overset{\|}{C}}-OH + SOCl_2 \longrightarrow$

b. $CH_3\underset{\underset{CH_3}{|}}{CH}\overset{O}{\overset{\|}{C}}-Cl + NH_3 \longrightarrow$

c. $CH_3\underset{\underset{CH_3}{|}}{CH}\overset{O}{\overset{\|}{C}}-Cl + HO\underset{\underset{CH_3}{|}}{CH}CH_3 \longrightarrow$

43. Complete these equations:

a. 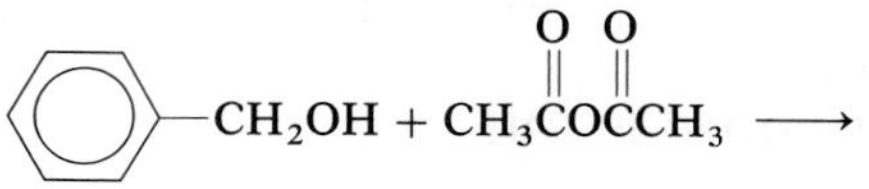 $\bigcirc-CH_2OH + CH_3\overset{O}{\overset{\|}{C}}O\overset{O}{\overset{\|}{C}}CH_3 \longrightarrow$

b. 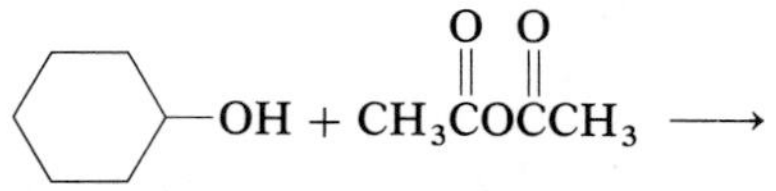 $\bigcirc-OH + CH_3\overset{O}{\overset{\|}{C}}O\overset{O}{\overset{\|}{C}}CH_3 \longrightarrow$

44. Complete these equations:

a. $CH_3\overset{O}{\overset{\|}{C}}-NH_2 + HCl + H_2O \longrightarrow$

b. 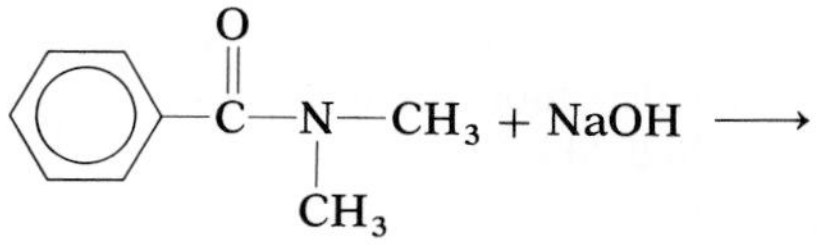 $\bigcirc-\overset{O}{\overset{\|}{C}}-\underset{\underset{CH_3}{|}}{N}-CH_3 + NaOH \longrightarrow$

45. List the reagents necessary to carry out these conversions:

a. $CH_3CH_2\underset{\underset{CH_3}{|}}{CH}CH_2OH \xrightarrow{?} CH_3CH_2\underset{\underset{CH_3}{|}}{CH}\overset{O}{\overset{\|}{C}}-OH$

b. $HOCH_2CH_2CH_2OH \xrightarrow{?} HO\overset{O}{\overset{\|}{C}}CH_2\overset{O}{\overset{\|}{C}}OH$

c. 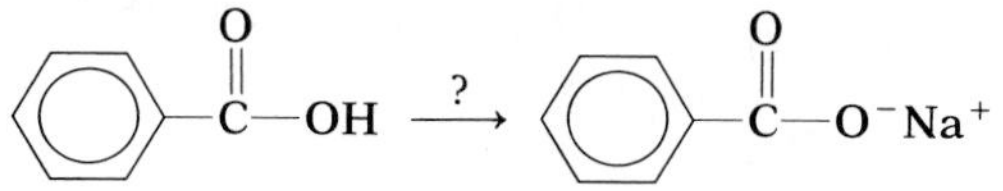 $\bigcirc-\overset{O}{\overset{\|}{C}}-OH \xrightarrow{?} \bigcirc-\overset{O}{\overset{\|}{C}}-O^-Na^+$

46. List the reagents necessary to carry out these conversions:

a. 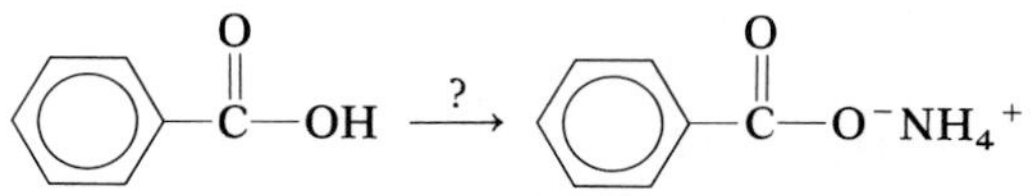 $\bigcirc-\overset{O}{\overset{\|}{C}}-OH \xrightarrow{?} \bigcirc-\overset{O}{\overset{\|}{C}}-O^-NH_4^+$

b. 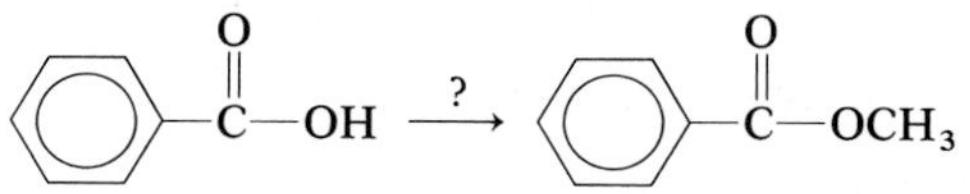 $\bigcirc-\overset{O}{\overset{\|}{C}}-OH \xrightarrow{?} \bigcirc-\overset{O}{\overset{\|}{C}}-OCH_3$

c. $CH_3CH_2CH_2CH_2OH \xrightarrow{?}$
$$CH_3CH_2CH_2CH_2O\overset{O}{\overset{\|}{C}}CH_3$$

d. $CH_3CH_2CH_2\overset{O}{\overset{\|}{C}}-OCH_3 \xrightarrow{?}$
$$CH_3CH_2CH_2C\overset{\overset{O}{\|}}{\underset{O^-}{\diagdown}} + CH_3OH$$

47. List the reagents necessary to carry out these conversions:

a. $CH_3\overset{O}{\overset{\|}{C}}-Cl \xrightarrow{?} CH_3\overset{O}{\overset{\|}{C}}-NH_2$

b. $CH_3\overset{O}{\overset{\|}{C}}-OH \xrightarrow{?} CH_3\overset{O}{\overset{\|}{C}}-Cl$

c. 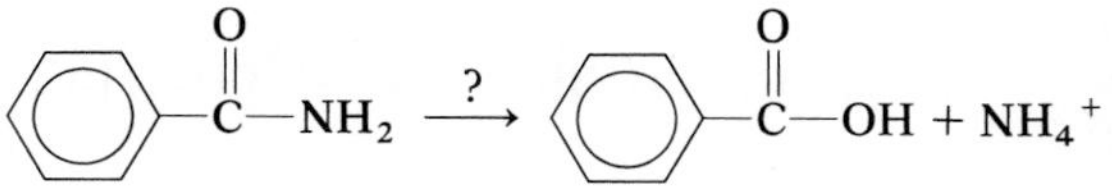 $\bigcirc-\overset{O}{\overset{\|}{C}}-NH_2 \xrightarrow{?} \bigcirc-\overset{O}{\overset{\|}{C}}-OH + NH_4^+$

d. $CH_3\underset{\underset{CH_3}{|}}{CH}-C\overset{\overset{O}{\|}}{\underset{NH_2}{\diagdown}} \xrightarrow{?} CH_3\underset{\underset{CH_3}{|}}{CH}-C\overset{\overset{O}{\|}}{\underset{O^-}{\diagdown}} + NH_3$

48. A lactone is a cyclic ester. What product is formed in each of the following reactions?

a. 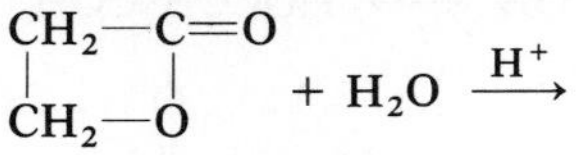
$$CH_2\text{---}C\text{==}O$$
$$|\qquad\quad | \qquad + H_2O \xrightarrow{\;H^+\;}$$
$$CH_2\text{---}O$$

b. 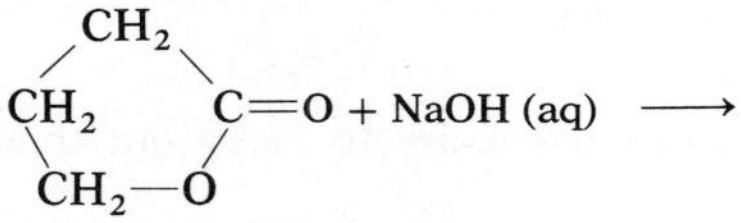
$$\begin{array}{c} CH_2 \\ \diagup \qquad \diagdown \\ CH_2 \qquad\quad C\text{==}O + NaOH\ (aq) \longrightarrow \\ \diagdown \qquad \diagup \\ CH_2\text{---}O \end{array}$$

49. A lactam is a cyclic amide. What product is formed in each of the following reactions?

a. 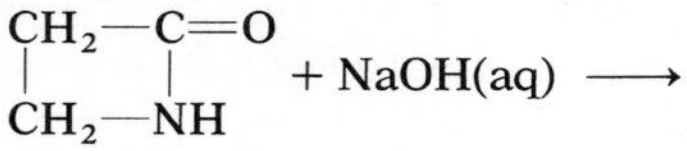
$$CH_2\text{---}C\text{==}O$$
$$|\qquad\quad |\qquad + NaOH(aq) \longrightarrow$$
$$CH_2\text{---}NH$$

b. 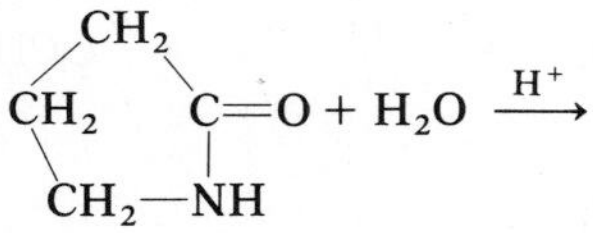
$$\begin{array}{c} CH_2 \\ \diagup \qquad \diagdown \\ CH_2 \qquad\quad C\text{==}O + H_2O \xrightarrow{\;H^+\;} \\ \diagdown \qquad \diagup \\ CH_2\text{---}NH \end{array}$$

50. Pellagra is a vitamin-deficiency disease. Corn contains the antipellagra factor nicotinamide, which is not readily absorbed in the digestive tract. When corn is treated with quicklime (calcium hydroxide) to make hominy, the vitamin is rendered more available. What sort of reaction takes place? Write the equation. Nicotinamide is

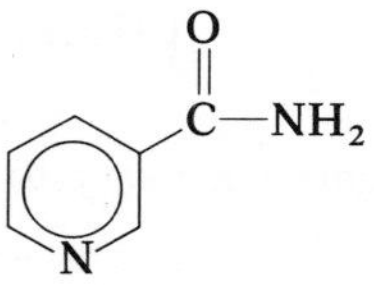

$$\underset{N}{\underset{\diagup\diagdown}{\bigcirc}}\overset{\overset{O}{\|}}{C}\text{---}NH_2$$

51. The following compounds are isomers. Circle and name the functional groups in each.

a. $CH_3CH_2CH_2\overset{\overset{O}{\|}}{C}OH$ b. $CH_3CH_2\overset{\overset{O}{\|}}{C}CH_2OH$

c. $CH_3\overset{\overset{O}{\|}}{C}CH_2CH_2OH$ d. $H\overset{\overset{O}{\|}}{C}CH_2CH_2CH_2OH$

e. $CH_3OCH_2CH_2\overset{\overset{O}{\|}}{C}H$ f. $CH_3CH_2OCH_2\overset{\overset{O}{\|}}{C}H$

g. $CH_3CH_2CH_2O\overset{\overset{O}{\|}}{C}H$ h. $CH_3OCH_2\overset{\overset{O}{\|}}{C}CH_3$

i. $CH_3CH_2O\overset{\overset{O}{\|}}{C}CH_3$ j. $CH_3CH_2\overset{\overset{O}{\|}}{C}OCH_3$

k. $CH_3\overset{\overset{O}{\|}}{C}\text{---}\overset{\overset{OH}{|}}{C}HCH_3$ l. $CH_3\overset{\underset{OH}{|}}{C}HCH_2\overset{\overset{O}{\|}}{C}H$

52. Draw structural formulas for the following:
   a. diethyl hydrogen phosphate
   b. methyl dihydrogen phosphate
   c. triphosphoric acid

53. Name the following compounds:

a. 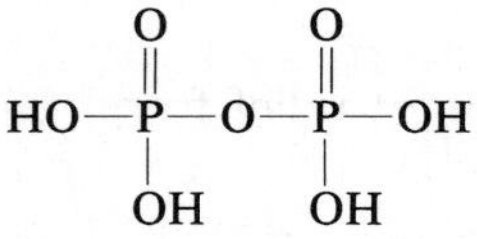
$$HO\text{---}\overset{\overset{O}{\|}}{\underset{\underset{OH}{|}}{P}}\text{---}O\text{---}\overset{\overset{O}{\|}}{\underset{\underset{OH}{|}}{P}}\text{---}OH$$

b. 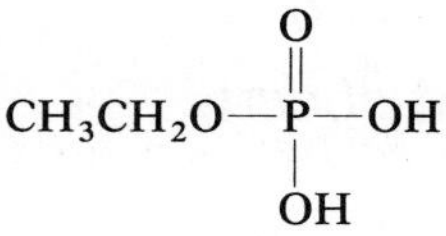
$$CH_3CH_2O\text{---}\overset{\overset{O}{\|}}{\underset{\underset{OH}{|}}{P}}\text{---}OH$$

54. Sulfuric acid, $HO\overset{\overset{O}{\|}}{\underset{\underset{O}{\|}}{S}}OH$, forms two esters with methanol. Give their structures and names.

55. Kan-Kleener, Inc., decided to add ethyl butyrate to its Snidey Bowl toilet cleaner to give it a pineapple fragrance. The principal active ingredient in the bowl cleaner was hydrochloric acid. After a short time, the product smelled like stale vomit rather than pineapple. Use the chemistry that you learned in this chapter to explain what happened.

F

Drugs: Some Esters and Amides

People have long sought relief from pain and from discomfort. Alcohol, opium, cocaine, and marijuana have been used as medicines for centuries. Often they were used for their pleasurable effects, not just for relief of pain.

In this special topic, we will discuss several important drugs that are either esters or amides, types of compounds discussed in the preceding chapter.

F.1 ASPIRIN AND OTHER SALICYLATES

Soon after acetic anhydride became available, in the nineteenth century, chemists began to acetylate a variety of physiologically active compounds. Such structural modifications often change the properties of drugs to enhance their effectiveness or to minimize undesirable side effects. Two such cases will be described here, that of aspirin and that of heroin.

The first successful synthetic pain relievers were derivatives of salicylic acid (Figure F.1) Salicylic acid was first isolated from willow bark in 1860, although an English clergyman named Edward Stone had reported to the Royal Society as early as 1763 that an extract of willow bark was useful in reducing fever. Salicylic acid is itself a good **analgesic** (pain reliever) and **antipyretic** (fever reducer), but it is sour and irritating when taken by mouth. Chemists sought to modify the structure of the molecule to remove this undesirable property while retaining (or even improving) the desirable properties.

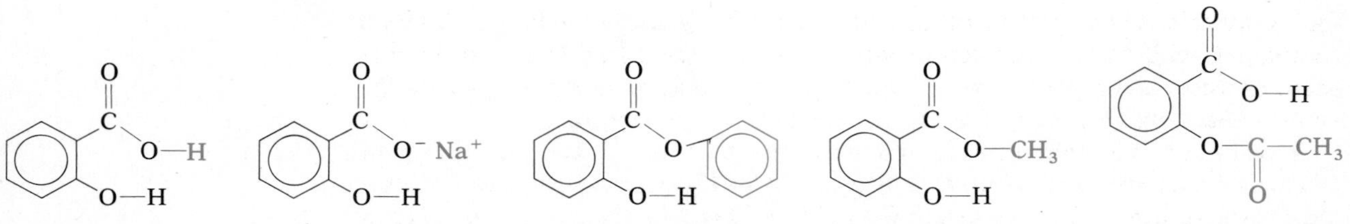

| Salicylic acid | Sodium salicylate | Phenyl salicylate (salol) | Methyl salicylate | Acetylsalicylic acid |

FIGURE F.1 *Salicylic acid and some of its derivatives.*

Sodium salicylate was first used in 1875. It was less unpleasant when taken by mouth than salicylic acid but proved to be highly irritating to the lining of the stomach. Phenyl salicylate (salol) was introduced in 1886. It passes unchanged through the stomach. In the small intestine it is hydrolyzed (Section 17.8) to the desired salicylic acid, but phenol, which is rather toxic, is also formed. Acetylsalicylic acid (aspirin) was first introduced in 1899 and soon became the largest-selling drug in the world. Over 50 billion tablets are produced annually in the United States (that's over 200 tablets for every person in the country).

Acetylsalicylic acid is made by treatment of salicylic acid with acetic anhydride. In this reaction, the hydroxyl group of the phenol reacts exactly like that of an alcohol.

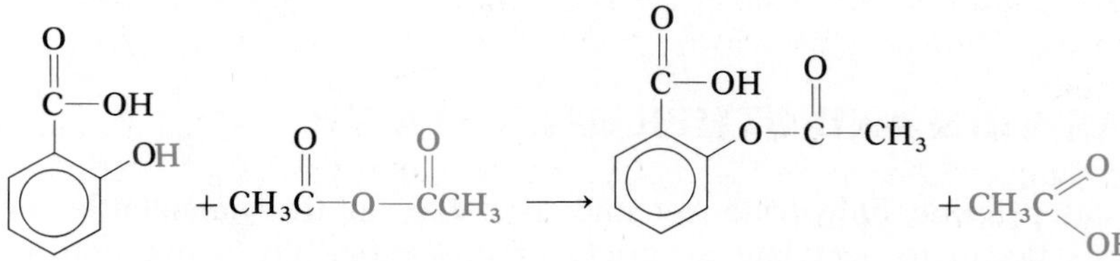

Aspirin relieves minor aches and pains, reduces fever, and suppresses inflammation. It works by reducing the number of pain impulses to the brain. It doesn't cure whatever is causing the pain; it merely kills the messenger. Recent evidence indicates that aspirin acts, at least in part, by inhibiting the synthesis of **prostaglandins,** compounds that are involved in inflammation, increased blood pressure, and contractions of smooth muscles.

Although probably one of the safest and most effective pain relievers known, aspirin is not without its hazards. Recent studies indicate that there is some intestinal bleeding every time aspirin is ingested. The blood loss is usually minor (0.5 mL to 2.0 mL per two-tablet dose), but it may be substantial in some cases. Because it inhibits the clotting of blood, aspirin should not be used by people facing surgery, childbirth, or other hazards involving the possible loss of blood (nonuse should start a week before the hazard). On the other hand, small doses seem to lower the risk of coronary heart attack and stroke, presumably by the same anticoagulant action that causes bleeding in the stomach.

Aspirin is a registered trade name of the German Bayer Company. In Germany, Canada, and other countries, *aspirin* still means the Bayer brand. Other brands are sold as acetylsalicylic acid or ASA. In the United States, Bayer has lost its rights to the trade name, and aspirin is used as a generic name for acetylsalicylic acid. Aspirin is a chemical compound, and, as with other compounds, its properties are invariant. Aspirin *tablets* usually contain 325 mg (5 grains) each of this compound, held together with some sort of inert binder. The latter may be starch, clay, or a sugar. Various brands of aspirin have been extensively tested. The conclusions of impartial studies are invariably the same: the only significant difference in brands is price.

Arthritis Pain Formula (APF) and similar "extra-strength" formulations simply have 500 mg of aspirin rather than the usual 325 mg. They have no other active ingredients. Simple arithmetic tells us that three plain aspirin tablets are equal to two APF tablets in dosage and are usually much lower in price.

F.2 THE OPIATES: MORPHINE AND HEROIN

Morphine, an opium alkaloid, is a **narcotic,** that is, a drug that produces both sedation (narcosis) and relief of pain (analgesia) (Figure F.2). It is also strongly addictive. Chemists, therefore, also sought to modify the properties of morphine.

In the laboratory, reaction of morphine (which is both an alcohol and a phenol) with acetic anhydride produces diacetylmorphine, a product in which both the alcohol and phenol hydroxyl groups have been esterified.

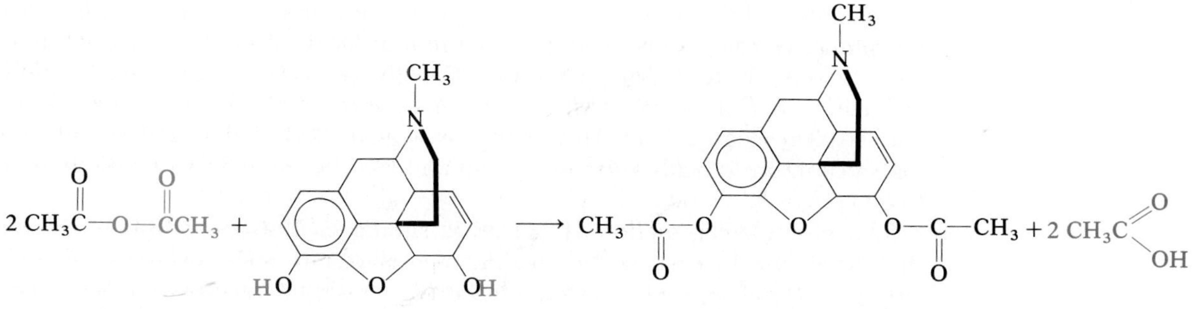

Morphine

Heroin
(diacetylmorphine)

FIGURE F.2
Morphine is the narcotic isolated from raw opium. Codeine and heroin are derivatives of morphine. (a) Scraping raw opium. (b) Opium gum. (c) Smoking opium, codeine, heroin, and morphine. (Photos courtesy of the U.S. Department of Justice, Drug Enforcement Administration, Washington, D.C.)

(a)

(b)

(c)

This morphine derivative was first prepared by chemists at the Bayer Company of Germany in 1874. Indeed, the name *heroin* is Bayer's trade name for diacetyl-morphine. Heroin received little attention until 1890, when it was proposed as an antidote for morphine addiction! Shortly thereafter, Bayer was widely advertising heroin as a sedative for coughs, often in the same advertisement with aspirin (Figure F.3). However, it was soon found that heroin induced addiction more quickly than morphine and that heroin addiction was harder to cure.

The physiological action of heroin is similar to that of morphine except that heroin produces a stronger feeling of euphoria for a longer period of time. Heroin is so strongly addictive that it seems one or two injections are sufficient to induce dependence in some individuals. Heroin is not legal in the United

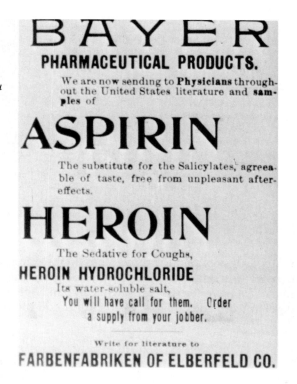

States, even by prescription. There are some individuals lobbying for legalization of the drugs for use in terminally ill people who are suffering great pain.

Chemists have synthesized thousands of morphine analogs. Only a few have shown significant analgesic activity. Most are addictive. Morphine acts by binding to receptors in the brain. Those molecules that have morphinelike action are called **agonists.** Morphine **antagonists** are drugs that block the action of morphine, most likely by blocking the receptors. Some molecules have both agonist and antagonist activity. These show great promise as analgesics. An example is pentazocine (Talwin). It is less addictive than morphine, and yet it is effective for relief of pain. There is some hope that an effective analgesic could be developed that is not addictive, but to date the two effects seem inseparable.

Pure antagonists such as naloxone are of value in treating opiate addicts. Overdosed addicts can be brought back from death's door by an injection with naloxone. Long-acting antagonists can block the action of heroin for as much as a month, thus aiding an addict in overcoming his or her addiction.

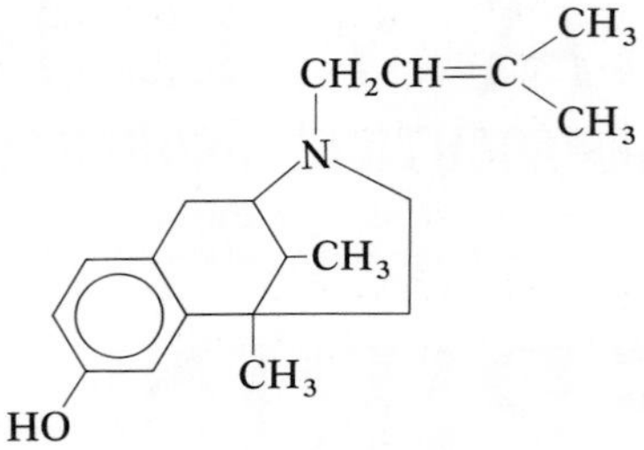

Pentazocine
(Talwin)

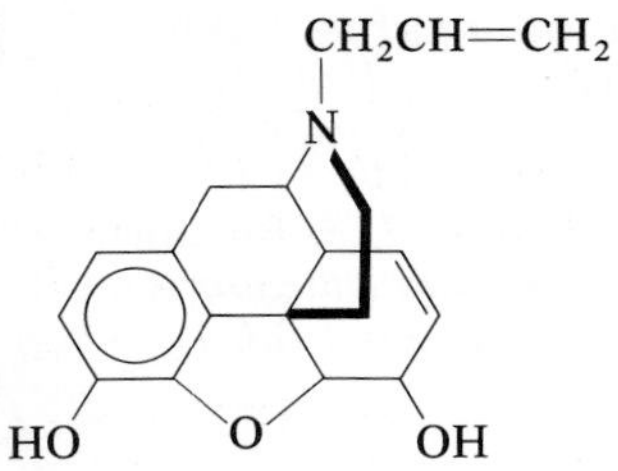

Naloxone

F.3 ASPIRIN SUBSTITUTES AND COMBINATION PAIN RELIEVERS

Many drugs contain the amide function. Acetanilide was once used as an anti-pyretic. It is quite toxic, however, and was largely replaced by aspirin and by two of its own derivatives, phenacetin and acetaminophen, both of which are derived from acetanilide by substitution on the benzene ring.

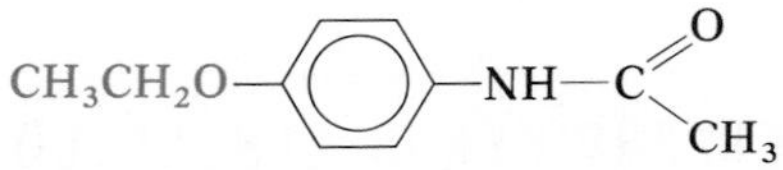

Phenacetin
(*p*-ethoxyacetanilide)

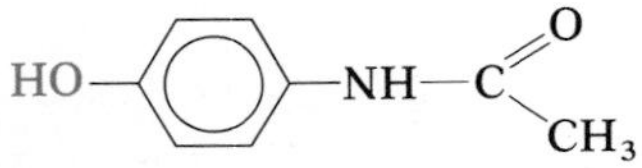

Acetaminophen
(*p*-hydroxyacetanilide)

Phenacetin has about the same effectiveness as aspirin in reducing fever and relieving minor aches and pains. It has been implicated in damage to the kidneys and in blood abnormalities. Phenacetin was banned from use in the United States in 1983.

For many years, the most familiar **combination pain reliever** was aspirin, phenacetin, and caffeine (APC). This combination was available under a variety of trade names, or it could be purchased as APC Tablets U.S.P., usually at a lower price than the proprietary medications.

Anacin, which, along with Empirin and Excedrin, was once an APC formulation, now contains only aspirin and caffeine. Caffeine is a mild stimulant found in coffee, tea, and cola syrup. There is no reliable evidence that caffeine enhances the effect of aspirin in any significant way. In fact, recent evidence indicates that for fever reduction, caffeine *counteracts* the action of aspirin. Combinations containing caffeine are therefore *less effective* than plain aspirin for this use. A tablet of Anacin does contain a little more aspirin (400 mg) than a regular (325-mg) aspirin tablet. The usual two-tablet dose of Anacin provides 800 mg of aspirin. You could get the same dose—at a lower price—from two and a half regular aspirin tablets.

Some people who are allergic to aspirin may safely take acetaminophen. This compound is comparable to aspirin in relief of pain and reduction of fever but usually costs more than aspirin. It does not reduce inflammation, nor does it interfere with the clotting of blood. It is available by itself under such trade names as Tylenol, Panadol, and Datril.

In Excedrin, acetaminophen has replaced the phenacetin that was dropped from the original formulation. This highly advertised product also contains aspirin and caffeine. The advertising claim was made that Excedrin is more effective than plain aspirin for pain other than headache. The claim failed to mention that the pain for which Excedrin is more effective is that suffered by women who have just had babies! Now, the relief of such pain is certainly a noble undertaking, but the implication that the effect is a general one might be regarded as somewhat self-serving.

In 1984 the U.S. FDA approved ibuprofen as an over-the-counter pain reliever.

$$CH_3CHCH_2 \underset{\underset{CH_3}{|}}{} \!\!\!\!\!- \!\!\! \bigcirc \!\!\! - \!\! CH\!\!\underset{\underset{CH_3}{|}}{}\!\! C\!\!\overset{\overset{O}{\|}}{O}H$$

Ibuprofen

Ibuprofen has long been available by prescription under the trade name Motrin in tablets of 300 mg, 400 mg, and 600 mg. The FDA action led to the introduction of ibuprofen under the trade names Advil and Nuprin, each of which consists of 200 mg tablets. Ibuprofen is anti-inflammatory and antipyretic. Unfortunately, people who are allergic to aspirin also are sensitive to ibuprofen, and the drug is considerably more expensive than aspirin.

To summarize, extensive studies have shown that of all the preparations on the market, plain aspirin is the cheapest, safest, and most effective for most people. Over-the-counter drugs must list their ingredients on the label. A concerned consumer can easily look up the properties—such as use, toxicity, and side effects—of these ingredients in a reference book such as *The Merck Index.*[*]

F.4 LSD AND THALIDOMIDE

One of the most interesting amides of all is the N,N-diethylamide of lysergic acid, better known as LSD (from the German *lysergsaure diethylamid*). The physiological properties of this compound were discovered quite accidentally

---

[*] Windholz, Martha, ed., *The Merck Index*, 10th ed., Rahway, N.J.: Merck, 1983.

by Albert Hofmann in 1943. Hofmann, a chemist at the Sandoz Laboratories in Switzerland, unintentionally ingested some LSD. He later took 250 μg, which he considered a small dose, to verify that LSD had caused the symptoms he had experienced. Hofmann had a very rough time for the next few hours, exhibiting such symptoms as visual disturbance and schizophrenic behavior.

Lysergic acid is obtained from ergot, a fungus that grows on rye. It can be converted chemically into LSD by treatment first with thionyl chloride and then with a compound called diethylamine.

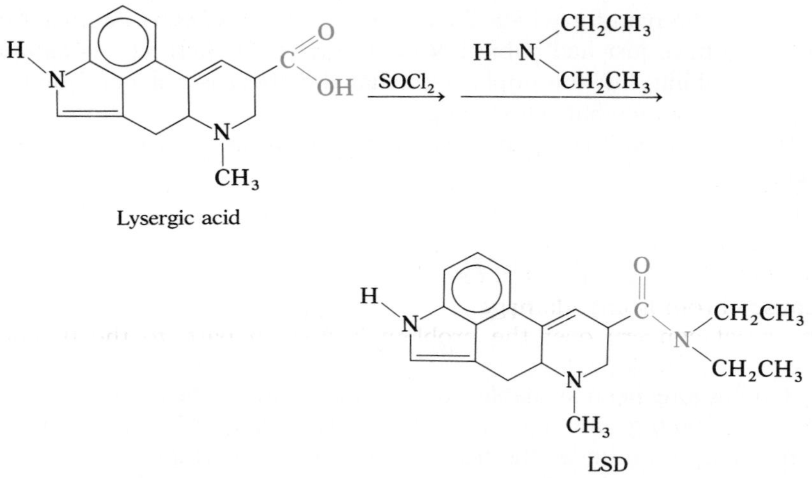

Lysergic acid

LSD

LSD is a member of a class of drugs that qualitatively change the way in which we perceive things. These drugs are called **hallucinogenic, psychotomimetic,** or **psychedelic** drugs, because they induce hallucinations, psychoses, and colorful visions. LSD is probably the most powerful hallucinogen of them all. Though mind-expanding qualities have been attributed to it, there is no sound evidence confirming this attribute, as illustrated by the story of the guy who took his notebook with him on a trip. While he was airborne, he had a great insight. When he came down, he told everyone of his marvelous discovery. He had met ultimate truth, face-to-face. Finally someone asked what he had learned. He opened his notebook and read, "Green grass is pretty." Now, no one but the most hardened mower of lawns would disagree with the statement, but most people reach that conclusion without chemically "enhancing" their perception. Enhanced creativity is just another drug-induced illusion.

LSD is a potent drug, as indicated by the small amount required for one to experience its fantastic effects. The usual dosage is probably about 10 to 100 μg. To give you an idea of how small 10 μg is, let us compare that amount of LSD to the amount of aspirin in one aspirin tablet. One aspirin tablet contains about

FIGURE F.4
Illicit dosage forms of
lysergic acid diethylamide
(LSD). (Courtesy of the U.S.
Bureau of Narcotics and
Dangerous Drugs,
Washington, D.C.)

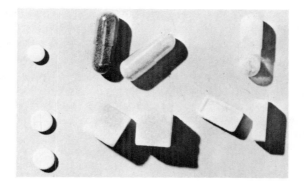

300 000 μg of aspirin. Some common illegal dosage forms of LSD are shown in Figure F.4.

Is LSD a dangerous drug? A few facts are known, but most are disputed. For example, in one study, LSD administered to pregnant hamsters caused gross fetal deformities. Other studies, however, seem to suggest that LSD does not cause chromosomal damage.

The great concern over the problem is due, in part, to the thalidomide tragedy of the late 1950s and early 1960s. Thalidomide was a completely legal, amidelike drug used as a tranquilizer.

Thalidomide

It was considered so safe, based on laboratory studies, that it was widely prescribed for pregnant women and, in Germany, was available without a prescription. It took several years for the human population for provide information that laboratory animals had not provided. The drug had a disastrous effect on developing human embryos. Children born to women who had taken the drug during the first 12 weeks of pregnancy suffered from phocomelia, a condition characterized by shortened or missing arms or legs and other physical defects. The drug was widely used in Germany and Great Britain, and these two countries bore the brunt of the tragedy. The United States escaped relatively unscathed because an official of the Food and Drug Administration had believed that there was evidence to doubt the drug's safety and did not, therefore, approve it for use in the United States (Figure F.5).

FIGURE F.5
Frances Kelsey, of the
U.S. Food and Drug
Administration, withheld
approval of the tranquilizer
thalidomide in the United
States. This action prevented
the use of this drug for
"morning sickness" and
saved many a mother from
the tragedy of giving birth to
a "thalidomide baby."
(Courtesy of the U.S. Food
and Drug Administration,
Washington, D.C.)

PROBLEMS

1. Define and give an example for each of the following:
   a. analgesic   c. narcotic
   b. antipyretic   d. hallucinogen
2. How does heroin differ in structure from morphine? In physiological properties?
3. Examine the labels of at least five "combination" pain relievers (e.g., Excedrin, Empirin, Anacin). Make a list of the ingredients in each. Look up the properties (medical use, dosage, side effects, toxicity, etc.) in a reference work such as *The Merck Index*.
4. Aspirin is a chemical compound.
   a. What is its structure and chemical name?
   b. In what ways may one brand of aspirin differ from another?
   c. In what ways must brands of aspirin be the same?
5. Do a cost analysis on at least five brands of plain aspirin, calculating the cost per grain. Also calculate the cost per gram (1 g = 15 grains).
6. To what family of organic compounds does ibuprofen belong?
7. What alkyl group is attached to the *para* position of the benzene ring of the ibuprofen molecule? Can you see how the generic name ibuprofen was derived?
8. List the reagents necessary to carry out the following conversions:

   a.

   b. CH_3CH_2O—⟨ ⟩—NH_2 $\xrightarrow{?}$

   CH_3CH_2O—⟨ ⟩—NH—$\overset{\overset{\displaystyle O}{\|}}{C}CH_3$

9. List the reagents necessary to carry out the following conversions:

   a.

   b.

10. List the reagents necessary to carry out the following conversions:

    a.

    b.

11. "New Maximum Strength" Bayer aspirin is "1000 mg of pure ... aspirin per [2 tablet] dose." How many "regular" aspirin tablets of 325 mg each would you take to get about the same amount of aspirin (probably at a much lower price)?

Amines and Derivatives

If carbon compounds are the basis of life, nitrogen compounds are the bases of life (pun intended). Recall from Chapter 10 that ammonia is a nitrogen-containing weak base. Amines are alkyl (and aryl) derivatives of ammonia, and they too are weak bases. These organic bases occur in living (and especially in once-living but now-decaying) organisms.

Nitrogen is an essential constituent of many physiologically active compounds. All enzymes—indeed, all proteins—contain nitrogen. Many vitamins and hormones contain nitrogen. Most drugs incorporate nitrogen atoms. And nitrogenous bases are part of the complex structure of the compounds that carry our genetic heritage, the nucleic acids DNA and RNA (bases in acids!). In this chapter, we discuss the amines generally. In Special Topic G, we discuss a number of related nitrogen-containing compounds that exhibit interesting physiological effects. The discussion of proteins and nucleic acids we shall save for later chapters.

We shall also save for subsequent chapters a consideration of the implications of the following facts: Plants can take inorganic nitrogen, usually in the form of nitrate or ammonium salts, and combine it with carbon compounds from photosynthesis to make all the organic nitrogen compounds they require. But animals are not quite so clever. They require some preformed organic nitrogen compounds in their diet, compounds that are essential to their health but that they themselves cannot synthesize.

18.1 STRUCTURE AND CLASSIFICATION OF AMINES

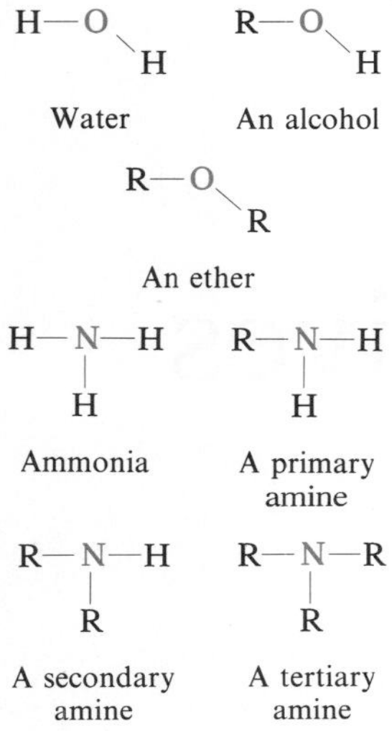

H—O⟍H   R—O⟍H

Water   An alcohol

R—O⟍R

An ether

H—N—H   R—N—H
  | |
  H   H

Ammonia   A primary
            amine

R—N—H   R—N—R
  | |
  R   R

A secondary   A tertiary
  amine   amine

FIGURE 18.1
Amines are derived from ammonia in a manner similar to that in which alcohols and ethers are derived from water.

$$NH_2$$
$$|$$
$$CH_3—CH—CH_3$$

I

$$OH$$
$$|$$
$$CH_3—CH—CH_3$$

II

$$CH_3—N—CH_3$$
$$|$$
$$H$$

III

$$CH_3—O—CH_3$$

IV

Chapter 15 shows that the alcohols and ethers can be considered derivatives of water. In a similar manner, the amines are derived from ammonia. Amines are classified according to the number of carbon atoms bonded directly to the nitrogen atom. A **primary amine** has one alkyl (or aryl) group on the nitrogen, a **secondary amine** has two, and a **tertiary amine** has three (Figure 18.1).

This use of the terms *primary*, *secondary*, and *tertiary* must be distinguished from our previous use of these terms in connection with the alcohols (Section 15.2). For example, consider structures I and II. Compound I is a primary amine because only one of nitrogen's bonds is attached to a carbon atom. But compound II is a *secondary* alcohol. When determining whether an alcohol is primary, secondary, or tertiary, one counts the number of carbons bonded not to the oxygen but to the carbon attached to the oxygen. Another difference can be seen in structures III and IV. Compound III is a secondary amine, but compound IV is an ether (*not* an alcohol, secondary or otherwise). When there is only one carbon group attached to oxygen, the compound is an alcohol. But if there are two, the compound is an ether. In contrast, whether there are one or two (or three) alkyl or aryl groups attached to nitrogen, the compounds are all classed as amines.

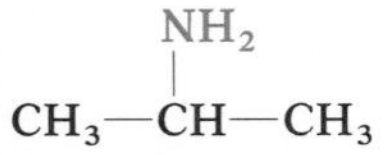

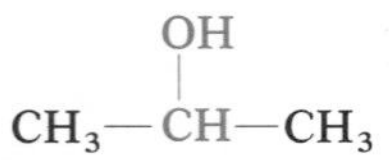

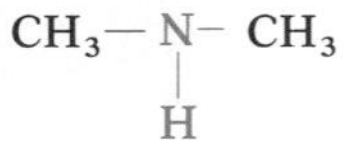

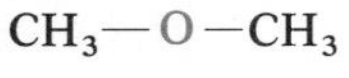

Let's look at the structure of ammonia and the amines a little more closely. In ammonia, three hydrogens are bonded to nitrogen. The nitrogen also has an unshared pair of electrons.

$$H:\overset{..}{\underset{..}{N}}:H \quad \left(or \; H—\overset{..}{\underset{|}{N}}—H \right)$$
$$\overset{\cdot\cdot}{H} \qquad\qquad\quad H$$

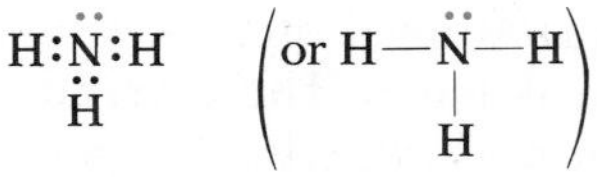

Ammonia can undergo a reaction in which it shares its normally nonbonding electrons. It can accept a proton and form the ammonium ion.

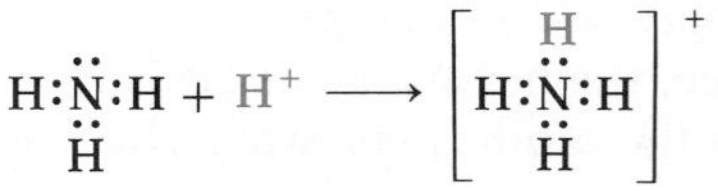

$$H:\overset{..}{\underset{..}{N}}:H + H^+ \longrightarrow \left[H:\overset{\overset{H}{..}}{\underset{..}{N}}:H \right]^+$$
$$\overset{\cdot\cdot}{H} \qquad\qquad\qquad \overset{\cdot\cdot}{H}$$

This, of course, is the reaction of ammonia as a base. An amine also has an unshared pair of electrons, and it too can act as a base. This reaction we shall consider in detail in Section 18.5. For the moment, let's concentrate on the fact that in the ammonium ion, nitrogen is bonded to four different hydrogens. Just as amines are derived from ammonia by replacement of one or several of the hydrogens with alkyl or aryl groups, so too can substituted ammonium ions be derived from the simple ammonium ion. Any or all of the hydrogens on the ammonium ion can be replaced by alkyl (or aryl) groups.

TABLE 18.1   *The Classification of Amines*

Class	Symbol	General Formula	Examples	
Primary	1°	R—NH$_2$	CH$_3$CH$_2$CH$_2$NH$_2$	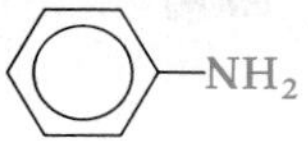 ⬡—NH$_2$
Secondary	2°	R—N—H 　　| 　　R	CH$_3$NHCH$_3$	⬡—NHCH$_3$
Tertiary	3°	R—N—R 　　| 　　R	CH$_3$—N—CH$_3$ 　　　| 　　　CH$_3$	⬡—N(CH$_3$)$_2$
Quaternary (salts)	4°	R—N$^+$—R 　　| 　　R	CH$_3$—N$^+$—CH$_3$ 　　　| 　　　CH$_3$	⬡—CH$_2$N$^+$(CH$_3$)$_3$

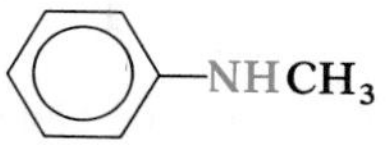

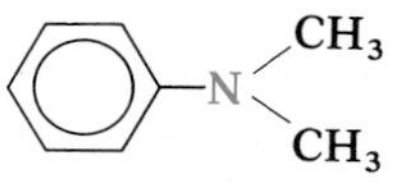

$$\left[\begin{matrix}H\\R-\underset{H}{\overset{|}{N}}-H\end{matrix}\right]^+ \left[\begin{matrix}R\\R-\underset{H}{\overset{|}{N}}-H\end{matrix}\right]^+ \left[\begin{matrix}R\\R-\underset{H}{\overset{|}{N}}-R\end{matrix}\right]^+ \left[\begin{matrix}R\\R-\underset{R}{\overset{|}{N}}-R\end{matrix}\right] ✗$$

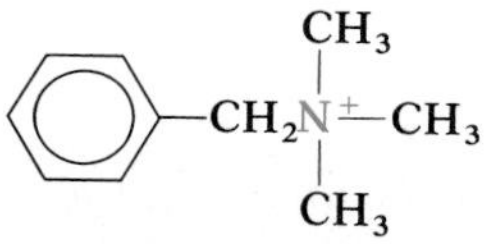

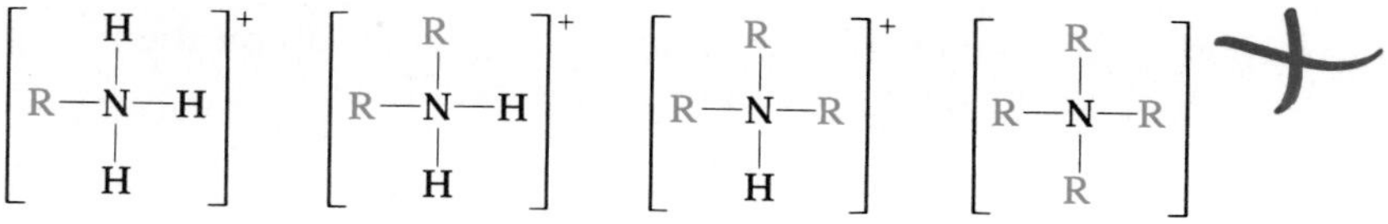

The ion in which all four hydrogens have been replaced by alkyl groups is termed a *quaternary* ammonium ion. Compounds that incorporate this type of ion are called **quarternary ammonium salts** (Table 18.1).

18.2 NAMING THE AMINES

To name simple aliphatic amines, we merely specify the alkyl groups attached to nitrogen and add the suffix **-amine.**

Name and classify

$$CH_3CH_2CH_2CH_2NH_2$$

example 18.1

The alkyl group attached to nitrogen is butyl; thus, the name is butylamine. There is only one alkyl group attached to nitrogen, so the amine is primary.

example 18.2

Name and classify

$$CH_3CH_2NHCH_2CH_3$$

There are two ethyl groups attached to the nitrogen. The compound is diethylamine, a secondary amine.

example 18.3

Name and classify

$$CH_3NHCH_2CH_2CH_3$$

There is a methyl group and a propyl group on nitrogen. The compound is methylpropylamine, a secondary amine.

example 18.4

Name and classify

$$CH_3CH_2\!-\!\overset{\displaystyle |}{\underset{\displaystyle CH_3}{N}}\!-\!CH_3$$

There are two methyl groups and one ethyl group on the nitrogen. The compound is ethyldimethylamine, a tertiary amine.

The primary amine in which the nitrogen is attached directly to a benzene ring is called **aniline.** Aryl amines are named as derivatives of this parent compound (Figure 18.2).

Compounds in which the nitrogen is attached to both a benzene ring and an alkyl group are also named as derivatives of aniline. The alkyl groups are named first, and their position of attachment (i.e., at the nitrogen atom) is indicated by the capital letter N.

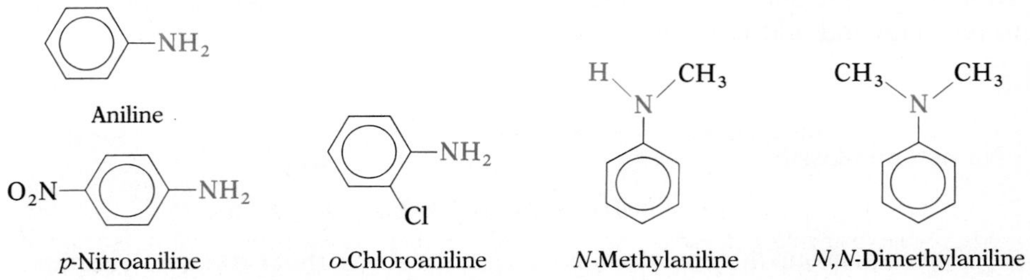

FIGURE 18.2 Aniline and some of its derivatives.

Name the compound

$$Br—\langle\bigcirc\rangle—NH_2$$

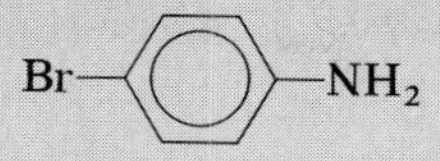

The compound is named as a derivative of aniline. It is *p*-bromoaniline.

example 18.5

Draw *p*-ethylaniline and N-ethylaniline.
    Both compounds are derivatives of aniline. The first compound is a primary amine with an ethyl group located *para* to the *amino* ($—NH_2$) group.

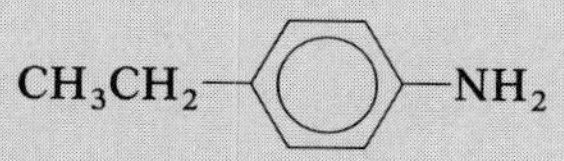

example 18.6

$$CH_3CH_2—\langle\bigcirc\rangle—NH_2$$

The second compound is a secondary amine with the ethyl group attached at the nitrogen.

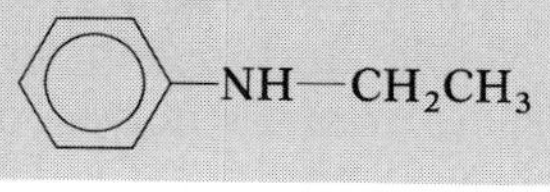

$$\langle\bigcirc\rangle—NH—CH_2CH_3$$

For amines that incorporate other functional groups or those in which the alkyl groups cannot be simply named, the amino group is named as a substituent (Figure 18.3).

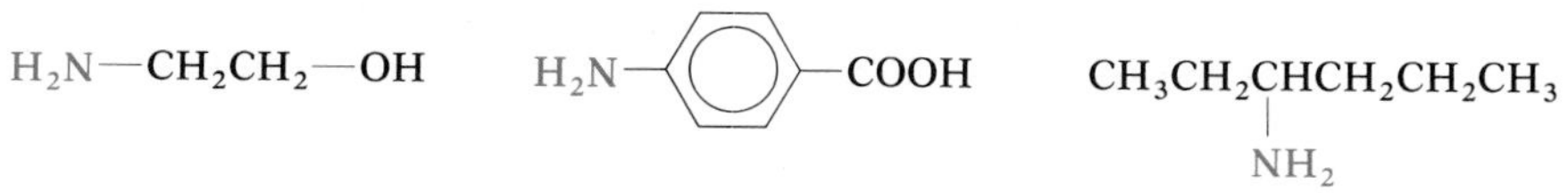

$$H_2N—CH_2CH_2—OH \qquad H_2N—\langle\bigcirc\rangle—COOH \qquad CH_3CH_2CHCH_2CH_2CH_3$$
$$\qquad\qquad\qquad\qquad\qquad\qquad\qquad\qquad\qquad\qquad\qquad\qquad | $$
$$\qquad\qquad\qquad\qquad\qquad\qquad\qquad\qquad\qquad\qquad\qquad\qquad NH_2$$

2-Aminoethanol   *p*-Aminobenzoic acid   3-Aminohexane
(ethanolamine)

FIGURE 18.3   *Compounds in which the amino group is named as a substituent. The 2-aminoethanol structure is included in the structure of many biologically active amines such as epinephrine and norepinephrine (Special Topic G) and acetylcholine (Chapter 23).*

example
18.7

Draw 2-amino-3-methylpentane.

Always start with the parent compound. First draw the five-carbon pentane chain. Then attach a methyl group at the third carbon atom and an amino group at the second.

$$\underset{\begin{array}{c} | \\ \text{NH}_2 \end{array}}{\text{CH}_3\text{CH}} - \underset{\begin{array}{c} | \\ \text{CH}_3 \end{array}}{\text{CH}} - \text{CH}_2\text{CH}_3$$

Ammonium ions in which one or more hydrogens have been replaced with alkyl groups are named in a manner analogous to that used for simple amines. The alkyl groups are named as substituents, and the parent species is regarded as the ammonium ion.

example
18.8

Name the following ions:

$$\overset{+}{\text{CH}_3\text{NH}_3} \qquad (\text{CH}_3)_2\overset{+}{\text{NH}_2} \qquad (\text{CH}_3)_3\overset{+}{\text{NH}} \qquad (\text{CH}_3)_4\overset{+}{\text{N}}$$

The structures are, in order, methylammonium, dimethylammonium, trimethylammonium, and tetramethylammonium ions.

Ions in which one of the ammonium ion's hydrogens has been replaced by a benzene ring are named as anilinium ions instead of ammonium ions. Such ions are prepared (Section 18.5) from the corresponding aniline, and the name reflects this fact.

Anilinium ion

18.3 PREPARATION OF AMINES

Organic chemists have a variety of methods of making amines. Among the more important is the reaction of an alkyl halide with ammonia or with another amine. In this reaction, the alkyl group of the halide is transferred to the nitrogen atom of the second reactant (ammonia in the equation below).

$$\text{RCl} + 2\,\text{NH}_3 \longrightarrow \text{RNH}_2 + \text{NH}_4\text{Cl}$$

A by-product of the initial reaction is hydrogen chloride, but this product immediately reacts with excess ammonia to form the ammonium chloride by-product shown in the equation. As is frequently done in organic chemistry, the equation may be written in a shorthand form that ignores this inorganic by-product.

$$\text{CH}_3\text{CH}_2\text{Cl} \xrightarrow{\text{NH}_3} \text{CH}_3\text{CH}_2\text{NH}_2$$

A second general method of preparing amines, called **reductive amination,** is a combination of two reactions we have encountered previously. A carbonyl compound is treated with ammonia and hydrogen in the presence of a catalyst. Initially, the carbonyl compound reacts with ammonia in a reaction similar to that described in Section 16.9.

$$R-\underset{\underset{R}{|}}{C}=\overbrace{O+H_2}NH \longrightarrow R-\underset{\underset{R}{|}}{C}=N-H + H_2O$$

The double bond of the intermediate product is then reduced by the hydrogen-catalyst combination used previously with alkenes (Section 14.11).

$$R-\underset{\underset{R}{|}}{C}=N-H \xrightarrow{\text{H}_2\text{, Ni catalyst}} R-\underset{\underset{R}{|}}{\overset{\overset{H}{|}}{C}}-\overset{\overset{H}{|}}{N}-H$$

Overall, the reaction involves the conversion of an aldehyde or ketone group to an amino group.

$$CH_3CH_2CH_2\overset{\overset{O}{||}}{C}CH_3 \xrightarrow{\text{NH}_3\text{, H}_2\text{, Ni}} CH_3CH_2CH_2\overset{\overset{NH_2}{|}}{C}HCH_3$$

The reaction is of great interest to chemists because it is versatile and frequently gives high yields of desired products. But we mention it particularly because living systems use a similar reaction sequence to convert carbonyl groups to amine groups. The reagents these systems use differ somewhat from those we've shown, and the catalysts are enzymes rather than metals, but the sequence of transformations is essentially the same as in reductive amination.

18.4 PHYSICAL PROPERTIES OF AMINES

Primary and secondary amines have hydrogen on nitrogen; thus, they are capable of intermolecular hydrogen bonding. These forces are not as strong as those between alcohol molecules (which have hydrogen on oxygen, a more electronegative element than nitrogen). Amines boil at higher temperatures than alkanes but at lower temperatures than alcohols of comparable molecular weight. Tertiary amines have no hydrogen on nitrogen and cannot form intermolecular hydrogen bonds. They have boiling points comparable to those of the ethers (Table 18.2).

All three classes of amines can hydrogen bond to water. Amines of low molecular weight are quite soluble in water, the borderline of water solubility coming at five or six carbon atoms.

Amines have interesting (!) odors. The simple ones smell very much like ammonia. Higher aliphatic amines smell like decaying fish. Or perhaps we

$$H_2NCH_2CH_2CH_2CH_2NH_2$$

1,4-Diaminobutane
(putrescine)

$$H_2NCH_2CH_2CH_2CH_2CH_2\ NH_2$$

1,5-Diaminopentane
(cadaverine)

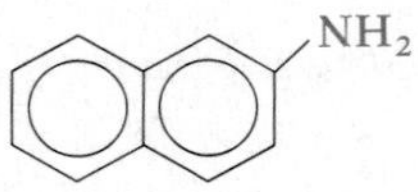

β-Naphthylamine

FIGURE 18.4
Some amines of interest.
Putrescine and cadaverine
have odors indicated by their
names. β-Naphthylamine is
a carcinogen.

TABLE 18.2 *Physical Properties of Some Amines and Comparable Oxygen-Containing Compounds*

Compound	Class	Molecular Weight	Boiling Point (in degrees Celsius)
Butylamine	1°	73	78
Diethylamine	2°	73	55
Butyl alcohol	—	74	118
Propylamine	1°	59	49
Trimethylamine	3°	59	3
Ethyl methyl ether	—	60	6

should put it the other way around: decaying fish give off odorous amines. The stench of rotting flesh is due in part to putrescine and cadaverine, two compounds that are diamines.

Aromatic amines generally are quite toxic. They are readily absorbed through the skin, and you must take care to prevent serious accidents when working with these compounds. Several aromatic amines, including especially β-naphthylamine, are potent carcinogens (cancer-inducing chemicals) (Figure 18.4).

18.5 THE AMINES AS BASES

As we noted in Section 18.1, ammonia is basic; it can accept a proton from water to form ammonium ions and hydroxide ions.

$$\text{:NH}_3 + H_2O \rightleftharpoons NH_4^+ + OH^-$$

Ammonia is a weak base; this equilibrium strongly favors the un-ionized forms. Similarly, the amines have an unshared electron pair on nitrogen and can accept protons.

$$CH_3\ddot{N}H_2 + H_2O \rightleftharpoons [CH_3NH_3]^+ + OH^-$$

Simple aliphatic amines are somewhat more basic than ammonia, although still much less basic than compounds such as sodium hydroxide. Aromatic amines, such as aniline, are much weaker bases than ammonia.

Amines react readily with strong acids, such as the mineral acids, to form salts.

$$
\begin{array}{c}
CH_3 \\
| \\
CH_3-N: \\
| \\
CH_3
\end{array}
+ HNO_3(aq) \longrightarrow
\left[
\begin{array}{c}
CH_3 \\
| \\
CH_3-N\,H \\
| \\
CH_3
\end{array}
\right]^+
NO_3^-
$$

Trimethylamine   Trimethylammonium
                                                        nitrate

Amine salts are named like other salts: the name of the cation is given first, followed by that of the anion. Remember that the ions formed from aliphatic amines are named as substituted ammonium ions (Example 18.8).

example 18.9

Name the salt

$$[CH_3NH_2CH_2CH_3]^+CH_3COO^-$$

The cation has a methyl and an ethyl group attached to nitrogen and is, therefore, the ethylmethylammonium ion. The anion is the acetate ion. The salt is therefore ethylmethylammonium acetate.

Salts of aniline are named as anilinium compounds (Figure 18.5). An older system, still in use for naming drugs, calls the salt of aniline and hydrochloric acid "aniline hydrochloride." By this older system, the formula of the compound is frequently drawn to correspond to the name. Keep in mind that these

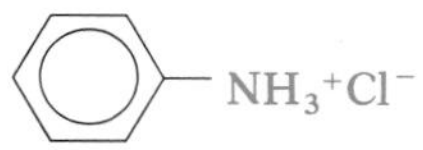

Anilinium chloride

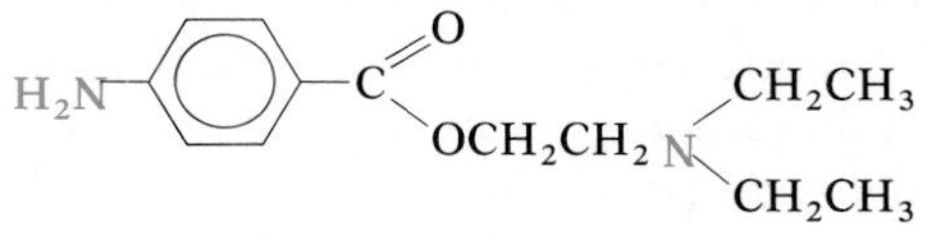

Procaine

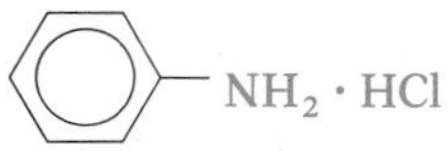

"Aniline hydrochloride"

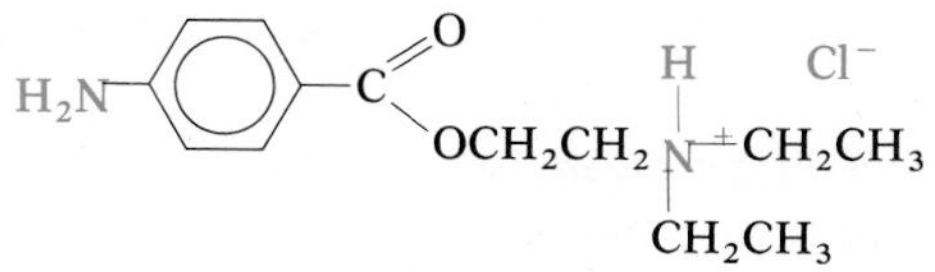

Procaine hydrochloride (Novocaine)

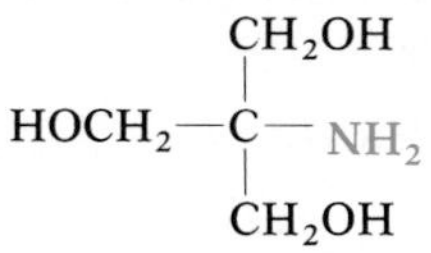

"Tris"

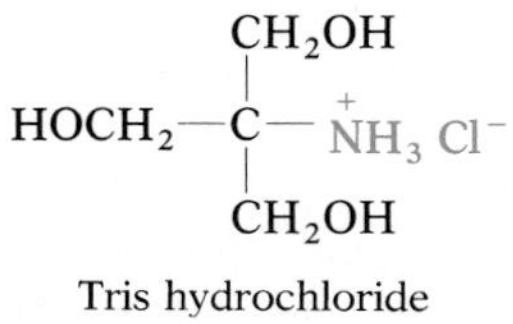

Tris hydrochloride

FIGURE 18.5 Amine salts are more soluble in water than the "free bases" from which they are derived. The salts of amines with hydrochloric acid are often named as hydrochlorides.

compounds are really ionic—they are salts—even though the name and formula seem to indicate a loose association of molecules. The properties of the compounds (solubility, for example) are those characteristic of salts.

Physiologically active amines are often converted to salts and thereby rendered water soluble. For instance, procaine is soluble only to the extent of 0.5 g in 100 g of water. The hydrochloride is soluble to the remarkable degree of 100 g in 100 g of water. Procaine hydrochloride, perhaps better known by the trade name Novocaine, is widely used as a local anesthetic (Figure 18.5).

We also make use of the chemistry of amines when we put lemon juice on fish. The unpleasant fishy odor is due to amines. The citric acid in the juice converts the amines to nonvolatile salts, thus eliminating the odor.

You know that a weak acid and its salt (e.g., acetic acid and sodium acetate) can be used in the preparation of buffer solutions (Section 11.5). Similarly, amines and their salts can also be used in the preparation of buffers. Whereas the acid-acid salt combination yields a solution that is buffered at acidic pH values, the amine-amine salt buffer will stabilize the pH in a basic range. An important example is tris(hydroxymethyl)aminomethane, often called simply "tris." This compound and its salt, tris hydrochloride, buffer in the range of 7 to 9 pH units. These compounds find wide use in the cosmetic and textile industries, in cleaning compounds, and in biochemical research. "Tris" is also used in the treatment of metabolic acidosis (Figure 18.5).

18.6 OTHER CHEMICAL PROPERTIES OF THE AMINES

In the preceding section, we saw that water-soluble amines give basic solutions and that amines generally react with mineral acids to form salts. In Chapter 17, the preparation of substituted amides from amines is mentioned. Primary or secondary amines can react with acid chlorides or acid anhydrides to form amides with substituents on the nitrogen atom.

Benzoyl chloride   Diethylamine   N,N-Diethylbenzamide

Acetic anhydride   Propylamine   N-Propylacetamide

Amides are also made by heating of ammonium salts of carboxylic acids.

$$CH_3C\underset{O^-NH_4^+}{\overset{O}{\diagup}} \xrightarrow{\text{heat}} CH_3C\underset{NH_2}{\overset{O}{\diagup}} + H_2O$$

Ammonium acetate   Acetamide

These and similar reactions are of considerable importance in the synthesis of proteins (Chapter 22), plastics (Chapter 19), and medicinals.

We shall consider two additional reactions of amines—reactions of considerable importance to those studying biological processes. First, amines react with nitrous acid. The nature of the products depends on the class of the amine involved. Primary amines give a quantitative yield of nitrogen gas.

$$R—NH_2 + HNO_2 \longrightarrow N_2(g) + \text{other products}$$

Nitrous acid is unstable, so it is usually made *in situ* (right in the reaction vessel) by addition of hydrochloric acid to sodium nitrite.

$$NaNO_2(aq) + HCl(aq) \longrightarrow HNO_2(aq) + NaCl(aq)$$

When a primary amine is added to this solution, bubbles of nitrogen gas can be seen escaping. These bubbles are an indication that the amine is a primary one, because secondary and tertiary amines do not release nitrogen gas. Therefore, this reaction serves as a qualitative test for primary amines. If the amount of the original amine is carefully measured and the nitrogen is collected and its volume measured and corrected to conditions of standard temperature and pressure, the reaction can be used for the quantitative determination of primary amino groups. One molecule of nitrogen (N_2) is liberated for each free amino group. The procedure is referred to as the **Van Slyke method** and is used especially with amino acids and proteins.

Secondary amines react with nitrous acid to form oily N-nitroso compounds.

$$R—N—[H + HO]—N{=}O \longrightarrow R—N—N{=}O + H_2O$$
$$\underset{R}{|} \qquad\qquad\qquad\qquad \underset{R}{|}$$

The appearance of an oil upon the addition of an amine to a solution of nitrous acid indicates that the amine is probably secondary. However, the test is no longer recommended, because most N-nitroso compounds are potent carcinogens. Since our diet contains sodium nitrite, and our stomachs contain hydrochloric acid, nitrosamines are formed from secondary amines in the breakdown products of the food we eat. Sodium nitrite is added to preserved meats to prevent botulism, a deadly affliction caused by a bacterial toxin. It has been postulated that the high rates of stomach cancer in countries that have prepared meats in the diet are due to the nitrites found in these products.

Tertiary amines also react with nitrous acid. Generally, the only product is a salt. Sometimes, though, the tertiary amine is cleaved to a secondary one that then can form a nitroso derivative.

A second reaction used to detect the presence of amines is that with the compound ninhydrin. Some amines react with ninhydrin to form a purple to blue anion.

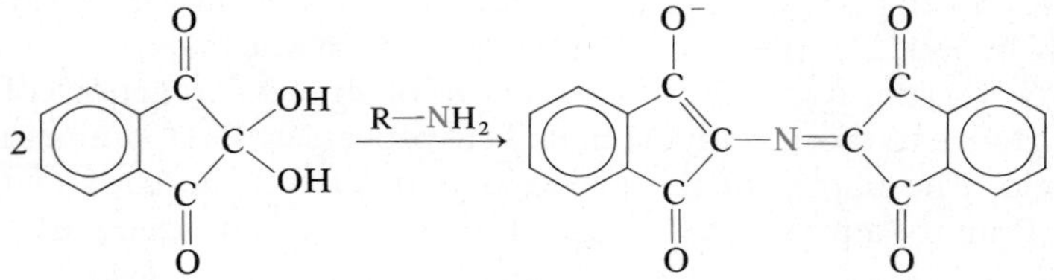

The ninhydrin test is especially suited to the detection of amino acids. Mixtures of amino acids, such as those that result from the hydrolysis of proteins, can be separated by a process called paper chromatography. At the conclusion of such a separation, the individual amino acids are scattered at different locations on the paper chromatogram. The amino acid cannot be seen, however, until the chromatogram is sprayed with ninhydrin and the colored ion becomes visible. Then the position of the constitutents of the original mixture can be compared with those of known amino acids, and an identification can be made on this basis.

18.7 HETEROCYCLIC AMINES

In Chapter 14, a variety of cyclic hydrocarbons were introduced. All the atoms in the rings of these compounds are carbon atoms. There are other cyclic compounds in which the molecules have nitrogen, oxygen, sulfur, or other elements in the rings. These are called **heterocyclic compounds;** those rings with only carbon atoms are called **carbocyclic compounds.** In this section, we will consider a variety of heterocyclic amines, compounds of considerable importance (Figure 18.6).

The compounds pyrrole and pyrrolidine each have four carbon atoms and one nitrogen atom in a ring. Pyrrole is an aromatic compound, having properties similar to those of benzene. Pyrrolidine, with four more hydrogen atoms than pyrrole, behaves like an aliphatic amine. Imidazole also has a five-membered ring, but it contains two nitrogen atoms and only three carbon atoms. Like pyrrole, imidazole has aromatic properties.

Pyridine and piperidine each have five carbon atoms and one nitrogen atom. Pyridine is aromatic; piperidine is aliphatic. Another six-membered heterocycle is pyrimidine, which has two nitrogen atoms and four carbon atoms. Pyrimidine is another aromatic compound.

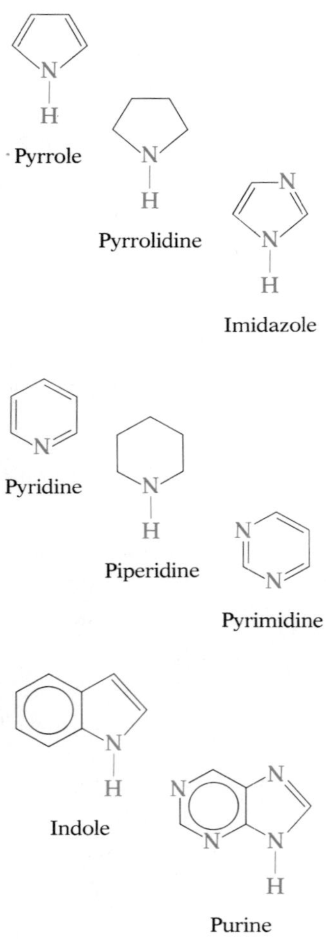

Pyrrole

Pyrrolidine

Imidazole

Pyridine

Piperidine

Pyrimidine

Indole

Purine

FIGURE 18.6
Some heterocyclic amines.

FIGURE 18.7 *Jacques Louis David's painting* The Death of Socrates *(1787) shows Socrates drinking the cup of hemlock to carry out the death sentence decreed by the rulers of Athens. (Metropolitan Museum of Art, New York. Wolfe Fund, 1931.)*

Other heterocyclic compounds have two rings that share a common side (a situation we encountered with naphthalene among the carbocyclic compounds). Indole has a benzene ring fused with a pyrrole ring. Purine has a pyrimidine ring sharing a side with an imidazole structure. Bases related to purine and pyrimidine make up a part of the structure of the nucleic acids, compounds that comprise the genetic material of cells and that direct protein synthesis. Nucleic acids are discussed in Chapter 23, in which we encounter one of the truly outstanding examples of the critical importance of the shape of molecules and of molecular structure in general.

Many amines, particularly heterocyclic ones, occur naturally in plants. Like most amines, these compounds are basic. They are called **alkaloids,** a name that means "like alkalis." Knowledge of many of these, at least in crude form, dates back to antiquity. Opium, which contains about 10% morphine (Special Topic F) as the principal alkaloid, has been used for thousands of years, although morphine was not isolated until 1805.

When the Greek philosopher Socrates was accused of corrupting the youth of Athens in 399 B.C., he was given the choice of exile or death. He chose the latter and implemented his decision by drinking a cup of hemlock (Figure 18.7). His hemlock was probably prepared from the fully grown but unripened fruit of *Conium maculatum,* or poison hemlock (Figure 18.8). The fruit would probably have been carefully dried and then brewed into a "tea." Hemlock contains several alkaloids, but the principal one is coniine. Coniine causes nausea, weakness, paralysis, and ultimately—as in the case of Socrates—death.

Hemlock tea, anyone?

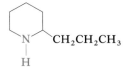

Coniine

FIGURE 18.8
Conium maculatum (poison hemlock).

■ PROBLEMS ■

1. Draw structural formulas for the eight isomeric amines that have the molecular formula $C_4H_{11}N$. Give each a common name and classify it as a primary, secondary, or tertiary amine.

2. Draw structural formulas for the five isomeric amines that have the molecular formula C_7H_9N and that contain a benzene ring. Name each compound and classify it as a primary, secondary, or tertiary amine.

3. Identify the following compounds as amines, alcohols, phenols, or ethers. In the case of amines and alcohols, classify the compounds as primary, secondary, or tertiary.

   a. $CH_3CH_2CH_2OH$

   b. $CH_3CH_2CH_2NH_2$

   c. $CH_3\overset{\underset{\displaystyle OH}{|}}{C}HCH_3$

   d. $CH_3\overset{\underset{\displaystyle NH_2}{|}}{C}HCH_3$

   e. (cyclic ether structure with O)

   f. (piperidine structure with N—H)

   g. $CH_3CH_2NHCH_2CH_3$

   h. $CH_3CH_2OCH_2CH_3$

   i. $CH_3\overset{\underset{\displaystyle CH_3}{|}}{N}CH_3$

   j. $CH_3\overset{\underset{\displaystyle CH_3}{|}}{\overset{\overset{\displaystyle OH}{|}}{C}}CH_3$

   k. (benzene ring)—NH_2

   l. (benzene ring)—OH

4. Distinguish between a carbocyclic compound and a heterocyclic compound.

5. Draw structural formulas for the following:
   a. dimethylamine   **c.** cyclohexylamine
   b. diethylmethylamine

6. Draw structural formulas for the following:
   a. 3-aminopentane   **b.** 1,6-diaminohexane

7. Draw the structural formula of 2-aminoethanol.

8. Draw structural formulas for the following:
   a. aniline   **c.** N,N-dimethylaniline
   b. *m*-bromoaniline

9. Draw structural formulas for the following:
   a. pyridine   **b.** purine   **c.** pyrimidine

10. Draw structural formulas for the following:
    a. anilinium bromide
    b. "aniline hydrochloride"
    c. tetramethylammonium chloride

11. Name the following compounds:
    a. $CH_3CH_2CH_2NH_2$

    b. $CH_3CH_2NHCH_3$

    c. $CH_3CH_2\overset{\underset{\displaystyle CH_2CH_3}{|}}{N}CH_2CH_3$

12. Name the following compounds:
    a. O_2N—(benzene ring)—NH_2

    b. (benzene ring)—$NHCH_2CH_3$

13. Name the following compound:
    $[CH_3CH_2NH_2CH_2CH_3]^+Br^-$

14. Draw the structural formula of, and name the amine formed in, each of these reactions.

    a. $CH_3\overset{\underset{\displaystyle CH_3}{|}}{C}HCH_2—Cl \xrightarrow{NH_3}$

    b. $CH_3CH_2CH_2CH_2CH_2—Cl \xrightarrow{NH_3}$

15. Draw the structural formula of, and name the amine formed in, each of these reactions.

    a. $CH_3CH_2—\overset{\displaystyle O}{\underset{\displaystyle H}{C}} \xrightarrow{NH_3,\ H_2,\ Ni}$

    b. $CH_3\overset{\underset{\displaystyle CH_3}{|}}{C}H—CH_2—\overset{\displaystyle \overset{\displaystyle O}{\|}}{C}—CH_3 \xrightarrow{NH_3,\ H_2,\ Ni}$

16. What is an alkaloid? Name several alkaloids.

17. Which compound has the higher boiling point: butylamine or pentane? Why?

18. Which compound has the higher boiling point: butylamine or butyl alcohol? Why?

19. Which compound has the higher boiling point: trimethylamine or propylamine? Why?

20. Which compound is more soluble in water: $CH_3CH_2CH_3$ or $CH_3CH_2NH_2$? Why?

21. Which compound is more soluble in water: $CH_3CH_2CH_2NH_2$ or $CH_3CH_2CH_2CH_2CH_2CH_2NH_2$? Why?

22. Which of the following compounds is more soluble in water? Why?

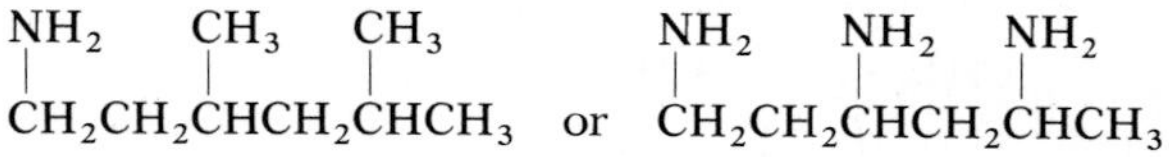

$$\underset{\underset{CH_2CH_2CHCH_2CHCH_3}{|\qquad\quad|\quad\;}}{NH_2\quad\;\;CH_3\;\;CH_3} \quad \text{or} \quad \underset{\underset{CH_2CH_2CHCH_2CHCH_3}{|\qquad\;\;|\qquad|}}{NH_2\;\;NH_2\;\;NH_2}$$

23. Amine X is insoluble in water, yet it dissolves readily in aqueous hydrochloric acid. Explain.

24. Draw the structural formula of the salt formed.

$$CH_3NH_2 + HBr \longrightarrow$$

25. Draw the structural formula of the salt formed.

—NHCH$_3$ + HNO$_3$ $\longrightarrow$

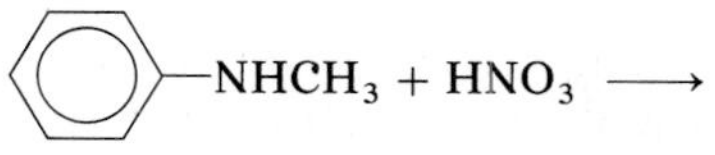

26. Draw the structural formula of the salt formed.

$$\underset{\underset{CH_3}{|}}{CH_3-N-CH_3} + H_2SO_4 \longrightarrow$$

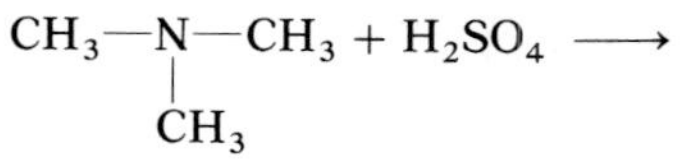

27. Draw the structural formula of the salt formed.

+ HCl $\longrightarrow$

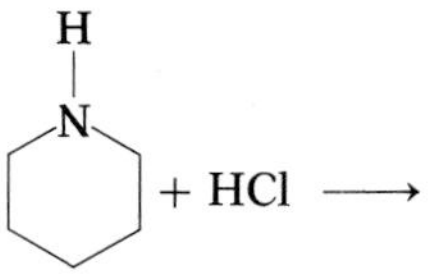

28. Draw the amide, if any, that is derived from hexanoic acid and butylamine.

29. Draw the amide, if any, that is derived from

$$\underset{O}{\overset{\displaystyle\parallel}{CH_3CH_2C}}-OH \quad \text{and} \quad \underset{\underset{CH_3}{}}{CH_3-\overset{\displaystyle H}{\overset{|}{N}}-CH_3}$$

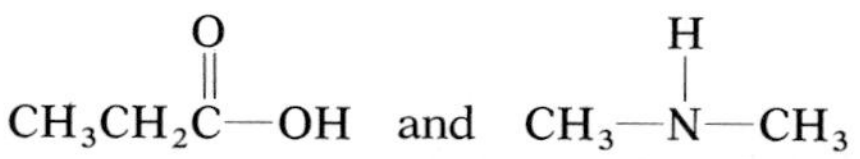

30. Draw the amide, if any, that is derived from benzoic acid and aniline.

31. Draw the amide, if any, that is derived from

$$\underset{O}{\overset{\displaystyle\parallel}{CH_3C}}-OH \quad \text{and} \quad \underset{\underset{CH_3}{|}}{CH_3-N-CH_3}$$

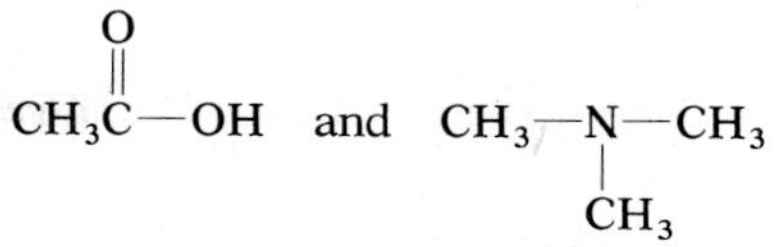

32. Draw the carboxylic acid and amine from which the following amide was formed:

$$\underset{\underset{CH_3}{|}}{CH_3CH_2N}-\overset{\displaystyle O}{\overset{\displaystyle\parallel}{C}}CH_2CH_3$$

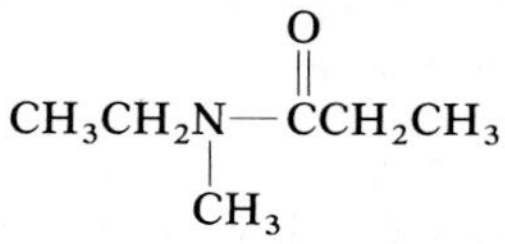

33. Draw the carboxylic acid and amine from which the following amide was formed:

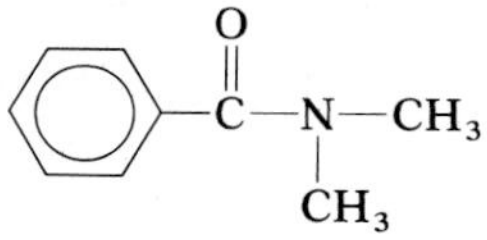

34. Draw the carboxylic acid and amine from which the following amide was formed:

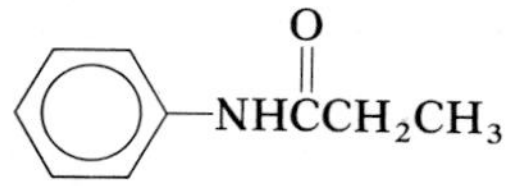

35. Draw the carboxylic acid and amine from which the following amide was formed:

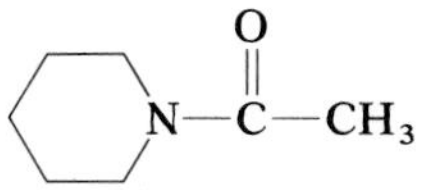

36. A carboxyl group and an amino group combined to form the amide functional group in the following compound. Draw the two starting materials for the reaction.

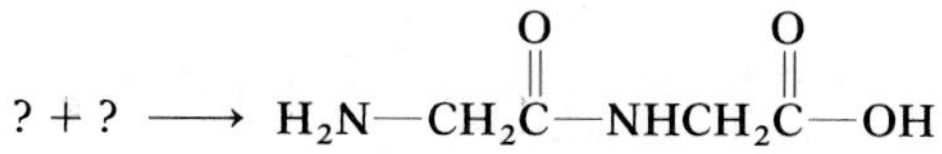

$$? + ? \longrightarrow H_2N-CH_2\overset{\displaystyle O}{\overset{\displaystyle\parallel}{C}}-NHCH_2\overset{\displaystyle O}{\overset{\displaystyle\parallel}{C}}-OH$$

37. Draw the structural formula for the principal organic product of each of these reactions.

a.

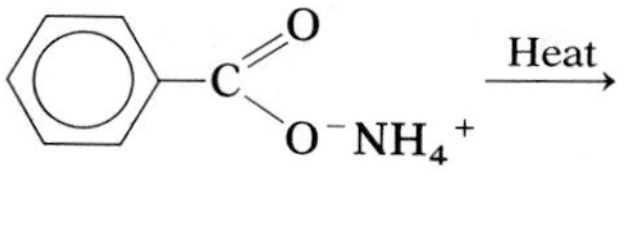

b. 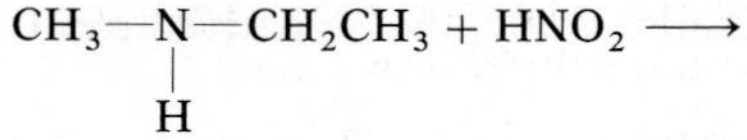 $CH_3\!-\!\overset{\displaystyle |}{\underset{\displaystyle H}{N}}\!-\!CH_2CH_3 + HNO_2 \longrightarrow$

38. Why are nitrate and nitrite food additives considered by some people to be health hazards?

39. List the reagents needed to carry out the following conversions:

a. $CH_3CH_2CH_2CH_2CH_2Br \overset{?}{\longrightarrow}$

$CH_3CH_2CH_2CH_2CH_2NH_2$

b. 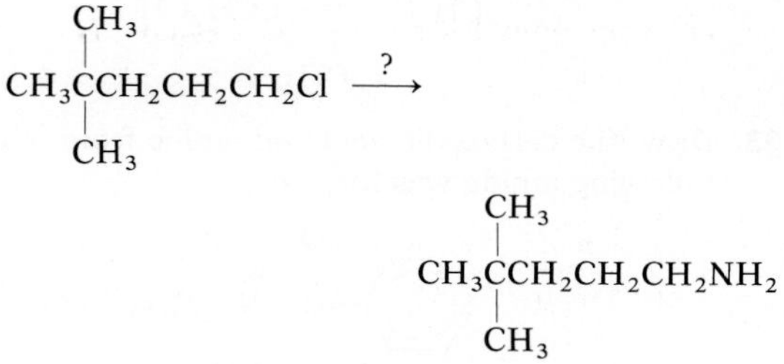 $CH_3\overset{\displaystyle CH_3}{\underset{\displaystyle CH_3}{\overset{\displaystyle |}{\underset{\displaystyle |}{C}}}}CH_2CH_2CH_2Cl \overset{?}{\longrightarrow}$

$CH_3\overset{\displaystyle CH_3}{\underset{\displaystyle CH_3}{\overset{\displaystyle |}{\underset{\displaystyle |}{C}}}}CH_2CH_2CH_2NH_2$

c. $CH_3\overset{\displaystyle }{\underset{\displaystyle O}{\overset{\displaystyle |}{C}}}CH_3 \overset{?}{\longrightarrow} CH_3\overset{\displaystyle }{\underset{\displaystyle NH_2}{\overset{\displaystyle |}{C}}HCH_3}$

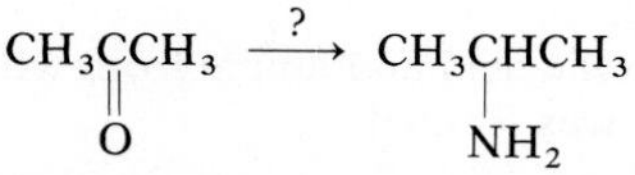

40. List the reagents needed to carry out the following conversions:

a. 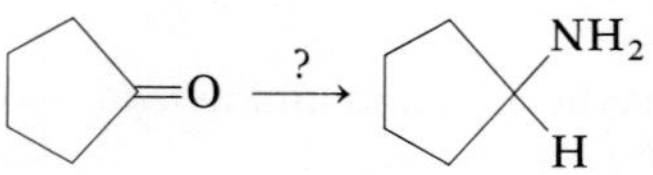 $\bigcirc\!=\!O \overset{?}{\longrightarrow} \bigcirc\!\!\overset{\displaystyle NH_2}{\underset{\displaystyle H}{}}$

b. 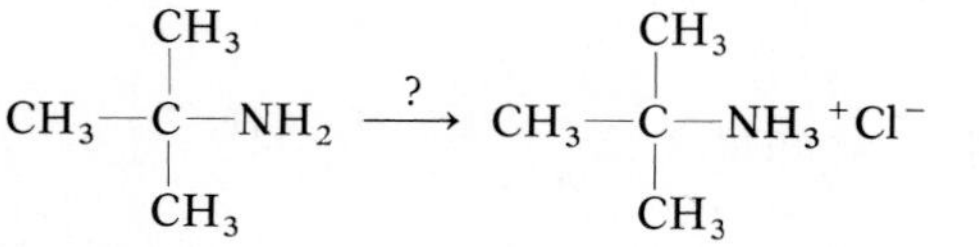 $CH_3\!-\!\overset{\displaystyle CH_3}{\underset{\displaystyle CH_3}{\overset{\displaystyle |}{\underset{\displaystyle |}{C}}}}\!-\!NH_2 \overset{?}{\longrightarrow} CH_3\!-\!\overset{\displaystyle CH_3}{\underset{\displaystyle CH_3}{\overset{\displaystyle |}{\underset{\displaystyle |}{C}}}}\!-\!NH_3{}^+Cl^-$

c. $\bigcirc\!\!-\!NH_2 \overset{?}{\longrightarrow} \bigcirc\!\!-\!NH_3{}^+NO_3{}^-$

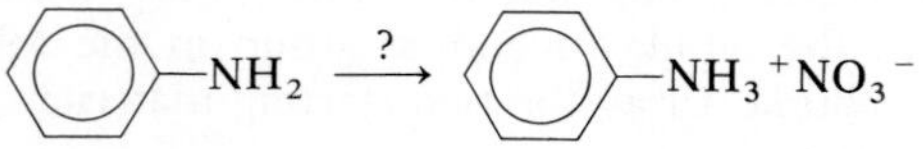

d. 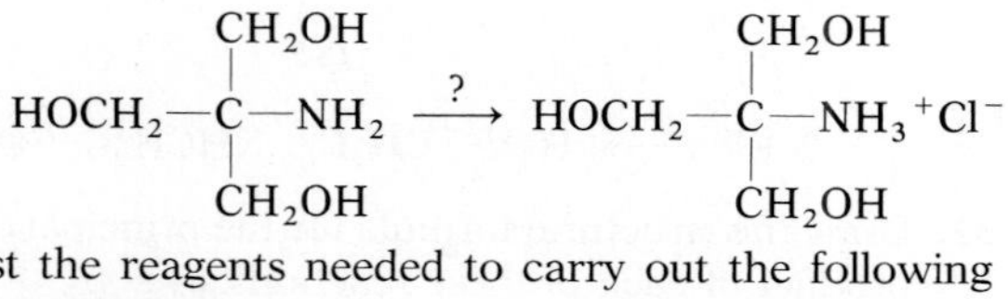 $HOCH_2\!-\!\overset{\displaystyle CH_2OH}{\underset{\displaystyle CH_2OH}{\overset{\displaystyle |}{\underset{\displaystyle |}{C}}}}\!-\!NH_2 \overset{?}{\longrightarrow} HOCH_2\!-\!\overset{\displaystyle CH_2OH}{\underset{\displaystyle CH_2OH}{\overset{\displaystyle |}{\underset{\displaystyle |}{C}}}}\!-\!NH_3{}^+Cl^-$

41. List the reagents needed to carry out the following conversions:

a. 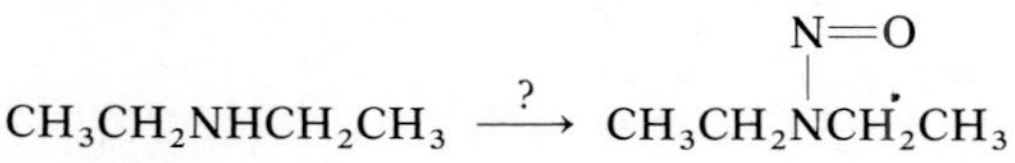 $CH_3CH_2NHCH_2CH_3 \overset{?}{\longrightarrow} CH_3CH_2\overset{\displaystyle N=O}{\overset{\displaystyle |}{N}}CH_2CH_3$

b. 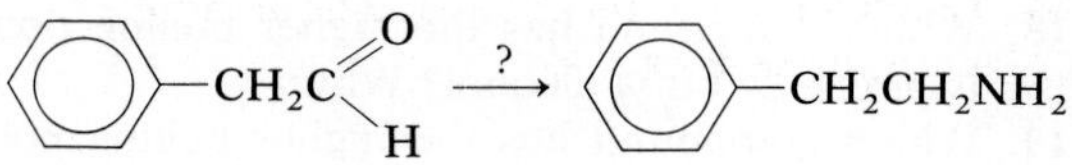 $\bigcirc\!\!-\!CH_2\overset{\displaystyle O}{\overset{\displaystyle \|}{C}}\overset{}{\underset{\displaystyle H}{}} \overset{?}{\longrightarrow} \bigcirc\!\!-\!CH_2CH_2NH_2$

c. $\bigcirc\!\!-\!\overset{\displaystyle O}{\overset{\displaystyle \|}{C}}\!-\!O^-NH_4{}^+ \overset{?}{\longrightarrow} \bigcirc\!\!-\!\overset{\displaystyle O}{\overset{\displaystyle \|}{C}}\!-\!NH_2$

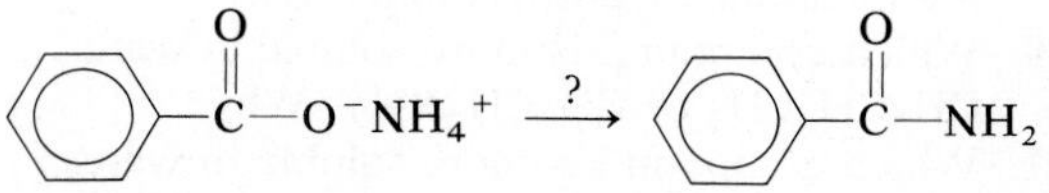

42. List the reagents necessary to carry out the following conversions:

a.

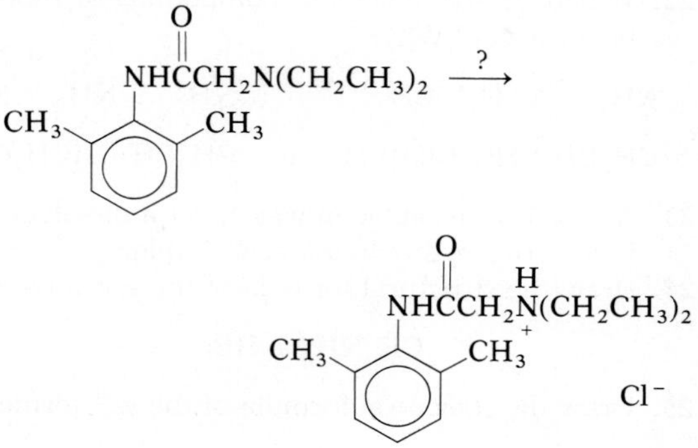

$CH_3\overset{\displaystyle \overset{O}{\|}}{\underset{\displaystyle CH_3}{\bigcirc\ NHCCH_2N(CH_2CH_3)_2}} \overset{?}{\longrightarrow}$

$CH_3\overset{\displaystyle \overset{O}{\|}\ \ \ \ \overset{H}{\underset{+}{}}}{\underset{\displaystyle CH_3}{\bigcirc\ NHCCH_2N(CH_2CH_3)_2}}\ \ Cl^-$

b. 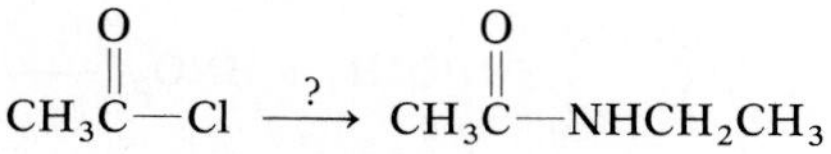 $CH_3\overset{\displaystyle O}{\overset{\displaystyle \|}{C}}\!-\!Cl \overset{?}{\longrightarrow} CH_3\overset{\displaystyle O}{\overset{\displaystyle \|}{C}}\!-\!NHCH_2CH_3$

c. $CH_3CH_2CH_2CH_2NH_2 \overset{?}{\longrightarrow}$

$CH_3CH_2CH_2CH_2NH\overset{\displaystyle O}{\overset{\displaystyle \|}{C}}CH_3$

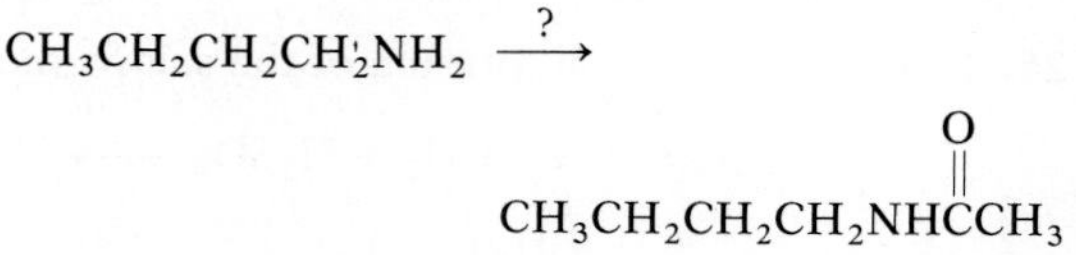

d.

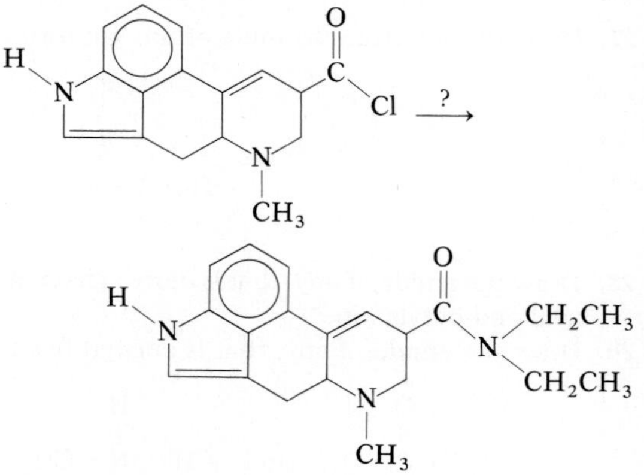

43. What chemical reaction occurs when lemon juice is added to fish? Why is this desirable?

44. Classify each of the following as an amine, an amide, both, or neither.

a.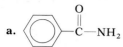

b. 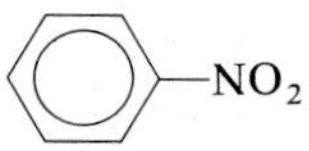 ⟨⟩—NO_2

c. $CH_3CH_2NHCH_2CH_3$

d. $CH_3CH_2NH\overset{\overset{\displaystyle O}{\|}}{C}CH_2CH_3$

e. $H_2NCH_2CH_2CH_2\overset{\overset{\displaystyle O}{\|}}{C}NH_2$

f. 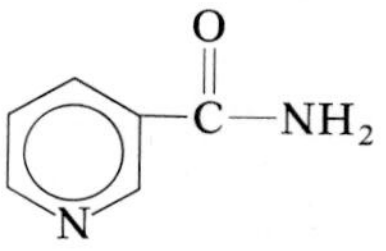 pyridine ring with $-\overset{\overset{\displaystyle O}{\|}}{C}-NH_2$

45. Tell whether the following compounds form acidic, basic, or neutral solutions in water:

a. $CH_3CH_2NH_2$

b. $CH_3\overset{\overset{\displaystyle O}{\|}}{C}NH_2$

c. $CH_3\overset{\overset{\displaystyle O}{\|}}{C}OH$

d. CH_3CH_2OH

46. Write equations for the reaction of anthranilic acid (*o*-aminobenzoic acid) with each of the following:

a. NaOH

b. HCl

c. acetic anhydride

d. ammonia

e. H_2SO_4

f. CH_3OH, H^+

G

Brain Amines and Related Drugs

S ome drugs are rather simple amines. Others, including some discussed in other chapters, are alkaloids (Section 18.7). In this special topic, we will take a look at some amines and related compounds that affect our mental state, render us insensitive to pain, put us to sleep, and calm our anxieties. First, though, let's take a look at nerve cells (**neurons**) and how they work.

G.1 SOME CHEMISTRY OF THE NERVOUS SYSTEM

The nervous system is made up of billions of neurons with 10^{15} connections between them. The brain operates with a power output of about 25 W and has capacity for about 10 trillion bits of information. Nerve cells vary a great deal in shape and size. One type is shown in Figure G.1. The essential parts of each cell are the cell body, the axon, and the dendrites. We discuss here only those nerves that make up the involuntary (autonomic) nervous system. These nerves carry messages between the organs and glands that act involuntarily (such as the heart, the digestive organs, and the lungs) and the brain and spinal column.

Although the axons on a given nerve cell may be up to 60 cm long, there is no continuous pathway from an organ to the central nervous system. Messages must be transmitted across tiny, fluid-filled gaps, or **synapses** (Figure G.2). When an electrical signal from the brain reaches the end of an axon, specific chemicals (called **neurotransmitters**) that carry the impulse across the synapse to the next cell are liberated. There are perhaps a few dozen neurotransmitters. Each has a specific function. Messages are carried to other nerve cells, to

FIGURE G.1
A human nerve cell.

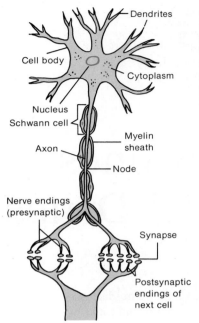

FIGURE G.2
Schematic diagram of the pathway by which messages are transmitted to and from an acceptor cell in a gland or an organ to the central nervous system.

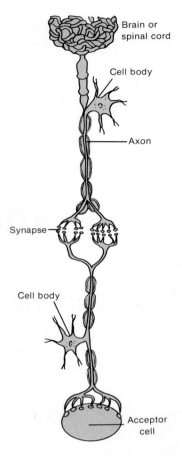

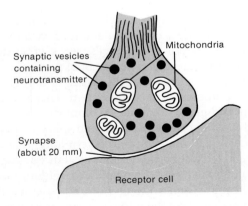

FIGURE G.3   Schematic diagram of a synapse. When an electrical signal reaches the nerve ending, neurotransmitter molecules are released from the vesicles. They then migrate across the narrow gap (synapse) and move to the receptor cell, where they fit specific sites.

muscles, and to the endocrine glands (such as the adrenal glands). Each neurotransmitter fits a specific receptor site on the receptor cell (Figure G.3). Many drugs (and some poisons) act either by blocking the receptor and preventing the neurotransmitter from acting on it, or by mimicking the action of the neurotransmitter. Several of the neurotransmitters are amines, as are some of the drugs that affect the chemistry of our brains.

G.2 BRAIN AMINES

We all have our ups and downs in life. These moods probably result from multiple causes, but it appears likely that a variety of chemical compounds formed in the brain are involved. Before we consider these ups and downs, though, let's take a look at epinephrine, an amine.

Commonly called adrenalin, epinephrine is secreted by the adrenal glands. A tiny amount of epinephrine causes a great increase in blood pressure. When a person is under stress or is frightened, the flow of adrenalin prepares the body for fight or flight. Because culturally imposed inhibitions prevent fighting or fleeing in most modern situations, the adrenalin-induced supercharge is not used. This sort of frustration has been implicated in some forms of mental illness.

Epinephrine
(adrenalin)

Norepinephrine

One widely held theory of a biochemical basis for mental illness involves a relative of epinephrine. Norepinephrine (NE) is a neurotransmitter formed in the brain. When NE is formed in excess, the person is in an elated, perhaps hyperactive, state. In large excess, NE could induce a manic state. A deficiency of NE, on the other hand, could cause depression.

Drugs that block the action of NE could also lead to depression, while those that enhance or mimic its action act as stimulants. Several such drugs are discussed in the sections that follow.

Another neurotransmitter, serotonin, also seems to play a role in mental illness. Serotonin is involved in sleep, sensory perception, and the regulation of body temperature. Its exact role in mental illness is not clear. A metabolite of serotonin, 5-hydroxyindoleacetic acid (5-HIAA), is found in unusually *low* levels in the spinal fluid of suicide victims. This indicates that abnormal serotonin metabolism may play a role in depression.

Richard Wurtman of the Massachusetts Institute of Technology has found a relationship between diet and serotonin levels in the brain. Serotonin is produced in the body from the amino acid tryptophan (Figure G.4). The synthesis

Serotonin

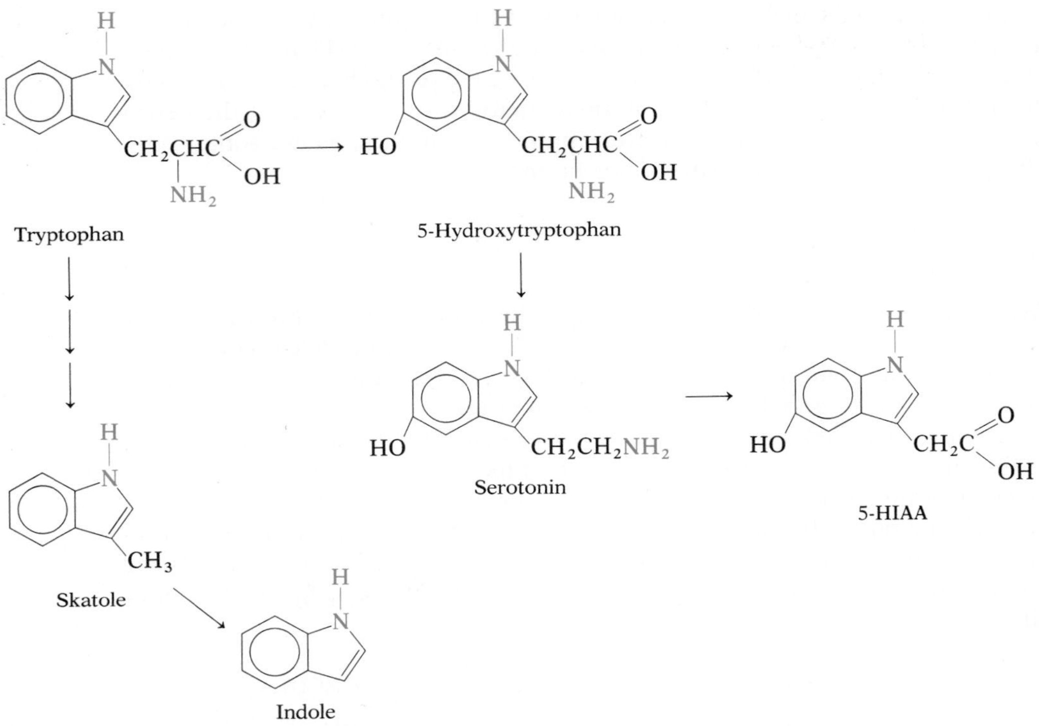

FIGURE G.4   *Serotonin is produced in the brain from tryptophan. A variety of other interesting compounds are also produced from tryptophan. One route leads (through several intermediates) to skatole, the principal odiferous ingredient in human feces, and to indole. Skatole and indole are used in perfumes.*

involves several steps, each catalyzed by an enzyme. Wurtman found that diets high in carbohydrates lead to high levels of serotonin. Lots of protein lowers the serotonin concentration. That may seem strange, because protein has lots of tryptophan and carbohydrates have little. But, Wurtman says, protein is only 1% tryptophan. In the presence of all those other amino acids, little tryptophan reaches the brain. With a carbohydrate meal, the hormone insulin lowers the level of the other amino acids in the blood, allowing relatively high levels of tryptophan to reach the brain.

Norepinephrine also is synthesized in the body from an amino acid. It is derived from tyrosine. The synthesis is complex and proceeds through several intermediates (Figure G.5). Each step is catalyzed by one or more enzymes. The intermediate compounds also have physiological activity: dopa has been used successfully in the treatment of Parkinson's disease, and dopamine has been employed to treat low blood pressure. Since tyrosine is also a component

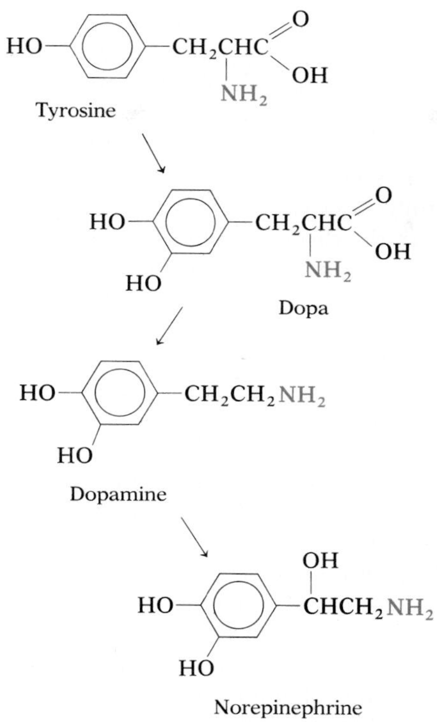

FIGURE G.5
The biosynthesis of norepinephrine from tyrosine.

Tyrosine

Dopa

Dopamine

Norepinephrine

of our diets, it may well be that our mental state depends to a fair degree on our diet.

It has been estimated that nearly 1 out of every 10 people in the United States suffers from mental illness. Over half the patients in hospitals are there because of mental problems. When the biochemistry of the brain is more fully understood, mental illness may be cured (or at least alleviated) by administration of drugs. In the remainder of this chapter, we see just how far we have already come in learning to control our moods with drugs. As is true for so many things, the potential for good that such compounds represent is matched by a potential for abuse.

G.3 STIMULANT DRUGS: AMPHETAMINES

Among the more widely known stimulant drugs are a variety of synthetic amines, all related to β-phenylethylamine (Figure G.6). Note the similarity in structure of these synthetic amines to the natural stimulants, epinephrine and norepinephrine. All are derived from the basic β-phenylethylamine structure. Perhaps the amphetamines act as stimulants by mimicking the natural brain amines.

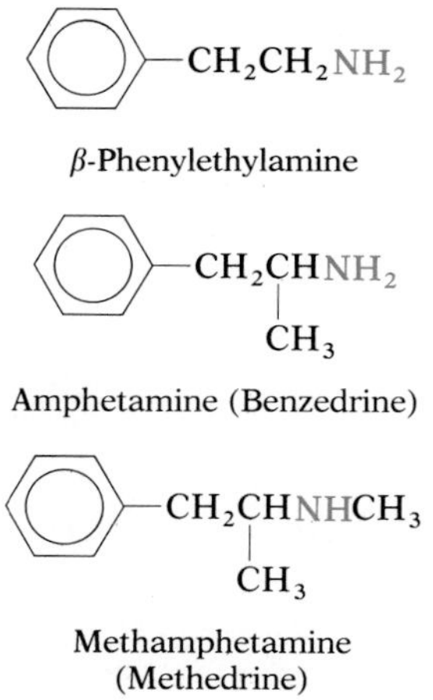

β-Phenylethylamine

Amphetamine (Benzedrine)

Methamphetamine (Methedrine)

FIGURE G.6
The stimulant drugs amphetamine and methamphetamine are chemical derivatives of β-phenylethylamine.

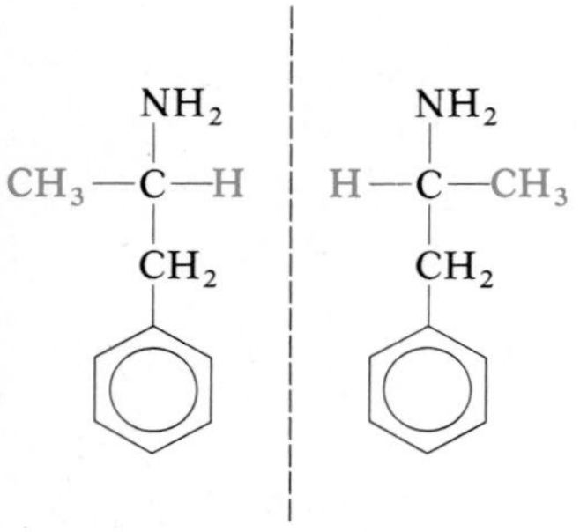

Amphetamine and methamphetamine have been widely abused. Amphetamine has been extensively used for weight reduction. It has also been employed for treating mild depression and narcolepsy, a rare form of sleeping sickness. Amphetamine induces excitability, restlessness, tremors, insomnia, dilated pupils, increased pulse rate and blood pressure, hallucinations, and psychoses. It is no longer recommended for weight reduction. It was found that, generally, any weight loss was only temporary. The greatest problem, however, was the diversion of vast quantities of amphetamines into the illegal drug market. Amphetamines are inexpensive. Armed forces personnel, truck drivers, and college students have been among the heavy users.

Methamphetamine has a more pronounced psychological effect. Generally, it is the "speed" that abusers inject into their veins. Such injections, at least initially, are said to give the individual a euphoric "rush." Shooting methamphetamine is quite dangerous, though, because this chemical is rather toxic.

Dextroamphetamine (Dexedrine) is another stimulant drug that has been widely abused. It is structurally related to amphetamine in a quite subtle way. Actually, amphetamine is not a single compound but a mixture of two isomers. These isomers have the same atoms and same groups of atoms, but they are oriented differently in space. One isomer is the mirror image of the other, so that if the dotted line in Figure G.7 were the surface of a mirror, then the reflection in the mirror of the right-hand compound would look like the left-hand compound. These isomers are not superimposable; therefore, they are not identical. In other chapters, we emphasize the fact that structural formulas that look different may actually represent the same compound. Now we have a set of formulas that look pretty much alike but that represent two different compounds. If you could build a model of each one of the mirror image isomers, you could prove to yourself that they cannot be made to coincide with one another exactly (cannot be superimposed). In lieu of that, we can at least show you for contrast a set of mirror images that are also superimposable, that is, which represent a single compound. Figure G.8 shows β-phenylethylamine itself and its mirror image. It takes only a cursory inspection to see that we have simply drawn exactly the same structure twice. β-Phenylethylamine does not exist as a mixture of isomers but as a single compound.

The two isomers that we have called amphetamine are related to one another in much the same way that your right hand is related to your left. You can't fit a right-hand glove on your left hand or vice versa. Similarly, the mirror image amphetamine isomers fit enzymes differently and thus have different physiological effects. The dextro ("right-handed") isomer is a stronger stimulant than the levo ("left-handed"). Dexedrine is the trade name for the pure dextro isomer. Benzedrine is the trade name for a mixture of the two isomers in equal amounts. Dexedrine is two to four times as active as Benzedrine.

Mirror image isomerism is quite common in organic chemicals of biological importance. This topic is discussed in detail in Special Topic I.

FIGURE G.7
Mirror image isomers of amphetamine.

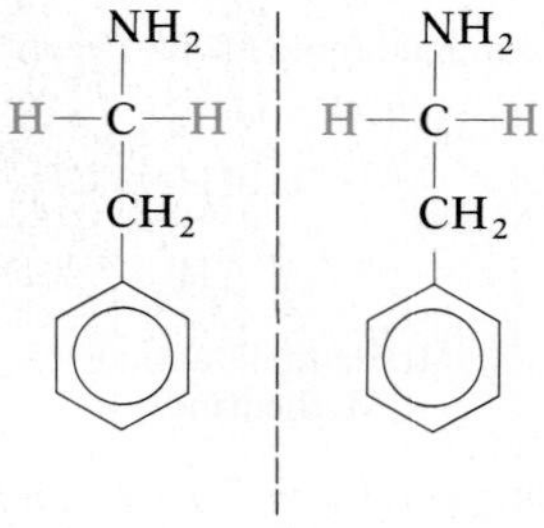

FIGURE G.8
Mirror images of β-phenylethylamine.

G.4 CAFFEINE, NICOTINE, AND COCAINE

Perhaps the most widely known stimulant of all is caffeine, an alkaloid found in coffee beans, tea leaves, and cola nuts. Caffeine is usually consumed in the form of beverages made from these plant materials. The effective dose of caffeine is about 200 mg, obtainable in about two cups of strong coffee or tea. It is also available in tablet form as a stay-awake or keep-alert type of drug. Two well-known brands of caffeine tablets are No-Doz and Vivarin. Each tablet of No-Doz contains 100 mg of caffeine—about the same amount of caffeine as in a cup of coffee. Vivarin tablets contain 200 mg each.

Is caffeine addictive? The "morning grouch" syndrome indicates that it may be mildly so. There is recent evidence that caffeine may be involved in chromosome damage. To be safe, people in their child-producing years should perhaps avoid large quantities of caffeine. Overall, the hazards of caffeine ingestion seem to be slight.

Another common alkaloid is nicotine, usually taken in the form of smoking tobacco. Nicotine is highly toxic to animals. It is especially deadly when injected. The lethal dose for a human is estimated to be about 50 mg. Nicotine is used in agriculture as a contact insecticide.

Nicotine seems to have a rather transient effect as a stimulant. This initial response is followed by depression.

Is nicotine addictive? Casual observation of a person who is trying to quit smoking seems to indicate that it is. There is also evidence of the development of tolerance. It is difficult, however, to separate all the social factors involved in smoking from the physiological effects.

Cocaine is another well-known alkaloid. This drug was first isolated in 1860 from the leaves of the coca plant (Figure G.10). Its structure was determined by Richard Willstätter in 1898. Cocaine is a powerful stimulant. Many of the Indians living on and around the eastern slopes of the Andes Mountains chew the leaves of the coca plant—mixed with lime and ashes—for their stimulant effect. Cocaine usually arrives in the United States as glistening white crystals; hence the slang name "snow." Some cocaine is legally imported for legitimate human and veterinary medical purposes. A far greater quantity is smuggled in for the illegal drug market.

Cocaine acts by blocking the uptake of NE, allowing high levels of the amine to develop. These high levels cause nerve cells to fire wildly, and the brain becomes like an overloaded telephone switchboard. Use of cocaine increases stamina and reduces fatigue. Moreover, it gives the user a feeling of confidence and power. It induces euphoria in some users. These stimulant effects are short-lived (about 1 hour) and are followed by depression. Overdose or repeated use can lead to paranoia, psychosis, and hallucinations. Overdose can cause death through heart attack or respiratory arrest.

Cocaine was once used as a local anesthetic, the first compound successfully

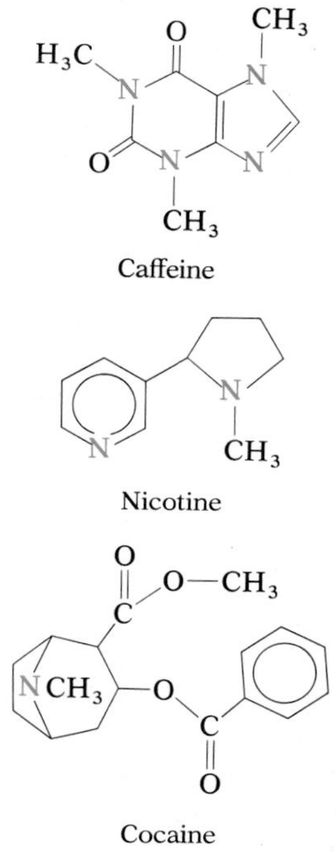

Caffeine

Nicotine

Cocaine

FIGURE G.9
Three stimulant drugs.

FIGURE G.10
Coca leaves and illicit forms of cocaine. (Courtesy of the U.S. Bureau of Narcotics and Dangerous Drugs, Washington, D.C.)

used as such. It is quite toxic, however. Even before Willstätter determined the structure of cocaine, the search was under way for synthetic compounds with similar properties.

G.5 LOCAL ANESTHETICS

Local anesthetics are drugs that block transmission of nerve signals when applied to nerve tissue. They act on all kinds of nerves and on all parts of the nervous system.

For dental work or minor surgery, it is usually desirable to deaden the pain in a part of the body only. After the initial work with cocaine, several other compounds were found that would perform that function. For example, certain esters of *p*-aminobenzoic acid act as local anesthetics (Figure G.11). The ethyl and butyl esters are used to relieve the pain of burns and open wounds. These are applied as salts, usually in the form of ointments.

More powerful in their anesthetic action are derivatives in which a second nitrogen atom is incorporated in the alkyl group of the ester. Perhaps the best known of these is procaine (Novocaine), first synthesized by Alfred Einhorn in 1905. Procaine can be injected as a local anesthetic. It can also be injected into the spinal column so that it deadens the entire lower portion of the body.

Lidocaine and mepivicaine are presently the most widely used local anesthetics. Dental use accounts for over 50 million dosages of each of the two

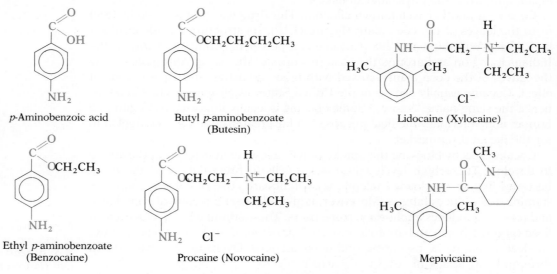

FIGURE G.11 *Some local anesthetics.*

drugs annually. They are used mainly for tooth extraction and root canal work.

Lidocaine is also used to treat cardiac arrhythmias. It probably is effective simply by decreasing the excitability of the cardiac cells.

Note that lidocaine is not a derivative of *p*-aminobenzoic acid but that it does share some structural features with the other compounds shown in Figure G.11.

A **general anesthetic** acts on the brain to produce unconsciousness as well as insensitivity to pain. Diethyl ether (Chapter 15) was the first general anesthetic. It was introduced into surgical practice in 1846 by a Boston dentist, William Morton. Inhalation of ether vapor produces unconsciousness by depressing the activity of the central nervous system. Ether is relatively safe. There is a fairly wide gap between the effective level for anesthesia and the lethal dose. The disadvantages are high flammability and a side effect of nausea.

Chloroform (Special Topic E) was introduced as a general anesthetic in 1847. It was widely used for years. It is nonflammable and produces effective anesthesia, but has a number of serious drawbacks. For one, it has a narrow safety margin: the effective dose is close to the lethal dose. It also causes liver damage. It must be protected from oxygen during storage to prevent the formation of deadly phosgene gas.

Modern anesthetics include fluorine-containing compounds such as halothane, enflurane, and methoxyflurane (Figure G.12). These compounds are nonflammable and relatively safe for the patient. Their safety, particularly that of halothane, for operating room personnel, however, has been questioned. For example, female workers suffer a higher rate of miscarriages than the general population.

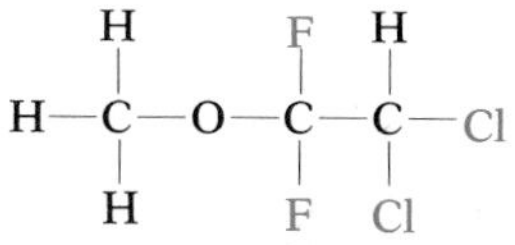

Modern surgical practice has moved away from the use of a single anesthetic. Generally a patient is given a strong sedative such as thiopental (Section G.6) by injection. This produces unconsciousness. The gaseous anesthetic then administered provides insensitivity to pain and keeps the patient unconscious. A relaxant, such as curare,* may also be employed. These compounds produce relaxation; thus, only light anesthesia is required. This avoids the hazards of deep anesthesia.

The potency of an anesthetic is related to its solubility in fat (a relatively nonpolar substance). General anesthetics seem to work by dissolving in the fatlike membranes of nerve cells. This changes the permeability of the membranes, and the conductivity of the nerve cells is depressed.

_____

* Curare is the arrow poison used by South American Indian tribes. Large doses of curare kill by causing a complete relaxation of all muscles. Death occurs because of respiratory failure.

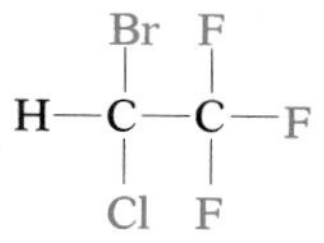

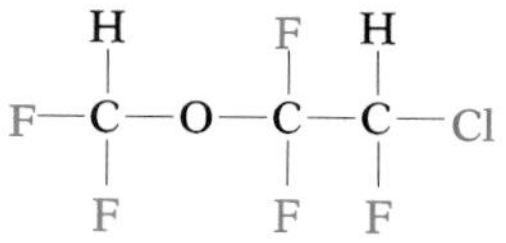

FIGURE G.12
Some general anesthetics.

G.6 BARBITURATES: SEDATION, SLEEP, AND SUICIDE

As a family of related compounds, the barbiturates display a wide variety of properties. They produce mild sedation, deep sleep, and even death.

Barbituric acid was first synthesized in 1864 by Adolph von Baeyer, a young student of August Kekulé (Chapter 14). Baeyer made it from urea and malonic acid. Urea occurs in urine; malonic acid is derived from apples.

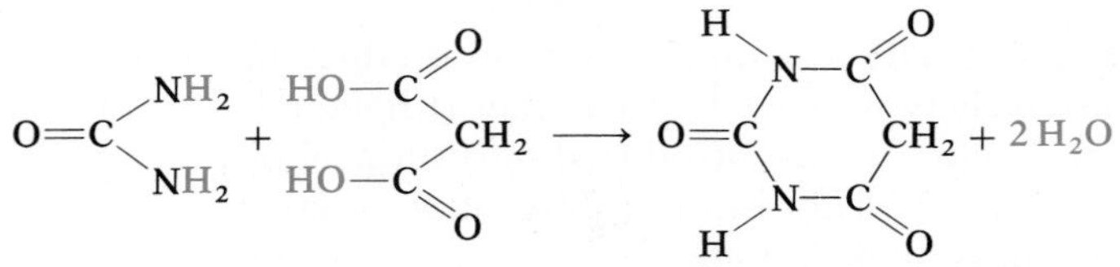

| Urea | Malonic acid | Barbituric acid |

The story goes that the compound was discovered on the day of Saint Barbara. The name is supposedly derived from the words *Barbara* and *urea*.

The medicinal value of the barbiturates was discovered in 1903 by Joseph von Mering. A derivative, called barbital, was found to be useful in putting dogs to sleep. The barbiturates (Figure G.13) are actually cyclic amides (not amines), but notice that the barbiturate ring resembles that of pyrimidine. Recent evidence indicates that the barbiturates may act by substituting for the pyrimidine bases in nucleic acids, thus interfering with protein synthesis.

Several thousand barbiturates have been synthesized through the years, but only a few have found widespread use in medicine.

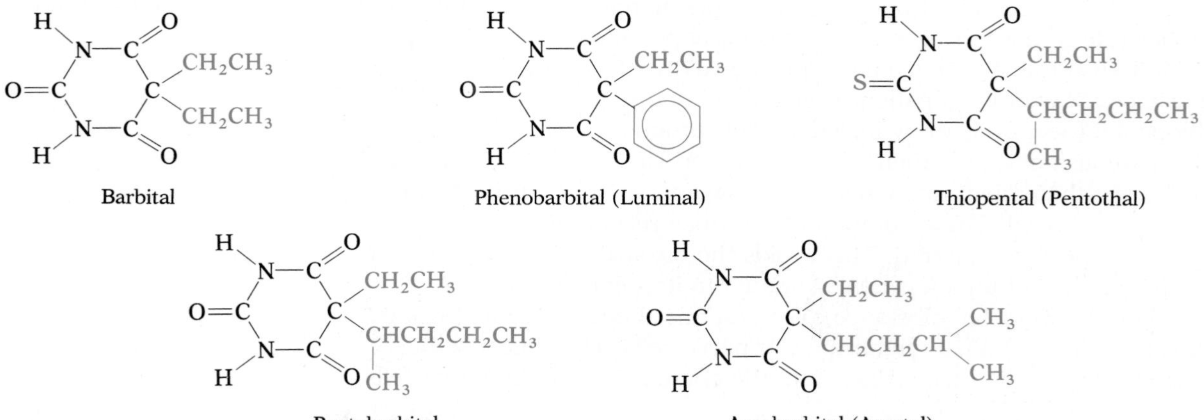

FIGURE G.13   *Some barbiturate drugs.*

Pentobarbital (Nembutal) is employed as a short-acting hypnotic drug. Before the discovery of modern tranquilizers, pentobarbital was widely used as a calmative against anxiety and other disorders of psychic origin.

Phenobarbital (Luminal) is a long-acting drug. It, too, is a hypnotic and can be used as a sedative. Phenobarbital is widely employed as an anticonvulsant for epileptics. The action of amobarbital (Amytal) is intermediate in duration relative to pentobarbital and phenobarbital.

Thiopental (Pentothal), a compound that differs from pentobarbital only in that an oxygen atom attached to the ring has been replaced by a sulfur atom, is widely used in anesthesia. Thiopental has been investigated as a possible "truth" drug. It does seem to aid psychiatric patients in the recall of traumatic experiences. It also helps uncommunicative individuals talk more freely. It does not, however, prevent one from withholding the truth or even from lying. No true "truth" drug exists.

The barbiturates were once used in small dosages as sedatives. The dosage for sedation was generally a few milligrams. In larger dosages (of about 100 mg), the barbiturates induce sleep. They are the sleeping pills once so widely used—and abused—by middle-class, often middle-aged, people. The lethal dose is in the vicinity of 1500 mg (1.5 g). Barbiturates are the drug of choice for most suicides. News reports list the cause of death as "an overdose of sleeping pills." There is also a potential for accidental overdose. After taking a couple of tablets, individuals may be groggy but not yet asleep. Unable to remember whether they took their sleeping pills or not, they take more.

The barbiturates are especially dangerous when ingested along with ethyl alcohol. This combination is far more potent than one would judge by simply summing the effects of the two depressants. The effect of the barbiturate is enhanced by factors of up to 200 when taken with alcoholic beverages. This **synergistic effect** has probably led to many deaths. We will never know how often alcohol is involved in deaths reported as due to "an overdose of sleeping pills." Synergistic effects are not limited to alcohol-barbiturate combinations. Fully one-half of the 100 most frequently prescribed drugs interact with alcohol. These drugs include antihistamines, tranquilizers, and medicines used to treat hypertension. Alcohol-drug interactions kill at least 2500 people per year in the United States. Most of these deaths are accidental. One should never take two drugs at the same time without competent medical supervision.

The barbiturates are strongly addictive. Habitual use leads to the development of a tolerance to the drug. One requires ever-larger doses to get the same degree of intoxication. Barbiturates are legally available by prescription only, but they are a part of the illegal drug scene also. Generally, they are known as "downers" because of their depressant, sleep-inducing effect.

Side effects of barbiturates are similar to those associated with alcohol. Abuse leads to hangovers, drowsiness, dizziness, and headaches. Withdrawal

A barbiturate

Thymine

Ketamine

Phencyclidine
(PCP)

symptoms are often severe and may be accompanied by convulsions and delirium. In fact, some medical authorities now say that withdrawal from barbiturates is more likely to cause death than withdrawal from heroin.

Barbiturates are cyclic amides (Chapter 17). Notice, however, that the barbiturate ring resembles that of thymine, one of the bases found in nucleic acids (Chapter 23). Recent evidence indicates that barbiturates may act by substituting for thymine (or cytosine or uracil) in nucleic acids, thus interfering with protein synthesis.

G.7 DISSOCIATIVE ANESTHETICS: KETAMINE AND PCP

As an anesthetic, thiopental is administered intravenously. Ketamine, another intravenous anesthetic, is called a **dissociative anesthetic:** it induces hallucinations similar to those reported by people who have had near-death experiences. They seem to remember observing their rescuers from a vantage point above it all, or moving through a dark tunnel toward a bright light. Unlike thiopental, ketamine seems to affect associative pathways before it hits the brain stem.

Little is known of the action of ketamine at the molecular level. If it acts by fitting receptors in the body, we can assume that our bodies produce their own chemicals that fit those receptors. These compounds may be synthesized or released only in extreme circumstances—such as in near-death experiences. Is it possible that we are on the threshold of the discovery of the chemistry of "life after death"?

Related to ketamine is phencyclidine (PCP), known on the street as "angel dust." PCP is soluble in fat and has no appreciable water solubility. It is stored in fatty issue and released when the fat is metabolized; this accounts for the "flashbacks" commonly experienced by users.

PCP has become an important part of the illegal drug scene. It is cheap and easily prepared, and though it was ruled much too dangerous for human use, it is available as an animal tranquilizer. Many users have had bad "trips" with PCP, but every few years a new crop of young people appears on the scene to be victimized by this hog tranquilizer.

G.8 ANTIANXIETY AGENTS

We will look at one other class of nitrogen-containing compounds before leaving the amine scene. Strictly speaking, not all the tranquilizers are amines; the carbamates are perhaps more closely related to the amides (Chapter 17). But the compounds in this section have similar physiological properties and will be grouped together on that basis.

The hectic pace of life in the modern world has driven people to seek rest and relaxation in chemicals. Ethyl alcohol is undoubtedly the most widely used tranquilizer. The drink before dinner—to "unwind" from the tensions of the day—is very much a part of the American way of life. Many people, however, seek their relief in other chemical forms.

Several over-the-counter drugs—Cope, Vanquish, and Compoz, among others—claim to be able to help us cope with or vanquish our problems or at least to compose ourselves in the face of minor adversity. Such products usually contain a little aspirin plus an antihistamine. The latter has a side effect of making one drowsy. These products have come under attack by consumer groups for being worthless at best—and perhaps even dangerous.

An antihistamine frequently used is diphenhydramine. When an **allergen** (a substance that triggers an allergic reaction) binds to the surface of certain cells, it triggers the release of histamine. The histamine causes the redness, swelling, and itching associated with allergies. Antihistamines block the release of histamine, but most also enter the brain and act upon the cells controlling sleep. Terfenadine (Seldane), an antihistamine approved for prescription use in 1985, blocks the release of histamine, but cannot enter the brain and thus does not cause drowsiness.

Diphenhydramine

Histamine

Terfenadine

Another group of drugs, available only by prescription, is widely employed to calm nervous tension. Several of these drugs are carbamates, compounds whose functional group resembles both an amide and an ester (Figure G.14). Simple derivatives, such as ethyl carbamate, act as mild soporifics (sleep-inducing agents). The best-known tranquilizer in this group is meprobamate (Equanil or Miltown). Proceeds from the sale of meprobamate amount to many millions of dollars per year. Another carbamate, carisoprodol (Soma), is employed as a muscle relaxant.

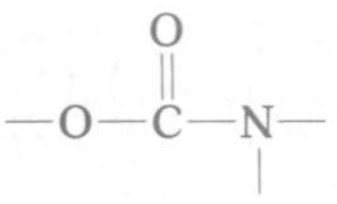

The carbamate functional group

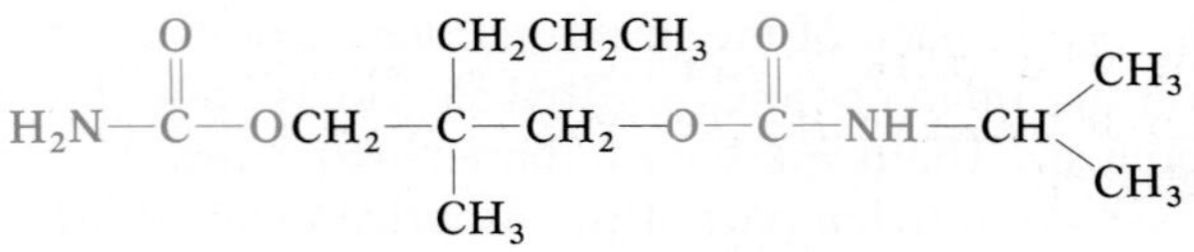

Carisoprodol (Soma)

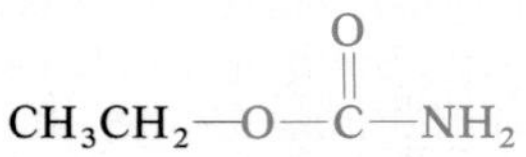

Ethyl carbamate

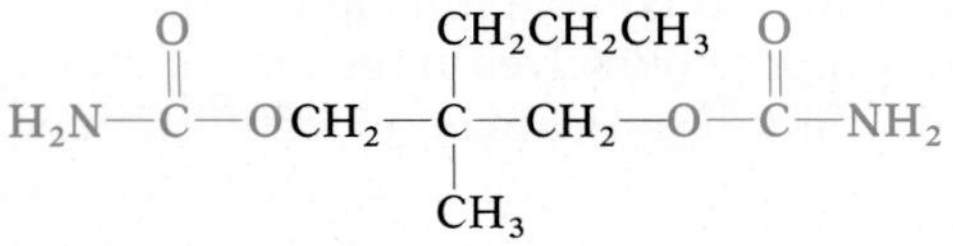

Meprobamate (Equanil, Miltown)

FIGURE G.14 Some carbamates.

Another class of widely used antianxiety drugs is the benzodiazepines, compounds that feature seven-membered heterocyclic rings (Figure G.15). The best known of these are diazepam (Valium) and chlordiazepoxide (Librium). For many years, Valium was the most widely prescribed drug in the United States.

A related drug, flurazepam (Dalmane), is widely used to treat insomnia. It has replaced the barbiturates as the ''sleeping pill'' of choice.

The benzodiazepine derivatives and the carbamates were formerly called ''minor tranquilizers.'' They are still widely used for treatment of anxiety. Some studies have shown them to be remarkably effective. Others have found the drugs to be little better than a placebo—a pill that looks like the drug tablet but that contains no active ingredient. The actual effect apparently depends upon the expectation of the patient. To the extent that these antianxiety agents really work, they do so simply by making people feel better by making them feel dull and insensitive. They do not solve any of the underlying problems that cause anxiety.

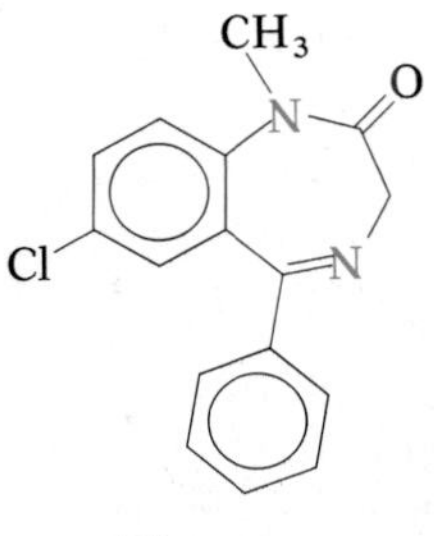

Diazepam (Valium)

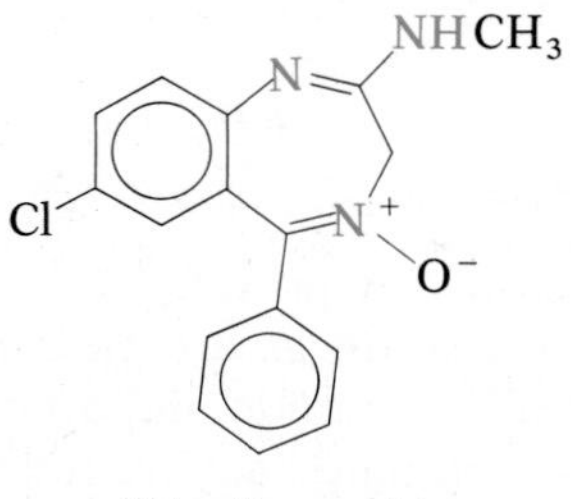

Chlordiazepoxide (Librium)

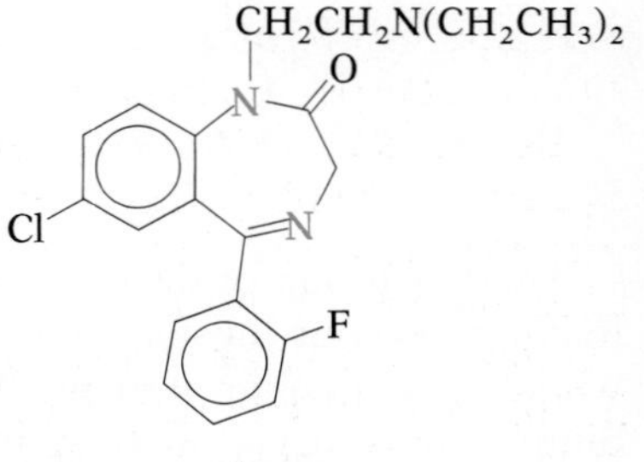

Flurazepam (Dalmane)

FIGURE G.15 Some benzodiazepine drugs.

And what price tranquility? After 20 years' use, it was finally found that Valium is addictive. People trying to go off it after prolonged use go into painful withdrawal.

G.9 ANTIPSYCHOTIC AGENTS

For centuries, the people of India used the snakeroot plant, *Rauwolfia serpentina*, to treat a variety of ailments including fever, snakebite, and other poisonings, and—most importantly—to treat maniacal forms of mental illness. Western scientists became interested in the plant near the middle of the twentieth century—after disdaining such remedies as quackery for many generations.

In 1952, rauwolfia was introduced into American medical practice as a hypotensive (blood-pressure-reducing) agent by Robert Wilkins, of Massachusetts General Hospital. In the same year, Emil Schlittler, of Switzerland, isolated an active alkaloid that he named reserpine and that has the following impressive (intimidating?) structure:

Reserpine

Rauwolfia was found not only to reduce blood pressure but also to bring about sedation. The latter finding attracted the interest of psychiatrists, who found reserpine so effective that by 1953 it had replaced electroshock therapy for 90% of their psychotic patients.

Also in 1952, chlorpromazine (Thorazine), which had been used in France as an antihistamine, was tried on psychotic patients in the United States as a tranquilizer. It was found to be extremely effective against the symptoms of schizophrenia.

Many compounds related to chlorpromazine have been synthesized (Figure G. 16). Several of these have been found to have interesting physiological properties. Promazine itself is a tranquilizer, but not as potent a one as chlorpromazine.

Thioridazine (Mellaril) is a potent tranquilizer, reputed to be without some of the undesirable side effects of chlorpromazine.

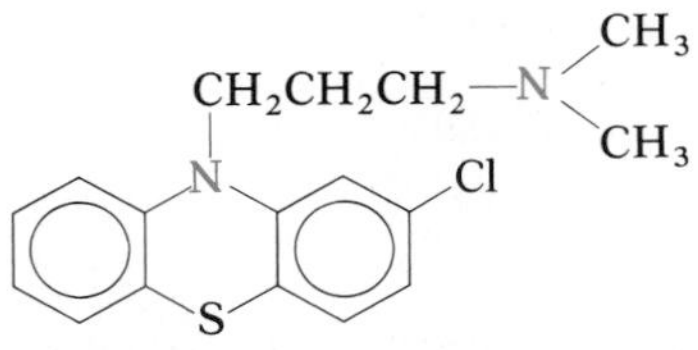

Chlorpromazine
(Thorazine)

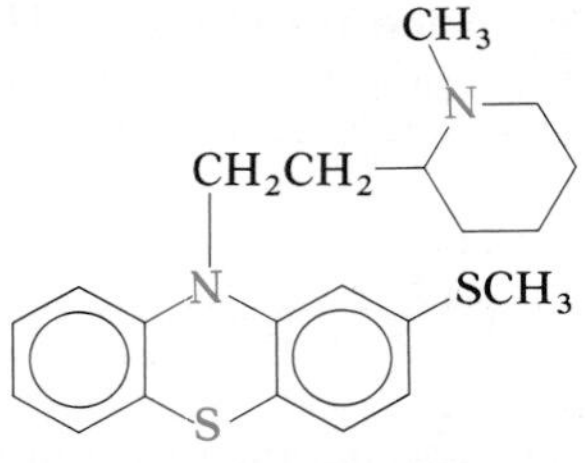

Thioridazine
(Mellaril)

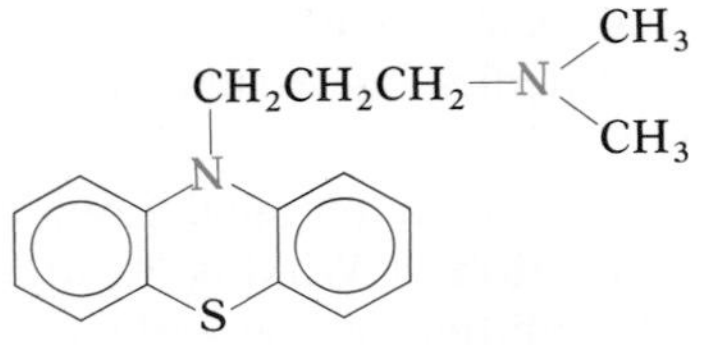

Promazine

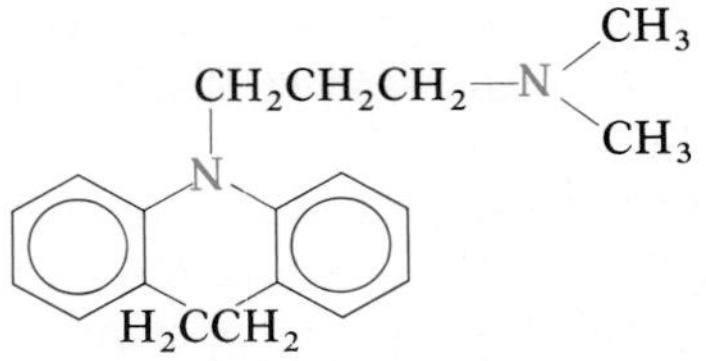

Imipramine

FIGURE G.16 Three tranquilizers and a psychic energizer.

It is worth noting that imipramine, which differs from promazine only in that the sulfur atom is replaced by a $—CH_2CH_2—$ group, is not a tranquilizer at all. Instead, it is a psychic energizer. This indicates that slight structural changes sometimes can result in profound changes in properties and that we have a long way to go in understanding why drugs act as they do.

The antipsychotic drugs—reserpine and the promazine derivatives—have been one of the real triumphs of chemical research. They have served to greatly reduce the number of patients confined to mental hospitals by controlling the symptoms of schizophrenia to the extent that 95% of all schizophrenics no longer need hospitalization. These drugs are not cures. We can only hope that continued research will ascertain the causes of schizophrenia. At that time perhaps a real cure—or, better yet, a preventative—can be found.

PROBLEMS

1. Define or identify each of the following terms:

a. neuron

b. synapse

c. neurotransmitter

d. allergen

2. Which two naturally occurring amines are presently considered to play major roles in the biochemistry of mental health? What are their proposed roles?

3. Which amino acids serve as precursors for the amines of Problem 2? How may our mental state in part be related to our diet?

4. How do amphetamines exert a stimulant effect?

5. How does cocaine exert a stimulant effect?

6. When administered intravenously to rats, the LD_{50} of procaine is 50 mg/kg and that of cocaine is 17.5 mg/kg of body weight. Which drug is more toxic?

7. What is a local anesthetic? How does a local anesthetic work?

8. What is a general anesthetic? How does a general anesthetic work?

9. Name two dissociative anesthetics. How do they work?

10. List three general anesthetics. What are the advantages and disadvantages of each?

11. How does curare kill? How can such a deadly poison be used on people in surgery without killing them?

12. Which of these anesthetics is dangerous because of its flammability?
a. diethyl ether   c. chloroform
b. halothane   d. cyclopropane

13. For each of the following anesthetics, describe a disadvantage associated with its use. Do not include flammability.
a. nitrous oxide   c. diethyl ether
b. halothane

14. What do we mean when we say amphetamines are "uppers"? Why are barbiturates called "downers"?

15. What is the basic structure common to all barbiturate molecules? How is the basic structure modified to change the properties of individual barbiturate drugs?

16. What is synergism?

17. Examine the structure of the reserpine molecule. Identify the following:
a. five ether functional groups
b. two amine functional groups
c. two ester functional groups

18. Acetbutolol has been proposed as a drug for the treatment of heart disease (angina and arrhythmias) and hypertension. There are five functional groups in the compound. Name the five families of organic compounds to which acetbutolol could be assigned.

19. Labetalol has been proposed as a drug for the treatment of angina and hypertension. Circle the four functional groups in the molecule and name the families of organic compounds that incorporate these functional groups.

20. If the minimum lethal dose (MLD) of amphetamine is 5 mg per kilogram of body weight, what would be the MLD for a 70-kg person? Can toxicity studies on animals always be extrapolated to humans?

21. Cocaine is usually administered as the salt cocaine hydrochloride that is sniffed up the nose, where it is readily absorbed through the watery mucous membranes. Some prefer to take their cocaine by smoking it (mixed with tobacco, for example). Before smoking, the cocaine hydrochloride must be converted back to the free base (that is, to the molecular form). Explain the choice of dosage form for each route of administration.

22. Which of the anesthetics in Figure G.12 are ethers? What halogen atoms are incorporated into each?

23. Overdoses of phencyclidine (PCP) are treated by intravenous administration of ammonium chloride. The ammonium ion presumably converts the PCP to a salt that is somewhat more water soluble and thus more readily excreted.

$$C_{17}H_{25}N + NH_4{}^+ \rightleftharpoons C_{17}H_{25}NH^+ + NH_3$$

Which is the stronger acid, $NH_4{}^+$ or $C_{17}H_{25}NH^+$? Which is the stronger base, $C_{17}H_{25}N$ or NH_3?

24. Drugs such as lithium carbonate and reserpine block the release of norepinephrine. How might these be useful for treating manic patients?

25. Electroconvulsive therapy (shock treatment) induces the release of norepinephrine. What sort of mental problems are treated with this therapy?

26. Haloperidol is one of the most widely prescribed antipsychotic drugs. What five functional groups are present in the molecule?

H

Organic Compounds of Sulfur

O rganic compounds containing sulfur make life possible. They also make life unpleasant (as when skunk and nature lover meet). This enormous range of effects makes organic sulfur compounds extremely interesting to people studying the chemistry of life. We now know, for example, that sulfur atoms in protein molecules play a major role in determining the shapes of these compounds. And the shapes of proteins, particularly enzymes, are critical to their function. Without active enzyme catalysts, biochemical reactions would not be fast or specific enough to sustain life.

As is our habit, we shall look at some of the simple organic compounds of sulfur first, saving for other chapters our consideration of the more complex compounds so vital to living cells.

H.1 THIOLS: THE SULFHYDRYL GROUP

Alcohols and ethers can be considered to be derived from water by replacement of one or both of its hydrogen atoms by alkyl groups. Some classes of sulfur-containing compounds can be derived from hydrogen sulfide in a similar way. The prefix *thio-* indicates the substitution of a sulfur atom for an oxygen atom in a compound. If ROH is an alcohol, RSH is a *thio*alcohol, or *thiol*.

The functional group (—SH) is called a **sulfhydryl group** (it contains sulfur and hydrogen). For many years the thiols were called **mercaptans,** a name derived from their ability to react with mercury (from the Latin *mercurium captans*, literally, "seizing mercury"). You may still encounter the name, though chemists seldom use it anymore.

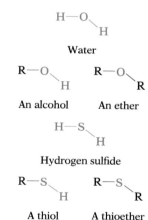

Water

An alcohol   An ether

Hydrogen sulfide

A thiol   A thioether

TABLE H.1   *A Comparison of the Physical Properties of an Alcohol and a Thiol*

Compound	Formula	Molecular Weight	Boiling Point (in degrees Celsius)	Solubility (in grams per 100 g of water)
Ethanol	CH_3CH_2OH	46	78	∞
Ethanethiol	CH_3CH_2SH	62	37	1.5

$CH_3CH_2CH_2CH_2OH$

1-Butanol
(butyl alcohol)

$CH_3CH_2CH_2CH_2SH$

1-Butanethiol
(butyl mercaptan)

The thiols are named in the same way that the alcohols are, except that in the IUPAC system the *-ol* becomes *-thiol* and the final *-e* of the parent hydrocarbon name is retained. Thus, it is butanol but butan*e*thiol. In the common system, *alcohol* is replaced by *mercaptan*.

Unlike oxygen, sulfur cannot participate in hydrogen bonding. Thus, the thiols have lower boiling points than the corresponding alcohols, even though they have greater molecular weights. Thiols do not hydrogen bond to water, either, and hence are much less soluble in water than alcohols with the same number of carbon atoms (Table H.1).

Perhaps the most notable and noticeable property of thiols is their offensive odor. The unforgettable odor of a skunk is due to two thiols and a compound called a disulfide (Figure H.1). Small amounts of thiols are added to natural gas in order that leaks of the otherwise odorless gas can be detected by smell. Someone who says, "I smell gas," is actually smelling the thiol in the gas mixture. Low concentrations of thiols may contribute desirable qualities to flavors and odors. Oysters and cheddar cheese contain small amounts of methanethiol. The onion is made piquant by the presence of 1-propanethiol. It is also true that not all thiols stink. Odor decreases with increasing molecular weight. With 10 or more carbon atoms per molecule, the odor may be pleasant.

trans-2-Butene-1-thiol

$CH_3CHCH_2CH_2SH$
  |
  CH_3

3-Methyl-1-butanethiol

CH_3—S—S—CH_2
C=C
H / CH_3

Methyl-1-(*trans*-2-butenyl)
disulfide

FIGURE H.1   *The odor of a skunk is made up of three sulfur compounds. trans-2-Butene-1-thiol accounts for about 40% of that odor. 3-Methyl-1-butanethiol and methyl-1-(trans-2-butenyl) disulfide account for about 30% each. (Photo by John Gerard.)*

The thiols of low molecular weight are somewhat toxic. Their foul odors, however, make it possible for us to detect levels much lower than those that are dangerous. For example, our noses can detect 0.02 ppb (by volume) of ethanethiol in air. The obnoxious odor would cause almost anyone to flee long before toxic levels were reached.

H.2 CHEMICAL PROPERTIES OF THE THIOLS

The thiols form insoluble compounds with heavy metal ions such as those of mercury and lead.

$$2\,RSH + Hg^{2+} \longrightarrow RS-Hg-SR + 2\,H^+$$
$$2\,RSH + Pb^{2+} \longrightarrow RS-Pb-SR + 2\,H^+$$

Most enzymes have amino acids with sulfhydryl groups at or near the active sites. Heavy metal ions tie up these groups, rendering the enzyme inactive.

$$\begin{array}{c} SH \\ | \\ \\ | \\ SH \end{array} + Hg^{2+} \longrightarrow \begin{array}{c} S \\ | \\ Hg \\ | \\ S \end{array} + 2\,H^+$$

It is this loss of enzyme activity that ultimately leads to the various symptoms of lead or mercury poisoning.

The antidote for mercury poisoning also contains sulfhydryl groups. British scientists, searching for an antidote for the arsenic-containing war gas, Lewisite, came up with a compound effective for heavy metal poisoning as well. The compound, a dithio analog of glycerol (Chapter 15), came to be known as **BAL** (British AntiLewisite).

$$\begin{array}{ccc} CH_2 & CH & CH_2 \\ | & | & | \\ OH & SH & SH \end{array}$$
BAL

BAL acts by picking up the mercury between the sulfhydryl groups in a sort of pincer action.

$$\begin{array}{ccc} CH_2 & CH & CH_2OH \\ | & | \\ S & & S \\ & Hg & \\ S & & S \\ | & | \\ HOCH_2 & CH & CH_2 \end{array}$$
Mercury chelated
by BAL

The BAL is said to chelate (from the Greek *chela,* ''claw'') the metal ion. Thus tied up, the mercury cannot attack the vital enzymes.

Now the bad news. The symptoms of mercury poisoning may not show up for several weeks. By the time the symptoms—loss of equilibrium, sight, feeling, hearing—are recognizable, extensive damage has already been done to the brain and nervous system. Such damage is largely irreversible. The BAL antidote is effective only when a person knows that poison has been ingested and seeks treatment right away.

Thiols are readily oxidized. Mild oxidizing agents bring about the removal of two hydrogen atoms from two thiol molecules. The remaining pieces of the thiols combine to form a new molecule, called a disulfide, with a covalent bond between two sulfur atoms.

$$R—S—\boxed{H\ \ H}—S—R + I_2 \longrightarrow RS—SR + 2\,HI$$

A disulfide

$$2\,RSH + H_2O_2 \longrightarrow RS—SR + 2\,H_2O$$

$$2\,CH_3CH_2SH + I_2 \longrightarrow CH_3CH_2S—SCH_2CH_3 + 2\,HI$$

Ethanethiol   Diethyl disulfide

Such mild oxidations are important in protein chemistry. Protein molecules consist of long, complex chains of atoms. One protein chain may be joined to another by oxidation involving one sulfhydryl group on each chain.

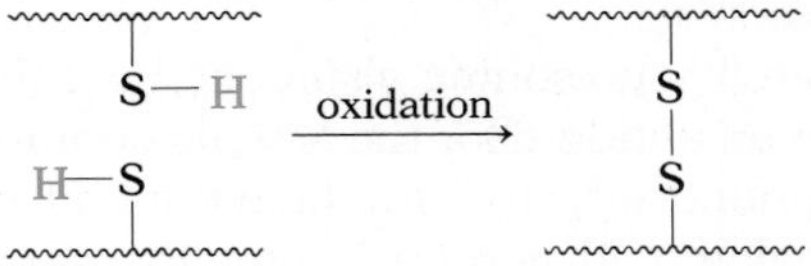

$$
\begin{array}{c}
S—H \\
H—S
\end{array}
\xrightarrow{\text{oxidation}}
\begin{array}{c}
S \\
| \\
S
\end{array}
$$

Or a loop may be formed in a single chain when two well-separated sulfhydryl groups are oxidized.

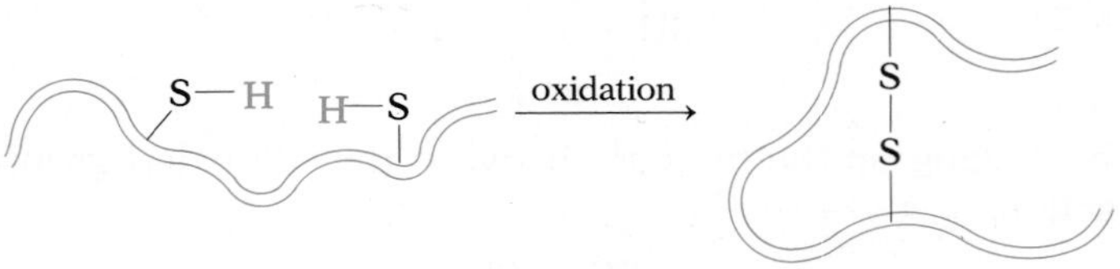

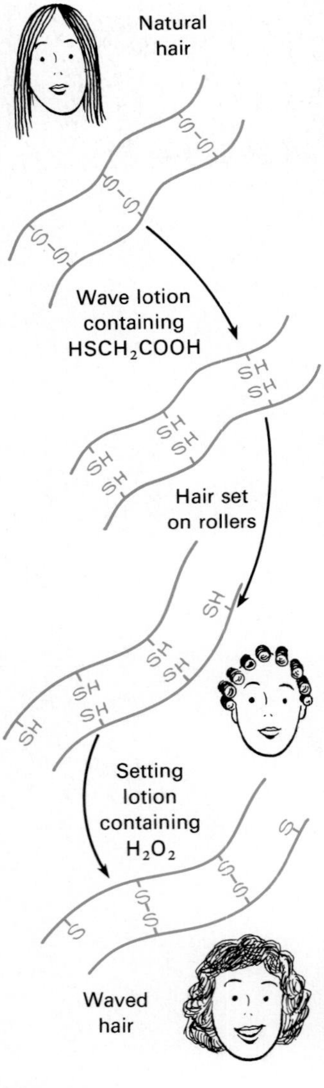

$$S—H \quad H—S \xrightarrow{\text{oxidation}} \begin{array}{c} S \\ | \\ S \end{array}$$

One practical application of this reaction is the permanent waving of hair (Figure H.2). Hair is protein, and it is held in shape by disulfide linkages between adjacent protein chains. The first step involves the use of a lotion containing a reducing agent such as thioglycolic acid.

$$HS—CH_2C\overset{\displaystyle O}{\underset{\displaystyle OH}{<}}$$

Natural hair

Wave lotion containing HSCH₂COOH

Hair set on rollers

Setting lotion containing H₂O₂

Waved hair

FIGURE H.2
The chemistry of the ''permanent'' waving of hair.

This wave lotion ruptures the disulfide linkages of the hair protein. The hair is then set on curlers or rollers and is treated with a mild oxidizing agent such as hydrogen peroxide (H_2O_2). Disulfide linkages are formed in new positions to give new shape to the hair. Exactly the same chemical process can be used to straighten naturally curly hair. The change in hair style depends only on how one arranges the hair after the disulfide bonds have been reduced and before reoxidation takes place.

H.3 MISCELLANEOUS ORGANIC SULFUR COMPOUNDS

There are a great variety of other classes of sulfur compounds. We will survey a few of them here.

The sulfur analogs of the ethers are called thioethers or, alternatively, dialkyl sulfides. (Note that dialkyl sulfides are not the same as the disulfides mentioned in Section H.2.) The thioethers have unpleasant odors, though generally not as bad as those of the thiols. They have low boiling points and are essentially insoluble in water. Dimethyl sulfide (CH_3SCH_3) is used as an odorant in natural gas. Methionine is an amino acid (Chapter 22) with a sulfide group.

$$CH_3SCH_2CHCOOH$$
$$| \\ NH_2$$

Methionine

Another important class of sulfur compounds is the sulfonic acids. Aromatic sulfonic acids can be made by direct sulfonation of arenes. For example, benzene reacts with fuming sulfuric acid (H_2SO_4 containing excess SO_3) to form benzenesulfonic acid.

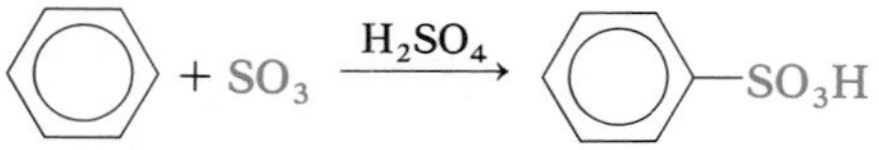

The sulfonic acids are strong acids, essentially completely ionized in aqueous solution.

$$RSO_3H + H_2O \longrightarrow RSO_3^- + H_3O^+$$

They react with bases to form salts.

$$RSO_3H + NaOH(aq) \longrightarrow RSO_3^-Na^+(aq) + H_2O$$

Modern **synthetic detergents** are often sodium salts of sulfonic acids. Those in use today are mainly linear alkylsulfonate (**LAS**) detergents.

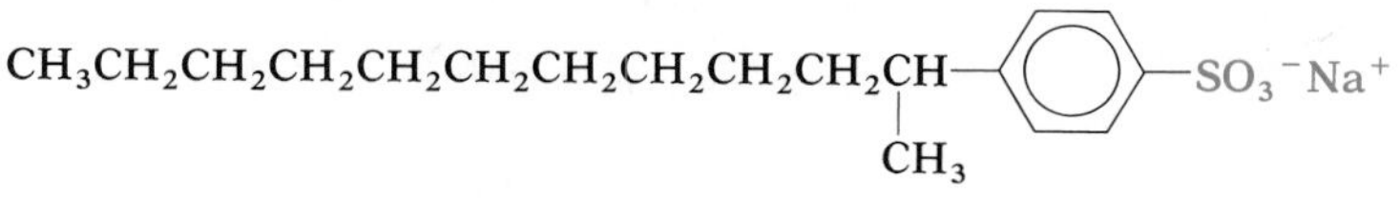

LAS

The cleansing action of synthetic detergents is quite similar to that of soap. The mechanism of cleaning is discussed in some detail in Chapter 21.

Derivatives of sulfonic acid analogous to those of the carboxylic acids (Chapter 17) can be formed. There are, for example, acid chlorides derived from the sulfonic acids. Benzenesulfonyl chloride is typical. This compound reacts with ammonia and the amines to form benzenesulfonamides.

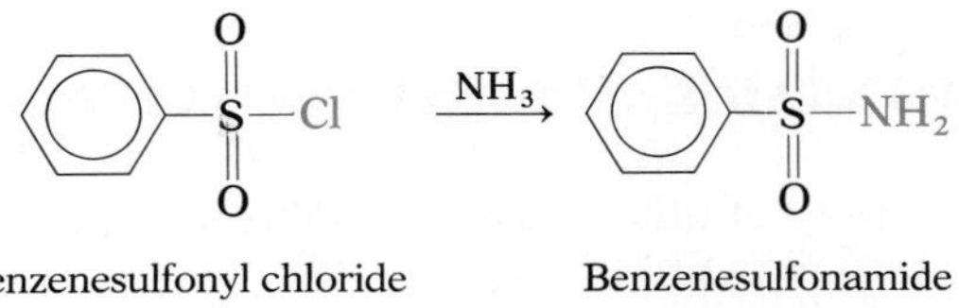

Benzenesulfonyl chloride   Benzenesulfonamide

The sulfa drugs, among the oldest of the antibacterial agents, are sulfonamides. The simplest sulfa drug, sulfanilamide, was discovered by the German chemist Gerhard Domagk in 1935. Domagk was awarded the Nobel Prize in 1939. The sulfa drugs inhibit the growth of bacteria by causing the bacteria to substitute it for *p*-aminobenzoic acid, a nutrient necessary for bacterial growth. The structures of three sulfa drugs and of *p*-aminobenzoic acid are given in Figure H.3. A summary of all the classes of organosulfur compounds that we have discussed is provided in Table H.2.

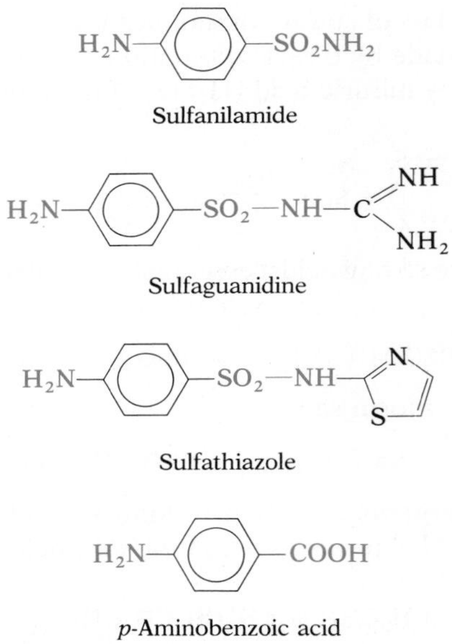

Sulfanilamide

Sulfaguanidine

Sulfathiazole

p-Aminobenzoic acid

FIGURE H.3   *Three sulfa drugs and* p-*aminobenzoic acid.*

TABLE H.2 *Some Families of Sulfur-Containing Organic Compounds*

Functional Group	Family Name	Example	Compound Name
R—S—H	Thiols (mercaptans)	CH_3SH	Methanethiol (methyl mercaptan)
R—S—R	Thioethers (sulfides)	CH_3—S—CH_3	Dimethyl sulfide
R—S—S—R	Disulfides	CH_3—S—S—CH_3	Dimethyl disulfide
R—S̈—OH (with two O double bonds)	Sulfonic acids	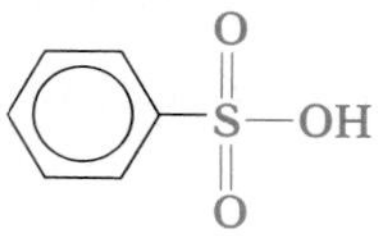 (benzene ring)—S̈—OH (with two O double bonds)	Benzenesulfonic acid
R—S̈—Cl (with two O double bonds)	Sulfonyl chlorides	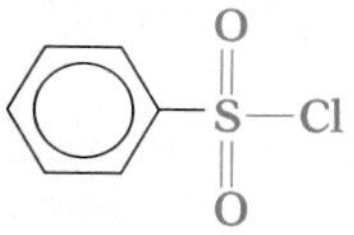 (benzene ring)—S̈—Cl (with two O double bonds)	Benzenesulfonyl chloride
R—S̈—NH_2 (with two O double bonds)	Sulfonamides	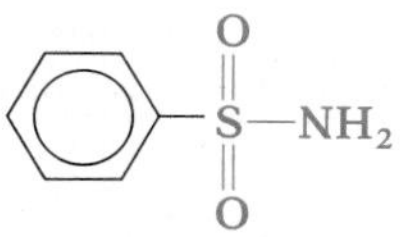 (benzene ring)—S̈—NH_2 (with two O double bonds)	Benzenesulfonamide

PROBLEMS

1. Give structural formulas for the following:
   a. ethanethiol   c. 2-methyl-2-butanethiol
   b. 2-propanethiol

2. Give structural formulas for the following:
   a. methyl mercaptan   b. *sec*-butyl mercaptan

3. Give IUPAC names for the following:
   a. $CH_3CH_2CH_2SH$

   b. $CH_3CH_2CHCH_2SH$
              |
              CH_3

   c. $CH_3CH_2CH_2CHCH_2CH_3$
                   |
                   SH

4. Which has the higher boiling point? Why?
   a. ethanol or ethanethiol
   b. $CH_3CH_2CH_2CH_2SH$ or $CH_3CH_2CH_2CH_2OH$

5. Give structural formulas for the following:
   a. dibutyl sulfide   b. diisopropyl sulfide

6. Give names for the following:
   a. $CH_3CH_2SCH_2CH_3$

   b. $CH_3CHCH_2SCH_2CHCH_3$
           | |
           CH_3 CH_3

7. Give structural formulas for the following:
   a. dimethyl disulfide   c. di-*tert*-butyl disulfide
   b. dipropyl disulfide

8. Give names for the following:
   a. $CH_3CH_2SSCH_2CH_3$

   b. $CH_3CHSSCHCH_3$
           | |
           CH_3 CH_3

c. CH$_3$CH$_2$CHSSCHCH$_2$CH$_3$
         | |
       CH$_3$ CH$_3$

9. Give structural formulas for the following:

  a. *o*-chlorobenzenesulfonic acid
  b. *m*-bromobenzenesulfonyl chloride
  c. *p*-fluorobenzenesulfonamide

10. Give names for the following:

  a. 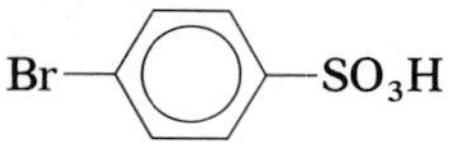 Br—⬡—SO$_3$H

  b. 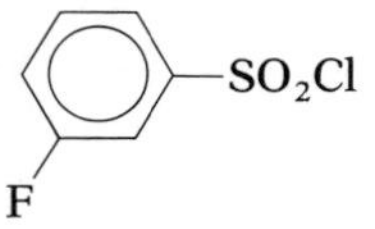 ⬡—SO$_2$Cl
     F

  c. 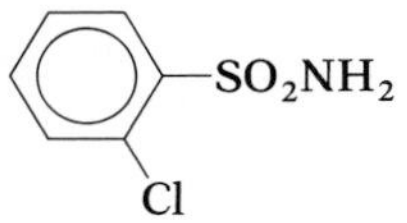 ⬡—SO$_2$NH$_2$
     Cl

  d. 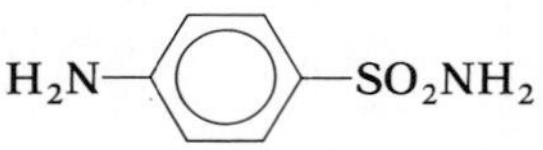 H$_2$N—⬡—SO$_2$NH$_2$

11. Give the structure of the product formed when ethanethiol is treated with a mild oxidizing agent such as iodine.

12. What products are formed by the action of a reducing agent on CH$_3$CH$_2$SSCH$_2$CH$_2$CH$_3$?

13. Give the structural formula for the product of the following reaction:

  ⬡ $\xrightarrow[\text{H}_2\text{SO}_4]{\text{SO}_3}$

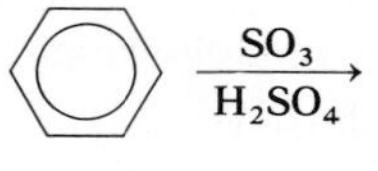

14. Give the structural formulas for the products of the following reactions:

  a. 2 CH$_3$SH $\xrightarrow{\text{Hg}^{2+}}$

  b. 2 CH$_3$CH$_2$SH $\xrightarrow{\text{Pb}^{2+}}$

15. Write a balanced equation for the reaction of benzenesulfonic acid with aqueous NaOH.

16. What type of linkages hold the protein molecules together in hair?

17. What chemical reactions are involved in curling hair? In straightening hair?

18. Write an equation for the ionization of benzenesulfonic acid in water. Is this compound a strong acid or a weak acid?

19. Give the structure of the product formed when benzenesulfonyl chloride reacts with each of the following:

  a. NH$_3$ **b.** CH$_3$NH$_2$

20. List the reagents necessary to carry out the following conversions:

  a. 2 CH$_3$CH$_2$CH$_2$SH $\xrightarrow{?}$
        CH$_3$CH$_2$CH$_2$—S—S—CH$_2$CH$_2$CH$_3$

  b. CH$_3$CH$_2$CH$_2$SH $\xrightarrow{?}$
        CH$_3$CH$_2$CH$_2$S—Hg—SCH$_2$CH$_2$CH$_3$

21. List the reagents necessary to carry out the following conversions:

  a. 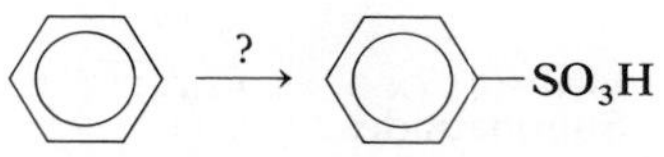 ⬡ $\xrightarrow{?}$ ⬡—SO$_3$H

  b. 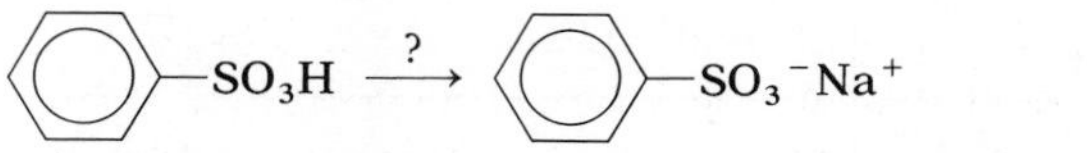 ⬡—SO$_3$H $\xrightarrow{?}$ ⬡—SO$_3^-$Na$^+$

22. List the reagents necessary to carry out the following conversions:

  a. 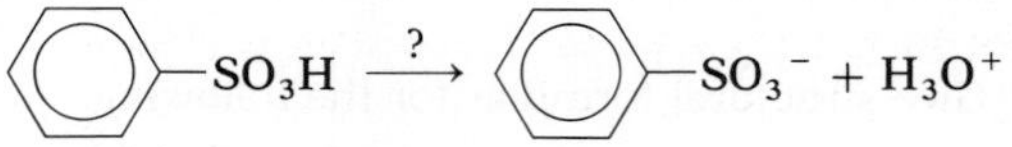 ⬡—SO$_3$H $\xrightarrow{?}$ ⬡—SO$_3^-$ + H$_3$O$^+$

  b. 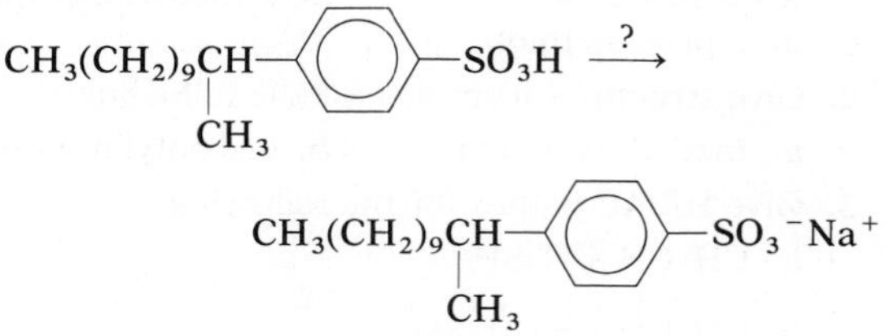 CH$_3$(CH$_2$)$_9$CH—⬡—SO$_3$H $\xrightarrow{?}$
           |
          CH$_3$

        CH$_3$(CH$_2$)$_9$CH—⬡—SO$_3^-$Na$^+$
                |
             CH$_3$

  c. 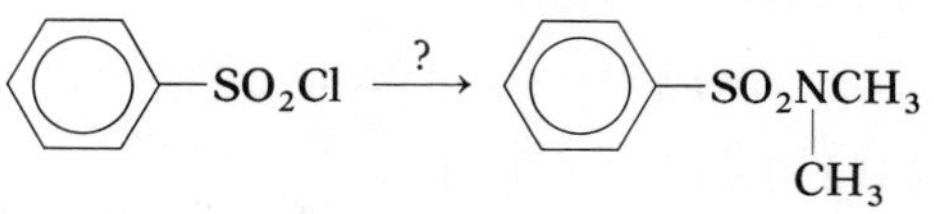 ⬡—SO$_2$Cl $\xrightarrow{?}$ ⬡—SO$_2$NCH$_3$
                           |
                          CH$_3$

Polymers

A variety of materials occur in nature that have rather remarkable properties. Some are hard and tough. Some have great tensile strength and flexibility. Others are edible. These materials have served humanity for centuries in the form of wood, wool, paper, cotton, and silk, as clothing and housing, and, in the form of starch and proteins, as food.

In Germany in 1920, Hermann Staudinger proposed that these materials were composed of huge molecules in the form of chains that are several thousand times as large as most ordinary molecules. Up to now we have considered molecules as small as that of hydrogen (H_2) and as complicated as that of reserpine ($C_{33}H_{40}N_2O_9$). You may think it strange, then, when we say that we are now going to look at some *large* molecules. If reserpine (Section G.9) with a molecular weight of 609 isn't large, what is? Well, starch and cellulose (Chapter 20) are, and so are proteins (Chapter 22) and nucleic acids (Chapter 23). The average molecular weights of these materials run into the thousands and even millions. As a class, these huge molecules and others of comparable size are referred to as **macromolecules** (from the Greek *makros*, "large" or "long").

It took many years for scientists to discover that some of the special properties of these substances were related to the enormous size of their constituent molecules. And it took even longer for them to determine the structure of these molecules. The major finding of these studies was that the giant molecules were constructed from "building blocks," smaller, repeating units like the bricks that make up a wall.

The temptation to improve on nature has always been great, and it has rarely been resisted. Chemists studying the structure and properties of the macromolecules of nature soon began producing their own macromolecules, designed for specific purposes and with certain properties enhanced to suit those purposes.

19.1 CHEMICAL MODIFICATION OF NATURAL POLYMERS

The oldest attempts to improve on nature simply involved chemical modification of the natural macromolecules. The synthetic material celluloid, as its name implies, was derived from natural cellulose. As Chapter 20 explains, cellulose contains many hydroxyl groups. Like any alcohol, cellulose reacts with acids to form esters. When the acid is nitric acid, the ester is cellulose nitrate (also called nitrocellulose or **gun cotton**). The reaction is represented diagrammatically here.

$$\text{Cellulose}-\text{OH} \xrightarrow{\text{HNO}_3} \text{Cellulose}-\text{ONO}_2$$

FIGURE 19.1
Celluloid, a nitrocellulose product, was the first synthetic plastic. Around the turn of the century, many commercial uses were found for celluloid. (Courtesy of E.I. du Pont de Nemours and Co., Inc., Wilmington, Del.)

When that product was dissolved in camphor and alcohol, the end product that resulted bore little resemblance to the original chips of wood or fibers of cotton from which the cellulose came. The new material, celluloid, could be molded into hard, smooth balls. It was, in fact, originally developed as a substitute for the ivory used in billiard balls. It was also used in the film of early movies and to make stiff collars that didn't require repeated starching (accepted male apparel in the early twentieth century—see Figure 19.1). As the name gun cotton indicates, cellulose nitrate is used in gun powder and other explosives. The dangerous flammability of celluloid led to its removal from the market.

Two other modifications of cellulose are still in use today. Cellulose can be treated with sodium hydroxide and carbon disulfide (CS_2) to produce a water-soluble intermediate, cellulose xanthate.

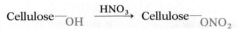

$$\text{Cellulose}-\text{OH} \xrightarrow[\text{NaOH}]{\text{S}=\text{C}=\text{S}} \text{Cellulose}-\overset{\overset{\text{S}}{\|}}{\underset{\text{O}}{}}\text{C}-\text{S}^-\text{Na}^+$$

A thick solution of this material is forced through fine holes into a dilute sulfuric acid solution that regenerates cellulose as fine, continuous, cylindrical threads called **rayon.**

$$\text{Cellulose}-\overset{\overset{\text{S}}{\|}}{\underset{\text{O}}{}}\text{C}-\text{S}^-\text{Na}^+ \xrightarrow{\text{H}^+} \text{Cellulose}-\text{OH}$$

If the solution is forced through a narrow slit, a thin transparent film or sheet of the same material is obtained. In this form, however, the product is called **cellophane.** In 1927, cellophane in the form of a transparent wrapping was quickly adopted by merchants who wanted their product to be visible. It is unlikely that anyone during those early years ever imagined that the same product would serve as the dialyzing membrane in artificial kidneys (Figure 19.2).

Cellulose acetate (also called **rayon acetate** or simply acetate) results when cellulose is treated with acetic anhydride. Like any alcohol, the cellulose reacts with acetic anhydride to form an acetate ester.

$$\text{Cellulose—OH} \xrightarrow{\text{CH}_3\overset{\overset{\text{O}}{\|}}{\text{C}}\text{O}\overset{\overset{\text{O}}{\|}}{\text{C}}\text{CH}_3} \text{Cellulose—O}\overset{\overset{\text{O}}{\|}}{\text{C}}\text{CH}_3$$

FIGURE 19.2   These workers are packaging disposable tubing for carrying blood during the hemodialysis process. (Courtesy of COBE Laboratories, Inc., Lakewood, Colo.)

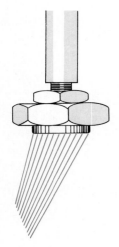

FIGURE 19.3
Formation of fibers by
extrusion through a
spinneret.

The ester is dissolved in acetone and forced through the tiny holes in a spinneret (Figure 19.3). Warm air evaporates the solvent, leaving fine, lustrous threads of rayon acetate. The difference in appearance of the natural cellulose in cotton fabrics and the semisynthetic modification is striking.

It didn't take long for the chemical industry to recognize the potential of synthetics. Scientists both in and out of industry shortly began making macromolecules from scratch (or, at least, from very small molecules).

19.2 POLYMERIZATION

Chemists call giant molecules assembled from much smaller building blocks **polymers** (from the Greek *poly*, "many," and *meros*, "parts"). In a laboratory (and in living systems, for that matter), the preparation of polymers involves the hooking together of many smaller molecules. The small molecules, the building blocks, are called **monomers** (from the Greek *monos*, "one" and *meros*, "parts"). The polymer is as different from the monomer as long strips of spaghetti are from the particles of flour that spaghetti is made of. For example, polyethylene, the familiar solid, waxy "plastic" of plastic bags, is a polymer prepared from a monomer (ethylene) that is a gas.

There are two general types of polymerization reactions—one is called addition polymerization, and the other is known as condensation polymerization.

In **addition polymerization**, the building blocks or monomers add to one another in such a way that the polymeric product contains all the atoms of the starting monomers. The polymerization of ethylene to form polyethylene can be represented by the reaction of a few monomer units.

$$\cdots + \;\overset{H}{\underset{H}{>}}C=C\overset{H}{\underset{H}{<}}\; + \;\overset{H}{\underset{H}{>}}C=C\overset{H}{\underset{H}{<}}\; + \;\overset{H}{\underset{H}{>}}C=C\overset{H}{\underset{H}{<}}\; + \cdots \longrightarrow$$

$$\cdots\overset{\displaystyle H}{\underset{\displaystyle H}{\overset{|}{\underset{|}{C}}}}-\overset{\displaystyle H}{\underset{\displaystyle H}{\overset{|}{\underset{|}{C}}}}-\overset{\displaystyle H}{\underset{\displaystyle H}{\overset{|}{\underset{|}{C}}}}-\overset{\displaystyle H}{\underset{\displaystyle H}{\overset{|}{\underset{|}{C}}}}-\overset{\displaystyle H}{\underset{\displaystyle H}{\overset{|}{\underset{|}{C}}}}-\overset{\displaystyle H}{\underset{\displaystyle H}{\overset{|}{\underset{|}{C}}}}\cdots$$

The dotted lines in the formula of the product are like etc.'s: they indicate that the structure is extended for many units in each direction. Notice that the two carbon atoms and four hydrogen atoms of each monomer molecule are incorporated into the polymer structure. Polyethylene is an addition polymer. A truer picture of the polymer is given in Figure 19.4, which shows a ball-and-stick model of a short segment. Still better is the space-filling representation of Figure 19.5. All of these models are deficient in that they contain too few atoms. Real polyethylene molecules have varying numbers of carbon atoms—from a few hundred to several thousand.

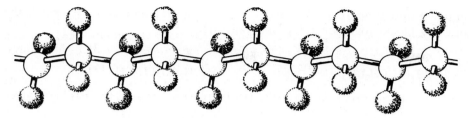

FIGURE 19.4   Ball-and-stick model of a segment of a polyethylene molecule.

In **condensation polymerization,** a portion of the monomer molecule is not incorporated in the final polymer. As an example, let's consider the formation of one type of nylon. (There are several different nylons, each prepared from a different monomer or set of monomers, but all share certain common structural features. We shall encounter another nylon later in this chapter.) In our present example, the monomer is 6-aminohexanoic acid. The polymerization involves the reaction of a carboxyl group of one monomer molecule with the

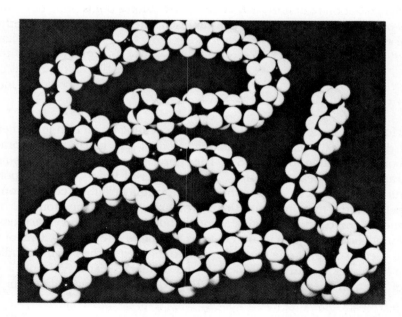

FIGURE 19.5   Space-filling model of a polyethylene molecule of 150 carbon atoms. (Reprinted with permission from Mandelkern, L., An Introduction to Macromolecules, Berlin: Springer-Verlag, 1973. Copyright © 1973 by Springer-Verlag, Inc.)

amine group of another. This reaction produces an amide bond that holds the building blocks together in the final polymer.

$$\cdots \overline{HO}\!-\!\underset{\underset{O}{\|}}{C}\!-\!CH_2CH_2CH_2CH_2CH_2\!-\!N\!-\!\overline{H + HO}\!-\!\underset{\underset{O}{\|}}{C}\!-\!CH_2CH_2CH_2CH_2CH_2\!-\!N\!-\!\overline{H} + \cdots \longrightarrow$$

$$\cdots \underset{\underset{O}{\|}}{C}\!-\!CH_2CH_2CH_2CH_2CH_2\!-\!\underset{\underset{H}{|}}{N}\!-\!\underset{\underset{O}{\|}}{C}\!-\!CH_2CH_2CH_2CH_2CH_2\!-\!\underset{\underset{H}{|}}{N} \cdots + x\,H_2O$$

Water molecules are formed as a by-product. It is this formation of a nonpolymeric by-product that distinguishes condensation polymerization from **addition polymerization.**

Alkenes, which tend to undergo addition reactions (Section 14.11), usually serve as the monomers in addition polymerizations. By substitution of various groups for one or more hydrogens of the simple ethylene molecule, a fantastic array of synthetic polymers can be obtained (Table 19.1). Condensation polymerization is used in the preparation of polyamides (such as nylon), polyesters (such as Dacron), and the polyurethanes (such as foam rubber).

The polymerization equations we have written above are quite cumbersome. There are other ways of writing the fomulas for polymers and the equations for polymerization reactions, ways that do not involve our writing out long segments of the polymer chain. We can represent the polymerization of ethylene by the equation

$$n \;\underset{H}{\overset{H}{>}}\!C\!=\!C\!\underset{H}{\overset{H}{<}} \longrightarrow \left[-\underset{\underset{H}{|}}{\overset{\overset{H}{|}}{C}}\!-\!\underset{\underset{H}{|}}{\overset{\overset{H}{|}}{C}}\!- \right]_n$$

In the formula for the polymer product, the repeating polymer unit (sometimes called the **segmer,** meaning "repeating segment") is placed within brackets with bonds extending to both sides. The subscript n indicates that this molecular fragment is repeated an unspecified number of times in the full polymer structure. The simplicity of the abbreviated formula facilitates certain comparisons between the monomer and the polymer. Notice that the monomer in the ethylene polymerization contains a double bond while the polymer does not. The double bond of the reactant contains two pairs of electrons. One of these pairs is used to connect one monomer unit to the next in the polymer (indicated by the lines sticking out to the sides in the segmer). That leaves only a single pair of electrons between the two carbons of the polymer segmer, in other words, a single bond.

It is also worth pointing out that the backbone of the polymer consists of the carbon atoms that were originally sharing the double bond. That probably sounds like a repeat of what we said in the preceding paragraph. Therefore, to make the point clearer, let's look at another polymerization reaction. This one involves propylene as the monomer.

TABLE 19.1 *A Selection of Addition Polymers*

Monomer	Polymer	Polymer Name	Some Uses
$H_2C=CH_2$	$\left[\begin{array}{cc} H & H \\ -C-C- \\ H & H \end{array}\right]_n$	Polyethylene	Plastic bags, bottles, toys, electrical insulation
$H_2C=CH-CH_3$	$\left[\begin{array}{cc} H & H \\ -C-C- \\ H & CH_3 \end{array}\right]_n$	Polypropylene	Indoor-outdoor carpeting, bottles
$H_2C=CH-\bigcirc$	$\left[\begin{array}{cc} H & H \\ -C-C- \\ H & \bigcirc \end{array}\right]_n$	Polystyrene	Simulated wood furniture, styrofoam insulation and packing materials
$H_2C=CH-Cl$	$\left[\begin{array}{cc} H & H \\ -C-C- \\ H & Cl \end{array}\right]_n$	Poly(vinyl chloride), PVC	Plastic wrap, simulated leather (Naugahyde), phonograph records, garden hoses
$H_2C=CCl_2$	$\left[\begin{array}{cc} H & Cl \\ -C-C- \\ H & Cl \end{array}\right]_n$	Poly(vinylidene chloride), Saran	Food wrap
$F_2C=CF_2$	$\left[\begin{array}{cc} F & F \\ -C-C- \\ F & F \end{array}\right]_n$	Polytetrafluoroethylene, Teflon	Nonstick coating for cooking utensils, electrical insulation
$H_2C=CH-CN$	$\left[\begin{array}{cc} H & H \\ -C-C- \\ H & CN \end{array}\right]_n$	Polyacrylonitrile, Orlon, Acrilan, Creslan, Dynel	Yarns, wigs
$H_2C=CH-O-\overset{\overset{O}{\|}}{C}-CH_3$	$\left[\begin{array}{cc} H & H \\ -C-C- \\ H & O-C-CH_3 \\ & \overset{\|}{O} \end{array}\right]_n$	Poly(vinyl acetate), PVA	Adhesives, textile coatings, chewing gum resin, paints
$H_2C=\overset{\overset{CH_3}{\|}}{\underset{\underset{O}{\|}}{C}}-C-O-CH_3$	$\left[\begin{array}{cc} H & CH_3 \\ -C-C- \\ H & C-O-CH_3 \\ & \overset{\|}{O} \end{array}\right]_n$	Poly(methyl methacrylate), Lucite, Plexiglas	Glass substitute, bowling balls

$$n \quad \underset{H}{\overset{H}{>}}C=C\underset{CH_3}{\overset{H}{<}} \quad \longrightarrow \quad \left[\begin{array}{c} H \quad H \\ -\underset{|}{\overset{|}{C}}-\underset{|}{\overset{|}{C}}- \\ H \quad CH_3 \end{array} \right]_n$$

Polypropylene

We hope that looks exactly as you expected it to look. But what if we had written the formula for propylene this way?

$$CH_2{=}CH{-}CH_3$$

Then many novice chemists would write the abbreviated formula for the polymer this way:

$$[-CH_2-CH-CH_3-]_n \qquad \text{(Incorrect structure)}$$

That formula is wrong. For one thing, it indicates that the right-hand carbon shares five bonds (a no-no). And if one ignores that problem, the formula still suggests that the polymer looks like a straight chain of carbon atoms, that is, like

$$-C-C-C-C-C-C-C-C-C-C-C-C-$$

The correct formula, shown in the original equation, indicates that the polymer is branched.

$$-\underset{\underset{C}{|}}{C}-C-\underset{\underset{C}{|}}{C}-C-\underset{\underset{C}{|}}{C}-C-\underset{\underset{C}{|}}{C}-C-$$

Including hydrogen atoms, the polymer would look like this:

$$\cdots C-C-C-C-C-C-C-C-C-C-C-C-C-C-C-C\cdots$$
(with H and CH_3 substituents)

The original doubly bonded carbon atoms form the main chain of the polymer, and any groups attached to these carbons are attached as substituents to the main polymer chain. Table 19.1 offers a number of examples.

19.3 THE RELATION OF PHYSICAL PROPERTIES TO STRUCTURE

Polymers are classified not only according to how they are made but also according to their reaction to heat. To the general public, synthetic polymers have come to be known as "plastics." However, *plastic* has a much more restricted meaning for chemists. The term **thermoplastic** is used to describe only those substances that can be softened by heat and then formed by pressure. In contrast, **thermosetting** polymers are those that are fusible at some stage in their production but that become permanently hard under the influence of heat and pressure. They cannot be softened and remolded. A major achievement of polymer chemistry has been the correlation of properties such as these with specific structural features of the giant molecules. Polyethylene offers an excellent example of how variations in structure can affect the properties of the resulting polymer. There are two kinds of polyethylene "plastics," those of high density and those of low density (Figure 19.6). The high-density material has a greater rigidity and strength. It is used for such things as threaded bottle caps, radio and television cabinets, toys, and large-diameter pipes. Low-density polyethylenes are waxy, semirigid, translucent materials with low melting points. They are used in plastic bags, refrigerator dishes, insulation for electrical wiring, squeeze bottles, and many other common household articles.

What structural differences can account for these variations in properties? **High-density polyethylene** consists primarily of linear molecules, that is, of long, unbranched chains. These linear molecules can assume a fairly ordered crystalline structure. In such an arrangement, chains can run alongside one another in close contact for relatively great distances. This permits maximum dispersion forces (Section 7.5) between chains. The overall effect of increased attraction between chains is to impart strength and rigidity to the polymeric material.

The **low-density polyethylenes** have branched chains, which prevent the molecules from assuming a crystalline structure. The branches get in the way when two chains come into close contact (or try to). This decreases the dispersion forces and weakens the attraction between chains, resulting in the more flexible material with a lower melting point. Figure 19.7 shows two arrangements of polymer molecules.

What if we want a really tough material, something very strong, very rigid, something that won't melt at all on heating? The answer might be one of the

FIGURE 19.6
These three bottles are made of polyethylene and were heated in the same oven for the same length of time. Those that melted have branched polyethylene molecules; the other is composed of unbranched chains. (Photo by Mari Ansari.)

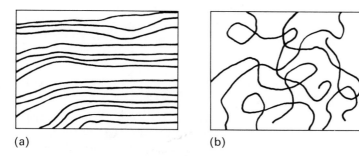

(a)   (b)

FIGURE 19.7   *Organization of polymer molecules. (a) Crystalline arrangement. (b) Amorphous arrangement.*

oldest synthetic thermosetting polymers, Bakelite. Bakelite is a condensation polymer. It is prepared from a combination of monomers, phenol and formaldehyde, and is therefore more generally referred to as a phenol-formaldehyde resin. (The use of two or more monomers in a single polymer is not an uncommon procedure, as some later examples will indicate.) The reaction proceeds stepwise, with formaldehyde adding first to the ortho and para positions of the phenol molecule.

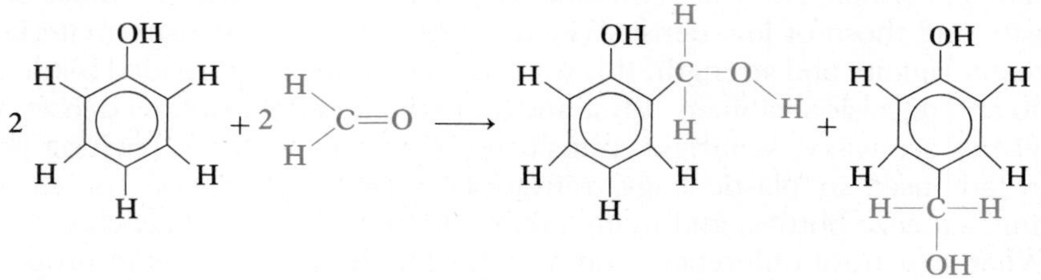

The substituted molecules then interact and split out water.

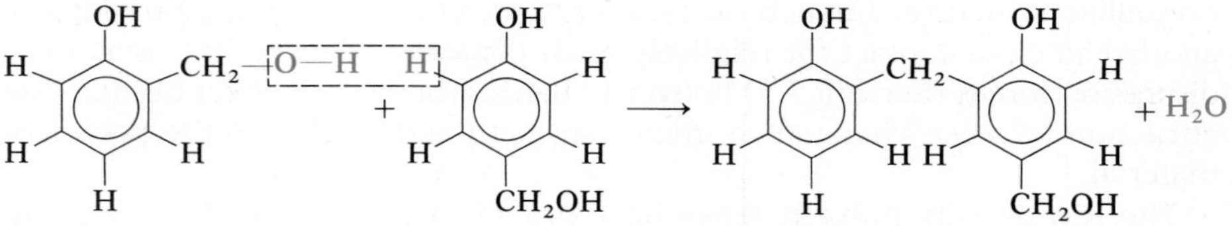

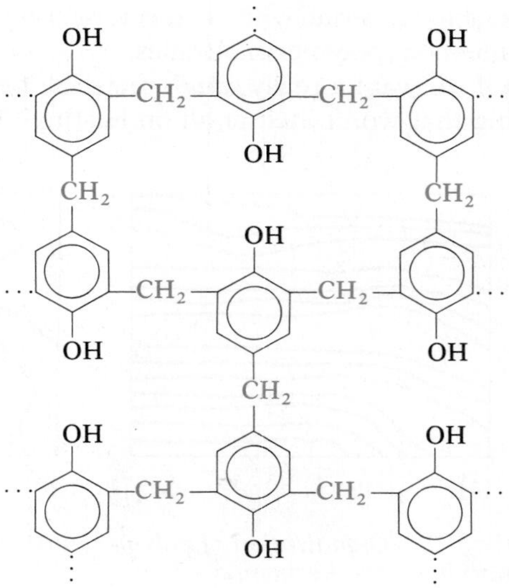

FIGURE 19.8
*Bakelite, a cross-linked
condensation polymer.*

The hookup of molecules continues until an extensive network is achieved. Water is driven off by heat as the polymer sets. The final polymer is a huge, complex, three-dimensional network whose structure cannot easily be conveyed in a drawing. Nonetheless, Figure 19.8 is our attempt to do just that. The extensive cross-linking results in rigidity. The final polymer has great strength without having great weight, a most useful combination of properties.

19.4 ELASTOMERS

Hardness and rigidity are not the only desirable properties for a polymer. Frequently flexibility and, more particularly, elasticity are what are needed. The natural polymer rubber is the prototype for this kind of material.

Natural rubber can be broken down into a simple hydrocarbon called isoprene.

$$CH_2{=}\underset{CH_3}{C}{-}\underset{H}{C}{=}CH_2$$

Isoprene

The molecules of the rubber polymer itself are now known to have the structure

$$\cdots\underset{CH_3}{CH_2}{>}C{=}C\underset{H}{<}\underset{CH_3}{CH_2{-}CH_2}{>}C{=}C\underset{H}{<}\underset{CH_3}{CH_2{-}CH_2}{>}C{=}C\underset{H}{<}\underset{CH_3}{CH_2{-}CH_2}{>}C{=}C\underset{H}{<}CH_2\cdots$$

or

$$\left[\underset{CH_3}{-CH_2}{>}C{=}C\underset{H}{<}CH_2{-}\right]_n$$

These long chains can be coiled and twisted and intertwined with one another. The stretching of rubber corresponds to the straightening of the coiled molecules. Natural rubber is soft and tacky when hot. It can be made harder by reaction with sulfur. This process, called **vulcanization,** cross-links the hydrocarbon chains with sulfur atoms (Figure 19.9). As with the phenol-formaldehyde

FIGURE 19.9
Vulcanized rubber. The subscript x indicates an indefinite number.

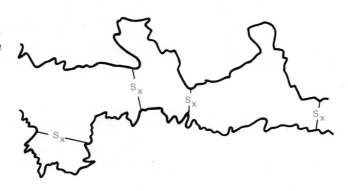

TABLE 19.2 *The Hardness of Rubber and Extent of Cross-Linking*

Item	Monomer Units Between Cross-Links
Surgical gloves	100–150
Kitchen gloves	50–80
Artificial heart membrane	30–40
Bike inner tube	20–30
Bike tire	10–20
Radio cabinet	5–10

resin, the three-dimensional cross-linked structure makes the vulcanized rubber a harder, stronger substance, suitable for automobile tires, for example (Table 19.2). Surprisingly, cross-linking also improves the elasticity of the rubber. With just the right degree of cross-linking, the individual chains are still relatively free to uncoil and stretch. But when a stretched piece of this material is released, the cross-links serve as a memory aid, pulling the chains back to their original prestretched arrangement (Figure 19.10). Rubber bands owe their snap to this sort of molecular structure.

When World War II cut supplies of natural rubber, a major search for rubber substitutes was undertaken. The resulting synthetics often bear a striking molecular resemblance to nature's own elastomer. For example, the synthetic material polychloroprene (Neoprene) is made from a monomer that is similar to isoprene but has a chlorine in place of isoprene's methyl group.

$$n \; CH_2{=}C{-}CH{=}CH_2 \longrightarrow \left[\begin{array}{c} {-}CH_2 \qquad\quad CH_2{-} \\ \diagdown \quad C{=}C \quad \diagup \\ Cl \qquad\qquad H \end{array}\right]_n$$

(a) (b) (c)

FIGURE 19.10 *Vulcanization of rubber cross-links the molecular chains. (a) In unvulcanized rubber, the chains slip past one another when the rubber is stretched. (b) Vulcanization involves the addition of sulfur cross-linkages between the chains. (c) When the vulcanized rubber is stretched, the sulfur cross-linkages prevent the chains from slipping past one another. Vulcanized rubber is stronger and more elastic than unvulcanized rubber.*

Another of the synthetic rubbers illustrates the principle of **copolymerization.** In this process, a mixture of two monomers forms a product in which the chain contains units of both building blocks. Such a material is called a **copolymer.** SBR rubber is a copolymer of styrene (25%) and butadiene (75%). A segment of an SBR molecule might look something like

| Butadiene unit | Butadiene unit | Styrene unit | Butadiene unit |

This synthetic is more resistant to oxidation and abrasion than natural rubber but has less satisfactory mechanical properties.

Chemists have even learned to make polyisoprene, a substance identical in every way to natural rubber except that the former compound is not harvested on plantations of rubber trees.

19.5 FIBERS AND FABRICS

Let's consider one other area in which synthetics have successfully mimicked nature's own work. Cotton, wool, and silk have long been spun and woven into fabrics for clothing and other uses. The fibrous nature and, particularly, the great tensile strength of these materials are perfectly suited to such purposes. Synthetic polymers with similar physical properties have revolutionized the clothing industry and have outdone the natural fibers in their resistance to stretching and shrinking (and to moths).

Both addition and condensation polymerization have yielded useful materials. Polyacrylonitriles (Orlon, Acrilan, Creslan, and the like) are addition polymers.

Polyesters (Dacron) and polyamides (nylon) are condensation polymers. Dacron polyester is made from the condensation of ethylene glycol with terephthalic acid.

$$n \; \overline{HO}-CH_2-CH_2-\overline{OH + n \; H}O-\overset{\overset{\displaystyle O}{\|}}{C}-\bigcirc-\overset{\overset{\displaystyle O}{\|}}{C}-O\overline{H} \longrightarrow$$

Ethylene glycol Terephthalic acid

$$\left[-O-CH_2-CH_2-O-\overset{\overset{\displaystyle O}{\|}}{C}-\bigcirc-\overset{\overset{\displaystyle O}{\|}}{C}-\right]_n + n \; H_2O$$

Dacron

The same polymer is marketed as Mylar, the "tape" that, when magnetically coated, is used in tape recorders and videotape machines.

The most common synthetic polyamide, Nylon 66 (for a description of another nylon, see Section 19.2), is also made by the condensation of two different monomers.

$$n \; \overline{H}\overset{\overset{\displaystyle H}{|}}{N}-CH_2CH_2CH_2CH_2CH_2CH_2-\overset{\overset{\displaystyle H}{|}}{N}-\overline{H + n \; HO}-\overset{\overset{\displaystyle O}{\|}}{C}-CH_2CH_2CH_2CH_2-\overset{\overset{\displaystyle O}{\|}}{C}-\overline{OH} \longrightarrow$$

1,6-Hexanediamine Adipic acid

$$\left[-\overset{\overset{\displaystyle H}{|}}{N}-CH_2CH_2CH_2CH_2CH_2CH_2-\overset{\overset{\displaystyle H}{|}}{N}-\overset{\overset{\displaystyle O}{\|}}{C}-CH_2CH_2CH_2CH_2-\overset{\overset{\displaystyle O}{\|}}{C}-\right]_n + n \; H_2O$$

Nylon 66

Nylon 66 was invented by Wallace H. Carothers and his associates at the Du Pont laboratories. Du Pont set up a basic research group whose purpose was to learn more about the nature of materials. Carothers led the investigation of marcromolecular materials. By 1928 his group had succeeded in synthesizing several polymers with molecular weights of 10 000 or more. It wasn't until 1935, however, that he developed nylon 66 into a useful fiber.

Silk and wool, natural protein fibers, are also polyamides. All these fibers, like high-density polyethylene, owe their strength to the ordered, relatively rigid arrangement of their long molecules. However, the polyamides and polyesters possess features not shared by polyethylene. The former compounds contain polar functional groups, and the interaction of these groups gives these polymers their unique tensile strength.

Design of materials has reached a point where a seemingly impossible combination of properties can be incorporated into a single substance. For example, spandex fibers, which are used for stretch fabrics (Lycra) in ski pants, girdles, and bathing suits, combine the elasticity of rubber (necessitating coiled, flaccid molecular chains) and the tensile strength of a fiber (necessitating a crystalline arrangement of fairly rigid, highly ordered chains). How can scientists make materials such as this? By grafting two molecular structures onto one polymer chain. Blocks of components with fiber character are alternated with blocks of elastomer in the same giant molecule, and the fabric made from the polymer exhibits both sets of properties, flexibility and rigidity!

19.6 BULK PROPERTIES OF POLYMERS

There are many other types of polymers, both synthetic and natural. The molecules are large, and variations in their structures—and consequent variations in their properties—are almost unlimited. The development of synthetic polymers has led to a whole new world of plastics, rubber substitutes, and synthetic fibers. There is no sharp line separating these types of polymers. The same basic material can be used in two (or even three) of the categories. Nylon, for example, can be molded to make plastic toys or television cabinets, or it can be spun into a fiber and woven into cloth. The major differences between polymer types are mainly in the organization of the big molecules into bulk materials. The molecules of elastic materials are randomly oriented and tangled with one another. On the other hand, molecules in fibers of high tensile strength are highly crystalline in arrangement (Table 19.3).

An important parameter of most polymers is the **glass transition temperature** (T_g). Above this temperature, the polymer is rubbery and tough; below it, the polymer is like glass—hard, stiff, and brittle. Each polymer has a characteristic T_g. We want automobile tires to be tough and elastic, so we use materials with a low T_g. On the other hand, we want plastic substitutes for glass to be glassy. Thus, they have T_g's well above room temperature. We make use of the T_g concept in everyday life. To remove chewing gum from clothing, we apply a piece of ice to lower the temperature of the gum below the T_g of the polyvinyl acetate resin that makes up the bulk of the gum. The cold, brittle resin then crumbles readily and can be removed.

TABLE 19.3   *The Hardness of Nylon and Its Degree of Crystallinity*

Polymeric Material	% Crystallinity
Soft shopping bag	15
Undergarment	20–30
Sweater	15–25
Hosiery	60–65
Cord for tires	75–90
Monofilament fishing line	90

19.7 PLASTICIZERS

It is difficult to process certain polymers, particularly the vinyl types (see Table 19.1). As formed, they may be rather hard and brittle. They can be made more flexible and easier to handle by addition of chemicals called **plasticizers.**

Ideally, plasticizers are liquids of low volatility. They act by lowering the T_g of the plastic to make it more flexible and less brittle. Undiluted polyvinyl chloride (PVC) cracks and breaks easily, but with proper plasticizers added, it is soft and pliable. Plastic raincoats, garden hoses, and seat covers for automobiles can be made from the modified PVC (Figure 19.11). Plasticizers are generally lost by diffusion and evaporation as a plastic article ages. The plastic becomes brittle, then cracks and breaks. Plasticizers in seat covers sometimes show up on car windows as a "fog" that is almost impossible to remove.

One type of plasticizer once used widely but now largely banned is the polychlorinated biphenyls (PCBs). These compounds are discussed in some detail in Special Topic E.

The phthalates comprise another class of plasticizer. These compounds are esters derived from phthalic acid. The parent acid and some of the common plasticizers derived from it are shown in Figure 19.12. Unlike the PCBs, which are used as a complex mixture, the phthalates are generally used as separate compounds, with the individual substance chosen to impart the desired flexibility to the plastic.

FIGURE 19.11 *Some "vinyl" objects. (Photo © Zeva Oelbaum, New York.)*

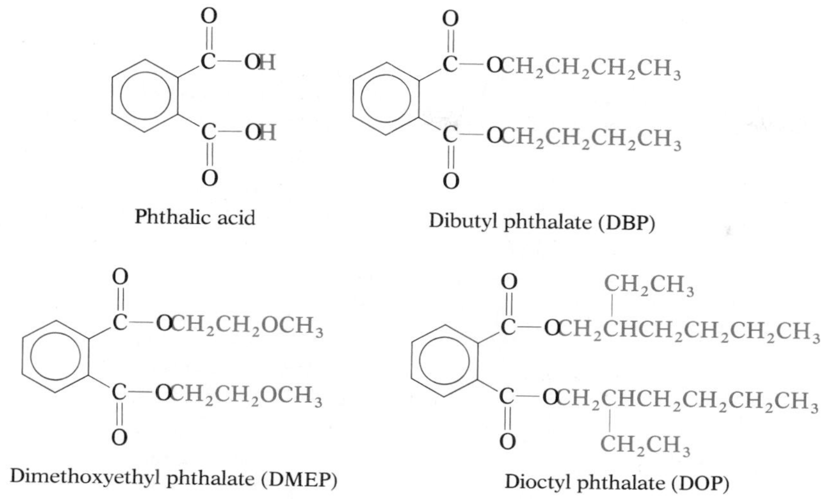

FIGURE 19.12 *Phthalic acid and some derivatives. Dioctyl phthalate is also called di-2-ethylhexyl phthalate (DEHP).*

Phthalate plasticizers, like the PCBs, appear to have low acute toxicity. Their long-term effects are generally unknown. Phthalates have been leached from the PVC bags in which whole blood has been stored. It has been proposed that these plasticizers may contribute to shock lung, a sometimes fatal condition observed in some patients after a blood transfusion. Phthalates have been found in the heart muscle of cattle, dogs, rabbits, and rats. Very high dosages can induce mutations and birth defects in laboratory animals. What implication, if any, these findings have for human health is still being debated.

19.8 PEOPLE OF PLASTIC PARTS BIOMEDICAL POLYMERS

One of the most interesting uses of polymers has been in replacements for diseased, worn out, or missing parts of the human body (Figure 19.13). Artificial ball-and-socket hip joints made of steel (the ball) and plastic (the socket) are now being installed at a rate of 25 000 a year. People crippled by arthritis are not only freed from pain but are given much more freedom of movement. Patients with heart and circulatory problems can enter a hospital for a "valve job" or the replacement of worn out or damaged parts. Pyrolytic carbon heart

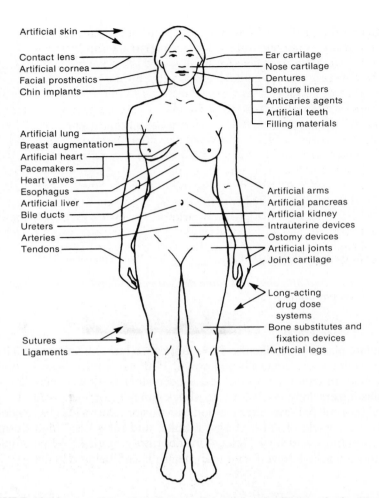

FIGURE 19.13 A variety of replacement parts for the human body. Many of the parts are made from synthetic polymers. (Courtesy of Southern Research Institute, Birmingham, Ala.)

valves derived from a polymer are widely used. Knitted Dacron tubes replace arteries blocked or damaged by atherosclerosis. Plastic implants reconstruct breasts removed because of cancer.

On contact with most foreign substances, blood begins to clot. To prevent such a reaction, most synthetics must be chemically treated and coated with heparin, a natural anticoagulant, before being used as implants. Dacron can be used for artificial arteries because it is relatively inert in this regard. Another approach to the problem of substituting synthetic materials for body parts is to use naturally occurring substances to construct the biomedical polymers. For example, polymers of glycolic and lactic acids have been used in synthetic films for covering burn wounds. Ordinarily, burns have to be covered with human skin from donors or with specially treated pigskin in order for infection and excessive loss of fluids to be prevented. Frequent changes of these covers are required because of the body's tendency to reject these foreign tissues. In contrast, the synthetic film is absorbed and metabolized rather than rejected.

The development of biomedical polymers has barely begun. In the future lies the prospect of the replacement of entire parts of the body. Artificial hearts (Figure 19.14) made of synthetic elastomers are now used to keep people alive who would otherwise die of heart failure. The first such patient, Barney Clark, was kept alive for four months. Such surgery is costly and controversial. Clark's treatment cost $200 000. The equipment is also cumbersome; the patient must remain permanently attached to an external pump or have the artificial heart replaced with a transplanted heart at a later date.

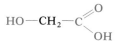

Glycolic acid

Lactic acid

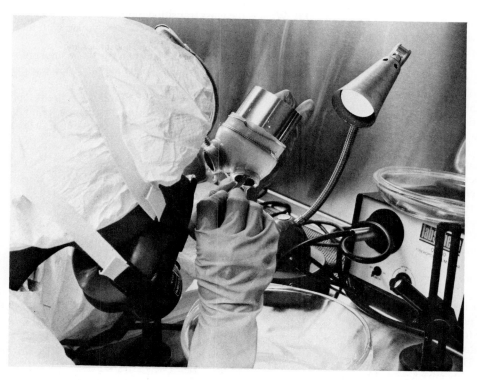

FIGURE 19.14   A worker casts the plastic diaphragm of the Jarvik-7 artificial heart. (Courtesy of Symbion, Inc., Salt Lake City.)

Polymer chemistry is a relatively new science. It has developed for the most part since World War II, yet has already produced an extraordinary array of products, ranging from Silly Putty to artificial hearts. Fully one-half of all the U.S. chemists employed in industry work in some way with polymers. They will no doubt continue to provide us with new materials that meet the needs and wants of society.

PROBLEMS

1. Define these terms:
   a. macromolecule
   b. polymer
   c. monomer
   d. segmer
   e. addition polymerization
   f. condensation polymerization
   g. vulcanization
   h. elastomer
   i. copolymer
   j. plasticizer

2. Make a list of plastic objects (or parts of objects) that you encounter in your daily life. Try to identify a few of the kinds of polymers used in making the items. Compare your list with those of some of your classmates.

3. How do thermosetting and thermoplastic polymers differ from each other in structure and properties?

4. What structural feature usually is found in molecules used as monomers for addition polymerization?

5. Why is rubber elastic? How does vulcanization improve its elasticity?

6. How do low-density and high-density polyethylenes differ in structure and properties?

7. In working this problem, refer to Table 19.1. Draw a section of polymer chain that is a minimum of six carbons long for each of the following:
   a. PVC   **b.** Teflon   **c.** Saran

8. What would the polymer formed from 1,2-dichloroethene look like?

9. Where will plastics come from when the Earth's supplies of coal and petroleum are exhausted?

10. Isobutylene polymerizes to form polyisobutylene, a sticky polymer used as an adhesive. Copolymerized with butadiene, isobutylene forms butyl rubber. Write a two-dimensional representation of a segment of the polyisobutylene molecule. Then write the structure of a segment of the butyl rubber molecule. In each case, show at least four monomer units incorporated into the polymer.

$$CH_2{=}\overset{\displaystyle CH_3}{\underset{\displaystyle CH_3}{C}}$$

Isobutylene

$$CH_2{=}CH{-}CH{=}CH_2$$

Butadiene

11. What is the glass transition temperature (T_g) of a polymer? For what uses do we want polymers with a low T_g? With a high T_g?

12. Dacron has a high T_g. How can the T_g be lowered so that a manufacturer can permanently crease a pair of Dacron slacks?

13. Kodel is a polyester fiber. The monomers are terephthalic acid and 1,4-cyclohexanedimethanol. Write the structure of a segment of Kodel containing at least one of each monomer unit.

Terephthalic acid

1,4-Cyclohexanedimethanol

14. From what monomers might the following copolymer be made?

15. Kevlar, a polyamide used to make bulletproof vests, is made from terephthalic acid and phenylene diamine. Write the structure for a segment of the Kevlar molecule.

$$H_2N-\langle\bigcirc\rangle-NH_2$$

Phenylene diamine

16. What is cellulose nitrate? How is it made?
17. What is rayon? How is it made?
18. What is cellophane? How is it made?
19. What is rayon acetate? How is it made?
20. Draw the structure of the polymer made from acrylonitrile ($CH_2{=}CH-C{\equiv}N$). Show at least four segmers.
21. Draw the structure of the polymer made from vinyl acetate ($CH_2{=}CH-O\overset{\displaystyle O}{\overset{\|}{C}}CH_3$). Show at least four segmers.
22. Draw the structure of the polymer made from 1-hexene ($CH_2{=}CHCH_2CH_2CH_2CH_3$). Show at least four segmers.
23. Draw the structure of the polymer made from styrene ($CH_2{=}CH-\langle\bigcirc\rangle$). Show at least four segmers.

24. Draw the structure of the polymer made from methyl methacrylate ($CH_2{=}\overset{\displaystyle CH_3}{\overset{|}{C}}-\overset{\displaystyle O}{\overset{\|}{C}}OCH_3$). Show at least four segmers.
25. Nylon 46 is made from 1,4-diaminobutane and adipic acid (hexanedioic acid). Draw the structure of nylon 46 showing at least two units from each monomer. Is nylon 46 an addition polymer or a condensation polymer? Explain.
26. The microorganism *Alcaligenes eutrophus* produces a natural macromolecular material called polyhydroxybutyrate. The material is a polymer of 3-hydroxybutanoic acid. Draw a structure for a segment of the polymer showing at least four monomer units. To what class of polymers does polyhydroxybutyrate belong?
27. Draw the structure of the polymer made from glycolic acid ($HOCH_2COOH$). Show at least four segmers.
28. Draw the structure of the polymer made from lactic acid (2-hydroxypropanoic acid). Show at least four segmers.
29. Which is more extensively cross-linked, the rubber in surgical gloves or that in an automobile tire?
30. Which has a higher degree of crystallinity, the nylon in a blouse or that in a 40-lb test fishing line?

Stereochemistry

olecules important to life, especially the carbohydrates and proteins, often exhibit a special kind of isomerism. They have a sort of "handedness" in that certain molecules are related to one another much like your left hand is related to your right. The isomerism is rather subtle, but extremely important. We consider it here in some detail.

The molecular geometry that results in handedness involves the same atoms or groups in the same order of attachment. The molecules differ only in the orientation of the groups in space. Such isomers are called **stereoisomers**, and their study is called **stereochemistry** (Figure I.1).

Much of what we know about stereoisomers comes from their effect on plane-polarized light. Let's take a look at the nature of this light before we examine the molecules that act upon it.

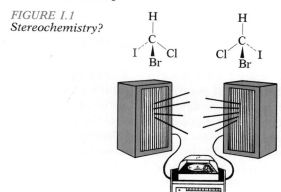

FIGURE I.1
Stereochemistry?

I.1 POLARIZED LIGHT AND OPTICAL ACTIVITY

First of all, let's establish a convention for drawing waves. A wave is something that goes up and down or increases and decreases or varies regularly in some such way. Here's a wave coming toward you and moving to the right and left across the page.

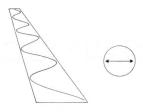

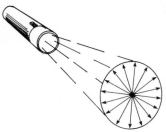

FIGURE I.2
The beam of light in this illustration is not polarized.

The arrow in the circle is our convention for indicating the same thing. When you see such an arrow, you are supposed to imagine a wave moving out of the page toward you and vibrating back and forth in the direction indicated by the arrow.

Here's another example.

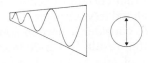

The wave is coming toward you in an up and down motion. The vertical double-headed arrow stands for the same motion.

Now what's all this about waves anyway? Well, light is a wave, that is, it has characteristics that are associated with wavelike motion. A beam of ordinary light can be pictured as a bundle of waves, some of which move up and down, some sideways, and others at all conceivable angles (Figure I.2). While such light can be described as ordinary, a more scientific term is **nonpolarized.** Which brings us to what we wanted to talk about all along—polarized light.

The waves of **polarized light** vibrate in a single plane. Both of the beams of light shown in Figure I.3 are polarized. The two polarized beams differ only in the angle of the plane of polarization (represented in our drawings by the different orientation of the double-headed arrows).

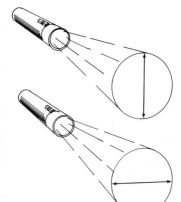

FIGURE I.3
Both of the illustrated light beams are polarized. The planes of polarization differ.

Sunlight, in general, is not polarized, nor is the light from an ordinary light bulb, nor the beam of light from an ordinary flashlight. One way to get polarized light is to pass ordinary light through Polaroid sheets, such as those used for the lenses of some sunglasses. These lenses are made by carefully orienting organic compounds in plastic, a material that permits only light vibrating in a single plane to pass through (Figure I.4). To the eye, polarized light doesn't

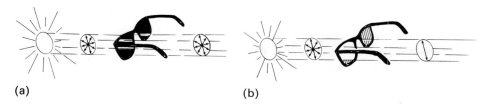

(a)   (b)

FIGURE I.4   Ordinary sunglasses (a) dim light by preventing some light from passing through the lenses. But these lenses do not discriminate among light waves vibrating at different angles—all are cut down. The light reaching the eyes through ordinary sunglasses is nonpolarized. Sunglasses with Polaroid lenses (b) selectively pass light waves vibrating in a single plane. Light waves vibrating in the other planes are rejected, i.e., not allowed to pass through. The light reaching the eyes through Polaroid sunglasses is plane polarized.

"look" any different than nonpolarized light. One can detect polarized light, however, by using a second sheet of polarizing material (Figure I.5).

Certain substances act on polarized light by rotating the plane of vibration. Such substances are said to be **optically active.** The extent of optical activity is measured in a device called a **polarimeter** (Figure I.6). The device consists of two polarizing lenses, called the **polarizer** and the **analyzer.** With the sample tube empty (Figure I.6a), maximum light reaches the observer's eye when both the polarizer and the analyzer are aligned so that both pass light vibrating in the same plane (the pointer on the analyzer indicates 0° difference between the alignment of the analyzer and the polarizer). When an optically active substance is placed in the sample tube (Figure I.6b), the plane of polarization is rotated, namely, the polarized light emerging from the sample tube is vibrating in a different direction than was the light that entered the tube. To see the maximum amount of light when the sample is in place, the observer must rotate the

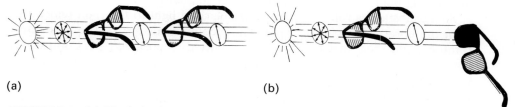

(a)   (b)

FIGURE I.5   (a) The light that passes through the first Polaroid lens is polarized. The second pair of glasses is oriented like the first, therefore its Polaroid lens passes the polarized light. (b) The second pair of glasses is oriented 90° to the first. Thus, the plane of polarization of the light that made it through the first lens is oriented incorrectly for the second lens. No light gets through. If the second pair of sunglasses were not made with Polaroid lenses, some light would get through no matter how they were oriented, because such glasses do not distinguish among various orientations of the plane of polarization.

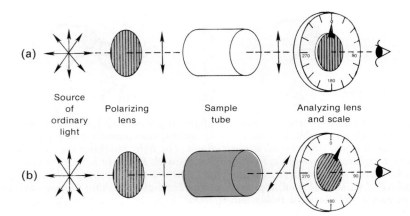

FIGURE I.6 (a) Schematic diagram of polarimeter with sample tube empty or containing a substance that is not optically active. (b) Sample tube containing an optically active substance.

analyzing lens to accommodate this change in the plane of polarization. The angle of rotation, indicated by the pointer on the analyzing lens, corresponds to the change in the plane of polarization caused by the sample.

The size of the angle depends not only on the structure of the optically active material but also on the length of the sample tube, the concentration of the solution, even the color of light used and the temperature of the experiment. However, scientists have agreed to certain standard conditions for reporting the angle of rotation. When a rotation is calculated and reported with these conditions taken into account, the value is referred to as the specific rotation (symbol = $[\alpha]$) and is a physical constant as characteristic of the material as the melting point, boiling point, density, or solubility.

Some substances rotate the plane of polarized light to the right (clockwise from the observer's point of view). These compounds are said to be **dextrorotatory,** and their specific rotation is reported as positive ($+$). Compounds that rotate the plane to the left (counterclockwise) are said to be **levorotatory** and are reported as negative ($-$).

I.2 CHIRAL CARBON ATOMS

Optical activity arises from the handedness, or **chirality** (Greek *cheir*, "hand"), of molecules. We all know that a glove for a right hand won't fit a left hand. In the familiar greeting of the Western world, the handshake, right hand grasps right hand. It is impossible to grasp a left hand with a right hand in the same

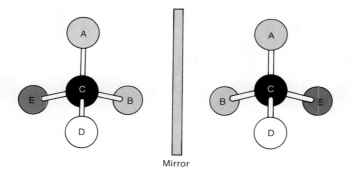

FIGURE I.7   *Molecules in which a carbon atom has four different groups attached may exist as two different arrangements. These are nonsuperimposable mirror images.*

way. Similarly, molecules may react differently with one another, depending on the handedness.

We shall limit our consideration of chirality to one type of chiral compound, one that contains *a carbon atom with four different groups attached.* At one time such a carbon atom was commonly referred to as an asymmetric carbon atom, but in modern terms it is called a **chiral center,** or sometimes, a chiral carbon atom.

With four different groups attached, a carbon atom may have *two different* arrangements in space (or configurations). One arrangement is the mirror image of the other (Figure I.7).* The two mirror image forms are nonsuperimposable, thus not identical (Figure I.8). Isomers that differ only in the orientation of the atoms in space are called **stereoisomers.** The particular isomers encountered here—a molecule and its nonsuperimposable mirror image—are called **enantiomers.** Stereoisomerism is a broader term that includes enantiomers, geometric (*cis-trans*) isomers (Chapter 14), and diastereomers (next section). (The other general classification is structural isomerism. Structural isomers differ in the order of attachment of the atoms, e.g., 1-chloropropane and 2-chloropropane are structural, not stereoisomers.)

---

* By arrangement or **configuration** we mean the position of the atoms relative to one another. These molecules, like any others, can be turned upside down or spun about or tipped forward or backward, just as you can stand on your feet or on your head or lie on your back, etc. If you had a mole on your right arm, however, no matter what position you assumed, the mole would still be on your right arm. Your mirror image would always have a left-arm mole. No amount of spinning or turning will cause the mole to change from your right to your left arm. The molecules pictured in Figure I.7 also have such a distinguishing arrangement. If, for these molecules, we call atom A the head and the side to which atom D is attached the front, then one compound always has atom E on its right side and the other always has E on its left side.

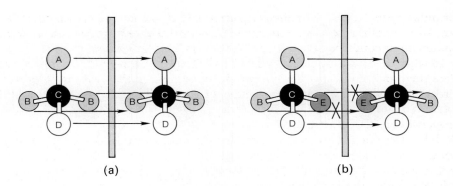

(a)   (b)

FIGURE I.8   (a) Molecules that do not have a chiral center are superimposable on their mirror images. (b) Molecules that have a chiral center (four different groups on carbon) are not superimposable on their mirror images.

CH₃ĊHC $\overset{O}{\underset{OH}{\diagup}}$
|
OH

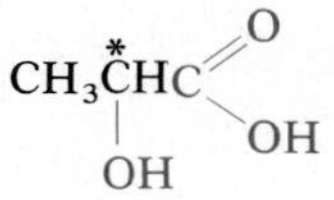

Lactic acid

Enough of this hypothetical stuff. Let's look at some real molecules that have chiral centers. A usual representation of lactic acid is given in the margin. Examination of the structure reveals that the second carbon atom (marked with an asterisk) has four different groups attached—a hydrogen atom, a carboxyl group, a hydroxyl group, and a methyl group. This carbon, then, is a chiral center. There should be *two* lactic acids. Are there? Yes! (Why, it's enough to give you faith in chemical theory!) One lactic acid is found in sour milk. It is levorotatory and may be designated as (−)-lactic acid. Another lactic acid is found in muscle tissue, particularly after exercise. It is dextrorotatory and is called (+)-lactic acid. How are the two related? They are enantiomers: one is the mirror image of the other. Using perspective formulas, the enantiomers may be represented as follows:

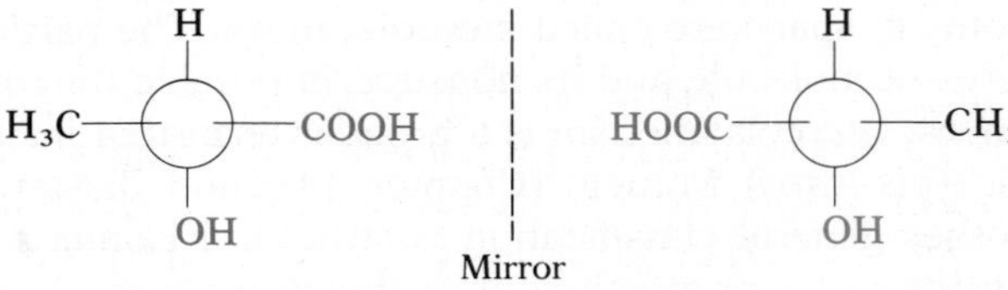

$$H_3C \overset{H}{\underset{OH}{\bigcirc}} COOH \qquad HOOC \overset{H}{\underset{OH}{\bigcirc}} CH_3$$

Mirror

Enantiomers can be drawn using "flat" formulas, for example,

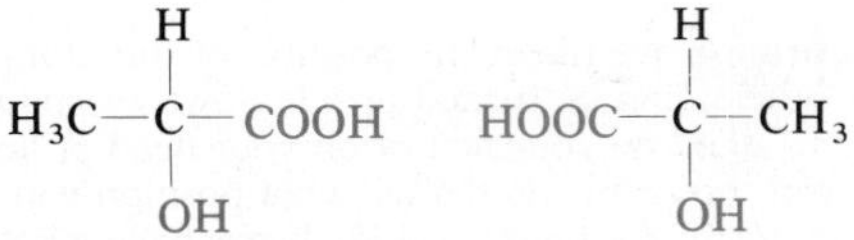

$$H_3C \overset{H}{\underset{OH}{-}C-} COOH \qquad HOOC \overset{H}{\underset{OH}{-}C-} CH_3$$

but such drawings require more of *you*. You have to remember that the horizontal bonds project above the plane of the paper and the vertical bonds project below. Because they are easier to draw, we shall eventually use these "flat" formulas (often called **Fischer projections**) to represent stereoisomers. For now, we'll continue to use perspective drawings while we introduce some of the fundamental concepts of stereochemistry.

In what ways do the actual isomers differ? In many respects, they seem more alike than different. All of their physical properties are identical save one, the direction in which they rotate the plane of polarized light. One has a specific rotation of $+2.6°$, the other $-2.6°$. Only the sign is different. Simple chemical properties are also the same. Both form acidic solutions. Both neutralize bases. Both form esters. Indeed, when reacting with *achiral* molecules, the two enantiomers exhibit identical chemical properties. Such common reagents as water, hydroxide ion, and ethyl alcohol do not contain chiral centers and are achiral. It is only when the lactic acid isomers react with other chiral molecules that they behave differently. They may react at different rates and to different extents. The products formed will have different properties. Are these differences really important? In living cells reactions are controlled by enzymes, and enzymes are chiral. That means that enantiomers may behave quite differently in living cells. Enantiomers may have different tastes and smells. One may be an effective drug and the other worthless. One may be essential to health and the other toxic. Quite literally, we may be talking about differences between life and death.

When lactic acid is made from propionic acid in the laboratory, it shows *no* optical activity.

$$CH_3CH_2C\underset{OH}{\overset{O}{\diagup}} \longrightarrow CH_3\overset{*}{C}HC\underset{\underset{OH}{|}\;OH}{\overset{O}{\diagup}}$$

Achiral   Chiral

How can this be? The lactic acid has a chiral center. The answer: in syntheses of this sort, the $(+)$ and $(-)$ forms are formed in exactly equal amounts. Such a mixture of enantiomers is called a **racemic modification.** It shows no optical activity because the mixture contains equal amounts of molecules with equal but opposite rotatory power. Everything cancels out. Racemic lactic acid is designated $(\pm)$. Notice in Table I.1 that the racemic mixture may exhibit different physical properties than the pure enantiomers.

TABLE I.1 *Properties of Lactic Acids*

Form	Melting Point (°C)	Specific Rotation	pKa
$(+)$	53	$+2.6$	3.8
$(-)$	53	-2.6	3.8
$(\pm)$	16.8	0	3.8

$$CH_3\overset{*}{C}H \overset{*}{-} \overset{*}{C}H-CH_2CH_3$$
$$||$$
$$OHOH$$

2,3-Pentanediol

I.3 MULTIPLE CHIRAL CENTERS

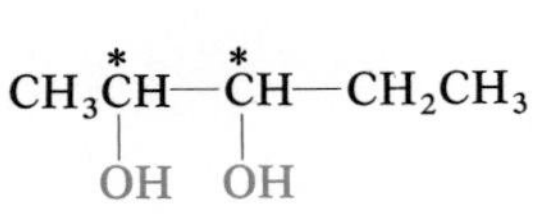

Molecules may have more than one chiral center. Indeed, simple carbohydrate molecules generally have several each. Giant molecules, such as starch, cellulose, and the proteins, may have several hundred or even several thousand chiral centers. Let us look first, though, at molecules with just two.

Consider 2,3-pentanediol (margin). There are *four ways* in which the groups can be arranged about the two chiral centers (Figure I.9). Note that structures I and II are enantiomers; they are nonsuperimposable mirror images of one another. Compounds III and IV make up another set of enantiomers. What is the relationship, though, between structures II and III? They are stereoisomers because they differ only in their spatial arrangement, but they are not enantiomers because they are not mirror images. Such sets of isomers are called **diastereomers**. Note that I and IV are diastereomers, as are II and III and II and IV. Diastereomers generally have *different* physical properties (like boiling point and solubility, as well as specific rotation). Unlike enantiomers, they may be separated by distillation or fractional crystallization.

Let us consider one last set of compounds before moving on to the next chapter. These compounds, the tartaric acids, were involved in the earliest studies

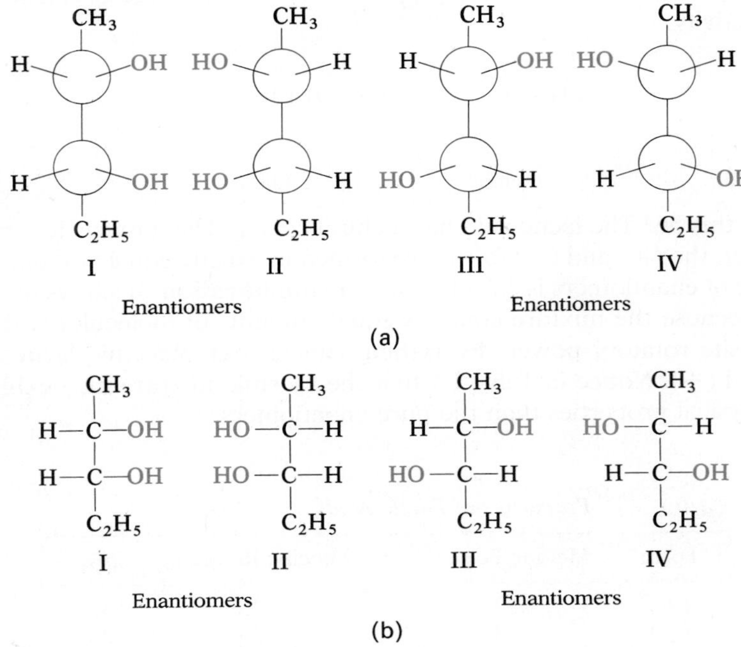

FIGURE I.9   The four stereoisomeric 2,3-pentanediols. (a) Perspective drawings. (b) Flat projection formulas.

relating structure and optical activity. The investigator was Louis Pasteur, a French chemist who invented the process, now called pasteurization, of heating milk to destroy pathogenic bacteria and slow the fermentative action. Pasteur's work also led to the germ theory of disease, to immunization procedures, and to the original discovery of the stereoisomerism associated with enantiomers. The compounds he studied also included a new type of stereoisomer.

The formula for tartaric acid, like that of 2,3-pentanediol, contains two chiral carbon atoms (margin). There is one notable difference between tartaric acid and 2,3-pentanediol. In tartaric acid, each of the two chiral carbons is attached to the same four different groups, namely, to —COOH, —OH, —H, and —CH—COOH. In 2,3-pentanediol, one chiral carbon was attached to a
    |
   OH
methyl group, whereas the other was attached to an ethyl group. This difference is significant, as we shall see.

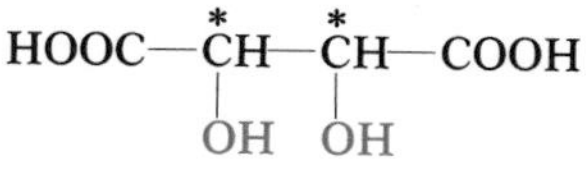

Writing out the perspective formulas of tartaric acid (Figure I.10), we see a pair of enantiomers (I and II). The other apparent pair (III and IV), though, are not enantiomers; they are not even isomers. They are, in fact, exactly the same compound. The structures may be superimposed by rotating one of them

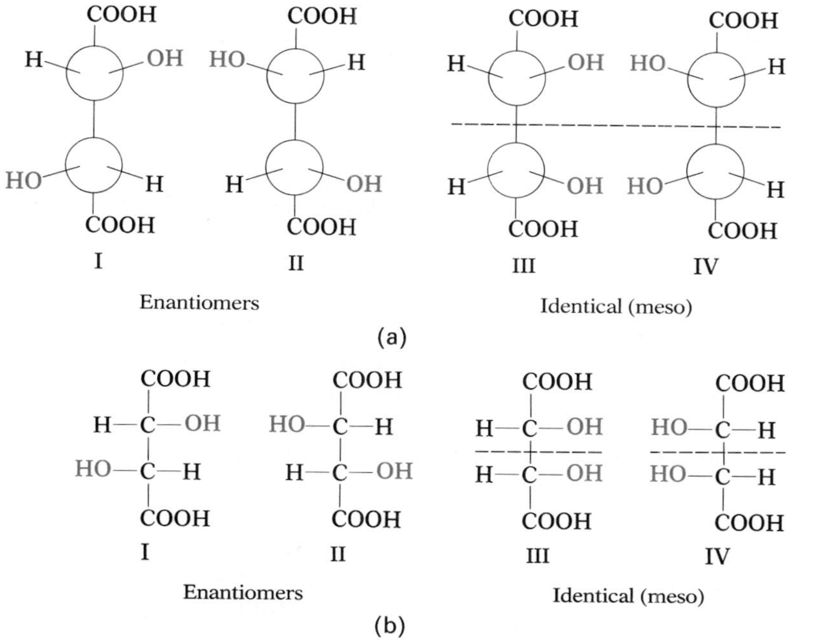

$$HOOC-\overset{*}{C}H-\overset{*}{C}H-COOH$$
$$\underset{OH}{|}\underset{OH}{|}$$

Tartaric acid

FIGURE I.10 *The three stereoisomeric tartaric acids. (a) Perspective drawings. (b) Flat projection formulas. Note that the meso form has an internal plane of symmetry.*

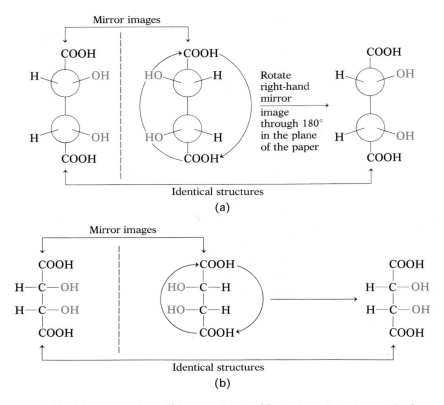

FIGURE I.11   *Meso tartaric acid is superimposable on its mirror image. It does not exist as a pair of enantiomers, but as a single compound only. (a) Perspective drawings. (b) Flat projection formulas.*

180° in the plane of the paper (Figure I.11). It is important to emphasize again that enantiomers are not simply mirror images of one another, but *nonsuperimposable* mirror images. Structures III and IV are mirror images, but they are also superimposable and, hence, identical. The corresponding structures for 2,3-pentanediol (Figure I.9) are not superimposable because the CH_3 group and the C_2H_5 group cannot be interchanged.

The single compound represented by both structures III and IV in Figure I.10 is termed a **meso** compound. It is a diastereomer of compound I and compound II. All meso compounds are optically inactive. Meso compounds contain at least two chiral centers, but have an internal symmetry plane; in Figure I.10 we have indicated this mirror plane in structure III and IV by a dashed line. The meso molecule as a whole is *not* chiral. Table I.2 lists all the forms of tartaric acid.

TABLE I.2 *Properties of the Tartaric Acids*

Form	Melting Point (°C)	Specific Rotation	pKa	Solubility (g/100 g H_2O)
(+)	168–170	+12.0	2.93	133
(−)	168–170	−12.0	2.93	133
(±)	206	0	2.96	21
Meso	140	0	3.11	125

Let us summarize here the various conditions that result in a lack of optical activity. A compound (like ethanol, CH_3CH_2OH) may contain no chiral center. It is not chiral (i.e., it is achiral), therefore, not optically active. A meso compound contains chiral centers, but is not chiral because it also contains an internal mirror plane. It, too, is optically inactive. A racemic modification contains chiral molecules, but is not optically active because the dextrorotatory molecules cancel out the effects of the levorotatory ones.

We've spent a long time preparing you for the subject of carbohydrates. Perhaps this will give you a hint as to why so much background is required: blood sugar is just one of 16 stereoisomeric carbohydrates with the structural formula shown.

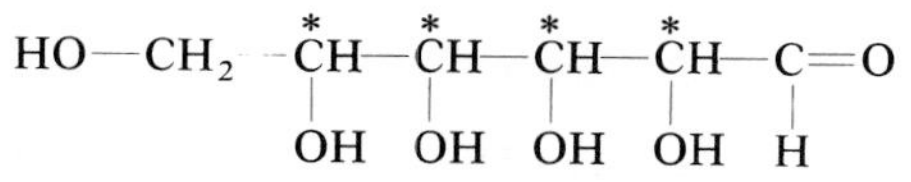

$$HO-CH_2-\overset{*}{C}H-\overset{*}{C}H-\overset{*}{C}H-\overset{*}{C}H-C=O$$
$$\qquad\qquad | \quad\; | \quad\; | \quad\; | \quad\; |$$
$$\qquad\quad OH \; OH \; OH \; OH \; H$$

Blood sugar is one of 16
compounds with this formula

You can calculate the maximum number of stereoisomers for a given structure by using the rule

Number of stereoisomers $= 2^n$

where n is the number of chiral carbon atoms. Those structures that have meso forms will have fewer than the calculated number. We have seen that 2,3-pentanediol (Section I.3) has two chiral centers and exists as

$$2^2 = 4$$

stereoisomers. On the other hand, tartaric acid also has two chiral centers, but exists in only three stereoisomeric forms, including one that is meso.

PROBLEMS

1. Define:
   a. chiral center
   b. enantiomers
   c. polarized light
   d. diastereomers
   e. stereoisomers
   f. specific rotation
   g. meso form
   h. racemic modification
   i. levorotatory
   j. dextrorotatory

2. Which of the carbon atoms shown in color are chiral?

a. ⬡—$CH_2$$\overset{|}{C}H$—$NH_2$
  $\overset{|}{C}H_3$

b. $CH_2CH_2CH_2$—$N\overset{\displaystyle CH_3}{\underset{\displaystyle CH_3}{}}$

(phenothiazine ring with N, S, Cl)

c. $H_2N\overset{O}{\overset{||}{C}}OCH_2\overset{\displaystyle CH_2CH_2CH_3}{\underset{\displaystyle CH_3}{C}}CH_2O\overset{O}{\overset{||}{C}}NH_2$

3. Are these structures mirror images? Are they superimposable?

$$H—\overset{\displaystyle CH_3}{\underset{\displaystyle CH_3}{C}}—OH \quad HO—\overset{\displaystyle CH_3}{\underset{\displaystyle CH_3}{C}}—H$$

4. Are these structures mirror images? Are they superimposable?

$$CH_3CH_2—\overset{\displaystyle H}{\underset{\displaystyle OH}{C}}—CH_3 \quad CH_3—\overset{\displaystyle H}{\underset{\displaystyle OH}{C}}—CH_2CH_3$$

5. Place an asterisk (*) beside each chiral carbon atom in the following molecules:

a. HO—⬡—$\overset{|}{C}HCH_2$—NH—CH_3
  $\underset{\displaystyle OH}{}$
  (HO on ring)

Epinephrine

b. $\overset{O}{\underset{HO}{\overset{\displaystyle \diagdown}{C}}}$—$CH_2$—$CH_2$—$\overset{|}{C}H$—$\overset{O}{\overset{\diagup}{C}}$$\diagdown$$O^-Na^+$
  $\underset{\displaystyle NH_2}{}$

MSG

6. Place an asterisk (*) beside each chiral carbon atom in the following molecules:

a. H_2N—$\overset{|}{C}H$—$\overset{O}{\overset{||}{C}}$—$NH$—$\overset{|}{C}H$—$\overset{O}{\overset{||}{C}}$—$OCH_3$
  $\overset{|}{C}H_2$ $\overset{|}{C}H_2$
  $\overset{|}{C}OOH$ ⬡

Aspartame

b. CH_3—$\overset{\displaystyle H}{\underset{\displaystyle CH_3}{N^+}}$—$\overset{\displaystyle CH_3}{C}HCH_2$—$\overset{O\diagdown \diagup CH_2CH_3}{\underset{}{C}}$
  Cl^- $\overset{|}{C}H_3$ (two phenyl rings)

Methadone

7. Circle each chiral carbon atom.

a. $CH_3\overset{|}{C}HCH_2OH$
  $\overset{|}{O}H$

b. $CH_3\overset{|}{C}HCOOH$
  $\overset{|}{N}H_2$

c. $C_6H_5CH_2\overset{|}{C}HCH_3$
  $\overset{|}{N}H_2$

d. $CH_3\overset{|}{C}HCH_2CH_3$
  $\overset{|}{B}r$

e. $CH_3\overset{|}{C}HCHO$
  $\overset{|}{O}H$

f. $CH_3\overset{|}{C}H$—$\overset{|}{C}HCH_3$
  $\overset{|}{O}H$ $\overset{|}{O}H$

8. Which can exist in meso forms?

a. $CH_3\overset{|}{C}H$—$\overset{|}{C}HCH_3$
  $\overset{|}{O}H$ $\overset{|}{O}H$

b. $CH_3\overset{|}{C}H$—$\overset{|}{C}HCH_2CH_3$
  $\overset{|}{B}r$ $\overset{|}{B}r$

c. $HOOC\overset{|}{C}H$—$\overset{|}{C}HCOOH$
  $\overset{|}{O}H$ $\overset{|}{O}H$

9. Which of the following can exist in meso forms?

a. CH$_2$OH   **b.** CHO   **c.** COOH   **d.** CH$_3$
| | | |
CHOH   CHOH   CHOH   CHCl
| | | |
CHOH   CHOH   CHOH   CHCl
| | | |
CH$_2$OH   CH$_2$OH   CH$_2$OH   CH$_3$

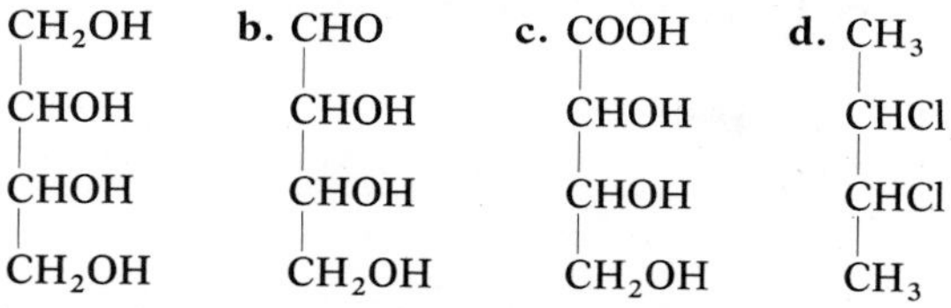

10. Indicate whether the compound is chiral.

a. 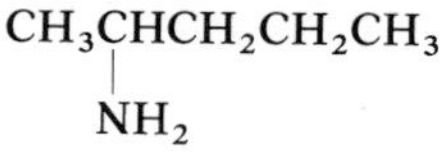 CH$_3$CHCH$_2$CH$_2$CH$_3$
    |
   NH$_2$

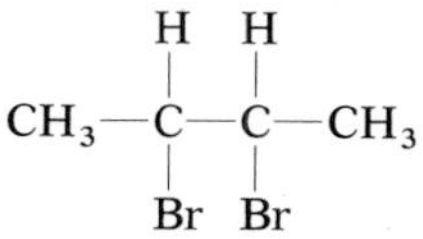

b. 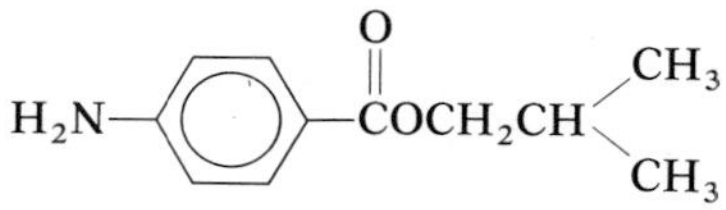 H$_2$N—⬡—COCH$_2$CH⟨CH$_3$, CH$_3$⟩ (with O double bond above C)

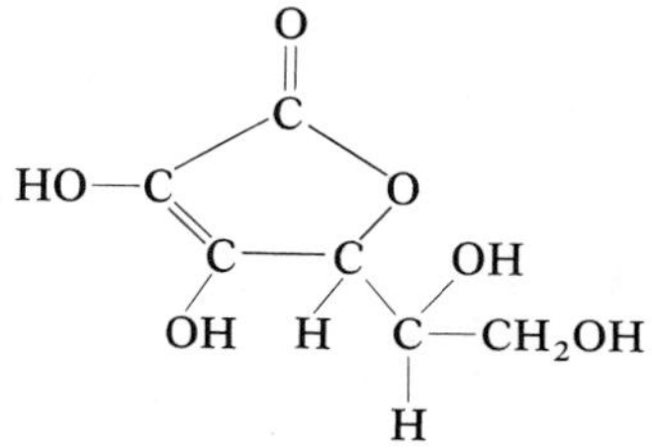

c. HO—C (with O double bond) —O ... C—C OH / OH H C—CH$_2$OH / H

d. 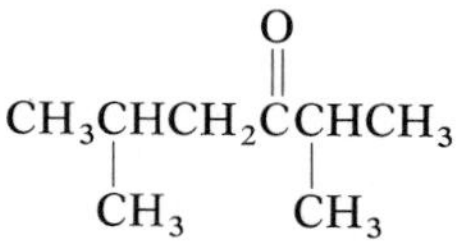 CH$_3$CHCH$_2$CCHCH$_3$ (with O double bond on middle C)
    | |
   CH$_3$ CH$_3$

11. Indicate whether the compound is chiral.

a. 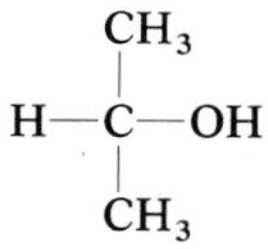
    CH$_3$
    |
H—C—OH
    |
    CH$_3$

b.
    H   H
    | |
CH$_3$—C—C—CH$_3$
    | |
    Br Br

c.
    H   Br
    | |
CH$_3$—C—C—CH$_3$
    | |
    Br H

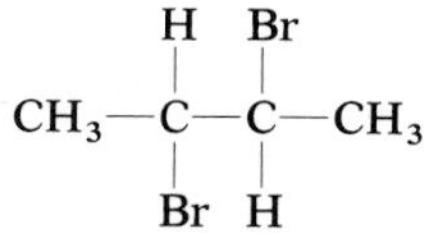

12. Compare (+)-lactic acid and (−)-lactic acid with respect to
  a. boiling point
  b. melting point
  c. specific rotation
  d. solubility in H$_2$O
  e. reaction with ethanol
  f. reaction with (+)-*sec*-butylamine

13. How does a meso compound differ from a racemic modification?

14. Draw projection formulas for (+), (−), and meso forms of 2,3-butanediol. Which is meso? Can you tell which is (+) and which is (−)?

15. In Chapter 15, Problem 3, you are asked to draw structures for the eight isomeric pentyl alcohols. Which of these could exist in enantiomeric forms? (That is, which molecules have chiral carbon atoms?) Draw projection formulas for each pair of enantiomers.

16. Draw Fischer projections (flat) for the stereoisomers of bromochlorofluoromethane.

17. Draw Fischer projections (flat) for 2,3-dichlorobutanal. Label pairs of enantiomers.

18. How many stereoisomers are there for each of the following?

a. 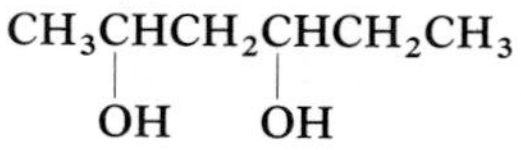 CH$_3$CHCH$_2$CHCH$_2$CH$_3$
        | |
       OH   OH

b. 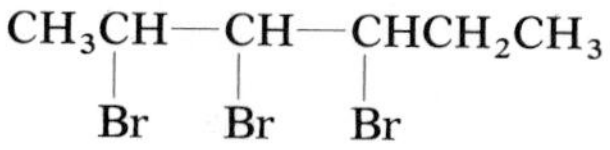 CH$_3$CH—CH—CHCH$_2$CH$_3$
        | | |
       Br   Br   Br

19. How many stereoisomers are there for the following structure?

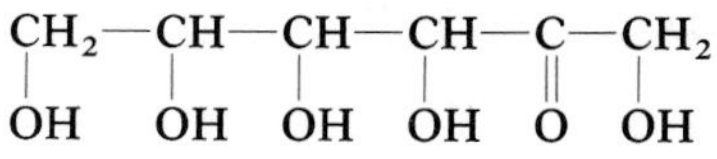

CH$_2$—CH—CH—CH—C—CH$_2$
 | | | | ‖ |
OH   OH   OH   OH O   OH

20. (−)-Menthol melts at 43 °C, boils at 212 °C, has a density of 0.890 g/cm^3, and a specific rotation of −50°. List the corresponding properties of (+)-menthol.

21. Would (+)-nicotine and (−)-nicotine react the same way with HCl? With lactic acid? With tartaric acid?

22. 1-Butene reacts with HCl to form 2-chlorobutane. Does the reactant have a chiral center? Does the product have a chiral center? Would 2-chlorobutane formed in this manner show optical activity? Why or why not?

Carbohydrates

Almost everyone knows what carbohydrates are: they're what you eat or don't eat depending on whose diet book you fancy. In fact, as any dietitian or nutritionist will tell you, carbohydrates must be included in any well-balanced diet. They are the body's primary source of energy. This energy is stored in the complex molecular structure of the carbohydrates when these compounds are synthesized by green plants from carbon dioxide and water. The entire biosphere, and that includes us, is dependent on the endothermic reaction in which simple compounds plus energy yield complex compounds. When we metabolize the complex compounds, the atoms rearrange themselves back into simple compounds and, in the process, release their stored energy for our use.

The photosynthetic reactions in green plants produce glucose. This simple sugar can be used by the plant for energy, or it can be converted to a more complex form—starch—to serve as a source of energy for later use, perhaps as nourishment for the plant's seeds. Some of the glucose is converted to cellulose, the structural material of plants. We can gather and eat the parts of plants that store energy—seeds, roots, tubers, fruits—and use some of that energy for ourselves. We can eat the cellulose, too, but we can't get any energy from it. Why? The answer lies in the chemical structure of cellulose.

We discuss the metabolism of carbohydrates in Chapter 27. For now, let's look at their structures and some simple reactions.

20.1 CARBOHYDRATES: DEFINITIONS AND CLASSIFICATIONS

Carbohydrates are compounds of carbon, hydrogen, and oxygen. They include the starches, the sweet-tasting compounds called sugars, and structural materials such as cellulose (Figure 20.1). The term **carbohydrate** has its origin in a misinterpretation of the molecular formulas of many of these substances. For example, the formula for blood sugar is $C_6H_{12}O_6$, but we could also represent it as that of a "carbon hydrate" $(C \cdot H_2O)_6$. These compounds are not hydrates of carbon, however. They are alcohols, all containing the hydroxyl (—OH) functional group. Most contain a real or latent carbonyl (C=O) group. By a **latent** carbonyl group we mean a functional group such as a hemiacetal or an acetal (Chapter 16) that can be more or less readily converted to a carbonyl group. Consequently, some carbohydrates give reactions as aldehydes or ketones even though the "carbonyl" may exist primarily in a hemiacetal form. Complex carbohydrates containing the acetal function can be hydrolyzed like other acetals to simpler compounds that give reactions typical of aldehydes or ketones.

Carbohydrates are often called saccharides (from the Latin *saccharon*, "sugar"). In fact, William Proust, an English physician who first recognized the three general classes of foodstuffs (now called carbohydrates, fats, and proteins), suggested that they be called the saccharine, the oily, and the albuminous. Simple carbohydrates, those that cannot be further hydrolyzed, are called **monosaccharides.** Carbohydrates that can be hydrolyzed to two mono-

FIGURE 20.1 Wood, lettuce, potatoes, and cotton are all substances that are rich in carbohydrates. Sugar is a pure carbohydrate. (Photo © The Terry Wild Studio.)

saccharide units are called **disaccharides,** and carbohydrates that can be hydrolyzed to many monosaccharide units are called **polysaccharides.**

Sugars are also classified experimentally into two groups. Those that reduce Tollens's reagent (or related reagents) are called **reducing sugars** (Section 20.7). Those that give a negative test, that is, that are not readily oxidized, are called **nonreducing sugars.**

20.2 PHYSICAL PROPERTIES

Carbohydrate molecules contain several —OH groups each. The molecules can form an extensive network of intermolecular hydrogen bonds. Most carbohydrates are crystalline solids at room temperature. They have relatively high melting points, and often will char before melting. Carbohydrate molecules also readily form hydrogen bonds to water molecules. The simpler ones are readily soluble in water. For example, 100 g of glucose will dissolve in 100 mL of water at 25 °C.

20.3 MONOSACCHARIDES: FURTHER CLASSIFICATIONS

Monosaccharides are further classified by the number of carbon atoms per molecule.

Number of Carbon Atoms	Class
3	Triose
4	Tetrose
5	Pentose
6	Hexose
7	Heptose

The **-ose** ending indicates a carbohydrate; the prefixes we have encountered previously.

Still another system classifies monosaccharides on the basis of the carbonyl group present. Those that have an aldehyde group are called aldoses, and those with a ketone group are called ketoses. The two systems are often combined. For example, an aldopentose is a monosaccharide with five carbon atoms and an aldehyde function. Similarly, a ketohexose has six carbons and a ketone function.

The simplest sugars are the trioses. Two of these are derived from glycerol by oxidation. These trioses are important intermediates in the metabolism in muscles. Dihydroxyacetone is a ketotriose. Glyceraldehyde is an aldotriose.

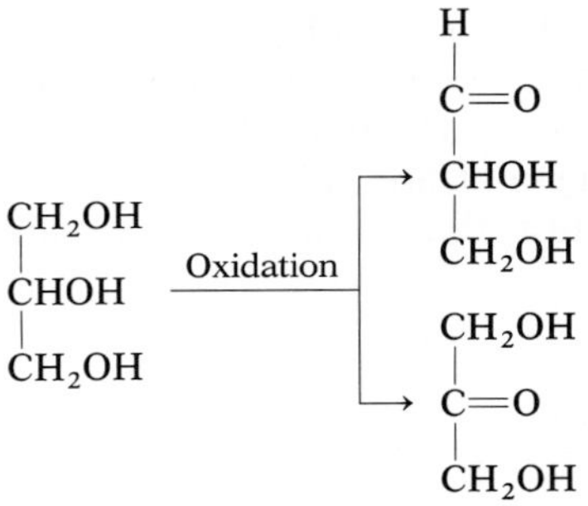

$$
\begin{array}{c}
CH_2OH \\
| \\
CHOH \\
| \\
CH_2OH
\end{array}
\quad \xrightarrow{\text{Oxidation}} \quad
\begin{array}{l}
\begin{array}{c}
H \\
| \\
C{=}O \\
| \\
CHOH \\
| \\
CH_2OH
\end{array}
\quad \text{Glyceraldehyde} \\[2em]
\begin{array}{c}
CH_2OH \\
| \\
C{=}O \\
| \\
CH_2OH
\end{array}
\quad \text{Dihydroxyacetone}
\end{array}
$$

$$
\begin{array}{c}
H \\
| \\
C{=}O \\
| \\
H{-}C{-}OH \\
| \\
CH_2OH
\end{array}
$$

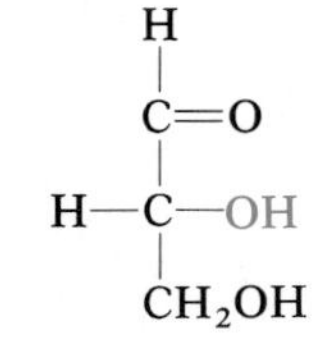

I

$$
\begin{array}{c}
H \\
| \\
C{=}O \\
| \\
HO{-}C{-}H \\
| \\
CH_2OH
\end{array}
$$

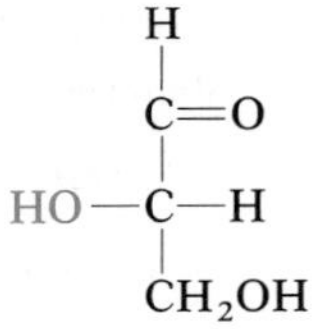

II

The glyceraldehyde molecule is **chiral;** it exhibits "handedness." It exists in two nonidentical mirror image forms (margin). Formula I, with the aldehyde group (CHO) at the top and the primary alcohol group (CH$_2$OH) at the bottom, has the hydrogen on the left and the hydroxyl on the right. Chemists refer to arrangement I as the D configuration. Arrangement II is said to have the L configuration.

All the important carbohydrates in nature are related to D-glyceraldehyde. They have the same handedness about the carbon atom just above the primary alcohol group. All these sugars are said to be members of the D series. Such handedness is of considerable importance. We cannot obtain energy from L carbohydrates (though beings on some other planet, with enzymes that are mirror images of ours, might be able to use only L forms). Since we will encounter only D carbohydrates, we will not emphasize this form of isomerism here. The interested reader is urged to read Special Topic I.

There are two aldopentoses of interest to us (Figure 20.2). One is D-ribose, the sugar unit that occurs in ribonucleic acids (RNA). Related to D-ribose, but not an isomer of it (its molecular formula is different), is D-2-deoxyribose, the sugar unit that occurs in deoxyribonucleic acids (DNA). As the prefix *deoxy*- implies, this sugar is "missing" an oxygen on the second carbon atom. We discuss these compounds in Chapter 24 also. For now, though, let's look at some hexoses. There are 16 isomeric aldohexoses, but we shall discuss only 3

FIGURE 20.2
Two aldopentoses.

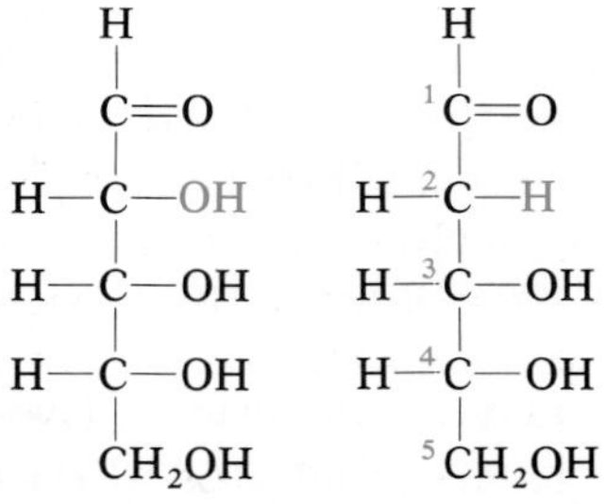

$$
\begin{array}{c}
H \\
| \\
C{=}O \\
| \\
H{-}C{-}OH \\
| \\
H{-}C{-}OH \\
| \\
H{-}C{-}OH \\
| \\
CH_2OH
\end{array}
\qquad
\begin{array}{c}
H \\
| \\
{}^1C{=}O \\
| \\
H{-}{}^2C{-}H \\
| \\
H{-}{}^3C{-}OH \\
| \\
H{-}{}^4C{-}OH \\
| \\
{}^5CH_2OH
\end{array}
$$

D-Ribose   D-2-Deoxyribose

of them here. There are quite a few ketohexoses, too, but we shall deal with only a single ketohexose here. All of the sugars we discuss in the remainder of this chapter belong to the D family. If no family designation is given, you can assume the compound is a D sugar.

20.4 D-GLUCOSE

D-Glucose is undoubtedly the most important hexose. This sugar is sometimes called **dextrose.** D-Glucose is the normal "blood sugar," making up about 0.065% to 0.110% (a concentration of 65 to 110 mg%) of our blood. It is essential to life, for it is the main sugar our cells use directly for the production of energy.

Glucose is readily available from the hydrolysis of starch and cellulose (Sections 20.9 and 20.10). It occurs, along with sucrose (Section 20.8) and fructose (Section 20.5), in the sap of plants. Honey is mainly a mixture of glucose and fructose. It has been estimated that half of the carbon atoms in the biosphere are tied up in glucose. Unfortunately for the hungry people of the world, most of it is in the form of cellulose (Section 20.10), which has little or no food value for humans.

Figure 20.3 illustrates D-glucose. This formula follows our convention of writing the aldehyde group at the top and the primary alcohol group at the bottom. Glucose is a D sugar because the hydroxyl group at the fifth carbon (the one just above the CH_2OH) is on the right. In fact all the hydroxyl groups except that at the third carbon are to the right. You should learn to draw this formula for glucose, as well as the analogous formulas for the other three monosaccharides in Figure 20.3.

20.5 MANNOSE, GALACTOSE, AND FRUCTOSE

Of the remaining aldohexoses, only two occur widely in nature. Mannose is a component of the polysaccharide mannan, found in some berries. A particularly good source is "vegetable ivory," the endosperm of palm nuts. Buttons were once widely made from this material, with the waste being hydrolyzed to mannose. A structure for mannose is presented in Figure 20.3. Note that the configuration differs from that of glucose only at the second carbon atom.

The other common aldohexose is galactose. This sugar, along with glucose, is obtained upon hydrolysis of the disaccharide lactose, or milk sugar (Section 20.8). It is also a component of the polysaccharides called galactans, found in the cell walls of plants. Galactose is found in the brain and nerves, where it is a constituent of compounds called cerebrosides and gangliosides. Figure 20.3 also shows the structure of galactose. Notice that the configuration differs from that of glucose only at the fourth carbon atom.

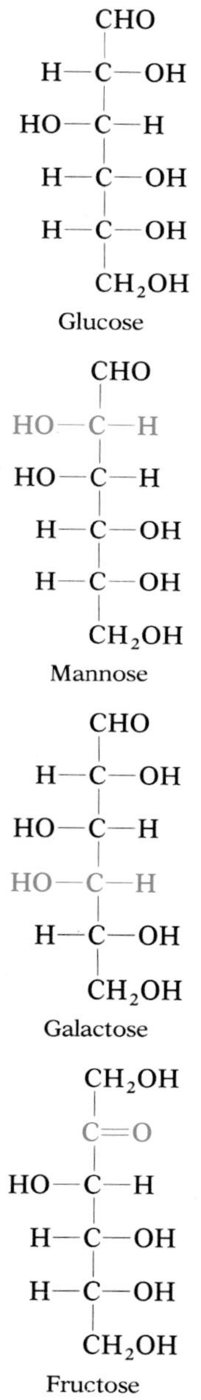

FIGURE 20.3
Structures of four important hexoses.

TABLE 20.1   *Sweetness of Some Compounds Relative to Sucrose as 100*

Compound	Class	Relative Sweetness
Glucose	Monosaccharide	74
Fructose	Monosaccharide	173
Galactose	Monosaccharide	70
Lactose	Disaccharide	16
Sucrose	Disaccharide	100
Maltose	Disaccharide	33
Aspartame	Dipeptide[†]	16 000
Saccharin*		30 000

* Note that saccharin is *not* a saccharide.
[†] Discussed in Chapter 22.

The only ketohexose we shall consider is D-fructose, whose structure is shown in Figure 20.3. It occurs, along with glucose and sucrose, in honey and fruit juices. It is the only sugar found in the semen of bulls and men. Fructose and glucose are obtained by the hydrolysis of the disaccharide sucrose, or table sugar (Section 20.8). Fructose is also called **levulose.**

Fructose is the sweetest of the common sugars (Table 20.1). Many nonsugars are several hundred or several thousand times as sweet, however.

Structurally, fructose is a 2-ketohexose. From the third through the sixth carbon atoms its structure is the same as that of glucose.

20.6 MONOSACCHARIDES: CYCLIC STRUCTURES

So far we have represented the monosaccharides as hydroxyaldehydes and ketones. These representations account for many of the properties (Section 20.7) of these simple sugars. However, Chapter 16 tells us that aldehydes and ketones react with alcohols to form hemiacetals and hemiketals. You should not be surprised, then, to find that hydroxyl groups and carbonyl groups conveniently located on the same molecule react with one another and that, consequently, the monosaccharides exist mainly as hemiacetals. The situation is further complicated by the fact that the ring can close in either of two ways, with the hydrogen on the left or with the hydrogen on the right.

Aldehyde group   Hydroxyl group   Hemiacetals

Thus, each hemiacetal exists in two forms.
Let's look at this reaction for glucose.

"Now wait a minute," you say. "Things are bad enough with all these complications, and now you ignore a hydroxyl group located right next to the carbonyl in order to play around with the hydroxyl at the fifth carbon. And you had to write a long, silly-looking bond to do it!" A reasonable objection—which is why we must introduce a different type of formula. In this new formula, we'll take into account approximately correct bond angles. Figure 20.4 shows a model of the free-aldehyde form of glucose. Notice how it folds around on itself. The center structure in Figure 20.5 is drawn to resemble the model. Note that the hydroxyl on the fifth carbon is quite near the carbonyl carbon, so it's not surprising that it is this hydroxyl group that reacts with the carbonyl. When the reaction occurs, the originally doubly bonded oxygen may be

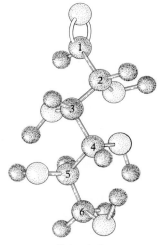

Extended

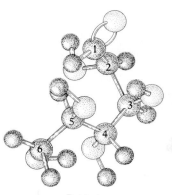

Folded

FIGURE 20.4
Models of the free-aldehyde form of glucose. Note that in the folded model the oxygen atom on the fifth carbon is near the carbonyl carbon.

α-Glucose   Open-chain form   β-Glucose

FIGURE 20.5 *In aqueous solution, glucose exists as an equilibrium mixture of these three forms. α-Glucose accounts for about 36%; the open-chain form, for about 0.02%; and β-glucose, for about 64%.*

pushed up or down, giving rise to two different hemiacetal forms. The structure on the left, with the hydroxyl on the first carbon projected downward, represents what is called the **alpha** (α) form. That on the right, with the hydroxyl on the first carbon pointed upward, is the **beta** (β) form.

Crystalline glucose may exist in either the alpha or the beta form. The two forms have different properties. The alpha form melts at 146 °C, and the beta form melts at 150 °C. In solution, one gets an equilibrium mixture. You can start out with either pure crystalline hemiacetal form, but as soon as these materials are dissolved in water, the unstable hemiacetal group opens to form the free carbonyl and then closes to either hemiacetal, reopening and reclosing in succession. This interconversion is referred to as **mutarotation** (from the Latin *mutare*, "to change").* At equilibrium, the mixture is about 36% alpha and 64% beta. There is less than 0.02% of the aldehyde form. Nevertheless, that is enough to give most of the characteristic reactions of aldehydes. As the small amount of free aldehyde is used up in a reaction, the hemiacetal forms open up to form more free aldehyde. Thus, *all* the molecules may eventually react as aldehyde entities, even though very little "free" aldehyde is present at any given time.

The difference in the alpha and beta forms may seem trivial, but keep in mind that such differences are often crucial in biochemical reactions. We shall encounter some examples of this principle later in this chapter.

20.7 PROPERTIES OF SOME MONOSACCHARIDES

Glucose, mannose, galactose, and fructose are crystalline solids at room temperature. With five hydroxyl groups per molecule, these sugars are quite soluble in water.

Chemically, these monosaccharides undergo the reactions to be expected from the functional groups present. The hydroxyl groups react to form esters and ethers. These reactions, though, are more important commercially for the polysaccharide cellulose (Section 20.10) than for the simpler sugars.

One important reaction is the oxidation of the aldehyde group, which is one of the most easily oxidized organic functional groups. This can be accomplished by any mild oxidizing agent. Ones frequently used are Tollens's, Benedict's, and Fehling's reagents. The Tollens test is based on the reduction of silver ions, and both Benedict's and Fehling's tests involve the reduction of copper complexes.

---

* The two forms of glucose also differ in the way they affect plane-polarized light (Special Topic I). α-D-Glucose has a specific rotation of $+112°$ and β-D-glucose has a specific rotation of $+18.7°$. If either is placed in solution, the observed rotation slowly changes (the substance undergoes **mutarotation**) to an equilibrium value of $+52.7°$.

These reactions aid in the detection of reducing sugars (in the urine of diabetics, for example).

$$\underset{\substack{\text{An aldose}}}{\text{CHO}} + \underset{\substack{\text{Tollens's reagent}\\\text{(clear solution)}}}{\text{Ag(NH}_3)_2{}^+} \longrightarrow \underset{\substack{\text{Carboxylate}\\\text{anion}}}{\text{COO}^-} + \underset{\substack{\text{Silver}\\\text{mirror}}}{\text{Ag}}$$

$$\text{CHO} + \underset{\substack{\text{Benedict's reagent}\\\text{(blue solution)}}}{\text{Cu(citrate)}_2{}^2} \longrightarrow \text{COO}^- + \underset{\substack{\text{Brick-red}\\\text{precipitate}}}{\text{Cu}_2\text{O}}$$

$$\text{CHO} + \underset{\substack{\text{Fehling's reagent}\\\text{(blue solution)}}}{\text{Cu(tartrate)}_2{}^{2-}} \longrightarrow \text{COO}^- + \underset{\substack{\text{Brick-red}\\\text{precipitate}}}{\text{Cu}_2\text{O}}$$

It should not be surprising that aldoses give positive tests, but ketoses do too. This is because, in basic solutions, ketoses (and aldoses) are subject to keto-enol tautomerism (Section 16.10).

$$\underset{\text{A ketose}}{\underset{\substack{|\\\text{C}=\text{O}}}{\overset{\text{CH}_2\text{OH}}{\overset{|}{}}}} \underset{\overrightarrow{\longleftarrow}}{\overset{\text{OH}^-}{}} \underset{\text{An enol}}{\underset{\substack{\|\\\text{C}-\text{OH}}}{\overset{\text{CH}-\text{OH}}{}}} \underset{\overrightarrow{\longleftarrow}}{\overset{\text{OH}^-}{}} \underset{\text{An aldose}}{\underset{\substack{|\\\text{H}-\text{C}-\text{OH}}}{\overset{\text{HC}=\text{O}}{}}}$$

In the resulting equilibrium, some aldose is always formed. Because of this, basic solutions of all monosaccharides exhibit some aldehyde character. Since the oxidizing reagents commonly used to detect reducing sugars are basic, *all* monosaccharides act as reducing sugars.

Hemiacetals react with alcohols to form acetals (Section 16.8). The cyclic hemiacetal forms of the monosaccharides do likewise.

An α-aldohexose + ROH $\xrightarrow{\text{HCl}}$ An α-glycoside (an acetal) + H_2O

The alcohol with which the hemiacetal reacts may be as simple as methanol or as complicated as another sugar molecule (Section 20.8). While hemiacetals are unstable and open up to form free aldehyde groups, the **glycosides** (acetals) are quite stable. They do not undergo mutarotation and do not reduce Tollens's, Benedict's, or Fehling's reagents. To recognize a reducing sugar, look for a *hemi*acetal group. In a hemiacetal, one carbon atom (shown explicitly below) is attached by single bonds to two different oxygen atoms. One of these oxygens is part of an ether group, and the other is in a hydroxyl group. Acetals, on the other hand, are 1,1-diethers.

Hemiacetal   Hemiacetal   Acetal

Reactions similar to these occur in living cells. It is not mere coincidence that the easily oxidized sugars are the "quick energy" foods in our diets. We devote an entire chapter (Chapter 27) to the reactions of the carbohydrates in our bodies.

20.8 MALTOSE, LACTOSE, AND SUCROSE: SOME DISACCHARIDES

There are three common disaccharides—maltose, lactose, and sucrose. Hydrolysis of 1 mol of disaccharide yields 2 mol of monosaccharide. Using word equations, we can write

$$\text{Maltose} + H_2O \longrightarrow 2\text{ Glucose}$$
$$\text{Lactose} + H_2O \longrightarrow \text{Glucose} + \text{Galactose}$$
$$\text{Sucrose} + H_2O \longrightarrow \text{Glucose} + \text{Fructose}$$

All three disaccharides are white crystalline solids. Sucrose is quite soluble in water (200 g in 100 mL) and lactose moderately soluble (20 g in 100 mL). All three molecules are too large to pass through cell membranes. Now let's look at each of these, in turn, in more detail.

Maltose occurs free, to a limited extent, in sprouting grain. Its major source, however, is the partial hydrolysis of starch. Lactose occurs in mammalian milk. Sucrose occurs widely in plant juices, but our principal sources are sugarcane and sugar beets. Each of these three disaccharides consists of two monosaccharide units joined together through an acetal linkage.

Maltose is composed of two D-glucose units, joined by an acetal linkage from the first carbon of one ring to the fourth carbon of the other.

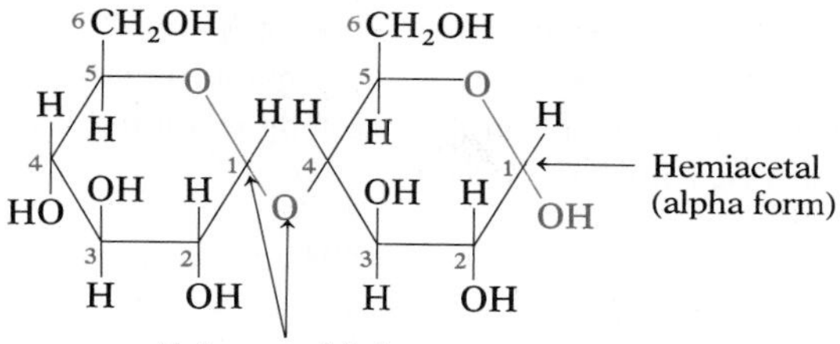

The acetal linkage is alpha. Notice that we have drawn the ring on the right in the alpha form of a hemiacetal. This right-hand ring can open to the free aldehyde and then reclose to the beta form.

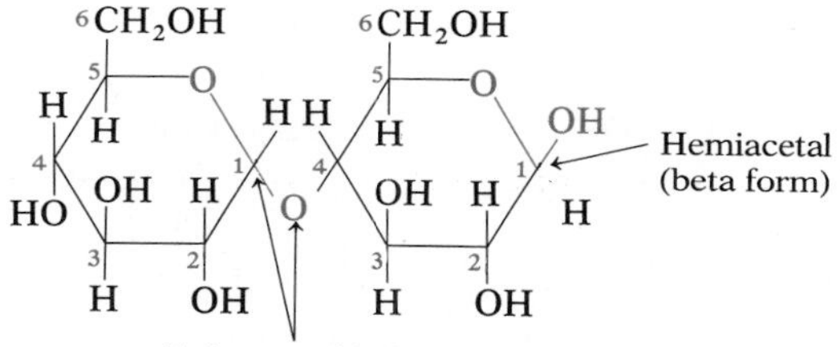

Thus, maltose is a reducing sugar. It can exist in two forms, differing in configuration only about the first carbon of the right-hand ring. The acetal link *between* the two rings is fixed, however. It must remain alpha. If two glucose units were joined by a beta acetal linkage, the compound would not be maltose.

Maltose is formed from starch by the action of enzymes known as **amylases,** which are present in malt, yeast, and saliva. Further hydrolysis of maltose by the enzyme *maltase* or by dilute acid opens the acetal link between the two rings and gives glucose as the sole product. Recall that simple acetals are hydrolyzed by aqueous acid.

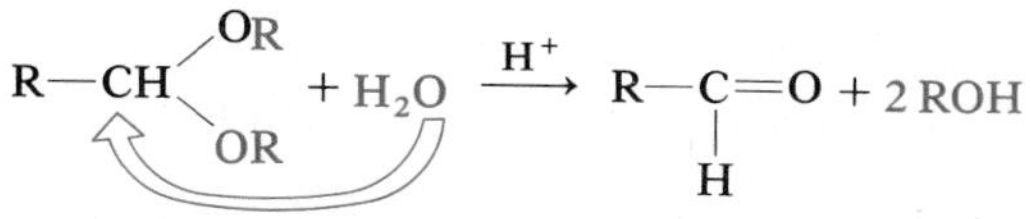

$$R-CH \underset{OR}{\overset{OR}{\diagup}} + H_2O \xrightarrow{H^+} R-\underset{H}{\overset{}{C}}=O + 2\,ROH$$

In a similar manner, the acetal linkage in maltose is hydrolyzed.

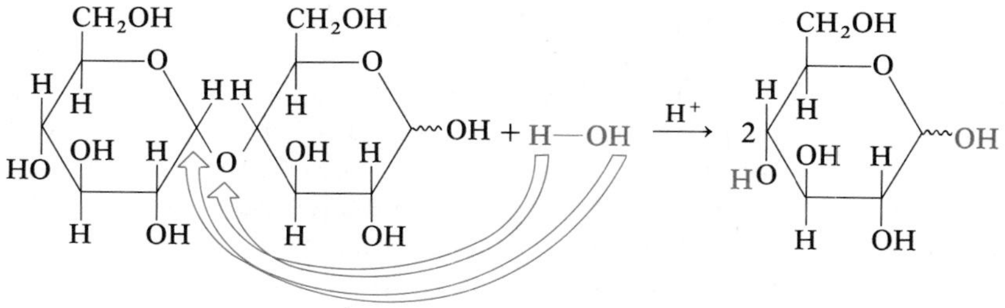

Maltose

Glucose

Lactose or milk sugar occurs to the extent of 5% to 7% in human milk and 4% to 6% in the milk of cattle. In lactose, a galactose unit is joined to a glucose unit. The first carbon of the galactose ring is joined by a beta acetal link to the fourth carbon of the glucose ring.

Note that lactose is also a reducing sugar because the first carbon of the glucose unit is in the hemiacetal form. Like maltose and the monosaccharides, lactose exists in two forms, alpha and beta. Lactose is hydrolyzed to one glucose unit and one galactose unit.

Sucrose (cane or beet sugar, commonly called table sugar) is the most important disaccharide. It is composed of a glucose unit joined through its first carbon atom to the second carbon of a fructose unit. The latent carbonyl groups of both units are involved in the connection between the two rings.

(Notice that the fructose ring contains five members, not six.) This is the only isomer of sucrose. Search as you may, you'll find no hemiacetal unit in sucrose. The first carbon of the glucose ring is an acetal carbon locked in the alpha arrangement; the second carbon of the fructose ring is a ketal carbon, fixed in the arrangement shown. We have now encountered our first nonreducing sugar. Neither ring of sucrose can undergo ring opening.

Sucrose can be hydrolyzed to give a 1:1 mixture of fructose and glucose.

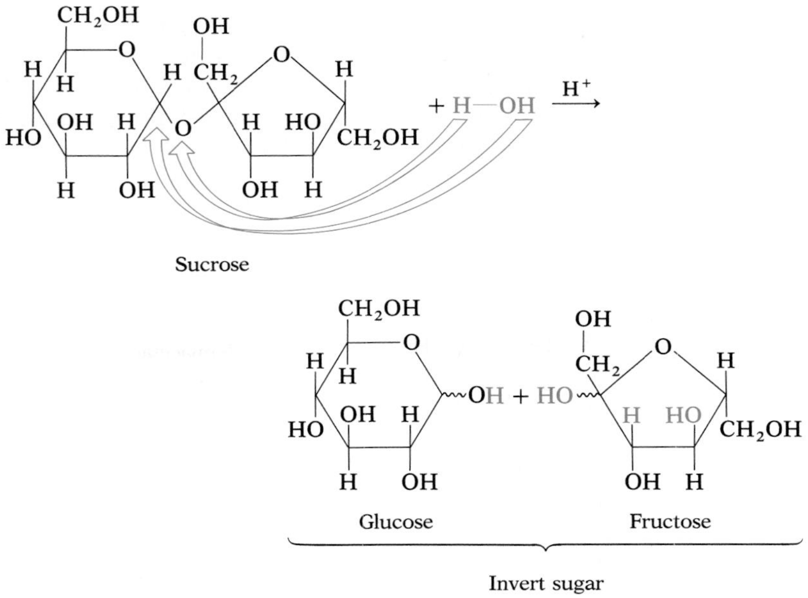

Sucrose

Glucose   Fructose

Invert sugar

This mixture is called *invert sugar*.* It is somewhat sweeter than sucrose. Honey is mostly invert sugar. Per capita consumption of sugars in the United States is about 45 kg per year. Much of it is consumed in soft drinks, presweetened cereals, and other highly processed foods with little or no nutritive value. These "empty calories" contribute greatly to tooth decay, obesity, and heart disease.

Some food faddists claim that raw sugar is much better for you than refined sugar. Raw sugar does contain a few trace minerals, but hardly enough to make it a lot more desirable than refined sugar. People in the United States probably consume much too much sugar—whether raw or refined.

---

* Sucrose is dextrorotatory with a specific rotation of $+66.5°$. During hydrolysis, the observed rotation drops off and eventually becomes negative. This is because fructose has a high negative specific rotation ($-92.4°$ at equilibrium) that more than balances the positive rotation of glucose ($+52.7°$ at equilibrium). Because the sign of rotation changes during the reaction, the process is known as **inversion** and the products as **invert sugar.**

20.9 STARCH

Plants make glucose by photosynthesis. They store it in the form of starch, a polymer of glucose. Plant seeds, especially the cereal grains, and tubers, such as potatoes, are particularly rich in starch. These are the energy reserves for the plant. They also serve as sources of energy for animals and people who eat them.

Starch can be separated into two fractions. One, called **amylose,** forms a colloidal dispersion in hot water. The other, **amylopectin,** is completely insoluble. On the average, plant starches are about 20% amylose and 80% amylopectin. Both forms are composed only of D-glucose units, but they differ in several ways. Amylose consists of an unbranched chain of glucose units joined from the first to the fourth carbons by alpha linkages (the same arrangement one finds in maltose). There may be from 60 to 300 glucose units per chain.

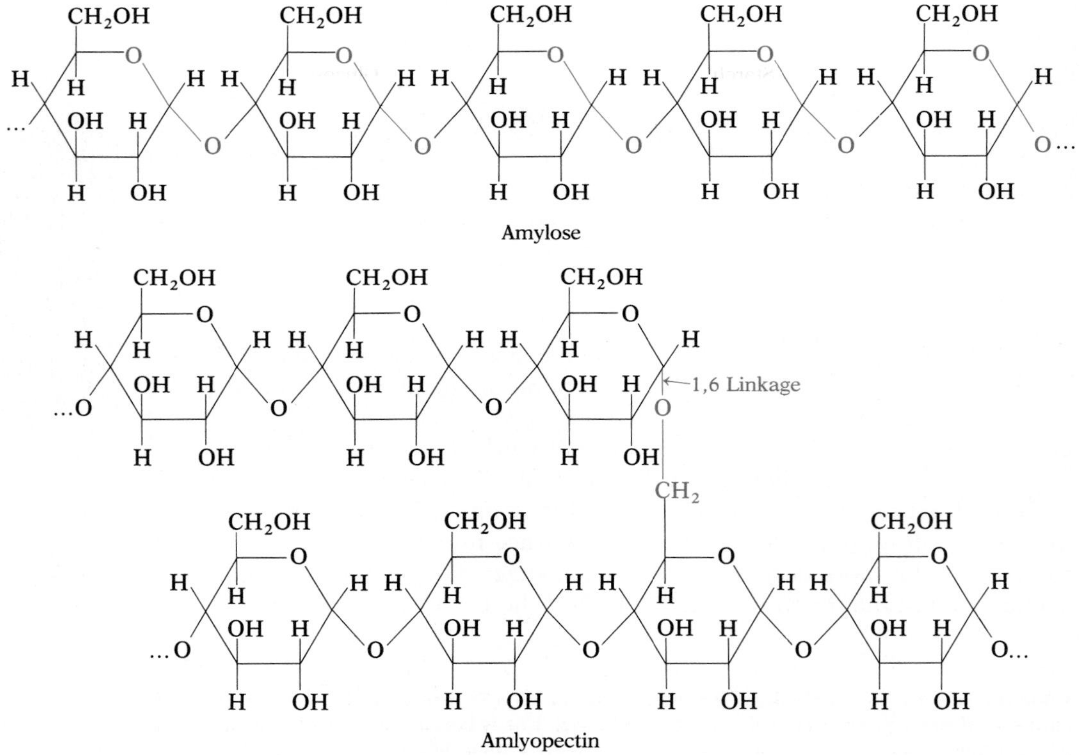

FIGURE 20.6   *Structures of small segments of amylose and amylopectin molecules.*

Amylopectin has similar chains, but it also has branches off the sixth carbon. There may be 300 to 6000 glucose units per amylopectin molecule. Partial structures of these molecules are shown in Figure 20.6.

Starch in plants occurs as very large granules. These granules rupture in boiling water to form a paste. On cooling, the paste gels. Potatoes and cereal grains, when cooked in water, form this type of "starchy" broth.

Commercial starch is a white powder. Under controlled conditions, starch can be partially hydrolyzed to products of intermediate molecular weight called **dextrins.** These are used in pastes, mucilages, and fabric sizing.

The acetal linkages of starch are also hydrolyzed when the compound is eaten. Hydrolysis is catalyzed by acids and by enzymes and leads ultimately to the formation of glucose.

$$\xrightarrow[\text{H}^+]{\text{H}_2\text{O}}$$

Starch   Glucose

This monosaccharide can be transported by the blood to all parts of the body. Whereas plants store carbohydrates as starch, animals convert glucose to a polymer called **glycogen** for storage in the liver and muscles. We can store only about 500 g of glycogen, enough for about a day's reserve of energy. Glycogen is similar to amylopectin in structure. In liver and muscle tissue, it is arranged in granules (Figure 20.7). These granules are clusters of small particles.

Starch is not a reducing sugar. Although there is a hemiacetal function at one end of each molecule, the molecules are so large that the effect of these end groups is too minor to give a visible Tollens, Benedict's, or Fehling's test. Starch is usually detected by its reaction with iodine. The two substances form a complex that is an intense blue-black in concentrated solutions. The test is sensitive enough to detect even minute amounts of starch in solution.

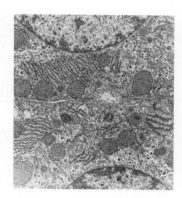

FIGURE 20.7
Electron micrograph of glycogen granules in a liver cell of a rat. (Photo by Patricia J. Schulz.)

20.10 CELLULOSE

Like starch, cellulose is a polymer of glucose. It differs, however, in that the glucose units are joined by beta acetal linkages (Figure 20.8). This may seem to be a minor difference, but its significance is tremendous. Most animals can digest and metabolize starch. People and some other animals get no food value from cellulose. We can eat potatoes, but we can't eat grass. Grazing animals

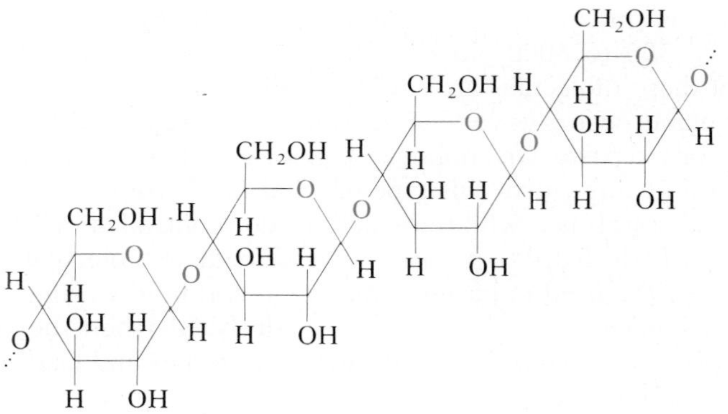

FIGURE 20.8   *Structure of a small segment of a cellulose molecule.*

and termites are able to use cellulose due to the action of microorganisms in their digestive tracts.

Although it has no nutritive value, cellulose makes up the greater part of dietary fiber. The fibrous portions of plants (stems, peels, and seeds) are rich in fiber. Bran, celery, beans, apples, raspberries, and figs are good sources of dietary fiber.

Some studies in the 1970s greatly increased public interest in fiber. First, it was noted that people in the developed countries are much more likely to get cancer of the colon than are those in the underdeveloped nations. Also, people in the developed countries eat diets rich in highly processed, low-bulk foods, while those in more "primitive" areas eat high-fiber diets. High-fiber diets lead to frequent and robust bowel movements. In addition to providing bulk, the fiber absorbs a lot of water, leading to softer stools. Low-bulk diets result in less frequent bowel action, with high retention times for feces in the colon.

Bacteria act upon the materials in the colon. With a high-fiber diet, the materials seldom remain in the colon for more than 1 day. With a low-bulk diet, the retention time can be as long as 3 days, allowing for prolonged bacterial activity that produces a high level of mutagenic chemicals. Chemicals that are mutagenic often are also carcinogenic. Thus, a possible chemical link between low-fiber diets and cancer of the colon has been established.

Different linkages between monosaccharide units result in different three-dimensional forms for cellulose and starch. For example, cellulose in the cell walls of plants is arranged in **fibrils**—bundles of parallel chains. The fibrils, in turn, are arranged parallel to each other in each layer of the cell wall (Figure 20.9). Alternate layers have the fibrils running in opposite directions, imparting great strength to the wall.

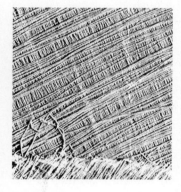

FIGURE 20.9
Electron micrograph of the cell wall of an alga. The wall consists of successive layers of cellulose fibers in parallel arrangement. (Photo by R.D. Preston, F.R.S., and E. Frei.)

Cotton fibrils are nearly pure cellulose. These and other cellulosic fibers can be used directly for the spinning of thread and the weaving of fabrics. Alternatively, the fibers can be modified by conversion of the alcohol functions to esters (Chapter 19).

Since so much of the carbon in the biosphere exists as cellulose, considerable research has gone into converting it to glucose or some other form of food for humans. There are enzymes that will degrade cellulose, but to do so efficiently requires some pretreatment of the fibers. Perhaps some day, through the ingenuity of research chemists, we will be able to eat grass and straw—after proper conversion, of course. For the moment we must continue to depend on reactions such as

$$\begin{array}{c}\text{Cellulose}\\\text{(grass)}\end{array} + \text{cow} \longrightarrow \begin{array}{c}\text{carbohydrates, proteins, and fats}\\\text{(milk, meat, butter)}\end{array}$$

PROBLEMS

1. Define these terms:
   a. triose
   b. aldose
   c. hexose
   d. disaccharide
   e. polysaccharide
   f. aldopentose
   g. ketotetrose
   h. invert sugar
   i. mutarotation
   j. glycoside
2. Draw formulas for D-glyceraldehyde and for L-glyceraldehyde. What do the prefixes mean?
3. Specify which is a D sugar and which is an L sugar.

   a.
   ```
        CHO
         |
   H —  C — OH
         |
   H —  C — OH
         |
   H —  C — OH
         |
        CH₂OH
   ```
   b.
   ```
         CHO
          |
   HO —  C — H
          |
   H —   C — OH
          |
   HO —  C — H
          |
         CH₂OH
   ```
   c.
   ```
         CHO
          |
   H —   C — OH
          |
   HO —  C — H
          |
         CH₂OH
   ```
   d.
   ```
         CH₂OH
          |
   H —   C — OH
          |
         CHO
   ```

4. Identify the sugars as aldoses or ketoses.
   a. D-glyceraldehyde
   b. D-ribose
   c. D-deoxyribose
   d. D-galactose
   e. D-glucose
   f. D-fructose
   g. L-fructose

5. Identify each of the following as a triose, tetrose, pentose, or hexose:
   a. L-glucose
   b. D-deoxyribose
   c. D-fructose
   d. L-glyceraldehyde
6. Draw a ketotetrose.
7. Draw an aldoheptose.
8. From memory, draw formulas for the open-chain forms of D-glucose, D-mannose, D-galactose, and D-fructose.
9. Draw the cyclic structure for α-D-glucose.
10. Knowing that mannose differs from glucose only in the configuration at the second carbon, draw the cyclic structure for α-D-mannose.
11. What is meant by a latent carbonyl group?
12. What is a reducing sugar?
13. For each of these abbreviated sugar formulas, indicate whether the glycosidic link is alpha or beta.

    a.
    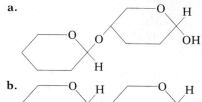

    b.

    c.
    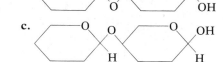

d.

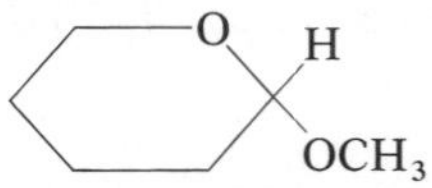

14. Which of the structures in Problem 13 contain a hemiacetal group? If present, is the hemiacetal group alpha or beta?

15. Which of the structures shown in Problem 13 is *not* a reducing sugar?

16. Indigo (the dye used to color blue jeans) occurs as the glucoside indican in *Indigofera tinctoria*. Is indican an alpha or beta glucoside?

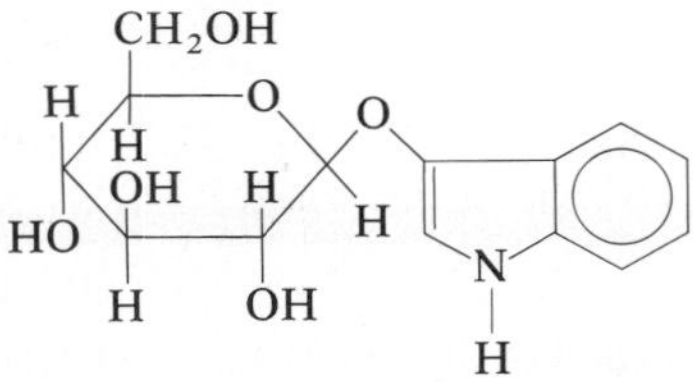

17. What monosaccharide is obtained from the hydrolysis of starch?

18. What monosaccharide is obtained from the hydrolysis of cellulose?

19. What monosaccharide is obtained from the hydrolysis of maltose?

20. What monosaccharides are obtained from the hydrolysis of lactose?

21. What monosaccharides are obtained from the hydrolysis of sucrose?

22. Identify these sugars by their proper names:
    a. blood sugar   **d.** levulose
    b. milk sugar   **e.** table sugar
    c. dextrose

23. Melibiose is a disaccharide that occurs in some plant juices. Its structure is

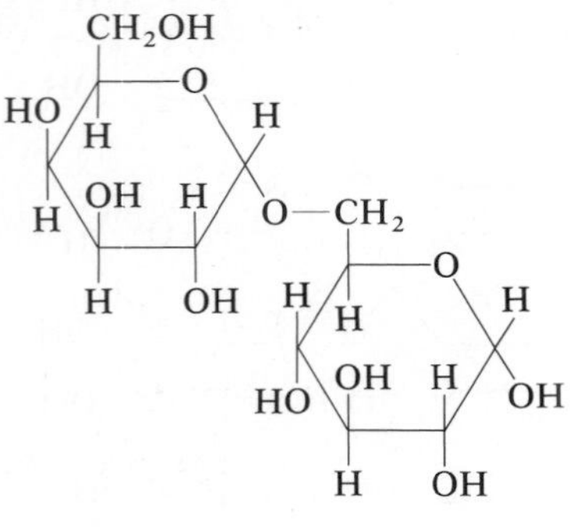

What monosaccharide units are incorporated in melibiose?

24. What type of linkage (alpha or beta) joins the two rings of melibiose (Problem 23)?

25. Is melibiose (Problem 23) a reducing sugar? If so, circle the hemiacetal function and indicate whether it is alpha or beta.

26. The structure of a methyl glycoside of glucose is

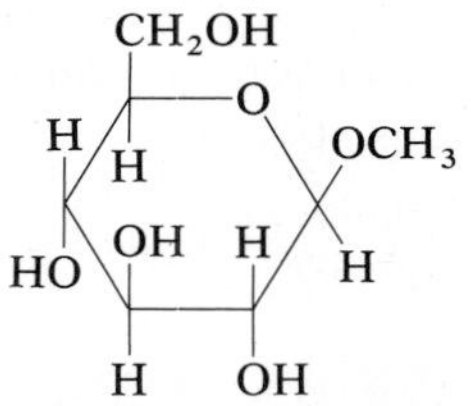

    a. Is carbon 1 in the alpha or the beta arrangement?
    b. Is the compound a reducing sugar?
    c. Will it give a positive test with Benedict's reagent?

27. The L form of β-hydroxybutyric acid occurs in the urine of diabetics in amounts up to 30 g per day. Draw the structure of this enantiomer. (Hint: Place the carbon atoms in a vertical column with the carboxyl group at the top.)

28. Draw structures for the two enantiomeric forms of 3-chloro-1,2-propanediol. (Hint: Place the carbon atoms in a vertical column with the CH_2OH at the bottom.) The L form has activity as a male antifertility drug. Label the D and L isomers.

29. Draw an enol tautomer for D-glyceraldehyde:

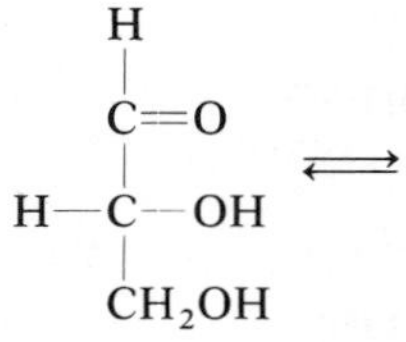

30. How do amylose and amylopectin differ? How are they similar?

31. How do amylose and cellulose differ? How are they similar?

32. How do amylopectin and glycogen differ? How are they similar?

33. List the reagents necessary for the following conversions:

a.

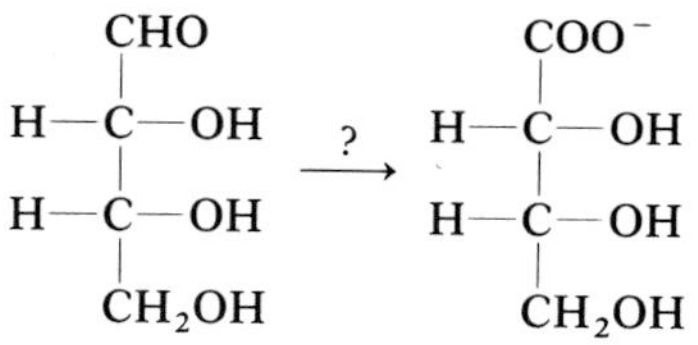

b.

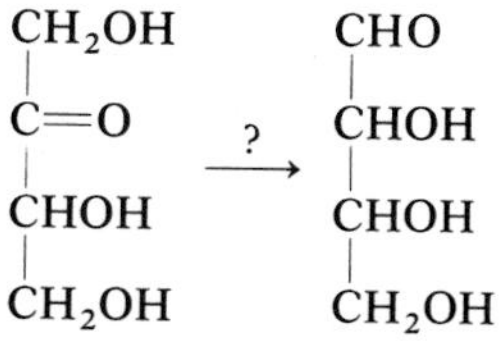

34. List the reagents necessary for the following conversions:

a.

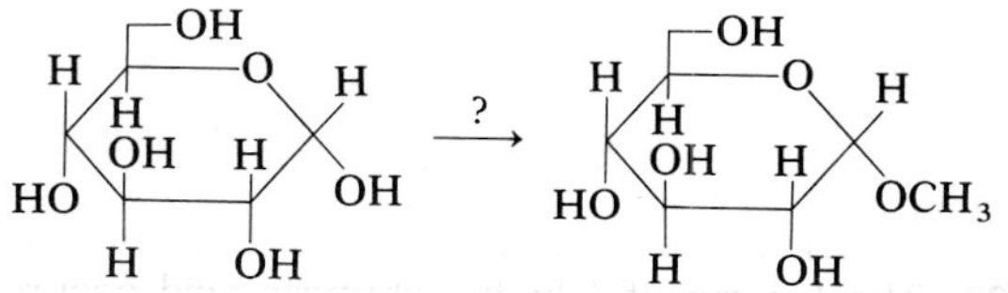

b.

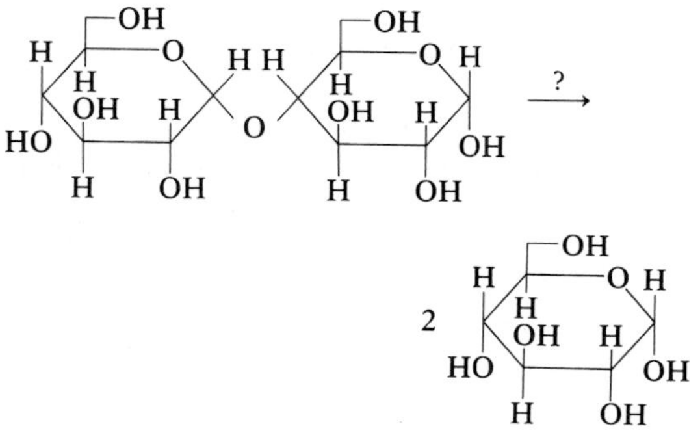

35. Draw the structure for β-D-glucose.
36. By reference to Problem 35, draw the structure for β-D-galactose.
37. What monosaccharide units make up the disaccharide lactulose?

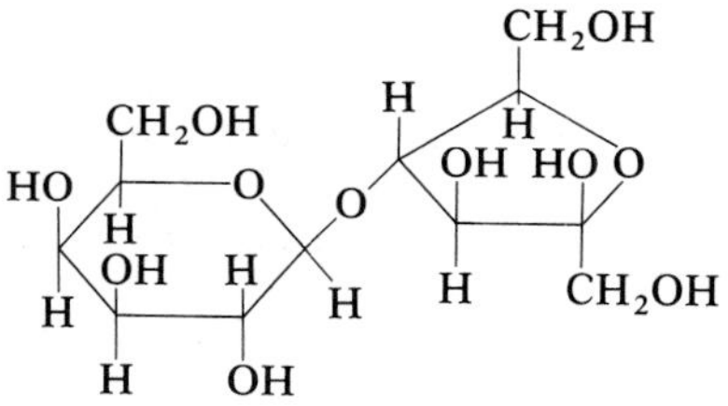

38. What monosaccharide units make up the disaccharide gentiobiose?

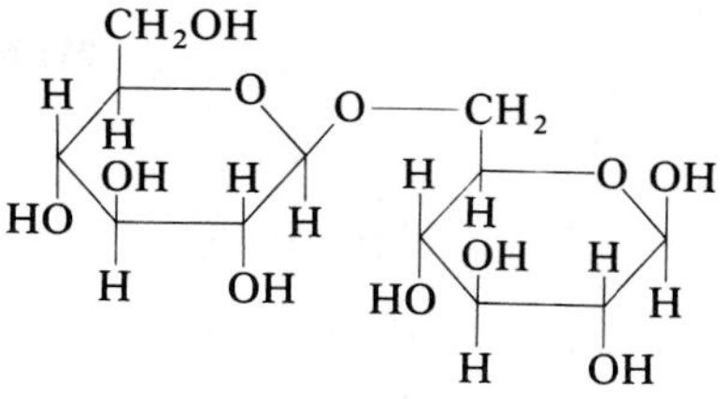

39. The disaccharide cellobiose has two D-glucose units joined by a beta linkage. Draw the alpha form of cellobiose.
40. In the schematic below, *Glc* represents glucose. What substance is indicated?

. . . . . Glc-Glc

. . . . . . Glc-Glc-Glc-Glc

. . . . . Glc-Glc-Glc-Glc-Glc-Glc-Glc-Glc-Glc

. . . . . Glc-Glc-Glc-Glc-Glc-Glc-Glc-Glc-Glc-Glc-Glc-Glc-Glc

. . . . . Glc-Glc-Glc-Glc-Glc-Glc-Glc-Glc

41. Fructose is found in many "natural foods" stores. (It even is promoted as a diet aid!) Fructose is made from sucrose by hydrolysis (reaction with water) and separation from the coproduct glucose, or by the catalytic isomerization of glucose (from corn syrup). Is fructose made in one of these ways "natural"? Explain your answer.
42. Ascorbic acid (vitamin C) is made from D-glucose. Its structure is shown below. What is the configuration of ascorbic acid (D or L)? Would the mirror image isomer (enantiomer) be expected to have the same vitamin activity as the isomer shown here?

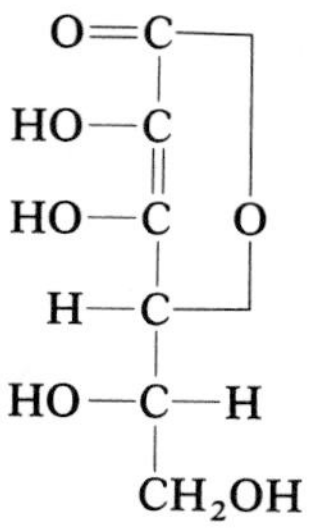

43. Xylulose is found in the urine of humans with pentosuria. Based on the structure below, classify xylulose as fully as possible.

$$
\begin{array}{c}
CH_2OH \\
| \\
C{=}O \\
| \\
H{-}C{-}OH \\
| \\
HO{-}C{-}H \\
| \\
CH_2OH
\end{array}
$$

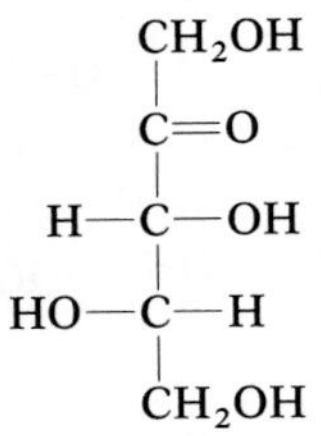

44. Erythrulose can be prepared from D-fructose. Based on the structure below, classify erythrulose as fully as possible.

$$
\begin{array}{c}
CH_2OH \\
| \\
C{=}O \\
| \\
HO{-}C{-}H \\
| \\
CH_2OH
\end{array}
$$

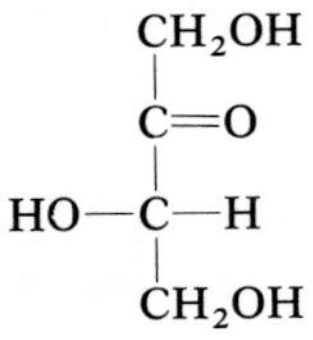

45. Tartaric acid from grapes has the structure shown. Is its configuration D or L? Draw the structure of its enantiomer.

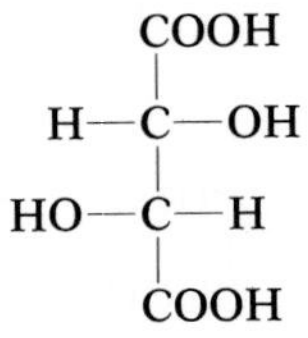

$$
\begin{array}{c}
COOH \\
| \\
H{-}C{-}OH \\
| \\
HO{-}C{-}H \\
| \\
COOH
\end{array}
$$

CHAPTER

21

Lipids

The food that we eat is divided into three primary groups: the carbohydrates (Chapter 20), the proteins (Chapter 22), and the lipids (which we shall discuss in this chapter). The best-known lipids are the fats. These compounds, in our affluent society, often occupy the lowest estate among the three classes of foods. People dieting to lose weight frequently try to eliminate fats from their diet. Gram for gram, fats pack about twice the caloric content of carbohydrates. While that may be bad news for the dieter, it also says something about the efficiency of nature's designs. The body has a limited capacity for storing carbohydrates. It can tuck away a bit of glycogen in the liver or in muscle tissue, but carbohydrates, primarily in the form of glucose, are meant to serve the body's immediate energy needs. If we intend to store energy reserves, then the more energy we can pack into a given space, the better off we are. The oxidation of fats supplies about 9 kcal/g, whereas the oxidation of carbohydrates supplies only 4 kcal/g. The body, an efficient organism, is geared to store fats, and its capacity for doing so is astounding. Most of us store enough energy as fats to last a month or so, but there is a recorded instance of a man weighing 486 kg. If all that energy were stored as carbohydrate, he would have weighed a ton or more.

The body's ability to store fat may elicit from you feelings of disgust or despair rather than awe. But a quick summary of the other functions of lipids in the body may provide a more positive picture of those essential compounds. They play an important role in brain and nervous tissue. Fats serve as protective padding and insulation for vital organs. Without fats in our diets, we'd be deficient in the fat-soluble vitamins, A, D, E, and K. Most important, lipids make up the major part of the membranes of each of the 10 trillion cells in our bodies.

21.1 WHAT'S A LIPID?

Of the three types of foodstuffs, two are classified by functional groups. As Chapter 20 states, carbohydrates are polyhydroxyaldehydes or ketones. The proteins, as we shall soon see, are polyamides. But lipids are not poly-anythings in particular. They tend to be esters or compounds that can form esters, but that takes in a lot of territory and doesn't even hint at the wide variation in structure found among the lipids.

What makes a lipid a lipid is its solubility. Considering that water is the major solvent in living systems and that reactions of physiological importance tend to take place in aqueous solutions, it is not surprising that insolubility in water should be considered a noteworthy feature when found in important body constituents. Lipids are soluble in relatively nonpolar organic solvents such as carbon tetrachloride, chloroform, and diethyl ether but are generally insoluble in water.

Compounds isolated from body tissues are classified as lipids if they are more soluble in organic solvents than in water. Included in this category are esters of glycerol (an alcohol) and the fatty acids (or phosphoric acid); steroids such as cholesterol; compounds that incorporate sugar units or a complicated aminoalcohol called sphingosine; and compounds called prostaglandins, which some regard as potential "miracle" drugs. Because of this broad variation in structure, we can't present a general formula for lipids. We shall, instead, consider one subclass at a time and try to point out similarities and differences in structure as we go along.

21.2 FATTY ACIDS

Fatty acids are so named because they are structural components of fats. Chemically, fatty acids are generally long-chain carboxylic acids. Nearly all contain an even number of carbon atoms. Few are branched. Some, the unsaturated fatty acids, contain one or more double bonds. Free fatty acids are rare, occurring in nature in only small amounts. The fats and other lipids, however, provide a reservoir from which the fatty acids can be obtained. Table 21.1 lists some common fatty acids and an important source of each of them.

The normal tetrahedral bond angles of carbon require that the chain of saturated fatty acid molecules assume a zigzag configuration (Figure 21.2a), but the molecule viewed as a whole is relatively straight. (Each angle in these zigzag formulas represents one carbon atom in the fatty acid chain.) Such molecules

FIGURE 21.1
Cooking oil, butter, birth control pills, and gallstones are rich in compounds that are classified as lipids.

fit rather nicely into a crystal lattice (Figure 21.2b), a capability that gives these acids and the fats derived from them relatively high melting points. Unsaturated fatty acids occur almost always in the *cis* configuration. This results in a bend in the molecules (Figure 21.2c). These molecules don't stack neatly (Figure 21.2d); hence, the forces between molecules are smaller. The unsaturated fatty acids and unsaturated fats, consequently, have lower melting points. Most are liquids at room temperature.

TABLE 21.1   Some Fatty Acids in Natural Fats

Number of Carbon Atoms	Condensed Structure	Melting Points (°C)	Name	Source
4	$CH_3CH_2CH_2COOH$	−8	Butyric acid	Butter
6	$CH_3(CH_2)_4COOH$	−3	Caproic acid	Butter
8	$CH_3(CH_2)_6COOH$	−17	Caprylic acid	Coconut oil
10	$CH_3(CH_2)_8COOH$	31	Capric acid	Coconut oil
12	$CH_3(CH_2)_{10}COOH$	44	Lauric acid	Palm kernel oil
14	$CH_3(CH_2)_{12}COOH$	54	Myristic acid	Oil of nutmeg
16	$CH_3(CH_2)_{14}COOH$	63	Palmitic acid	Palm oil
18	$CH_3(CH_2)_{16}COOH$	70	Stearic acid	Beef tallow
18	$CH_3(CH_2)_7CH{=}CH(CH_2)_7COOH$	13	Oleic acid	Olive oil
18	$CH_3(CH_2)_4CH{=}CHCH_2CH{=}CH(CH_2)_7COOH$	−5	Linoleic acid	Soybean oil
18	$CH_3CH_2(CH{=}CHCH_2)_3(CH_2)_6COOH$	−11	Linolenic acid	Fish oil

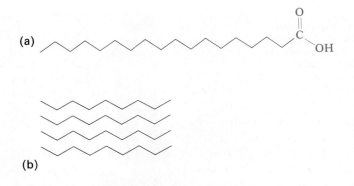

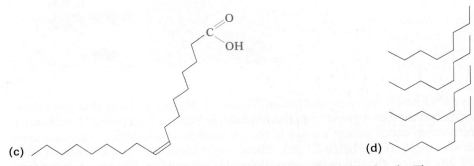

FIGURE 21.2 (a) A schematic representation of a stearic acid molecule. These saturated acids stack nicely in a crystal lattice (b). Oleic acid (c), with its cis *double bond, will not fit neatly into a crystalline arrangement (d).*

21.3 SATURATED FATS

Chemically, **fats** are esters of the fatty acids and the trihydroxy alcohol glycerol. Three acid units are combined with the three hydroxyl groups in the glycerol molecule. The fatty acid units can be identical, as in tristearin.

$$
\begin{array}{c}
\underset{\text{Glycerol}}{
\begin{array}{l}
CH_2O\!-\!\boxed{H \ \ HO}\!-\!\overset{\displaystyle O}{\overset{\|}{C}}\!-\!C_{17}H_{35}\\[6pt]
CH\!-\!O\!-\!\boxed{H \ \ HO}\!-\!\overset{\displaystyle O}{\overset{\|}{C}}\!-\!C_{17}H_{35}\\[6pt]
CH_2O\!-\!\boxed{H \ \ HO}\!-\!\overset{\displaystyle O}{\overset{\|}{C}}\!-\!C_{17}H_{35}
\end{array}}
\quad\longrightarrow\quad
\underset{\text{Tristearin}}{
\begin{array}{l}
CH_2O\!-\!\overset{\displaystyle O}{\overset{\|}{C}}\!-\!C_{17}H_{35}\\[6pt]
CH\!-\!O\!-\!\overset{\displaystyle O}{\overset{\|}{C}}\!-\!C_{17}H_{35} \; + \; 3\,H_2O\\[6pt]
CH_2O\!-\!\overset{\displaystyle O}{\overset{\|}{C}}\!-\!C_{17}H_{35}
\end{array}}
\end{array}
$$

Glycerol   Stearic acid   Tristearin

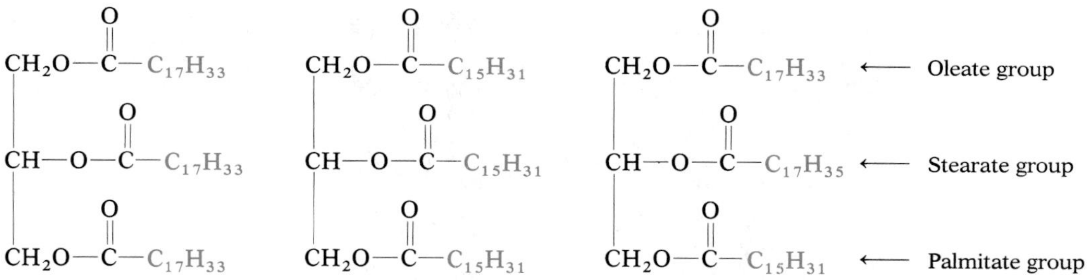

$$\begin{array}{ccc}
\text{(a)} & \text{(b)} & \text{(c)} \\
\text{Triolein} & \text{Tripalmitin} & \text{A mixed triglyceride}
\end{array}$$

FIGURE 21.3   *Triolein (a) and tripalmitin (b) are examples of simple triglycerides. A mixed triglyceride (c) has more than one kind of fatty acid incorporated.*

Fat molecules such as tristearin are called **simple triglycerides.** Other examples are triolein and tripalmitin (Figure .21.3). Most naturally occurring fats contain different fatty acid units in the same molecule. These are called **mixed triglycerides.** For example, there may be an oleate, a stearate, and a palmitate group present (Figure 21.3c).

Animal fats generally contain both saturated and unsaturated fats, but the former predominate. At room temperature, animal fats are usually solids. At body temperature in warm-blooded creatures, though, these fats are likely to exist in the liquid state. The fatty acid composition of several typical fats is given in Table 21.2.

Fats undergo a variety of chemical reactions; the most important is hydrolysis. Fats are esters. They can be hydrolyzed in either acidic or basic media. Acid hydrolysis is of little importance, however, because it is difficult to dissolve fats in acidic media. Basic hydrolysis is of considerable importance in the

TABLE 21.2   *Fatty Acid Distribution Ranges (as Percentages) in Several Fats and Oils*

Fat or Oil	Lauric	Myristic	Palmitic	Stearic	Oleic	Linoleic	Others
Animal fats							
Beef tallow	0.2	2–3	25–30	21–26	39–42	1–3	1.2–1.8*
Lard	—	1	25–30	12–16	41–51	3–8	4.2–8.2*
Butter	1–4	8–13	25–32	8–13	22–29	3	6.6–15.5†
Vegetable oils							
Palm	—	1–6	32–47	1–6	40–52	2–11	—
Corn	—	0–2	8–10	1–4	30–50	34–56	1–4*
Soybean	—	0.3	7–11	2–5	22–34	50–60	3–14‡
Cottonseed	—	0–3	17–23	1–3	23–44	34–55	0–1

* Mainly the unsaturated fatty acid having 16 carbon atoms per molecule (palmitoleic acid).
† Mainly the acids having 4, 6, 8, or 10 carbon atoms per molecule.
‡ Mainly linolenic acid.

making of soap and will be discussed in detail in Section 21.5. When we eat fats, they are hydrolyzed by enzymes. This process is discussed in Chapter 28. When fats are heated, their glycerol portions are dehydrated to acrolein.

$$\underset{\underset{\text{Glycerol}}{}}{CH_2-CH-CH_2} \;\;\overset{\text{Heat}}{\longrightarrow}\;\; \underset{\underset{\text{Acrolein}}{}}{CH_2{=}CH-C{\overset{O}{\underset{H}{\diagdown}}}} \;\; + \; 2\,H_2O$$

This unsaturated aldehyde is partially responsible for the acrid odor of burning fat.

21.4 UNSATURATED FATS: OILS

The solid fats are derived mainly from animals. Liquid fats, called **oils,** are obtained principally from vegetable sources. Structurally, oils are identical to fats except that they incorporate a higher proportion of unsaturated acid units (Table 21.2). Coconut and palm oils, which are highly saturated, and fish oils, which are relatively unsaturated, are notable exceptions to the general rule.

Fats are often classified according to the degree of unsaturation of the fatty acids they incorporate. **Saturated fatty acids** contain no double bonds, **monounsaturated fatty acids** contain one double bond per molecule, and **polyunsaturated fatty acids** are those that have two or more double bonds. **Saturated fats** contain a high proportion of saturated fatty acids; the fat molecules have relatively few double bonds. **Polyunsaturated fats** (oils) incorporate mainly unsaturated fatty acids; these fat molecules have many double bonds.

Saturated fats have been implicated, along with cholesterol, a steroid (Section 21.9), in one type of **arteriosclerosis** (hardening of the arteries). As this condition develops, lipids deposit on the walls of arteries. Eventually these deposits become calcified (harden), robbing the vessels of their elasticity (Figure 21.4). There is a strong correlation between diets rich in saturated fats and incidence of the disease. It is this correlation that has led to a concern over the relative amounts of saturated and unsaturated fats in our diets. Advertisers who recommend that you buy corn oil margarine (prepared from relatively unsaturated vegetable oil) rather than butter (a relatively saturated animal fat) take advantage of this concern.

There is also statistical evidence that fish oils can prevent heart disease. Researchers at the University of Leiden in the Netherlands have found that Greenlanders who eat a lot of fish have a low risk of heart disease, even though their diet is high in total fat and cholesterol. The probable effective agents are the unsaturated fatty acids such as linolenic acid. Other studies have shown

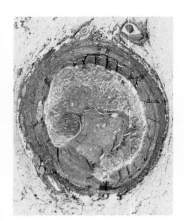

FIGURE 21.4
Photomicrograph of a cross section of a "hardened" artery, showing deposits of plaque. The deposits contain cholesterol. (Courtesy of Biomedical Graphics, University of Minnesota Hospitals, Minneapolis.)

that diets with added fish oil lead to lower cholesterol and triglyceride levels in the blood.

The degree of unsaturation of a fat or an oil is usually measured in terms of the **iodine number.** Recall that halogens such as chlorine and bromine add readily to carbon-carbon double bonds.

$$\underset{/}{\overset{\backslash}{}}C{=}C\underset{\backslash}{\overset{/}{}} + Br_2 \longrightarrow -\overset{|}{\underset{|}{C}}-\overset{|}{\underset{|}{C}}- \atop Br \quad Br$$

Iodine also adds, but less readily. The iodine number of a fat is the number of grams of iodine that will be consumed by 100 g of fat or oil. The more double bonds a fat contains, the more iodine is required for the addition reaction; thus, a high iodine number means a high degree of unsaturation. Representative iodine numbers are given in Table 21.3. Notice the generally lower values for the animal fats (butter, tallow, lard) compared with those for the vegetable oils.

Vegetable oils can be converted to solids or semisolids by hydrogenation. Margarine, a butter substitute, and vegetable shortening, a lard substitute, consist of vegetable oils that have been partially hydrogenated. Generally, because of the advertising appeal of polyunsaturated products, the hydrogenation is limited to the degree necessary to give the products the proper consistency. The consumer would get much greater unsaturation by using the oils directly, but most people would rather spread margarine than pour oil on their toast. Table 21.4 lists iodine numbers for butter and various kinds of margarines.

On standing at room temperature in contact with moist air, fats and oils soon turn rancid. This rancidity, characterized by disagreeable odors, results

TABLE 21.3 Typical Iodine Numbers for Some Fats and Oils

Fat or Oil	Iodine Number
Butter	25–40
Beef tallow	30–45
Palm oil	37–54
Lard	45–70
Olive oil	75–95
Peanut oil	85–100
Cottonseed oil	100–117
Corn oil	115–130
Fish oils	120–180
Soybean oil	125–140
Safflower oil	130–140
Sunflower oil	130–145
Linseed oil	170–205

TABLE 21.4 *Iodine Values and Comparative Unsaturation Ratings of Various Margarines and Butter*

Food Product	Iodine Value	Comparative Unsaturation*
Butter	27	100
Margarines		
Hard type A	68	252
Hard type B	72	267
Hard type C	77	285
Soft type D	84	311
Soft type E	88	326
Liquid type F	90	333
Liquid type G	93	344

Reprinted with permission from Mancott, Anatol, and Tietjen, John, "Polyunsaturation in Food Products," *Chemistry*, November 1974, p. 29. Copyright © 1974 by the American Chemical Society.
* Calculated by dividing the iodine number of the substance by the iodine number of butter and multiplying the result by 100.

from two reactions. Hydrolysis produces volatile fatty acids. Butter, for example, yields foul-smelling butyric, caprylic, and capric acids. Oxidation of the unsaturated fatty acid components by oxygen in the air also produces a variety of volatile, odorous compounds. The structural unit

$$\sim\sim\sim CH=CH-CH_2-CH=CH\sim\sim\sim$$

in linoleic and linolenic acids is readily oxidized. One particularly offensive product, formed by the cleavage of both double bonds, is a compound called malonaldehyde. The stale, sweaty odor of the unwashed skin is due in large part to the oxidation of fats and oils excreted by the body.

Malonaldehyde

Margarine is a water in oil dispersion; that is, an emulsion (Section 9.10). The emulsifying agents are lecithin (Section 21.7) and monoglycerides and diglycerides (glycerol esterified by one and two fatty acid units, respectively). Margarine is colored by annato, a vegetable dye, or β-carotene, a vitamin A precursor (Special Topic J). A lot of chemistry goes into the preparation of this butter substitute.

21.5 SOAP

Animal fats are available in large quantities as a by-product of the meat-packing industry. The fatty acids and many other long-chain organic compounds are derived from these fats. The most important derivatives are soaps

FIGURE 21.5
Soap is made in large reaction vessels such as these, which are three stories deep. Such containers are stocked with 59 000 kg of fat and 34 000 kg of lye solution. The yield is about 68 000 kg of soap and 6400 kg of glycerol. (Courtesy of Procter & Gamble Co., Cincinnati, Ohio.)

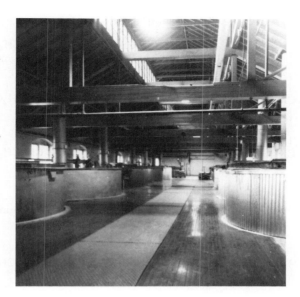

(Figure 21.5). The reaction that converts animal fats to soaps is called, logically, saponification, which we discussed originally as one of the reactions of esters in general (Section 17.8). Animal fat is cooked with lye (sodium hydroxide). Glycerol and the sodium salts of the fatty acids are the products. It is the mixture of sodium salts that is called **soap.** The reaction is illustrated, with tristearin as the fat, in Figure 21.6.

Soap produced from fat and lye usually contains excess sodium hydroxide, which is removed by washing. The soap is then molded and cut into bars. Potassium hydroxide is sometimes employed instead of sodium hydroxide. Potassium soaps are softer and produce a finer lather. They are used in shaving creams. Most soaps contain a number of additives such as perfumes, dyes, and germicides. Scouring soaps contain abrasives to (quite literally) cut grime. Floating soaps have air beaten in before the soap solidifies. This process lowers the density of the bar of soap to less than that of water.

Dirt and grime usually adhere to skin, clothing, and other surfaces because they are combined with greases and oils—body oils, cooking fats, lubricating greases, or a variety of similar substances—that act a little like sticky glues. Since oils are not miscible with water, washing with water alone does little good. Soap molecules have a "split personality." One end is ionic and dissolves in water. The other end is like a hydrocarbon and dissolves in oils (Figure 21.7). If we represent the ionic end of the molecule as a circle and the hydrocarbon end as a zigzag line, we can illustrate the cleansing action of soap schematically (Figure 21.8). The hydrocarbon "tails" stick into the oil. The ionic "heads" remain in the aqueous phase. In this manner, the oil is broken

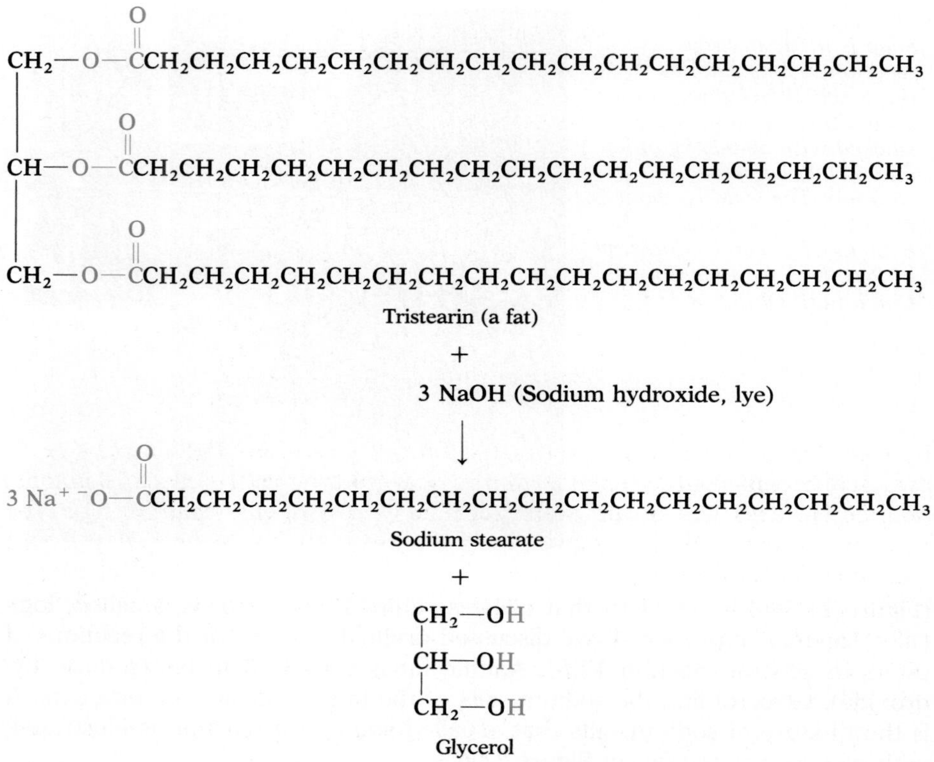

$$CH_2-O-\overset{\overset{O}{\|}}{C}CH_2CH_2CH_2CH_2CH_2CH_2CH_2CH_2CH_2CH_2CH_2CH_2CH_2CH_2CH_2CH_2CH_3$$

$$CH-O-\overset{\overset{O}{\|}}{C}CH_2CH_2CH_2CH_2CH_2CH_2CH_2CH_2CH_2CH_2CH_2CH_2CH_2CH_2CH_2CH_2CH_3$$

$$CH_2-O-\overset{\overset{O}{\|}}{C}CH_2CH_2CH_2CH_2CH_2CH_2CH_2CH_2CH_2CH_2CH_2CH_2CH_2CH_2CH_2CH_2CH_3$$

Tristearin (a fat)

+

3 NaOH (Sodium hydroxide, lye)

↓

$$3\ Na^+\ {}^-O-\overset{\overset{O}{\|}}{C}CH_2CH_2CH_2CH_2CH_2CH_2CH_2CH_2CH_2CH_2CH_2CH_2CH_2CH_2CH_2CH_2CH_3$$

Sodium stearate

+

$$CH_2-OH$$
$$CH-OH$$
$$CH_2-OH$$

Glycerol

FIGURE 21.6 Soap is produced by the reaction of animal fat with sodium hydroxide. Tristearin is a fat, and sodium stearate is a soap.

Hydrocarbon end (dissolves in oils)   Ionic end (dissolves in water)

(a)

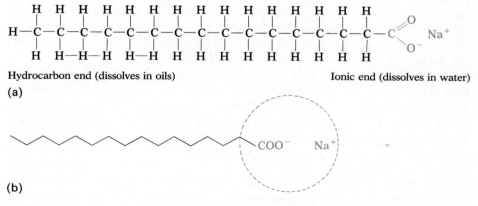

(b)

FIGURE 21.7 Sodium palmitate, a soap. (a) Structural formula. (b) A schematic representation.

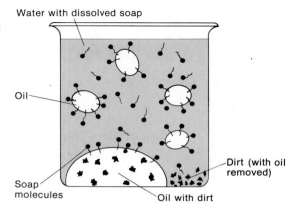

FIGURE 21.8
The action of soap in removing dirt.

into tiny droplets and dispersed throughout the solution. The droplets don't coalesce, because of the repulsion of the charged groups (the carboxyl anions) on their surfaces. The oil and water form an emulsion, with soap acting as the emulsifying agent. With the oil no longer "gluing" it to the surface, the dirt can be easily removed.

For cleaning clothes and for many other purposes, soap has been largely replaced by synthetic detergents because soaps have two rather serious shortcomings. One of these is that in acidic solutions, soaps are converted to free fatty acids.

$$CH_3CH_2CH_2CH_2CH_2CH_2CH_2CH_2CH_2CH_2CH_2CH_2CH_2COO^-Na^+ + H^+ \longrightarrow$$

A soap

$$CH_3CH_2CH_2CH_2CH_2CH_2CH_2CH_2CH_2CH_2CH_2CH_2CH_2COOH$$

A fatty acid

The fatty acids, unlike soap, don't have an ionic end. Lacking the necessary "split personality," they can't emulsify the oil and dirt; that is, they do not exhibit any detergent action. What is more, these fatty acids are insoluble in water and would thus precipitate as a greasy scum. To counteract this lack of detergent action in acidic solution, various alkaline substances are added to laundry soap formulations to keep the pH high. These basic compounds include carbonates, phosphates, and silicates.

The second serious disadvantage of soap is that it doesn't work very well in hard water. Hard water is just water that contains certain metal ions, particularly magnesium, calcium, and iron ions. The soap anions react with these metal ions to form greasy, insoluble curds.

$$2\,CH_3CH_2CH_2CH_2CH_2CH_2CH_2CH_2CH_2CH_2CH_2CH_2CH_2COO^-\,Na^+ + Ca^{2+} \longrightarrow$$

Soap (soluble)

$$(CH_3CH_2CH_2CH_2CH_2CH_2CH_2CH_2CH_2CH_2CH_2CH_2CH_2COO^-)_2Ca^{2+} + 2\,Na^+$$

Bathtub ring (insoluble)

(a) Reduction:

$$\begin{matrix} CH_2OCO(CH_2)_{10}CH_3 \\ | \\ CHOCO(CH_2)_{10}CH_3 + 6\,H_2 \\ | \\ CH_2OCO(CH_2)_{10}CH_3 \end{matrix} \xrightarrow{\text{Ni (catalyst)}} 3\,CH_3(CH_2)_{10}CH_2OH + \begin{matrix} CH_2OH \\ | \\ CHOH \\ | \\ CH_2OH \end{matrix}$$

Trilaurin   Lauryl alcohol   Glycerol
(dodecyl alcohol)

(b) Reaction with sulfuric acid:

$$CH_3(CH_2)_{10}CH_2OH + H_2SO_4 \longrightarrow CH_3(CH_2)_{10}CH_2OSO_3H + H_2O$$

(c) Neutralization:

$$CH_3(CH_2)_{10}CH_2OSO_3H + NaOH \longrightarrow CH_3(CH_2)_{10}CH_2OSO_3{}^-Na^+ + H_2O$$

FIGURE 21.9   Sodium lauryl sulfate is synthesized in a three-step process. First trilaurin is reduced (a) to lauryl alcohol by hydrogen over a nickel catalyst. The alcohol then reacts with sulfuric acid (b) to form an ester. Finally, the lauryl hydrogen sulfate, which still has an acidic hydrogen, is neutralized (c) with sodium hydroxide (or, alternatively, with sodium carbonate).

These deposits make up the familiar bathtub ring. They leave freshly washed hair sticky and are responsible for the "telltale gray" of the family wash.

Synthetic detergents also react with the ions in hard water to form calcium, magnesium, or iron salts. These salts, however, are relatively soluble, and clean about as effectively as the original sodium salts.

Although most synthetic detergents are now made from petroleum (Special Topic H), early ones were made from fats. Some, such as sodium lauryl sulfate (sodium dodecyl sulfate) still are. Sodium lauryl sulfate is widely used in specialty products such as toothpastes and shampoos. The synthesis of this detergent from trilaurin is shown in Figure 21.9.

21.6 WAXES

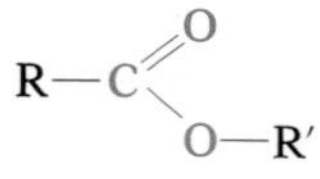

$$R-C\overset{\displaystyle O}{\underset{\displaystyle O-R'}{\diagup}}$$

A wax
(A simple ester)

In everyday usage, the word *wax* refers to a substance that is hard when cold yet easily molded when warm. Familiar waxes include the mixture of alkanes called paraffin wax, synthetic polymers such as Carbowax, and carnauba wax, a mixture of esters. In chemistry, however, the term **wax** refers to esters formed from long-chain fatty acids and long-chain *monohydric* alcohols (i.e., alcohols having one hydroxyl group). The general formula for a wax, then, is the same as that of a simple ester. For a wax, however, R and R' are limited to alkyl groups containing a large number of carbon atoms.

Most natural waxes are mixtures of such esters. Many also contain free alcohols, hydrocarbons, and esters of dibasic acids, hydroxy acids, and diols.

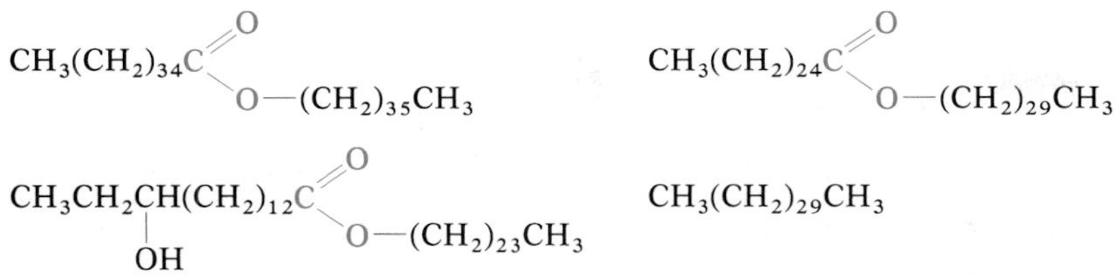

$$CH_3(CH_2)_{34}C \overset{O}{\underset{O-(CH_2)_{35}CH_3}{<}}$$

$$CH_3(CH_2)_{24}C \overset{O}{\underset{O-(CH_2)_{29}CH_3}{<}}$$

$$CH_3CH_2CH(CH_2)_{12}C \overset{O}{\underset{O-(CH_2)_{23}CH_3}{<}}$$
$$\underset{OH}{|}$$

$$CH_3(CH_2)_{29}CH_3$$

FIGURE 21.10   *Three esters and an alkane found in beeswax.*

All have similar properties; they feel "waxy," are insoluble in water, and melt at temperatures above body temperature (37 °C) and below the boiling point of water (100 °C).

Beeswax is the material from which bees construct honeycombs. Upon saponification, beeswax yields alcohols and fatty acid salts with even numbers of carbon atoms. The alcohols generally have 24 to 36 carbon atoms. The fatty acids have up to 36 carbon atoms; about one-fourth are hydroxy acids. About 20% of beeswax is hydrocarbons. These have odd numbers of carbon atoms from 21 to 33. Figure 21.10 shows some of the typical molecules found in beeswax.

Beeswax is a by-product of the production of honey. Commercial operations can always leave behind enough honey to sustain the colony of bees and ensure future production. Obtaining the wax called spermaceti, however, is a little harder on the creature that produces it. Spermaceti crystallizes when the oil from the head of the sperm whale (*Cetacea*) is cooled. Whales must be killed before the product can be obtained. The principal constituent of spermaceti is cetyl palmitate.

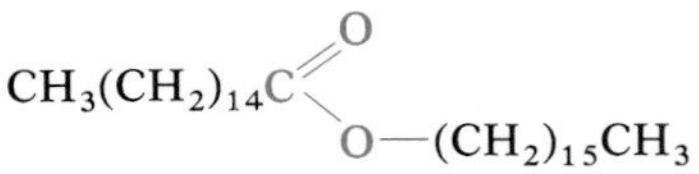

$$CH_3(CH_2)_{14}C \overset{O}{\underset{O-(CH_2)_{15}CH_3}{<}}$$

Esters of lauric, myristic, and palmitic acids are also present, as are esters of higher alcohols.

Spermaceti melts at temperatures from 42 to 50 °C. It was once widely used in ointments, cosmetics, soaps, and candles and for laundering. Demand for spermaceti, sperm oil, and whale meat for human and pet food has led to the near extinction of several species of whale.

Lanolin is a wax from sheep's wool. It is a mixture of esters and polyesters of 33 alcohols and 36 fatty acids. Some of the alcohols are steroids (similar to cholesterol). Lanolin is used as a base for ointments and cosmetic lotions.

An important vegetable wax is obtained from Brazilian palms of the genus *Copernicia*. The exudate from the leaves is called carnauba wax. This wax has a high melting point (82 to 86 °C) and is quite hard. It is used extensively in floor waxes, automobile waxes, and shoe polishes. Carnauba wax is thought to be a polyester, the monomer of which has the structure $HO(CH_2)_nCOOH$, where n is an odd number from 17 to 29.

Many natural waxes have been replaced by synthetic materials, mainly polymers. By careful control of the molecular weight, the properties of natural waxes can be closely duplicated. For example, Carbowax, a polymer of ethylene glycol ($HO-CH_2CH_2-OH$), is available in a range of average molecular weights. Synthetic waxes are used in cosmetics and ointments and in certain industrial processes.

21.7 PHOSPHOLIPIDS

Glycerol is an alcohol. It forms esters with acids. When the acids are carboxylic acids, the esters formed are fats. But glycerol can also form ester with inorganic acids, such as phosphoric acid. The **phosphatides,** a most important class of **phospholipids** (phosphorus-containing lipids), are esters of glycerol in which there are two fatty acid groups and one phosphoric acid residue. Further, the phosphoric acid is also esterified with another alcohol molecule, usually an amino alcohol (Figure 21.11). Since these compounds are getting quite complicated, it is useful to relate their structure to those of other lipids we have studied. Notice that the molecule is identical to the triglycerides (fats and oils) up to the phosphoric acid part. Let's simplify things by representing those parts

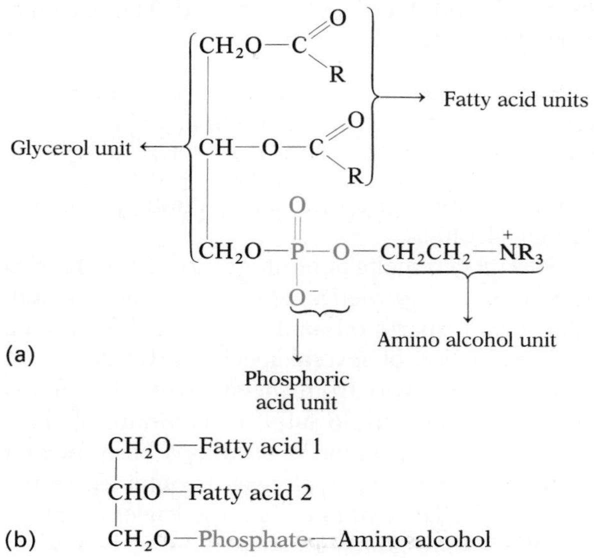

FIGURE 21.11
Structural formula (a) and schematic representation (b) of a phosphatide.

(a)

(b)
$$CH_2O-\text{Fatty acid 1}$$
$$CHO-\text{Fatty acid 2}$$
$$CH_2O-\text{Phosphate}-\text{Amino alcohol}$$

FIGURE 21.12
Two common phosphatides.

$$CH_2O—\text{Fatty acid 1}$$
$$CHO—\text{Fatty acid 2}$$
$$CH_2O—\text{Phosphate}—OCH_2CH_2\overset{+}{N}H_3$$

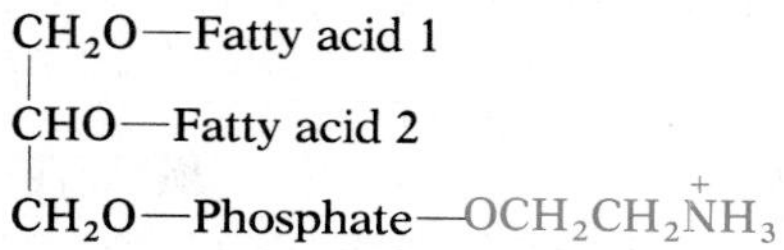

Phosphatidylethanolamine
(A cephalin)

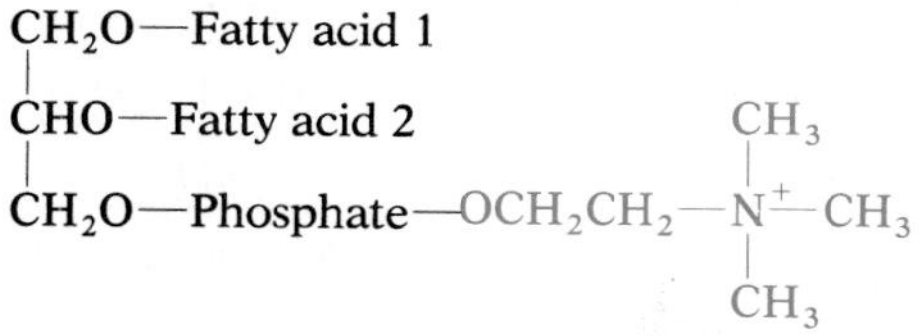

$$CH_2O—\text{Fatty acid 1}$$
$$CHO—\text{Fatty acid 2}$$
$$CH_2O—\text{Phosphate}—OCH_2CH_2—\overset{CH_3}{\underset{CH_3}{\overset{|}{\underset{|}{N^+}}}}—CH_3$$

Phosphatidylcholine
(A lecithin)

schematically (as in Figure 21.11b). We will then write out the structure of the amino alcohol to emphasize the parts that are different.

Two common phosphatides are shown in Figure 21.12. When the phosphatide contains the *ethanolamine* structural group (we've shown the amine group ionized), the compounds are called **cephalins.** The cephalins are found is brain tissue and nerves. They are also involved in blood clotting.

When *choline* is the amino alcohol unit, the compounds are called **lecithins.** Lecithins occur in all living organisms. They, too, are important constituents of nerve and brain tissue. Eggs are especially rich in lecithins. Commercial grades of lecithins are available from soybeans. These lecithins are widely used in foods as emulsifying agents. Many candy bars list lecithin among their ingredients.

Some snake venoms contain an enzyme that can catalyze the hydrolysis of one of the fatty acid ester units of lecithins. The remaining portion of the phospholipid then causes a breakdown of red blood cells.

Like the phosphatides, **sphingolipids** contain a phosphoric acid unit. They are classified separately, however, because they are based on the unsaturated amino alcohol, sphingosine, rather than glycerol. Sphingomyelin is the ''simplest'' sphingolipid. It contains fatty acid, phosphoric acid, sphingosine, and choline units (Figure 21.13).

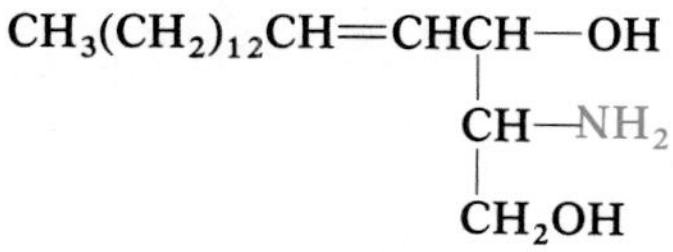

$$HOCH_2CH_2NH_2$$

Ethanolamine
(2-aminoethanol)

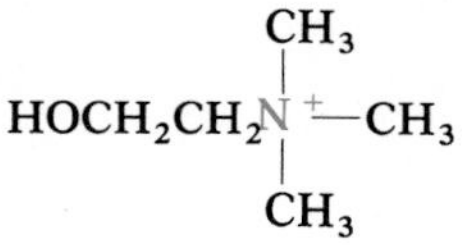

$$HOCH_2CH_2\overset{CH_3}{\underset{CH_3}{\overset{|}{\underset{|}{N^+}}}}—CH_3$$

Choline

$$CH_3(CH_2)_{12}CH=CHCH—OH$$
$$\underset{|}{CH—NH_2}$$
$$CH_2OH$$

Sphingosine

Sphingomyelins are important constituents of the myelin sheath that surrounds the axon of a nerve cell. Multiple sclerosis is one of several diseases related to a fault in the myelin sheath.

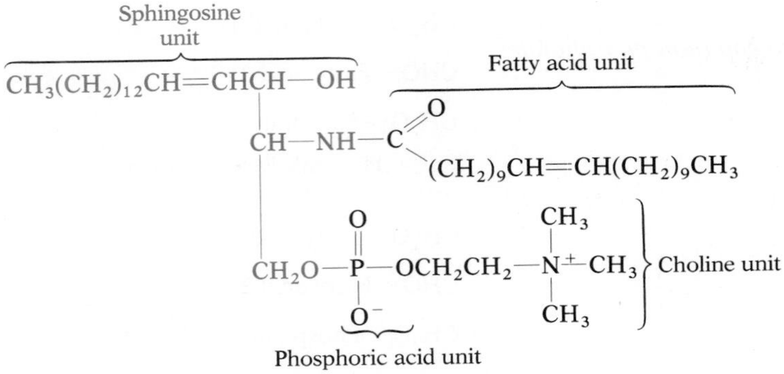

Sphingosine
unit

Fatty acid unit

$$CH_3(CH_2)_{12}CH=CHCH-OH$$

$$CH-NH-C\underset{(CH_2)_9CH=CH(CH_2)_9CH_3}{\overset{O}{\big\|}}$$

$$CH_2O-P\underset{O^-}{\overset{O}{\underset{\big|}{\overset{\big\|}{-}}}}OCH_2CH_2-\underset{CH_3}{\overset{CH_3}{\underset{\big|}{\overset{\big|}{N^+}}}}CH_3 \Big\} \text{ Choline unit}$$

Phosphoric acid unit

FIGURE 21.13   A sphingomyelin.

21.8 GLYCOLIPIDS

Glycolipids may incorporate glycerol or sphingosine, fatty acids, and, always, a sugar unit. They contain no phosphoric acid. The simplest contain two fatty acid units and one sugar unit, such as a galactose unit, combined with glycerol (Figure 21.14). Again notice the similarity between this molecule and a fat molecule. They're the same except that the glycolipid attaches a sugar unit where the fat has its third fatty acid unit. These simple glycolipids occur in

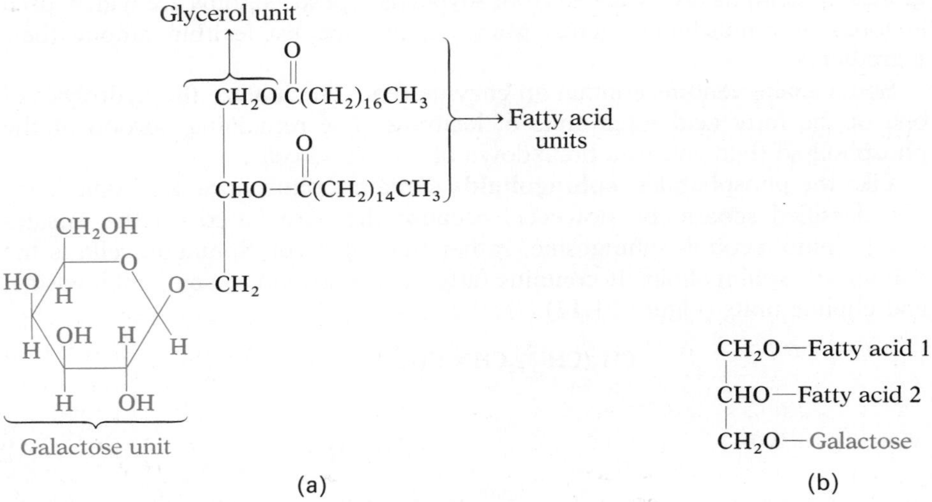

Glycerol unit

$$CH_2O\overset{O}{\overset{\big\|}{C}}(CH_2)_{16}CH_3$$

$$CHO-\overset{O}{\overset{\big\|}{C}}(CH_2)_{14}CH_3$$

Fatty acid
units

CH$_2$OH

Galactose unit

CH$_2$O—Fatty acid 1

CHO—Fatty acid 2

CH$_2$O—Galactose

(a)

(b)

FIGURE 21.14   Structural formula (a) and schematic representation (b) of a simple glycolipid.

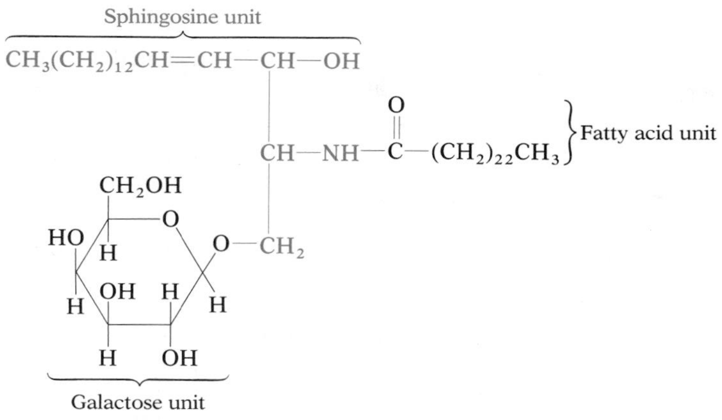

Sphingosine unit

$$CH_3(CH_2)_{12}CH=CH-CH-OH$$

$$CH-NH-\overset{\overset{\displaystyle O}{\|}}{C}-(CH_2)_{22}CH_3 \Big\} \text{Fatty acid unit}$$

Galactose unit

FIGURE 21.15 *Cerebrosides are both* glycolipids *and* sphingolipids.

microorganisms and plants. Another group of glycolipids are called **cerebrosides** (Figure 21.15). Cerebrosides contain a galactose unit, a fatty acid unit, and a sphingosine unit. The cerebrosides qualify not only as glycolipids but also as sphingolipids. The structure resembles that of a sphingomyelin, except that a sugar unit is connected where the sphingomyelin has a choline phosphate group. The cerebrosides are important constituents of the membranes of nerve and brain cells. Gaucher's disease, a hereditary affliction, results from the substitution of glucose for galactose in the cerebroside. Large amounts of these abnormal cerebrosides accumulate, causing enlargement of the liver and the spleen.

Related compounds, called gangliosides, are also found in cell membranes. The structures of these compounds are even more complex than those of the cerebrosides.

21.9 STEROIDS: CHOLESTEROL AND BILE SALTS

All the lipids discussed so far are **saponifiable;** they react with aqueous alkali to yield simpler components such as glycerol, fatty acids, amino alcohols, or sugars. As extracted from cellular material, however, the lipids contain a small but important fraction that does not react with alkali. The most important **nonsaponifiable** lipids obtained from animal cells belong to a class of compounds called **steroids.** These compounds include the bile salts, cholesterol and related compounds, hormones such as cortisone, and the all-important sex hormones. Those steroids with hormonal activity are discussed in detail in Special Topic K.

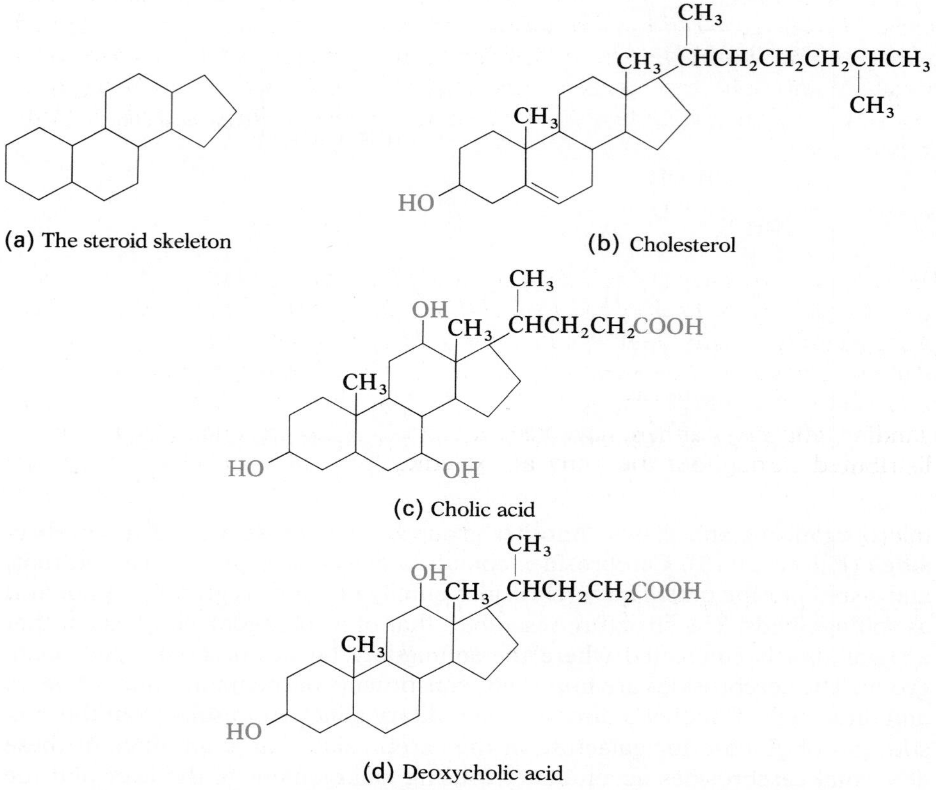

(a) The steroid skeleton

(b) Cholesterol

(c) Cholic acid

(d) Deoxycholic acid

FIGURE 21.16   The steroid skeleton (a) and three important steroids. Cholesterol (b) is a component of all animal tissues. Bile salts are derived from cholic acid (c) and deoxycholic acid (d).

FIGURE 21.17
Gallstones are often composed largely of cholesterol. (Courtesy of Biomedical Graphics, University of Minnesota Hospitals, Minneapolis.)

All steroids have a common structural unit—a system of four fused rings (Figure 21.16). Three rings have six members each; the fourth has only five carbons. The most abundant of the steroids is cholesterol. As its name indicates and its structure confirms, cholesterol is an alcohol. However, its 27 carbon atoms are more than enough to establish its hydrocarbon character, including its solubility in nonpolar solvents. Cholesterol is a common component of all animal tissues. The brain is about 10% cholesterol, but the function of cholesterol there is unknown. Cholesterol is a major component of certain types of gallstones (Figure 21.17). It is also found in the lipid deposits in hardened arteries, although it has not been established whether it is there as a cause of hardening of the arteries or as an effect from some other process.

Bile is a fluid that is formed in the liver and stored in the gallbladder. It is released from time to time into the digestive tract, where it aids in the digestion of fats (Special Topic L). The bile salts are the principal active ingredients of bile. They are derived from cholic acid and deoxycholic acid, which are steroids. The bile salts derived from these compounds have an ionic portion and a less polar portion. They are emulsifying agents, acting much like soap as they break up fats into tiny particles and suspend them in the body's aqueous solutions. The emulsified fats are much more readily digested and more easily transported.

21.10 PROSTAGLANDINS

Perhaps no compounds since birth-control pills have stimulated as much activity in pharmaceutical companies as have the group of compounds called **prostaglandins.** Over 6000 papers are published each year on these compounds. Biochemically, prostaglandins are derived from arachidonic acid, a fatty acid with 20 carbon atoms (Figure 21.18). There are six primary prostaglandins, and many others have been identified. These compounds are widely distributed throughout the body at extremely low concentrations. They are

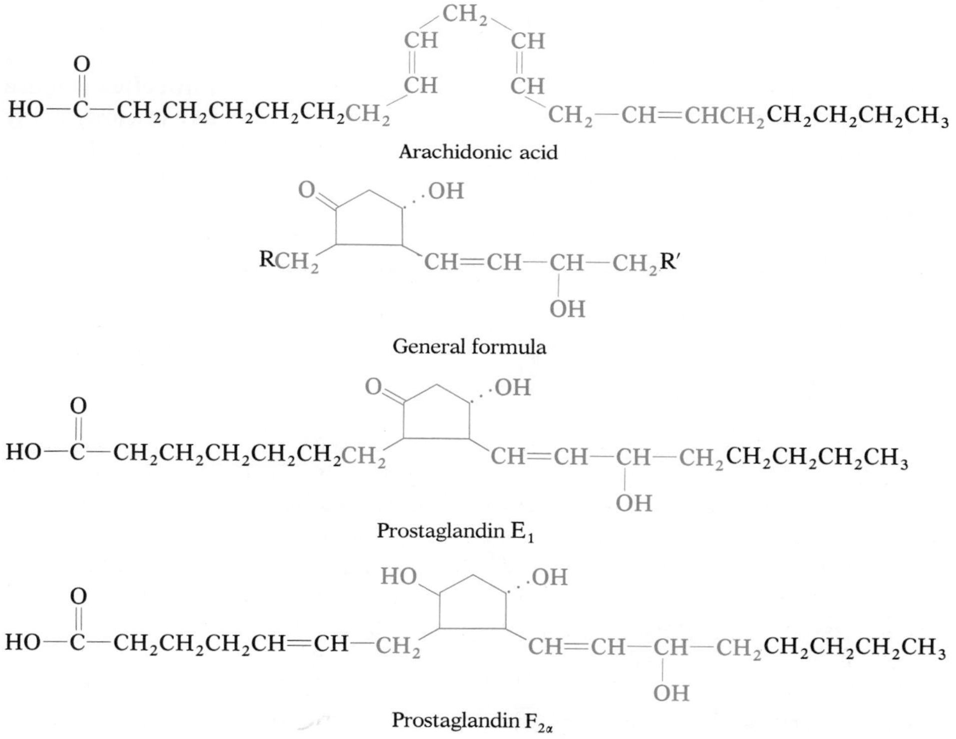

FIGURE 21.18   *Prostaglandins, the latest of the "miracle" drugs. The prostaglandins are derived from an unsaturated carboxylic acid having 20 carbon atoms.*

among the most potent biological chemicals, and extremely small doses can elicit marked changes.

In some respects, the prostaglandins act much as hormones do, regulating such things as smooth-muscle activity, blood flow, and secretion of various substances. This range of physiological activity led to the synthesis of hundreds of prostaglandins and their analogs. Two such derivatives are now in use in the United States to induce labor. Others have been employed clinically to lower or increase blood pressure, to inhibit stomach secretions, to relieve nasal congestion, to provide relief from asthma, and to prevent the formation of the blood clots associated with heart attacks and strokes. Several of these drugs will likely be on the market soon, after further testing has demonstrated their relative safety and confirmed their therapeutic value.

21.11 CELL MEMBRANES

All the components of a living cell are enclosed within a membrane. Plant cells have rigid walls of cellulose to protect and surround the cell (Figure 21.19), but animal cells have only the cell membrane (Figure 21.20).

When polar lipids such as soaps, phospholipids, and glycolipids are placed in water, they disperse, forming clusters of molecules called **micelles** (Figure 21.21a). The hydrocarbon "tails" of these lipids are directed inward, away from the polar solvent; the hydrophilic (water-loving) heads are directed outward into the water. Each micelle may contain thousands of the polar lipid molecules.

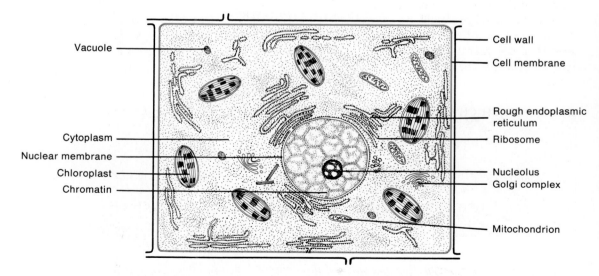

FIGURE 21.19 An idealized "typical" plant cell.

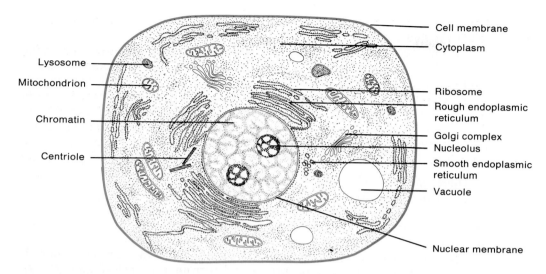

FIGURE 21.20 *An idealized "typical" animal cell.*

Polar lipids also form **monolayers,** one molecule thick, on the surface of water (Figure 21.21b). The polar heads stick into the water, and the nonpolar tails stick up in the air. In addition, **bilayers** may be formed (Figure 21.21c), with the hydrophobic (water-fearing) tails sandwiched between the hydrophilic heads sticking outward into the water. These bilayers have properties quite similar to those of cell membranes.

In addition to lipids, cell membranes contain protein molecules. The proteins move about in a "sea" of lipids. This description is called the **fluid mosaic model** of cell membranes. The lipid molecules also move laterally within the bilayer. Indeed, they move much more rapidly than the larger protein molecules.

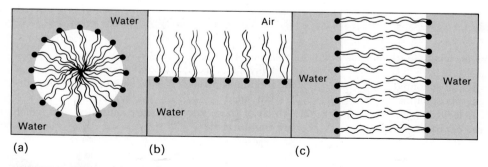

FIGURE 21.21 *(a) Micelle, (b) monolayer, and (c) bilayer formed when polar lipids are added to water.*

FIGURE 21.22
A cell membrane model
showing proteins in a lipid
bilayer. (Adapted from
Green, D., BioScience
21:409.)

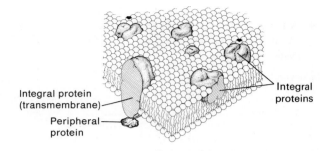

Integral protein
(transmembrane)

Peripheral
protein

Integral
proteins

There are two classes of proteins in the cell membrane (Figure 21.22). **Integral proteins** extend into the lipid bilayer. Those integral proteins that extend completely across the membrane are called **transmembrane proteins.** **Peripheral proteins,** the other class of proteins, seem to bond only loosely to the lipid bilayer. They appear to be attached to integral proteins by hydrogen bonds and electrostatic forces.

The lipid part of the membrane serves as a barrier to the passage of ions and polar molecules. Nonpolar molecules, such as the general anesthetics (Special Topic G), can infiltrate the bilayer and change its permeability. Polar and ionic substances enter by way of channels through the integral proteins. It appears that there are special proteins to facilitate the passage of certain molecules. A specific interaction occurs between the carrier protein and the molecule. The complex then moves to the other surface of the membrane and discharges its passenger. If this occurs in the direction dictated by the concentration (high to low), it requires no energy and is called **facilitated diffusion.** If the movement takes place against the concentration gradient, it requires energy and is called **active transport.** (The source of this energy, cellular ATP, is discussed in Chapter 27.)

This model of cell membranes is a tentative one. It will be modified as research continues in this exciting area.

PROBLEMS

1. Define and illustrate or give an example of each of the following:
   a. lipid
   b. fatty acid
   c. simple triglyceride
   d. mixed triglyceride
   e. fat
   f. oil
   g. polyunsaturated fat
   h. iodine number
   i. soap
   j. phospholipid
   k. phosphatide
   l. lecithin
   m. cephalin
   n. steroid
   o. prostaglandin
   p. saponifiable lipid
   q. nonsaponifiable lipid
   r. micelle
   s. monolayer
   t. bilayer
   u. monounsaturated fatty acid
   v. wax
   w. glycolipid
   x. sphingolipid
   y. cerebroside

2. Which of these fatty acids are saturated and which are unsaturated?
   a. caproic acid
   b. oleic acid
   c. stearic acid
   d. palmitic acid
   e. linolenic acid
   f. caprylic acid

3. How many carbon atoms are there in a molecule of each of the compounds listed in Problem 2?

4. Arrange these fatty acids in order of increasing levels of unsaturation: linoleic acid, linolenic acid, oleic acid.
5. Write a structural formula for lauric acid.
6. Write a structural formula for oleic acid.
7. Write a structural formula for glyceryl tristearate (tristearin).
8. Write a structural formula for sodium oleate.
9. Write a structural formula for calcium myristate.
10. In the determination of the iodine number of a fat, what functional group in the fat molecule reacts with the reagent?
11. Which would have the higher iodine number—tristearin or triolein?
12. Which would you expect to have a higher iodine number—corn oil or beef tallow? Explain your reasoning.
13. Which would you expect to have the higher iodine number—hard or liquid margarine? Explain your reasoning.
14. What fat would be formed by the complete hydrogenation of triolein? Of trilinolein?
15. Draw the structural formulas of the products of this saponification reaction:

$$
\begin{array}{l}
CH_2-O-\overset{\displaystyle O}{\overset{\|}{C}}-CH_2CH_2CH_2CH_2CH_2CH_2CH_3 \\[2ex]
CH-O-\overset{\displaystyle O}{\overset{\|}{C}}-CH_2CH_2CH_2CH_2CH_3 \quad\xrightarrow{\;NaOH\;} \\[2ex]
CH_2-O-\overset{\displaystyle O}{\overset{\|}{C}}-CH_2CH_2CH_2CH_2CH_2CH_2CH_2CH_3
\end{array}
$$

16. Write the equation for the saponification of glyceryl trilaurate.
17. How would the equation for Problem 16 differ if the compound being saponified were triolein?

18. What general structural features make soaps good cleaning agents?
19. Describe the process by which soap acts to remove oily dirt.
20. The structures of synthetic detergents are designed to give these compounds the same kind of cleaning action as soap. Which of the compounds at the bottom of the page would be effective cleaning agents?
21. Why does soap lose its cleaning efficiency in hard water?
22. Synthetic detergents work well in hard water that deactivates soap. Why?
23. Classify the following as saponifiable or nonsaponifiable lipids:
    a. tristearin   c. a lecithin
    b. cholesterol   d. a prostaglandin
24. Which are phospholipids?
    a. tristearin   d. prostaglandin E_1
    b. cholesterol   e. cephalin
    c. phosphatides   f. triolein
25. Which compounds are esters of fatty acids?
    a. cephalin   c. vegetable oil
    b. cholesterol   d. prostaglandins
26. Both lecithins and cephalins are classed as phosphatides. How do lecithins differ from cephalins?
27. What general structural feature do phosphatides share with soaps and detergents?
28. Using a circle to represent the ionic group and zigzag lines to represent the two fatty acid units, draw the arrangement of phosphatide molecules in a micelle and in a bilayer.
29. Draw the basic steroid skeleton.
30. Which compounds are classified as steroids?
    a. tristearin   d. prostaglandins
    b. cholesterol   e. cephalin
    c. lecithin   f. cholic acid
31. Which of the following are derived from glycerol, which from sphingosine, and which from neither?
    a. fats   b. oils

a. $CH_3CHCH_2CHCH_2CHCH_2CHCH_2CHCH_2CHCH_2CHCH_2CHCH_2CHCH_2-O-\overset{\displaystyle O}{\underset{\displaystyle O}{\overset{\|}{\underset{\|}{S}}}}-O^-K^+$
    $\underset{CH_3}{|}\ \underset{CH_3}{|}\ \underset{CH_3}{|}\ \underset{CH_3}{|}\ \underset{CH_3}{|}\ \underset{CH_3}{|}\ \underset{CH_3}{|}$

b. $\overset{OH}{\underset{|}{}}\ \overset{OH}{\underset{|}{}}\ \overset{OH}{\underset{|}{}}\ \overset{OH}{\underset{|}{}}\ \overset{OH}{\underset{|}{}}\ \overset{OH}{\underset{|}{}}\ \overset{OH}{\underset{|}{}}\ \overset{OH}{\underset{|}{}}$
   $CH_2CH_2CHCH_2CHCH_2CHCH_2CHCH_2CHCH_2CHCH_2CHCH_2CHCH_2\overset{\displaystyle O}{\overset{\|}{C}}-O^-Na^+$

c. $\overset{SO_3{}^-Na^+}{\underset{|}{}}\qquad \overset{SO_3{}^-Na^+}{\underset{|}{}}$
   $CH_2CH_2CH_2CHCH_2CH_2CH_2$
   $\qquad\qquad\underset{SO_3{}^-Na^+}{|}$

c. waxes **f.** cerebrosides
d. phosphatides **g.** prostaglandins
e. steroids

32. Describe two ways in which butter can go rancid.

33. Why doesn't soap work in acidic water? Write the equation.

34. What fat would be formed by the complete hydrogenation of triolein? Of trilinolein?

35. List some possible future uses of prostaglandins as drugs.

36. Salts of the bile acids are involved in the digestion of lipids (Chapter 26). Relate the action of the bile salts to that of soap.

37. What reagents or catalysts or both are needed to carry out the following conversions?

a. $CH_3(CH_2)_7CH{=}CH(\dot{C}H_2)_7COOH \longrightarrow$
$$CH_3(CH_2)_{16}COOH$$

b.
$$
\begin{array}{l}
CH_2O\overset{\displaystyle O}{\overset{\|}{C}}{-}R \\[4pt]
CHO\overset{\displaystyle O}{\overset{\|}{C}}{-}R \\[4pt]
CH_2O\overset{\displaystyle O}{\overset{\|}{C}}{-}R
\end{array}
\longrightarrow
3\ RCO^- +
\begin{array}{l}
CH_2OH \\[4pt]
CHOH \\[4pt]
CH_2OH
\end{array}
$$

c.
$$
\begin{array}{l}
CH_2OH \\
CHOH \\
CH_2OH
\end{array}
\longrightarrow
\begin{array}{l}
H{-}C\overset{\displaystyle O}{\diagup} \\
CH \\
CH_2
\end{array}
$$

38. How many molecules of I_2 would be taken up by one molecule of
a. triolein? **b.** trilinolein?

39. Saponification of lipid **X** yields glycerol and sodium salts of lauric, oleic, and stearic acid. Give a structure of **X**.

40. How is sodium lauryl sulfate made?

41. Palmitic acid (hexadecanoic acid) melts at 63 °C while palmitoleic acid (*cis*-9-hexadecenoic acid) melts at −1 °C. Account for the difference in melting points.

42. Describe the fluid mosaic model of cell membranes.

43. What is an integral protein? What is its role in the function of a cell membrane?

44. What is a peripheral protein? Where is it located on the cell membrane?

45. What is facilitated diffusion?

46. What is active transport?

47. Draw the structure of the cerebroside that has palmitic acid as its fatty acid and glucose as its sugar.

48. A principal wax in spermaceti is an ester of palmitic acid (hexadecanoic acid) and cetyl alcohol (1-hexadecanol). Draw the structure of the ester.

49. Which of the following can diffuse through the lipid portion of a cell membrane?
a. glucose **c.** $CH_3CH_2OCH_2CH_3$
b. NaCl

50. Discuss the role of cholesterol, saturated fats, and fish oils in arteriosclerosis.

51. The system that transports Na^+ across a cell membrane is called the "sodium pump." Sodium ions are pumped out of a cell in which the concentration of Na^+ is maintained at 0.10 M and into the extracellular fluid where the Na^+ concentration is 0.14 M. Is this an example of facilitated diffusion or of active transport? Explain.

52. A carrier protein complexes a molecule of glucose at the outer surface of a cell membrane. The complex then moves to the other side of the membrane and releases the glucose molecule. The concentration of glucose is higher outside the cell than inside. Is this facilitated diffusion or active transport? Explain.

53. Identify each component of the following phospholipid:

$$
\begin{array}{l}
CH_2O\overset{\displaystyle O}{\overset{\|}{C}}(CH_2)_{16}CH_3 \\[6pt]
CHO\overset{\displaystyle O}{\overset{\|}{C}}(CH_2)_7CH{=}CH(CH_2)_7CH_3 \\[6pt]
CH_2O\overset{\displaystyle O}{\underset{\displaystyle O}{\overset{\|}{P}}}{-}OCH_2CH_2\overset{+}{N}\begin{array}{l}CH_3\\[-2pt]{-}CH_3\\[-2pt]CH_3\end{array}
\end{array}
$$

54. Serine is an amino acid with the structure
$$HOCH_2\underset{\underset{\textstyle NH_2}{|}}{CH}COOH.$$
Give the structure for phosphatidyl serine.

Proteins

Carbohydrates, lipids, and proteins—these are the three classes of foods. All are essential to life, but proteins perhaps are closest to the stuff of life itself. No living part of the human body—nor of any other organism, for that matter—is completely without protein. There is protein in the blood, in the muscles, in the brain, and even in tooth enamel. The smallest cellular organisms, the bacteria, contain protein. And the viruses, so small that they make the bacteria look like giants, are nothing but large molecules of conjugated proteins—nucleoproteins. **Nucleoproteins** are combinations of proteins with nucleic acids. Nucleic acids (Chapter 23) cause proteins to be formed, and through these proteins they control the structure and function of cells.

Each type of cell makes its own specific kinds of proteins. Proteins serve as structural material for animals, much as cellulose does for plants. Muscle tissue is largely protein. So are skin and hair. Proteins are made in different forms in different animals: silk, wool, nails, claws, feathers, horns, and hoofs are all proteins (Figure 22.1).

In an overcrowded, hungry world, protein is of increasing importance. The rich nations have it—in the form of beef steak, fish, fowl, soybeans. The poor nations need it but can't afford it. A nation's consumption of sulfuric acid has long been considered an indication of its industrial development. Perhaps its consumption of protein is a better indication of the quality of life of its people, for without protein no nation can have the healthy, vigorous people who are vital to progress. The last major wars were fought over land and mineral resources. The next wars may well be fought over proteins—or over the nitrates, phosphates, and energy needed to produce them.

FIGURE 22.1
Leather, silk, wool, and meat tenderizer are all rich in protein. Insulin is a protein hormone. (Photo © The Terry Wild Studio.)

22.1 AMINO ACIDS: STRUCTURE AND PHYSICAL PROPERTIES

$H_2N-\overset{\alpha}{\underset{\quad}{C}}-C\overset{O}{\underset{OH}{}}$

An alpha amino acid (incorrect structure)

$H_3\overset{+}{N}-\overset{R}{\underset{H}{C}}-C\overset{O}{\underset{O^-}{}}$

A zwitterion (correct structure)

$H_2\overset{+}{N}-\overset{H}{\underset{}{C}}-COO^-$
$H_2C\quad CH_2$
CH_2

Proline

OH
$CH-CH_2$
$CH_2\quad CH-COO^-$
$\overset{+}{N}H_2$

Hydroxyproline

The genetic code specifies 20 amino acids for incorporation into proteins. Some of these are modified after incorporation, but we can define **proteins** fairly accurately as copolymers of 20 or so amino acids. As the name indicates, these building blocks contain an amino group and a carboxylic acid group. Further, the amino group is always on the alpha carbon; all are α-amino acids (margin). The partial formula indicates the proper placement of these groups, but the structure is not really correct. Acids react with bases to form salts; the carboxyl group is acidic, and the amino group is basic. Therefore, these two functional groups interact, the acid transferring a proton to the base. The resulting product is an inner salt, or **zwitterion,** a compound in which the anion and cation are part of the same molecule (margin).

All the 20 amino acids specified by the genetic code fit this general structure except proline. While the other 19 have primary amine groups, proline, in which there is a ring incorporating the nitrogen atom, has a secondary amine function. (Proline is also one of the amino acids frequently modified after its incorporation into proteins; a hydroxyl group is added to change proline to hydroxyproline.)

The physical properties of amino acids confirm the zwitterion structure. All are crystalline solids at room temperature. Attempts to determine melting points result in decomposition rather than melting because the covalent bonds break before the ionic ones. Charring generally occurs above 200 °C. Water solubility varies, but is generally greater than solubility in less polar solvents such as chloroform and benzene. These properties are consistent with a zwitterion structure. We shall, therefore, ordinarily draw amino acids as zwitterions.

What makes one alpha amino acid different from another is the identity of the "R" group, which is also attached at the alpha position. If this R group is anything other than hydrogen, the amino acid exists in the form of mirror image isomers (Special Topic I). Nearly all naturally occurring amino acids have the L configuration. Note the similarity in the configurations of L-glyceraldehyde (Section 20.2) and the amino acid L-serine (margin).

CHO
|
HO—C—H
|
CH₂OH

L-Glyceraldehyde

COO⁻
|
H₃N⁺—C—H
|
CH₂OH

L-Serine

H₃N⁺—CH₂—C(=O)(O⁻)

Glycine

22.2 SOME INTERESTING AMINO ACIDS

The simplest alpha amino acid is glycine, in which the R group is a hydrogen.

Several amino acids have simple hydrocarbon side chains. These include alanine (in which the R group is methyl), phenylalanine (in which the R group is $C_6H_5CH_2$), valine (in which the R group is isopropyl), leucine (in which the R group is isobutyl), and isoleucine (in which the R group is *sec*-butyl). Structures of these and other amino acids are given in Table 22.1.

Several amino acids contain polar side chains. The hydroxyl groups in serine, threonine, hydroxyproline, and tyrosine molecules, for example, may be involved in hydrogen bonding. Such bonding may affect the shape of protein molecules in important ways, as we shall see.

Perhaps even more important to the shape of protein molecules is the presence of ionizable groups in the side chains of certain amino acids. Aspartic and glutamic acids have acidic groups in their side chains. Three others—histidine, lysine, and arginine—have basic groups in their side chains. These groups are ionizable. If the amino acids containing them are incorporated into a protein, a proton transfer can occur when the groups are brought into close contact.

—COOH   H₂N— ⟶ —COO⁻   H₃N⁺—

This is exactly the same acid-base reaction given previously. Here, however, it is the side chains that are interacting and not the amino and acid groups that give the amino acids, as a class, their name. The ionic charges that result from side chain interactions, because of their strong attraction for one another, contribute markedly to the structure of some proteins. Such interactions are called **salt bridges.**

In addition to the 20 amino acids specified by the genetic code, there are many other amino acids of physiological importance that are not derived from proteins. Most have been isolated from plants, but some are found in animals.

TABLE 22.1   *Amino Acids Specified by the Genetic Code*

Name	Abbreviation	Essential	Structure
Nonpolar side chains			
Glycine	Gly	No	CH_2-COO^- $^+NH_3$
Alanine	Ala	No	$CH_3-CH-COO^-$ $^+NH_3$
Phenylalanine	Phe	Yes	⬡$-CH_2-CH-COO^-$ $^+NH_3$
Valine	Val	Yes	$CH_3-CH-CH-COO^-$ CH_3 $^+NH_3$
Leucine	Leu	Yes	$CH_3CHCH_2-CH-COO^-$ CH_3 $^+NH_3$
Isoleucine	Ile	Yes	$CH_3CH_2CH-CH-COO^-$ CH_3 $^+NH_3$
Proline	Pro	No	CH_2-CH_2 COO^- CH_2 C $^+NH_2$ H
Methionine	Met	Yes	$CH_3-S-CH_2CH_2-CH-COO^-$ $^+NH_3$
Polar, but not ionizable, side chains			
Serine	Ser	No	$HO-CH_2-CH-COO^-$ $^+NH_3$
Threonine	Thr	Yes	$CH_3CH-CH-COO^-$ OH $^+NH_3$
Asparagine	Asn	No	$\overset{O}{\overset{\|}{H_2N-C}}-CH_2-CH-COO^-$ $^+NH_3$

(continued)

TABLE 22.1 *(continued)*

Name	Abbreviation	Essential	Structure
Glutamine	Gln	No	$H_2N-\overset{\displaystyle O}{\overset{\|}{C}}-CH_2CH_2-\underset{+NH_3}{CH}-COO^-$
Cysteine	Cys	No	$HS-CH_2-\underset{+NH_3}{CH}-COO^-$
Tyrosine	Tyr	No	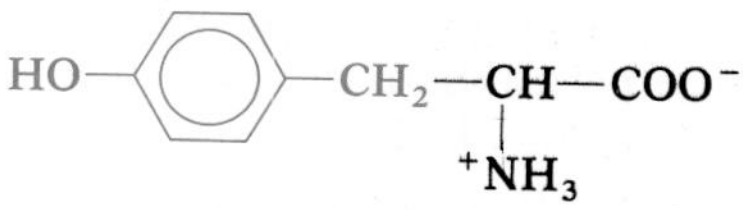 $HO-\langle\bigcirc\rangle-CH_2-\underset{+NH_3}{CH}-COO^-$
Tryptophan	Trp	Yes	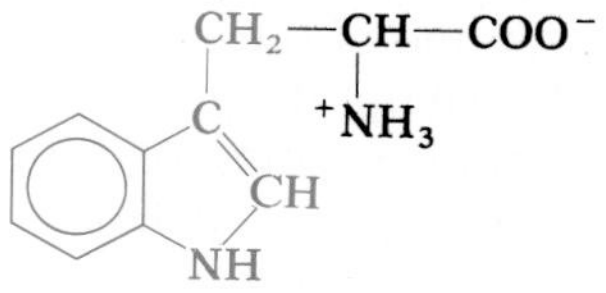 $CH_2-\underset{+NH_3}{CH}-COO^-$ (indole ring)

Basic side chains, ionizable

Name	Abbreviation	Essential	Structure
Lysine	Lys	Yes	$\overset{+}{H_3N}CH_2CH_2CH_2CH_2-\underset{NH_2}{CH}-COO^-$
Arginine	Arg	—*	$H_2N-\underset{+NH_2}{\overset{\|}{C}}-NHCH_2CH_2CH_2-\underset{NH_2}{CH}-COO^-$
Histidine	His	—†	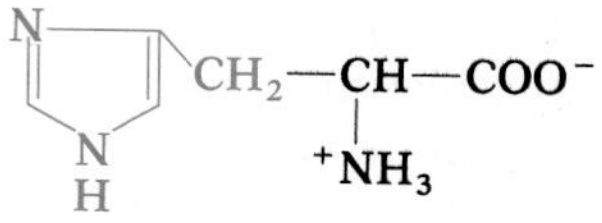 $CH_2-\underset{+NH_3}{CH}-COO^-$ (imidazole ring)

Acidic side chains, ionizable

Name	Abbreviation	Essential	Structure
Aspartic acid	Asp	No	$HOOC-CH_2-\underset{+NH_3}{CH}-COO^-$
Glutamic acid	Glu	No	$HOOC-CH_2CH_2-\underset{+NH_3}{CH}-COO^-$

* Essential to growing children but not to adult humans.
† Essential to rats but not to humans.

$$\overset{\gamma}{C}H_2\overset{\beta}{C}H_2\overset{\alpha}{C}H_2-COO^-$$
$$|$$
$$^+NH_3$$

γ-Aminobutyric acid

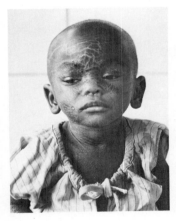

$$H_2N-\langle\bigcirc\rangle-COOH$$

p-Aminobenzoic acid

For example, γ-aminobutyric acid is a chemical neurotransmitter found free in the brain. Amino acids that incorporate the anilino group are also of physiological importance. *p*-Aminobenzoic acid is required for the synthesis of the essential vitamin folic acid (Special Topic J).

22.3 THE ESSENTIAL AMINO ACIDS

The adult human body can synthesize all but eight of the amino acids needed for making proteins. Those eight (they are indicated in Table 22.1) are called **essential amino acids.** They must be included in the diet. We eat proteins, break them down in our bodies to their constituent amino acids, and then use some of these amino acids to build other protein structures essential for health. Each of the essential amino acids is a **limiting reagent.** When the body runs out of one of them, synthesis of protein goes awry.

An **adequate protein** source supplies all the essential amino acids in the quantities needed for growth and repair of body tissues. Most proteins from plant sources are deficient in one or more amino acids. Corn protein is lacking in lysine and tryptophan. People whose diet consists chiefly of corn may suffer from malnutrition even though the amount of calories supplied by the food is adequate. Protein from rice is short of lysine and threonine. Wheat protein is lacking in lysine. Even soy protein, a good nonanimal protein, is lacking in the essential amino acid methionine.

Most proteins from animal sources contain all the essential amino acids in adequate amounts. Lean meat, milk, fish, eggs, and cheese are adequate proteins. In fact, gelatin is about the only inadequate animal protein. It contains almost no tryptophan and has only small amounts of threonine, methionine, and isoleucine. A proper diet must include all the essential amino acids in sufficient quantity.

Inadequate protein is already a problem in much of the world. A protein-deficiency disease called kwashiorkor (Figure 22.2) is common in parts of Africa where corn is the major food. As population increases, animal protein will become scarcer in the developed countries.

Plants, such as grasses or grains, trap a small fraction of the solar energy that falls upon them. They use this energy to convert carbon dioxide, water, and mineral nutrients such as nitrates, phosphates, and sulfates into protein. Cattle eat the plant protein, digest it, and convert a small portion of it into animal protein. People eat this animal protein, digest it, and reassemble some of the amino acids into human protein. Some of the energy originally trapped by the green plants is lost at every step. If people ate the plant protein directly, one highly inefficient step would be skipped. Extreme vegetarianism is dangerous, however. One would have to eat a wide variety of plant materials to be sure of getting enough of all the essential amino

FIGURE 22.2
A lack of proteins and vitamins causes the deficiency disease known as kwashiorkor, which can be recognized by retarded growth, discoloration of the skin and hair, bloating, swollen belly, and mental apathy. (Courtesy of the World Health Organization, New York. Photo by Paul Almasy.)

acids. Even then, an all-vegetable diet is likely to be lacking in vitamin B_{12}, for this nutrient is not found in plants. Other nutrients scarce in all-plant diets include calcium, iron, riboflavin, and (for children not exposed to sunlight) vitamin D. A modified vegetarian diet that includes milk, eggs, cheese, and fish can provide excellent nutrition.

22.4 REACTIONS OF AMINO ACIDS

Amino acids can act either as acids or as bases. Indeed, they (and the proteins) act as buffers in living organisms. In the presence of added acid, the carboxylate group of the zwitterion captures protons.

$$\overset{+}{H_3N}-CH-COO^- + H^+ \longrightarrow \overset{+}{H_3N}-CH-COOH$$
$$\quad\quad\; | \quad\quad\quad\quad\quad\quad\quad\quad\quad\; |$$
$$\quad\quad\; R \quad\quad\quad\quad\quad\quad\quad\quad\quad R$$

Note that the product is a positive ion. If base is added, protons are removed from the amino group of the zwitterion.

$$\overset{+}{H_3N}-CH-COO^- + OH^- \longrightarrow H_2N-CH-COO^- + H_2O$$
$$\quad\quad | \quad\quad\quad\quad\quad\quad\quad\quad\quad\quad\quad |$$
$$\quad\quad R \quad\quad\quad\quad\quad\quad\quad\quad\quad\quad\quad R$$

In this case, a negative ion is formed. In both instances, the amino acid acts to maintain the pH of the system, that is, to tie up added acid and base.

At some intermediate pH value, an amino acid exists almost entirely as the zwitterion. That particular pH is called the **isoelectric point.** At the isoelectric point the positive and negative charges on an amino acid (or protein) just balance, and the molecule *as a whole* is electrically neutral. Note that the isoelectric point does *not* necessarily coincide with pH 7 (the value we ordinarily associate with neutrality). At its isoelectric point an amino acid (zwitterion) behaves very much like the salt of a weak acid and a weak base. As we noted in Section 11.4, the solution of such a salt can be slightly acidic, slightly basic, or neutral. It depends on whether the acid or the base from which the salt was formed is stronger. Each amino acid has its own characteristic isoelectric point. The simple amino acids with un-ionizable side chains have isoelectric points ranging from pH 5.0 to pH 6.5. Basic amino acids (those in which the side chain incorporates a basic group) have isoelectric points at relatively high pH values. Acidic amino acids have isoelectric points at quite low pH values (Table 22.2). By adjusting the pH of a solution, one can vary the net charge on an amino acid, thereby causing amino acids (or proteins) to migrate at different rates in an electric field. This process of separating mixtures of amino acids is called **electrophoresis.**

TABLE 22.2 *Isoelectric Points of Representative Amino Acids*

Amino Acid	Class	pH
Aspartic acid	Acidic	2.77
Glutamic acid	Acidic	3.22
Cysteine	Simple	5.02
Methionine	Simple	5.75
Tryptophan	Simple	5.89
Glycine	Simple	5.97
Alanine	Simple	6.11
Lysine	Basic	9.74

In a typical experiment, a paper strip saturated with a buffer solution at a particular pH is suspended between two reservoirs of the buffer (Figure 22.3). A sample of the solution of amino acids is applied to the center of the paper, and an electrical potential is applied between the two buffer solutions. If an amino acid has an isoelectric point equal to the pH of the buffer, it will be mainly in the zwitterion form and will not migrate. An amino acid that is

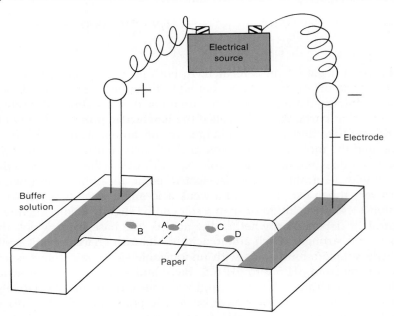

FIGURE 22.3 *An apparatus for electrophoresis. Amino acid A is in the zwitterion form and has not migrated. B exists as an anion and has moved toward the positive electrode. C and D are in cationic forms and have migrated toward the negative electrode.*

primarily in the anionic form will move toward the positive electrode. Those mainly in the cationic form will migrate toward the negative electrode. Amino acids of different sizes will move at different rates, even if both have the same electrical charge. After a period of time, the various amino acids will have separated into individual spots on the paper.

Proteins also have characteristic isoelectric points (Section 22.5) and can be separated by electrophoresis. Variations of the technique use *gel* media instead of paper. Electrophoresis is an important tool in clinical laboratories. Blood proteins can be separated by this method. Detection of unusual proteins (or unusual amounts of a common one) can be an important indicator of a disease (Chapter 29). Gel electrophoresis is also employed in the sequencing of nucleic acids (Chapter 23).

Simple chemical tests also can be used to identify amino acids. These compounds undergo characteristic reactions of carboxylic acids and amines. One of possible interest is that with nitrous acid. This reagent reacts with free amino groups, and nitrogen gas is formed.

$$R—\underset{\underset{+NH_3}{|}}{CH}—COO^- + HNO_2 \longrightarrow R—\underset{\underset{OH}{|}}{CH}—COOH + N_2(g) + H_2O$$

This reaction is the basis of the Van Slyke method for determining the number of free amino groups in a protein (Chapter 18). The ninhydrin test (Chapter 18) gives a purple color when amino acids are present and serves as a qualitative test for amino acids.

The preceding reaction and other similar ones are quite important in the detection and separation of amino acids. The most important reaction of all, though, is the polymerization reaction that forms peptides and proteins (Sections 22.5 and 22.6).

22.5 THE PEPTIDE BOND

Proteins are polymers of amino acids. How are the monomer units held together? In Chapter 18 we discussed the reaction of amines with carboxylic acid derivatives to form amides. If we heat the salt of an amine and a carboxylic acid, an amide is formed.

$$CH_3—C\underset{O^-\ H_3\overset{+}{N}—CH_3}{\overset{\diagup O}{\diagdown}} \xrightarrow{\text{Heat}} CH_3—C\underset{NH—CH_3}{\overset{\diagup O}{\diagdown}} + H_2O$$

Methylammonium acetate   N-Methylacetamide

Similarly, the amino group on one amino acid molecule can react with the carboxyl group on another.

$$H_3\overset{+}{N}-CH-C\overset{O}{\underset{O^-}{}} + H_3\overset{+}{N}-CH-C\overset{O}{\underset{O^-}{}} \longrightarrow$$
$$RR$$

$$H_3\overset{+}{N}-CH-C\overset{O}{\underset{NH-CH-C}{}} \overset{O}{\underset{O^-}{}} + H_2O$$

Peptide bond

The amide linkage (—CONH—) is called a **peptide bond** when it joins two amino acid units. Note that there is still a reactive amino group on the left and a carboxyl group on the right. Each of these can react further to join more amino acid units. This process can continue until thousands of units have joined to form a giant molecule—a polymer called a **protein.** (Synthesis of proteins in living organisms is quite complex and is discussed in more detail in Chapter 23.)

$$\cdots CH-\overset{O}{\overset{\|}{C}}-NH-CH-\overset{O}{\overset{\|}{C}}-NH-CH-\overset{O}{\overset{\|}{C}}-NH-CH-\overset{O}{\overset{\|}{C}}-NH-CH-\overset{O}{\overset{\|}{C}}-NH\cdots$$
$$RRRRR$$

When only two amino acids are joined, the product is called a **dipeptide.**

$$H_3\overset{+}{N}-CH_2-\overset{O}{\overset{\|}{C}}-NH-CH-C\overset{O}{\underset{O^-}{}}$$
$$CH_2$$

Glycylphenylalanine (a dipeptide)

When three amino acids are combined, the substance is a **tripeptide.**

$$H_3\overset{+}{N}-CH-\overset{O}{\overset{\|}{C}}-NH-CH-\overset{O}{\overset{\|}{C}}-NH-CH-C\overset{O}{\underset{O^-}{}}$$
$$CH_2OHCH_3CH_2SH$$

Serylalanylcysteine (a tripeptide)

Note how these substances are named. Other combinations are named in a similar manner. Those with more than 10 amino acid units are often simply called **polypeptides.** When the molecular weight of a compound exceeds 10 000, it is called a protein. The distinction between polypeptides and proteins

TABLE 22.3   *Isoelectric Points of Selected Proteins*

Protein	Source	pH
Silk fibroin	Silk	2.2
Casein	Cow's milk	4.6
Gelatin	Pig's feet	4.8
Serum albumin	Human blood serum	4.9
Albumin	Egg white	4.9
Lactoglobin	Cow's milk	5.2
Insulin	Cow's pancreas	5.3
Hemoglobin	Human red blood cells	6.7
Ribonuclease	Pancreas	9.5
Lysozyme	Egg white	10.7

is an arbitrary one, and it is not always precisely applied. Nor need it be, for names are merely labels and do not change the properties of substances.

As with amino acids, every protein has a characteristic isoelectric point (Table 22.3). In proteins, however, the alpha carboxyl and amino groups are tied up in the peptide bonds that hold the molecule together. Therefore, it is the ionizable side chains of the constituent amino acids that establish the isoelectric point of proteins. At its isoelectric point, the protein molecule as a whole is electrically neutral; it may contain many ionized groups, but the positively charged side chains are exactly balanced by negatively charged ones.

Proteins are usually *least* soluble at their isoelectric points. The size of many proteins places them in the category of colloids (Section 9.13). At pH values other than the isoelectric point, the molecules carry a net charge. These charges on the surface of the colloidal proteins repel the other colloidal particles and keep them from coalescing. Thus, they form colloidal dispersions. At the isoelectric point, however, the colloidal protein molecules are electrically neutral and no longer repel one another. Therefore they come together to form larger aggregates that eventually precipitate from solution.

22.6 THE SEQUENCE OF AMINO ACIDS

For peptides and proteins to be physiologically active, it is not enough that they incorporate certain amounts of specific amino acids. The order or *sequence* in which the amino acids are connected is also of critical importance. Glycylalanine is different from alanylglycine.

$$\overset{H}{\underset{+}{H_3N}}—\overset{H}{\underset{|}{CH}}—\overset{O}{\overset{\|}{C}}—NH—\overset{CH_3}{\underset{|}{CH}}—COO^-$$

Glycylalanine

$$\overset{+}{H_3N}—\overset{CH_3}{\underset{|}{CH}}—\overset{O}{\overset{\|}{C}}—NH—\overset{H}{\underset{|}{CH}}—COO^-$$

Alanylglycine

Although the difference seems minor, the two substances behave differently in the body.

When chemists describe peptides (and proteins), they find it much simpler to indicate the amino acid sequence by using the abbreviations for the amino acids (Table 22.1). The sequence for glycylalanine is written Gly-Ala, and that for alanylglycine is Ala-Gly. It is understood from this shorthand that the peptide is arranged with the free amino group to the left and the free carboxyl group to the right.

As the length of a peptide chain increases, the possible sequential variations become almost infinite. And this potential for many different arrangements is exactly what one needs in a material that will make up such diverse things as hair and skin and eyeballs and toenails and a thousand different enzymes. Just as we can make millions of different words with our 26-letter English alphabet, we can make millions of different proteins with the 20 or so different amino acids. Just as one can write gibberish with the English alphabet, one can make nonfunctioning proteins by putting together the *wrong sequence* of amino acids. Yet while the correct sequence is ordinarily of utmost importance, it is not always absolutely required. Just as you can sometimes make sense of incorrectly spelled English words, a protein with a small percentage of ''incorrect'' amino acids may continue to function. It may not function as well, however, as a protein with the correct sequence. And sometimes a seemingly minor change can have a disastrous effect. Some people have hemoglobin with a single incorrect amino acid unit in about 300. That ''minor'' error is responsible for sickle cell anemia, an inherited condition that ordinarily proves fatal (Chapter 26).

22.7 SOME PEPTIDES OF INTEREST

The functional diversity of various peptides produced in nature is absolutely amazing. The tripeptide glutathione, for example, is found in all cells. Its structure is

We have not used our shorthand method for drawing the structure of this compound because of an interesting structural feature. The glutamic acid residue is connected to the rest of the peptide not through its alpha carboxyl group but through the carboxyl group of its side chain. Glutathione plays an important role in plant respiration. It also functions in animal cells to keep certain enzymes active. Since the amino acid cysteine as such is not present in large quantities in animal tissue, glutathione represents the most abundant sulfhydryl (—SH) compound in cells. Many enzymes also contain sulfhydryl groups (Chapter 24). The oxidation of the groups to disulfides inactivates the enzymes.

$$2 \text{ Enzyme}-\text{SH} \longrightarrow \text{enzyme}-\text{S}-\text{S}-\text{enzyme}$$

$$\underset{\text{Active}}{\phantom{2 \text{ Enzyme}-\text{SH}}} \qquad \underset{\text{Inactive}}{\phantom{\text{enzyme}-\text{S}-\text{S}-\text{enzyme}}}$$

Glutathione reduces the enzymes back to free (and active) sulfhydryl compounds.

$$\text{Enzyme}-\text{S}-\text{S}-\text{enzyme} + 2 \text{ glutathione}-\text{SH} \longrightarrow$$

$$\text{glutathione}-\text{S}-\text{S}-\text{glutathione} + 2 \text{ enzyme}-\text{SH}$$

The brain produces a variety of peptides, several of which act like morphine to relieve pain. Rather, we should say that morphine acts like these peptides, because the brain was making peptides long before people discovered the pain-killing effect of the juice of the opium poppy.

Morphine acts by fitting specific receptor sites in the brain. These receptors were discovered in 1973 by Solomon Snyder and Candace Pert at Johns Hopkins University. Why should the human brain have receptors for a plant alkaloid? There seemed to be no good reason, so scientists started a search for morphinelike substances produced in the body. Several soon were found. All were peptides. Two such peptides contained five amino acid units each; they differed only in the terminal amino acid. The one with leucine at the carboxyl end is called *leu*-enkephalin and that with methionine at that end is called *met*-enkephalin. Both are potent pain relievers, but their use in medicine so far is quite limited because they are rapidly broken down by enzymes that hydrolyze proteins. It is hoped, though, that analogs more resistant to hydrolysis can be employed as morphine substitutes for relief of pain. Unfortunately, it appears that both the natural enkephalins and the analogs, like morphine, are addictive.

Tyr-Gly-Gly-Phe-Leu

Leu-enkephalin

Tyr-Gly-Gly-Phe-Met

Met-enkephalin

It appears that enkephalins and related peptides are released as a response to pain deep in the body. Bruce Pomeranz of the University of Toronto has collected evidence that indicates that acupuncture anesthetizes by stimulating the release of the brain "opiates." The long needles stimulate deep sensory nerves that cause the release of the peptides that then block the pain signals.

Enkephalin release has also been used to explain other phenomena, once thought to be largely psychological. A soldier, wounded in battle, feels no pain until the skirmish is over. His body has secreted its own painkiller. An athlete can perform with an injury and not feel pain until the contest is over.

Many hormones (Special Topic K) are peptides or proteins. Falling within the weight limits we have set for peptides are insulin and glucagon, pancreatic

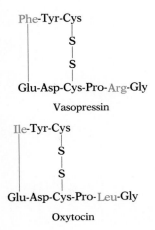

Phe-Tyr-Cys
S
S
Glu-Asp-Cys-Pro-Arg-Gly

Vasopressin

Ile-Tyr-Cys
S
S
Glu-Asp-Cys-Pro-Leu-Gly

Oxytocin

hormones that play important roles in the metabolism of carbohydrates (Chapter 27).

Two smaller peptide hormones, vasopressin and oxytocin, are isolated from extracts of the posterior lobe of the pituitary gland. Both molecules are cyclic structures that incorporate cysteine units with disulfide bonds. Note that the two differ in only two amino acid units. Vasopressin acts to increase the amount of water reabsorbed by the kidneys. It is sometimes called an anti-diuretic hormone (ADH). It also causes an increase in blood pressure and has been used medically to combat the low blood pressure characteristic of post-surgical shock. Oxytocin governs smooth-muscle contraction. It is involved in the uterine contractions that accompany childbirth and in the ejection of milk by lactating females. Table 22.4 offers a comparison of the effects of vasopressin and oxytocin. Remember the great similarity in the structure of these compounds as you look at the table.

Bradykinin, a nonapeptide produced in the blood by the cleavage of larger protein molecules, has this sequence of amino acids.

Arg-Pro-Pro-Gly-Phe-Ser-Pro-Phe-Arg

In terms of activity, bradykinin, a peptide, sounds almost like the prostaglandins, which are lipids (Chapter 21). It is a potent biochemical that causes a lowering of blood pressure, stimulation of smooth-muscle tissue, an increase in capillary permeability, and pain. The reverse peptide has been synthesized.

Arg-Phe-Pro-Ser-Phe-Gly-Pro-Pro-Arg

It shows none of the activity of bradykinin.

The octapeptide angiotensin II is produced in the kidneys.

Asp-Arg-Val-Tyr-Ile-His-Pro-Phe

TABLE 22.4   *A Comparison of the Effects of Oxytocin and Vasopressin*

Structure or Function Affected	Vasopressin	Oxytocin
Water diuresis	Inhibits	Has no effect on
Blood pressure	Elevates	Slightly lowers
Coronary arteries	Constricts	Slightly dilates
Intestinal contractions	Stimulates	Has questionable effect on
Uterine contractions*	Stimulates	Stimulates
Ejection of milk	Slightly stimulates	Stimulates

Reprinted with permission from White, A., Handler, P., and Smith, E. L., *Principles of Biochemistry*, 5th ed., New York: McGraw-Hill, 1973.

* Response varies with species, as well as with the stage of the normal and the reproductive cycle.

This substance is the most powerful vasoconstrictor known. It acts to maintain blood pressure. Some forms of hypertension probably involve overproduction of angiotensin II. Drugs that act to suppress its production are important in the control of hypertension.

As a final example we shall consider a hormone that regulates the coloring in animals, melanophore-stimulating hormone (MSH). Melanophores are cells that contain the pigments called melanins. MSH is involved in establishing protective coloration in animals. An injection of MSH will cause human skin to darken. There are two MSH groups, one designated alpha and one beta. The α-MSH from all species is identical. The β-MSH differs from one source to another (Figure 22.4). Such variation from species to species is the rule rather than the exception.

There are many other physiologically important peptides. Some of nature's most potent toxins, such as snake venom and bacterial toxins, are peptides.

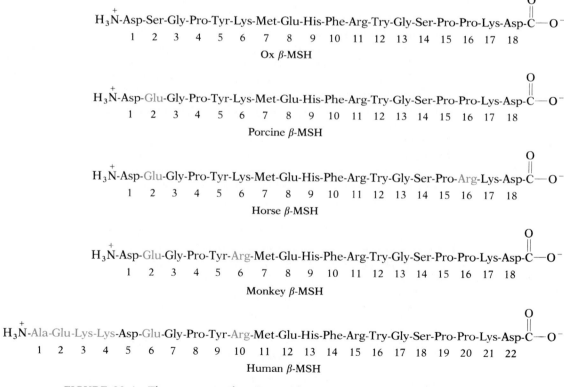

FIGURE 22.4 *The sequence of amino acids in proteins varies from one species to another. The variation illustrated here is for a hormone involved in establishing protective coloration in animals.*

Additional peptides are discussed in Chapter 24 and Special Topics J and K. Before leaving the topic, however, we do want to point out that the amino acid sequences of hundreds of peptides have been determined. Some of these peptides contain hundreds of amino acid units. Further, many of these have been synthesized in laboratories around the world. The first polypeptide hormone synthesized was oxytocin, in 1953. Synthetic peptides have the same physiological properties as the corresponding natural ones.

22.8   CLASSIFICATION OF PROTEINS

Proteins are big polypeptides, those with molecular weights of 10 000 or more. Some have molecular weights ranging into the millions. There are many kinds of proteins. Each has its own characteristic composition, amino acid sequence, and three-dimensional shape.

One way to classify proteins is based on solubility. Some proteins, such as those that make up hair, skin, muscles, and connective tissue, are fiberlike in nature. These **fibrous proteins** are insoluble in water. They usually serve structural, connective, and protective functions. Examples of fibrous proteins are keratins, collagens, myosins, and elastins. Hair and the outer layer of skin are composed of *keratin*. Connective tissues contain *collagen*. *Myosins* are muscle proteins and are involved in contraction and extension of muscles. *Elastins* are found in the elastic tissue of artery walls and in ligaments. We will discuss collagen, a typical fibrous protein, in Section 22.12. Myosin is discussed in Chapter 27.

Globular proteins, the other major class, are soluble in aqueous media. The protein chains of globular proteins are folded so that the molecule as a whole is roughly spherical in shape. The mixtures of globular proteins and water are actually colloidal dispersions rather than true solutions. Familiar examples are the *albumins*. Egg albumin is obtained from egg whites. Serum albumin is present in blood, and it plays a major role in maintaining a proper balance of osmotic pressures in the body (Chapter 26). *Globulins* are a second group of globular proteins. Hemoglobin and myoglobin are two examples. Another is the serum globulins, which are a part of our defense against disease. We will discuss one example, myoglobin, in some detail in Section 22.13. Its characteristics are rather typical of globular proteins.

22.9   STRUCTURE OF PROTEINS

The structure of proteins is generally discussed at four organizational levels. The **primary structure** of a protein molecule is its amino acid sequence. To specify the primary structure, one merely writes out the sequence of amino

acids in the long-chain molecule. What "holds" the primary structure together
are the peptide links between the amino acid units.

The sequence of amino acids in small peptides is determined by use of a
reagent, phenylisothiocyanate, which reacts with the amino-terminal end of
the peptide. (Larger proteins are sequenced by cleaving them into smaller units,
then sequencing the units.) Reaction of phenylisothiocyanate with the amino
group forms a phenylthiocarbamoyl (PTC) derivative (Figure 22.5). The PTC
derivative can be hydrolyzed under such mild conditions that the peptide bonds
are not broken. The end amino acid winds up in a cyclic compound called a
phenylthiohydantoin (PTH) derivative. The PTH derivative can be identified,
and the remaining peptide can be subjected to another round of reactions.
Peptides of up to about 40 amino acids can be sequenced in this manner.

Molecules of proteins aren't just arranged at random as tangled threads.
The chains are held together in unique configurations. The term **secondary
structure** refers to the arrangement of chains about an axis. This arrangement
may be a pleated sheet, as in silk (Section 22.10), a helix, as in wool (Section
22.11), or whatever. The term **tertiary structure** refers to the spatial relationships of amino acid units that are relatively far apart in the protein chain. In
describing tertiary structure, we frequently will talk about how the molecule is

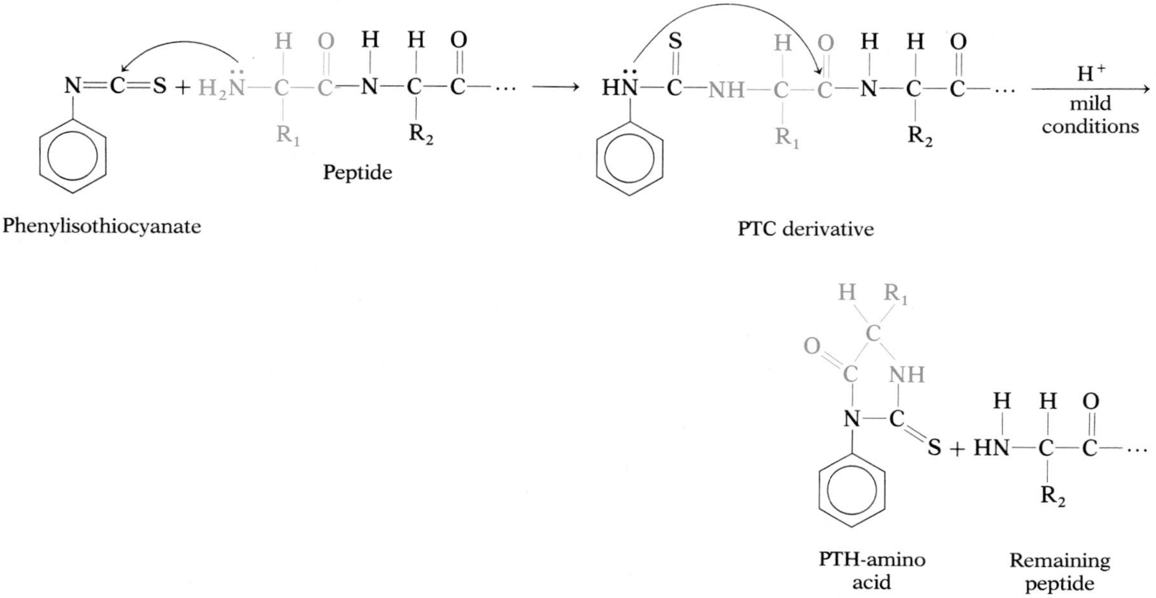

*FIGURE 22.5   The reactions involved in determining the primary structure of a
protein.*

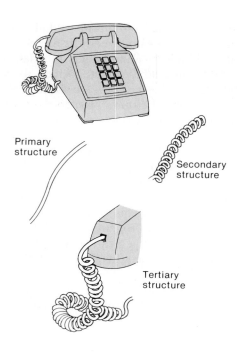

FIGURE 22.6
*Three levels of structure of a
telephone cord.*

FIGURE 22.6
*Three levels of structure of a
telephone cord.*

Primary
structure

Secondary
structure

Tertiary
structure

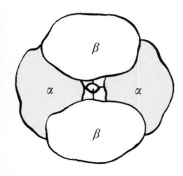

FIGURE 22.7
*The quaternary structure of
hemoglobin has the four
coiled chains stacked in a
nearly tetrahedral
arrangement.*

folded. An example is the protein chain in globular proteins (Section 22.8),
which is folded into a compact, spherical shape.

We can relate these three levels of organization to a more familiar object.
Think of the coiled cord on a telephone receiver. The cord starts out as a long,
straight wire (Figure 22.6). We'll call that the primary structure. The wire is
coiled into a helical arrangement. That's its secondary structure. When the
receiver is hung up, the coiled cord folds in a particular pattern. That would
be its tertiary structure.

Some proteins contain more than one polypeptide chain; that is, a protein
molecule can be an aggregate of subunits. Hemoglobin is the most familiar
example. A single hemoglobin molecule contains four polypeptide units. *Each
unit* is roughly comparable to a myoglobin molecule. The four units are arranged in a specific pattern (Figure 22.7). When we describe the **quaternary
structure** of hemoglobin, we describe in detail the way in which the four units
are packed together in the hemoglobin molecule. We consider hemoglobin in
much greater detail in Chapter 26. The next sections of this chapter will help
you gain some insight into secondary and tertiary organization. With that
background, a second look at the picture of the hemoglobin structure will be
more worthwhile. (To continue our analogy of a telephone cord, we would
have to consider several such cords, stacked in a particular way. That might

stretch our analogy more than we could stretch a cord.) A schematic representation of the four levels of protein structure is shown in Figure 22.8.

Peptide bonds fix the primary structure of proteins. What forces hold proteins in the structural arrangements we have referred to as secondary, tertiary, and quaternary? There are four kinds of forces—hydrogen bonds, ionic forces called salt bridges, covalent disulfide linkages, and dispersion forces. The last

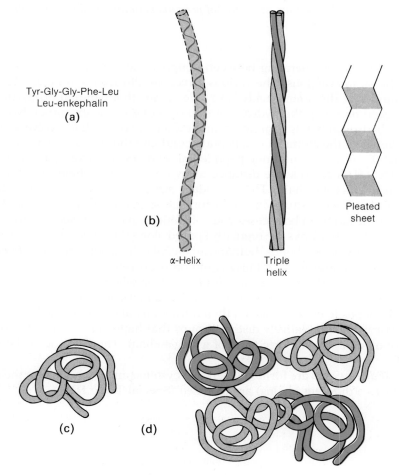

Tyr-Gly-Gly-Phe-Leu
Leu-enkephalin
(a)

(b)

α-Helix

Triple
helix

Pleated
sheet

(c)   (d)

FIGURE 22.8 *A schematic representation of the four levels of structure in proteins: (a) primary, (b) secondary, (c) tertiary, and (d) quaternary.*

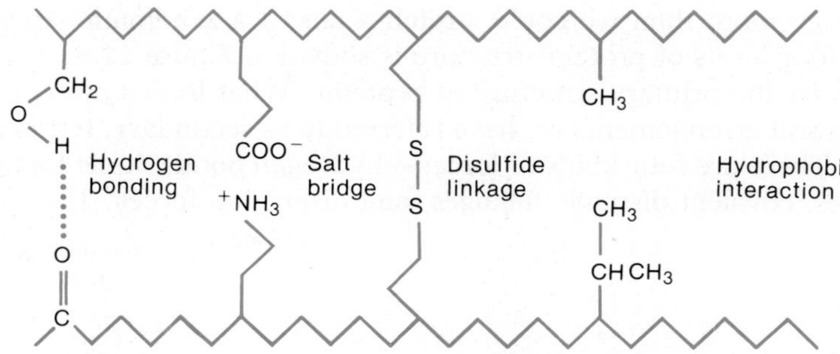

FIGURE 22.9   The tertiary structure of proteins is maintained by four different types of interactions.

are the only forces operating between nonpolar side chains; these are called hydrophobic interactions. The various forces are illustrated in Figure 22.9.

A number of the amino acids have side chains that can and do participate in *hydrogen bonding* (the hydroxyl group of serine is one example). Nonetheless, the hydrogen bonds of greatest importance are those that involve an interaction between the atoms of one peptide bond and those of another. Thus, the carbonyl (C=O) oxygen of one peptide link may form a hydrogen bond to an amide (N—H) located some distance away on the same chain or located on an entirely different chain. The secondary structure of wool (Section 22.11) illustrates the former situation; the secondary structure of silk (Section 22.10) illustrates the latter. In both cases you will notice a pattern of such interactions. Because the peptide links are regularly spaced along the chain, there is a regular repetition of hydrogen bonds, both intramolecular (in wool) and intermolecular (in silk). Such patterns are a fairly common occurrence.

As we saw in Section 22.2, *salt bridges* occur when an amino acid with a basic side chain appears opposite one with an acidic side chain. Proton transfer results in opposite charges, which then attract one another. These interactions can occur between relatively distant groups that happen to come into contact because of some folding or coiling of a single chain. They also occur between chains.

Disulfide linkages are formed when two cysteine units (whether on the same chain or two different chains) are oxidized (Special Topic H).

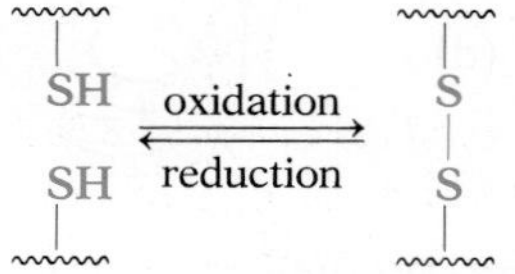

$$\begin{array}{ccc}
\text{SH} & & \text{S} \\
| & \xrightarrow{\text{oxidation}} & | \\
& \xleftarrow{\text{reduction}} & \\
\text{SH} & & \text{S} \\
| & & | \\
\end{array}$$

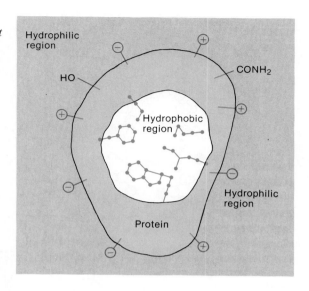

FIGURE 22.10
A folded protein chain might have a hydrophobic region on the inside and a hydrophilic region on the outside.

The disulfide bond is a covalent bond. The strength of this bond is much greater than that of a hydrogen bond. The much larger number of hydrogen bonds does compensate somewhat for their individual weakness.

Still weaker are the **hydrophobic interactions** between nonpolar side chains. These can be important, however, when other types of interactions are missing or are minimized. The hydrophobic interactions are made stronger by the cohesiveness of the water molecules surrounding the protein. Nonpolar side chains minimize their exposure to water by clustering together on the inside folds of the protein in close contact with one another (Figure 22.10). Hydrophobic interactions become fairly significant in structures such as that of silk, in which a high proportion of amino acids in the protein have nonpolar side chains.

22.10 SILK

Silk is a fine, soft, shiny fiber produced by the silkworm as it spins its cocoon. It was once widely used in clothing for the well-to-do. The expressions "silk hat" and "silk-stocking district" are relics of another age, for silk has been largely replaced by synthetics such as nylon (Chapter 19).

Silk consists of **fibroin**, a fibrous protein. The polypeptide chains exist in an extended zigzag arrangement (Figure 22.11). Unlike most proteins, silk is

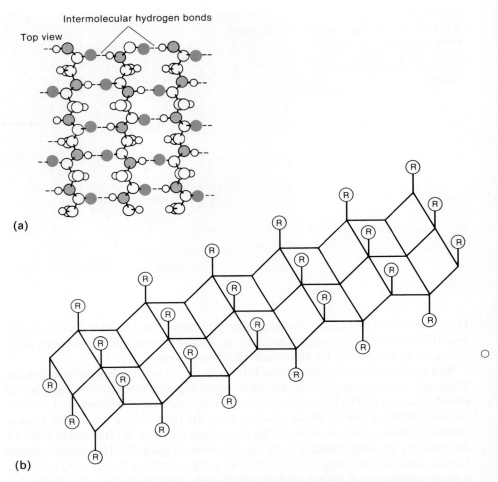

FIGURE 22.11   *Pleated sheet conformation of polypeptide chains. (a) Ball-and-stick model. (b) Schematic drawing emphasizing the pleats. (Redrawn from Gutsche, C. David, and Pasto, Daniel J.,* Fundamentals of Organic Chemistry, *© 1975. Reprinted by permission of Prentice-Hall, Inc., Englewood Cliffs, N.J.)*

composed primarily of just three amino acids, glycine (45%), alanine (30%) and serine (12%), with the remaining 13% being made up of various other amino acids. Indeed, partial degradation shows that much of each chain consists of a repetition of six units.

· · · Gly-Ser-Gly-Ala-Gly-Ala · · ·

In this arrangement, every other amino acid unit is glycine. Most of the remaining side chains, such as the CH_3 in alanine and the CH_2OH in serine, are small. The molecules are stacked in extended arrays, with hydrogen bonds holding adjacent chains together. The appearance of this arrangement (Figure 22.11) has led to its designation as the **pleated sheet conformation.** Silk fibers have these hydrogen-bonded layers arranged one over the other.

The properties of silk—strength, flexibility, and resistance to stretching—are a consequence of its structure. To break the fibers would involve rupturing thousands of hydrogen bonds or breaking covalent bonds. The interaction of small side chains allows for great flexibility, and since the chains are already fully extended, the fibers cannot be stretched easily.

There is an old saying that "you can't make a silk purse out of a sow's ear." Like many other clichés, this one has its limits. A chemist has proven that you can indeed make silklike fibers from the aural appendages of female hogs. More important commercially, though, is the ability of chemists to make substitutes for silk—such as nylon—out of petroleum, coal, and air.

22.11 WOOL

Wool is another natural protein material, but its composition and properties are quite different from those of silk. Wool has a greater variety of amino acids than silk, including many with large side chains. The structure of proteins such as wool, hair, and muscle was first determined by Linus Pauling and co-workers in 1950. Pauling showed that the amino acid units were arranged in a right-handed helix, or **alpha helix** (Figure 22.12). Each turn of the helix requires 3.6 amino acid units. The NH groups in one turn form hydrogen bonds to the carbonyl groups in the next (the dotted lines in the center model of Figure 22.12).

Unlike silk, wool can be stretched. (Think of how the coiled wire on the telephone can be stretched.) Such stretching results in an elongation of the hydrogen bonds joining the turns. When the stretching is discontinued, the helix will return to its original configuration, unless, of course, you have radically disrupted the structure by treating the wool with very hot water (Section 22.14).

Hair, horn, nails, and muscles also have the alpha-helical structure. This helix extends along the axis of the fiber. Adjacent helices are not precisely parallel but are wound about one another like the strands of a rope. These ropes, called protofibrils, may have three or seven alpha helices as strands. Horns, nails, and claws have less flexibility than wool and hair, due to more extensive cross-linking by disulfide bridges.

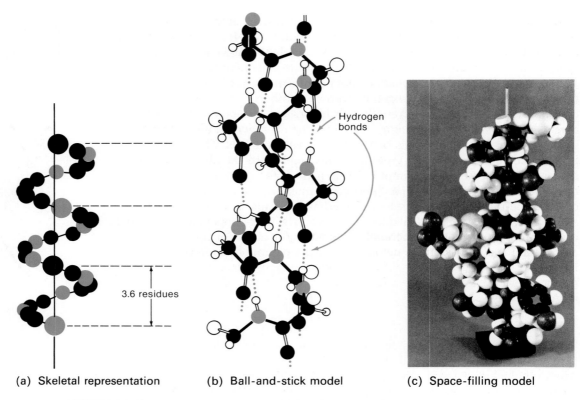

(a) Skeletal representation   (b) Ball-and-stick model   (c) Space-filling model

FIGURE 22.12   Three representations of the alpha-helical conformation of a polypeptide chain. (a) The skeletal representation best shows the helix. (b) Hydrogen bonding between turns of the helix is shown in the ball-and-stick model. (c) The space-filling model shows the actual shape of a short segment of the chain. (Skeletal representation and ball-and-stick model redrawn from Gutsche, C. David, and Pasto, Daniel J., Fundamentals of Organic Chemistry, *© 1975. Reprinted by permission of Prentice-Hall, Inc., Englewood Cliffs, N.J. Space-filling model courtesy of Science Related Materials Inc., Janesville, Wis.)*

22.12 COLLAGEN: THE PROTEIN OF CONNECTIVE TISSUES

The principal protein of connective tissues is **collagen.** As much as 60% of all mammalian protein is of this type. Most of the organic portion of skin, bones, tendons, and teeth is collagen. It also occurs in most other parts of the body as fibrous inclusions. In all, collagen makes up about 25% of all the protein in the human body.

Like other fibrous proteins, collagen is not readily digestible. Treatment with boiling water converts collagen to **gelatin.** Gelatin is not only water soluble but digestible. The cooking of meat converts part of the tough connective tissue to gelatin, making the meat more tender.

In composition, collagen is about 33% glycine, with another 20% to 25% consisting of the heterocyclic compounds proline and hydroxyproline. Collagen also contains acidic and basic amino acids. It does not contain enough of the essential amino acids; the gelatin derived from it is a poor-quality protein.

Structurally, collagen consists of three protein chains, each wound about its own axis in a left-handed helix (Figure 22.13). Unlike wool, collagen does not have hydrogen bonding between coils. Indeed, its structure is much more open than that of the tightly coiled alpha helix. Hydrogen bonds between chains do hold the three chains of collagen together. Collagen chains are also somewhat cross-linked by covalent bonds. As an animal grows older, the extent of cross-linking increases and the meat gets tougher.

Collagen is of considerable commercial importance. The process of tanning increases the degree of cross-linking, converting skin to leather. The soluble gelatin derived from collagen is used in food, film emulsions, and glue and in many other ways.

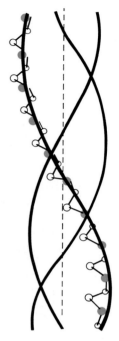

FIGURE 22.13
The collagen molecule is a triple helix.

22.13 MYOGLOBIN

Myoglobin is a typical globular protein. Myoglobin molecules consist of a single chain of 153 amino acid residues. This protein chain is connected to a *heme* unit. The structure of heme, also found in hemoglobin, is

It is a relatively flat molecule with an iron atom at its center. The heme unit is called a **prosthetic** group, and it is capable of combining with oxygen under

FIGURE 22.14
A skeletal representation of the tertiary structure of myoglobin. (Reprinted with permission from Neurath, H., ed., The Proteins, *vol. 2, New York: Academic Press, 1964. Copyright © 1964 by Academic Press, Inc.)*

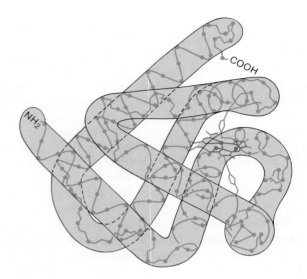

proper conditions. Whereas hemoglobin is involved in oxygen transport in the blood, myoglobin serves to store oxygen in muscle tissue.

The polypeptide chain has eight segments with the alpha-helical structure (Figure 22.14). These are folded back upon one another in such a way as to give the molecule an overall shape like that of a ball or globe. The shape of the final molecule includes a hole that nicely accommodates the heme unit. The tertiary structure also brings two amino acid side chains into position to anchor the heme unit to the protein portion of the myoglobin molecule.

There are many other proteins of vital importance. We will encounter several more, including the all-important enzymes, in subsequent chapters.

22.14 DENATURATION OF PROTEINS

In many ways, proteins are remarkable compounds. Their highly organized structures are truly masterworks of chemical architecture. But highly organized structures tend to have a certain delicacy, and this is true of many proteins. We shall define **denaturation** as the process in which a protein is rendered incapable of performing its assigned function. If the protein can't do its job, we say it has been **denatured.** (Sometimes denaturation is equated with the precipitation or coagulation of a protein. Our definition is a bit broader.) The process is sometimes reversible, but usually it is not. You have certainly observed the denaturation of egg albumin. The clear egg "white" turns to an opaque white "white" when the egg is boiled or fried. What you are observing is the denaturation and coagulation of the albumin. No one has yet reversed that process!

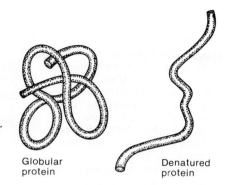

FIGURE 22.15
Denaturation of a protein.
The globular protein is
folded into the tertiary
conformation necessary for
its functioning. The
denatured protein can
assume various random
conformations. It is not
active but may, under proper
conditions, refold to the
active conformation.

Globular
protein

Denatured
protein

The primary structure of proteins is quite sturdy. In general, it takes fairly vigorous conditions for peptide bonds to be hydrolyzed (though chemists have devoted much effort to developing gentler methods and enzymes manage to hydrolyze proteins with remarkable ease). At the secondary and tertiary levels, however, proteins are quite vulnerable to attack (Figure 22.15). Disulfide bonds are a bit resistant, but the critical hydrogen bonds can be disrupted in many ways.

How can you denature a protein? Obviously heat is one way. It works with eggs and also with bacteria, which, like you, need active protein to live. Autoclaving of surgical instruments sterilizes them through the action of heat on bacterial protein.

Other forms of energy can be used; ultraviolet light, for example, will denature protein and is occasionally used for sterilization.

Chemicals can be used. Ethyl alcohol will badly disrupt the hydrogen bonds of proteins. It sterilizes the skin (by killing bacteria present) when injections are to be given. A 70% aqueous solution is used rather than the more commonly available 95% because the higher concentration is so effective at coagulating protein that it forms a surface "crust" that prevents the interior of the bacteria from being attacked (Figure 22.16). There is also a group of acidic compounds referred to as alkaloidal reagents (they react with alkaloids) that will react with proteins. The reagents disrupt salt bridges and hydrogen bonds and cause precipitation of the protein. One, tannic acid, is sometimes applied to severe burns. The precipitated protein forms a crust over the wounds and thus slows the loss of body fluids.

Heavy metals (lead and mercury, for example) denature proteins and bring about their coagulation, probably by interacting with sulfhydryl groups or ionizable side chains. That can be bad when the protein is an essential enzyme (Chapter 24). Mercury poisoning and lead poisoning are debilitating diseases that are potentially fatal. There is a positive side, however. An antidote can take advantage of the same phenomenon. A person who had ingested mercury could be fed raw eggs immediately. The mercury would then react with egg protein

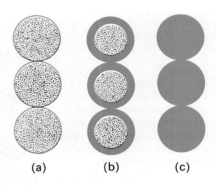

FIGURE 22.16
*Effect of alcohol on bacteria.
Dark areas represent
coagulated protein. (a)
Bacteria before application
of alcohol. (b) After
application of 100% alcohol.
(c) After application of 70%
alcohol, which is more
effective than 100% alcohol.
(Used with permission from
Arnow, L. Earle,*
Introduction to Physiological
and Pathological Chemistry,
*9th ed. St. Louis: C. V.
Mosby Co., 1976.)*

(a)   (b)   (c)

in the stomach rather than with other, more essential proteins. This treatment would require that the stomach contents be pumped out or vomited to prevent the ultimate digestion of the egg protein and the consequent release in the body of mercury ions. Quite clearly, the technique works only for acute poisonings and not for the far more common chronic mercury poisoning (Chapter 24).

There are many other ways of denaturing proteins that we have not discussed (introduction of radical changes in pH, for example). The point should be clear, however. The very complexity that makes proteins so versatile also makes them vulnerable. There is a considerable range of vulnerability. The delicately folded globular proteins are much more readily denatured than are the tough, fibrous proteins of hair and skin.

On the other hand, there is increasing evidence that an unfolded protein, given the proper conditions and enough time, will refold and may again exhibit biological activity. Such evidence suggests that, for these molecules, the primary structure determines the secondary and tertiary structures. A given sequence of amino acids seems naturally to adopt its particular three-dimensional arrangement if conditions are right.

We have emphasized structure in this chapter. In other chapters we concentrate on the function of several kinds of proteins, particularly the enzymes.

PROBLEMS

1. What is the general structure for an alpha amino acid?

2. Draw the side chains of the following amino acids:
   a. alanine
   b. phenylalanine
   c. serine
   d. aspartic acid
   e. cysteine
   f. lysine

3. Write structural formulas for the following:
   a. glycine
   b. alanine
   c. phenylalanine

4. Write structural formulas for the following:
   a. glycylalanine
   b. alanylglycine

5. Write the structural formula for phenylalanylglycyl-alanine.

6. Write a structural formula for an amino acid with an acidic side chain and give the name of the compound.

7. Write a structural formula for an amino acid with a basic side chain and give the name for the compound.

8. Write structural formulas for the following:
   a. cysteine   **b.** methionine

9. Write a structural formula for the anion formed when glycine reacts with a base.

10. Write a structural formula for the cation formed when glycine reacts with an acid.

11. To which family of mirror image isomers do almost all naturally occurring amino acids belong?

12. What is the difference between a polypeptide and a protein?

13. Define, describe, or illustrate each of the following:
   a. peptide bond
   b. sequence of amino acids
   c. essential amino acid
   d. disulfide linkage
   e. salt bridge
   f. hydrophobic interaction
   g. primary structure
   h. secondary structure
   i. tertiary structure
   j. quaternary structure
   k. zwitterion
   l. isoelectric point
   m. an adequate protein
   n. globular protein
   o. fibrous protein
   p. electrophoresis
   q. prosthetic group

14. Amino acid units in a protein are connected by peptide bonds. What is another name for the functional group linking the units of proteins?

15. Translate the following abbreviated form into the structural formula for the peptide: Ser-Ala-Gly.

16. Write the abbreviated version of the following structural formula:

17. In general, what are the potential problems associated with a strict vegetarian diet?

18. Describe the structure of silk and explain how its properties reflect its structure. What name is given to the secondary structure of silk?

19. Describe the structure of wool and relate this to wool's elasticity. What name is given to the secondary structure of wool protein?

20. Describe the structure of collagen.

21. Name the four kinds of interactions that maintain the tertiary structure of protein.

22. The following sets of amino acids are involved in maintaining the tertiary structure of a peptide. In each case, identify the type of interaction (see Problem 21) involved.
   a. aspartic acid and lysine
   b. phenylalanine and alanine
   c. serine and lysine
   d. 2 cysteines

23. Classify these proteins as fibrous or globular:
a. albumin	**d.** hemoglobin
b. myosins	**e.** keratins
c. collagen	**f.** myoglobin

24. Which class of proteins shows greater solubility in aqueous solution, fibrous or globular?

25. Which class of proteins is more easily denatured, fibrous or globular?

26. Describe some ways of denaturing a protein.

27. What level(s) of structure is(are) ordinarily disrupted in denaturation?

28. Is denaturation of protein usually reversible?

29. What triggered the search that lead to the discovery of the endorphins and enkephalins?

30. Why are the endorphins and enkephalins compared to morphine?

31. What is the difference between *leu*-enkephalin and *met*-enkephalin?

32. Using the formula given in Section 22.7, write the shorthand structure of the product that would result from the reduction of the disulfide bond in oxytocin. Do you think the product would show the same biological activity as oxytocin?

$$\underset{H_3\overset{+}{N}}{}-\underset{\underset{CH_3}{|}}{CH}-\underset{\overset{O}{\|}}{C}-NH-\underset{\underset{\underset{OH}{|}}{CH_2}}{CH}-\underset{\overset{O}{\|}}{C}-NH-\underset{\underset{\underset{SH}{|}}{CH_2}}{CH}-\underset{\overset{O}{\|}}{C}-NH-\underset{\underset{\underset{C_6H_5}{|}}{CH_2}}{CH}-\underset{\overset{O}{\|}}{C}-NH-\underset{\underset{H}{|}}{CH}-\underset{\overset{O}{\|}}{C}-NH-\underset{\underset{H}{|}}{CH}-\underset{\overset{O}{\|}}{C}-O^-$$

33. What do we mean when we say that hemoglobin shows species variation?
34. Bacteria synthesize D-alanine and use it in the biosynthesis of cell walls. Draw the structure for D-alanine. A deutrated derivative (one in which ordinary hydrogen, ^1_1H, has been replaced by deuterium, ^2_1H or D), 2-deutero-3-fluoro-D-alanine, acts as an antibiotic by inhibiting cell wall synthesis. Draw the structure for the derivative.
35. How many amino acids are specified by the genetic code for the synthesis of proteins?
36. Which amino acid is not chiral?
37. Which amino acid specified by the genetic code does *not* contain a primary amine group? Give its structure.
38. The genetic code does not specify hydroxyproline, yet this amino acid is common in collagen. Explain.
39. Two cysteine units joined through a disulfide linkage are sometimes considered a different amino acid called *cystine*. Draw the structure for cystine.
40. Explain how the physical properties of amino acids reflect their zwitterion structure.
41. Describe how the primary structure of a peptide is determined.
42. How are hydrophobic interactions reinforced by the water molecules surrounding the proteins?
43. For each of the following amino acids, state whether it is more likely to be on the inside or the outside of a globular protein:
    a. phenylalanine   d. lysine
    b. aspartic acid   e. leucine
    c. serine   f. glutamic acid

44. Which amino acid is more likely to be in a pleated sheet protein, alanine or phenylalanine?
45. What is unusual about the structure of the tripeptide glutathione?
46. Draw the structure for the tripeptide Ala-Ser-Val. How many chiral carbon atoms are there in the molecule? How many stereoisomers are possible? How many of these stereoisomers occur in the naturally occurring tripeptide?
47. Draw the structure for γ-aminobutyric acid (GABA). Is GABA found in proteins? What is its role in the body?
48. The isoelectric point of silk fibroin is 2.2. Which of the following amino acids is likely to be present in large amounts?
    a. aspartic acid
    b. histidine
    c. lysine
49. Write equations to show how alanine can act as a buffer.
50. Suggest a cure for kwashiorkor (other than including animal protein in the diet).
51. What happens to milk proteins when they reach the stomach and contact the hydrochloric acid there?
52. Silver nitrate is put into the eyes of newborn babies as a disinfectant against gonorrhea. What is the effect of Ag^+ on bacterial protein?
53. How do oxytocin and vasopressin differ in structure? Which has the lower isoelectric point?
54. Carbohydrates are incorporated into *glycoproteins*. How does the incorporation of sugar units affect the solubility of a protein?

CHAPTER 2

Nucleic Acids and Protein Synthesis

In contrast to what many people believe, the complexity of the sciences increases as one proceeds from physics to chemistry to biology. Because the language of physics is mathematics, most people regard physics as the most difficult of the sciences. Yet, physical phenomena can be described with mathematical precision because the relationships involved are comparatively simple. We can write an equation that accurately describes the behavior of gases or of subatomic particles. A functioning cell defies such analysis.

Nonetheless, the cell is slowly yielding its secrets. One of these secrets, perhaps the most important one, is the method by which the cell stores and transmits information on how to reproduce itself. Nucleic acids are the molecules that store the patterns of life. It is through nucleic acids that these patterns are passed from one generation to the next. Nucleic acids also control the synthesis of proteins, including the enzymes that mediate those biochemical reactions that make an organism what it is.

As we've noted so often, with understanding comes control. The biochemists and molecular biologists who are unraveling these mechanisms are also learning how to manipulate the structure of living matter. The repair of defective genes, the design of precise molecular medicines, and control—for better or worse—of our heredity may lie in the future. While the twentieth century may be remembered as the nuclear age, the twenty-first century could well become the age of molecular biology.

23.1 THE BUILDING BLOCKS: SUGARS, PHOSPHATES, AND BASES

Nucleoproteins are found in every living cell. They are exceedingly complex, as might be expected from the role they play: they are the information and control centers of the cell. More about that later in the chapter. Let's look first at what nucleoproteins are made of. One way to find out is to take them apart.

Working carefully, chemists can separate nucleoproteins into a nucleic acid portion and a protein portion.

$$\text{Nucleoprotein} \longrightarrow \text{nucleic acid} + \text{protein}$$

The protein is highly basic. Hydrolysis reveals that it contains many units of the amino acids lysine and arginine.

$$\text{Protein} \xrightarrow{\text{H}_2\text{O}} \text{lysine} + \text{arginine} + \text{other amino acids}$$

The nucleic acids can also be hydrolyzed. Controlled hydrolysis gives units called **nucleotides.** These can be further hydrolyzed to phosphoric acid and compounds called **nucleosides.** Nucleosides can be hydrolyzed to the ultimate molecular constituents that are purine and pyrimidine bases (Chapter 18) and a pentose sugar (Chapter 20).

$$\text{Nucleic acids} \xrightarrow{\text{H}_2\text{O}} \text{nucleotides} \xrightarrow{\text{H}_2\text{O}} \begin{cases} \text{nucleosides} \\ + \\ \text{H}_3\text{PO}_4 \end{cases} \xrightarrow{\text{H}_2\text{O}} \begin{cases} \text{two purine bases} \\ + \\ \text{two pyrimidine bases} \\ + \\ \text{a pentose sugar} \end{cases}$$

There are actually two kinds of nucleic acids. Each is a gigantic polymer with nucleotides as the repeating units. Deoxyribonucleic acid (DNA) occurs in the cell nucleus. Ribonucleic acid (RNA) is found in all parts of the cell. The two nucleic acids differ only slightly in composition. Complete hydrolysis of DNA ultimately gives two purine bases, adenine and guanine, and two pyrimidine bases, cytosine and thymine (Figure 23.1).

Also obtained from DNA are phosphoric acid and the simple sugar 2-deoxyribose. Ribonucleic acid (RNA) differs only in that one of the pyrimidine bases, thymine, is replaced by another, uracil, and the sugar, deoxyribose, is replaced by ribose. These differences in composition are summarized in Table 23.1.

(We present a somewhat oversimplified analysis here. Actually, the DNA bases from many sources are methylated, that is, the molecules have a hydrogen atom replaced by a methyl group. Substitutions in RNA are even more varied. Methylation of uracil can produce thymine, so some RNAs do contain thymine. We will stick with the simpler view in order to emphasize the main features of nucleic acid structures.)

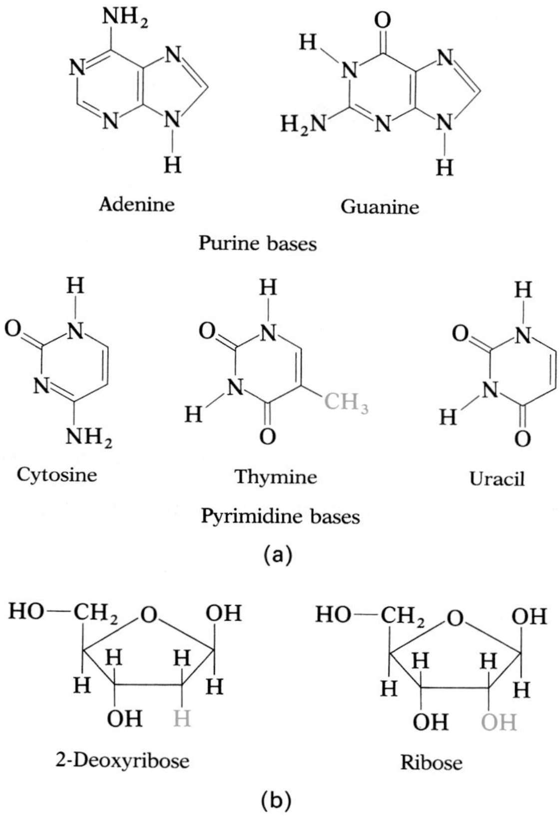

FIGURE 23.1   *The principal organic bases (a) and the two pentose sugars (b) obtained upon complete hydrolysis of nucleic acids.*

TABLE 23.1   *Ultimate Hydrolysis Products of DNA and RNA*

	DNA	RNA
Purine bases	Adenine Guanine	Adenine Guanine
Pyrimidine bases	Cytosine Thymine	Cytosine Uracil
Pentose sugar	2-Deoxyribose	Ribose
Inorganic acid	Phosphoric acid	Phosphoric acid

23.2 NUCLEOSIDES: A SUGAR AND A BASE

When a purine or pyrimidine base is combined with one of the pentose sugars, a compound called a **nucleoside** is formed. If the sugar is ribose, the compound is a **ribonucleoside.** If 2-deoxyribose is the sugar involved, the product is a **deoxyribonucleoside.**

Several representative nucleosides are shown in Figure 23.2. The bases in these molecules are attached at the first carbon of the pentose unit, replacing a hydroxyl group that ordinarily appears there. Note that *adenosine* is derived from adenine and ribose. Uridine is similarly derived from uracil and ribose. Deoxythymidine is made up of the base thymine and the pentose

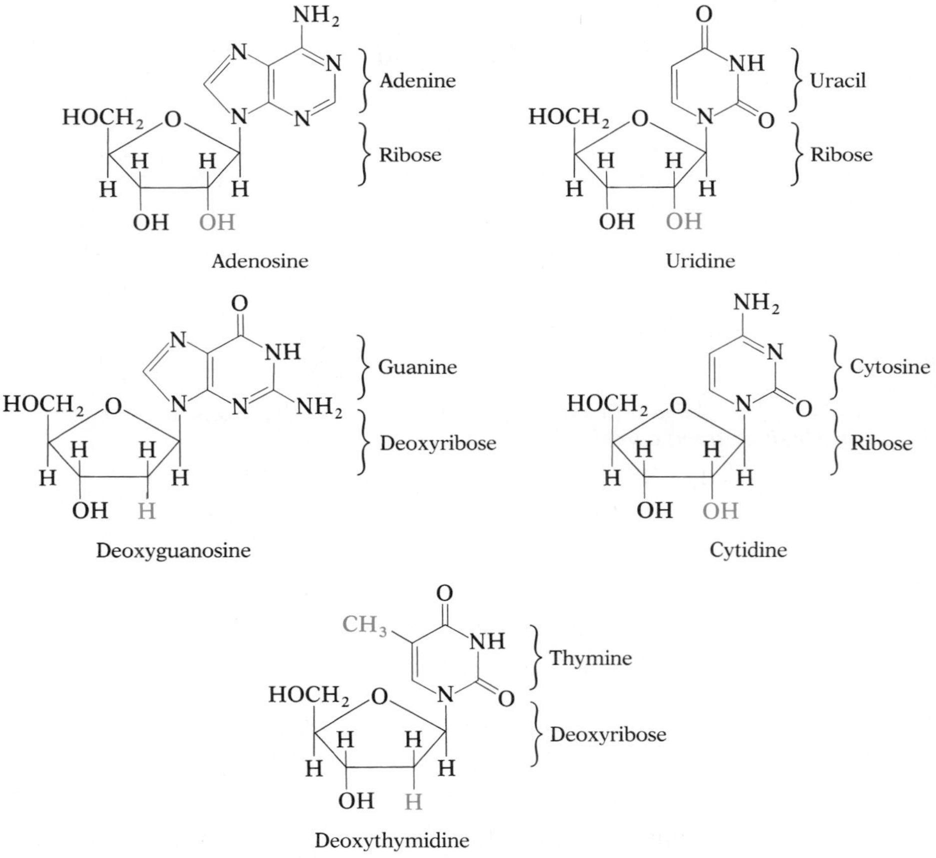

FIGURE 23.2   *Structures of some nucleosides.*

2-deoxyribose. Since there are 5 bases and 2 sugars, 10 nucleosides are possible. Half of these would be deoxynucleosides and half would be ribonucleosides. Only 8 of the 10 usually are obtained by hydrolysis of natural nucleic acids. The thymine-ribose combination and the uracil-deoxyribose combination are rare.

Adenosine may serve as a chemical regulator throughout the body. Receptor sites for it have been identified. It appears that adenosine may regulate the function of neurons in the brain, dilate blood vessels in the heart, constrict bronchial tubes, and inhibit the aggregation of platelets. Caffeine may act as a stimulant by blocking adenosine receptors.

Several nucleoside derivatives have been used in medicine. One, puromycin (Figure 23.3a), is derived from adenosine. It is an antibiotic. First obtained from cultures of the fungus *Streptomyces alboniger*, puromycin is effective against protozoa and has shown some antitumor activity.

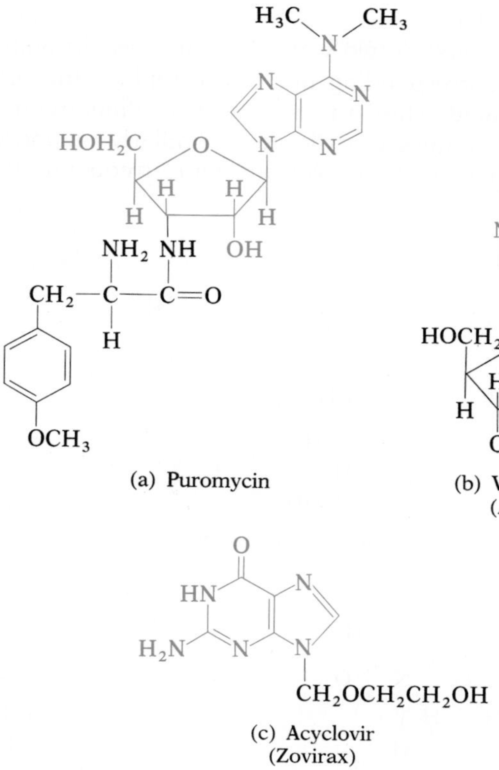

(a) Puromycin

(b) Vidarabine (Ara-A)

(c) Acyclovir (Zovirax)

FIGURE 23.3 *Puromycin (a) and vidarabine (b), two nucleoside derivatives used in medicine. The antiviral drug acyclovir (c) is probably converted to a nucleoside derivative in the body.*

Vidarabine (Figure 23.3b) is an antiviral drug. The base in vidarabine is adenine, but the sugar, arabinose, is an isomer of ribose. Another antiviral drug, acyclovir (Figure 23.3c) has shown promise against the herpes viruses that cause genital sores, chicken pox, shingles, mononucleosis, and cold sores. Acyclovir, which contains guanine, is probably converted to a nucleoside derivative in the body.

23.3 NUCLEOTIDES: A SUGAR, A BASE, AND A PHOSPHATE

Nucleotides are phosphate esters of nucleosides. A hydroxyl group on the pentose molecule serves as the alcohol in the esterification process. The primary alcohol group ($-CH_2OH$) is generally the one involved. (Compare the nucleoside adenosine, in Figure 23.2, with the nucleotide adenosine monophosphate, in Figure 23.4.)

The nucleotides are named in two ways. One involves dropping the ending from the name of the corresponding nucleoside (either **-ine** or **-osine**) and adding the ending **-ylic acid.** Thus, the nucleoside uridine, upon esterification with phosphoric acid, becomes uridylic acid. Similarly, guanosine becomes guanylic acid. The other system, simpler and more frequently used, involves

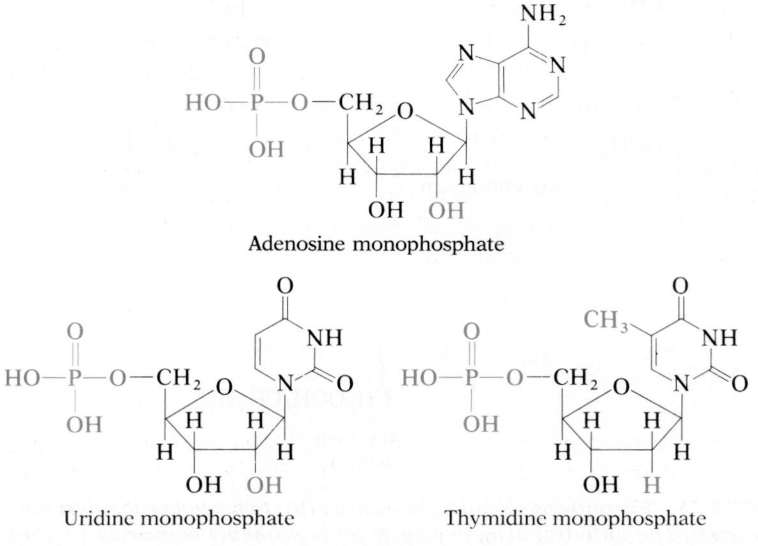

Adenosine monophosphate

Uridine monophosphate

Thymidine monophosphate

FIGURE 23.4   *Structures of three important nucleotides.*

TABLE 23.2   *Naming the Nucleotides*

| Base | Nucleoside | Nucleotide | | Abbreviation |
		As an Acid	As a Monophosphate	
Adenine	Adenosine	Adenylic acid	Adenosine monophosphate	AMP
Guanine	Guanosine	Guanylic acid	Guanosine monophosphate	GMP
Cytosine	Cytidine	Cytidylic acid	Cytidine monophosphate	CMP
Thymine	Thymidine	Thymidylic acid	Thymidine monophosphate	TMP
Uracil	Uridine	Uridylic acid	Uridine monophosphate	UMP

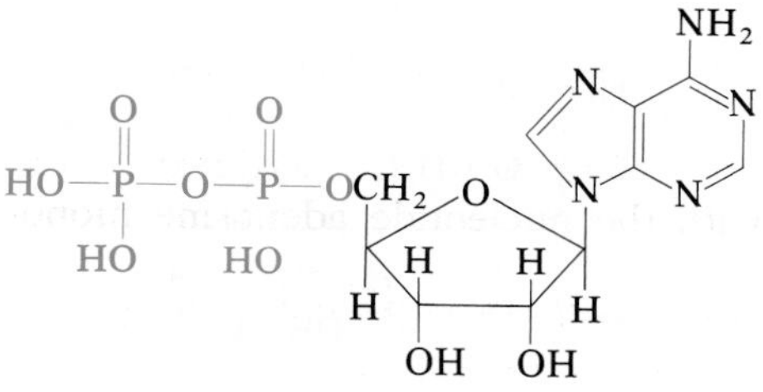

FIGURE 23.5   *Structures of two important nucleotide derivatives.*

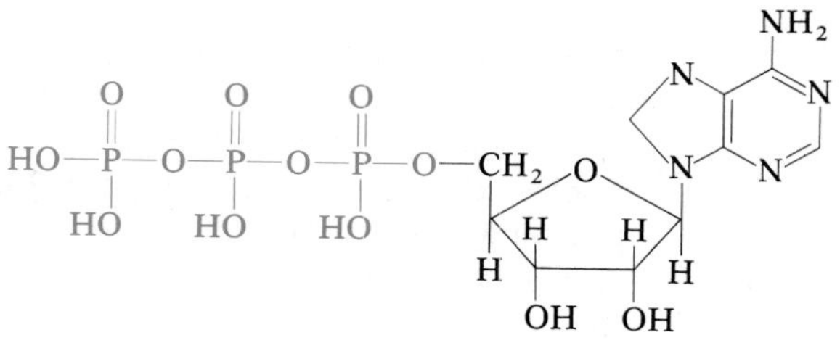

using the nucleoside name as is and adding the word **monophosphate.** Thus, adenosine, upon esterification, becomes adenosine monophosphate, often abbreviated AMP. The prefix **deoxy-** indicates that the sugar involved is deoxyribose rather than ribose. These naming systems are summarized in Table 23.2.

The structures of three representative nucleotides are given in Figure 23.4. Nucleotides are the monomers from which DNA and RNA are synthesized. In addition, the nucleotides and some of their derivatives perform a variety of other functions in the cell. Adenosine diphosphate (**ADP**) and adenosine triphosphate (**ATP**) are involved in many metabolic and biosynthetic processes. We encounter them often in other chapters. Structures of these nucleotide derivatives are shown in Figure 23.5.

23.4 THE BASE SEQUENCE: PRIMARY STRUCTURE OF NUCLEIC ACIDS

The nucleic acids are polymers of nucleotides. The backbone of the polymer is a polyester chain. This backbone involves the phosphoric acid and the pentose sugar portions of the nucleotides. The connection between one nucleotide unit and the next is made when a phosphate hydroxyl group in one unit reacts with a sugar hydroxyl on the next (Figure 23.6).

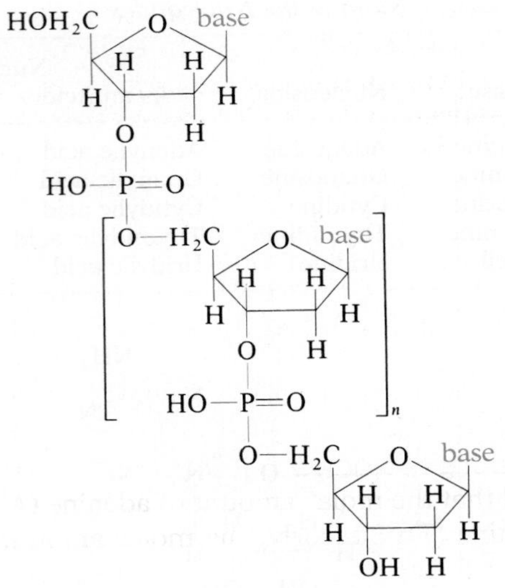

FIGURE 23.6
The polymeric backbone of a nucleic acid, shown here for deoxyribonucleic acid.

A base is attached to each sugar unit. The polymer can be represented schematically by

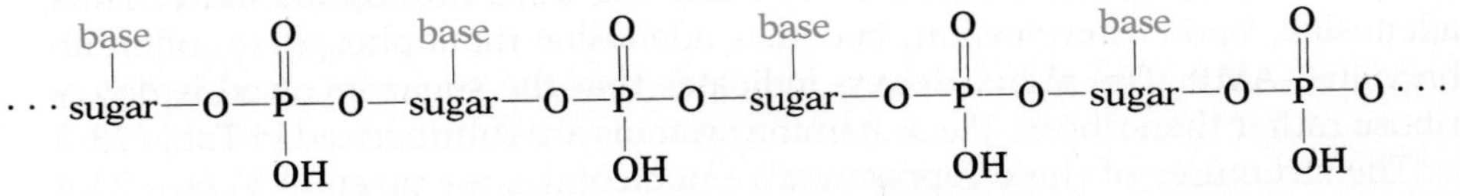

As we have seen, the sugar in DNA is 2-deoxyribose, and that in RNA is ribose. The bases in DNA are adenine, guanine, cytosine, and thymine. Those in RNA are adenine, guanine, cytosine, and uracil. Note the one ionizable hydrogen on each phosphate unit. That is what makes these compounds nucleic *acids*. However, in solution or combined with basic proteins as nucleoproteins, the acid may be ionized (Figure 23.6).

Nucleic acids resemble proteins in one respect: to completely specify the primary structure of a nucleic acid, one must specify the **sequence** of bases. Unlike the proteins, which have 20 different amino acids, there are only 4 different bases in a nucleic acid. However, the molecular weight of nucleic acids is often much greater than that of proteins, ranging into the billions for mammalian DNA. In the 1960s, sequencing was extremely laborious. For example, the primary structure of the nucleic acid called alanine transfer RNA, a molecule with 77 nucleotide units, has been determined. The work, done by Robert W. Holley and co-workers at Cornell University, took 7 years. Holley was rewarded with a share of the Nobel Prize in 1968 for his part in the project.

The sequence of bases in short strands of nucleic acids is determined using gel electrophoresis (Chapter 22). Enzymes (Chapter 24) are used to cleave the nucleic acids at specific base sequences. The primary structure of each fragment is determined. Then overlapping parts are matched to give the base sequence of the whole strand. Now the work is largely automated, and sequences of several thousand base units per molecule have been determined.

23.5 THE SECONDARY STRUCTURE OF NUCLEIC ACIDS: BASE PAIRING AND THE DOUBLE HELIX

The shape of the giant DNA molecules was long a mystery. Early studies revealed no more than the fact that the structure exhibited a periodic pattern. A real breakthrough occurred in 1950, when Erwin Chargoff, of Columbia University, showed that the molar amount of adenine (A) in DNA was always equal to that of thymine (T). Similarly, the molar amount of guanine (G) was the same as that of cytosine (C). The bases must be paired, A to T and G to C. But how? The race was on, with an almost certain Nobel Prize for the winner. Many illustrious scientists, including Linus Pauling, were working on the problem. However, in 1953, two relative unknowns in the world of science announced that they had worked out the structure of DNA. Using data from X-ray studies, which involved quite sophisticated chemistry, physics, and mathematics, and working with models not unlike a child's construction set, James D. Watson and Francis Crick determined that DNA must be composed of two helices wound about one another, to form a **double helix.** The phosphate and sugar groups (the backbone of the nucleic acid polymer) form the outside of the structure, which is rather like a spiral staircase. The heterocyclic amines are paired on the inside—with guanine always opposite cytosine and adenine always opposite thymine. In our staircase analogy, these base pairs are the stairsteps (Figure 23.7). This structure can explain how cells are able to divide and go on functioning, how genetic data are passed on to new generations, and even how proteins are built to required specifications. It all depends on the base pairing. Figure 23.8 shows a space-filling model of the DNA double helix.

Which brings up the still unanswered question "Why do the bases pair in that precise pattern, always A to T and T to A, always G to C and C to G?" The answer is hydrogen bonding and a truly elegant molecular design. Figure 23.9 shows the two sets of base pairs. You should notice two things. First, a pyrimidine is paired with a purine in each case, and the long dimensions of both pairs are identical (1.085 nm). If two pyrimidines were paired or two purines were paired, the two pyrimidines would take up less space than a purine and a pyrimidine, and the two purines would take up more space, as is illustrated in Figure 23.10. If this were the situation, the structure of DNA

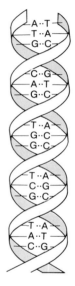

FIGURE 23.7
The DNA double helix as portrayed by Watson and Crick.

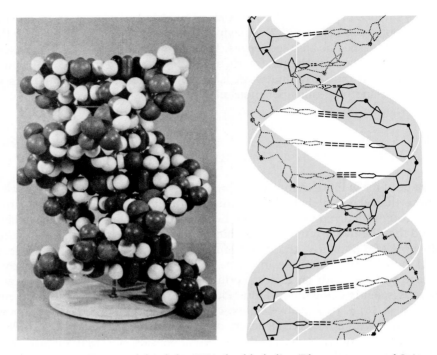

FIGURE 23.8 Two models of the DNA double helix. (Photo courtesy of Science Related Materials Inc., Janesville, Wis.)

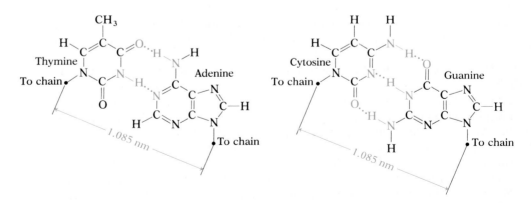

FIGURE 23.9 Base pairing of thymine to adenine and of cytosine to guanine.

FIGURE 23.10
Difference in widths of possible base pairs.

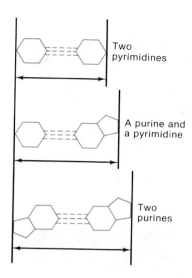

Two pyrimidines

A purine and a pyrimidine

Two purines

would be like a staircase made with stairs of different widths. In order for the two strands of the double helix to fit neatly, a pyrimidine must always be paired with a purine.

The other thing you should notice from Figure 23.9 is that when guanine is paired with cytosine, three hydrogen bonds can be drawn between the bases. No other pyrimidine-purine pairing will permit such extensive interaction. Indeed, in the pairing shown in the figure, both pairs of bases fit like lock and key.

There are about 10 base pairs per turn of the double helix. The acidic phosphate units are on the outside. In nucleoproteins, the highly basic proteins probably wrap around the double helix. Proton transfers from the phosphate units of DNA to the lysine and arginine side chains of the protein result in ionic charges. The protein is held to the nucleic acid, at least in part, by these salt bridges.

The molecules of RNA consist of single strands of the nucleic acid. Some internal (intramolecular) base pairing may occur in sections where the molecule folds back on itself. Because of this, portions of the molecule may exist in a double-helical form (Figure 23.11). The importance of base pairing to the proper functioning of the various types of RNA will be shown in later sections.

Watson and Crick received the Nobel Prize in 1962 for discovering, as Crick put it, "the secret of life" (Figure 23.12). It was not long after the development of the models that DNA was synthesized in the laboratory. In 1967, Arthur Kornberg, of Stanford University, carried out a test-tube synthesis of a single strand of DNA that was able to reproduce itself. Kornberg, of course, had to

FIGURE 23.11
RNA occurs as single strands that can form double-helical portions by internal base pairing.

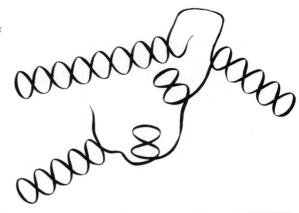

FIGURE 23.12
James D. Watson and Francis Crick, who proposed the double helix model of DNA. (Courtesy of Harvard University Biological Laboratories, Cambridge, Mass.)

add the appropriate precursors, and he added the enzymes and cofactors (Chapter 24) essential to the process. Synthesis of life in a test tube? Hardly. A strand of DNA is still a long way from even the simplest functioning cell. Synthesis of human life? Human DNA consists of about five billion base pairs. It is unlikely that the overall base sequence of such a complex DNA will ever be determined, and synthesis would be even more difficult than sequence determination.

23.6 DNA: SELF-REPLICATION

Cats have kittens that grow up to be cats. Bears have cubs that grow up to be bears. Why is it always so that each species reproduces after its own kind? How does a fertilized egg "know" that it should develop as a kangaroo and not as a koala?

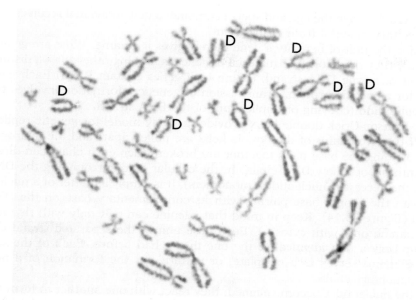

FIGURE 23.13 *Human chromosomes in the metaphase stage, just before division. Note that there are 47 chromosomes instead of the normal 46. This cell came from a boy who had an extra chromosome 13. Since numbers 13, 14, and 15 are indistinguishable in this photograph, they are all labeled D, and there are 7 D chromosomes instead of the expected 6. (Courtesy of Eeva Therman, University of Wisconsin, Madison.)*

The physical basis of heredity has been known for a long time. Most higher organisms reproduce sexually. A sperm cell from the male unites with an egg cell from the female. The fertilized egg so formed must carry all the information needed to make all the various cells, tissues, and organs necessary for the functioning of the new individual. For humans, that single cell must carrry the information for the making of legs, liver, and lungs and heart, head, hair, and hands—in short, all the instructions ever needed for growth and maintenance of the individual. In addition, if the species is to survive, information must be set aside in germ cells—either sperm or eggs—for the production of new individuals.

The basic hereditary material is found in the nuclei of all cells, concentrated in elongated, threadlike bodies called **chromosomes** (Figure 23.13). The number of chromosomes varies with the species. Human body cells have 46. The basic units of heredity, called **genes,** are arranged along the chromosomes in a linear fashion. During cell division, each chromosome produces an exact duplicate of itself. Germ cells carry only half the chromosomes of the body cells. Thus, in sexual reproduction, the entire complement of chromosomes is achieved only when the egg and sperm combine; a new individual receives half its hereditary material from each parent.

Calling the unit of heredity a gene merely gives it a name. What are genes? What are they made of? The material of genes is nothing other than a distinct segment of a long DNA strand. (Some viral genes contain RNA.) Each gene codes for a specific protein. Transmission of genetic information involves the **self-replication** (copying or duplication) of the DNA strand.

The Watson-Crick double helix provides a ready model for genetic replication. If the two chains of the double helix are pulled apart and the hydrogen bonds holding the base pairs together are broken, then each chain can direct the synthesis of a new DNA chain. In the cellular fluid surrounding the DNA are all the necessary nucleotides (monomers). It is simply a matter of a nucleotide with the proper base pairing with its complementary base on the DNA strand (Figure 23.14). Keep in mind that adenine can pair only with thymine and guanine only with cytosine. Each base unit in the separated strand can pick up only a unit identical to the one that it had before. Each of the separating chains serves as a template, or pattern, for the formation of a new complementary chain.

As the nucleotides become aligned, they react with one another to form the polyester backbone of the new chain. In this way, each strand of the original DNA molecule forms a duplicate of its former partner. Whatever information was encoded in the original DNA double helix is now contained in each of the

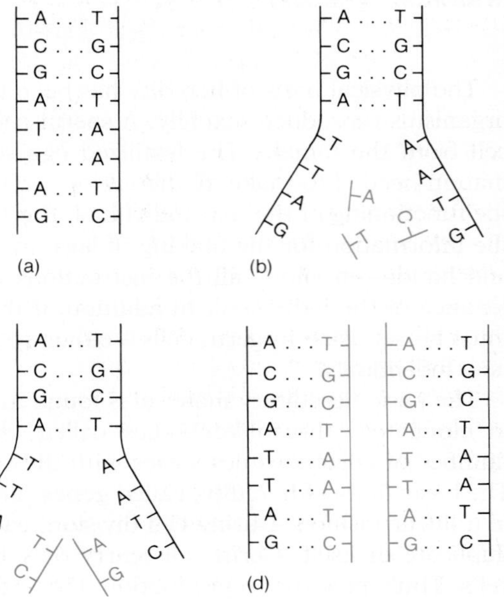

FIGURE 23.14
DNA replication. (a) The original double helix, flattened out here for clarity. (b) The helix beginning to split. Some free nucleotides from the cell are shown. (c) Nucleotides from the cell beginning to pair with bases on each original strand. (d) The two new double helixes, each identical to the original. (This representation is simplified. The "nucleotides" are actually triphosphate derivatives, and the process is catalyzed by an enzyme called DNA polymerase.)

replicates. When the cell divides, each daughter cell gets one of the DNA molecules and all of the information that was available to the parent cell.

We keep saying that there is information encoded in the DNA molecule. DNA is often compared to a set of directions for putting together a model airplane or for knitting a sweater. Knitting directions store information as words on paper. Letters of the alphabet are arranged in a certain way (e.g., ''knit one, purl two''), and these words direct the knitter to perform a certain operation with needles and yarn. If all of the directions are correctly followed, the ball of yarn becomes a sweater.

How is information stored in DNA? It is the particular arrangement of bases along the DNA chain that encodes the directions for building an organism. Just as *saw* means one thing in English and *was* means another, the sequence of bases C-G-T means something, and G-C-T means something else. Although there are only four ''letters''—the four bases—in the genetic code of DNA, their sequence along the long strands can vary so widely that there is an essentially unlimited information storage system. Even a tiny bacterium, 2 μm long and 1 μm in diameter, has 3 million base pairs. The genetic material of a human cell consists of 5 billion base pairs, and these can specify 20 billion bits of information—enough information to fill 1000 books of 2000 pages each. Thus each cell can carry all the information it needs to determine all the hereditary characteristics of even the most complex organisms.

Some viruses carry genetic information in RNA. A virus is just a bundle of genetic material (DNA or RNA) wrapped in a protein package. When a virus invades a cell, it takes over and operates that cell's machinery to its own advantage. A small virus can have as few as 5000 base units.

23.7 RNA: PROTEIN SYNTHESIS AND THE GENETIC CODE

Even if we accept the fact that DNA carries a message, there is still the problem of how this message is read. How is a particular sequence of bases along the DNA chain translated into a complex organism? The answer is that DNA directs the synthesis of proteins. Proteins serve as building materials and, most importantly, as enzymes (Chapter 24). Enzymes are catalysts that control all of the metabolic reactions that occur in the complex chemical factory we call a living organism. The mechanism by which the DNA blueprint is translated into protein molecules involves the intermediacy of RNA molecules. The process is called **transcription.** It is catalyzed by an enzyme called RNA polymerase.

In the first step, a portion of the DNA message is transcribed to a special kind of RNA, called messenger RNA (mRNA). (This is a little like xeroxing the directions for making a door from a library book that describes how to build an entire house.) One can envision a DNA double helix partially unzipping and

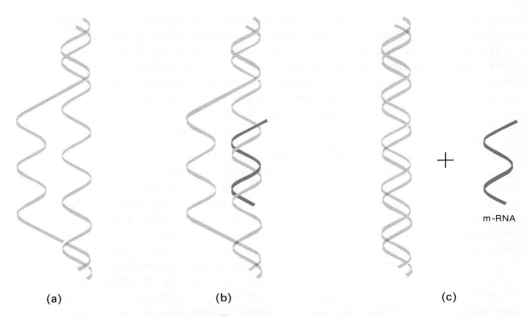

(a)   (b)   (c)

FIGURE 23.15   The DNA double helix is partly unwound (a), messenger RNA is formed on the separated portion (b), and released (c).

then a limited portion of one DNA strand directing the synthesis of the single-stranded mRNA molecule (Figure 23.15). Once again, the process involves a buildup of complementary nucleotides along the single DNA strand. The base sequence of DNA specifies the base sequence of mRNA. Thymine in DNA calls for adenine in mRNA, cytosine specifies guanine, guanine calls for cytosine, and adenine requires uracil. Recall from Section 23.1 that, in RNA molecules, uracil replaces DNA's thymine. Notice the similarity between the structures of these two bases (Figure 23.1).

DNA Base	*Complementary RNA Base*
Adenine	Uracil
Thymine	Adenine
Cytosine	Guanine
Guanine	Cytosine

The next step in protein synthesis involves the reading of the blueprint now encoded in the mRNA molecule and its **translation** into a protein structure. The mRNA travels from the nucleus to the cytoplasm of the cell, where nucleoprotein granules called ribosomes are located. The mRNA becomes attached to the ribosomes, and it is here that the genetic code is deciphered.

In the cytoplasm is another kind of RNA, called transfer RNA (tRNA). The tRNA is responsible for translating the specific base sequence of an mRNA

FIGURE 23.16
A given transfer RNA can carry only one kind of amino acid, which is specified by the base triplet in the anticodon.

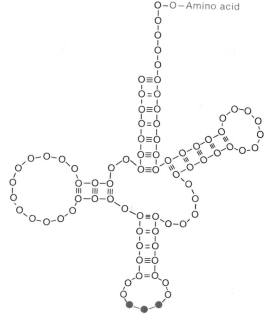

Anticodon base triplet

molecule into the specific amino acid sequence of a protein molecule. A tRNA molecule has a looped structure (Figure 23.16). On one of these loops is a critical set of three bases (called the triplet). One of the trailing ends of the tRNA is attached to an amino acid. A given tRNA molecule can carry only one of the different amino acids. And all tRNA molecules with a particular base triplet carry the same kind of amino acid. For example, all tRNA molecules whose base triplet is CCU carry the amino acid glycine. A tRNA molecule with the base triplet GCG always has an arginine molecule attached. There are many more ways of sequencing three bases than there are kinds of amino acids. Thus, one kind of amino acid is usually associated with more than one triplet sequence, so that a tRNA molecule with a CCG triplet, as well as one with a CCU triplet, may carry glycine. What is important is that a specific pattern of bases is associated with a specific pattern of amino acids.

We are now ready to build proteins. Strung out along some ribosomes in the cytoplasm of a cell is an mRNA molecule. The sequence of bases along this molecule was determined by the DNA of a gene in the cell nucleus. Floating around in the cytoplasm surrounding the mRNA are tRNA molecules, each carrying its own amino acid. Let us suppose that a portion of the mRNA base sequence reads · · · C-G-C-G-G-A-G-G-C · · · . The first three bases (C-G-C) could pair, through hydrogen bonding, with a nucleic acid that had a G-C-G sequence.

There, floating around in the cytoplasm, is just such a nucleic acid, a tRNA molecule with the amino acid arginine attached. This tRNA pairs up with the first three bases of the mRNA.

$$\cdots \text{C—G—C—G—G—A—G—G—C} \cdots$$
$$\text{G—C—G}$$
$$\text{Arg}$$

The next three bases of the mRNA, G-G-A, will pair up with a C-C-U tRNA molecule, which carries the amino acid glycine.

$$\cdots \text{C—G—C—G—G—A—G—G—C} \cdots$$
$$\text{G—C—G} \quad \text{C—C—U}$$
$$\text{Arg} \qquad \text{Gly}$$

When the two tRNA molecules are appropriately lined up along the mRNA, a peptide bond forms betwen the two amino acids.

$$\cdots \text{C—G—C—G—G—A—G—G—C} \cdots$$
$$\text{G—C—G} \quad \text{C—C—U}$$
$$\text{Arg—Gly}$$

The next three bases on the mRNA (G-G-C) pair with a C-C-G tRNA molecule.

$$\cdots \text{C—G—C—G—G—A—G—G—C} \cdots$$
$$\text{G—C—G} \quad \text{C—C—U} \quad \text{C—C—G}$$
$$\text{Arg—Gly} \qquad \text{Gly}$$

The amino acid carried by the tRNA is then joined by a peptide bond to the previously linked amino acids.

$$\cdots \text{C—G—C—G—G—A—G—G—C} \cdots$$
$$\text{G—C—G} \quad \text{C—C—U} \quad \text{C—C—G}$$
$$\text{Arg—Gly—Gly}$$

In this way the protein chain is gradually built up. The chain is released from the tRNA and mRNA as it is formed (Figure 23.17).

Each base triplet strung consecutively along the *mRNA* molecule is called a **codon.** The base triplets of the *tRNA* molecules are called **anticodons.** A codon on mRNA always pairs with its complementary tRNA anticodon. A complete dictionary of the genetic code has been compiled. Figure 23.18 shows which amino acids are called for by all the possible mRNA codons. The amino acids serine and leucine are each specified by six different codons. Two amino acids, tryptophan and methionine, have only one codon each. Note that three

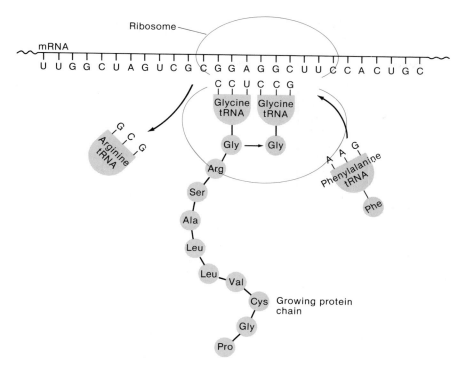

FIGURE 23.17 *The synthesis of a protein molecule.(Adapted with permission from Lipman, F. "Polypeptide Chain Elongation in Protein Biosynthesis."* Science *164(1969):1024. Copyright © AAAS.)*

codons, UAA, UAG, and UGA, are stop signals, calling for termination of the protein chain.

All proteins for muscle, hair, skin, and the all-important enzymes are formed by the mechanism just outlined. Next time you look in a mirror, remember that what you see depends ultimately on something as fragile as a hydrogen bond.

23.8 ONCOGENES

Some of the messages carried by our genes may not be friendly ones. In the 1970s, scientists began identifying genes that seemed to trigger or sustain the processes that convert normal cells to cancerous ones. So far about 20 or 30 of these genes have been identified. They have been named **oncogenes.**

First base	Second base: U		Second base: C		Second base: A		Second base: G		Third base
U	UUU	Phe	UCU	Ser	UAU	Tyr	UGU	Cys	U
	UUC	Phe	UCC	Ser	UAC	Tyr	UGC	Cys	C
	UUA	Leu	UCA	Ser	UAA	Stop	UGA	Stop	A
	UUG	Leu	UCG	Ser	UAG	Stop	UGG	Trp	G
C	CUU	Leu	CCU	Pro	CAU	His	CGU	Arg	U
	CUC	Leu	CCC	Pro	CAC	His	CGC	Arg	C
	CUA	Leu	CCA	Pro	CAA	Gln	CGA	Arg	A
	CUG	Leu	CCG	Pro	CAG	Gln	CGG	Arg	G
A	AUU	Ile	ACU	Thr	AAU	Asn	AGU	Ser	U
	AUC	Ile	ACC	Thr	AAC	Asn	AGC	Ser	C
	AUA	Ile	ACA	Thr	AAA	Lys	AGA	Arg	A
	AUG	Met	ACG	Thr	AAG	Lys	AGG	Arg	G
G	GUU	Val	GCU	Ala	GAU	Asp	GGU	Gly	U
	GUC	Val	GCC	Ala	GAC	Asp	GGC	Gly	C
	GUA	Val	GCA	Ala	GAA	Glu	GGA	Gly	A
	GUG	Val	GCG	Ala	GAG	Glu	GGG	Gly	G

FIGURE 23.18 The genetic code for protein synthesis.

(Oncology is the branch of medicine that treats tumors; the term comes from the Greek *onkos,* meaning "mass.")

Oncogenes are believed to arise from ordinary benign genes, but more research is needed to find out exactly how they develop. Some evidence points to mutations caused by chemicals or other environmental factors. A suppressed gene may be relocated to a place where it is activated. It appears that more than one oncogene must be turned on, perhaps at different stages of the process, before a cancer develops.

The discovery of oncogenes holds some promise for medicine. Since oncogenes, like all other genes, code for proteins, diagnostic tests for cancer may be available in a few years. Drugs that suppress the oncogenes are another possibility. Antibodies to the cancer proteins can presumably be manufactured.

It may be a long time before you can go to a drugstore and buy an oncogene diagnostic kit, but the discovery of oncogenes may be an important step toward understanding and eventually conquering cancer.

FIGURE 23.19
A drop of the first human insulin produced through recombinant DNA technology. (Courtesy of Eli Lilly and Co.)

23.9 GENETIC ENGINEERING

Only a few years ago, millions of diabetics were threatened with a shortage of life-sustaining insulin. The supply of pig and cow insulin, by-products of the meatpacking industry, was increasing slowly while the number of diabetics who needed it was increasing rapidly. Now human insulin (Figure 23.19), a protein coded for by human DNA, is being produced by the cell machinery of bacteria. A plentiful supply is assured. Human growth hormone, once available only from cadavers and at great cost, is also being made by bacteria. It is being used to treat children with **dwarfism,** a condition that results from inadequate production of this hormone by the pituitary gland. Even the complex interferons, proteins produced as a part of the body's defense mechanism, are being made by yeast and bacteria. Large quantities of these materials are now available for testing as antiviral and anticancer agents.

All these accomplishments result from **genetic engineering.** It has become possible to manipulate the genes of organisms. Scientists have developed what is called **recombinant-DNA** technology, that is, methods for splicing together DNA from different species. Once the base sequence of a gene has been determined, the gene can be synthesized. The synthetic gene can then be inserted into **plasmids,** circular pieces of double-stranded DNA that occur in microorganisms. The modified plasmids do all the things DNA does; they replicate, and they control the synthesis of proteins. Microorganisms multiply rapidly. Soon vast vats of these modified creatures are producing a protein specified by a human gene.

Not all engineered organisms are designed to produce medicines. Scientists have developed bacteria that will "eat" the oil released in an oil spill. Bacteria have been redesigned in an attempt to reduce frost damage to crops. (The unmodified bacteria seem to act as "seeds" on which ice crystals form.) Nor have the young people of the world been left out. Bacteria have had genes inserted that enable them to produce indigo, the dye used for blue jeans. Designer genes for your designer jeans!

Concerns have been expressed over potential harm from recombinant-DNA organisms. Initially, scientists worried about the possibility of producing a deadly "artificial" organism. What if a gene that causes cancer were spliced into the DNA of a bacterium that normally inhabits our intestine? We would have no natural immunity against such an artificial organism. To protect against such a development, strict guidelines for recombinant-DNA research have been instituted.

The new molecular genetics has already resulted in some impressive achievements. Its possibilities are mind-boggling—elimination of genetic defects, a cure for cancer, a race of geniuses, and who knows what else. Knowledge gives power. It does not necessarily give wisdom. Who will decide what sort of creatures the human species should be? The greatest problem we shall be likely to face in our use of bioengineering is that of choosing who is to play "God" with the new "secret of life."

PROBLEMS

1. Name the two kinds of nucleic acids.
2. Which of the two kinds of nucleic acids is concentrated in the nucleus of the cell?
3. What sugar is incorporated in the RNA polymer? What is the sugar unit in DNA?
4. Compare DNA and RNA with respect to the major bases present in each type of nucleic acid.
5. How do DNA and RNA differ in secondary structure?
6. For each of the following, indicate whether the compound is a nucleoside, a nucleotide, or neither:

a.

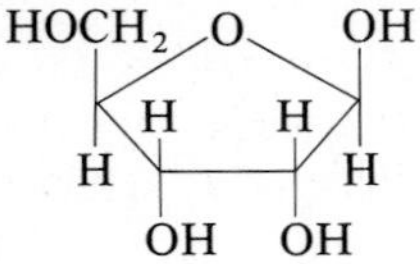

b.

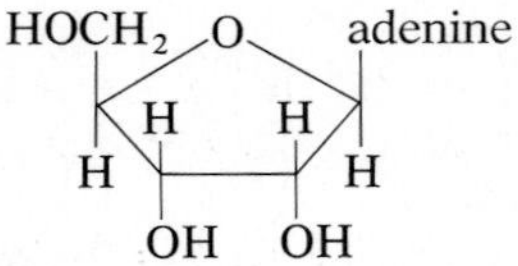

c.

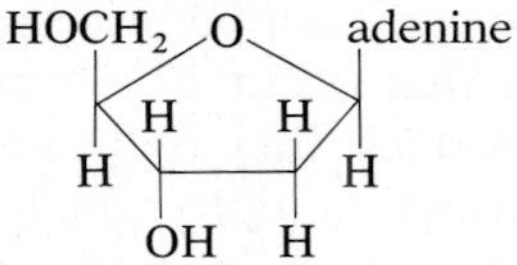

d.

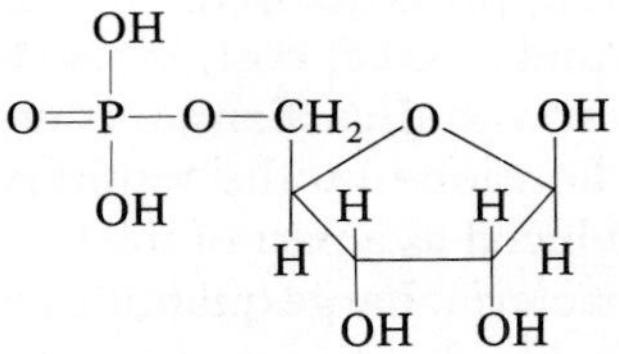

e.

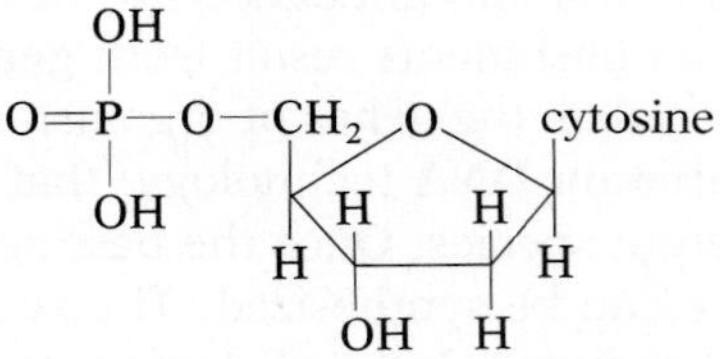

f.

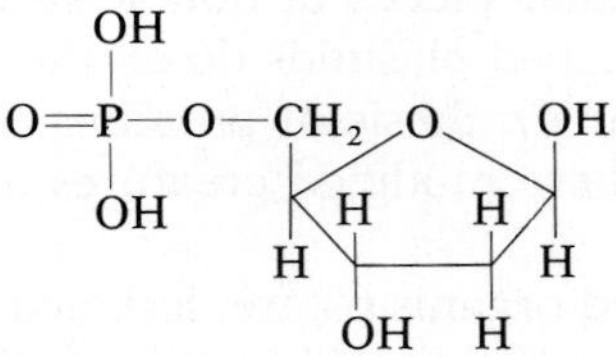

7. For each of the structures shown in Problem 6, indicate whether the sugar unit is ribose or deoxyribose.

8. Would the nucleoside shown commonly be incorporated in DNA or RNA or neither?

a. HOCH$_2$ O   adenine

b. HOCH$_2$ O   adenine

c. HOCH$_2$ O   cytosine

d. HOCH$_2$ O   thymine

e. HOCH$_2$ O   uracil

f. HOCH$_2$ O   guanine

9. Which base is purine and which is pyrimidine?

a.

b.

10. For each compound in Problem 8, indicate whether the base is a purine or a pyrimidine.

11. Answer the questions for the molecule shown.

CH$_3$

a. Is the compound a nucleic acid, a nucleotide, or a nucleoside?
b. Is the base a pyrimidine or a purine?
c. Would the compound be incorporated in DNA or RNA?

12. Answer the same questions posed in Problem 11 for the following compound.

13. Using a schematic representation, show a length of nucleic acid polymer chain, indicating the positions of the sugar, phosphoric acid, and base units.

14. The primary structure of a protein is defined by the sequence of amino acids. What defines the primary structure of nucleic acids?

15. With the same sort of schematic representation used in Problem 13, show the overall design of the double helix.

16. Why is it structurally important in the DNA double helix that a purine base always pair with a pyrimidine base?

17. What kind of intermolecular force is involved in base pairing?

18. In DNA, which base would be paired with the base listed?

a. cytosine	**c.** guanine
b. adenine	**d.** thymine

19. In an RNA molecule, which base would pair with the base listed?

a. adenine	**c.** uracil
b. guanine	**d.** cytosine

20. Describe the process of replication.

21. In replication, a parent DNA molecule produces two daughter molecules. What is the fate of each strand of the parent DNA double helix?

22. We say DNA controls protein synthesis, yet most DNA resides within the cell nucleus while protein synthesis occurs outside of the nucleus. How does DNA exercise its control?

23. Explain the role of messenger RNA in protein synthesis.

24. Explain the role of transfer RNA in protein synthesis.

25. Which nucleic acid(s) is(are) involved in the process referred to as transcription?

26. Which nucleic acid(s) is(are) involved in the process referred to as translation?

27. Which nucleic acid contains the codon?
28. Which nucleic acid contains the anticodon?
29. The base sequence along one strand of DNA is · · · A-T-T-C-G · · · . What would be the sequence of the complementary strand of DNA?
30. What sequence of bases would appear in the messenger RNA molecule copied from the original DNA strand shown in Problem 29?
31. If the sequence of bases along a messenger RNA strand is · · · U-C-C-G-A-U · · · , what was the sequence along the DNA template?
32. What are the complementary triplets on tRNA for the following triplets on mRNA?
    a. U-U-U   c. A-G-C
    b. C-A-U   d. C-C-G
33. What are the complementary triplets on mRNA for the following triplets on tRNA molecules?
    a. U-U-G   c. U-C-C
    b. G-A-A   d. C-A-C
34. Using Figure 23.18, identify the amino acids carried by the tRNA molecules in Problem 32.
35. Using Figure 23.18, identify the amino acids carried by the tRNA molecules in Problem 33. Remember that Figure 23.18 lists the mRNA codons.
36. Refer to Figure 23.18. What amino acid sequence would result if the base sequence on mRNA is
    a. · · · U-U-A-C-C-U-C-G-A · · ·
    b. · · · G-C-G-U-C-A-U-A-A · · ·
    c. · · · C-C-C-C-C-C-C-C-C · · ·
37. If this DNA base sequence, · · · T-T-A-C-T-C-T-C-A · · · , acts as a template for mRNA formation, what amino acid sequence would eventually be produced from the mRNA?
38. What is the relationship between the cell parts called chromosomes, the units of heredity called genes, and the nucleic acid DNA?
39. What is the basic process in recombinant-DNA technology?
40. Discuss some applications of genetic engineering.
41. Chemical bonds can be broken by ultraviolet radiation, gamma rays, certain chemical substances, etc. What are the biological implications of these facts? (Hint: What effect might the breaking of chemical bonds have on DNA?)
42. Certain genes are implicated in cancer. Some of these *oncogenes* can be activated by single point mutations. Ordinarily the DNA triplet GGT codes for the amino

acid glycine. What amino acid is formed when the DNA triplet is changed by mutation to TGT? to GTT? to CGT?
43. What is the difference between AMP, ADP, and ATP?
44. Using Figure 23.9 as a guide, show the hydrogen bonding between uracil and adenine.
45. Using Figure 23.9 as a guide, pair cytosine with adenine and thymine with guanine and evaluate the possibility for hydrogen bonding. Compare these interactions with those for the correct C-G and A-T pairings.
46. If a DNA segment is 432 base units long and all the units code for protein, how many amino acid units will there be in the protein?
47. From what base and what sugar is the following anticancer drug derived? (Fluorine has been substituted for hydrogen.)

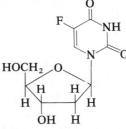

48. From what base is the following anticancer drug derived? (Assume SH has replaced NH_2.)

49. Name the following compound. What is its three-letter abbreviation?

50. Following are the results of two different point mutations. Which is likely to be more serious?
    a. valine is substituted for leucine
    b. glutamic acid is substituted for leucine

24

Enzymes

nzymes are a highly specialized class of proteins. They control and direct the myriad of chemical reactions that occur in living cells. Enzymes are biological catalysts produced by cells but able to act independently of the cells. For example, an enzyme isolated from the cell that made it will still catalyze reactions, even in a test tube or a laundry tub.

Enzymes have enormous catalytic power. They speed up reactions by factors of 1 million or more. Most reactions in living cells would be imperceptibly slow without enzymes. Even a simple reaction like the conversion of carbon dioxide to carbonic acid is catalyzed by an enzyme.

$$CO_2 + H_2O \xrightleftharpoons[\text{}]{\text{enzyme}} H_2CO_3$$

Without the enzyme, the body tissues couldn't dump carbon dioxide into the blood fast enough. The enzyme, called carbonic anhydrase, speeds up the reaction by a factor of 10 million! Each enzyme molecule can convert 100 000 molecules of carbon dioxide per second.

Several thousand enzymes are known. Many are available commercially in varying grades of purity. Some have even been isolated as highly purified crystals.

24.1 THE ANATOMY OF ENZYMES

All known enzymes are proteins. As such, they can be considered according to their primary structure (their amino acid sequence) and their secondary,

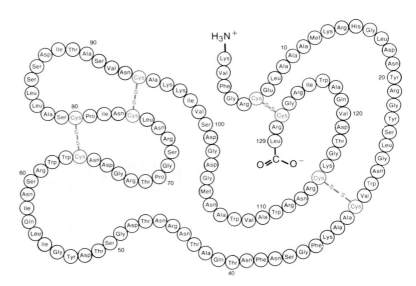

FIGURE 24.1   *The enzyme lysozyme has four disulfide cross-links. (From* Biochemistry, *2nd ed., by Lubert Stryer, W. H. Freeman and Co. Copyright © 1981. After Canfield, R. E., and Liu, A. K.* Journal of Biological Chemistry *240 (1965): 2000; Phillips, D. C.* Scientific American *(5) 215(1966):79.)*

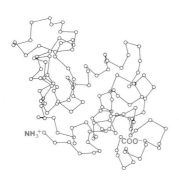

FIGURE 24.2
The three-dimensional structure of lysozyme, showing only the C—C—N skeleton. (Courtesy of David C. Phillips, Oxford University, Oxford.)

tertiary, and quaternary features. The primary structure of the protein hormone insulin (Special Topic K) was determined by Frederick Sanger in 1953. Shortly thereafter many researchers began to determine the amino acid sequences of proteins, including some enzymes.

The first enzyme to have its three-dimensional structure determined was lysozyme. The amino acid sequence of this protein was determined independently by Pierre Jolles and co-workers at the University of Paris and by Robert Canfield, of Columbia University, in 1963. Lysozyme, obtained from egg white, is composed of a single polypeptide chain of 129 amino acid units. The chain is cross-linked at four locations by disulfide bonds (Figure 24.1).

Lysozyme breaks down the cell walls of bacteria. Its mode of action, however, cannot be understood from its primary structure alone. The three-dimensional structure, determined by the folds in the molecule, must be known in order for the properties of the molecule to be understood. This three-dimensional structure was worked out by David C. Phillips and co-workers at the Royal Institution in London in 1965. The skeleton of this structure is shown in Figure 24.2. What looks like an apparently random orientation of the chain is actually a precise folding that is exactly repeated in every molecule of the active enzyme.

Lysozyme is activated by **phosphorylation** of a serine residue, that is, the alcohol group in serine is converted to an ester of phosphoric acid.

$$\ldots NHCHCO \ldots \qquad \ldots NHCHCO \ldots$$
$$CH_2OH \qquad \longrightarrow \qquad CH_2OPO_3{}^{2-}$$

The phosphate comes from adenosine triphosphate (ATP). The reaction is catalyzed by an enzyme called a protein kinase.

With the structure of lysozyme known, it was possible to determine how it functions. The bacterial cell wall is composed of a polysaccharide. A portion of the polysaccharide attaches itself to a specific portion of the lysozyme molecule. That portion of the enzyme has glutamic acid and aspartic acid residues. These interact with the acetal linkage of the polysaccharide, breaking the bond. When enough of these acetal linkages cleave, the bacterial cell wall ruptures.

Three-dimensional structures of relatively few enzymes have been determined. Detailed knowledge of how those few work, however, has enabled us to understand better the action of enzymes in general.

24.2 CLASSIFICATION AND NAMING OF ENZYMES

The first enzymes to be discovered were named according to their source or method of discovery. The enzyme pepsin, which aids in the hydrolysis of proteins, is found in the digestive juices of the stomach (from the Greek *pepsis*, "digestion"). Ptyalin, found in saliva (from the Greek *ptyalon*, "spittle"), starts the hydrolysis of starches.

As more enzymes were discovered, a more systematic nomenclature developed. A **substrate** is the compound or type of compound that an enzyme acts upon. You name an enzyme, then, by dropping the ending of the name of the substrate and adding the suffix *-ase*. Maltase is an enzyme that catalyzes the hydrolysis of the disaccharide maltose. A lipase is an enzyme that catalyzes the cleavage of lipids. A dipeptidase aids in the splitting of dipeptides to amino acids. A still more systematic nomenclature has been devised by the International Union of Biochemistry. Used mainly by research workers, that system will not be discussed further here.

Enzymes are generally classified according to the *type* of chemical reaction that they catalyze. There are six major types.

1. **Oxidoreductases** catalyze all the reactions in which one compound is oxidized and another is reduced. **Oxidases** operate in physiological

oxidation reactions. They are essential in cells that use oxygen to produce energy. This group also includes the **dehydrogenases,** enzymes that oxidize by the removal of hydrogen. For example, alcohol dehydrogenase speeds the oxidation of ethanol to acetaldehyde.

$$CH_3CH_2OH \longrightarrow CH_3CHO + ''2\,H''$$

2. **Transferases** facilitate the transfer of groups such as methyl, amino, and acetyl from one molecule to another. **Transaminase** catalyzes the transfer of an amino group from one molecule to another. This reaction is involved in the removal of the amino group during the metabolism of amino acids (Chapter 29). **Kinases** are involved in the transfer of phosphate groups.
3. **Hydrolases** are enzymes that catalyze hydrolysis reactions. These include **lipases,** which act on fat and other lipids, **carbohydrases,** which speed up the hydrolysis of carbohydrates to monosaccharides, and **proteases** and **peptidases,** which catalyze the hydrolysis of proteins and peptides.
4. **Lyases** aid in the removal of certain groups without hydrolysis. Examples are the **decarboxylases,** which catalyze the removal of carboxyl groups.
5. **Isomerases,** as their name implies, catalyze the conversion of a compound into another that is isomeric with it.
6. **Ligases** or **synthetases** are involved in the formation of new bonds from carbon to a nitrogen, an oxygen, a sulfur, or another carbon atom.

We encounter additional examples, particularly of the first four classes, in other chapters.

24.3 PRECISION PROTEINS: ENZYME SPECIFICITY

Enzymes are highly specific. They catalyze only one reaction or a group of closely related reactions. For example, the enzyme *urease* catalyzes a single reaction, the hydrolysis of urea.

$$\overset{\displaystyle O}{\underset{\displaystyle \|}{NH_2-C-NH_2}} + H_2O \xrightarrow{\text{urease}} 2\,NH_3 + CO_2$$

Urease has no effect on other compounds, even closely related ones such as the amides. Such **absolute specificity** is rather rare among enzymes characterized to date.

Many enzymes show **stereospecificity.** These react with only one of a pair of mirror image isomers. *Arginase,* for example, catalyzes the hydrolysis of L-arginine to ornithine and urea.

$$
\begin{array}{ccccc}
\text{COO}^- & & & \text{COO}^- & \\
| & & & | & \\
\overset{+}{\text{H}_3\text{N}}-\text{C}-\text{H} & & & \overset{+}{\text{H}_3\text{N}}-\text{C}-\text{H} & \\
| & & & | & \\
\text{CH}_2 & & & \text{CH}_2 & \\
| & & & | & \quad\;\; \text{O} \\
\text{CH}_2 & & & \text{CH}_2 & \quad\;\; \| \\
| & +\,\text{H}_2\text{O} \xrightarrow{\;\text{arginase}\;} & & | & +\,\text{H}_2\text{N}-\text{C}-\text{NH}_2 \\
\text{CH}_2 & & & \text{CH}_2 & \\
| & & & | & \quad\;\; \text{Urea} \\
\text{NH} & & & \text{NH}_2 & \\
| & & & & \\
\text{C}{=}\text{NH} & & & \text{Ornithine} & \\
\diagup & & & & \\
\text{H}_2\text{N} & & & & \\
\text{L-Arginine} & & & &
\end{array}
$$

On the other hand, arginase has no effect on the rate of hydrolysis of D-arginine.

Other enzymes are **reaction specific.** Those that catalyze the hydrolysis of peptide linkages might also catalyze the hydrolysis of closely related ester linkages but would do so with less efficiency. Some enzymes are highly **linkage specific.** *Trypsin,* for example, splits only those peptide bonds on the carboxyl side of lysine and arginine units in proteins.

$$
\cdots \text{Arg}-\text{xxx}\cdots + \text{H}_2\text{O} \xrightarrow{\;\text{trypsin}\;} \cdots \text{Arg}-\text{COO}^- + \overset{+}{\text{H}_3\text{N}}-\text{xxx}\cdots
$$

$$
\cdots \text{Lys}-\text{xxx}\cdots + \text{H}_2\text{O} \xrightarrow{\;\text{trypsin}\;} \cdots \text{Lys}-\text{COO}^- + \overset{+}{\text{H}_3\text{N}}-\text{xxx}\cdots
$$

Thrombin is even more specific. It cleaves only those bonds between arginine and glycine.

$$
\cdots \text{Arg}-\text{Gly}\cdots + \text{H}_2\text{O} \xrightarrow{\;\text{thrombin}\;} \cdots \text{Arg}-\text{COO}^- + \overset{+}{\text{H}_3\text{N}}-\text{Gly}\cdots
$$

Thrombin has no effect on the hydrolysis of other peptide linkages. It won't even attack the bonds between glycine and arginine if the order is reversed.

$$
\cdots \text{Gly}-\text{Arg}\cdots + \text{H}_2\text{O} \xrightarrow[\;\;\times\;\;]{\;\text{thrombin}\;} \text{no reaction}
$$

Another kind of specificity is **group specificity.** *Carboxypeptidase,* for example, cleaves only those terminal amino acids at the end of a protein chain that has a free carboxyl group. *Aminopeptidase,* on the other hand, cleaves those at the other end of the chain, those with a free amino group.

$$
\overset{+}{\text{H}_3\text{N}}-\text{CH}-\overset{\displaystyle\text{O}}{\overset{\displaystyle\|}{\text{C}}}\underset{\overset{|}{\text{R}}}{}\!\!\!\cdots\cdots\cdots\cdots\cdots\cdots\cdots\cdots\cdots-\text{NH}-\underset{\overset{|}{\text{R}}}{\text{CH}}-\text{COO}^-
$$

(Aminopeptidase) (Carboxypeptidase)

Enzyme specificity is crucial in chemical reactions in the cell. It ensures, for the most part, that the proper reactions occur in the proper place at the proper time.

24.4 HOW ENZYMES WORK: ACTIVE SITES

Enzymes act by lowering the activation energy for a chemical reaction. This they do by changing the reaction path (Chapter 5). The new route involves the formation of bonds between the enzyme and substrate to form an **enzyme-substrate complex.** The complex then separates into products and the regenerated enzyme. Schematically, this may be written

Enzyme + substrate $\rightleftarrows$ enzyme-substrate complex $\rightleftarrows$ enzyme + product

Note that the steps are reversible. The enzyme catalyzes both the forward and reverse reactions. The enzyme does not change the extent of a reaction;

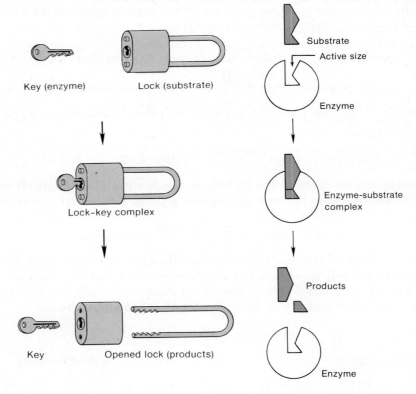

FIGURE 24.3   *The lock-and-key model for enzyme action.*

it merely changes the rate at which the reaction comes to equilibrium. An equilibrium may be shifted left or right by the usual factors (Chapter 5), such as temperature, concentration of reactant and product, or removal of a reactant or product.

Enzyme action is often explained in a simple way by the use of a **lock-and-key model.** The substrate must fit a portion of the enzyme, called the **active site,** quite precisely (Figure 24.3), just as a key must fit a tumbler lock in order to open it.

Not only must the enzyme and substrate fit precisely, but they are probably held together by electrical attraction. That requires that certain charged groups on the enzyme complement certain charged or partially charged groups on the substrate. Formation of new bonds to the enzyme by the substrate probably weakens bonds within the substrate. The bonds in the substrate can then be easily broken.

The lock-and-key analogy works well for some systems, but it must be modified to fit other data. We know from experimental data that the active site of an enzyme changes shape when the substrate binds. Daniel Koshland, an American biochemist, has proposed the **induced-fit model** to account for these data. He compares the changes in the shape of the active site to the changes that occur in a glove when a hand is inserted (Figure 24.4).

The active site on an enzyme is usually rather small, and only that small part comes into direct contact with the substrate. Side chains of amino acid units, usually those containing amino, carboxyl, or sulfhydryl groups, are often involved. The amino acid units involved need not be next to one another in the primary sequence. They must, however, be brought close by the cross-linking and folding of the protein chain. The active site of *chymotrypsin,* a protease, involves the amino terminal of the chain, a cross-linked loop incorporating histidine units, and an aspartic acid side chain (Figure 24.5). The polar groups at this site no doubt interact strongly with the peptide linkage that is to be hydrolyzed.

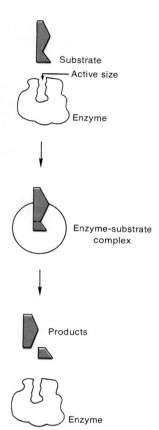

FIGURE 24.4
The induced-fit model for enzyme action.

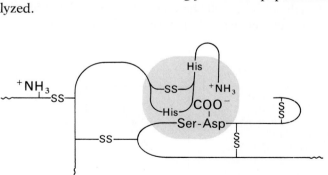

FIGURE 24.5 *A diagram of a portion of a chymotrypsin molecule, showing the active site.*

TABLE 24.1 *Some Important Enzymes*

Enzyme	Reaction Catalyzed	Turnover Number (per minute)
Carbonic anhydrase	$H_2O + CO_2 \rightleftharpoons H_2CO_3$	36 000 000
Catalase	$2 H_2O_2 \rightleftharpoons 2 H_2O + O_2$	5 600 000
Fumarase	Fumaric acid $\rightleftharpoons$ malic acid	1 200 000
Lactate dehydrogenase	Pyruvic acid $\rightleftharpoons$ lactic acid	60 000
Succinate dehydrogenase	Succinic acid $\rightleftharpoons$ fumaric acid	1 150
DNA polymerase I	Addition of nucleotides to DNA chain	900
Lysozyme	Hydrolysis of specific polysaccharide bonds	30

Portions of the enzyme molecule other than the active site may be involved in the catalytic process. The rearrangement of these other parts of the molecule during the reaction may cause essential changes at the active site.

Table 24.1 lists some important enzymes and gives an indication of how rapidly some of them work. The **turnover number** is the number of substrate molecules converted to product in one minute by one enzyme molecule. Typical turnover numbers are about 1000 reactions per minute. Some, like carbonic anhydrase, are much faster. Others, such as lysozyme, are much slower.

24.5 FACTORS INFLUENCING ENZYME ACTIVITY: CONCENTRATION, TEMPERATURE, AND pH

The rates of reactions involving enzymes are influenced by a variety of factors, including concentration of substrate, concentration of enzymes, temperature, and the pH of the medium. The enzyme and substrate must come together for the reaction to take place. The chance of this meeting depends on the concentration of each substance. If the concentration of substrate, [S], is varied, the reaction rate will increase as [S] increases (Figure 24.6). Eventually, a con-

FIGURE 24.6
The effect of substrate concentration [S] on reaction rate with enzyme concentration held constant. When [S] is low, the reaction rate increases with [S]. At high values of [S], the reaction rate is independent of [S], and a maximal rate is observed.

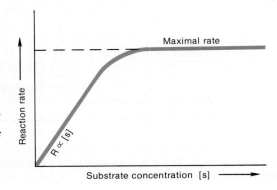

FIGURE 24.7
*The effect of enzyme
concentration on reaction
rate, with the concentration
of substrate constant and
in excess.*

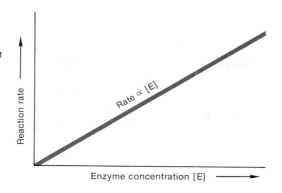

centration is reached that saturates all the active sites on the enzymes. The rate levels off; further increase in concentration of substrate leaves the rate unchanged.

It is as if you had 10 taxis (enzymes) waiting to take people (substrates) to a certain destination. If there were only 3 people at the taxi stand and the trip took 5 minutes, the rate at which they arrived at their destination would be 3 people every 5 minutes. If the concentration of people at the stand were increased to 5, the rate would increase to 5 arrivals in 5 minutes. With 10 people, you would have 10 arrivals every 5 minutes. With 20 people at the stand, the rate would still be 10 arrivals in 5 minutes. The taxis have been saturated (in our analogy, each taxi can carry only one passenger). If the taxis could carry 2 or 3 passengers each, the same principle would apply. The rate would simply be higher (20 or 30 people in 5 minutes) before it leveled off.

If the concentration of substrate is held constant and remains in excess (if we always have more people than taxis), the rate of a reaction is proportional to the concentration of enzyme. The more enzyme, the faster the reaction (the more taxis, the more people can be transferred). This relationship holds over a wide range of enzyme concentrations (Figure 24.7).

Enzymes are proteins with acidic and basic groups. It is not surprising, therefore, that enzyme activity is influenced by the concentration of hydronium ions. Each enzyme has its own **optimum pH.** If a graph is plotted of reaction rate versus pH, the curve will show a sharp decline on both sides of some optimum value (Figure 24.8). The optimum pH is generally close to the physiological pH of 7.4 for enzymes in the human body. Pepsin, which operates in the acidic media of the stomach, has an optimum pH of 1.6. It has no activity in basic media. Glycine oxidase operates best at a pH of 8.8. It has no activity below a pH of 5.

Let's illustrate the effect of pH on enzyme activity with an example. Our old friend lysozyme (Section 24.1) hydrolyzes certain bonds in polysaccharides that occur in the cell walls of bacteria. Its optimum pH is 5. Activity declines

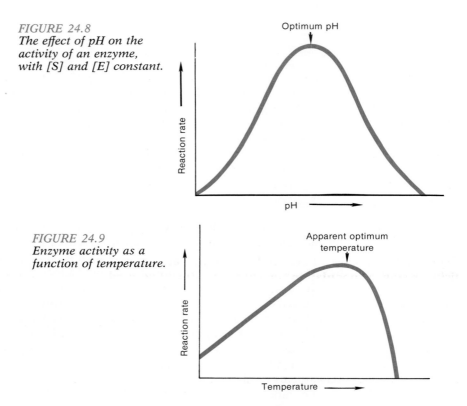

FIGURE 24.8
The effect of pH on the
activity of an enzyme,
with [S] and [E] constant.

Optimum pH

Reaction rate

pH

FIGURE 24.9
Enzyme activity as a
function of temperature.

Apparent optimum
temperature

Reaction rate

Temperature

rapidly at both higher and lower pH. Lysozyme is active only when an aspartic acid side chain is ionized (as COO^-) and a glutamic acid side chain is not (as COOH). At low pH both are protonated. At high pH both are ionized. In either case, the enzyme doesn't function.

Reaction rates generally are affected by temperature. The higher the temperature, the faster the reaction. This is also true for reactions catalyzed by enzymes—up to a point. Enzymes are proteins. Proteins are denatured by higher temperatures, and their enzymatic activity is destroyed. The temperature range for activity of enzymes generally is 10 to 50 °C. The **optimum temperature** for many enzymes in the human body is 37 °C—body temperature (Figure 24.9).

24.6 ENZYME REGULATION AND ALLOSTERISM

Most biochemical processes involve several steps, each catalyzed by an enzyme. The product of each step becomes the substrate for the next enzyme. A reaction product of one enzyme may control the activity of another enzyme. Consider the system

$$A \xrightarrow{E_1} B \xrightarrow{E_2} C \xrightarrow{E_3} D \xrightarrow{E_4} F$$

where the Es are various enzymes controlling the steps of a four-stage reaction. A through D represent substrates of the different enzymes. The last product in the chain, F, may inhibit the activity of enzyme E_1. (This inhibition may be competitive or noncompetitive; see Section 24.9.) When the concentration of F is low, all the reactions proceed rapidly. As the concentration of F rises, though, the action of E_1 is slowed and eventually stops. The buildup of the concentration of F signals E_1 to quit because the cell has enough F for a while. When E_1 shuts down, the whole chain of reactions stops. This mechanism for regulating enzyme activity is called **feedback control.**

Sometimes enzyme regulation takes place by a mechanism involving a site removed from the active site. An enzyme so regulated is called an **allosteric enzyme,** and the process is called **allosterism.** The substance binding at the regulatory site is called a **modulator** or a **regulator.** It may inhibit enzyme activity (**negative modulation**) or enhance it (**positive modulation**). Allosteric enzymes are often large and contain two or more subunits. Often the regulatory site is on one protein chain and the active site on another.

Figure 24.10 shows a simple model for negative modulation of an allosteric enzyme with two subunits. When an inhibitor binds to the regulatory site, the conformations of the proteins change, and the active site is less accessible to the substrate. The reaction rate drops. In the absence of the inhibitor, the substrate binds readily, and reaction proceeds rapidly.

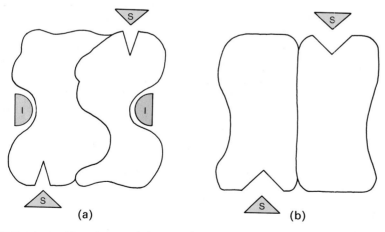

FIGURE 24.10   *Negative modulation of an allosteric enzyme. With an inhibitor attached to the regulatory site (a), the substrate cannot bind effectively, and the reaction slows. Conformations are different without the inhibitor present (b), the substrate binds readily, and the reaction proceeds rapidly.*

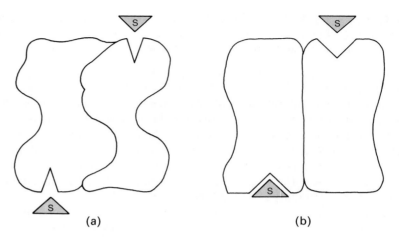

(a)   (b)

FIGURE 24.11 Positive modulation of an allosteric enzyme. With no substrate bound (a), the active sites are relatively inaccessible. When one molecule of substrate is bound (b), though, both protein units change conformation, and a second unit of substrate binds readily.

A simple model for positive modulation is shown in Figure 24.11. When a regulator molecule binds to a site on one protein subunit, the conformation of that subunit and a neighboring subunit both change. The neighbor can then bind substrate more effectively, and the reaction rate is enhanced.

Allosteric regulation occurs in nonenzyme proteins, too. Positive modulation explains how hemoglobin's ability to transport oxygen is enhanced by the binding of an oxygen molecule to one subunit (Chapter 26).

24.7 PROENZYMES: ACTIVATION OF ENZYME PRECURSORS

A protein called *pepsinogen* is secreted from the cells of the stomach. The acid in the stomach converts pepsinogen into pepsin, an active protease. The reaction is also **autocatalytic:** it is catalyzed by the product pepsin. That is, once some pepsin is formed, it speeds the reaction, which produces more pepsin. Activation involves removal of 42 amino acid residues, as small peptides, from the pepsinogen molecule.

$$\text{Pepsinogen} + H_3O^+ \xrightarrow{\text{pepsin}} \text{pepsin} + \text{small peptides}$$

Substances such as pepsinogen, which are not enzymes but which are converted into enzymes by other substances, are called **proenzymes,** or (sometimes) **zymogens.**

Another proenzyme is *trypsinogen,* formed in the pancreas. It is converted into *trypsin* by the removal of a hexapeptide from the amino end. This activation is itself catalyzed by an enzyme, enterokinase.

$$\text{Trypsinogen} \xrightarrow{\text{enterokinase}} \text{trypsin} + \text{a hexapeptide}$$

Activation of proenzymes is quite important. We wouldn't want an enzyme breaking down the proteins of our pancreas. Better to have a proenzyme made there that can be activated in the small intestine, where it can hydrolyze the proteins in our food.

24.8   COFACTORS: COENZYMES, VITAMINS, AND METALS

Some enzymes consist entirely of protein chains. Others contain some other chemical component necessary to the proper function of the enzyme. Such a component is called a **cofactor.** The cofactor may be a metal ion such as zinc (Zn^{2+}), manganese (Mn^{2+}), magnesium (Mg^{2+}), iron(II) (Fe^{2+}), or copper(II) (Cu^{2+}). For example, the enzyme alcohol dehydrogenase requires zinc ion as a cofactor, and arginase requires manganese ion. If the cofactor is organic in nature, it is called a **coenzyme.** Coenzymes are nonprotein. The pure protein part of an enzyme is called an **apoenzyme.** Neither the coenzyme nor the apoenzyme has enzymatic activity alone. The catalytically active enzyme-cofactor complex is sometimes called the holoenzyme.

$$\text{Coenzyme} + \text{apoenzyme} \rightleftharpoons \text{holoenzyme}$$

Nonprotein Protein   (Active)
(inactive) (inactive)

Many coenzymes are vitamins or are derived from vitamin molecules. These important chemical substances and their relationships to vitamins are discussed in detail in Special Topic J.

24.9   ENZYME INHIBITION: POISONS AND DRUGS

Any substance that makes an enzyme less active or that renders it inactive is called an **inhibitor.** Some inhibitors act by binding at the active site and blocking access to the site by substrate. These are called **competitive inhibitors.** Other substances, the **noncompetitive inhibitors,** bind to some other site on the enzyme molecule. They hinder the action of the enzyme by changing its conformation. This may change the shape of the active site or hinder access to it, thus inhibiting the action of the enzyme.

Inhibition also can be either reversible or irreversible. Heavy metal ions usually act irreversibly. Let's look at some examples of inhibitors that serve as poisons that kill and some that act as drugs to sustain life.

Enzymes contain sulfhydryl groups (—SH) on cysteine residues. Special Topic H discusses how mercury ions can deactivate enzymes by reacting with sulfhydryl groups. Lead is another heavy metal poison. Recall that lead(II) ions react with hydrogen sulfide to form insoluble lead sulfide.

$$Pb^{2+} + H_2S + 2\,H_2O \longrightarrow PbS + 2\,H_3O^+$$

Lead (II) ion   Lead sulfide
              (insoluble)

Similarly, lead ions react with the sulfhydryl groups of enzymes, rendering them inactive. This can occur at a position removed from the active site (Figure 24.12). Poisoning by lead or mercury ions is an example of noncompetitive, irreversible inhibition.

The action of organic phosphorus compounds can be used to illustrate competitive inhibition. These compounds act on an enzyme that is an essential part of the chemistry of the nervous system (Special Topic G). Acetylcholine

FIGURE 24.12
Mercury poisoning is an example of irreversible, noncompetitive inhibition of an enzyme. Mercury ions react with sulfhydryl groups to change the conformation of the enzyme and destroy the active site.

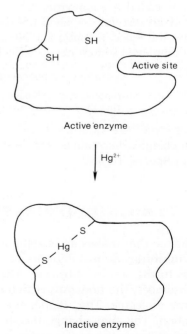

is a neurotransmitter. When an electrical signal from the brain reaches the end of an axon (nerve fiber), acetylcholine is liberated and migrates across the synapse to the receptor cell (refer to Figure G.3). It is thought to activate the next cell by changing the permeability of the cell membrane, perhaps allowing ions to flow through the membrane more readily.

Once acetylcholine has carried the impulse across the synapse, it is rapidly hydrolyzed to acetic acid and choline. This reaction is catalyzed by an enzyme called *cholinesterase.*

$$CH_3C\diagdown \substack{O \\ OCH_2CH_2-N^+-CH_3} + H_2O \xrightarrow{\text{cholinesterase}}$$

Acetylcholine

$$CH_3C\diagdown \substack{O \\ OH} + HOCH_2CH_2-N^+-CH_3$$

Acetic acid   Choline

Choline is relatively inactive. In essence, the breakdown of acetylcholine to acetic acid and choline resets the receptor to the "off" position, making it ready to receive another impulse. Other enzymes, such as acetylase, convert the acetic acid and choline back to acetylcholine, completing the cycle (Figure 24.13).

Anticholinesterase poisons block the action of cholinesterase. Organic compounds of phosphorus, such as the insecticides malathion and parathion and the war gases tabun and sarin (Figure 24.14), are well-known nerve poisons. The polar phosphorus-oxygen bond attaches tightly to cholinesterase (Figure

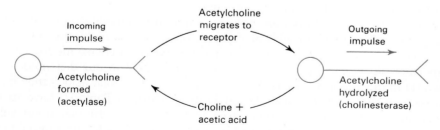

FIGURE 24.13   *The acetylcholine cycle.*

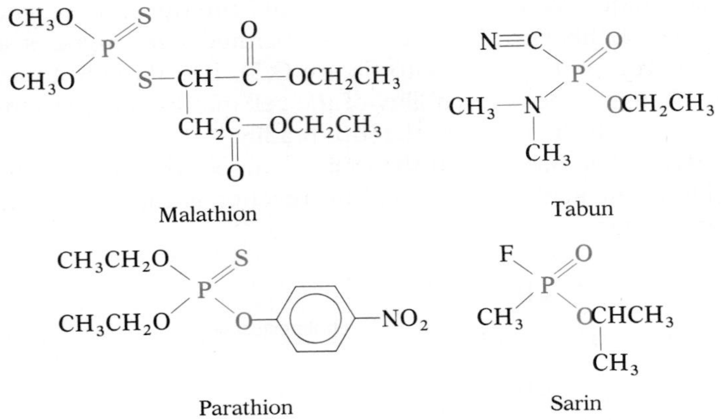

FIGURE 24.14 *Some organic phosphorus compounds. Malathion and parathion are insecticides. Tabun and sarin are nerve gases for use in chemical warfare.*

24.15), preventing the enzyme from performing its normal function. If the breakdown of acetylcholine is blocked, then this messenger compound builds up, causing the receptor nerves to "fire" repeatedly, to be continuously "on." This overstimulates the muscles, glands, and organs. The heart beats wildly and irregularly. The victim goes into convulsions and dies quickly.

Although the organic phosphorus compounds bind tightly to acetylcholinesterase enzymes, some can be displaced. Pralidoxime (2-PAM) is one antidote for organic phosphorus poisons.

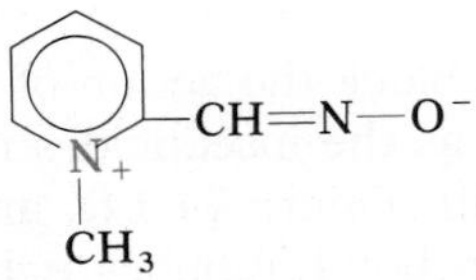

It is thought that the positive charge on nitrogen enables 2-PAM to bind to the site normally occupied by the quaternary nitrogen of acetylcholine, thus displacing the organic phosphorus compound.

Not all enzyme inhibitors are poisons; some are valuable drugs. The sulfa drugs (Special Topic H) provide an example of competitive inhibitors that save lives. The vitamin folic acid (Special Topic J) serves as a coenzyme for several important biochemical processes. We obtain folic acid in our diets and from bacteria in our digestive tracts. Bacteria can synthesize folic acid *if* they have access to *p*-aminobenzoic acid. The sulfa drugs (Special Topic H) have structures similar to that of *p*-aminobenzoic acid. When bacteria encounter sulfa compounds, a bacterial enzyme readily incorporates the sulfa drug into a false folic acid. This altered folic acid not only cannot function as a proper coenzyme,

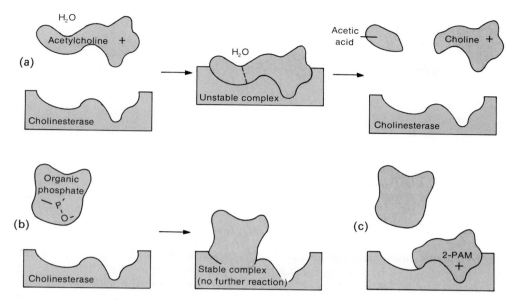

FIGURE 24.15 (a) *Cholinesterase catalyzes the hydrolysis of acetylcholine to acetic acid and choline. (b) An organic phosphate ties up cholinesterase, preventing it from breaking down acetylcholine. (c) 2-PAM displaces the organic phosphorus compound.*

but it serves as a competitive inhibitor of the enzyme. The bacteria are unable to make compounds such as certain amino acids and nucleotides, so they die. The human patient lives because humans don't make folic acid, false or otherwise.

Penicillin also acts by inhibiting an enzyme, a transpeptidase, that bacteria need for making their cell walls. With the enzyme inhibited by penicillin, the cell walls are flimsy, and the bacteria die. Unfortunately, resistant bacteria have developed an enzyme, penicillinase, that deactivates penicillin by hydrolyzing the amide linkage in a four-membered ring (called a β-lactam).

Penicillin

Scientists have attacked the resistant bacteria two ways. One approach is simply to change the groups about the β-lactam ring to prevent a fit with the enzyme. Another way is to use an inhibitor, such as clavulinic acid,

$$CH_2-CH \quad C=CH-CH_2OH$$

which forms a covalent bond to the active site of penicillinase, irreversibly inhibiting the enzyme. When given along with penicillin, clavulinic acid deactivates the enzyme that would otherwise deactivate the penicillin. The penicillin can then destroy the bacteria in the usual way.

24.10 USES OF ENZYMES

Scientists have taken advantage of the catalytic power of enzymes. Indeed, many enzymes have been isolated from living organisms and purified for use in the home and in industries. In the 1950s, ways were found to induce bacteria to produce commercial quantities of some kinds of enzymes. Enzymes are widely used in the food industry. Meat tenderizers contain *papain*, an enzyme obtained from the milky juice of unripe papayas. Papain is a proteolytic enzyme. It cuts the large protein molecules responsible for toughness down to smaller ones, making the meat more tender. Other proteolytic enzymes help predigest foods for babies. Tough, fibrous protein foods are broken down into "soups" that are more readily digested. Interestingly enough, protein hair conditioners are made in much the same way. Collagen, a by-product of the meat-packing industry, is broken down to a "soup" for the hair.

Enzymes are also used in medicine. One such use is in the treatment of one of several forms of hemophilia. The clotting of blood is a complicated process. (This should not be surprising: it is clear that nature must take some precautions to prevent such clotting while the blood is journeying through the blood vessels.) Many of the preliminary steps in the clotting process involve the conversion of proenzymes to their active forms. We shall detail only the final steps. First, prothrombin must be converted to thrombin. This process itself requires a proteolytic enzyme called factor X_a (read "factor ten-A").

$$\text{Prothrombin} \xrightarrow{\text{factor } X_a} \text{thrombin}$$

Thrombin, in turn, serves as an enzyme for the conversion of fibrinogen to fibrin.

$$\text{Fibrinogen} \xrightarrow{\text{thrombin}} \text{fibrin}$$

Fibrin is the actual material of the blood clot.

In the form of hemophilia mentioned here, the factor X_a is missing. The absence of this enzyme is hereditary. It has been found, however, that the venom of a snake called Russell's viper is effective in converting prothrombin to thrombin. This venom has been used as a treatment for some cases of hemophilia.

In the classic form of hemophilia, a substance called factor VIII is lacking. Bleeding crises are stanched with transfusions of concentrated clotting factor made from thousands of pints of donated blood and costing a patient thousands of dollars a year. Now scientists have isolated the gene for factor VIII, duplicated (cloned) it, and inserted it into bacteria. These bacteria produce tiny amounts of the factor. After testing in animals to ensure safety, trials in humans will begin. In a few years, factor VIII should be readily available for treating classic hemophilia.

A far less exotic but far more widespread medical application of enzymes is in the area of diagnosis. Clinical analysis for enzymes in body fluids or tissues is a common diagnostic technique. For example, elevated blood levels of creatine kinase (CK) and glutamic-oxaloacetic transaminase (GOT) accompany some forms of severe heart disease. A blood analysis that shows high levels of CK may indicate that the heart muscle has suffered serious damage. On the other hand, many forms of strenuous (and healthful) physical activity will also result in elevated CK levels. The enzyme mediates the reaction that serves as one source of energy for muscle contraction. Indeed, it is even possible for the CK level to rise simply because someone who hates needles has tensed up while waiting for the blood sample to be taken. Nonetheless, as Figure 24.16 suggests, analysis for specific enzymes is considered one valuable source of data on which to base a medical diagnosis.

☐ ROUTINE	LAB NO._____			IF IMPRINT PLATE NOT USED / PRINT HERE	PATIENT NO.		ROOM NO.	
☐ PRE-OP	SPECIMEN: ☐ BLOOD ☐ OTHER				PATIENT NAME		AGE	
☐ ASAP	DATE_____ TIME_____				ADDRESS			
☐ STAT	TECH_____				DOCTOR			

✔	TEST	NORM	RESULT	✔	TEST	NORM	RESULT	✔	TEST	NORM	RESULT	✔	TEST	NORM	RESULT
	LDH				Amylase				Albumin				Calcium		
	LDH-1				Acid Phos.				Globulin				Phosphorus		
	CPK				Cholesterol				BUN				Magnesium		
	CPK-MB				Triglyceride				Creat				Ethanol		
	Alk. Phos.				HDL-Chol.				Total Bili				Hgb AIC		
	SGOT (AST)				Iron				Direct Bili				Ammonia		
	SGPT (ALT)				TIBC				Indirect Bili				Lithium		
	GGPT				Total Prot.				Uric Acid				TECH:	DATE:	
BRIGGS, DES MOINES, IA 50306				CHEMISTRY I							PRINTED IN U.S.A				

FIGURE 24.16 A hospital form for clinical analysis of a blood sample. (Courtesy of the Briggs Corporation, Des Moines, Iowa.)

As we learn more about the structure and function of enzymes, we will no doubt find many more uses for them in industry, foods, and medicine.

PROBLEMS

1. Define and, where appropriate, give an example for each of the following:
   a. enzyme
   b. substrate
   c. optimum pH
   d. optimum temperature
   e. active site
   f. regulatory site
   g. cofactor
   h. coenzyme
   i. proenzyme
   j. feedback control
   k. apoenzyme
   l. holoenzyme

2. What is the substrate for each of the following enzymes?
   a. maltase
   b. cellulase
   c. peptidase
   d. lipase

3. Which enzyme is more specific, urease or carboxy-peptidase?

4. To which of the six major types of enzymes does each of the following belong?
   a. decarboxylase
   b. peptidase
   c. transaminase
   d. dehydrogenase
   e. kinase
   f. lipase

5. Do enzymes change the *extent* of a reaction? Explain fully.

6. What is the chemical nature of enzymes?

7. Describe the lock-and-key model of enzyme action.

8. Describe the induced-fit model of enzyme action.

9. Enzymes are said to exhibit specificity. Define or illustrate these terms:
   a. absolute specificity
   b. stereospecificity
   c. reaction specificity
   d. linkage specificity

10. Why does the body synthesize trypsin as the proenzyme trypsinogen rather than make it directly?

11. How does the body activate trypsinogen, that is, how is it converted to trypsin?

12. Alcohol dehydrogenase is an enzyme that catalyzes the conversion of ethanol to acetaldehyde. The active enzyme consists of a protein molecule and zinc ion. Identify the following components of this reaction system:
   a. substrate
   b. cofactor
   c. apoenzyme

13. Can the zinc ion mentioned in Problem 12 be called a coenzyme? Explain your answer.

14. Succinate dehydrogenase is an enzyme that is only active in combination with a nonprotein organic molecule called flavin adenine dinucleotide (FAD). Is FAD a cofactor? Is it a coenzyme?

15. The concentration of substrate X is low. What happens to the rate of the enzyme-catalyzed reaction if the concentration of X is doubled?

16. An enzyme has an optimum pH of 7.4. What happens to the activity of the enzyme if the pH drops to 6.8? If the pH rises to 8.0?

17. A bacterial enzyme has an optimum temperature of 35 °C. Will the enzyme be more or less active at normal body temperature? Will it be more or less active if the patient has a fever of 40 °C?

18. What is an enzyme inhibitor?

19. How does competitive inhibition work?

20. How does noncompetitive inhibition work?

21. What is allosterism?

22. Why should enzymes consist of 100 or more amino acid units when only a few amino acid units are involved in the active site?

23. What is the turnover number for an enzyme?

24. What is negative modulation? Give an example.

25. What is positive modulation? Give an example.

26. How do mercury ions act as poisons? Is their action reversible or irreversible?

27. What is acetylcholine? Describe the acetylcholine cycle.

28. How do organic phosphorus compounds act as poisons?

29. How does 2-PAM act as an antidote for poisoning by organic phosphorus compounds?

30. How do sulfa drugs kill bacteria?

31. How does penicillin kill bacteria?

32. How do resistant bacteria deactivate penicillin?

33. Describe two ways to get around the problem of bacterial resistance.

34. What is the active ingredient in meat tenderizer? How does it work?
35. Which has the longer polypeptide chain, prothrombin or thrombin?
36. What enzymes are monitored in the blood of a patient who is suspected of having a heart attack? Why? Why is the test not always reliable?
37. L-Aspartic acid is readily converted to fumaric acid in the presence of the enzyme aspartase, but D-aspartic acid does not react. Explain.
38. What is the name of the enzyme that is involved in the conversion of lactose to galactose and glucose? To what class of enzymes does it belong?
39. In a commercial process, glucose (in corn syrup) is converted enzymatically to fructose (to form high-fructose corn syrup). What type of enzyme is involved? (Recall that both glucose and fructose have the formula $C_6H_{12}O_6$.)
40. What type of enzyme is involved in the conversion

$$\text{of pyruvic acid } (CH_3\overset{\overset{\displaystyle O}{\|}}{C}\text{—COOH}) \text{ to lactic acid}$$

$$(CH_3\overset{\overset{\displaystyle OH}{|}}{CH}\text{—COOH})?$$

41. What type of enzyme is involved in the conversion of pyruvic acid to CH_3CHO and CO_2?
42. What type of side chains on amino acid residues at the active site of an enzyme might bind the COO^- group of a substrate?
43. How does increased enzyme concentration affect the rate of a reaction?
44. The active site of acetylcholinesterase has aspartic acid, histidine, and serine residues at the site. In acid solution, the enzyme is inactive, but activity increases as the pH rises. Explain.
45. The enzyme pepsin is active in the stomach but not in the intestine, where the pH is slightly basic. Estimate an optimum pH for pepsin. Would it more likely be around 2, 5, 8, or 12?
46. Sketch the general shape of a curve showing how the rate of a reaction varies with the concentration of substrate (with the concentration of enzyme constant).
47. Alcohol dehydrogenase is involved in the oxidation of both methanol and ethanol. How might ethanol work as an antidote for methanol poisoning?

Energy and Life

L ife requires energy. Living cells are inherently unstable and avoid falling
apart only because of a continued input of energy. Living organisms
are restricted to using certain forms of energy. Supplying a plant with
heat energy by holding it in a flame will do little to prolong its life. On the
other hand, a green plant is uniquely able to tap the richest source of energy
on Earth, sunlight. A green plant captures the radiant energy of the sun and
converts it to chemical energy, which it stores in carbohydrate molecules. The
plant can also convert those carbohydrate molecules to fat molecules and, with
proper inorganic nutrients, to protein molecules (Figure 25.1).

Animals cannot directly use the energy of sunlight. They must eat plants, or
other animals that eat plants, in order to get carbohydrates, fats, and proteins
with their stored chemical energy. Once digested and transported to the cell, a
food molecule can be used in either of two ways. It can be used as a building
block to make new cell parts or to repair old ones, or it can be "burned" for
energy.

The entire series of coordinated chemical reactions that keep cells alive is
called **metabolism.** In general, metabolic reactions are divided into two classes.
The degrading of molecules to provide energy is called **catabolism.** The process
of building up the molecules of living systems is called **anabolism.** In this
chapter, we focus on the energy transformations that accompany catabolic
chemical processes. We leave for other chapters a detailed discussion of the
chemical reactions involved in metabolism.

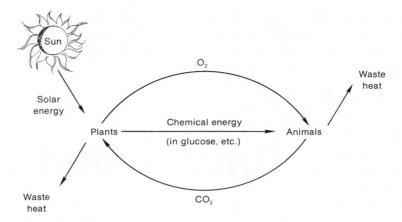

FIGURE 25.1   *Some energy transformations in living systems.*

25.1 LIFE AND THE LAWS OF THERMODYNAMICS

Thermodynamics is the study of the relationships between different forms of energy. Included within its scope are considerations of energy changes associated with chemical reactions. It is a quantitative science, and its laws can be rigorously derived using higher mathematics. That need not concern us here, for those aspects that are of interest in the chemistry of life can be rather simply stated and intuitively grasped. In fact, you had your first encounter with thermodynamics back in Chapter 5 when we discussed exothermic and endothermic reactions. At that time we concentrated on the transfer of *heat* energy. Our present discussion will not be so restricted.

The **first law of thermodynamics** states that energy is neither created nor destroyed. It can, however, be changed from one form to another. Life, for us, is a process that changes potential energy (the stored energy in our food) into the kinetic energy associated with breathing, the beating of our hearts, and general movement, whether it be walking, running, or twiddling our thumbs.

The **second law of thermodynamics** states that when energy is converted from one form to another, some of it is "lost." This is not a contradiction of the first law. Energy has not been destroyed. It has merely been lost in the sense that it is no longer useful. When you drive a car, for example, only about a quarter of the energy stored in the gasoline goes into moving the car down the road. The remainder is "lost" as heat. It may warm up the atmosphere around the car a bit, but it cannot be captured and put to work. We say that the car is about 25% efficient. The **efficiency** of a system, then, compares the amount of useful work accomplished to the amount of energy put into the system. Another statement of the second law is that no machine—or organism—can be 100% efficient. Neither technological improvement nor scientific progress will change this assessment. The efficiency of some processes can be increased, but it is theoretically impossible to achieve 100% efficiency.

The energy content of chemical compounds is changed when some chemical bonds are broken and new ones are formed. If, in this process, there is a net *release* of energy by the compounds, the reaction is said to be **exergonic.** If, on the other hand, a net *input* of energy is required, the reaction is **endergonic.** To survive, all living organisms must carry out exergonic reactions that release sufficient energy to maintain life processes. There must be endergonic reactions to counterbalance the energy-releasing processes. Reactions that store energy (such as the photosynthesis carried on by green plants) are endergonic. Because no process is 100% efficient, the endergonic reactions must store *more* energy than is required by organisms for survival. Some of the energy released by subsequent exergonic reactions will always be wasted. It's a point of law—the second law.

25.2 PREDICTING SPONTANEOUS CHANGE: FREE ENERGY

If you were holding a book and then released it, the book would drop to the floor spontaneously. If a book were lying on the floor, you would not stand patiently waiting for it to leap into your hands. You know that, no matter how long you wait, the book will not jump from the floor back into your hands. The movement of book from hand to floor could be described as natural, and the movement of book from floor to hand as unnatural.

A chemical reaction can also be described as having a natural direction. Thermodynamicists have defined a term, **change in free energy** or ΔG, that indicates the natural direction of reactions. A negative value for ΔG (reported in kilocalories per mole) means that the reaction can occur naturally or spontaneously. Note that *spontaneously* does not necessarily mean *immediately*. A ΔG value does not indicate *when* the reaction will proceed but *whether* it can proceed naturally, that is, of its own accord. A positive value for ΔG means that the direction of the reaction is not natural. If the value for ΔG is zero, then the system being considered is at equilibrium.

25.3 COUPLED REACTIONS

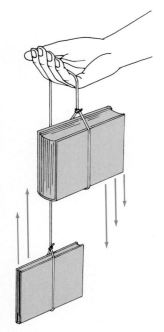

Many reactions essential to life are nonspontaneous. How can these reactions be forced to proceed in directions that are not natural? How can you bring a book from the floor to your hand? Obviously, one way is to bend over and pick it up. Another way can take advantage of the natural process of a book falling to the floor. Suppose a book on the floor is connected by a taut rope to a larger book in your hand. All you have to do is hold on to the rope and let the book in your hand drop to the floor. What you have is a primitive pulley. As the larger book drops, the smaller book on the floor is lifted to your hand (Figure 25.2). The two processes have been coupled, and the spontaneous one drives the nonspontaneous one.

In a living cell, an endergonic reaction with a positive ΔG can be made to proceed if it is *coupled* with an exergonic reaction with a negative ΔG of larger absolute value. By **coupling** we mean that the two reactions share a common intermediate that transfers energy from one reaction to the other. (In the book analogy, the rope serves as the intermediate for transferring energy from one book to the other.)

Let's look at some specific chemistry to demonstrate this principle of coupled reactions. Consider the reaction of glucose and fructose to form sucrose.

FIGURE 25.2
The coupling of a spontaneous and a nonspontaneous action.

Glucose + Fructose ⟶ Sucrose

This reaction has a positive ΔG; it is not spontaneous. The reaction must be coupled with another reaction that has a large negative ΔG. In some cells, the exergonic reaction is the hydrolysis of the compound ATP to the compound ADP. We shall discuss the structure of these two compounds in the next section. For the moment, you need only know that the conversion of ATP to ADP is an exergonic reaction.

$$\text{Glucose + fructose} \longrightarrow \text{sucrose} + H_2O \qquad \Delta G = +7.0$$
$$\text{ATP} + H_2O \longrightarrow \text{ADP} + HPO_4{}^{2-} \qquad \Delta G = -7.3$$

The sum of the ΔGs for these reactions is negative, so if the two reactions could be coupled, the formation of sucrose would be thermodynamically favored. The reactions are coupled through the common intermediate, glucose 1-phosphate. When ATP reacts with glucose to form this intermediate, some of the energy stored in the bonds of ATP is trapped in the bonds of the intermediate:

$$\text{Glucose} + \text{ATP} \longrightarrow \text{glucose 1-phosphate} + \text{ADP}$$

Then the intermediate reacts with fructose, and this time, some energy is trapped in the sucrose molecule.

$$\text{Glucose 1-phosphate} + \text{fructose} \longrightarrow \text{sucrose} + \text{HPO}_4{}^{2-}$$

Because glucose 1-phosphate is a product of the first reaction and a reactant in the second, it is not included in the equation for the overall reaction.

$$\text{Glucose} + \text{fructose} + \text{ATP} \longrightarrow \text{sucrose} + \text{ADP} + \text{HPO}_4{}^{2-}$$

25.4 ATP: UNIVERSAL ENERGY CURRENCY

Many cellular reactions, otherwise unfavored thermodynamically, can proceed if they are coupled with a high-energy reaction such as the hydrolysis of ATP or ADP. The structures of ATP (adenosine triphosphate), ADP (adenosine diphosphate), and AMP (adenosine monophosphate) are presented in Figure 25.3. These compounds were mentioned briefly in Chapter 23. They are also encountered frequently in other chapters. For the moment, we can focus on the phosphate linkages. ATP is hydrolyzed in two steps, each of which releases considerable energy. The standard free energy changes ($\Delta G°$)* are

$$\text{ATP} + \text{H}_2\text{O} \longrightarrow \text{ADP} + \text{HPO}_4{}^{2-} \qquad \Delta G° = -7.3 \text{ kcal/mol}$$
$$\text{ADP} + \text{H}_2\text{O} \longrightarrow \text{AMP} + \text{HPO}_4{}^{2-} \qquad \Delta G° = -6.5 \text{ kcal/mol}$$
$$\text{AMP} + \text{H}_2\text{O} \longrightarrow \text{Adenosine} + \text{HPO}_4{}^{2-} \qquad \Delta G° = -2.2 \text{ kcal/mol}$$

The free energy changes associated with the first two transformations (involving the cleavage of a P—O—P linkages) are significantly greater than that for the third transformation (cleavage of a P—O—C linkage). Because ATP and

---

* The **standard free energy change** is the change in free energy measured under standard conditions of temperature, pressure, and concentration. Standard conditions, as defined by most thermodynamicists, do not coincide with conditions inside cells, where the reactions in which we are interested take place. Attempts to take into account intracellular conditions (which are not constant throughout the body) might yield values for the three hydrolysis steps of 8.8, 8.6, and 3.0 kcal/mol respectively. The important point to note is that the first two values are higher than the third.

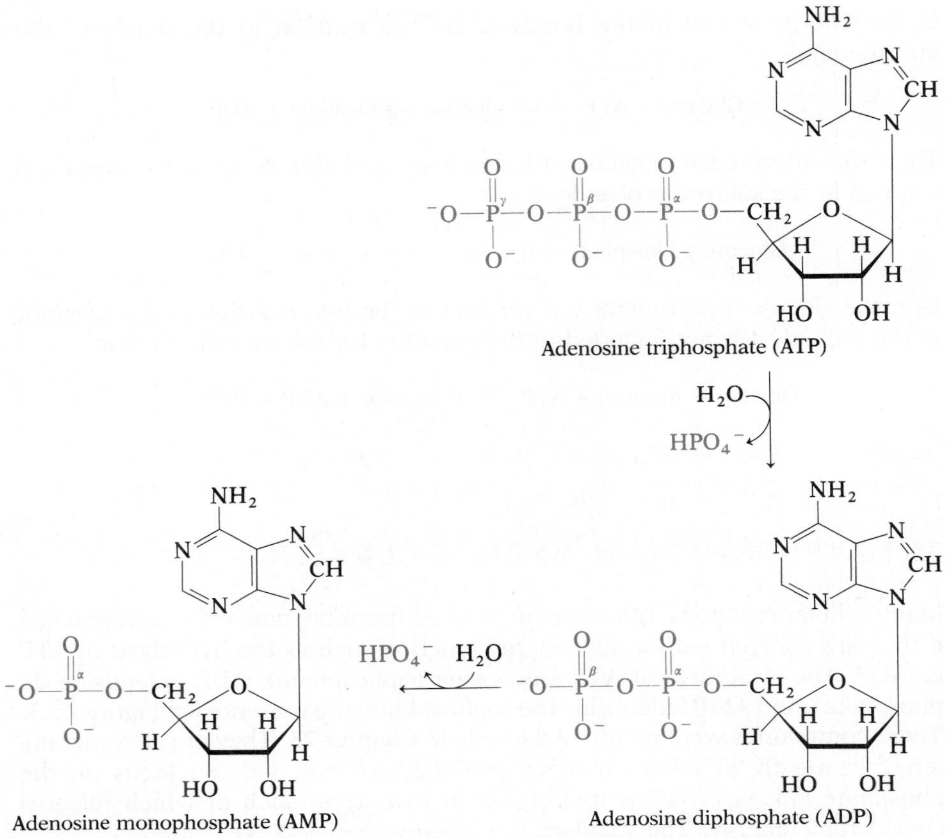

FIGURE 25.3 Stages in the hydrolysis of adenosine triphosphate.

ADP release large amounts of energy upon hydrolysis, they are called **high-energy compounds.** AMP, which yields considerably less energy upon hydrolysis, is not regarded as a high-energy substance. The cutoff for the high-energy designation is rather arbitrarily set at about -5 kcal/mol.

ATP is not the only high-energy compound in cells, but it is the most important one and is referred to as the *energy currency* of the cell. It acts as a general repository of energy, trapping energy released when the fuel molecules of the cell are metabolized and then releasing that energy to drive otherwise unfavored reactions (Figure 25.4).

The standard free energies of hydrolysis for a number of important organic phosphates are given in Table 25.1.

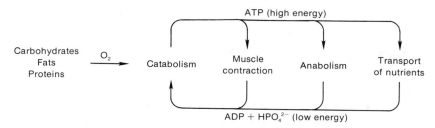

FIGURE 25.4 *ATP is called the energy currency of the cell. Energy stored in ATP by catabolic reactions can be used for mechanical work such as muscle contraction, for chemical synthesis (anabolism), and to transport nutrients.*

TABLE 25.1   *Free Energy of Hydrolysis of Some Phosphates*

Compound	Hydrolysis Products	Approximate $\Delta G°$ (kcal/mol)
Phosphoenolpyruvate (PEP)	Ketopyruvic acid + HPO_4^{2-}	-12.8
1,3-Diphosphoglyceric acid	3-Phosphoglyceric acid + HPO_4^{2-}	-12.0
Creatine phosphate	Creatine + HPO_4^{2-}	-10.5
Acetyl phosphate	Acetic acid + HPO_4^{2-}	-10.0
ATP	ADP + HPO_4^{2-}	-7.3
ADP	AMP + HPO_4^{2-}	-6.5
Glucose 1-phosphate	Glucose + HPO_4^{2-}	-5.0
Fructose 6-phosphate	Fructose + HPO_4^{2-}	-3.8
Glucose 6-phosphate	Glucose + HPO_4^{2-}	-3.3
3-Phosphoglyceric acid	Glyceric acid + HPO_4^{2-}	-2.4
Glycerol 3-phosphate	Glycerol + HPO_4^{2-}	-2.2
AMP	Adenosine + HPO_4^{2-}	-2.2

25.5 SYNTHESIS OF ATP

When glucose is oxidized, either by burning in air or by catabolism in living cells, the free energy change is -686 kcal/mol.

$$C_6H_{12}O_6 + 6\,O_2 \longrightarrow 6\,CO_2 + 6\,H_2O \qquad \Delta G° = -686 \text{ kcal/mol}$$

Similar measurements can be made for other food substances. When the energy is released by burning in air, its release is rather sudden, and the reaction is accompanied by a hot flame. If the same sudden release occurred in a living cell, the cell would be destroyed. Oxidation in cells of necessity proceeds through a series of small steps, and energy is released in small increments. Chemical intermediates, called **metabolites,** are formed as the multistep reaction proceeds. Part of the energy is released as heat and serves to maintain our body temperatures. Some of the many steps involve energy storage in bonds of ATP or other high-energy compounds.

We examine the *chemistry* of these metabolic processes in some detail in Chapters 27, 28, and 29. For now, let's just pick out some of the steps in those reaction sequences in which ATP is synthesized. For example, when glucose is metabolized by a process called anaerobic glycolysis (Section 27.4), one of the intermediate compounds formed is 1,3-diphosphoglyceric acid. Note from Table 25.1 that this compound has a higher (more negative) $\Delta G°$ for hydrolysis than ATP. That means it can transfer a phosphate group to ADP to form ATP. Here is how you assess the likelihood of that transfer. If the hydrolysis of ATP to ADP has a $\Delta G°$ of -7.3 kcal/mol, then the reverse reaction has a $\Delta G°$ of $+7.3$ kcal/mol.

$$\text{ADP} + \text{HPO}_4{}^{2-} \longrightarrow \text{ATP} + \text{H}_2\text{O} \qquad \Delta G° = +7.3 \text{ kcal/mol}$$

The hydrolysis of 1,3-diphosphoglyceric acid, according to Table 25.1, has a $\Delta G°$ of -12.0 kcal/mol.

$$\text{1,3-Diphosphoglyceric acid} + \text{H}_2\text{O} \longrightarrow \text{3-phosphoglyceric acid} + \text{HPO}_4{}^{2-}$$
$$\Delta G° = -12.0 \text{ kcal/mol}$$

If the two equations are added together to yield a net equation, the result is

$$\text{1,3-Diphosphoglyceric acid} + \text{ADP} \longrightarrow \text{3-phosphoglyceric acid} + \text{ATP}$$

The $\Delta G°$ of the coupled reactions is $-12.0 + 7.3 = -4.7$ kcal/mol. It is negative, and the coupled reactions are thermodynamically favored.

In another step of the anaerobic glycolysis process, the conversion of phosphoenolpyruvic acid (PEP) to pyruvic acid is coupled with the formation of ATP from ADP.

$$\text{PEP} + \text{ADP} \longrightarrow \text{pyruvic acid} + \text{ATP}$$

According to Table 25.1, the hydrolysis of PEP has a $\Delta G°$ of -12.8 kcal/mol. Since this value is higher than that for the hydrolysis of ATP (-7.3 kcal/mol), the coupled reactions are thermodynamically favored, that is, PEP (higher negative $\Delta G°$) can transfer its phosphate group to form ATP (lower negative $\Delta G°$).

In contrast to the preceding examples, the first step of the glycolysis process, the conversion of glucose to glucose 6-phosphate, is coupled to the conversion of ATP to ADP. In this step, instead of forming ATP, the hydrolysis of ATP is used to drive the reaction.

$$\text{Glucose} + \text{ATP} \longrightarrow \text{glucose 6-phosphate} + \text{ADP}$$

We could have predicted this from an examination of Table 25.1. The standard free energy of hydrolysis for glucose 6-phosphate is -3.3 kcal/mol. That is lower than the value for ATP. Therefore, we should expect the phosphate group to be transferred from ATP to form glucose 6-phosphate.

The third step of the anaerobic glycolysis pathway also requires ATP to convert fructose 6-phosphate to fructose 1,6-diphosphate.

$$\text{Fructose 6-phosphate} + \text{ATP} \longrightarrow \text{fructose 1,6-diphosphate} + \text{ADP}$$

So in glycolysis, some ATP is used up, and some is formed. In fact, the progression of one molecule of glucose through this metabolic pathway to form two molecules of lactic acid results in a net yield of two molecules of ATP. Some of the energy in the original glucose molecule is trapped in ATP molecules that can be used to drive other life-sustaining processes. What kind of processes? They range from the contraction of muscles to the transmission of nerve impulses to a whole spectrum of metabolic reactions. Unfortunately, two molecules of ATP for every molecule of glucose is a rather poor return on investment. Fortunately, ATP production associated directly with the conversion of glucose to lactic acid represents only a minor source of the body's energy currency. A much more productive route to ATP molecules is discussed in Section 25.7.

25.6 THREE IMPORTANT COMPOUNDS IN CATABOLISM

So far in our discussion we have used word *equations* with little reliance on complicated chemical formulas. You probably have noted that we have little reluctance to use big structures in other chapters, and you may be wondering why we are not using them here. The simple answer is that we now want to focus on energy changes, not on complicated structures.

However, before we proceed further, we must introduce a few more important chemicals. Two are coenzymes (Section 24.8). Their structures are given in Figure 25.5. Nicotinamide adenine dinucleotide (NAD^+) is derived from ADP, ribose, and nicotinamide, a B vitamin (Special Topic J). Flavin adenine dinucleotide (FAD) is derived from ADP, ribitol (a reduced form of ribose), and a tricyclic nitrogen-containing compound called flavin. The ribitol and flavin parts together make up riboflavin, another B vitamin. The business part of NAD^+ is the nicotinamide portion. (The $+$ on NAD^+ represents the positive charge on the nitrogen atom.) Similarly, the active portion of FAD is flavin. Both coenzymes attach to their respective apoenzymes through the ADP end, leaving the other end sticking out to participate in reactions.

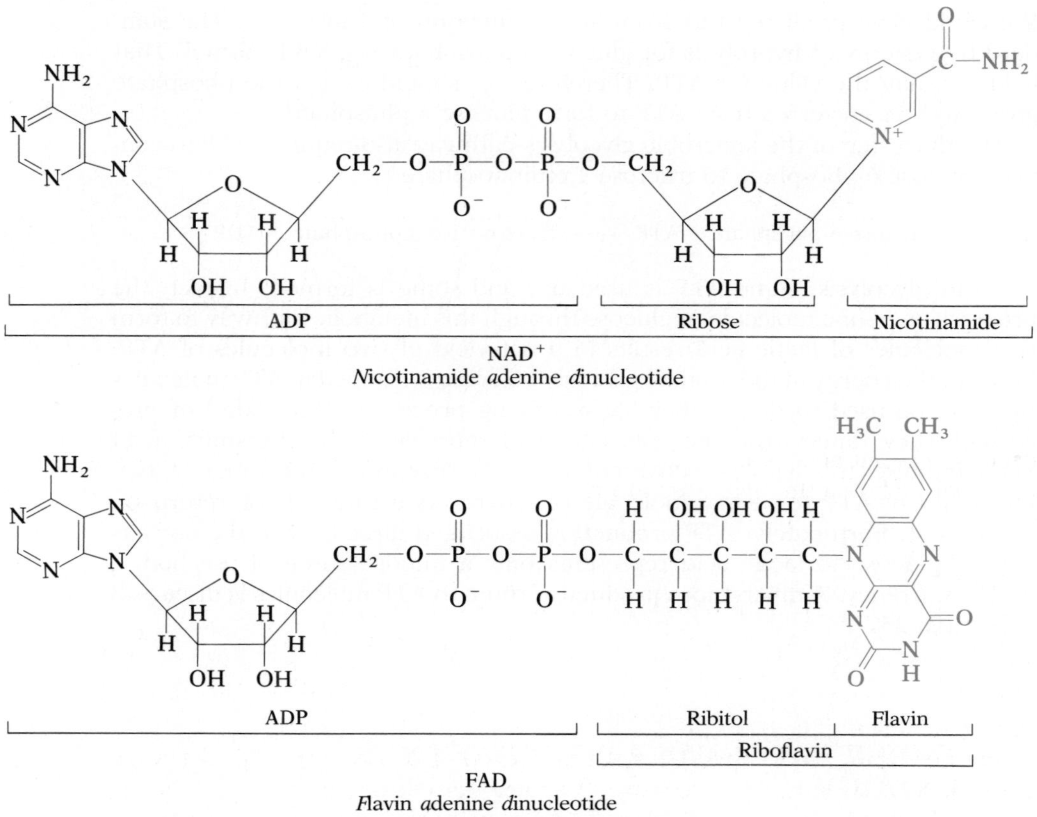

FIGURE 25.5 *The structures of NAD$^+$ and FAD.*

Both NAD$^+$ and FAD are oxidizing agents. When NAD$^+$ reacts, it is the nicotinamide portion of the molecule that is reduced by picking up a hydrogen atom.

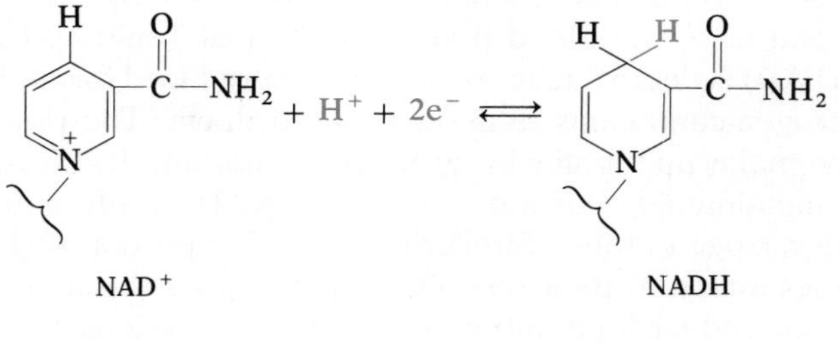

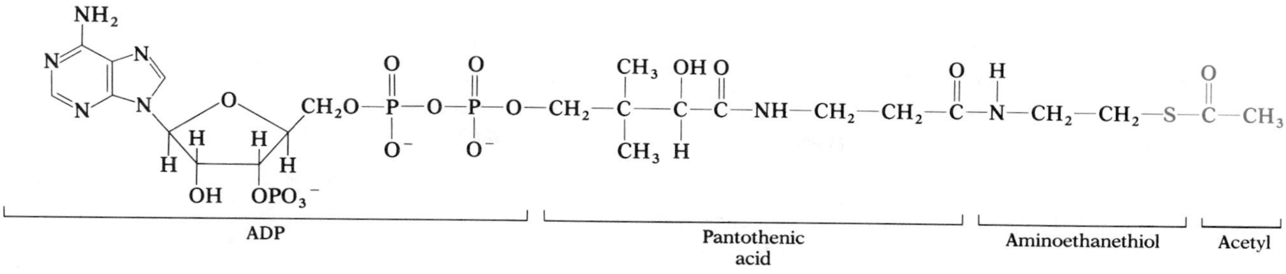

FIGURE 25.6 *The structure of acetyl coenzyme A (acetyl CoA).*

The reduced form of NAD^+ is called NADH. Similarly, when FAD is reduced (that is, when it acts as an oxidizing agent), it is the flavin portion of the molecule that picks up the hydrogen atoms.

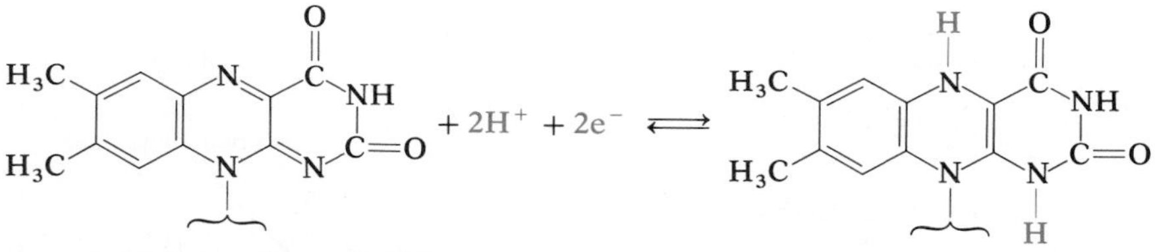

Reduced FAD is called $FADH_2$.

We need one other actor on stage to complete our cast for the oxidative phosphorylation show (next section). When carbohydrates, lipids, and proteins are metabolized (Chapters 27, 28, and 29, respectively), the metabolites are all interconnected through a common intermediate called acetyl coenzyme A (acetyl CoA). Figure 25.6 shows that this molecule is composed of an ADP unit, pantothenic acid (another B vitamin), an aminoethanethiol unit, and an acetyl group. Acetyl CoA is the starting point for the citric acid cycle (Section 27.5), an important part of the process for aerobic oxidation of carbohydrates.

25.7 OXIDATIVE PHOSPHORYLATION

The reactions of the citric acid cycle (also called the Krebs cycle) involve oxidation in 4 of the 10 steps (Figure 25.7). The net reaction of the cycle is the conversion of pyruvic acid to carbon dioxide and water.

$$CH_3\overset{O}{\underset{\|}{C}}-\overset{O}{\underset{\|}{C}}-OH + \tfrac{5}{2}O_2 \longrightarrow 3CO_2 + 2H_2O$$

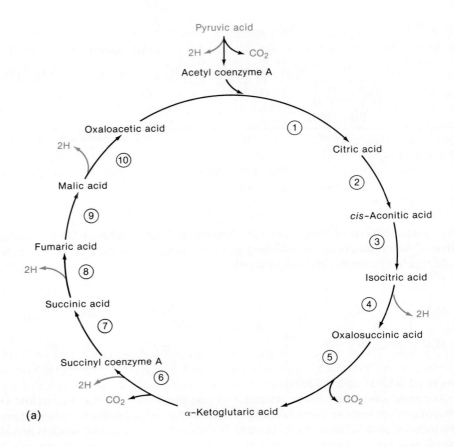

(a)

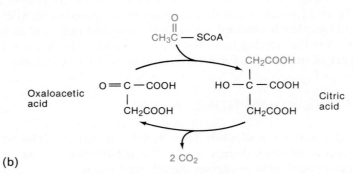

(b)

FIGURE 25.7 The Krebs cycle. (a) The conversion of pyruvic acid to carbon dioxide involves a large number of transformations. (b) This greatly condensed version of the cycle emphasizes the fate of the two carbon atoms in acetyl coenzyme A.

The process is called aerobic oxidation. Nowhere in the cycle, however, is oxygen directly involved. Nor do any of the steps result directly in the production of ATP. Let's look now at the four steps involving oxidation. We will take a detailed look at the chemistry of the cycle in Section 27.5. For now, we focus on energy changes.

Recall that oxidation can be defined as a removal of hydrogen atoms. In Steps 4, 6, 8, and 10 of the citric acid cycle, the intermediates are undergoing oxidation, indicated in the diagram (Figure 25.7) by the release of two hydrogen atoms at each of these steps. As is also indicated in the diagram, when pyruvic acid feeds into the cycle, it also undergoes oxidation as it is converted to acetyl coenzyme A. The hydrogen atoms we show as by-products when a metabolite is oxidized must actually be transferred to an oxidizing agent. Oxygen is a wonderful oxidizing agent, but it does not pick up these hydrogens directly. Instead, the hydrogen atoms are initially transferred to NAD^+ and FAD. These molecules serve as the oxidizing agents for the Krebs cycle. They accept the hydrogen atoms released by the Krebs cycle intermediates.

$$NAD^+ + 2H \longrightarrow NADH + H^+$$
$$FAD + 2H \qquad FADH_2$$

The supplies of these oxidizing agents are limited. In order for the Krebs cycle to keep on turning, it is necessary that the reduced forms of these compounds, NADH and $FADH_2$ be reoxidized. The ultimate oxidizing agent is—you guessed it—oxygen. The transfer of the hydrogens from NADH and $FADH_2$ to oxygen (to form water) is coupled with the synthesis of ATP. The process is referred to as **oxidative phosphorylation.** The overall reaction for NADH can be written

$$NADH + H^+ + 3\,ADP + 3\,HPO_4{}^{2-} + \tfrac{1}{2}O_2 \longrightarrow NAD^+ + 3\,ATP + 4\,H_2O$$

As in most biochemical processes, the actual oxidation involves a whole series of reactions (Figure 25.8), which make up what is called the **respiratory chain.** What travel down this chain are electrons. (Remember oxidation-reduction can be regarded as the transfer of hydrogens *or* of electrons.) The **cytochromes** (Figure 25.8) are iron-containing proteins. They are globular proteins, most of which contain one or more heme units. The iron may be in either the $+2$ or $+3$ oxidation state. The oxidized form (with Fe^{3+}) can accept an electron; the reduced form (Fe^{2+}) can donate one. Thus, electrons are transported from one cytochrome to another. Eventually, electrons are passed to molecules of oxygen (the reagent we replenish in our body with each breath), reducing the oxygen to water. The electrons originate ultimately in the metabolites undergoing oxidation in the Krebs cycle. For example, when isocitric acid is

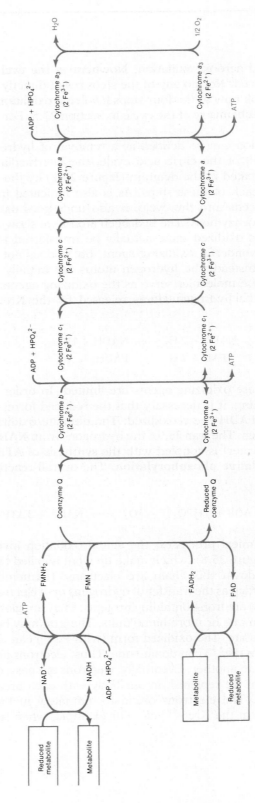

FIGURE 25.8   The electron transport system in oxidative phosphorylation.

oxidized to oxalosuccinic acid by NAD^+, its electrons are passed *via* NADH through the respiratory chain to oxygen.

Hydrogen cyanide and its salts have long been recognized as lethal poisons. The gas released in execution chambers is hydrogen cyanide. A number of potent pesticides incorporate cyanide salts. Sodium cyanide (NaCN) is used in the extraction of gold and silver from their ores and in electroplating baths. It is interesting to note that the first convictions (in 1985) for "corporate murder" were obtained against plant managers who allowed unsafe practices in the handling of sodium cyanide solutions that were being used to extract silver from used X-ray films. One worker died, and several others were made ill by excessive exposure to cyanide. The poison used in the mass suicide that occurred at Jonestown in Guyana was cyanide. So too was the lethal contaminant found in capsules of Tylenol that resulted in the withdrawal of Tylenol capsules from the market. The cyanide ion (CN^-), because of its small size, quickly makes its way into cells, where it binds to the iron in the heme of one of the cytochromes. This immediately shuts down the respiratory chain by inhibiting electron transfer. Cell respiration then ceases. The period from ingestion to death is only a few minutes, which is what makes this poison so notoriously lethal. There is little time in which to institute lifesaving measures.

25.8 MITOCHONDRIA

In humans, oxidative phosphorylation takes place in cell organelles called **mitochondria** (Figures 21.20 and 25.9). The number of mitochondria in a particular cell reflects the energy requirements of the cell, and the mitochondria can reproduce themselves if the energy requirements of the cell increase. Within the mitochondria are contained the enzymes of the respiratory chain

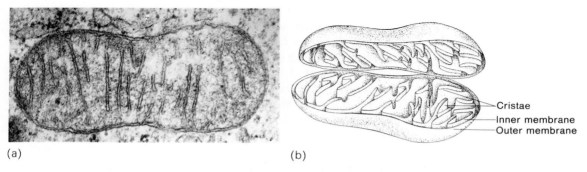

Cristae
Inner membrane
Outer membrane

(a)   (b)

FIGURE 25.9   (a) Electron micrograph of a mitochondrion. (b) Three-dimensional representation. (Photo courtesy of E. A. Munn.)

and the enzymes required for the various metabolic pathways. Fuel molecules move into the mitochondria, where they are oxidized and release their energy to ATP. The ATP molecules, originally formed within, make their way out of the organelle and diffuse throughout the cell, where they can release their stored energy to drive necessary endergonic reactions.

The mitochondria are exquisitely designed energy factories. In recognition of the central role that this organelle plays in cellular energy production, the mitochondria are called the *powerhouses of the cell.*

25.9 EFFICIENCY OF CELLULAR ENERGY TRANSFORMATIONS

The overall oxidation of NADH by oxygen is accompanied by the release of a large amount of energy, and much of that energy is trapped in ATP molecules. Carbohydrates, fats, and proteins can ultimately fuel the Krebs cycle. Here is a comparison of the overall yield of ATP from the complete oxidation of an average fat molecule and of glucose. Remember that the vast majority of ATP molecules are produced during oxidative phosphorylation.

$$C_{55}H_{102}O_6 + 77\tfrac{1}{2}O_2 + 437\,ADP + 437\,HPO_4{}^{2-} \longrightarrow 55\,CO_2 + 437\,ATP + 488\,H_2O$$

Fat

$$C_6H_{12}O_6 + 6\,O_2 + 36\,ADP + 36\,HPO_4{}^{2-} \longrightarrow 6\,CO_2 + 36\,ATP + 42\,H_2O$$

Glucose

A single fat molecule produces over ten times as much ATP as a single glucose molecule. Of course, the fat molecule contains 55 carbons and the glucose molecule only 6. The fat molecule produces about 8 ATP molecules per carbon oxidized, and the glucose molecule yields 6 per carbon. The glucose gives a lower per carbon yield of ATP because the carbon in glucose is in a higher oxidation state. The oxidation number of carbon in fat is about -1.6. (See if you can calculate this number from the formula for the fat, $C_{55}H_{102}O_6$.) The oxidation number of carbon in glucose is 0. Both fats and carbohydrates are oxidized to carbon dioxide (CO_2), in which carbon has an oxidation number of $+4$. Thus, the carbons in glucose go from 0 to $+4$, whereas the carbons in the fat change from -1.6 to $+4$. More ATP is formed from the oxidation of a fat carbon because the fat carbon has a longer way to go.

How efficiently does ATP trap the energy released in these oxidation reactions? If standard free energy values are used as the basis for the calculation, the efficiency of the process is about 38%. If actual intracellular conditions of temperature, pressure, and concentration are taken into account, efficiencies of 60% or better have been calculated.

We know from the second law of thermodynamics that the process cannot be 100% efficient. What happens to the energy that is not trapped? It is released as heat to the surroundings, that is, to the cell. It is this heat that maintains body temperature. If we are exercising strenuously and our metabolism speeds up to provide the necessary energy for muscle contraction, then more heat is also produced. We begin to sweat to dissipate some of that heat. As the sweat evaporates, the excess heat is carried from the body by the departing water vapor.

25.10 MUSCLE POWER

The stimulation of muscle causes it to contract. That contraction is work and requires energy. The immediate source of energy for muscle contraction is ATP. It is the energy stored in this molecule that is used to bring about the physical movement of muscle tissue. Two proteins, **actin** and **myosin,** play important roles in this process. Together actin and myosin form a loose complex called **actomyosin,** the contractile protein of which muscles are made (Figure 25.10). When ATP is added to isolated actomyosin, the protein fibers contract. It seems likely that the same process occurs *in vivo,* that is, in muscle in living animals. Not only does myosin serve as part of the structural complex in muscles, it also acts as an enzyme for the removal of a phosphate group from ATP. Thus, it is directly involved in liberating the energy required for the contraction.

In the resting person, muscle activity (including that of the heart muscle) accounts for only about 15–30% of the energy requirements of the body.

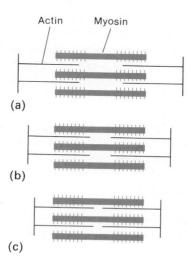

FIGURE 25.10
Diagram of actomyosin complex in muscle.
(a) Extended muscle.
(b) Resting muscle.
(c) Partially contracted muscle.

Other activities, such as cell repair or the transmission of nerve impulses or even the maintenance of body temperature, account for the remaining energy needs. During intense physical activity, the energy requirements of muscle may be more than 200 times the resting level.

Fats are the major source of energy for sustained, low- or moderate-intensity activity. Prolonged, heavy work is fueled primarily by aerobic oxidation of both fats and carbohydrates. Anaerobic metabolism of carbohydrates supplies a significant proportion of the energy required for short bursts of high-intensity activity. The carbohydrate tapped for such all-out efforts is the glycogen stored in muscle tissue. To rapidly generate the ATP required for strenuous work, muscle glycogen is hydrolyzed, and the resulting glucose is converted to lactic acid via anaerobic glycolysis. The generation of energy by this pathway is self-limiting. As lactic *acid* builds up in muscle tissue, the pH drops and deactivates enzymes required for glycolysis. The muscle's response to stimuli becomes weaker, and in extreme cases there may be no response at all. In this state, the muscle is described as fatigued.

When muscles use anaerobic pathways, they incur an **oxygen debt.** It is as if the body regards oxidation as the only proper source of energy for muscular activity and uses anaerobic metabolism as a temporary expedient. As soon as it can, the body oxidizes some of the resulting lactic acid back to pyruvic acid and ultimately to carbon dioxide and water. The energy released in the process is used to convert the rest of the lactic acid back to glycogen, which is restored for future use.

When is this oxygen debt repaid? Just as soon as the very high level of muscular activity ceases. Sprinters, while running a 100-m dash, breathe in oxygen but still obtain only a fraction of the energy required for their intense muscular activity through aerobic processes. When the race is over, the sprinters continue to take in great gulps of air. This air is used to repay the oxygen debt incurred during the race (Figure 25.11). We continue to breathe hard even after we stop vigorous activity because our body chemistry is still catching up and needs some more of a critical reagent.

On the other hand, long-distance runners derive only a small percentage of their energy needs from glycolysis. Aerobic oxidation (the Krebs cycle coupled to oxidative phosphorylation) provides most of the required ATP. Some glycogenolysis (hydrolysis of glycogen) and glycolysis (anaerobic metabolism of glucose) does occur, and after very long periods of this moderate muscle activity, lactic acid does build up and muscles do become fatigued. But since fats and some carbohydrate are supplying most of the energy through aerobic oxidation, the buildup of lactic acid takes much longer.

Muscle tissue seems to have been designed to provide for both short, intense bursts of activity and sustained, moderate levels of activity. Muscle fibers are divided into two categories, described as *fast twitch* and *slow twitch*. Table 25.2 lists some characteristics of these different types of muscle fibers. The

FIGURE 25.11
The race is over, but the metabolism that fueled the effort continues. (Photo © Jim Anderson, Black Star.)

Type I fibers described in Table 25.2 are called on during activity of light or moderate intensity. The respiratory capacity of these fibers is high, which means they can provide much energy via aerobic pathways, or, to put it another way, they are geared to oxidative phosphorylation. Notice that for Type I fibers, myoglobin levels are also high. Myoglobin is the heme-containing protein in

TABLE 25.2 *A Comparison of Types of Muscle Fiber*

	Type I	Type IIB*
Category	Slow twitch	Fast twitch
Color	Red	White
Respiratory capacity	High	Low
Myoglobin level	High	Low
Catalytic activity of actomyosin	Low	High
Capacity for glycogenolysis	Low	High
Number of mitochondria	High	Low

* There is a Type IIA fiber that resembles Type I in some respects and Type IIB in others. We will discuss only the two types described in the table.

muscle that transports oxygen (as hemoglobin does in the blood). Aerobic oxidation requires oxygen, and this muscle tissue is geared to supply high levels of oxygen. The capacity of Type I muscle fibers for glycogenolysis is low. This tissue is not geared to anaerobic generation of energy and does not require the hydrolysis of glycogen. The number of mitochondria in Type I muscle tissue is high, as we would expect since oxidative phosphorylation takes place in the mitochondria. The catalytic activity of the actomyosin complex is low. Remember that actomyosin is not only the structural unit in muscle that actually undergoes contraction—it is also responsible for catalyzing the hydrolysis of ATP to provide energy for the contraction. Low catalytic activity means that the energy is parceled out more slowly, which is not good if you want to lift 200 kg but is perfect for a jog of five miles.

The Type IIB fibers described in Table 25.2 have characteristics just the opposite of Type I fibers. Low respiratory capacity, low myoglobin levels, and fewer mitochondria all argue against aerobic oxidation, whereas a high capacity for glycogenolysis and high catalytic activity of actomyosin allow this tissue to generate ATP rapidly via glycolysis and also to hydrolyze that ATP rapidly in intense muscle activity. Thus, this type of muscle tissue gives you the capacity to do short bursts of vigorous work. We say *bursts* because this type of muscle fatigues relatively quickly. A period of recovery in which lactic acid is cleared from the muscle is required between brief periods of activity.

The fields of sports medicine and exercise physiology have done much to increase our understanding of muscle action. Endurance exercise training (for example, jogging for long distances) increases the size and number of mitochondria in muscle. There is an increase in the level of enzymes required for the transport and oxidation of fatty acids, for the Krebs cycle, and for oxidative phosphorylation. The increase in the mitochondrial enzymes is much greater for Type I fibers (used in prolonged, moderate-intensity activity) than for Type IIB fibers (brief, intense activity). Endurance training also increases myoglobin levels in skeletal muscles, providing for faster oxygen transport (Figure 25.12). These changes can be observed shortly after training begins, that is, within a week or two. Muscle changes resulting from endurance training do not necessarily include a significant increase in the size of the muscle, in contrast to the effect of strength exercises such as weight lifting. And weight lifting (presumably fueled primarily by anaerobic glycolysis) does not result in the mitochondrial changes we have just described (Figure 25.13). The mitochondria of heart muscle, which is working constantly anyway, also undergo no change during endurance training.

When we say that energy for muscle contraction is supplied by oxidative (aerobic) or glycolytic (anaerobic) processes or both, we mean that the ATP used up by muscular activity is replaced through these pathways. In addition, muscles contain a relatively large concentration of creatine phosphate. As ATP is used up, creatine phosphate reacts with ADP to form more ATP (compare

FIGURE 25.12 *The Boston Marathon, a test of the respiratory capacity of muscle. (Photo courtesy Stock, Boston; © Arthur Grace.)*

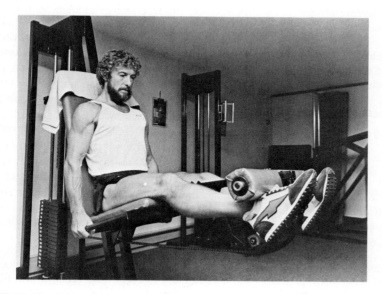

FIGURE 25.13 *Strength exercises build muscle mass but do not increase respiratory capacity of muscles. (Photo courtesy Stock, Boston; © Gale Zucker.)*

the standard free energies of hydrolysis for creatine phosphate and ATP, Table 25.1). Thus, creatine phosphate serves as a ready reserve for the quick restoration of ATP levels.

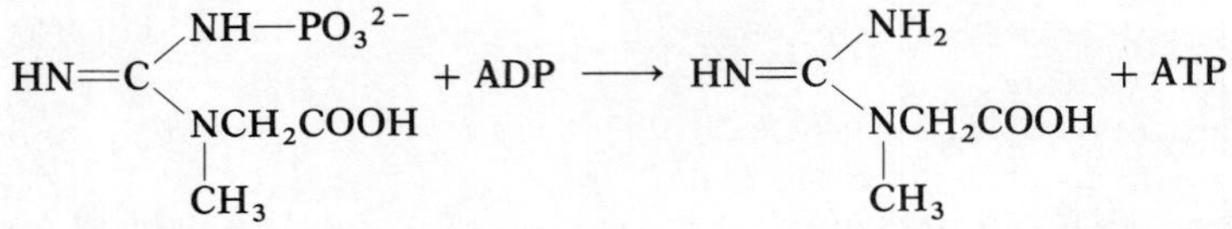

$$HN\!=\!C\!\!\begin{array}{c}NH\!-\!PO_3{}^{2-}\\NCH_2COOH\\|\\CH_3\end{array} + ADP \longrightarrow HN\!=\!C\!\!\begin{array}{c}NH_2\\NCH_2COOH\\|\\CH_3\end{array} + ATP$$

Creatine phosphate   Creatine

A standard clinical test performed on people suspected of having a myocardial infarction involves the measurement of creatine kinase (CK) levels in the blood (Section 24.10). Creatine kinase is the enzyme that catalyzes the phosphate transfer shown above.

We hope we have made clear the central role ATP plays in many life-sustaining metabolic reactions. Our cells can make ATP, using the energy stored in foods. But only green plants can make the food. They also replenish the oxygen we breathe. Have you thanked a green plant lately?

PROBLEMS

1. State the first law of thermodynamics.
2. Energy is stored in a flashlight battery. Eventually the battery runs down. Has the energy been destroyed? Explain your reasoning.
3. State the second law of thermodynamics.
4. Which is more efficient: a vehicle that travels 10 km on 1 L of gasoline or a similar vehicle that travels 30 km on 5 L of gasoline?
5. Lightning can cause the following reaction in air:

$$N_2 + O_2 + \text{lightning} \longrightarrow 2\,NO$$

In a firefly, the compound luciferin is converted to dehydroluciferin with the production of light.

$$\text{Luciferin} \longrightarrow \text{dehydroluciferin} + \text{light}$$

Which reaction is exergonic and which is endergonic?
6. What is the sign of ΔG for a spontaneous reaction? When a reaction reaches equilibrium, what is the sign of ΔG?
7. What is meant by $\Delta G°$ (*standard* free energy change)?
8. The free energy change for a reaction is negative. Is the rate of the reaction fast or slow? Explain.
9. What is a coupled reaction?

10. In a coupled reaction, which absolute value should be greater, the free energy change associated with the exergonic reaction or the free energy change associated with the endergonic reaction? Why?
11. Using Table 25.1, calculate the $\Delta G°$ for these reactions:

   a. PEP + acetic acid $\longrightarrow$
                  ketopyruvic acid + acetyl phosphate
   b. AMP + acetyl phosphate $\longrightarrow$
                       ADP + acetic acid
   c. Creatine phosphate + ADP $\longrightarrow$ ATP + creatine
   d. ATP + glycerol $\longrightarrow$
                  glycerol 3-phosphate + ADP
12. What is the structural difference among ATP, ADP, and AMP?
13. Why is ATP referred to as the energy currency of the cell?
14. Referring to Table 25.1, indicate which of these compounds are classified as high-energy phosphates:
   a. ATP   **d.** creatine phosphate
   b. AMP   **e.** glucose 1-phosphate
   c. PEP   **f.** glucose 6-phosphate

15. Is there a net consumption or synthesis of ATP in anaerobic glycolysis?

16. Is there a net consumption or synthesis of ATP in oxidative phosphorylation?

17. Is there a net consumption or synthesis of ATP during muscle contraction?

18. What are the oxidizing agents used in the Krebs cycle?

19. What is the electron acceptor at the end of the respiratory chain? To what product is this compound reduced?

20. What is the function of the cytochromes in the respiratory chain?

21. Summarize the movement of electrons from Krebs cycle through the oxidative phosphorylation chain by ordering the following compounds involved in the transport. Start with the compound in which the electrons originate and end with the final compound to which the electrons are transferred: O_2, FAD, pyruvic acid, cytochromes.

22. Why is the cyanide ion so toxic?

23. Which provides the greater number of ATP molecules: glycolysis of a glucose molecule or aerobic oxidation of a glucose molecule?

24. A fat yields more ATP molecules per carbon oxidized than does a carbohydrate. Why?

25. What is the function of the mitochondria?

26. What are the two functions of the actomyosin protein complex?

27. Which energy reserves (that is, carbohydrates or fats) are tapped in intense bursts of vigorous activity?

28. Which energy reserves (carbohydrates or fats) are mobilized to fuel prolonged low levels of activity?

29. Which type of metabolism (aerobic or anaerobic) is primarily responsible for providing energy for intense bursts of vigorous activity?

30. Which type of metabolism (aerobic or anaerobic) is primarily responsible for providing energy for prolonged low levels of activity?

31. What is meant by *oxygen debt*?

32. Identify Type I and Type IIB muscle fibers as
  a. fast twitch or slow twitch
  b. suited to aerobic oxidation or to anaerobic glycolysis

33. Explain why high levels of myoglobin and mitochondria are appropriate for muscle tissue that is geared to aerobic oxidation.

34. Why does the high catalytic activity of actomyosin in Type IIB fibers suggest that these are the muscle fibers engaged in brief, intense physical activity?

35. Why can the muscle tissue that utilizes anaerobic glycolysis for its primary source of energy be called on only for *brief* periods of intense activity?

36. Which type of muscle fiber is most affected by endurance training exercises? What changes occur in the muscle tissue?

37. Birds use large, well-developed breast muscles for flying. Pheasants can fly 80 km/hr, but only for short distances. Great blue herons can fly only about 35 km/hr, but can cruise great distances. What kind of fibers would each have in its breast muscles?

38. What role does creatine phosphate play in supplying energy for muscle contraction?

39. Why could elevated levels of creatine kinase (CK) in the blood be taken as a possible indication of a heart attack?

40. A person collapses while shoveling snow, and a blood analysis reveals elevated levels of creatine kinase. Why would it be unwise to conclude that the individual had suffered a heart attack on this evidence alone?

41. Define each of the following:
  a. anabolism   **b.** catabolism   **c.** metabolism

42. What are metabolites?

43. List the three component parts of the NAD^+ molecule. Which part is a B vitamin?

44. List the three components of the FAD molecule. Which two parts combined make up a B vitamin? What is the name of that vitamin?

45. What part of the coenzymes NAD^+ and FAD attach to the apoenzyme?

46. When NAD^+ and FAD act as oxidizing agents, which portion of each is reduced?

47. What substance serves to interconnect carbohydrate, lipid, and protein metabolisms?

48. What substance is the starting material for the citric acid cycle?

49. Use Table 25.1 to show that PEP should spontaneously transfer a phosphate group to ADP.

50. The following reactions occur in anaerobic glycolysis. Which is an oxidation reaction?
  a. Glucose ($C_6H_{12}O_6$) $\longrightarrow$ lactic acid ($C_3H_6O_3$)
  b. Glucose $\longrightarrow$ pyruvic acid ($C_3H_4O_3$)

Vitamins

Carbohydrates, fats, and proteins are the three major classes of foods. To remain healthy we must take in relatively large amounts of each of these substances. They are not, however, the only nutrients we require. Some of our needs are satisfied only by vitamins and minerals. The minerals, inorganic ions of critical importance to our health, are discussed as a group in Chapter 12. Vitamins are considered in this special topic.

We normally eat three meals a day to satisfy our need for carbohydrates, fats, and proteins, but all the necessary vitamins can be packed into a single small pill. So small are the required amounts that not even a vitamin pill is necessary. We can get all we need simply by eating a balanced diet. Nature has thoughtfully incorporated minute but adequate amounts of vitamins into our various foodstuffs.

J.1 WHAT ARE VITAMINS?

Vitamins are specific organic compounds that our bodies need for proper functioning. Our bodies can't synthesize these compounds; they must be included in the diet. Absence or shortage of a vitamin results in a vitamin-deficiency disease.

One such disease, called scurvy, had plagued seamen since early times. In 1747, British navy captain James Lind showed that the disease could be prevented by the inclusion of fresh fruit or vegetables in the diet. A convenient fresh fruit to carry on long voyages was the lime. British ships put to sea with barrels of limes aboard. The sailors ate a lime or two every day and remained free of scurvy. British sailors came to be known as "lime eaters" or simply "limeys."

In 1897, the Dutch scientist Christiaan Eijkman showed that polished rice lacked something found in the hulls of whole-grain rice. Lack of that "something" caused the disease beriberi, which was quite a problem in the Dutch East Indies at that time.

A British scientist, F. G. Hopkins, fed a synthetic diet of carbohydrates, fats, proteins, and minerals to a group of rats. The rats were unable to sustain healthy growth. Again, something was missing.

In 1912, Casimir Funk, a Polish biochemist, coined the word *vitamine* (from the Latin *vita*, "life") for these missing factors. Funk thought that all contained the amino group. The final *e* was dropped after it was found that not all the factors were amines. The generic term became *vitamin*. Eijkman and Hopkins shared the 1929 Nobel Prize in medicine and physiology for their important discoveries.

The vitamins are divided into two broad categories, the **fat-soluble** group, including A, D, E, and K, and the **water-soluble** group, made up of the B complex and vitamin C.

J.2 VITAMIN A AND THE CHEMISTRY OF VISION

Plants make a class of colored compounds called carotenoid pigments. Four of these can be converted into vitamin A by animals. These are **provitamins**—vitamin precursors. A typical carotenoid pigment, β-carotene, has a number of double bonds, all of which have the *trans* configuration.

This molecule can be cleaved at the central double bond to give two molecules of vitamin A. Vitamin A is an unsaturated alcohol called retinol.

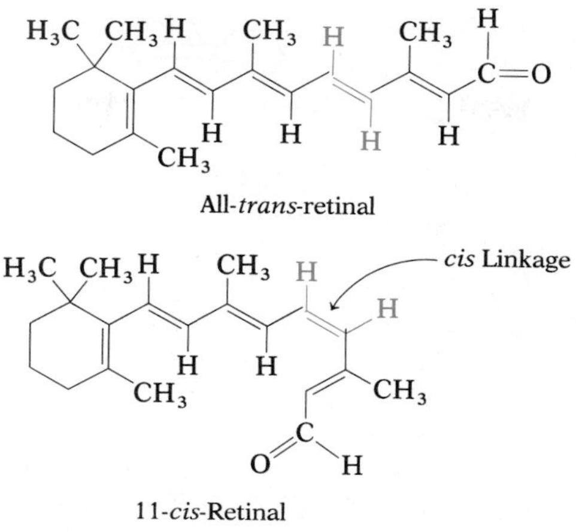

FIGURE J.1   *Retinal isomers.*

All-*trans*-retinal

11-*cis*-Retinal

As would be expected from its structure, vitamin A is insoluble in water but soluble in fats and fat solvents. The pure compound is a crystalline solid that melts at 64 °C.

The relationship between vitamin A and vision has been worked out in considerable detail. In the body, some vitamin A is isomerized and oxidized. The isomerization converts the all-*trans* compound to a molecule in which one of the double bonds has a *cis* arrangement. The oxidation converts the alcohol group to an aldehyde group. The product that results is called 11-*cis*-retinal (Figure J.1).

The retina of the human eye contains two kinds of receptor cells, rods and cones. 11-*cis*-Retinal forms a complex with various proteins, and these complexes are the photosensitive chemicals found in the rods and cones. Rhodopsin, a complex of 11-*cis*-retinal and a protein called opsin, is the visual pigment found in the cones. When light strikes rhodopsin, 11-*cis*-retinal is isomerized to the all-*trans* isomer (Figure J.1). This change in structure is accompanied by a change in electrical potential. The photoisomerization of the retinal triggers a nerve impulse—we see. The rhodopsin complex then splits into opsin and the free aldehyde. The all-*trans* retinal is converted by enzymes back to 11-*cis*-retinal, which again complexes opsin, completing the visual cycle (Figure J.2). Some retinal is lost during the regeneration of rhodopsin. It must be replaced by vitamin A from the bloodstream.

It is interesting to note that of all the tissues in the body, the retina has the highest respiration rate. You may think of seeing as a rather passive activity, but vision is metabolically demanding.

The role of vitamin A in vision is something we understand fairly well. In addition, we also know that a deficiency of this vitamin affects most of the

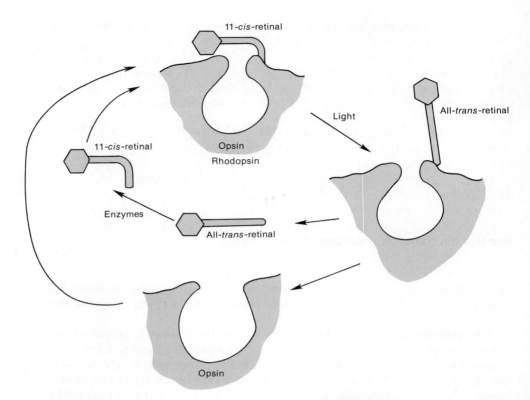

11-*cis*-retinal

11-*cis*-retinal

Light

All-*trans*-retinal

Opsin

Rhodopsin

Enzymes

All-*trans*-retinal

Opsin

FIGURE J.2 *A schematic diagram of the visual cycle.*

body's organs. What we do not know is the detailed biochemistry of vitamin A as it relates to all these other organs. We can, however, describe some of the effects of vitamin A deficiency.

Vitamin A is required for normal growth. Young animals fed a diet lacking in the vitamin simply fail to grow. One of the earliest manifestations of vitamin A deficiency is a loss of night vision (a function of the rod cells in the retina). Mucous membranes may harden, dry, and crack. In cases of severe deprivation, victims may exhibit xerophthalmia, a condition characterized by inflammation of the eyes and eyelids, leading ultimately to infection and blindness. Many children in developing countries are permanently blinded by vitamin A deficiency. Health workers in those countries often carry injectible solutions of the vitamin for emergency treatment of such cases. Vitamin A (an alcohol) can be stored as an ester in the liver. Adults, while on an adequate diet, can store several years' supply in this form. Small children, without such reserves, are particularly susceptible to vitamin A deprivation.

Vitamin A is found in high concentration in fish liver oils. Liver, eggs, fish, butter, and cheese are also good sources. It is interesting to note that polar bear liver has so much vitamin A that it is toxic. These large animals eat seals that eat fish. Fat-soluble vitamin A is concentrated in each step of the food chain. Eskimos who kill a polar bear know better than to eat its liver. If taken in excess, the vitamin is stored rather than excreted. Large excesses cause irritability, dry skin, and a feeling of pressure inside the head. Massive doses of vitamin A administered to pregnant rats result in malformed offspring.

The provitamin forms are found in carrots, tomatoes, and green vegetables. These forms appear to be much less toxic than the active vitamin. Studies have shown a statistical correlation between diets rich in carotenes and a decreased incidence of lung cancer. The effect was noted even in smokers. More research into this phenomenon is certainly to be expected.

J.3 VITAMIN D: ACTIVATION BY SUNLIGHT

Several chemical compounds have vitamin D activity. Only two commonly occur in foods or are used in drugs and food supplements. Each is formed from a precursor by the action of ultraviolet light (Figure J.3). Vitamin D_2 (calciferol) is synthesized by irradiation of ergosterol, a compound found in yeast and other molds. Vitamin D_3 is formed in the skin of animals by the action of sunlight on 7-dehydrocholesterol. The two vitamins differ only in the structure of their side chains; D_2 contains an extra carbon and a double bond. (There is no vitamin D_1. The material that was originally given this designation proved to be a mixture of compounds that included calciferol.)

Vitamin D increases the utilization of calcium and phosphorus by the body. Deficiency in infants and growing children results in abnormal bone formation, a condition known as **rickets** (Figure J.4). The condition is characterized by bowed legs, knobby bone growths where the ribs join the breastbone (called a "rachitic rosary"), pigeon breast, and poor tooth development. In adults, with completed bone growth, rickets does not develop. Women may develop osteomalacia, a condition characterized by fragile bone structure, if deficient in vitamin D. This condition is rare but may occur after several pregnancies.

The fish liver oils are a good source of vitamin D as well as of vitamin A. Irradiated ergosterol is added to milk as a supplemental source. Vitamin D is the "sunshine vitamin." Children who are outside a lot seldom suffer a deficiency, for sunlight converts 7-dehydrocholesterol in the skin to vitamin D.

Vitamin D, like vitamin A, is fat soluble. Amounts taken in excess are stored in body fat. The effects of large overdoses are even more severe than with

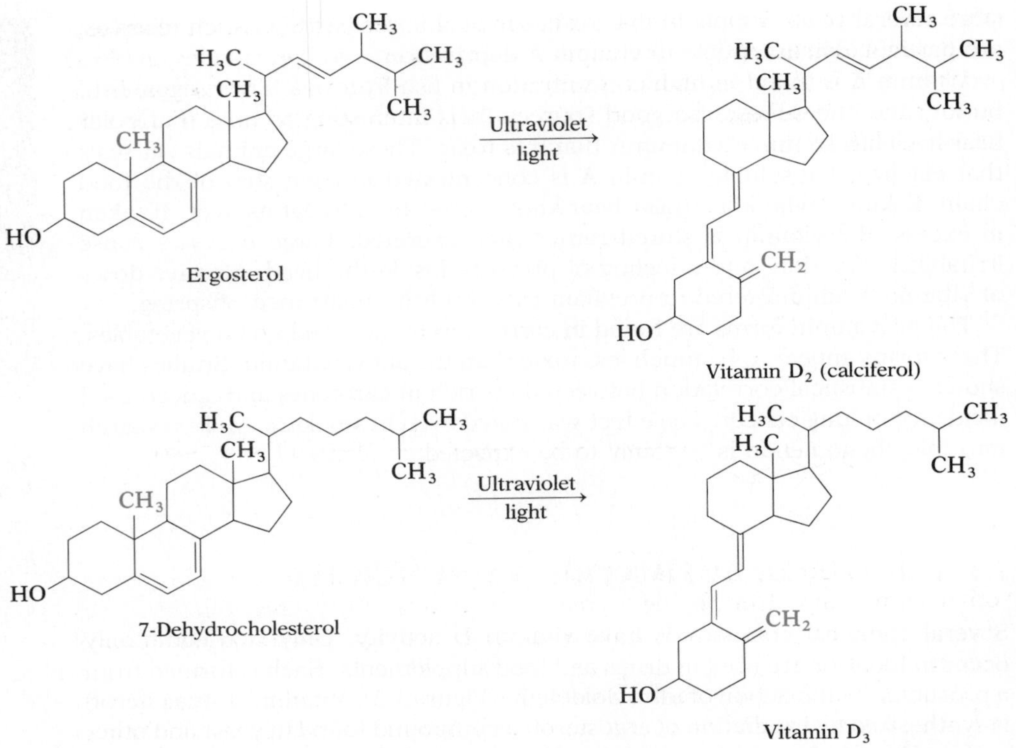

Ergosterol
Ultraviolet light
Vitamin D_2 (calciferol)
7-Dehydrocholesterol
Ultraviolet light
Vitamin D_3

FIGURE J.3 Formation of two forms of vitamin D by action of ultraviolet light on corresponding provitamins.

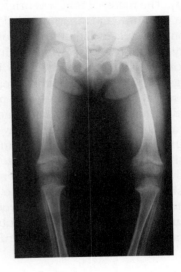

FIGURE J.4
This X ray shows the typical distortion of leg bones in rickets, a disease caused by a deficiency of vitamin D. (Photo by Joseph J. Mentrikoski, Department of Medical Photography, Geisinger Medical Center.)

vitamin A. Too much vitamin D can cause pain in the bones, nausea, diarrhea, and weight loss. Bonelike material may be deposited in kidney tubules, in blood vessels, and in heart, stomach, and lung tissue (Figure J.5). Like vitamin A, amounts of vitamin D in nonprescription capsules are regulated by the U.S. Food and Drug Administration.

J.4 VITAMIN E

As was true for vitamin D, there are several compounds with vitamin E activity. The compounds are called tocopherols; the most potent of these is α-tocopherol.

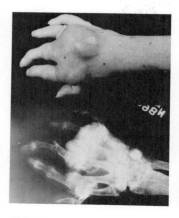

FIGURE J.5
Too much vitamin D causes the deposit of bonelike material in joints. (Courtesy of Marc Moldawer, M.D., The Methodist Hospital, Texas Medical Center, Houston.)

The tocopherols are phenols. Many phenols are antioxidants; they protect other compounds from being oxidized by reacting themselves with the oxidizing agent (oxygen in the atmosphere, for example). Vitamin E exhibits similar properties.

Rats deprived of vitamin E develop a scaly skin, muscular weakness, and sterility. Vitamin E deficiency can also lead to vitamin A deficiency. Vitamin A can be oxidized to a nonactive form when vitamin E is no longer present to act as an antioxidant. Indeed, the loss of vitamin E's antioxidant effect is believed to be responsible for all of the symptoms of vitamin E deficiency. Polyunsaturated fatty acids, especially, are oxidized at increased rates. The muscular dystrophy, sterility, and other symptoms manifested by animals deficient in vitamin E are believed to result from this "simple" change in body chemistry.

Because of its effect on fertility, vitamin E is called the antisterility vitamin. Some food faddists promote the ingestion of large amounts of vitamin E to increase sexual prowess, combat wrinked skin, and prevent heart attacks. There is no unambiguous evidence for any of these claims. However, Lester Packer and James R. Smith, of the University of California at Berkeley, have shown that vitamin E slows the aging of cultured human cells. Just how this might be extrapolated to the entire human organism is not clear.

Vitamin E is available in wheat germ oil, green vegetables, vegetable oils, egg yolk, and meat. Most nutritionists contend that it would be nearly impossible to eat a diet deficient in vitamin E. Like vitamins A and D, vitamin E is fat soluble and thus is stored in the body. Large doses waste money, but they seem not to be as harmful as excesses of vitamins A and D. Some people who take large daily doses do develop blurred vision, nausea, and intestinal problems. Further, a belief in the efficacy of vitamin E may lead to the postponement of needed medical treatment.

J.5 VITAMIN K AND BLOOD CLOTTING

Chemically, vitamin K has a fused ring system related to the structure of naphthalene (Chapter 14). One of the rings contains two carbonyl groups, an arrangement that has the special name **quinone.** Attached to the quinone ring are alkyl groups. One of these is usually methyl. The other has 20 or more carbon atoms. Many compounds have vitamin K activity. The structure of vitamin K_1 is given here.

$$\text{CH}_2-\text{CH}=\text{C}-(\text{CH}_2)_3-\text{CH}-(\text{CH}_2)_3-\text{CH}-(\text{CH}_2)_3-\text{CH}$$

Vitamin K, like vitamins A, D, and E, is insoluble in water but soluble in fats and fat solvents. Vitamin K is necessary for the formation of prothrombin (Chapter 24), one of the enzyme precursors involved in blood clotting. Deficiency will increase the time required for clotting of the blood. Symptoms are bleeding under the skin and in muscles, leading to ugly "bruises" from what would otherwise be minor blows. Infants lacking in vitamin K may die from hemorrhaging in the brain. Increased vitamin K intake by pregnant women has lowered the incidence of this disease in newborn infants. Good sources of vitamin K are spinach and other green leafy vegetables. Synthetic compounds with K activity are readily available.

FIGURE J.6
Inflammation and abnormal pigmentation characterize pellagra, caused by niacin deficiency. (Courtesy of the World Health Organization, New York.)

J.6 THE B COMPLEX

In Chapter 24 we discuss how some enzymes require cofactors, nonprotein components, for proper function. Organic cofactors are called coenzymes, and many coenzymes are vitamin B derivatives.

There is really no vitamin B. What was once called vitamin B has long since been recognized as a complicated mixture of factors. The term **B complex** is now used to designate a group of water-soluble vitamins found together in many food sources. Table J.1 lists the members of the B complex, their structures and sources, and the diseases associated with their deficiencies.

TABLE J.1   Members of the B Complex of Vitamins

Name and Structure	Sources	Deficiency Symptoms
B₁ (thiamine) Thiamine chloride	Germ of cereal grains, legumes, nuts, milk, and brewers yeast	Beriberi—polyneuritis resulting in muscle paralysis, enlargement of heart, and ultimately heart failure
B₂ (riboflavin)	Milk, red meat, liver, egg white, green vegetables, whole wheat flour (or fortified white flour), and fish	Dermatitis, glossitis (tongue inflammation)
Niacin (nicotinic acid and nicotinamide) Nicotinic acid Nicotinamide	Red meat, liver, collards, turnip greens, yeast, and tomato juice	Pellagra—skin lesions, swollen and discolored tongue, loss of appetite, diarrhea, various mental disorders (Figure J.6)

(continued)

TABLE J.1 (continued)

Name and Structure	Sources	Deficiency Symptoms
B₆ (pyridoxol, pyridoxal, and pyridoxamine) Pyridoxal Pyridoxol (pyridoxine) Pyridoxamine	Eggs, liver, yeast, peas, beans, and milk	Dermatitis, apathy, irritability, and increased susceptibility to infections; convulsions in infants
Pantothenic acid	Liver, eggs, yeast, and milk	(Possibly) emotional problems and gastro-intestinal disturbances
Biotin	Beef liver, yeast, peanuts, chocolate, and eggs (although this vitamin cannot be synthesized by humans, it is a product of their intestinal bacteria)	Dermatitis

TABLE J.1 (continued)

Name and Structure	Sources	Deficiency Symptoms
Folic acid 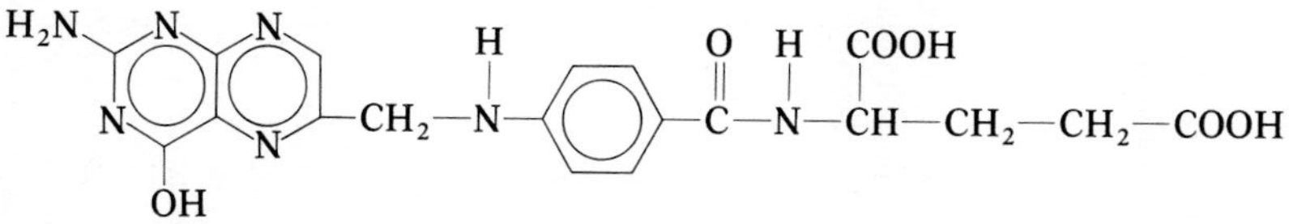	Liver, kidney, mushroom, yeast, and green leafy vegetables	Anemias (folic acid is used in the treatment of megaloblastic anemia, a condition characterized by giant red blood cells)
B_{12} (cyanocobalamine) 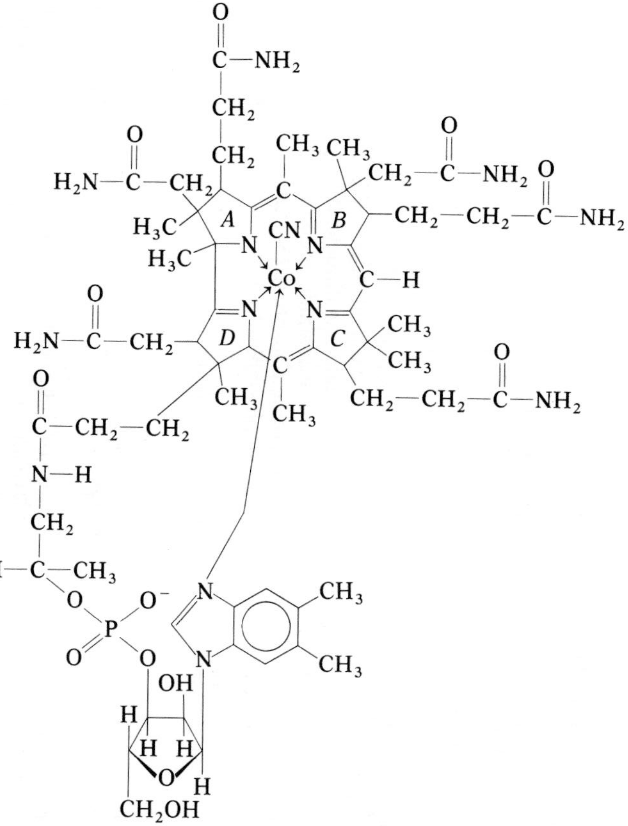	Liver, meat, eggs, and fish (not found in plants)	Pernicious anemia

The B complex vitamins are water soluble. The body has a limited capacity to store water-soluble vitamins. It will excrete anything over the amount that it can immediately use. Thus, water-soluble vitamins must be taken in at frequent intervals, whereas a single large dose of a fat-soluble vitamin can be used by the body over a period of several weeks or longer.

The structures of the B complex members (Table J.1) range from the relatively simple one of niacin to the very complex one of B_{12}. Each of the B vitamins is incorporated in coenzyme molecules. In some instances, the coenzymes are simple derivatives of the vitamins. In the coenzyme thiamine pyrophosphate, for example, the primary alcohol group of thiamine (Table J.1) is esterfied with pyrophosphoric acid.

You may recall that nicotinamide is a part of the coenzyme NAD^+ and that riboflavin is a component of FAD (see Figure 25.5). Pantothenic acid occurs

TABLE J.2 *The B Complex Coenzymes*

Vitamin	Name of Coenzyme	Some Biochemical Processes Involving the Coenzyme
B_1 (thiamine)	Thiamine pyrophosphate	Decarboxylation of pyruvic acid during carbohydrate metabolism (Chapter 27)
B_2 (riboflavin)	Flavin mononucleotide (FMN), flavin adenine dinucleotide (FAD)	Oxidation-reduction reactions
Niacin	Nicotinamide adenine dinucleotide (NAD^+), nicotinamide adenine dinucleotide phosphate ($NADP^+$)	Oxidation-reduction reactions
B_6 (pyridoxine group)	Pyridoxal phosphate	Decarboxylation and transamination of amino acids (Chapter 29)
Pantothenic acid	Coenzyme A (CoA)	Transfer of acetyl, succinyl, and benzoyl groups in carbohydrate synthesis (Chapter 27) and lipid synthesis (Chapter 28)
Biotin	Biotin	Carboxylation-decarboxylation reactions
Folic acid	Tetrahydrofolic acid (FH_4) and derivatives	One-carbon transfer reactions (purine and amino acid syntheses)
B_{12} (cyanocobalamine)	Cobamide coenzymes	Isomerization reactions, dehydrations, and methyl group biosynthesis

in coenzyme A (see Figure 25.6). The names and functions of these and other B complex coenzymes are listed in Table J.2.

J.7. VITAMIN C

Chemically, vitamin C is ascorbic acid. It is a white, crystalline solid that is quite soluble in water. Severe deficiency results in **scurvy,** a condition characterized by thin, porous bones, sore and bleeding gums, and a pronounced muscular weakness (Figures J.7 and J.8).

Vitamin C
(projection formula)

Vitamin C
(cyclic structure)

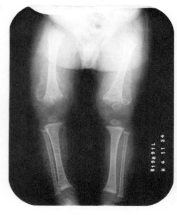

FIGURE J.7
An X-ray photograph of the legs of a patient with infantile scurvy, caused by a deficiency of vitamin C. (Reprinted with permission from Arnow, L. Earle, Introduction to Physiological and Pathological Chemistry, *9th ed., St. Louis: C. V. Mosby, 1976.) (Courtesy of Dr. Walter Burden, Babies Hospital, New York.)*

Citrus fruits are rich in vitamin C. Tomatoes, green peppers, and strawberries are also good sources.

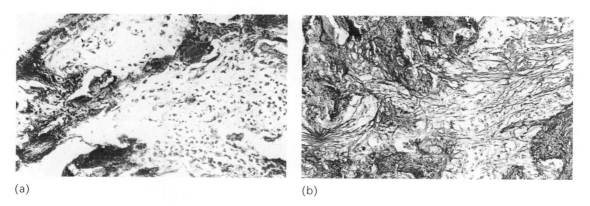

(a)

(b)

FIGURE J.8   *The effect of ascorbic acid on bone formation. (a) Cross section of bone-forming cells from a guinea pig dying of scurvy. Note that there are no connective tissue fibers. (b) Cross section of bone-forming cells from another guinea pig with scurvy after 72 hours of treatment with ascorbic acid. Note that connective tissue fibers have appeared. (Courtesy of Upjohn Co., Kalamazoo, Mich.)*

The function of ascorbic acid in the body is somewhat obscure. It seems to be a necessary cofactor in some hydroxylation reactions important to the synthesis of collagen. It may also have other important, but as yet unknown, functions.

Only about 40 to 75 mg of ascorbic acid per day are necessary to prevent scurvy. Daily intake of over 100 mg was thought to be excreted and thus wasted.

Linus Pauling (Figure J.9), winner of two Nobel prizes (for chemistry in 1954 and for peace in 1962), has proposed the use of massive doses of vitamin C for the prevention and cure of the common cold and a variety of other ailments. He recommends daily doses of 250 to 15 000 mg depending on the person and the circumstances.

Pauling theorized that humans are actually mutated mammals because, unlike most mammals, humans cannot synthesize ascorbic acid (vitamin C). A 70-kg goat can produce 13 g of the compound daily, yet the recommended daily dietary allowance for humans is 45 mg. This "abnormally" low level, according to Pauling, may be responsible for sudden infant death syndrome and for human vulnerability to viral diseases (including colds), cancer, heart and vascular diseases, and even drug addiction. Ascorbic acid does prevent melanoma (a kind of tumor) in tissue cultured cells, and it promotes the production of the antiviral protein interferon. Clinical tests of vitamin C therapy generally have not substantiated Pauling's claims, however. Studies indicate that with few exceptions people excrete all but 100 mg or so of the massive doses. The controversy continues. So does research, by Pauling and other investigators.

FIGURE J.9
Linus Pauling, winner of two Nobel Prizes. (Courtesy of Linus Pauling.)

PROBLEMS

1. Define and given an example of each of the following:
   a. vitamin   **b.** provitamin   **c.** coenzyme
2. Compare and contrast vitamins and minerals (Chapter 12) with regard to the following:
   a. inorganic or organic
   b. essentiality in the diet
   c. amounts needed by the body
3. Match the compound with its designation as a vitamin.

Compound	Designation
Ascorbic acid	Vitamin A
Calciferol	Vitamin B_{12}
Cyanocobalamine	Vitamin C
Retinol	Vitamin D
Tocopherol	Vitamin E

4. Which of the following are B vitamins?
   a. folic acid   **c.** niacin   **e.** thiamine
   b. insulin   **d.** riboflavin
5. Identify the vitamin deficiency associated with these diseases.
   a. scurvy   **c.** night blindness
   b. rickets
6. In each case, identify the deficiency disease associated with a diet lacking in the indicated vitamin.
   a. vitamin B_1 (thiamine)
   b. niacin
   c. vitamin B_{12} (cyanocobalamine)
7. What is the structural difference between water-soluble and fat-soluble vitamins?
8. Identify the following vitamins as water-soluble or fat-soluble:
   a. vitamin A   **c.** vitamin B_{12} **e.** vitamin K
   b. vitamin B_6 **d.** vitamin C
9. Identify the following vitamins as water-soluble or fat-soluble:
   a. calciferol   **c.** riboflavin
   b. niacin   **d.** tocopherol
10. Identify the following vitamins as water-soluble or fat-soluble:

a.
$$\underset{\text{CH}_3}{\underset{|}{\text{HOCH}_2\text{C}}}\text{---}\underset{\text{OH}}{\underset{|}{\text{CH}}}\text{---}\underset{}{\overset{\text{O}}{\overset{||}{\text{C}}}}\text{---NHCH}_2\text{CH}_2\text{COOH}$$

with CH_3 groups on the quaternary carbon.

b. H_3C CH_3 ... CH=CH—C(CH$_3$)=CH—CH=CH—C(CH$_3$)=CHCH$_2$OH (with cyclohexene ring bearing CH_3 groups)

11. Is an excess of a water-soluble vitamin or a fat-soluble vitamin more likely to be dangerous? Why?
12. Could a one-a-month vitamin pill satisfy all human requirements? Explain your answer.
13. On what basis did some scientists object to the vitamin C therapy for colds proposed by Linus Pauling? Can you offer any supporting arguments for Pauling's position?
14. How are vitamins related to coenzymes?
15. Does synthetic vitamin C differ from natural vitamin C?
16. If boiled vegetables are served as part of a meal, would the water-soluble or fat-soluble vitamins originally present be lost? Why?
17. Explain the role of vitamin A in the chemistry of vision.
18. Why is vitamin D called the "sunshine vitamin"?
19. What biochemicals are protected by vitamin E's antioxidant effect?
20. What is meant by the term *B complex?*
21. Which vitamin is a part of the coenzyme NAD^+?
22. Which vitamin is a part of the coenzyme FAD?
23. Which vitamin is a part of coenzyme A?
24. Vitamin B_{12} has the formula $C_{63}H_{88}CoN_{14}O_{14}P$ and a molecular weight of 1355 amu, yet it is soluble in water. Explain.

K

Hormones

L ike the vitamins (Special Topic J), hormones are organic compounds. Unlike vitamins, hormones can be synthesized in the body. Both vitamins and hormones play critical biochemical roles. The most striking similarity, however, is in their effective amounts. Even in very low concentrations, hormones produce dramatic changes in the body. This special topic, then, deals with another class of compounds, which, like the vitamins, illustrates the old adage that good things come in small packages.

K.1 THE ENDOCRINE SYSTEM

The hormones are synthesized in the endocrine glands (Figure K.1) and then discharged directly into the circulatory system. They serve as "chemical messengers." Hormones released in one part of the body signal profound physiological changes in other parts of the body. They cause reactions to speed up or slow down. In this way they control growth, metabolism, reproduction, and many other functions of body and mind.

If we consider hormones as messengers, then the pituitary gland (hypophysis) must be viewed as the central dispatcher or control. Many of the pituitary hormones control the production of hormones by other endocrine glands. The removal of portions of the pituitary gland results in atrophy of other endocrine glands. Shut down the central control, and you ultimately shut down much of the endocrine system. The pituitary itself responds to hormone signals from the hypothalamus. The hypothalamus secretes hormones, called releasing

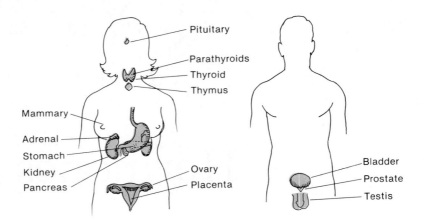

FIGURE K.1 *Approximate locations of endocrine glands in the human body.*

factors, that trigger the production of the pituitary hormones. The hypothalamus is triggered by nerve impulses. The sequence is this. A nerve impulse signals the hypothalamus, the hypothalamus signals the pituitary, the pituitary signals some target endocrine gland, the gland signals some target tissue, and the tissue responds in a specific way (Figure K.2).

Let us take just a moment to consider the extraordinary complexity of this system. For example, in response to neural stimulation, the hypothalamus produces thyrotropin-releasing factor (TRF). The TRF reaches the pituitary and causes that gland to produce thyroid-stimulating hormone (TSH). In the presence of TSH, the thyroid gland releases the hormone thyroxine. Thyroxine signals the cells to increase their metabolic rate. As the level of thyroxine builds up, a feedback mechanism causes the pituitary to slow down its production of thyroid-stimulating hormone. This, in turn, slows the production of thyroxine by the thyroid, which results in a slowing of the metabolic rate, which gets us back to where we started. The cycle includes the synthesis of proteins and peptides (through the complex process described in Chapter 23). It includes the multistep synthesis of the amino acid thyroxine and, ultimately, the speeding up of that complicated combination of reactions we lump together under the heading metabolism.

In a healthy individual all of this happens routinely, in perfect balance, and with no conscious direction. And that's just one of the myriad, interrelated biochemical processes that life requires. That it all works, and works so well, is nothing short of miraculous.

Many important human hormones and their physiological effects are listed in Table K.1. We discussed the hormones vasopressin and oxytocin in Chapter 22. The pancreatic hormones insulin and glucagon are considered in Chapter 27. In

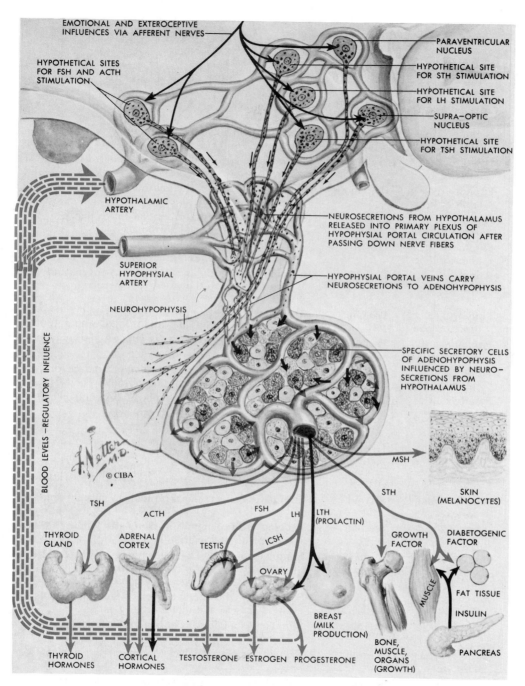

FIGURE K.2 *Schematic diagram illustrating the interrelationship of the hypothalamus, the pituitary, and the organs upon which they work. (Reproduced with permission from the CIBA Collection of Medical Illustrations by Frank Netter, M.D. Copyright © 1965 by the CIBA Pharmaceutical Co., Division of CIBA-Geigy Corporation. All rights reserved.)*

TABLE K.1   Some Human Hormones and Their Physiological Effects

Name	Gland and Tissue	Chemical Nature	Effect
Various releasing and inhibitory factors	Hypothalamus	Peptide	Triggers or inhibits release of pituitary hormones
Human growth hormone (HGH)	Pituitary, anterior lobe	Protein	Controls the general body growth; controls bone growth
Thyroid-stimulating hormone (TSH)	Pituitary, anterior lobe	Protein	Stimulates growth of the thyroid gland and production of thyroxine
Adrenocorticotrophic hormone (ACTH)	Pituitary, anterior lobe	Protein	Stimulates growth of the adrenal cortex and production of cortical hormones
Follicle-stimulating hormone (FSH)	Pituitary, anterior lobe	Protein	Stimulates growth of follicles in ovaries of females, sperm cells in testes of males
Luteinizing hormone (LH)	Pituitary, anterior lobe	Protein	Controls production and release of estrogens and progesterone from ovaries, testosterone from testes
Prolactin	Pituitary, anterior lobe	Protein	Maintains the production of estrogens and progesterone, stimulates the formation of milk
Vasopressin	Pituitary, posterior lobe	Protein	Stimulates contractions of smooth muscle; regulates water uptake by the kidneys
Oxytocin	Pituitary, posterior lobe	Protein	Stimulates contraction of the smooth muscle of the uterus; stimulates secretion of milk
Parathyroid	Parathyroid	Protein	Controls the metabolism of phosphorus and calcium
Thyroxine	Thyroid	Amino acid derivative	Increases rate of cellular metabolism
Insulin	Pancreas, beta cells	Protein	Increases cell usage of glucose; increases glycogen storage
Glucagon	Pancreas, alpha cells	Protein	Stimulates conversion of liver glycogen to glucose
Cortisol	Adrenal gland, cortex	Steroid	Stimulates conversion of proteins to carbohydrates
Aldosterone	Adrenal gland, cortex	Steroid	Regulates salt metabolism; stimulates kidneys to retain Na^+ and excrete K^+
Epinephrine (adrenalin)	Adrenal gland, medulla	Amino acid derivative	Stimulates a variety of mechanisms to prepare the body for emergency action including the conversion of glycogen to glucose
Norepinephrine (noradrenalin)	Adrenal gland, medulla	Amino acid derivative	Stimulates sympathetic nervous system; constricts blood vessels, stimulates other glands
Estradiol	Ovary, follicle	Steroid	Stimulates female sex characteristics; regulates changes during menstrual cycle
Progesterone	Ovary, corpus luteum	Steroid	Regulates menstrual cycle; maintains pregnancy
Testosterone	Testis	Steroid	Stimulates and maintains male sex characteristics

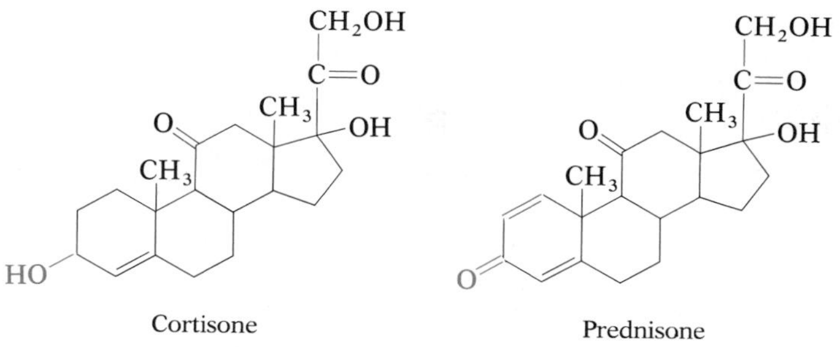

FIGURE K.3   *Cortisone and prednisone.*

the remaining sections of this special topic, we'll focus our attention on some steroid hormones.

K.2 CORTISONE: AN ADRENOCORTICAL HORMONE

The hormone cortisone is a steriod produced in the outer layer, or cortex, of the adrenal gland. Cortisone is involved in carbohydrate metabolism (Chapter 27). It is also used as a medicine. Applied topically or injected, cortisone acts to reduce inflammation. It was once widely used in the treatment of arthritis, but relief was transitory and repeated application caused serious side effects, including the disturbance of the delicately balanced endocrine system. Cortisone has been largely replaced by the related compound prednisone. The latter is effective in much smaller doses; thus side effects are greatly reduced. Structures of cortisone and prednisone are shown in Figure K.3.

K.3 SEX HORMONES

Closely related in structure to cortisone and prednisone are the sex hormones (Figure K.4). It is interesting to note that male sex hormones differ only slightly in structure from female hormones. In fact, the female hormone progesterone can be converted by a simple biochemical reaction into the male hormone testosterone. The physiological action of these structurally similar compounds is obviously markedly different.

Male sex hormones are called **androgens.** They are secreted by the testes. These hormones are responsible for development of the sex organs and for secondary sex characteristics such as the pitch of the voice and hair distribution.

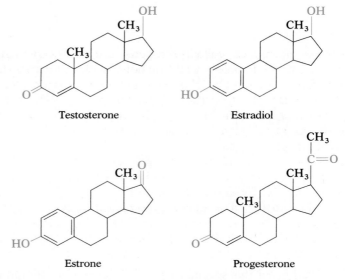

Testosterone

Estradiol

Estrone

Progesterone

FIGURE K.4   *Some important sex hormones.*

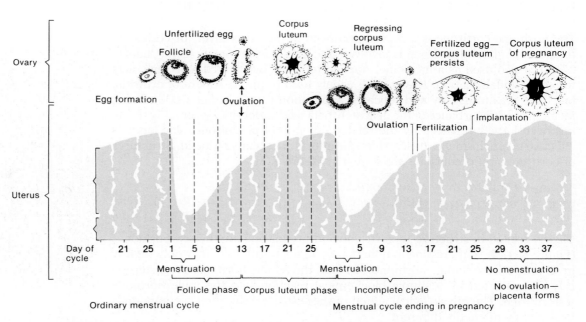

FIGURE K.5   *Changes in the ovary and the uterus during the menstrual cycle. Pregnancy—or the pseudopregnancy caused by the birth-control pill—prevents ovulation. (Adapted with permission from Sparks, Philip D.; Nord, Richard P.; Unbehaun, Laraine M.; Weeks, Thomas F.; and Hughes, Eileen G.,* Student Study Guide for the Biological Sciences, *3d ed., Minneapolis, Minn.: Burgess, 1973.)*

The female sex hormones, called **estrogens,** are produced mainly in the ovaries. They control the female sexual functions, such as the menstrual cycle (Figure K.5) and the development of breasts and other secondary sexual characteristics. Two important estrogens are estradiol and estrone.

Another important hormone produced in the ovaries is progesterone. It prepares the uterus for pregnancy and also prevents the further release of eggs from the ovaries during pregnancy.

Sex hormones—both natural and synthetic—are sometimes used therapeutically. For example, a woman who has had her ovaries removed may be given female hormones to compensate for those no longer produced by the ovaries. Some of the earliest chemical compounds employed in cancer chemotherapy were sex hormones. The male hormone testosterone was used to treat carcinoma of the breast in females; estrogens, female sex hormones, were given to males to treat carcinoma of the prostate. Sex hormones are also important in sex change operations. Before corrective surgery, hormones are administered to promote the development of the proper secondary sexual characteristics.

K.4 THE BIRTH CONTROL PILL

When administered by injection, progesterone serves as an effective birth control drug. This knowledge led to attempts by chemists to design a contraceptive that would be effective when taken orally.

The structure of progesterone was determined in 1934 by Adolf Butenandt. Just 4 years later, Hans Inhoffen synthesized the first effective oral contraceptive, ethisterone. Ethisterone had to be taken in large doses to be effective. It never became important as an oral contraceptive.

The next breakthrough came in 1951 when Carl Djerassi (Figure K.6) synthesized 19-norprogesterone, that is, progesterone with one of its methyl groups missing (Figure K.7). This compound was four to eight times as effective as progesterone as a birth control agent. Like progesterone itself, it had to be given by injection, an undesirable property for obvious reasons. Djerassi then put it all together. Removal of a methyl group made the drug more effective. The ethynyl group ($-C\equiv CH$) allowed oral administration. He then synthesized norethindrone. This drug proved effective when taken orally in small doses. Djerassi's work was published in 1954, and a patent was issued to him in 1956. At about the same time, Frank Colton (Figure K.8) synthesized norethynodrel. Note that the two substances differ only in the position of a double bond (see Figure K.7). Patents were issued to Colton's employer, G. D. Searle, in 1954 and 1955. Searle produced the first approved contraceptive, Enovid, in 1960. The pill contained 9.85 mg norethynodrel and 150 μg mestranol. Norethynodrel, norethindrone, and related compounds are called **progestins** because they

FIGURE K.6
Carl Djerassi, professor of chemistry at Stanford University and president of Zoecon Corporation, Palo Alto, California.

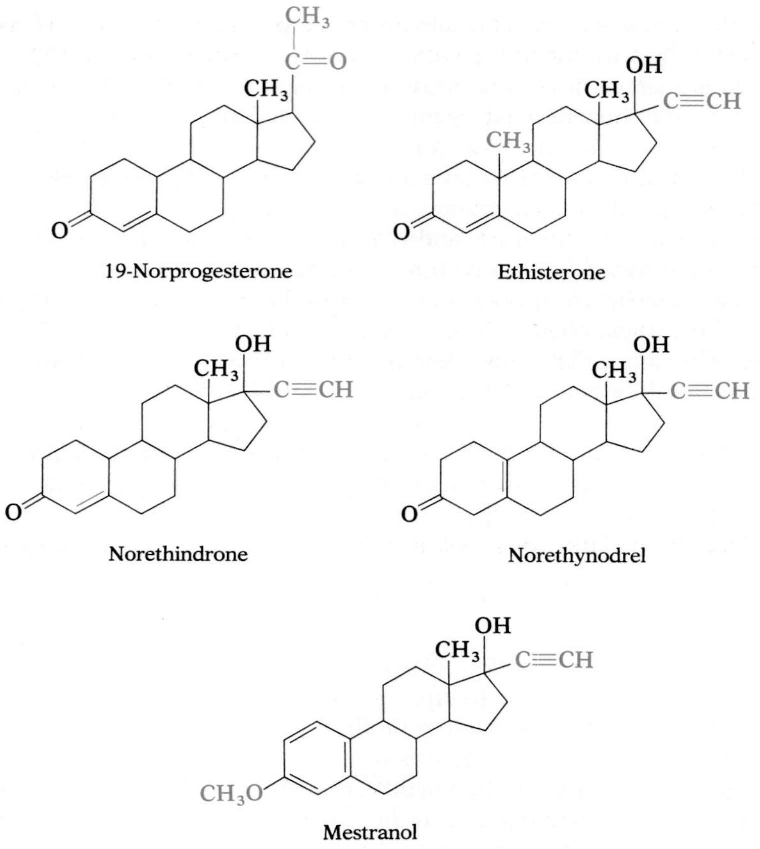

19-Norprogesterone

Ethisterone

Norethindrone

Norethynodrel

Mestranol

FIGURE K.7   *Some synthetic steroids.*

FIGURE K.8
Frank Colton, chemist at
G. D. Searle Co., first
synthesized norethynodrel, a
progestin used in Enovid, the
first birth control pills.
(Courtesy of Frank Colton.)

mimic the action of progesterone. Mestranol is a synthetic estrogen, added to regulate the menstrual cycle. The progestin acts by establishing a state of false pregnancy. A woman does not ovulate when she is pregnant (or in the state of false pregnancy established by the progestin). Since the woman does not ovulate, she cannot conceive.

The ultimate effects of prolonged tampering with the reproductive biochemistry of the human female remain to be seen. In use in the United States since 1960 by millions of women—with about 10 million still using it—the pill appears to be relatively safe. There are undesirable side effects in some women. Among these are hypertension, acne, and abnormal bleeding. It also causes clotting of the blood in some women—but so does pregnancy. Such blood clots can clog a blood vessel and cause death by stroke or coronary heart attack.

The death rate associated with the use of the pill is about 3 in 100 000. This is only one-tenth that associated with childbirth—about 30 in 100 000. For the general population, the pill is probably as safe as aspirin. Women who have blood with an abnormal tendency to clot should not take the pill, nor should those who experience any serious side effects.

In recent years, a minipill containing only about 150 μg of progestin and 30 μg of estrogen has been marketed. It supposedly avoids many of the problems associated with the early pills. Most of the undesirable side effects are thought to be caused by the estrogen.

Barring unforeseen developments, some form of chemical contraceptive seems to be with us to stay. The pill has already brought revolutionary social changes. The impact of science on society is perhaps nowhere more evident.

K.5 CYCLIC AMP: HOW HORMONES WORK

Now you have an idea of what hormones do. In this section, let's take a look at how they do it. First, let's look at a nucleotide derivative, cyclic AMP.

Cyclic AMP

AMP stands for adenosine monophosphate, one of the nucleotides discussed in Chapter 23. The molecule above is called cyclic AMP because the phosphate group has formed an intramolecular double ester. The hydroxyl groups at the third and fifth carbons of the sugar ring have both reacted to form ester links with the phosphate group, thus creating an additional ring in this nucleotide.

In many cases, when a hormone is released, the target tissue shows a sudden increase in the level of cyclic AMP. Table K.2 shows the range of hormones and tissues in which this effect has been noted. It has also been noted that many hormones never make their way into the cells of the target tissue but are stopped by the membranes surrounding the cells. Still the hormone triggers profound changes in the chemistry of the target cells. How?

TABLE K.2   *Cyclic AMP Levels Are Increased by Hormonal Action*

Tissue	Hormone	Principal Response
Bone	Parathyroid hormone	Calcium resorption
Muscle	Epinephrine	Glycogenolysis
Fat	Epinephrine	Lipolysis
	Adrenocorticotrophic hormone	Lipolysis
	Glucagon	Lipolysis
Brain	Norepinephrine	Discharge of Purkinje cells
Thyroid	Thyroid-stimulating hormone	Thyroxine secretion
Heart	Epinephrine	Increased contractility
Liver	Epinephrine	Glycogenolysis
Kidney	Parathyroid hormone	Phosphate excretion
	Vasopressin	Water reabsorption
Adrenal	Adrenocorticotrophic hormone	Hydrocortisone secretion
Ovary	Luteinizing hormone	Progesterone secretion

Modified from "Cyclic AMP," Ira Pastan. Copyright © 1972 by Scientific American, Inc. All rights reserved.

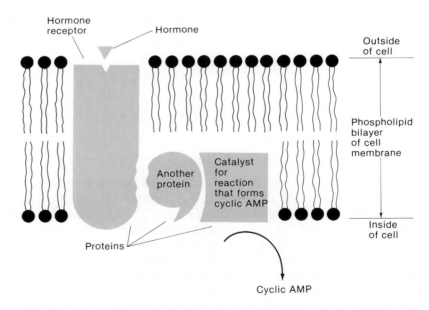

FIGURE K.9   *The binding of a hormone to a receptor site on the outside of a cell results in the activation of an enzyme that controls synthesis of cyclic AMP on the inside of the cell.*

The structure of cell membranes is discussed in Section 21.11. They consist of phospholipid bilayers in which proteins are imbedded. In the target cells, one of the protein particles has a receptor site for the appropriate hormone, that is, a portion of the protein is able to bind the hormone when the hormone arrives at the cell. The binding site is located on the outside of the cell membrane. When the hormone is bound to the protein, the conformation of the protein changes. (Section 24.6 discusses the action of an inhibitor on an enzyme.) As the shape of the binding protein changes, adjacent proteins, also imbedded in the bilayer, change their shape, too. One of these adjacent proteins is an enzyme responsible for the formation of cyclic AMP. A change in its conformation activates the enzyme, and cyclic AMP is produced on the inside of the cell (Figure K.9).

And what does cyclic AMP do? In the cytoplasm of the cell, cyclic AMP locates a protein to which it can bind. The binding of the cyclic AMP changes the conformation of this protein, which adjusts by splitting off a subunit of itself (Figure K.10). This subunit, once released, is an active enzyme ready to catalyze the reactions that produce the effects we observe. Via this mechanism,

FIGURE K.10

When cyclic AMP binds to a protein within a cell, the protein dissociates into subunits, some with catalytic activity.

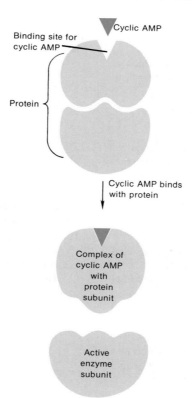

Cyclic AMP

Binding site for cyclic AMP

Protein

Cyclic AMP binds with protein

Complex of cyclic AMP with protein subunit

Active enzyme subunit

the hormone epinephrine ultimately causes the polysaccharide glycogen to convert to the monosaccharide glucose, which the cells oxidize for energy (Chapter 27).

PROBLEMS

1. In what respects are vitamins and hormones alike? In what ways do they differ?
2. Define and give an example of each of the following:
   a. hormone   c. estrogen
   b. androgen   d. progestin
3. Match the hormone to the gland that produces it.

Hormone	Gland
Thyroxine	Corpus luteum
FSH	Pancreas
Progesterone	Thyroid
Insulin	Pituitary
Epinephrine	Adrenal

4. What gland produces releasing factors?
5. What gland is the target of releasing factors?
6. Describe the general sequence of events by which a nerve impulse is translated by the endocrine system into a changed physiological state.
7. Match the hormone with the effect.

Hormone	Effect
FSH	stimulates male sex characteristics
Prolactin	stimulates female sex characteristics
Thyroxine	regulates cell uptake of glucose
Testosterone	stimulates milk production
Estradiol	increases rate of cellular metabolism
Insulin	stimulates growth of follicle in ovary

8. Answer the following questions about the menstrual cycle:
   a. What are the roles of the pituitary hormones FSH and LH?
   b. What is the relationship of the ovarian follicle and the corpus luteum?
   c. Which ovarian hormone affects the lining of the uterus in the early stages of the cycle? Which produces the changes in this lining in the later stages of the cycle?
   d. What triggers the start of a new cycle?

9. What is the general structural classification of the compounds incorporated in birth control pills?
10. How does the birth control pill work?
11. What structural feature renders a synthetic steroid sex hormone effective orally?
12. Which of the two categories of synthetic hormones, estrogens or progestins, is associated with undesirable side effects? Name some of these side effects.
13. The hormone aldosterone causes the kidney to reabsorb sodium, chloride, and bicarbonate ions. Name all the functional groups on aldosterone.

14. Compare the structural features of the androgen testosterone and the estrogen estradiol. In what ways are the structures similar? In what ways are the structures different?
15. Androstenedione is an androgen produced in the adrenal cortex. Name its functional groups. How does its structure differ from that of testosterone? How could it be made from testosterone?

16. The birth control pill Ovulen contains mestranol as the estrogen and the following compound as the progestin. Name all the functional groups in the compound.

17. Draw the structure of cyclic AMP.
18. When a hormone causes an increase in cyclic AMP levels within the cells, where is the binding site for the hormone located?
19. Explain how the binding of a hormone at its receptor site can release cyclic AMP within a target cell.
20. Outline a mechanism by which cyclic AMP may increase the rates of cellular reactions.
21. Guanosine monophosphate also exists in cyclic form. By analogy to cyclic AMP, Problem 17, draw the structure of cyclic AMP.

26

Body Fluids

The processes of life occur for the most part in solution. The solvent for life processes is water, and the solutes are many—simple ions, small molecules, large molecules, and colloidal aggregates. These are found in a variety of body fluids, and each is essential in its own way to life.

A 70-kg person has about 5 L of blood, including about 3 L of plasma. That person has about 10 L of interstitial fluid, the fluid that fills the space between cells. Fluid within the cells amounts to about 35 L. Each day the individual excretes an average of 1.5 L of fluid in urine, 0.2 L of water in the feces, and 1 L of water through the skin and lungs. All of these losses are balanced by water taken into or synthesized during metabolism within the body.

You are indeed a solution—or, rather, several solutions. But the solutions of which you are composed are not static; they constantly renew themselves. Because of this, their composition is quite dependent on the state of your health. Analysis of various body fluids represents one of the most powerful diagnostic techniques available to medical personnel.

26.1 BLOOD: AN INTRODUCTION

Blood is the principal transport system of the human body. It moves through a 100 000-km-long network of blood vessels. Some of these are so small that blood cells have to line up to pass through. Blood carries (1) oxygen from the lungs to the tissues, (2) carbon dioxide from the tissues to the lungs, (3) nutrients from the intestines to the tissues, (4) metabolic wastes from the tissues

TABLE 26.1 *Analysis of the Blood of a Typical Healthy Adult*

Component	Amount
Whole blood	
Blood volume	5L
Plasma volume	3L
Formed-cell volume (red cells, white cells, and platelets)	2L
Red cells	4 000 000/mm^3
White cells	7 000/mm^3
Platelets	300 000/mm^3
pH	7.4
Density	1.06 g/cm^3
Plasma	
Density	1.03 g/cm^3
Sodium ions (Na$^+$)	143 meq/L
Potassium ions (K$^+$)	5 meq/L
Calcium ions (Ca^{2+})	5 meq/L
Magnesium ions (Mg^{2+})	2.5 meq/L
Chloride ions (Cl$^-$)	103 meq/L
Bicarbonate ions (HCO$_3{}^-$)	27 meq/L
Hydrogen phosphate ions (HPO$_4{}^{2-}$)	3 meq/L
Sulfate ions (SO$_4{}^{2-}$)	1 meq/L
Organic acids	2 meq/L
Proteins (albumin, globulin)	20 meq/L

to the excretory organs, (5) hormones from the endocrine glands to their target tissues, and (6) three major kinds of blood cells. In addition, blood helps maintain a fairly constant body temperature, an acid-base balance, an electrolyte balance, and a water balance. A rather remarkable substance, blood. And we've only mentioned a few highlights.

Blood makes up about 8% of the body's weight. Moderate amounts lost through bleeding or blood donation are readily replaced. Table 26.1 contains a typical analysis of the blood of an average healthy adult.

Blood is a suspension of about 40% **formed elements** (red blood cells, white blood cells, and platelets) and about 60% fluid, called **plasma.** Normally, there are about 4 000 000 red blood cells, or **erythrocytes,** in each cubic millimeter of blood. If the number of red blood cells per cubic millimeter of blood is a good deal higher, a condition known as *polycythemia* exists. If the number is appreciably lower, a condition called *anemia* exists. Red blood cells are small and have no nucleus. They carry the respiratory protein hemoglobin. White blood cells, or **leukocytes,** are larger than the red blood cells (Figure 26.1).

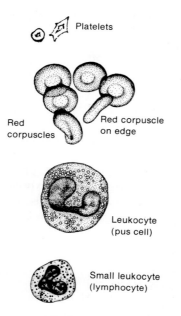

Platelets

Red corpuscles

Red corpuscle on edge

Leukocyte (pus cell)

Small leukocyte (lymphocyte)

FIGURE 26.1   The formed elements of blood

They contain nuclei and exist in a variety of forms. Leukocytes attack infectious bacteria. Normally there are about 7000 white cells in each cubic millimeter of blood. The number rises sharply during some acute infections. An increase in white cell count characterizes the condition known as leukemia. In some cases of this disease, the leukocyte count may reach 500 000 per cubic millimeter of blood.

The third type of formed element is the **platelet,** or **thrombocyte.** These are even smaller than the red cells (Figure 26.1). Like the red cells, they have no nucleus. There are usually about 300 000 platelets per cubic millimeter of blood. They are involved in the clotting of blood.

The fluid portion of the blood in which the formed elements are suspended is called **plasma.** Unless the blood is chemically treated, it is not possible to separate the plasma from the formed elements by filtration or any other technique. This is because freshly drawn blood tends to form clots on standing. Clotting is not simply the precipitation of the formed elements, but a series of complicated chemical reactions (Chapter 24). The fluid remaining after clotting occurs is called **serum** (not plasma). If one wishes to store blood plasma, an anticoagulant must be added to the freshly drawn blood to prevent clotting.

$$\text{Fresh blood (+anticoagulant)} \longrightarrow \text{formed elements + plasma}$$
$$\text{Fresh blood} \longrightarrow \text{fibrinogen (clot) + formed elements + serum}$$

Plasma or serum can be separated from the solid materials by a centrifuge.

26.2 ELECTROLYTES IN PLASMA AND ERYTHROCYTES

Plasma is about 90% water. Even red blood cells are about 65% water. A variety of ions are found in solution in both plasma and blood cells. These electrolytes play important roles in several life processes.

The concentration of electrolytes is often expressed as milliequivalents per liter (meq/L) of fluid. Equivalent weights are discussed in Chapter 11 for acids and bases. The milliequivalent weight of an ion is its atomic weight, expressed in milligrams, divided by the charge on the ion.

example
26.1

What is the milliequivalent weight of Na^+?
    The atomic weight of sodium is 23. The charge on the ion is 1 (the sign of the charge is ignored in these calculations).

$$\frac{23 \text{ mg}}{1} = 23 \text{ mg}$$

example
26.2

What is the milliequivalent weight of Ca^{2+}?
    The atomic weight of calcium is 40.

$$\frac{40 \text{ mg}}{2} = 20 \text{ mg}$$

example
26.3

What is the milliequivalent weight of SO_4^{2-}?
    The formula weight of the sulfate ion is 96.

$$\frac{96 \text{ mg}}{2} = 48 \text{ mg}$$

The number of milliequivalents in a certain amount of a substance is given by that amount (in milligrams) divided by the milliequivalent weight (i.e., the number of milligrams in a milliequivalent).

How many milliequivalents are there in 30 mg of Ca^{2+}?
    The milliequivalent weight of Ca^{2+} is 20 mg (Example 26.2).

example
26.4

$$\frac{30 \text{ mg}}{20 \text{ mg/meq}} = 1.5 \text{ meq}$$

What is the concentration of Ca^{2+} in milliequivalents per liter if there are 12 mg of Ca^{2+} per 100 mL of serum?
    First convert milliliters to liters.

example
26.5

$$\frac{12 \text{ mg}}{100 \text{ mL}} \times \frac{1000 \text{ mL}}{1 \text{ L}} = 120 \text{ mg/L}$$

The milliequivalent weight of Ca^{2+} is 20 mg. Therefore,

$$\frac{120 \text{ mg/L}}{20 \text{ mg/meq}} = 6.0 \text{ meq/L}$$

Typical electrolyte concentrations (in milliequivalents per liter) are given in Table 26.1. Sometimes concentrations are expressed in milligram percents (milligrams per 100 mL) of plasma. To convert milligram percents to milliequivalents per liter, multiply the number of milligrams by 10 and then divide the result by the milliequivalent weight of the ion (Example 26.5).

Sodium ions are found mainly in the plasma, and potassium ions are found chiefly in the erythrocytes. These positive ions do not migrate readily across cell membranes. Calcium and magnesium ions are the main dipositive ions in blood. They are usually determined in serum, because many anticoagulants added to preserve plasma tie up these ions. (Calcium ions are necessary for blood clotting; if they are removed, the blood will not clot.) Calcium ions are not found in erythrocytes, but magnesium ions are. The metabolism of calcium and the metabolism of phosphorus are closely related, and levels of the two usually vary in a reciprocal manner. Magnesium levels, on the other hand, are found sometimes to parallel the variations in calcium levels and at other times to parallel the variations of phosphorus levels.

The principal negative ions present in the blood, in addition to inorganic phosphorus (as HPO_4^{2-}), are chloride, bicarbonate, and sulfate. Bicarbonate and hydrogen phosphate are involved in the acid-base balance of the blood. These anions will be discussed in that regard in Section 26.5.

Table 26.2 lists some conditions that influence plasma or serum levels of certain electrolytes.

TABLE 26.2   *Pathological Conditions Affecting Electrolyte Levels*

| Electrolyte | Diseases or Conditions Associated With Changes in Concentration | |
	Elevated Levels	Depressed Levels
Sodium (Na^+)	Overactive adrenal glands	Severe kidney inflammation, Addison's disease (underactive adrenal cortex), pregnancy
Potassium (K^+)	Acute bronchial asthma, Addison's disease, uncontrolled diabetes, uremia	Hyperinsulinism (overproduction of insulin)
Calcium (Ca^{2+})	Overactive parathyroid	Underactive parathyroid, variety of kidney ailments, vitamin D deficiency
Magnesium (Mg^{2+})	Uremia, dehydration, Addison's disease, diabetic acidosis	Acute pancreatitis, chronic alcoholism
Chloride (Cl^-)	Kidney inflammation, heart disease	Burns, fever, pneumonia, vomiting, Addison's disease, diabetes
Hydrogen phosphate ($HPO_4{}^{2-}$)	Uremia, nephritis, underactive parathyroid, overdose of vitamin D	Rickets, overactive parathyroid

26.3 PROTEINS IN THE PLASMA

Proteins make up about 6.5% of the plasma. Dozens of different ones are present in plasma; the most important ones are **albumin,** the **globulins, fibrinogen,** and **prothrombin.** Fibrinogen and prothrombin are involved in the mechanism of blood clotting (Chapter 24). The globulins and their role in the body's immune response will be discussed in Section 26.4. All the proteins, but particularly albumin, are important in regulating the osmotic pressure of blood. To appreciate the significance of this function, it will be necessary for us to consider first a few facts about the movement of blood and its constituents.

How does material get from the blood to the cells? Diffusion of water, electrolytes, glucose, amino acids, and other materials occurs at a rapid rate through pores in the capillary walls. It occurs in both directions—from the blood into the interstitial fluid outside the capillary and from the interstitial fluid back into the blood. Diffusion through capillary walls occurs at a fantastic rate, estimated at equivalent to 1500 L of water per minute for someone weighing 70 kg. Ultimately material makes its way from the interstitial space through cell walls and into (or out of) the cells themselves.

The blood doesn't just sit in the capillaries while the diffusion takes place. It circulates, that is, continues to move through the vascular system. What keeps the blood moving is the heart, a pump that pushes the blood around the system. The pressure imparted to the blood by this pumping action falls off as the blood travels from the heart through the arteries, the capillaries, and the veins and finally back to the heart, where it gets a fresh push. Blood moves from the high-pressure arterial end of the capillaries to the low-pressure venous end. As it circulates through the capillaries, the diffusion process described above occurs. The pressure that keeps the blood moving around the system also has a tendency to push the blood plasma out of the porous capillaries into the interstitial space. In the absence of any countereffect, the rapid diffusion of material back and forth between blood and interstitial fluid would be accompanied by a slow but steady net loss of fluid from the blood to the interstitial space. Blood volume would drop, and tissue would swell with the extra fluid. There is a countereffect, though—osmotic pressure.

Because blood and interstitial fluid have all sorts of things dissolved in them, both fluids have characteristic osmotic pressures. Blood plasma has a higher one than interstitial fluid. Plasma contains proteins, molecules of colloidal size that cannot pass easily through the capillary pores into the interstitial compartment. Thus, whereas both plasma and interstitial fluid contain roughly equal concentrations of total electrolytes, the concentration of proteins (chiefly albumin) is higher in plasma than in interstitial fluid. The proteins make a relatively small contribution to the osmotic pressure of plasma compared with the effect of the electrolytes. However, it is this smaller contribution that tips the scales in favor of the plasma.

Remember that the higher the osmotic pressure of a system, the greater tendency it has to accumulate water (Chapter 9). This means that in the capillaries there is a competition between two pressures—the **hydrostatic pressure** (the pressure that the heart's pumping action gives the blood) and the **colloid osmotic pressure** (the osmotic pressure due to the colloidal proteins in the blood, sometimes called **oncotic** pressure). The hydrostatic pressure pushes fluid out of the capillaries; the colloid osmotic pressure pulls liquid into the capillaries. Who wins? They both do.

At the arterial end of the capillaries, the hydrostatic pressure (32 torr) exceeds the colloid osmotic pressure (about 25 torr), and there is a net loss of liquid to the interstitial space. At the venous end, the hydrostatic pressure is lower (15 torr), but the colloid osmotic pressure is essentially unchanged (still about 25 torr). At the venous end, then, the colloid osmotic pressure wins, and the net flow of fluid is into the capillary (Figure 26.2). If we are in good health, everything balances rather nicely. To be sure, nutrients have diffused from blood to interstitial fluid and wastes have diffused from interstitial fluid to blood, but the volume of fluid in the two systems has remained constant.

If the concentration of plasma protein drops, the lowering of osmotic pressure results in a net flow of fluid to the interstitial space. The tissues become waterlogged and swollen, a condition known as **edema.** A lowering of the plasma protein level (and consequent edema) may be caused by liver disease, malnutrition, or nephrosis. Liver disease affects the protein level because albumin is made in the liver. In malnutrition, albumin production also drops off. Nephrosis is a condition in which albumin is lost into the urine (Section 26.8).

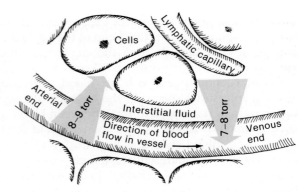

FIGURE 26.2
Fluid filters out of the arterial end of the capillary and back into the venous end. An imbalance in these processes results in a net loss of fluid to the interstitial space.

The medical condition called **shock** also results from a loss of fluid from the vascular system. Capillary permeability increases, and albumin is lost to the interstitial fluid. The resulting outflow of fluid from the vascular system reduces blood volume enough to cause a dramatic decrease in blood pressure. There is a consequent decrease in oxygen-transporting capability with potentially fatal results. When someone goes into shock following a traumatic injury, it means that the delicately balanced transport system in the body has gone awry and could fail completely. Shock is a physiological, not a psychological, state. Treatment of the condition involves bringing the blood volume back up to normal levels.

26.4 GAMMA GLOBULINS: THE IMMUNE RESPONSE

Some of the plasma proteins, the **gamma globulins,** are associated with a sophisticated defense mechanism in the body termed the **immune response.** When the body is invaded by foreign macromolecules (such as proteins), it responds by synthesizing its own specialized proteins called **antibodies** (also called immunoglobulins). Since the foreign substance triggers the synthesis of antibodies, the invader is called an **antigen** (it causes antibody generation). Each distinctive antigen causes a unique set of antibodies to be produced. The antibodies are designed to lock onto the particular antigen. They bind to it and, to oversimplify a bit, incapacitate it. In some instances the incapacitation involves the formation of a precipitate, with the antigen being tied up as part of an antibody-antigen clot.

When the body is invaded by pathogenic antigens (disease-causing bacteria or viruses, for example), its ability to produce specific antibodies represents an effective defense mechanism. It is even possible to prime this mechanism. A **vaccine** containing a weakened form of the antigen (dead bacteria, for example) will cause an individual to build up a certain level of antibodies for this particular antigen. If the same individual is then subjected to attack by a more virulent form of the antigen, the body responds much more rapidly and effectively to its presence. Having practiced on the dead "bug," the system can easily handle the live one; it has gained an immunity to that particular organism (Figure 26.3).

Unfortunately, the body can't tell "good" antigens from "bad" antigens. We want it to respond to a viral infection, but we wish it wouldn't attack purposely transplanted tissue. Nonetheless, the host body will respond to a donor skin graft, a kidney transplant, or a heart transplant as it does to a virus, by generating antibodies to attack it. When the body responds to foreign tissues in this way, the process is called **rejection.** This is a major problem associated with organ transplants. There are drugs that suppress the immune response, but when these are used on a transplant recipient to protect the transplanted organ, the patient also becomes more susceptible to infection. If you turn off the immune response, you turn off both the good and bad aspects of it.

There are a variety of diseases associated with the immune system. An overactive or misguided immune system can attack its own body organs, causing an **autoimmune** disease. It is thought that multiple sclerosis, some forms of arthritis, and some cases of diabetes are autoimmune diseases. The system senses something as foreign even though it isn't. It attacks and destroys the myelin sheath of nerves (multiple sclerosis), the connective tissues in the joints (arthritis), or the insulin-producing cells of the pancreas (diabetes).

At the opposite end of the scale is **acquired immune deficiency syndrome (AIDS).** In this deadly affliction, the immune system is destroyed, presumably by a virus. The victims then succumb to a variety of infections and to rare forms of cancer.

Scientists have sought to enhance certain immune responses as weapons against diseases. By genetic engineering, researchers have cloned the genes for specific antibodies and placed them in cells that then produce **monoclonal antibodies.** In theory, monoclonal antibodies should attack and overwhelm specific antigens. In practice, they have been a bit disappointing, but they still hold considerable promise.

FIGURE 26.3
Smallpox, which once killed and disfigured millions, has been eradicated by the use of vaccines. (Courtesy of the U.S. Agency for International Development, Washington, D.C.)

26.5 ACIDS AND BASES IN THE BLOOD

In Chapter 11, we examined buffer systems in general and blood buffers in particular. You will recall that the pH of the blood must be maintained within a very narrow range around 7.4. If it falls below 7.35, **acidosis** sets in. If it rises above 7.45, a condition called **alkalosis** exists. The life-sustaining range of blood pH covers only 1 pH unit—from about 6.8 to about 7.8.

The primary blood buffer is the bicarbonate-carbonic acid system (HCO_3^-/H_2CO_3). The buffering power of this system is enhanced by the interconversion of carbonic acid and carbon dioxide, so that the entire system can be represented as

$$H_3O^+ + HCO_3^- \rightleftharpoons H_2CO_3 \rightleftharpoons H_2O + CO_2$$
$$+$$
$$H_2O$$

Addition of acid shifts the equilibrium to the right, tying up hydronium ions. Added base will cause a shift to the left, releasing hydronium ions.

In certain situations, the buffers can be overwhelmed or may fail to function properly, and acidosis or alkalosis can result. If a person hyperventilates (breathes in and out very rapidly), too much carbon dioxide can be exhausted. This ultimately results in a shift in the equilibrium to the right. As hydronium ions are tied up, the pH increases and respiratory alkalosis occurs. If an individual is not ventilating adequately (because of some lung obstruction or disease, for example), carbon dioxide builds up in the blood, the equilibrium shifts to the left, more hydronium ions are released, the pH falls, and respiratory acidosis occurs. Metabolic disorders are more likely to produce acidosis than alkalosis because metabolism produces a variety of acids, including lactic, pyruvic, and phosphoric acids, as well as carbonic acid. We discuss acidosis further in Chapter 28.

Other plasma buffers are the monohydrogen phosphate-dihydrogen phosphate system and the plasma proteins. The latter, which have both acidic and basic groups on the side chains of certain amino acid residues, can give up protons if the blood starts to get alkaline or accept protons if the blood starts to get acidic.

The bicarbonate-carbonic acid buffer system also operates within the red blood cells. Here it is coordinated with two important hemoglobin buffers also found in the erythrocytes. We will look more closely at hemoglobin and its role in oxygen transport shortly. For now let's consider the interplay of the hemoglobin and bicarbonate-carbonic acid buffers. One of the hemoglobin buffers can be represented as HHb for the acidic form and Hb⁻ for the basic form.

Carbon dioxide produced by cell respiration diffuses into red blood cells. In the red blood cells, the carbon dioxide is rapidly converted to carbonic acid.

$$CO_2 + H_2O \rightleftharpoons H_2CO_3$$

The carbonic acid reacts with the basic form of hemoglobin, yielding the acidic form of hemoglobin and bicarbonate ion.

$$H_2CO_3 + Hb^- \rightleftharpoons HCO_3^- + HHb$$

Bicarbonate builds up in the cell. It has a tendency to diffuse out, but electrical neutrality must be preserved. Positive potassium ions in the erythrocytes won't cross the cell membrane readily, so neutrality is maintained by another negative ion, chloride, which diffuses in as bicarbonate moves out. This phenomenon is called the **chloride shift.** At equilibrium, about 60% of the bicarbonate ion is in the plasma, and 40% is in the erythrocytes. Bulimia is a condition in which a person gorges on food and then induces vomiting to keep from gaining weight. This leads also to a loss of chloride ion as hydrochloric acid from the stomach is lost in the vomit. There is then less chloride in the body, and the chloride shift is suppressed. Bicarbonate ion builds up, leading to alkalosis.

When the blood reaches the lungs, hemoglobin picks up fresh oxygen to form oxyhemoglobin, which we'll write for now as $HHbO_2$. This molecule is the acidic form of the second hemoglobin buffer. The basic form we'll write as HbO_2^-. In the capillaries of the lungs, the acidic oxyhemoglobin reacts with the bicarbonate ion accumulated by the erythrocytes.

$$HHbO_2 + HCO_3^- \longrightarrow HbO_2^- + H_2CO_3$$

The carbonic acid thus produced is rapidly converted to carbon dioxide,

$$H_2CO_3 \longrightarrow H_2O + CO_2$$

and the carbon dioxide is exhausted through the lungs to the atmosphere.

Through these acid-base reactions, hemoglobin plays a role in the transport of carbon dioxide from cells to the lungs. Its role as an oxygen carrier is discussed in the next section.

26.6 HEMOGLOBIN: OXYGEN TRANSPORT

One of the main functions of blood is the transport of oxygen. The oxygen-carrying component is a conjugated protein called hemoglobin. The hemoglobin molecule has a protein portion and a nonprotein portion, or prosthetic group. The prosthetic group is heme.

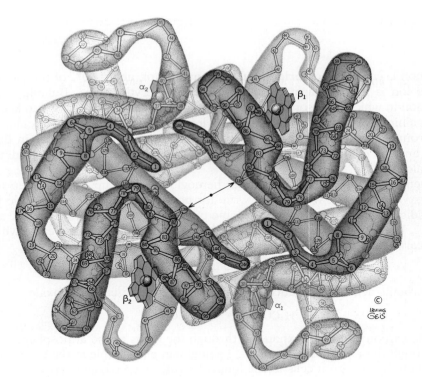

Oxyhemoglobin

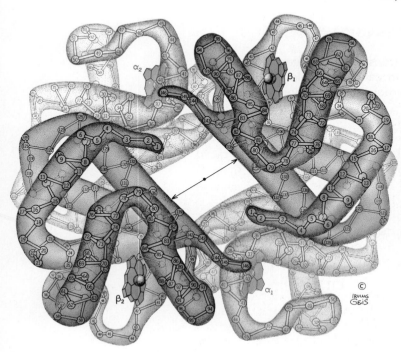

Deoxyhemoglobin

FIGURE 26.4
The quaternary structure of hemoglobin. Above: hemoglobin with oxygen (oxyhemoglobin); left: hemoglobin without oxygen (deoxyhemoglobin). The four protein subunits are designated α_1, α_2, β_1, and β_2. (Illustrations copyright by Irving Geis.)

A hemoglobin molecule contains four heme units, each with a central iron atom. This iron atom in each heme unit is attached to the nitrogen atoms of four pyrrole rings (Chapter 18). It is in the $+2$ oxidation state and is capable of forming two additional bonds.

The protein portion of the hemoglobin molecule has four polypeptide chains, which are not covalently bonded to one another. There are two identical alpha chains, each with 141 amino acid residues, and two beta chains, each with 146 amino acid units. Each chain contains sections that are coiled into alpha-helical conformations. Each chain is then folded into a globular tertiary structure. Finally, the quaternary structure of this protein has the four folded peptide subunits arranged in a roughly tetrahedral pattern (Figure 26.4). The heme units, indicated by rectangular shapes, are located in cavities in the folds.

The protein chains play an active role in the transport of oxygen. The fifth bonding site on the iron atoms is occupied by a nitrogen atom of a histidine residue that is the 87th amino acid unit in the alpha chain or the 92nd in the beta chain (margin).

Each hemoglobin molecule is capable of transporting four molecules of oxygen; each oxygen molecule occupies the sixth bonding site of an iron in one of the heme units. Binding of one oxygen molecule facilitates the attachment of another, because oxygenation changes the conformation of the chain. This seems also to induce changes in a neighboring heme-peptide unit, giving it a greater affinity for oxygen. Conversely, release of oxygen from one site facilitates release from a neighboring site. These factors enable hemoglobin to load and unload oxygen quite rapidly.

Normally, there are about 15 g of hemoglobin per 100 mL of blood. This amount of hemoglobin can combine with about 20 mL of gaseous oxygen (at STP). Without the hemoglobin, only 0.3 mL of gaseous oxygen could physically dissolve in 100 mL of plasma.

Various chemical substances act as poisons by interfering with the transport of oxygen by hemoglobin. The action of carbon monoxide was described in Chapter 6. Nitrite salts oxidize the iron in hemoglobin from the $+2$ to the $+3$ oxidation state. The resulting methemoglobin (Section 13.11) is unable to transport oxygen. Nitrate ions are reduced to nitrite ions by microorganisms in the digestive tract. Concern exists over the high level of nitrates from fertilizer in the groundwater in some areas. Babies are particularly sensitive. Nitrites or nitrates cause methemoglobinemia and result in the blue baby syndrome.

Red blood cells are manufactured in the bone marrow. The life span of a red blood cell is about 120 days. After that amount of time, the hemoglobin in the cell breaks down. The protein portion (globin) is metabolized like other proteins. The heme portion breaks down into bilirubin, a yellow-brown pigment, and other products. Bilirubin is the chief bile pigment. It is further changed to other colored compounds that are excreted. These give urine and

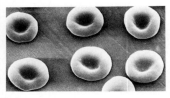

(a)

(b)

FIGURE 26.5

(a) Normal red blood cells. (b) Sickled cell. (Courtesy of Richard F. Baker, University of Southern California Medical School, Los Angeles.)

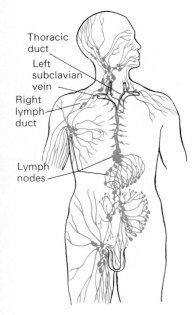

FIGURE 26.6
The lymphatic system.

feces their characteristic colors. The sometimes spectacular color changes observed in a bruise have a related cause. A bruise consists of blood released and trapped beneath the skin. As the blood is gradually broken down, a series of colored products is formed.

There are a variety of abnormal hemoglobins, each responsible for a particular disease. Perhaps the most notorious of the abnormal hemoglobins is that responsible for sickle cell anemia. This hemoglobin, called hemoglobin S (HbS) differs from ordinary hemoglobin in only one amino acid: at the sixth position of the beta chains, hemoglobin S has valine rather than glutamic acid. This seemingly minor variation not only changes the conformation of oxyhemoglobin but causes the red blood cells themselves to change from a round to a sickle shape (Figure 26.5). These sickled cells are rapidly destroyed. The bone marrow can't replace them fast enough, and anemia results. In addition, the sickled cells clog capillaries, particularly in the spleen, causing excruciating pain.

Sickle cell anemia is an inherited trait. The genetic code has one error in its message, calling for valine instead of glutamic acid. What a difference one minor "mistake" in the code can make.

26.7 LYMPH: A SECONDARY TRANSPORTATION SYSTEM

The lymphatic system also plays a role in the transportation of materials from one part of the body to another. It seems to return components of the interstitial fluid to the bloodstream. The lymphatic system (Figure 26.6) is composed of veins and capillaries but no arteries. Lymph capillaries are closed at one end. Lymph flow is quite slow compared with blood circulation. Interstitial fluid is absorbed into the lymph capillaries, in which it is called **lymph.** This lymph flows into larger and larger lymph veins. Eventually, two large lymph veins empty into veins of the blood circulatory system.

Lymph serves other functions. Fat absorbed in the intestine is picked up by lymph capillaries rather than blood capillaries. **Lymph nodes** serve as filters and as factories for the production of some forms of white blood cells. The white blood cells located there remove dead cells, bacteria, and other foreign elements from the lymph. It is in the nodes that antibodies are synthesized. These nodes are lumpy enlargements in the lymph veins. Sometimes the nodes are so effective at filtering out bacteria from an infected area that they become swollen. Someone suffering from a sore throat may also exhibit swollen and tender lymph nodes in the neck area.

The role of lymph, though secondary to that of blood, is nonetheless an important one.

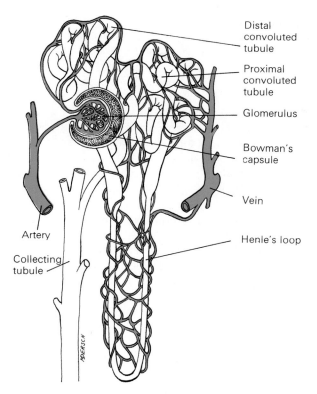

Distal convoluted tubule

Proximal convoluted tubule

Glomerulus

Bowman's capsule

Vein

Henle's loop

Artery

Collecting tubule

MDERSCH

FIGURE 26.7 *Nephrons are the functional units of kidneys, where metabolic wastes are removed from the blood.*

26.8 URINE: FORMATION AND COMPOSITION

The kidneys operate to remove metabolic waste products from the blood. The functional units of the kidneys are called **nephrons** (Figure 26.7). Each nephron has a bulbous **Bowman's capsule,** which tails into a long, highly convoluted urinary tubule. The capsule surrounds a network of arterial capillaries called the **glomerulus.** These capillaries rejoin to form a small artery (arteriole) and then divide into another network of capillaries that surrounds the tubule of the nephron. Finally these capillaries join again and form a small vein. Blood flows from the artery into the glomerulus (capillaries), then through the arteriole and the second capillary network surrounding the tubule, and finally out the vein (Figure 26.8).

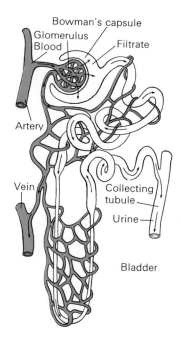

Bowman's capsule

Glomerulus
Blood

Filtrate

Artery

Vein

Collecting tubule

Urine

Bladder

FIGURE 26.8
This view of a nephron indicates the path of fluids through the unit.

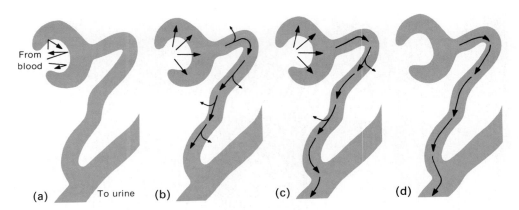

FIGURE 26.9   The fates of various substances in the kidney tubule. (a) Particles of colloidal size (proteins) or larger (blood cells) never filter into Bowman's capsule and, therefore, do not ordinarily appear in urine. (b) Substances with a high threshold (glucose, etc.) filter into the tubules but are then reabsorbed by the blood. They do not ordinarily appear in urine. (c) Substances with intermediate thresholds (inorganic ions such as Na^+ or Cl^-) filter into Bowman's capsule and are then partially reabsorbed by the blood. Dietary levels determine how much finally remains in urine. (d) Substances with low thresholds filter into the tubule and remain there. Urine is rich in urea.

Most of the constituents of the blood, except the formed bodies and large protein molecules, are filtered out through the capillary walls of the glomerulus. The glomerulus seems to act as a simple filter, with particle size being the main factor that decides what goes and what stays (Figure 26.9). The fluid that enters Bowman's capsule and eventually collects in the tubule contains many valuable components as well as wastes. Consider water alone. About 170 L per day is filtered into the tubules. Most of this water—and valuable constituents such as glucose, amino acids, and salt—is reabsorbed by the blood through the walls of the tubules. Waste products, such as urea, uric acid, and excess salts, are passed on into collecting tubules and eventually excreted as **urine.** Knowing that healthy adults pass 1.1 to 1.5 L of urine each day, you can calculate that over 99% of the water that is filtered through the glomerulus is reabsorbed through the tubules. We'd have quite a drinking problem if all that water weren't conserved.

The amount and composition of the urine is quite variable. Its volume depends on liquid intake, amount of perspiration, presence of fever or diarrhea, and other factors. The composition varies with diet and state of health. Most substances have a kidney threshold level. If the concentration of a substance in the blood exceeds this threshold value, the excess is not reabsorbed through the tubule but appears in the urine. The threshold level for glucose (blood sugar), for example, is quite high. Normally, nearly all glucose is reabsorbed.

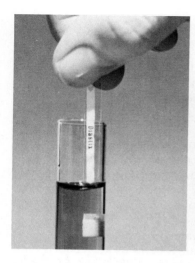 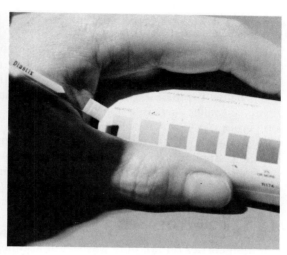

FIGURE 26.10   A simple test for sugar in urine. The reagent strip is dipped into the urine sample and then compared to a color chart that indicates sugar content. (DIASTIX® is a registered trademark of Ames Division, Miles Laboratories, Inc., Elkhart, IN 46515. Reprinted with permission.)

If the glucose level in the blood is too high, however, glucose shows up in the urine (as happens in cases of uncontrolled diabetes) (Figure 26.10).

Some components of urine have a relatively low solubility. When some condition (such as an increase in concentration or a change in pH) leads to the precipitation of these materials in the kidney, a stone is formed. Kidney stones (or **renal calculi**) usually consist of calcium phosphate [$Ca_3(PO_4)_2$], magnesium ammonium phosphate ($MgNH_4PO_4$), calcium carbonate ($CaCO_3$), calcium oxalate (CaC_2O_4), or a mixture of these. Stone formation may accompany increased concentration of calcium ion caused by disease or increased ingestion of calcium ion. People who eat foods rich in oxalates (such as spinach) have a relatively high incidence of oxalate kidney stones. (What a great excuse for kids who don't want to eat their spinach!)

The kidney does far more than just get rid of wastes. It plays a vital role in maintaining the balance of water, electrolytes, acids and bases, and other components of the body fluid. When the kidneys malfunction, the body is in trouble. Analysis of the urine can often give a good indication of the health of the individual.

Wilhelm Kolff invented the artificial kidney at the Cleveland Clinic in the 1950s. Before that time, kidney failure meant death. There are now 60 000 people on dialysis at an annual cost of $1.8 billion. This constitutes 3.6% of all costs of Medicare and Medicaid, which pay for dialysis. The number of patients and the costs are rising rapidly.

26.9 SWEAT

The skin is also an organ of excretion. Through it we lose water, electrolytes, nitrogenous wastes, and lipids. Water is lost directly through the skin and through the respiratory tract at a rate of about 700 mL per day. This water loss is called **insensible perspiration.** Perspiration from the 2.5 million sweat glands is called **sensible perspiration.** It is activated by a rise in blood temperature.

Sweat is about 99% water. It contains sodium ions, chloride ions, calcium ions, and smaller amounts of other minerals. A person working in a hot environment might lose 12 L of sweat per day, including 70 g of salt. Organic constituents of sweat include urea, lipids (body oils), creatinine, lactic acid,* and pyruvic acid. Drugs such as morphine, nicotine, and alcohol will also appear in sweat.

If the air around us is not too humid, the water in sweat evaporates. Perspiration carries off not only wastes but also heat. It helps us keep cool on hot days. Each gram of water that evaporates absorbs from the body 540 cal of heat (Chapter 7).

26.10 TEARS: THE CHEMISTRY OF CRYING

If ever you should cry over your chemistry grade, take consolation from the fact that this lacrimal fluid is responsible for maintaining the health of your eyes. Tears keep the eyes moist. The eyelids sweep the secretions of the lacrimal glands over the surface of the eye at regular intervals.

Tears are actually three-layered. There is an inner layer of mucus, then a layer of lacrimal secretions, and finally an outer layer of oily film that retards evaporation from the watery middle layer. Total normal secretion is about 1 g per day.

Chemically, tears have about the same composition as other body fluids. They have approximately the same salt content as blood plasma. Tears contain **lysozyme,** an enzyme that ruptures bacterial cell walls. This bactericidal action helps prevent eye infections.

Copious flows of tears may be caused by irritants, such as pepper, acid fumes, or a variety of chemical lachrymators. Tear gas, usually α-chloroacetophenone, is a specially designed eye irritant. The flow of tears may also be triggered by emotional upsets. Such crying is undoubtedly controlled by hormones. And if you feel like crying, go right ahead. Most psychologists say it's

α-Chloroacetophenone

---

* It is interesting to note that female mosquitoes, seeking their meal of blood, find us by following a warm stream of air laden with carbon dioxide and lactic acid. We could foil them by not sweating and not breathing!

good for you. Indeed, many people have long believed crying to be beneficial. Richard Crashaw, an English poet of the seventeenth century, called tears "the ease of woe."

26.11 THE CHEMISTRY OF MOTHER'S MILK

Newborn mammals are nourished by a secretion of the mammary glands of their mothers. Indeed, the presence of such glands in females characterizes the class Mammalia. The composition of milk varies from one species to another (Table 26.3). Amounts of fat and protein tend to be greater in marine mammals and those that live in cold climates.

Colloidal proteins and emulsified fats give milk its characteristic milky appearance. **Casein** is the precipitated protein from cow's milk. Humans beyond infancy often include some milk or milk products in their diet. This food is an excellent source of most nutrients. It includes all the essential amino acids. It also contains most vitamins and minerals. Cow's milk, though, is a poor source of iron, copper, and vitamin C. It is also short of vitamin D as it comes from the cow, but this vitamin is often added as a supplement to cow's milk intended for human consumption.

Milk sugar is lactose, a disaccharide incorporating a glucose and a galactose unit. Some infants lack enzymes necessary for the metabolism of galactose and suffer from a condition called **galactosemia.** Children with this condition may develop cataracts and some mental disorders. By removal of the sources of galactose (chiefly milk) from the diet of such infants, the condition can be controlled.

Cow's milk is sometimes given to infant humans as a substitute for their mother's milk. Evidence now indicates that this may not always be a wise substitution. Human milk not only more closely matches the nutritional needs of human infants, but it may also add to the infant's immunological defenses against disease. Newborn infants are unable to make antibodies. They have *temporary passive immunity* because their blood contains antibodies made by the mother and passed across the placenta. Mother's milk produced within the

TABLE 26.3 *Composition of Milks (in Grams per 100 g)*

Mammal	Fat	Protein	Lactose
Cow	4.4	3.8	4.9
Human	3.8	1.6	7.0
Goat	4.1	3.7	4.2
Reindeer	22.5	10.3	2.5
Porpoise	49.0	11.0	1.3

first few days after childbirth is a rich source of gamma globulins. There is clear evidence that these immunoglobulins protect some newborn animals (ungulates such as cows, horses, and sheep). A similar function for the immunoglobins in mother's milk in human infants is presumed.

PROBLEMS

1. List five functions of the blood.
2. How are plasma and serum obtained from whole blood?
3. What blood electrolyte is required for clotting?
4. What is a formed element of the blood?
5. What is an erythrocyte? What is its function?
6. What is a leukocyte? What is its function?
7. What is a thrombocyte? What is its function?
8. Name four protein fractions in blood plasma.
9. Name four cations in blood plasma.
10. Name four anions in blood plasma.
11. How many equivalents are there in each of the following?
    a. 1 mol of $H_2PO_4^-$ c. 2 mol of $C_2O_4^{2-}$
    b. 1 mol of Al^{3+}
12. How many equivalents are there in each of the following?
    a. 40 g of Ca^{2+} b. 9.6 g of SO_4^{2-}
13. How many moles are there in each of the following?
    a. 4.0 eq of Ca^{2+} c. 110 meq of Cl^-
    b. 0.050 eq of SO_4^{2-}
14. How many grams are there in each of the following?
    a. 2.0 eq of Na^+ c. 250 meq of SO_4^{2-}
    b. 0.250 eq of Mg^{2+}
15. The concentration of electrolytes in the spinal fluid of a patient was reported in milligram percent. Calculate the corresponding values in milliequivalents per liter.

$$Cl^- = 426 \text{ mg } \%$$
$$Ca^{2+} = 2.5 \text{ mg } \%$$
$$Na^+ = 322 \text{ mg } \%$$

16. The normal values for these ions in spinal fluid fall within the following ranges:

$$Cl^-: 118-132 \text{ meq/L}$$
$$Ca^{2+}: 2.1-2.9 \text{ meq/L}$$
$$Na^+: 138-158 \text{ meq/L}$$

Does the analysis of the patient's spinal fluid (given in Problem 15) reveal any abnormalities?
17. What is interstitial fluid?
18. What is lymph?
19. Explain the role of diffusion, blood pressure (hydrostatic pressure), and oncotic pressure (colloid osmotic pressure) in transferring material from capillaries to interstitial fluid and vice versa.
20. What is edema? What causes it?
21. What is shock?
22. What are the three main buffers in blood?
23. Using equations, explain how the bicarbonate-carbonic acid buffer would combat acidosis.
24. Using equations, show how the bicarbonate-carbonic acid buffer would combat alkalosis.
25. What condition results in respiratory acidosis?
26. What condition results in respiratory alkalosis?
27. What condition results in metabolic acidosis?
28. Explain the role of the hemoglobin buffers in carbon dioxide transport.
29. What is the chloride shift? Why does it occur?
30. What are antibodies?
31. What are antigens?
32. Describe the immune response.
33. What is an autoimmune disease?
34. What is a monoclonal antibody?
35. What proteins are involved in the immune response?
36. Where are antibodies formed?
37. In what blood protein fraction are antibodies found?
38. How does bulimia affect the acid-base balance of the body?
39. Acquired immune deficiency syndrome (AIDS) is a lethal disease. How does it kill? Relate the effects to the name of the disease.
40. What is a vaccine?
41. How are immunity and tissue rejection related?
42. Which of the body fluids functions to cool the body? By what mechanism does it work?

43. Give a general description of the hemoglobin molecule.
44. What is the source of bilirubin?
45. List three functions of the lymphatic system.
46. How is urine formed in the kidney?
47. List four factors that affect the volume of urine excreted.
48. List three organic components of urine.
49. What is the difference between sensible and insensible perspiration?
50. List three inorganic and three organic constituents of sweat.
51. What are the three layers of tears?
52. What is the function of lysozyme in tears?
53. Is cow's milk a "perfect" food? Does it contain a complete protein?
54. What advantage does human milk possess over cow's milk as a food for a human infant?

55. Which process results in the transfer of more solutes among fluids of the body: filtration or diffusion?
56. If the amount of colloidal protein in the interstitial compartment matched the amount in the blood plasma, would more or less fluid be likely to filter from the capillaries to the interstitial space?
57. When pneumonia blocks respiratory passageways and limits the transfer of carbon dioxide from blood to the atmosphere, blood CO_2 levels build up. Is the pH of the blood higher or lower than normal in this situation?
58. A common cause of death in young children is acidosis resulting from severe diarrhea. In severe diarrhea, large amounts of bicarbonate ion are excreted from the body. Why should this lead to acidosis?

L

Digestion

There on the table are some carbohydrates, lipids, and proteins, small portions of vitamins and minerals, and a generous dollop of water. What we want to do is to get all of these nutrients from the table to our cells. So we cook our meat and potatoes, cut them into small pieces, and proceed to eat.

The cooking and cutting have not prepared the food adequately for absorption into our bodies, however. Our cells demand a much more highly refined form of the basic nutrients. The huge complex molecules that constitute proteins, carbohydrates, and, to some extent, lipids must be broken down to simpler, more soluble substances. This process, called **digestion,** must occur before food can really be taken *into* the body. It is carried out in the **digestive tract,** a tunnel that runs *through* the body. Food in the digestive tract is in the tunnel, not in the body. The alternative name for the digestive tract, the **alimentary canal,** perhaps conveys this image more clearly.

Food is sluiced through the canal by a flow of digestive juices and by physical pushes imparted by sections of the canal. Compounds that are changed during this journey into suitable forms are absorbed through the walls of the canal into the circulatory systems of the body. Those materials that can't be absorbed make their way through the entire length of the canal and out again. Without the process of digestion, very little of the food we eat would nourish us.

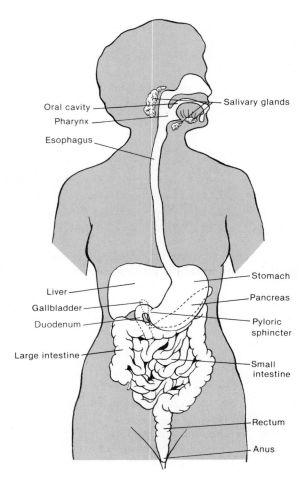

FIGURE L.1
The human digestive system, showing the organs somewhat displaced for clearer viewing. The small intestine is much longer than indicated here.

L.1 THE DIGESTIVE TRACT AND THE NATURE OF DIGESTION

The structures that make up the digestive tract are the mouth, pharynx (throat), esophagus, stomach, small intestine, large intestine, rectum, and anus. Other structures associated with the tract itself are the salivary glands, liver, gallbladder, and pancreas. Figure L.1 shows these structures somewhat schematically. Note that the first section of the small intestine, where it joins the stomach, is called the duodenum. Pancreatic juice and bile enter the digestive tract here.

Digestion is hydrolysis. Fats (esters) are hydrolyzed to glycerol and fatty acids. Proteins (polyamides) are hydrolyzed to amino acids. Polysaccharides and disaccharides (polyacetals) are hydrolyzed to monosaccharides. Nucleic acids (polynucleotides) are hydrolyzed to nucleotides and then to nucleosides and phosphoric acid.

Water for these hydrolysis reactions is present in saliva and the other **digestive juices.** Each step is catalyzed by specific enzymes. Each enzyme operates at or near its optimum pH and at body temperature (37 °C). Without the enzymes, no appreciable hydrolysis would occur.

L.2 THE BEGINNING OF DIGESTION: THE MOUTH

Enzymatic digestion starts in the mouth. Here the food encounters **saliva,** a digestive fluid secreted by the salivary glands. Saliva is over 99% water. It contains several inorganic ions and a variety of organic molecules typical of other body fluids. In the mouth, food is chewed, that is, torn or crushed to a finer consistency. Reduction of food to smaller particles facilitates digestion by providing greater surface area for enzymes to work on. Chewing also coats the food particles with **mucin,** a glycoprotein constituent of saliva. The mucin lubricates the food and makes it easier to swallow.

The principal digestive enzyme in the mouth is α-**amylase,** sometimes called ptyalin. This enzyme attacks the alpha acetal linkages in starch more or less at random. The pH of saliva is about 6.8, the optimum pH for α-amylase. Cleavage of the acetal linkages produces a mixture of dextrins, maltose, and glucose (Figure L.2). Digestion of starch in the mouth is probably not necessary, except perhaps as a means of removing particles of starchy food lodged between the teeth. Digestion of starch is completed in the small intestine (Section L.4). No appreciable digestion of fats or proteins occurs in the mouth.

The secretion of saliva can be triggered by the sight, taste, smell, or even the thought of food. An average person produces about 1.5 L of saliva a day. Excessive flow may be caused by certain pathological conditions, such as mercury poisoning.

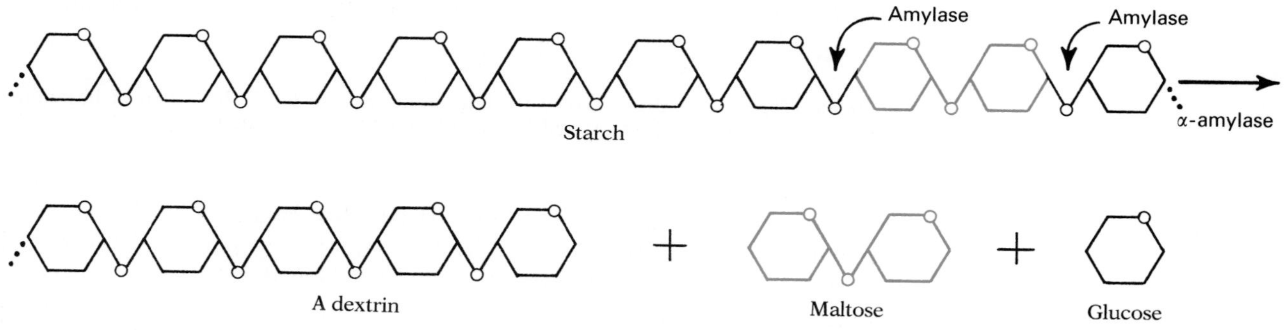

FIGURE L.2 *A schematic representation of the hydrolysis of starch to dextrins, maltose, and glucose. See Chapter 20 for details of structure.*

L.3 GASTRIC DIGESTION: CHYME

Histamine

Gastric juice is a mixture of secretions of the stomach. The chief components are water (more than 99%), mucin, the usual inorganic ions, hydrochloric acid, and some enzymes. A protein hormone called **gastrin,** produced in the stomach, starts the flow of gastric juices. Flow can also be started by histamine. Indeed, it may well be that gastrin acts by releasing histamine, which in turn stimulates the secretion of gastric juices.*

Hydrochloric acid is secreted by certain glands in the lining of the stomach. The pH of freshly secreted gastric juice is about 1.0, but contents of the stomach may partially neutralize it, raising the pH to from 1.5 to 2.5. (The pain of a gastric ulcer is at least partially due to the irritation of the ulcerated tissue by the acidic gastric fluid.) Hydrochloric acid is involved in the denaturation of food protein; it opens up the folds to expose the chains to more efficient enzyme action. Hydrochloric acid is also involved in the activation of **pepsin,** a protein-splitting enzyme.

Pepsinogen is a proenzyme secreted from certain cells in the wall of the stomach. It is activated (converted to pepsin) by hydrochloric acid and by pepsin that is already present. The activation by pepsin is called autocatalysis (Section 24.7). Pepsin catalyzes the hydrolysis of proteins into intermediate products called proteoses and peptones (Figure L.3). It acts only on peptide linkages adjacent to tyrosine or phenylalanine units. The optimum pH for pepsin is about 2, which makes the gastric fluid, with its pH of between 1.5 and 2.5, an ideal environment.

A lipase enzyme is also found in the gastric juice. It may be there only accidentally, having worked its way back up from the small intestine. At any rate, gastric lipase does very little in the way of digesting fats in the stomach, for its optimum pH is near 7, far from the pH of the stomach contents.

Another product of the stomach factory is a mucoprotein called **intrinsic factor.** This substance is essential to the absorption of vitamin B_{12}. *Pernicious anemia* is characterized by the failure of the stomach lining to produce intrinsic factor. Treatment of choice for pernicious anemia is injection of vitamin B_{12}.

---

* Cimetidine (Tagamet), a drug used in treating ulcers, acts to block the action of histamine. Cimetidine is now one of the most widely prescribed drugs in the United States.

Cimetidine

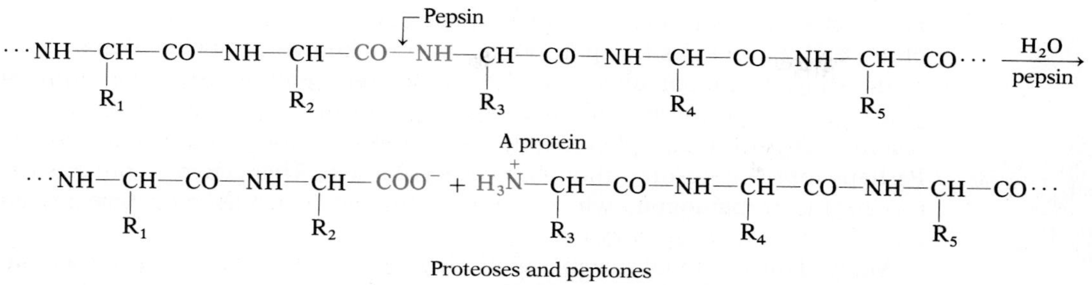

$$\cdots NH-CH-CO-NH-CH-CO \overset{\text{Pepsin}}{\longrightarrow} NH-CH-CO-NH-CH-CO-NH-CH-CO\cdots \xrightarrow[\text{pepsin}]{H_2O}$$

$$\underset{R_1}{} \quad \underset{R_2}{} \quad \underset{R_3}{} \quad \underset{R_4}{} \quad \underset{R_5}{}$$

A protein

$$\cdots NH-CH-CO-NH-CH-COO^- + H_3\overset{+}{N}-CH-CO-NH-CH-CO-NH-CH-CO\cdots$$

$$\underset{R_1}{} \quad \underset{R_2}{} \quad \underset{R_3}{} \quad \underset{R_4}{} \quad \underset{R_5}{}$$

Proteoses and peptones

FIGURE L.3 Pepsin breaks down proteins into soluble fragments called proteoses and peptones.

Food stays in the stomach for from 2 to 5 hours. It is broken down by pepsin and by the mechanical churning of the stomach into a thin, watery liquid called **chyme.** This material then passes, in small portions, into the **duodenum,** the first 30 cm of the small intestine.

L.4 INTESTINAL JUICES

As far as digestion is concerned, the small intestine is where the action is. The process, barely started in the mouth and carried forward only slightly in the stomach, is completed in the small intestine. Much coiled, the small intestine is about 7 m long in adult humans. Most of the enzymes of the intestinal juice are secreted from the duodenum.

Intestinal juice is involved in hydrolysis of all classes of foodstuffs. Several protein splitters complete the hydrolysis of proteins to amino acids. The disaccharides sucrose, lactose, and maltose are hydrolyzed to monosaccharides by enzymes secreted by the intestinal mucosa, as summarized by these word equations.

$$\text{Sucrose} + H_2O \xrightarrow{\text{sucrase}} \text{glucose} + \text{fructose}$$

$$\text{Lactose} + H_2O \xrightarrow{\text{lactase*}} \text{glucose} + \text{galactose}$$

$$\text{Maltose} + H_2O \xrightarrow{\text{maltase}} 2 \text{ glucose}$$

* It is interesting that nearly all human babies have the enzyme lactase but most adults do not. If that statement seems strange to you, then you are probably a Northern European or are of Northern European descent. People who have had their origins in Northern Europe comprise the principal populations—a definite minority—whose adults possess lactase. People who lack the enzyme get sick from drinking milk, due to the buildup of indigestible lactose, a condition called **lactose intolerance.** When milk is cooked or fermented, the lactose is at least partially hydrolyzed. People with lactose intolerance may still be able to enjoy cheese, yogurt, or cooked foods containing milk with little or no problem.

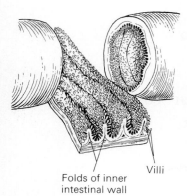

FIGURE L.4
A section of the small intestine opened to reveal the folds of the inner wall and the lining covered with vilii.

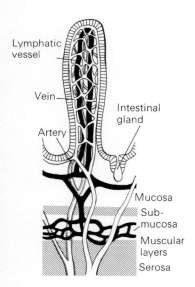

FIGURE L.5
Diagram of one of the intestinal villi. Each villus contains arteries, veins, and lymphatic vessels.

Fats are acted upon by lipase. Even nucleic acids are broken down to nucleotides and nucleosides by appropriate enzymes known as nucleotidases.

Intestinal juice contains a variety of protein-splitting enzymes. **Aminopeptidase** aids in the hydrolysis of polypeptides at linkages involving leucine residues. **Dipeptidases** split dipeptides containing a variety of amino acid units. Perhaps most interesting, though, is **enterokinase.** This enzyme converts the proenzyme **trypsinogen,** which comes from the pancreas, into **trypsin,** an active protein-splitting enzyme.

Many of the intestinal enzymes work right in the cells of the intestinal mucosa, not out in the hollow tube itself. Thus, final hydrolysis may be completed in the cells of the intestinal wall as the nutrients diffuse through.

Most of the absorption of digested foods takes place in the lower digestive tract, particularly in the small intestine. (Alcohol is an exception to this rule: it can be absorbed while still in the stomach. That means it gets into the bloodstream faster than most foods.) The inner walls of the intestines are lined with millions of microscopic projections called **villi** (Figure L.4). The presence of the villi increases the effective surface area of the intestine to 8 m², an area greater than that of the skin. This provides much more contact between the digested foodstuffs and the absorbing surface of the intestinal wall. Figure L.5 shows a detailed view of one of the villi. The structure is richly supplied with blood and lymph vessels. Nutrients pass through the walls of the villi and are absorbed by blood (or lymph, in the case of fats). The transport across cell walls involves energy, which requires cell activity. Thus, the absorption of nutrients from the intestines is not simply a matter of passive diffusion but of active transport in which cells participate in moving the material through.

We are being a bit premature in describing absorption of nutrients. The process does occur in the small intestine for the most part, but only after two other organs, the pancreas and the liver, have made their contributions to the digestive process.

L.5 PANCREATIC JUICES

The pancreas is a large organ lying just below the stomach (see Figure L.1). It produces two kinds of secretions. One kind includes insulin, a hormone of protein nature that is emptied into the bloodstream, where it influences carbohydrate metabolism (Chapter 27). The other, called pancreatic juice, passes through the pancreatic duct into the duodenum.

Pancreatic juice is rich in bicarbonate and hence is slightly alkaline, having a pH of from 7.7 to 8.0. When highly acidic chyme from the stomach reaches the duodenum, the intestinal cells release a hormone called **secretin.** This hormone reaches the pancreas through the circulatory system. It stimulates the secretion of pancreatic juice.

The pancreas produces enzymes for the hydrolysis of all three major food types. Three proteolytic enzymes are formed in the pancreas as proenzymes. Trypsinogen is converted to trypsin by the intestinal enzyme enterokinase (Section L.4). Trypsin, in turn, converts the pancreatic proenzymes **chymotrypsinogen** and **procarboxypeptidase** to **chymotrypsin** and **carboxypeptidase,** respectively.

$$\text{Trypsinogen} \xrightarrow{\text{enterokinase}} \text{trypsin}$$

$$\text{Chymotrypsinogen} \xrightarrow{\text{trypsin}} \text{chymotrypsin}$$

$$\text{Procarboxypeptidase} \xrightarrow{\text{trypsin}} \text{carboxypeptidase}$$

Each active enzyme works on a particular type of peptide linkage. Trypsin splits bonds of the type

$$\ldots \text{CO}-\text{NH}-\underset{\underset{\text{R}}{|}}{\text{CH}}-\text{CO}\!\!\!\diagup\!\!\!\text{NH}\ldots$$

where R represents a group that makes the amino acid arginine or lysine. Chymotrypsin acts similarly when R is such that the amino acid residue is tyrosine or phenylalanine. Carboxypeptidase lops off the terminal amino acid located at the carboxyl end of the peptide chain.

$$\ldots \text{CO}\!\!\!\diagup\!\!\!\text{NH}-\underset{\underset{\text{R}}{|}}{\text{CH}}-\text{COO}^-$$

These enzymes, with the dipeptidases, tripeptidases, and aminopeptidases from the intestinal mucosa, complete the job of protein digestion. The free amino acids are absorbed as rapidly as they are formed. The absorption involves active transport through cell membranes and requires energy. This energy is obtained by oxidative metabolism in the cells. Absorbed amino acids are ready to be incorporated into new protein or to be converted and used for energy (Chapter 29).

Pancreatic juice also contains a lipase sometimes called **steapsin.** Pancreatic lipase catalyzes the hydrolysis of fats to monoglycerides and fatty acids. A monoglyceride is a molecule in which glycerol is esterified with only one fatty acid. These components are absorbed and rapidly converted back to triglycerides. There is some evidence that the conversion back to triglycerides occurs during the absorption process. In any case, the digested lipids are absorbed by the villi lining the intestine and enter the lymphatic capillary. In this respect,

$$\underset{\text{A monoglyceride}}{\begin{array}{l} \text{CH}_2-\text{O}-\overset{\displaystyle \overset{\text{O}}{\|}}{\text{C}}-\text{R} \\ \text{CH}-\text{OH} \\ \text{CH}_2\text{OH} \end{array}}$$

the absorption of lipid differs from that of the other nutrients, which are absorbed directly into the blood. The fats, too, reach the blood, but only after traveling through the lymphatic network.

The hydrolysis of fats is extremely slow without the emulsifying action of bile salts (Section L.6).

Carbohydrate digestion by pancreatic juice involves the action of an α-amylase. This enzyme acts on the starch that escaped degradation in the mouth and upon the partially degraded dextrins produced by salivary amylase. Pancreatic amylase completes the breakdown of starch, giving maltose (87%) and some glucose (13%). The maltose is then cleaved by intestinal maltase to glucose (Section L.4).

L.6 BILE: AN INTERNAL EMULSIFYING AGENT

Bile, the third fluid that enters the duodenum, is formed in the liver and then stored temporarily in the gallbladder. Bile contains no digestive enzymes, yet it plays a number of vital roles. The composition of typical bile is given in Table L.1.

A normal adult produces 500 to 600 mL of bile each day. The fluid is slightly alkaline, with a pH of between 7 and 8. When food enters the duodenum, a hormone called cholecystokinin is released and enters the bloodstream. This hormone causes the gallbladder to empty its contents through the gall duct into the duodenum. This alkaline fluid, along with the pancreatic juice, helps neutralize the acidic chyme from the stomach, providing a favorable environment for the enzymes that act in the intestine. As each load of chyme is neutralized, the stomach releases a new portion into the duodenum.

Lipids are by definition insoluble in water. The digestive juices are aqueous solutions. To bring the lipids into contact with the enzymes dissolved in the digestive juices, it is necessary to emulsify the fats. This is done by the bile salts. The principal bile salts are sodium glycocholate and sodium

TABLE L.1   Composition of Bile

Component	Percentage
Water	97
Bile salts	0.7
Inorganic salts	0.7
Bile pigments	0.2
Fatty acids	0.15
Lecithin	0.1
Fat	0.1
Cholesterol	0.06

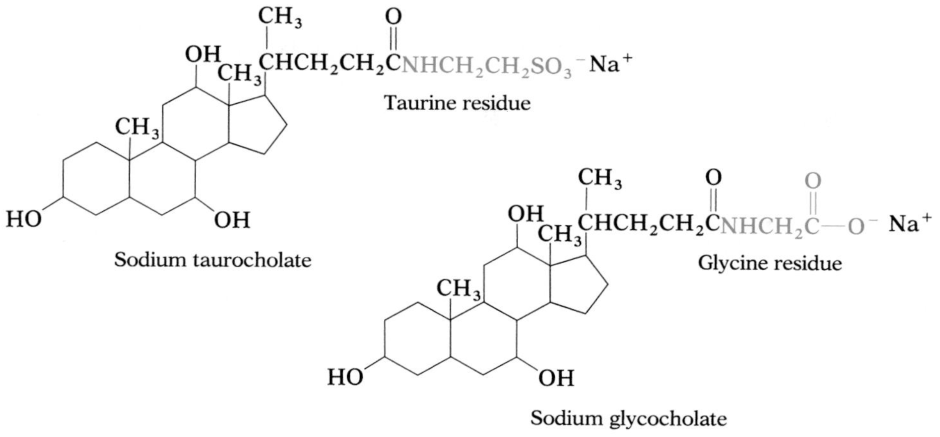

FIGURE L.6 *The principal bile salts.*

taurocholate (Figure L.6). (Taurine is an aminosulfonic acid with the formula $H_2NCH_2CH_2SO_3H$). The bile salts act in much the same way that soap does (Chapter 21). They break down large fat globules into smaller ones and keep the smaller globules suspended in the aqueous digestive medium. The greatly increased surface area of the fat particles and the opportunity afforded for more intimate contact with the lipase enzymes results in a much more rapid digestion of the fats.

Bile salts also aid in the absorption of fat-soluble vitamins and cholesterol. The bile serves as a route for the excretion of drugs, end products from the breakdown of hemoglobin (Section L.6), and heavy metal ions.

On occasion, cholesterol may precipitate in the gallbladder as **gallstones.** These may be exceedingly painful, making it necessary to remove the gallbladder.

L.7 BACTERIAL ACTION IN OUR INTESTINES

Soon after a child is born, bacteria invade its digestive tract. Thriving colonies of bacteria are established, mainly in the colon (large intestine) but also in the small intestine. Most of these bacteria are harmless. Many are beneficial. Bacteria in ruminants (grazing animals) break down cellulose, providing nourishment for their hosts as well as for themselves.

Certain bacteria act on carbohydrates in the intestine to form organic acids and gases such as methane, carbon dioxide, and hydrogen. Action of intestinal bacteria on proteins produces amines, organic acids, phenols, and ammonia. Sulfur-containing amino acids yield thiols and hydrogen sulfide.

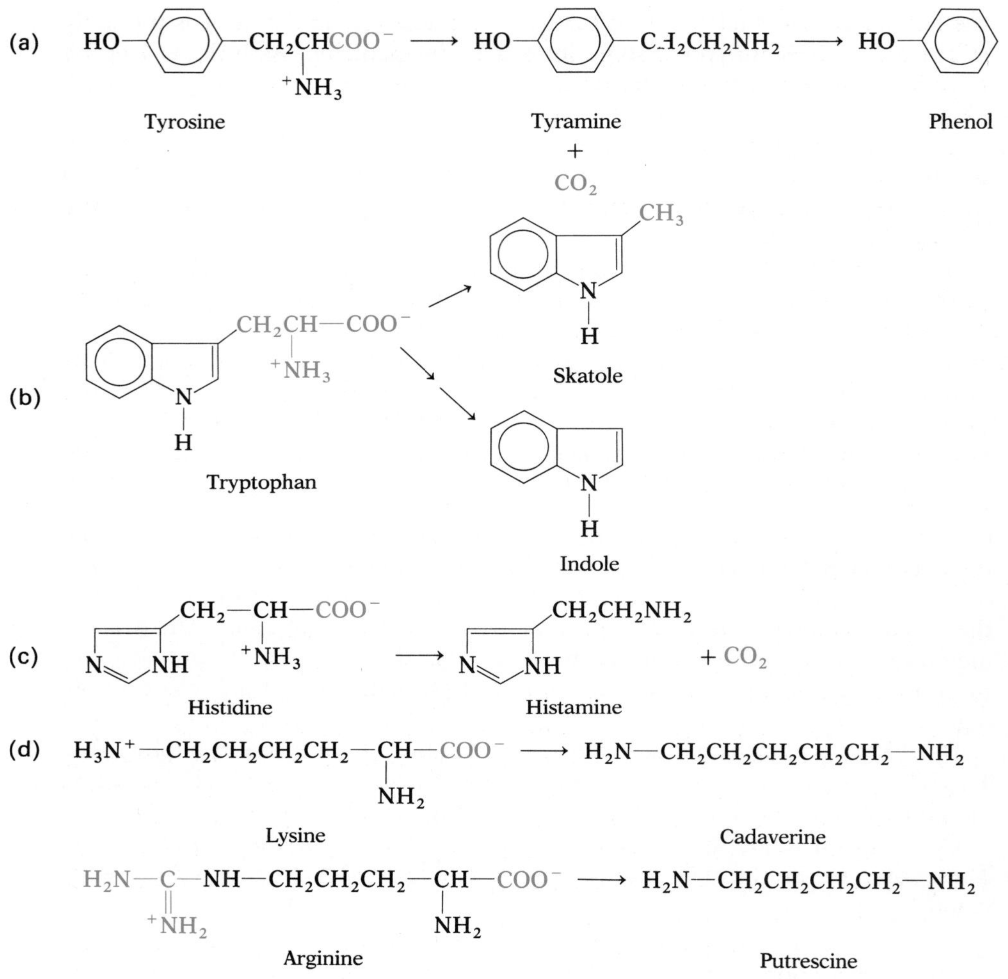

FIGURE L.7   *Bacterial action on selected amino acids.*

Let's look at a few specific cases. Tyrosine is converted to tyramine, and eventually to phenol, by bacterial action (Figure L.7a). Tyramine has a physiological action similar to, but weaker than, that of norepinephrine (Special Topic G), which it resembles structurally.

Tryptophan is converted by bacterial action to indole and skatole, two compounds largely responsible for the characteristic odor of feces (Figure L.7b).

Histidine is decarboxylated by bacterial action to histamine (Figure L.7c). Histamine is a powerful vasodilator. It is most likely involved in stimulating the flow of gastric juice (Section L.3). Its complete role in the body is probably still unknown. Histamine is sometimes used in medicine. As little as 1 μg by injection causes a pronounced drop in blood pressure, by dilating the blood vessels.

Lysine and arginine, basic amino acids, are decarboxylated by bacterial action to cadaverine and putrescine, respectively (Figure L.7d). These compounds have odors as foul as their names imply. They are also toxic, but not in the amounts usually found.

The organic acids formed by bacterial action stimulate the muscular action of the colon. If formed in large amounts, these acids produce a laxative effect.

Bacteria undoubtedly act upon many other substances in the intestines. Intestinal bacteria are also believed to produce the vitamin biotin. While bacteria within the intestines can prove beneficial to their host, their release into the peritoneal (abdominal) cavity can be quite dangerous. If the gastrointestinal tract is perforated (as may happen with certain ulcers), the bacteria released may cause a potentially fatal generalized infection, peritonitis.

L.8 FECES: FORMATION AND COMPOSITION

The contents of the small intestine enter the colon in a semiliquid state. The solid portion consists of undigested carbohydrates (largely cellulose), proteins (largely connective tissue), and fats. In addition there are dead mucosal cells, traces of unused digestive fluids, and large quantities of bacteria. Indeed, bacteria (both dead and living) make up about a third of the dry weight of feces. There are about 100 billion bacteria in each gram of stool. The main function of the colon is the reabsorption of water, which leaves behind a more solid, compact material. An average adult on a normal diet produces about 100 g of feces per day. This material is 60% to 70% water, and the remainder (about 30 to 40 g) is solid.

The normal brown color of feces is due to bilifuscins, produced by the bacterial reduction of bilirubin (Chapter 26). Abnormal colors sometimes indicate pathological conditions. Black, tarry stools indicate bleeding in the upper digestive tract. The color is due to methemoglobin, an oxidation product of hemoglobin. Blood in the stool that is still red indicates bleeding in the lower part of the tract.

Stools the color of clay indicate that little or no bile is entering the digestive tract. Such a condition indicates that an obstruction, such as a gallstone, is blocking the flow of bile. This condition, in which the patient's skin turns yellow due to a buildup of bilirubin, is called obstructive jaundice.

Certain drugs and foods also color the feces. Iron and bismuth compounds turn the stool black due to the formation of sulfides. Rhubarb may cause a yellowish stool.

Material may remain in the colon for a day or longer. Normal bowel movements may occur as often as three times a day or as seldom as once every three days. Laxatives, despite much advertising on television, are seldom needed. There is evidence, however, that the highly processed, low-bulk diet of the average person in the developed countries is harmful. The high-bulk diet, containing lots of fibers, peels, seeds, and the like, of more "primitive" people may be better. Such diets yield robust stools, with low retention times in the colon. Cancer of the colon has been correlated with the low-bulk diets of people in the industrialized nations. (Section 20.10).

PROBLEMS

1. Give the location of action and the function of each digestive enzyme:
   a. salivary amylase (ptyalin)
   b. pepsin
   c. trypsin
   d. chymotrypsin
   e. pancreatic amylase
   f. sucrase
   g. lactase
   h. pancreatic lipase (steapsin)
   i. carboxypeptidase
   j. dipeptidase
   k. enterokinase
   l. nucleotidase
   m. maltase
2. What is mucin? What is its function in saliva?
3. What is pepsinogen? How is it converted to pepsin?
4. What is intrinsic factor?
5. What ion causes pancreatic juice to be slightly alkaline?
6. If the optimum pH for an enzyme were 8, would it be more active in the stomach or in the small intestine?
7. How is chyme neutralized in the duodenum?
8. Describe the emulsifying action of bile salts. What function does emulsification serve?
9. What are gallstones? What is their chief chemical component?
10. What are the end products of carbohydrate digestion?
11. What are the end products of fat digestion?
12. What are the products of protein digestion?

13. Draw the products of the hydrolysis of alanylglycine.
14. Which class of foods is absorbed into the lymph instead of directly into the blood?
15. What causes the flow of gastric juices?
16. What causes the flow of pancreatic juices?
17. What causes the flow of bile?
18. What is lactose intolerance?
19. Write an equation for the complete hydrolysis of the tetrapeptide Arg-Tyr-Gly-Ser.
20. Write an equation for the hydrolysis of Arg-Tyr-Gly-Ser by the enzyme trypsin.
21. Write an equation for the hydrolysis of Arg-Tyr-Gly-Ser by the enzyme chymotrypsin.
22. Write an equation for the hydrolysis of Arg-Tyr-Gly-Ser by the enzyme carboxypeptidase.
23. Write an equation for the hydrolysis of lactose by lactase.
24. Write an equation for the complete hydrolysis of glyceryl tristearate.
25. Write an equation for the hydrolysis of glyceryl tristearate by the lipase steapsin.
26. What is the composition of feces?
27. What is indicated by a black, tarry stool? By a clay-colored stool?
28. Describe three functions of bacteria in the digestive tract. Is there a difference between bacteria in the digestive tract and in the abdominal cavity? Explain.
29. Explain the action of cimetidine in the treatment of ulcers.
30. Some people take laxatives to help them lose weight. Comment on the wisdom of this in the light of your

knowledge that laxatives act on the colon and nearly all the nutrients from food are absorbed in the small intestine.

31. Aspartame, an artificial sweetener about 160 times as sweet as sucrose, is the methyl ester of a simple dipeptide. Draw the products of complete hydrolysis of aspartame.

$$
\underset{\text{Aspartame}}{
\overset{\displaystyle COO^{-}}{\underset{\displaystyle \overset{+}{H_3N}}{|}}
}
$$

Aspartame

32. A can of pop can be sweetened with about 0.25 g of aspartame. This amount of aspartame would release about 0.027 g of methanol on hydrolysis. The lethal dose of methanol for humans is about 25 g. How many cans of aspartame-sweetened pop would you have to drink to ingest a theoretically lethal dose of methanol?

33. The enzyme superoxide dismutase (SOD) is thought to have an antiaging effect in cells. Some health food stores sell SOD for oral ingestion with the idea that it will retard aging. Do you think it will work? Explain.

34. What is the general reaction type used in digestion?

35. What are the products of the digestion of monosaccharides? disaccharides? polysaccharides?

36. In what section of the digestive tract does most of the digestion of carbohydrates take place?

37. If a cracker, which is rich in starch, is chewed for a long time, it begins to develop a sweet, sugary taste. Why?

Carbohydrate Metabolism

The energy changes associated with metabolism are discussed in Chapter 25. In this chapter and the next two, we focus on the chemical reactions that make up metabolic processes.

Green plants, using energy from the sun, make glucose. Some of the glucose is converted to other sugars, to starch for storage, and to the structural material, cellulose.

What photosynthesis has joined together, the digestive and metabolic processes take apart. Plants make glucose; plant and animal cells "burn" glucose for energy (Chapter 25). We can store a bit of glucose as glycogen, but carbohydrates are used mainly to meet rather short-term energy needs.

We obtain carbohydrates mainly from plant sources. We eat them primarily as starches, but also as disaccharides and even as monosaccharides. Special Topic L discusses how starches and disaccharides are broken down by the digestive juices into the monosaccharides glucose, fructose, and galactose. These are absorbed directly into the bloodstream through capillaries in the villi, the fingerlike projections on the walls of the small intestine.

Once absorbed into the bloodstream, the monosaccharides are interconverted and enter a variety of different reaction schemes called **pathways.** Some pathways are designed to liberate the energy stored in the monosaccharides. Others yield starchlike products that can be stored until the energy they contain is needed. Some of the schemes are cyclic; they are like merry-go-rounds, with passengers getting on and off. The passengers are changed as the cycle goes around. The cycle releases energy with each turn.

We shall travel several of these different pathways in this chapter. We shall learn how our cells get the energy we need for work and play, for breathing, and even for thinking.

27.1 SUGAR IN THE BLOOD

What happens to the monosaccharides after they leave the digestive tract and enter the bloodstream? First they are carried to the liver. There they are phosphorylated. The fructose product, fructose 6-phosphate, may immediately enter

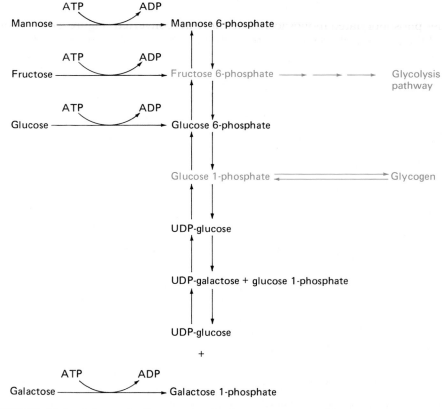

FIGURE 27.1   Schematic diagram for the interconversion of hexose sugars. UDP is uridine diphosphate, a nucleoside derivative that serves as a carrier of sugar molecules.

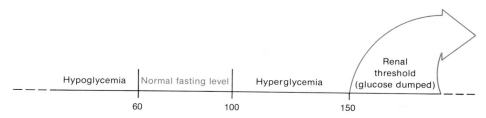

Blood sugar level, in milligrams of glucose per 100 mL of blood

FIGURE 27.2   *If the blood sugar level exceeds the renal threshold, glucose is found in the urine.*

the glycolysis pathway (Section 27.4), which provides energy to the cell. The fructose product can also enter a series of reversible reactions in which it and other phosphorylated monosaccharides are interconverted (Figure 27.1). Thus, all of the monosaccharides can ultimately enter the glycolysis pathway through the fructose 6-phosphate intermediate. Similarly, all the monosaccharides can be converted to glucose 1-phosphate. This intermediate can, in turn, be converted to glycogen, a compound that serves as an energy reserve in animals (Section 27.3).

The sugar in the blood is mainly glucose. Normal concentrations for a fasting person range from 60 to 100 mg of glucose per 100 mL of blood. There is considerable controversy over just what "normal" levels should be, with some medical professionals accepting lower or higher values than those we have given as normal. After a meal rich in carbohydrates, the level may be temporarily much higher. The condition of high blood sugar is called **hyperglycemia.** After severe starvation or vigorous exercise, the blood sugar concentration may fall below normal, leading to the condition called **hypoglycemia.*** Neither condition, if temporary, is necessarily pathological, because the body has several methods of regulating the level of glucose in the blood.

The kidneys may excrete excess glucose (Special Topic L). The **renal threshold** value is fairly high, however, ranging from 150 to 170 mg of glucose per 100 mL of blood (Figure 27.2). In general, the kidneys are designed to conserve the glucose in the blood. Only when the blood glucose level goes well above "normal" will the kidneys shunt some of the glucose into the urine. The liver also helps to regulate blood glucose level by converting excess sugar to glycogen

---

* There is a great deal of controversy regarding hypoglycemia. There are some clear-cut cases in which easily recognized symptoms are evident. These include general weakness, trembling, and rapid heartbeat. Severe cases can lead to delirium, coma, and even death. The treatment for hypoglycemia is simple. Give the patient some sugar. If the person is unconscious, glucose solution can be administered intravenously. A common cause of hypoglycemia is an overdose of insulin. Diabetics generally carry candy to counteract excess insulin.

(Section 27.3). Excess glucose may also be converted to fats for storage. And, of course, the glucose may be oxidized in the cells, producing energy.

> Of the various interconversions shown in Figure 27.1, that in which galactose is converted to a glucose derivative is of particular interest. Some babies are born lacking an enzyme necessary for this conversion. This condition, called **galactosemia,** is hereditary. Infants with galactosemia cannot tolerate milk, because the hydrolysis of milk sugar (lactose) to galactose and glucose in the intestine results in the buildup of galactose and eventually in a lack of appetite, weight loss, diarrhea, and jaundice. These infants may die of cirrhosis of the liver. Milk and other sources of galactose must be replaced in the diet of an infant with galactosemia.

27.2 HORMONES AND BLOOD SUGAR LEVELS

There are a variety of hormones that affect the blood sugar level. Perhaps the most important and certainly the best known of these is **insulin,** a protein-type hormone formed in the pancreas. Its primary structure was determined by Frederick Sanger, who won the Nobel Prize for chemistry in 1958. Insulin is made up of two chains (called A and B) cross-linked by two disulfide bonds (Figure 27.3). Another disulfide group forms a loop in the A chain.

Insulin secretion is stimulated by a rise in the level of blood sugar. The hormone then serves to increase the rate at which tissues and organs absorb the circulating glucose for storage, oxidation, or transformation. Insulin acts by increasing the transport of glucose across cell membranes. Under its influence the glucose content of the blood is rapidly diminished. If glucose is to be stored, it is converted to glycogen in muscles and to fat in adipose (fatty) tissues. Glucose can be oxidized to carbon dioxide and water in both kinds of tissue. Insulin also causes increased glycogen storage in the liver, reducing glucose output by that organ. Although it influences carbohydrate metabolism in muscles, adipose tissue, and the liver, insulin has no effect on carbohydrate metabolism in the brain or kidneys. Insulin also exhibits some activity in amino acid metabolism and protein production.

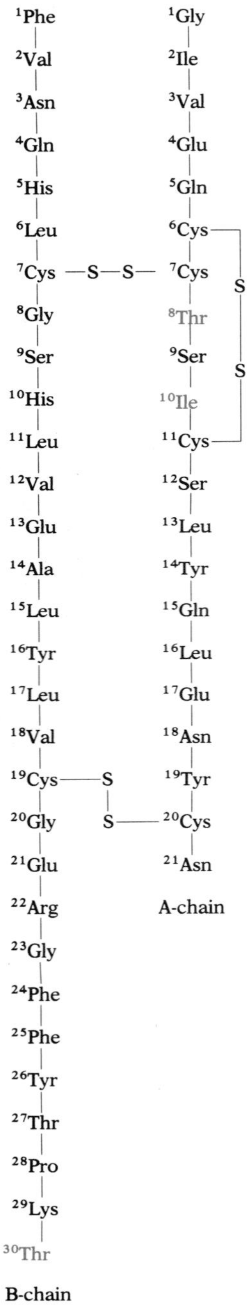

B-chain:
^{1}Phe — ^{2}Val — ^{3}Asn — ^{4}Gln — ^{5}His — ^{6}Leu — ^{7}Cys — ^{8}Gly — ^{9}Ser — ^{10}His — ^{11}Leu — ^{12}Val — ^{13}Glu — ^{14}Ala — ^{15}Leu — ^{16}Tyr — ^{17}Leu — ^{18}Val — ^{19}Cys — ^{20}Gly — ^{21}Glu — ^{22}Arg — ^{23}Gly — ^{24}Phe — ^{25}Phe — ^{26}Tyr — ^{27}Thr — ^{28}Pro — ^{29}Lys — ^{30}Thr

A-chain:
^{1}Gly — ^{2}Ile — ^{3}Val — ^{4}Glu — ^{5}Gln — ^{6}Cys — ^{7}Cys — ^{8}Thr — ^{9}Ser — ^{10}Ile — ^{11}Cys — ^{12}Ser — ^{13}Leu — ^{14}Tyr — ^{15}Gln — ^{16}Leu — ^{17}Glu — ^{18}Asn — ^{19}Tyr — ^{20}Cys — ^{21}Asn

FIGURE 27.3   The primary structure of human insulin. Porcine (pork) insulin differs only in that Ala replaces Thr at position 30 on the B chain. Bovine (cattle) insulin also has Ala in that position; in addition it has Ala in place of Thr at position 8 and Val in place of Ile at position 10 on the A chain.

If the pancreas is removed surgically, or if the cells that produce insulin are destroyed, the carbohydrate storage mechanisms don't work properly. The result is an increase in blood sugar level. The renal threshold is exceeded, and glucose is found in the urine. The production of large quantities of sugar-containing urine is characteristic of persons suffering from **diabetes mellitus,** the disease that results from faulty insulin production. Carbohydrate metabolism is impaired, so the liver produces keto acids by the metabolism of fatty acids (Chapter 28). Severe acidosis and **diabetic coma** may ensue. This condition is fatal if not reversed.

Human insulin is now available from genetically engineered bacteria (Chapter 23). Beef and pork insulin, which differ slightly in primary structure (see Figure 27.3), are available as by-products of the meat-packing industry. Since it is protein in nature, insulin cannot be administered orally. It would be deactivated by the same digestive processes that break down the protein we eat for food. It is usually injected subcutaneously. Administered at the proper level, insulin keeps carbohydrate metabolism regulated, enabling diabetics to live nearly normal lives.

The oral drugs used to control mild cases of diabetes are not protein in nature; rather, they are synthetic chemical substances such as tolbutamide (Orinase).

$$CH_3 - \bigcirc - SO_2NHCNHCH_2CH_2CH_2CH_3$$

These compounds stimulate the release of insulin from the pancreas. Therefore, they are only effective for persons who manufacture their own insulin but who fail to release the insulin in response to increased blood sugar level. The drugs have been shown to increase the risk of heart disease. Their use has become rather controversial. Second-generation drugs, such as glyburide, are now available. They stimulate the release of insulin, as do the first-generation drugs, but also increase the sensitivity of the cell receptors to the insulin. Like the earlier drugs, they must carry warnings about the increased risk of cardiovascular disease.

$$\underset{\underset{OCH_3}{|}}{\overset{Cl}{\bigcirc}} - \overset{O}{\overset{||}{C}} - NH - CH_2CH_2 - \bigcirc - SO_2NHCNH - \bigcirc$$

A **glucose tolerance test** is used to diagnose diabetes mellitus. A patient's blood glucose level is determined after a fast of several hours. Then a certain amount (25 to 100 g) of glucose is administered, and the patient's blood glucose level is measured at intervals. In normal persons, the level will rise initially but

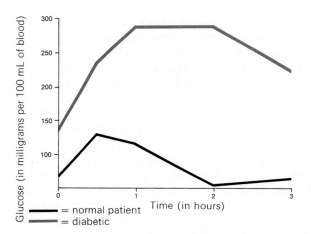

FIGURE 27.4   Glucose tolerance curves for a normal patient and for a diabetic.

will return to normal or slightly less than normal in about 2 hours. With diabetic patients, the level will be high at the start, will go even higher, and will remain at the elevated levels for several hours (Figure 27.4).

Glucagon is another pancreatic hormone. It is a polypeptide with 29 amino acid residues.

His-Ser-Gln-Gly-Thr-Phe-Thr-Ser-Asp-Tyr-Ser-Lys-Tyr-Leu-Asp-Ser-Arg-Arg ⎤

⎿ Ala-Gln-Asp-Phe-Val-Gln-Trp-Leu-Met-Asn-Thr

It raises the blood sugar level by increasing the rate of breakdown of glycogen to glucose in the liver. The overall effect of glucagon, then, is opposite to that of insulin.

Another hormone affecting blood sugar level is **epinephrine** (adrenalin).

$$HO-\bigcirc-CHCH_2NHCH_3$$
$$HO \qquad OH$$

This hormone is produced by the medulla of the adrenal glands. Its action as a "fight or flight" hormone is mentioned briefly in Special Topic G. Like glucagon, epinephrine acts to raise the blood glucose level by increasing the breakdown of glycogen to glucose. It also stimulates the conversion of muscle glycogen to lactic acid, raising the level of that metabolite in the blood. Fright may be such that enough epinephrine is secreted to raise the blood glucose level above the renal threshold. Glucose will appear in the urine. Sometimes this condition, called **emotional glycosuria,** will appear during a routine physical examination—if the examinee is sufficiently frightened.

TABLE 27.1 Hormones and Blood Glucose Levels

Hormone	Effect on Blood Glucose Level	Mechanism
Insulin	Lowers	Enhances transport across cell membranes
Glucagon	Raises	Enhances breakdown of glycogen to glucose in the liver
Epinephrine	Raises	Enhances glycogen breakdown to glucose (liver) and to lactic acid (muscles)
Cortisone and cortisol	Raises	Enhances conversions of amino acids to glucose in the liver
Human growth hormone	Raises	Stimulates glucagon secretion

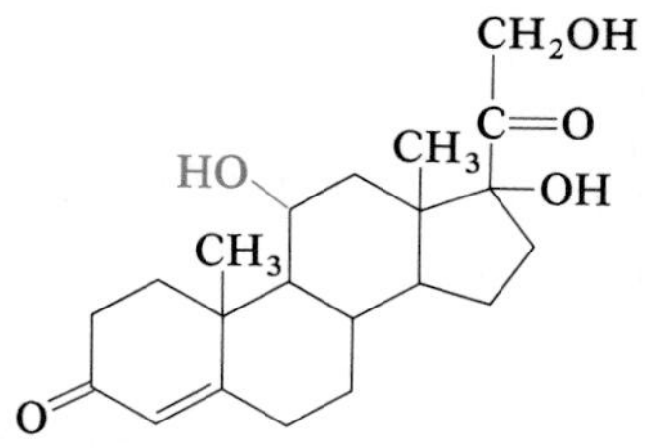

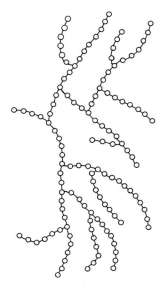

Cortisol

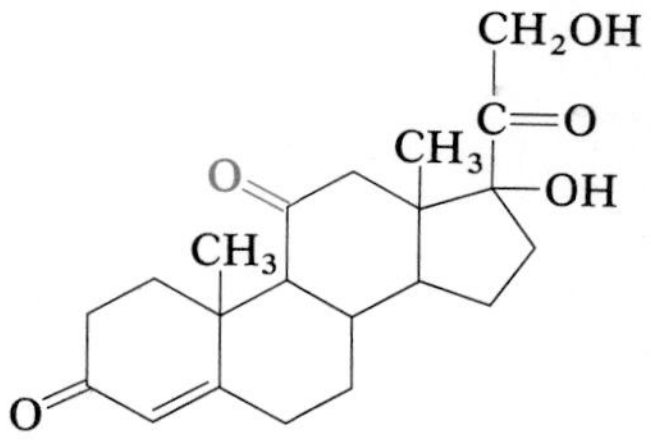

Cortisone

FIGURE 27.5 *Two hormones of the adrenal cortex.*

Hormones of the adrenal cortex also affect blood sugar level. *Cortisone* and *cortisol* are two such compounds (Figure 27.5). These hormones stimulate the synthesis of glucose in the liver from amino acids. They, too, are antagonistic to insulin. The adrenocortical hormones are controlled in turn by the adrenocorticotrophic hormone (ACTH). ACTH is produced in the anterior pituitary gland. Another pituitary hormone that influences the blood glucose level is **human growth hormone.** This hormone stimulates glucagon secretion and is antagonistic to insulin. The action of these various hormones is summarized in Table 27.1.

27.3 CARBOHYDRATE STORAGE: GLYCOGEN

The structure of glycogen is much like that of amylopectin (Chapter 20), the branched component of starch. Glycogen consists of a chain of glucose units, joined by an alpha linkage from the first carbon of one glucose unit to the fourth carbon of the next. In addition, there are branches off the sixth carbon. Indeed, glycogen is even more branched than amylopectin. A schematic representation is given in Figure 27.6. (For a more detailed structural formula see Figure 20.5.) In addition, glycogen shows a wider range of molecular weights than amylopectin. Molecular weight varies with the nutritional state of the organism and with species.

Two enzymes are involved in glycogen synthesis. **Glycogen synthetase** catalyzes the formation of straight chains of glucose units. The chain is extended when a glucose unit is transferred from a uridine diphosphate (UDP)

FIGURE 27.6
Schematic drawing of a glycogen molecule.

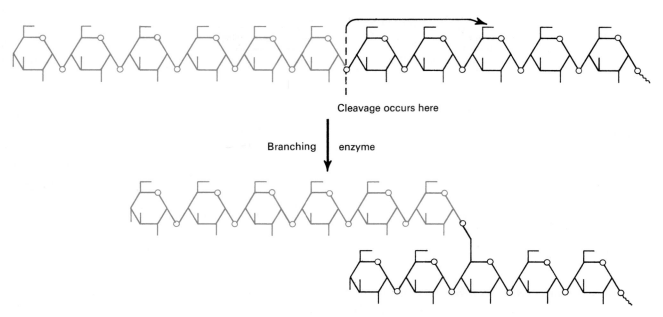

FIGURE 27.7   *Formation of a glycogen branch.*

molecule to a growing glycogen chain. The reactant glycogen chain has n glucose units, and the product chain has $n + 1$ units.

$$(\text{Glucose})_n + \text{UDP-glucose} \xrightarrow{\text{glycogen synthetase}} (\text{glucose})_{n+1} + \text{UDP}$$

Reactant
glycogen

Product
glycogen

The glycogen synthetase enzyme is not involved in the formation of the 1,6 bonds found in the branched glycogen structure. Another enzyme, called the **branching enzyme,** is responsible for breaking off a piece of the straight chain at one of the 1,4 linkages and repositioning it at the sixth carbon (Figure 27.7).

Glycogen serves as an energy reserve in animals. Our livers normally store about 100 g of glycogen. In exceptional circumstances, as much as 400 g may be stored. Muscles also have the capability of making and storing glycogen.

When the blood sugar level drops, glycogen is converted into free glucose. Muscle glycogen releases glucose to supply energy during muscular contraction (Section 27.6). In the liver, the enzyme **phosphorylase** catalyzes the reaction.

$$\text{Glucose 1-phosphate} + (\text{glucose})_n \rightleftharpoons (\text{glucose})_{n+1} + \text{HPO}_4{}^{2-}$$

As indicated, the process is reversible. When the blood glucose level drops, the equilibrium shifts left, resulting in the formation of glucose 1-phosphate. Other enzymes convert glucose 1-phosphate to free glucose. When blood glucose

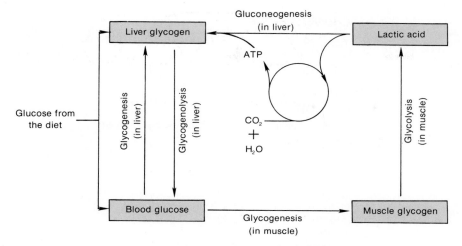

FIGURE 27.8   The Cori cycle.

levels rise, the equilibrium shifts to the right, forming glycogen of higher molecular weight.

There are other, more complicated processes by which glycogen can be converted to glucose. The details of these processes need not concern us here. The relationships between glycogen formation (**glycogenesis**) and glycogen breakdown (**glycogenolysis**) are summarized in Figure 27.8. The diagram is called the Cori cycle, in honor of Gerty and Carl Cori, who first described it. The Coris won a Nobel Prize in 1947, the third husband-wife team to achieve this distinction (Marie and Pierre Curie were the first; Irène and Frédéric Joliot-Curie, the second).

27.4 ANAEROBIC GLYCOLYSIS

The metabolism of monosaccharides, which occurs in much the same way in most plants and animals, can take any one of several interconnected pathways. We will discuss two of those pathways: (1) anaerobic glycolysis, called the Embden-Meyerhof pathway, and (2) aerobic oxidation, variously called the Krebs cycle, the citric acid cycle, or the tricarboxylic acid cycle. Let's look first at anaerobic glycolysis.

The overall result of **anaerobic glycolysis** is the conversion of a glucose molecule into two molecules of lactic acid.

$$C_6H_{12}O_6 \longrightarrow 2\ C_3H_6O_3$$

Glucose   Lactic acid

The word **glycolysis** means "splitting of a sugar." Note that glucose is neither oxidized nor reduced in the reaction. (The oxidation number of carbon in each

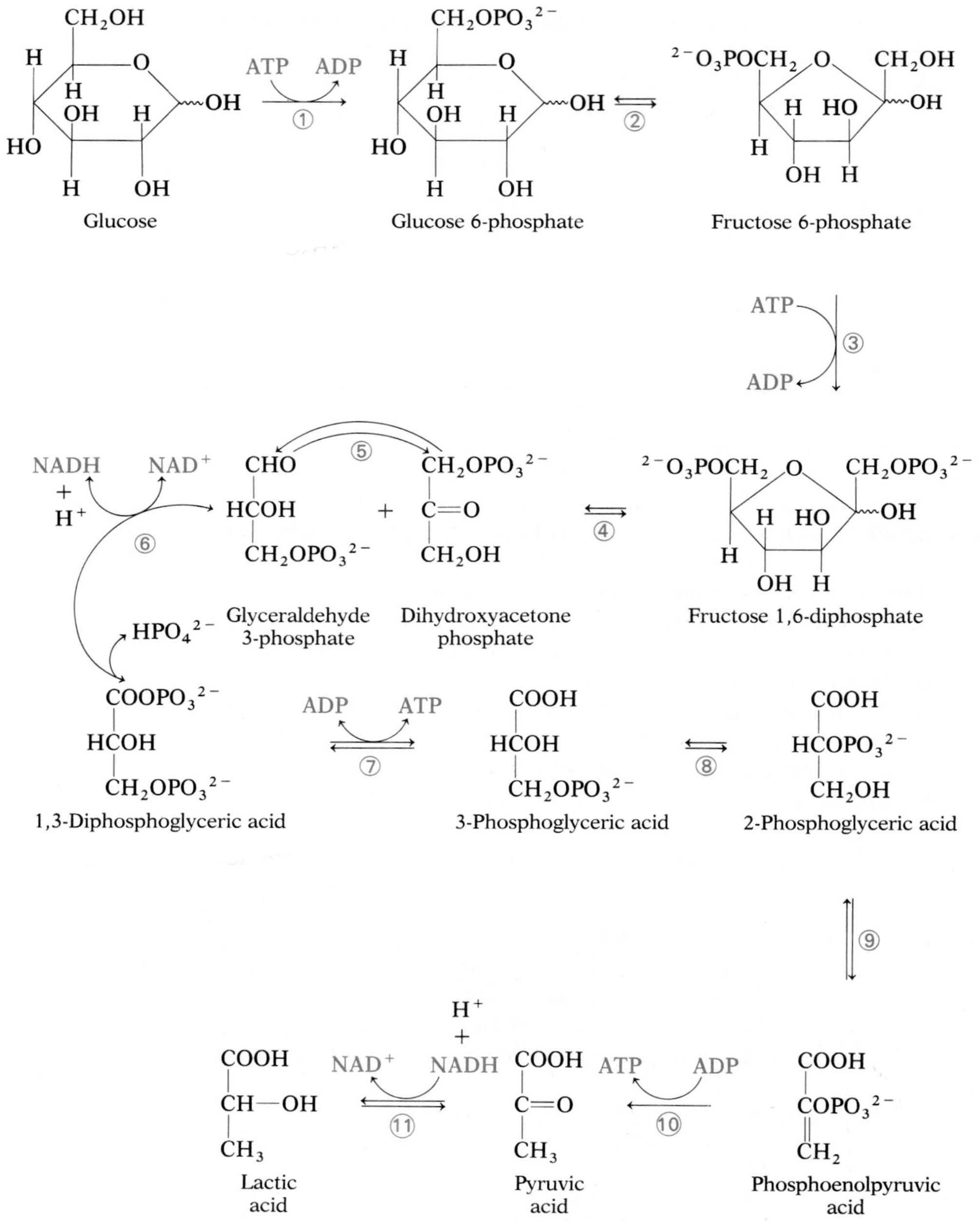

FIGURE 27.9   *The Embden-Meyerhof pathway of glycolysis.*

compound is 0.) There are, however, oxidation and reduction processes involved along the way.

Glycolysis releases energy. Part of that energy goes to convert inorganic phosphate and ADP to ATP (Chapter 25).

$$2\,ADP + 2\,HPO_4{}^{2-} \longrightarrow 2\,ATP + 2\,H_2O$$

The overall process that results from the two coupled reactions can be written

$$C_6H_{12}O_6 + 2\,ADP + 2\,HPO_4{}^{2-} \longrightarrow 2\,CH_3\underset{\underset{\displaystyle OH}{|}}{C}HCOOH + 2\,ATP + 2\,H_2O$$

Glucose   Lactic acid

So far glycolysis sounds simple enough, but the details of the pathway took many years for scientists to work out. The reaction scheme (Figure 27.9) is called the **Embden-Meyerhof pathway** in honor of Gustav Embden and Otto Meyerhof, German biochemists who made important contributions to the elucidation of the steps in the 1930s. Let's take a quick walk down the pathway, concentrating on the chemistry involved. You will have to refer to Figure 27.9 for the chemical structures involved in some of the steps.

Notice that step 1 uses ATP to convert glucose to glucose 6-phosphate. This step illustrates **substrate level phosphorylation,** that is, the use of a high-energy phosphate such as ATP to phosphorylate a low-energy compound such as glucose, converting it to high-energy glucose 6-phosphate.

Step 2 involves an isomerization, probably a keto-enol tautomerism (Chapter 16).

Step 3 is another substrate level phosphorylation. A molecule of ATP is consumed in the conversion of fructose 6-phosphate to fructose 1,6-diphosphate.

Step 3 is a crucial one in the process; it determines whether the reactions proceed to lactic acid or go back to glycogen. This step is catalyzed by phosphofructokinase. This enzyme is inhibited by high concentrations of ATP and of citric acid (an intermediate in the Krebs cycle, Section 27.5), and is activated by ADP and AMP. When the cell has adequate supplies of ATP, there is no need for glycolysis. On the other hand, when the cell has used up most of its ATP, the resulting ADP and AMP stimulate the enzyme to carry forward with the anaerobic breakdown of glucose, producing more ATP.

Step 4 is a reversible reaction in which a six-carbon structure is broken into two three-carbon units. The reverse of this step is an aldol condensation (Chapter 16). Both the forward and reverse processes are catalyzed by the enzyme aldolase.

The two three-carbon compounds formed in step 4, glyceraldehyde 3-phosphate and dihydroxyacetone phosphate, are in equilibrium with one another. Either can be changed into the other by a keto-enol hydrogen shift (step 5).

Step 6 is an **oxidative phosphorylation.** NAD^+ oxidizes the aldehyde group of glyceraldehyde 3-phosphate and phosphorylates it to 1,3-diphosphoglyceric acid. The overall process can be summarized as

$$
\begin{array}{c}
\text{CHO} \\
| \\
\text{H—C—OH} \\
| \\
\text{CH}_2\text{OPO}_3{}^{2-}
\end{array}
+ NAD^+ + HPO_4{}^{2-} \longrightarrow
\begin{array}{c}
\text{COOPO}_3{}^{2-} \\
| \\
\text{H—C—OH} \\
| \\
\text{CH}_2\text{OPO}_3{}^{2-}
\end{array}
+ NADH + H^+
$$

Glyceraldehyde   1,3-Diphosphoglyceric
3-phosphate   acid

In oxidative phosphorylation, high-energy compounds are *produced.* Note that the new group is a mixed anhydride of a carboxylic acid and phosphoric acid; it is not an ester. It is quite different from the group on the third (bottom) carbon atom; that group is an ordinary ester of phosphoric acid with an alcohol group.

Step 7 uses the high-energy compound 1,3-diphosphoglyceric acid to produce ATP.

$$
\begin{array}{c}
\text{COOPO}_3{}^{2-} \\
| \\
\text{H—C—OH} \\
| \\
\text{CH}_2\text{OPO}_3{}^{2-}
\end{array}
+ ADP \longrightarrow
\begin{array}{c}
\text{COOH} \\
| \\
\text{H—C—OH} \\
| \\
\text{CH}_2\text{OPO}_3{}^{2-}
\end{array}
+ ATP
$$

Since there are two molecules of 1,3-diphosphoglyceric acid produced for each molecule of glucose entering step 1, two molecules of **ATP** are produced per glucose unit. We are now back to even in ATP consumption (steps 1 and 3 each consumed one) and production (step 7 produced two).

Step 8 is a conversion of 3-phosphoglyceric acid to its isomer, which has the phosphate in the second position. Step 9 is a dehydration of the 2-phosphate to phosphoenolpyruvic acid (PEP), which, like ATP, is another high-energy compound.

$$
\begin{array}{c}
\text{COOH} \\
| \\
\text{H—C—OPO}_3{}^{2-} \\
| \\
\text{H—C—OH} \\
| \\
\text{H}
\end{array}
\rightleftharpoons
\begin{array}{c}
\text{COOH} \\
| \\
\text{C—OPO}_3{}^{2-} \\
\| \\
\text{H—C} \\
| \\
\text{H}
\end{array}
+ H_2O
$$

2-Phosphoglyceric   Phosphoenolpyruvic
acid   acid (PEP)

In step 10, **PEP** transfers its phosphate to ADP, converting the latter to ATP.

$$
PEP + ADP \longrightarrow ATP + \text{Pyruvic acid}
$$

Since there are two PEP molecules for each molecule of glucose entering the pathway, two ATPs are formed through this step. Thus, glycolysis produces a *net gain* of two ATP molecules.

The final step of this anaerobic process is the reduction of pyruvic acid to lactic acid by NADH.

$$CH_3\overset{\overset{\displaystyle O}{\|}}{C}-COOH + NADH + H^+ \longrightarrow CH_3\underset{\underset{\displaystyle H}{|}}{\overset{\overset{\displaystyle OH}{|}}{C}}-COOH + NAD^+$$

The NAD$^+$ consumed in step 6 is restored in step 11.

Another anaerobic process, the fermentation by yeast of glucose to ethyl alcohol, follows the same pathway through the first 10 steps. There are 2 additional steps in fermentation. Step 11 is a decarboxylation of pyruvic acid to acetaldehyde.

$$CH_3-\overset{\overset{\displaystyle O}{\|}}{C}-\overset{\overset{\displaystyle O}{\|}}{C}-O-H \longrightarrow CH_3-\overset{\nearrow O}{\underset{\searrow H}{C}} + CO_2$$

The final step, step 12 in the fermentation process, involves the reduction of acetaldehyde to ethyl alcohol by NADH.

$$CH_3\overset{\nearrow O}{\underset{\searrow H}{C}} + NADH + H^+ \longrightarrow CH_3CH_2OH + NAD^+$$

27.5 AEROBIC OXIDATION AND THE KREBS CYCLE

In Chapter 25, we look at the production of energy by a sequence of reactions called the **Krebs cycle.** Let us now examine the chemistry of the cycle in some detail. The Krebs cycle is also known by two other names. It is called the **citric acid cycle** because citric acid is a most important intermediate. It is also known as the **tricarboxylic acid cycle** because four of the intermediate compounds have three carboxyl groups each. It is best known, though, as the Krebs cycle, in honor of Hans Adolf Krebs, the German-born British biochemist who correlated a lot of data by postulating the cycle in 1937. Krebs won the Nobel Prize for physiology and medicine in 1953 in recognition of his work.

The cycle is fed by pyruvic acid from the anaerobic glycolysis pathway (Section 27.4) and by lactic acid from that source or from the Cori cycle (Section 27.3). Before they can enter the Krebs cycle, these compounds must be converted to acetyl coenzyme A. The process involves several steps, with each step

$$CH_3\overset{\overset{\displaystyle O}{\|}}{C}-S-CoA$$

Acetyl coenzyme A

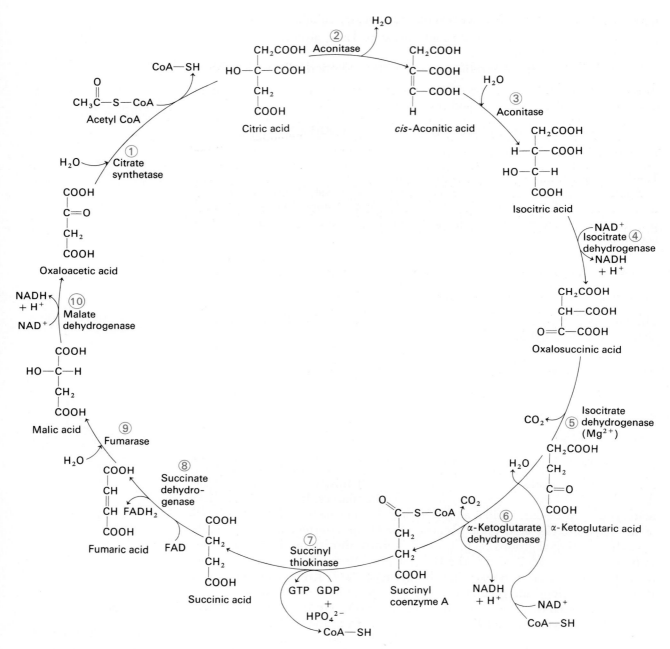

FIGURE 27.10 *The Krebs cycle.*

catalyzed by an enzyme. The net reaction for pyruvic acid is

$$CH_3\overset{O}{\underset{\|}{C}}-COOH + NAD^+ + CoASH \longrightarrow CH_3\overset{O}{\underset{\|}{C}}-S-CoA + NADH + H^+ + CO_2$$

Acetyl coenzyme A then transfers its acetyl group to oxaloacetic acid, forming citric acid, and the cycle is off and rolling (Figure 27.10).

Acetyl coenzyme A is a most versatile intermediate. Its structure was given in Figure 25.6. Acetyl coenzyme A is also formed in the metabolism of fats (Chapter 28) and of certain amino acids. In addition to being the "starter" for the Krebs cycle, it is used in the biosynthesis of a variety of larger molecules.

The aerobic oxidation of the Krebs cycle is much more efficient than the anaerobic processes of the Embden-Meyerhof pathway (Section 27.4). The net reaction shows the conversion of pyruvic acid to carbon dioxide and water with the production of 15 molecules of ATP per molecule of pyruvic acid.

$$CH_3\overset{O}{\underset{\|}{C}}-COOH + \tfrac{5}{2}O_2 + 15\,ADP + 15\,HPO_4{}^{2-} \longrightarrow 3\,CO_2 + 2\,H_2O + 15\,ATP$$

Let's look at some of the individual steps.

Step 1 is an aldol-type condensation involving acetyl coenzyme A and the ketone group of oxaloacetic acid.

Citric acid

The Krebs cycle is regulated in part at this first step. The reaction is catalyzed by citrate synthetase, an enzyme that is inhibited by ATP and NADH. When the cell has sufficient ATP for its immediate needs, ATP molecules interact with the enzyme to reduce its affinity for acetyl CoA. The latter is then shunted to the biosynthesis of fatty acids (Section 28.4).

The citric acid product is then dehydrated to *cis*-aconitic acid in step 2, a reaction catalyzed by the enzyme aconitase.

Citric acid   *cis*-Aconitic acid

In step 3 of the Krebs cycle, *cis*-aconitic acid is rehydrated to isocitric acid.

$$
H-OH \; + \; \begin{array}{l} CH_2COOH \\ | \\ C-COOH \\ \| \\ CH-COOH \end{array} \xrightarrow{\text{Aconitase}} \begin{array}{l} CH_2COOH \\ | \\ H-C-COOH \\ | \\ HO-CH-COOH \end{array}
$$

Isocitric acid

Fluoroacetic acid is a potent poison. It is readily converted to fluoroacetyl coenzyme A. Thus activated, the molecular fragment is transferred to oxaloacetic acid, forming fluorocitric acid. The latter is a powerful inhibitor of aconitase, binding tightly and hanging on. The energy-producing Krebs cycle grinds to a halt, and death follows rapidly. Sodium fluoroacetate (Compound 1080) is widely used to poison rats and predators such as the coyote. It is not selective; it will kill humans, pets, domestic animals—or any other organisms that live by aerobic oxidation involving the Krebs cycle.

Isocitric acid is oxidized by NAD^+ to oxalosuccinic acid in step 4.

$$
\begin{array}{l} CH_2COOH \\ | \\ CH-COOH \\ | \\ HO-CH-COOH \end{array} + NAD^+ \longrightarrow \begin{array}{l} CH_2COOH \\ | \\ CH-COOH \\ | \\ O=C-COOH \end{array} + NADH + H^+
$$

Oxalosuccinic acid

Note that this reaction involves the oxidation of a secondary alcohol function to a ketone function. This step is catalyzed by isocitrate dehydrogenase. This enzyme, like citrate synthetase, is inhibited by ATP and NADH. Unlike citrate synthetase, however, isocitrate dehydrogenase is activated by ADP. This step serves as the primary control point of the Krebs cycle.

Oxalosuccinic acid is readily decarboxylated, giving α-ketoglutaric acid (step 5).

$$
\begin{array}{l} CH_2COOH \\ | \\ CH-COO-H \\ | \\ O=C-COOH \end{array} \longrightarrow \begin{array}{l} CH_2COOH \\ | \\ CH_2 \\ | \\ O=C-COOH \end{array} + CO_2
$$

α-Ketoglutaric acid

In step 6, a rather complicated one, α-ketoglutaric acid is decarboxylated, oxidized by NAD^+, and coupled with coenzyme A to give succinyl coenzyme A.

$$
\begin{array}{l}
\text{CH}_2\text{COOH} \\
|\\
\text{CH}_2 \quad + \text{NAD}^+ + \text{CoA}-\text{SH} \longrightarrow \\
|\\
\text{O}{=}\text{C}-\text{COOH}
\end{array}
\qquad
\begin{array}{l}
\text{CH}_2\text{COOH} \\
|\\
\text{CH}_2 \quad + \text{CO}_2 + \text{NADH} + \text{H}^+ \\
|\\
\text{O}{=}\text{C}-\text{S}-\text{CoA}
\end{array}
$$

Succinyl coenzyme A

This thioester is then hydrolyzed to succinic acid (step 7) in a reaction coupled with the conversion of guanosine diphosphate (GDP) to guanosine triphosphate (GTP). This is the same type of oxidative decarboxylation as that involved in the conversion of pyruvic acid to acetyl CoA and carbon dioxide (p. 825).

$$
\begin{array}{l}
\text{CH}_2\text{COOH} \\
|\\
\text{CH}_2 \quad + \text{H}_2\text{O} + \text{HPO}_4{}^{2-} + \text{GDP} \longrightarrow \\
|\\
\text{O}{=}\text{C}-\text{S}-\text{CoA}
\end{array}
\qquad
\begin{array}{l}
\text{CH}_2\text{COOH} \\
|\\
\text{CH}_2 \quad + \text{CoA}-\text{SH} + \text{GTP} \\
|\\
\text{O}{=}\text{C}-\text{OH}
\end{array}
$$

Succinic acid

The succinic acid is oxidized (dehydrogenated) by flavin adenine dinucleotide (FAD) to fumaric acid (step 8).

$$
\begin{array}{l}
\quad\ \text{H} \\
\quad\ |\\
\text{H}-\text{C}-\text{COOH} \\
\quad\ |\\
\text{HOOC}-\text{C}-\text{H} \quad + \text{FAD} \longrightarrow \\
\quad\ |\\
\quad\ \text{H}
\end{array}
\qquad
\begin{array}{l}
\text{H}\quad\ \ \text{COOH} \\
\ \ \diagdown\ \diagup \\
\quad\ \text{C} \\
\quad\ \| \\
\quad\ \text{C} \quad + \text{FADH}_2 \\
\ \ \diagup\ \diagdown \\
\text{HOOC}\quad\ \text{H}
\end{array}
$$

Fumaric acid

Fumaric acid is hydrated to malic acid (step 9).

$$
\begin{array}{l}
\text{H}\quad\ \ \text{COOH} \\
\ \ \diagdown\ \diagup \\
\quad\ \text{C} \\
\quad\ \| \\
\quad\ \text{C} \quad + \text{H}-\text{OH} \xrightarrow{\ \text{fumarase}\ } \\
\ \ \diagup\ \diagdown \\
\text{HOOC}\quad\ \text{H}
\end{array}
\qquad
\begin{array}{l}
\quad\ \text{COOH} \\
\quad\ |\\
\text{H}-\text{C}-\text{OH} \\
\quad\ |\\
\text{H}-\text{C}-\text{H} \\
\quad\ |\\
\quad\ \text{COOH}
\end{array}
$$

Malic acid

Then the secondary alcohol group of malic acid is oxidized by NAD^+ to a ketone function in oxaloacetic acid (step 10).

$$
\begin{array}{l}
\text{HO}-\text{CH}-\text{COOH} \\
\qquad\ |\\
\qquad\ \text{CH}_2\text{COOH}
\end{array}
+ \text{NAD}^+ \longrightarrow
\begin{array}{l}
\text{O}{=}\text{C}-\text{COOH} \\
\quad\ |\\
\quad\ \text{CH}_2\text{COOH}
\end{array}
+ \text{NADH} + \text{H}^+
$$

Oxaloacetic acid

And the cycle is complete!

Oxaloacetic acid can accept an acetyl group from acetyl coenzyme A, and the cycle is ready for another spin. Every time we go around the cycle, two carbons are fed into the system as acetyl coenzyme A, and two carbons are kicked out as carbon dioxide molecules.

Taken as a whole, the Krebs cycle seems rather complex. All the reactions, though, are familiar types from organic chemistry: condensations, dehydrations, hydrations, oxidations, decarboxylations, and hydrolyses. The difference here is one of *conditions*. These reactions take place at constant temperature (37 °C in humans) and constant pH. Each is catalyzed by an enzyme.

The description of the Krebs cycle presented here describes the fate of carbon atoms. The energy flow in the cycle is described in Chapter 25. Recall that although we called the process aerobic oxidation, the oxidizing agents employed in the cycle are NAD^+ and FAD, not elemental oxygen. The process by which the electrons produced in the oxidative steps of the Krebs cycle are eventually transferred to oxygen is described in Section 25.7. For now, let us simply tabulate the number of ATP molecules formed by each turn of the citric acid cycle.

27.6 ATP SYNTHESIS AND AEROBIC OXIDATION

The Krebs cycle results in the formation of three NADH molecules and one FADH molecule. In addition, one molecule of guanosine triphosphate (GTP), a high-energy compound equivalent to ATP, is formed. Each NADH molecule, upon oxidation in the respiratory chain, yields three ATP molecules.

$$\text{NADH} + \text{H}^+ + 3\,\text{ADP} + 3\,\text{HPO}_4^- + \tfrac{1}{2}\text{O}_2 \longrightarrow \text{NAD}^+ + 3\,\text{ATP} + 4\,\text{H}_2\text{O}$$

The three NADH molecules, then, yield nine ATP molecules.

Oxidation of an $FADH_2$ molecule yields two ATP molecules.

$$\text{FADH}_2 + 2\,\text{H}^+ + 2\,\text{ADP} + 2\,\text{HPO}_4^- + \tfrac{1}{2}\text{O}_2 \longrightarrow \text{FAD} + 2\,\text{ATP} + 3\,\text{H}_2\text{O}$$

Since GTP is equivalent to ATP,

$$\text{GTP} + \text{ADP} \rightleftharpoons \text{GDP} + \text{ATP}$$

We can count a total of $(9 + 2 + 1) = 12$ molecules of ATP for each turn around the Krebs cycle. Two carbon atoms are kicked out as carbon dioxide. This amounts, then, to the conversion of the acetyl group (often called ''acetate'') of acetyl coenzyme A to carbon dioxide with a part of the energy going into the high-energy bonds of 12 ATP molecules.

If we go back to the starting material, glucose, we can calculate the total yield of ATP from one molecule of glucose. Let us assume that the glucose molecule follows the glycolysis route to pyruvic acid, that is, through the first ten steps of the Embden-Meyerhof pathway (Figure 27.9). Two molecules of

TABLE 27.2   *Yield of ATP From the Aerobic Oxidation of Glucose*

Source	Other Compounds Formed	ATP Formed per Glucose Molecule
1. Glycolysis	2 NADH	+2
2. Pyruvic acid → acetyl CoA	2 NADH	0
3. Citric acid cycle	2 GTP	+2
	6 NADH	
	2 FADH$_2$	
4. Oxidative phosphorylation		
2 NADH from 1		+6
Transport across		−2
mitochondrial membrane		
2 NADH from 2		+6
6 NADH from 3		+18
2 FADH$_2$ from 3		+4
		Total = +36

NADH are produced in step 6 of that pathway (one for each of the two glyceraldehyde 3-phosphate molecules that come from glucose). These two NADH, if transported from the cytoplasm where glycolysis occurs to the mitochondria, would yield six ATP. However, it takes one ATP each to transport the NADH across the mitochondrial membrane, so the net yield from these NADH is four ATP.

The pyruvate decarboxylation (p. 825), which leads into the Krebs cycle, also produces NADH. Since there are two pyruvic acid molecules, two NADH are formed in this step, leading to six ATP molecules. All the ATP molecules from the various sources are tabulated in Table 27.2. For each glucose molecule, 36 ATP molecules are formed.

$$C_6H_{12}O_6 + 6\,O_2 + 36\,ADP + 36\,HPO_4^- + 36\,H^+ \longrightarrow 6\,CO_2 + 36\,ATP + 42\,H_2O$$

Note that the important product from the aerobic oxidation of glucose is not carbon dioxide or water, but ATP.

PROBLEMS

1. Define these terms:
   a. blood sugar
   b. hypoglycemia
   c. hyperglycemia
   d. renal threshold
   e. glycolysis
   f. glycogenolysis
   g. glycogenesis
   h. glucose tolerance test
   i. galactosemia
2. What is the role of insulin in regulating blood glucose levels?
3. How does glucagon act to raise blood glucose levels?

4. How does epinephrine raise blood glucose levels?
5. How do cortisol and cortisone affect blood glucose levels?
6. How does human growth hormone affect blood glucose levels?
7. Glucose appears in the urine of a diabetic. Does glucose in the urine always indicate diabetes mellitus? Explain.
8. Why cannot insulin be taken orally?

9. In structure and purpose, how do the oral drugs such as Orinase differ from insulin?
10. What is meant when glyburide is called a second-generation drug?
11. What is the storage form of carbohydrate in the body?
12. In what tissues or organs are carbohydrates stored?
13. Draw a schematic diagram of the storage form of carbohydrates.
14. What roles do glycogen synthetase and the branching enzyme play in glycogen formation?
15. What is the ultimate product obtained from a glucose molecule that has entered the glycolysis pathway?
16. Which step in the Embden-Meyerhof pathway involves oxidation of a glucose metabolite? What is the oxidizing agent?
17. What critical role is played by both 1,3-diphosphoglyceric acid and phosphoenolpyruvic acid (PEP) in the Embden-Meyerhof pathway?
18. How does fermentation deviate from the Embden-Meyerhof pathway of glycolysis?
19. What is the main function of the Krebs cycle?
20. Two carbon atoms are fed into the Krebs cycle as acetyl coenzyme A. In what form are two carbon atoms removed from the cycle?
21. What are the oxidizing agents most immediately involved in the Krebs cycle?
22. How many molecules of ATP are formed by the oxidation of NADH in the respiratory chain?

23. How many molecules of ATP are formed by the oxidation of $FADH_2$ in the respiratory chain?
24. What is GTP?
25. Is the compound $R-\overset{\overset{\displaystyle O}{\|}}{C}-O-PO_3^{2-}$ an ester or an anhydride?
26. Is the compound $R-CH_2O-PO_3^{2-}$ as ester or an anhydride?
27. Which, if either, of the compounds in Problems 25 and 26 is a high-energy phosphate?
28. Which coenzyme acts as the oxidizing agent in glycolysis?
29. Through what set of reactions is NADH oxidized under aerobic conditions?
30. How many molecules of pyruvic acid are produced from one molecule of glucose in glycolysis?
31. What is substrate level phosphorylation?
32. What is oxidative phosphorylation?
33. What is the first step in the Embden-Meyerhof pathway?
34. Is the reaction in Problem 33 an example of substrate level phosphorylation or of oxidative phosphorylation?
35. What is the Cori cycle?
36. What happens to the monosaccharide galactose after it reaches the liver?
37. What happens to fructose in the liver?

28

Lipid Metabolism

I n Chapter 27, we see that carbohydrates are sources of energy for immediate use and reserve energy (in the form of glycogen). Carbohydrates
are good sources of energy, but in many respects fats are better. The
"burning" of 1 mol (180 g) of glucose produces 673 kcal of energy, a yield of
about 4 kcal/g.

$$C_6H_{12}O_6 + 6\,O_2 \longrightarrow 6\,CO_2 + 6\,H_2O + 673 \text{ kcal/mol}$$

Glucose

The "burning" of 1 mol (256 g) of palmitic acid, a typical fatty acid, gives
2340 kcal of energy.

$$CH_3(CH_2)_{14}COOH + 23\,O_2 \longrightarrow 16\,CO_2 + 16\,H_2O + 2340 \text{ kcal/mol}$$

This amounts to about 9 kcal/g. Gram for gram, fats are a much richer energy
source than carbohydrates. Fats are also more readily stored. The body can
accommodate only a limited amount of glycogen, but considerable fat reserves
can be stashed away. The glycogen reserves are mobilized more quickly, but
the fat reserves last longer.

In this chapter, we shall look at how our bodies store and metabolize fats.
We'll also consider some of the problems associated with lipid storage and
metabolism. Lipids offer one of the best illustrations of an old saying. Too much
of a good thing can be bad!

28.1 BLOOD LIPIDS: NORMAL FASTING LEVELS

Fats are emulsified by bile salts and are digested in the small intestine (Special Topic L). Pancreatic **lipase** splits the fats into fatty acids, glycerol, soaps (salts of fatty acids), and mono- and diglycerides (Figure 28.1). As these components pass through the intestinal wall, they are reassembled as triglycerides. The reassembled products first enter the lymph system, which eventually introduces them into the blood circulation. Phospholipids are broken down and reassembled in much the same way. Some small fatty acid molecules can be directly absorbed into the bloodstream.

Lipids are, by definition, relatively insoluble in water. Since blood is an aqueous solution, fats as such are not soluble in blood. But fats can be efficiently transported by the blood if they are first complexed with water-soluble proteins in the plasma. Such complexes are called **lipoproteins.** Fatty acids,

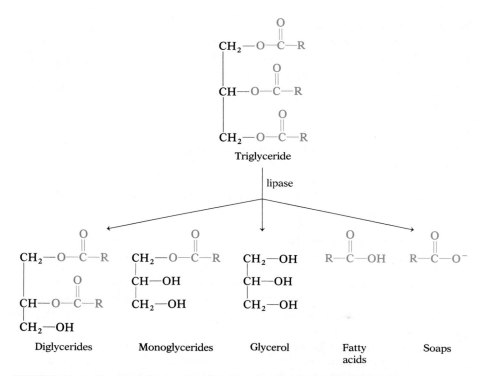

FIGURE 28.1 *Products formed in the digestion (hydrolysis) of fats. Triglyceride (fat) is hydrolyzed into diglycerides, monoglycerides, glycerol, fatty acids, and soaps.*

TABLE 28.1 *Lipoproteins in the Blood*

Class	Abbreviation	% Protein	Density (g/mL)	Main Function
Very-low-density	VLDL	5	1.006–1.019	Transport triglycerides
Low-density	LDL	25	1.019–1.063	Transport cholesterol to the cells for use
High-density	HDL	50	1.063–1.210	Transport cholesterol to the liver for processing for excretion

too, are generally complexed with proteins while being transported in the blood.

Classification of lipoproteins in the blood is based upon their density (Table 28.1). Very-low-density lipoproteins (VLDL) serve mainly to transport triglycerides. Low-density lipoproteins (LDL) are the main carriers of cholesterol. The LDL carry cholesterol to the cells for use. It is these lipoproteins that are thought to deposit cholesterol in arteries, leading to cardiovascular disease. The high-density lipoproteins (HDL) also carry cholesterol, but they carry it to the liver for processing and excretion. Exercise is thought to increase the levels of HDL, the "good cholesterol." High levels of LDL, the "bad cholesterol," put one at increased risk of heart attack and stroke.

There is one advantage to the low solubility of fats, and that is that these compounds can be separated from other components of the blood by extraction with fat solvents such as benzene or chloroform. Such an extraction is always one of the first steps in any clinical procedure aimed at determining lipid levels.

The concentration of lipids in the blood changes constantly. As lipids are absorbed from the digestive tract after a meal, the level in the blood increases. As blood lipids are removed to storage or are oxidized in certain tissues, blood lipid level falls. On demand, the body can synthesize lipids from other foods or can remove already-formed lipids from storage. Both activities would result in an increase in blood lipid level. Finally, the body excretes some lipids as a normal component of feces. These excreted lipids may be in the form of fats or soaps or fatty acids. Their removal via the intestine would tend to lower the blood lipid level.

So that the conditions under which lipid concentrations are determined are somewhat standard, lipids in the plasma are measured after fasting. This is the same procedure followed in the measurement of blood sugar levels. Typical *normal fasting levels* of lipids are given in Table 28.2. Abnormally high levels of triglycerides and cholesterol are thought by many medical practitioners to be involved in hardening of the arteries, a condition that may lead to rupture or blockage of vessels in the brain (a stroke) or in the heart (a heart attack, or coronary).

TABLE 28.2   *Range of Normal Levels of Lipids in Blood Plasma of a Fasting Person*

Constituent	Concentration (in milligrams per 100 mL)
Free cholesterol	30–60
Cholesterol esters	75–150
Total cholesterol	120–250
Triglycerides	25–260
Lecithin	100–225
Sphingomyelin	10–47
Total phospholipids	150–250
Total lipids	400–700

28.2 FAT DEPOTS

Fats are stored throughout the body. Principally, though, they are deposited in a special kind of connective tissue called **adipose tissue.** Storage places are called **fat depots.** Considerable fat is stored around vital organs such as the heart, liver, kidneys, and spleen. There it serves as a protective cushion, helping prevent injury to the organs. Fat is also stored under the skin, where it helps insulate against sudden temperature changes. The fat acts just like the insulation in the walls of a house, trapping body heat and preventing it from escaping to the surroundings. In some ways, the fatty tissue also acts like a furnace of a house. When the outside temperature drops, metabolic activity in the cells generates heat to compensate for that lost to the environment.

Some lipids, particularly phospholipids, are found as highly organized functional units of cells, for example, in bilayers that serve as cell membranes. The triglycerides in adipose tissue, on the other hand, are stored in the form of little fat globules. Each cell contains a single droplet of fat that can occupy as much as 90% of the cell's volume. The cells are all held more or less in place by the connective tissue. This adipose tissue can be quite active metabolically, and it is richly supplied with blood. When the body needs energy, the cells of the adipose tissue release fatty acids from storage. Normally this occurs only after glycogen reserves have been depleted, a few hours after eating. When the glycogen is gone, the body first uses the small amount of fat stored in the liver and then calls on the depots.

When food is taken into the body in excess of immediate needs, it is shunted to the storage areas. Fats are tucked away in the adipose tissue. Carbohydrates are converted to glycogen. When no more glycogen storage area is available, carbohydrates are converted to fats and stored in the depots. There is a continuous, dynamic change as molecules come and go from storage. If your weight is constant, there is an "equilibrium" in which the number of molecules arriving for storage is equal to the number being removed from the depot. We will discuss two situations in which there is an imbalance—obesity and starvation—in subsequent sections.

28.3 OXIDATION OF FATTY ACIDS

Fat reserves are metabolized in a multistep process that produces energy. The basic outline of this process is given in Figure 28.2. First, the fat molecules are hydrolyzed to glycerol and free fatty acids. The glycerol is phosphorylated in the liver to glycerol phosphate and then is oxidized to dihydroxyacetone phosphate.

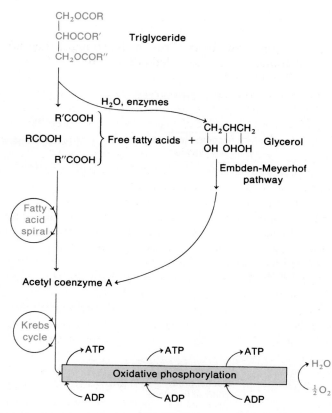

FIGURE 28.2   *An outline showing how fat reserves (triglyceride) are used for the production of energy.*

$$
\begin{array}{ccc}
\text{CH}_2\text{OH} & \text{ATP} \quad \text{ADP} & \text{CH}_2\text{OPO}_3\text{H}_2 \quad \text{NAD}^+ \quad \overset{\text{H}^+}{\underset{\text{NADH}}{+}} \quad \text{CH}_2\text{OPO}_3\text{H}_2 \\
| & & | \\
\text{CH}\!-\!\text{OH} & \longrightarrow & \text{CH}\!-\!\text{OH} \longrightarrow \quad \text{C}\!=\!\text{O} \\
| & & | \\
\text{CH}_2\text{OH} & & \text{CH}_2\text{OH} \qquad\qquad\qquad \text{CH}_2\text{OH}
\end{array}
$$

Dihydroxyacetone phosphate is an intermediate in the Embden-Meyerhof glycolytic pathway (Chapter 27), so fats as well as carbohydrates contribute fuel to that energy-producing process.

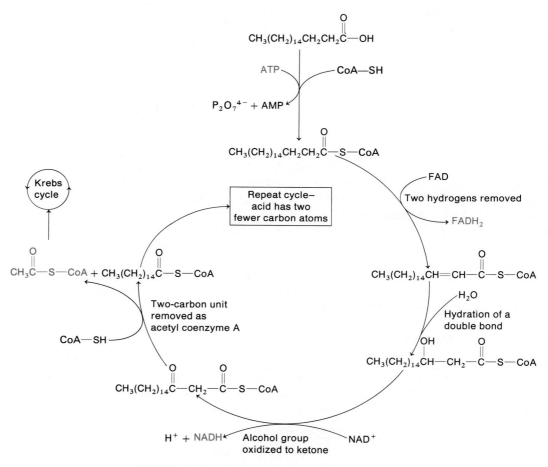

FIGURE 28.3 *The fatty acid spiral for stearic acid.*

FIGURE 28.4
The fatty acid "cycle" is really a spiral, with two carbon atoms chopped off for each swing around the cycle.

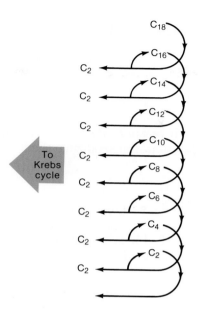

The free fatty acids formed by the hydrolysis of fats are metabolized mainly in the liver. The cells of heart and skeletal muscle are also able to obtain energy by oxidizing fatty acids. The oxidation proceeds by a series of steps called the **fatty acid spiral** (Figure 28.3). As a fatty acid makes a swing around the cycle, two carbon atoms are chopped off (Figure 28.4). The acid entering the next swing (as its CoA thioester) is smaller by two carbon atoms. On its last swing through the cycle, the acid yields two molecules of acetyl coenzyme A.

The first reaction of the cycle involves activation of the fatty acid. The acid reacts with ATP and coenzyme A in the presence of magnesium ions to form the fatty acid thioester of coenzyme A along with adenosine monophosphate (AMP) and inorganic pyrophosphate.

$$R-CH_2-CH_2-\overset{O}{\overset{\|}{C}}-OH + ATP + CoA-SH \xrightarrow{Mg^{2+}} R-CH_2-CH_2-\overset{O}{\overset{\|}{C}}-S-CoA + AMP + P_2O_7^{4}$$

The second reaction is an oxidation (dehydrogenation) reaction in which a double bond is introduced between the alpha and beta carbon atoms of the acid unit of the thioester. The oxidizing agent is flavin adenine dinucleotide (FAD).

$$R-\overset{\beta}{CH_2}-\overset{\alpha}{CH_2}-\overset{O}{\overset{\|}{C}}-S-CoA + FAD \longrightarrow R-\overset{\beta}{CH}=\overset{\alpha}{CH}-\overset{O}{\overset{\|}{C}}-S-CoA + FADH_2$$

The third reaction involves the addition of water to the double bond.

$$R-CH=CH-\overset{\overset{\displaystyle O}{\|}}{C}-S-CoA + H-OH \longrightarrow R-\underset{\beta}{\overset{\overset{\displaystyle OH}{|}}{CH}}-\overset{\alpha}{CH_2}-\overset{\overset{\displaystyle O}{\|}}{C}-S-CoA$$

In the fourth reaction, the secondary alcohol group is oxidized to a ketone group. This time the oxidizing agent is nicotinamide adenine dinucleotide (NAD$^+$).

$$R-\underset{\beta}{\overset{\overset{\displaystyle OH}{|}}{CH}}-\overset{\alpha}{CH_2}-\overset{\overset{\displaystyle O}{\|}}{C}-S-CoA + NAD^+ \longrightarrow R-\underset{\beta}{\overset{\overset{\displaystyle O}{\|}}{C}}-\overset{\alpha}{CH_2}-\overset{\overset{\displaystyle O}{\|}}{C}-S-CoA + NADH + H^+$$

This step is called **beta oxidation.**

The fifth step involves the cleavage of the β-keto ester by coenzyme A.

$$R-\underset{\beta}{\overset{\overset{\displaystyle O}{\|}}{C}}-\overset{\alpha}{CH_2}-\overset{\overset{\displaystyle O}{\|}}{C}-S-CoA + CoA-S-H \longrightarrow R-\overset{\overset{\displaystyle O}{\|}}{C}-S-CoA + CH_3-\overset{\overset{\displaystyle O}{\|}}{C}-S-CoA$$

The products are acetyl coenzyme A and a second thioester. The acetyl coenzyme A incorporates the carboxyl group and the alpha carbon of the original acid, and the other thioester contains a fatty acid unit with two carbon atoms fewer than the original. Acetyl coenzyme A enters the Krebs cycle, and the other thioester takes another spin around the fatty acid spiral.

The oxidation of stearic acid (which has 18 carbon atoms) requires eight turns of the fatty acid spiral (see Figure 28.4) and results in the formation of nine molecules of acetyl coenzyme A. During the oxidation, eight molecules each of FADH$_2$ and NADH are formed. These enter the respiratory chain (Chapter 27), resulting in the formation of ATP.

Each FADH molecule yields two ATP molecules, and each NADH molecule produces three, for a total of $16 + 24 = 40$ ATP. However, in the first step, a molecule of ATP was hydrolyzed to *AMP*. That is equivalent to *two* being hydrolyzed to ADP. So the net ATP production is $40 - 2 = 38$ ATP.

Each of the nine molecules of acetyl coenzyme A that enters the Krebs cycle gives rise to 12 ATP molecules. That means a total of 108 ATP from this source. The overall yield from one molecule of stearic acid is $108 + 38 = 146$ ATP. These calculations are summarized in Table 28.3.

Not all the acetyl coenzyme A formed by oxidation of fatty acids enters the Krebs cycle. A small portion is used to synthesize new fatty acids (Section

TABLE 28.3   *ATP Formed From the Oxidation of Stearic Acid*

Source	ATP Formed per Stearic Acid Molecule
8 $FADH_2$	$+16$
8 NADH	$+24$
Hydrolysis of ATP to AMP (step 1)	-2
9 Acetyl coenzyme A	$+108$
Total $=$	$+146$

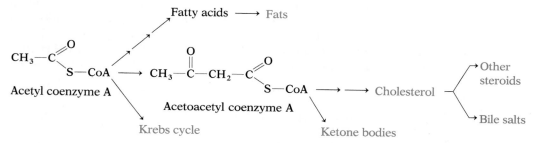

FIGURE 28.5   *Acetyl coenzyme A plays a variety of roles in cellular chemistry.*

28.4). Some of it goes to produce cholesterol, bile salts, and other steroids (Section 28.11). Still other acetyl coenzyme A is used in the formation of the ketone bodies (Section 28.7). The various routes available to acetyl coenzyme A are summarized in Figure 28.5.

28.4 BIOSYNTHESIS OF FATTY ACIDS

Some fatty acids are made by a reversal of the fatty acid spiral. The main pathway for fatty acid synthesis, however, is somewhat different, though it too starts with acetyl coenzyme A. In this sequence, acetyl coenzyme A is carboxylated in the presence of ATP, manganese ions, and the B vitamin biotin to form malonyl coenzyme A.

$$\overset{\alpha}{C}H_3 - \overset{O}{\overset{\|}{C}} - S - CoA + CO_2 + ATP \xrightarrow{Mn^{2+}} \underset{COOH}{\overset{\alpha}{C}H_2} - \overset{O}{\overset{\|}{C}} - S - CoA + ADP + HPO_4^{2-}$$

Malonyl coenzyme A

Malonyl coenzyme A then condenses with another molecule of acetyl coenzyme A to form an acetoacetyl group. This group is then transferred to the fatty acid synthetase enzyme complex with the release of coenzyme A.

$$
\underset{\substack{O \\ \parallel}}{CH_3CSCoA} + \underset{\substack{O \\ \parallel}}{\underset{\alpha}{CH_2}CSCoA} + \text{enzyme complex} \longrightarrow \underset{\substack{O \quad O \\ \parallel \quad \parallel}}{CH_3C\underset{\alpha}{C}H_2C}\text{—enzyme complex} + CO_2 + \text{CoASH}
$$
$$
\underset{\substack{| \\ COOH}}{}
$$

Notice that in the first step, a molecule of carbon dioxide is incorporated in the product. In the second step, the same carbon dioxide molecule is eliminated. The addition of the carbon dioxide unit serves only to activate the alpha position of the original acetyl coenzyme A molecule toward condensation. As the desired condensation takes place, the group is eliminated.

The keto group of the condensation product is reduced by NADPH to a secondary alcohol group.

$$
CH_3\underset{\substack{O \\ \parallel}}{—C}—CH_2\underset{\substack{O \\ \parallel}}{—C}—S\text{—enzyme} + NADPH + H^+ \longrightarrow CH_3\underset{\substack{OH \\ |}}{—CH}—CH_2\underset{\substack{O \\ \parallel}}{—C}—S\text{—enzyme} + NADP^+
$$

The hydroxy compound is dehydrated, and then the unsaturated compound is reduced (hydrogenated) by NADPH.

$$
CH_3\underset{\substack{OH \\ |}}{—CH}—CH_2\underset{\substack{O \\ \parallel}}{—C}—S\text{—enzyme} \longrightarrow CH_3—CH{=}CH\underset{\substack{O \\ \parallel}}{—C}—S\text{—enzyme} \quad \xrightarrow{\text{NADPH} + H^+ \quad NADP^+}
$$
$$
\underset{\substack{+ \\ H_2O}}{}
$$

$$
CH_3—CH_2—CH_2\underset{\substack{O \\ \parallel}}{—C}—S\text{—enzyme}
$$

The resulting four-carbon product begins a second round of the cycle by undergoing condensation with another molecule of malonyl coenzyme A. Again this condensation is accompanied by the loss of a carbon dioxide molecule, so the chain length increases by only two carbon atoms. Each repetition of the cycle adds another two carbon atoms to the chain. The cycle is diagrammed in Figure 28.6.

Fatty acid synthesis through reversal of the fatty acid spiral possesses some features in common with the sequence just outlined. In the former case, acetyl coenzyme A rather than malonyl coenzyme A is directly involved in the condensation steps, and different reducing agents and enzyme systems are used. But, in general, the sequence of reactions is the same: condensation is followed by reduction, which is followed by dehydration, which is followed by another

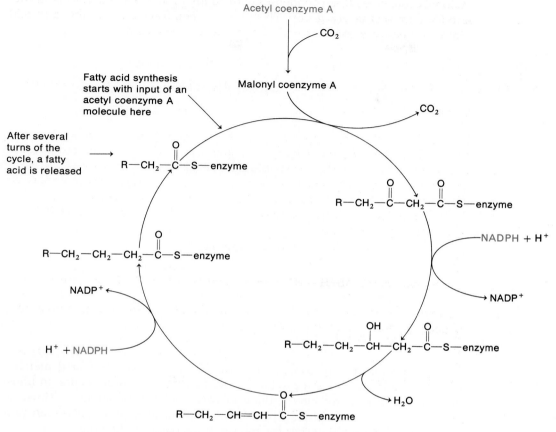

FIGURE 28.6 *The synthesis of fatty acids.*

reduction. Recall from Chapter 17 that most naturally occurring fatty acids have an even number of carbon atoms. Now you know why. They are put together two atoms at a time.

28.5 BIOSYNTHESIS OF FATS

The body can manufacture fats as well as metabolize them. Fats are synthesized in the body from fatty acids and glycerol. These two starting materials may be available from the hydrolysis of other fats. The fatty acids can also be synthesized from scratch, or rather from acetyl coenzyme A (Section 28.4). The

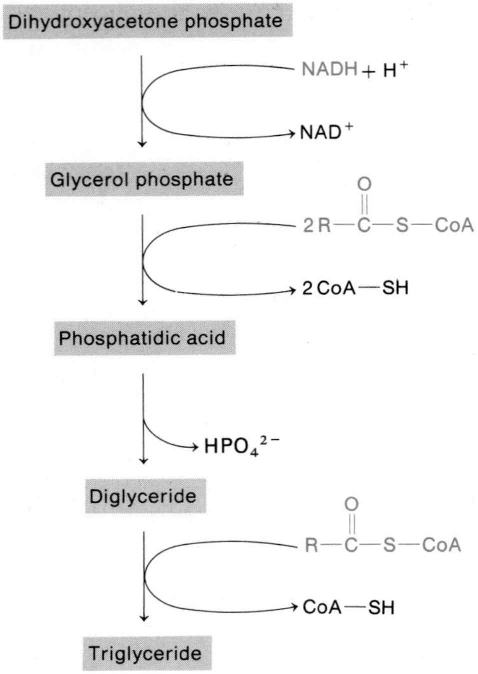

FIGURE 28.7
The biosynthesis
of triglyceride.

Dihydroxyacetone phosphate

$NADH + H^+$

NAD^+

Glycerol phosphate

$2 R—\overset{\displaystyle O}{\overset{\|}{C}}—S—CoA$

$2 CoA—SH$

Phosphatidic acid

HPO_4^{2-}

Diglyceride

$R—\overset{\displaystyle O}{\overset{\|}{C}}—S—CoA$

$CoA—SH$

Triglyceride

acetyl coenzyme A may come from carbohydrate, protein, or lipid metabolism. In addition to being a product of fat digestion, glycerol occurs, in phosphorylated form, as an intermediate in carbohydrate metabolism. Thus, all three categories of foodstuffs—proteins, carbohydrates, and lipids—can provide the needed raw materials for the synthesis of fat in the body.

Before fatty acids and glycerol can be combined to make a fat molecule, both must be activated. Fatty acids react as coenzyme A derivatives; glycerol reacts as a glycerol phosphate. Fat biosynthesis is outlined in Figure 28.7. Glycerol phosphate reacts with two fatty acid coenzyme A molecules to form phosphatidic acid, the phosphate ester of a diglyceride.

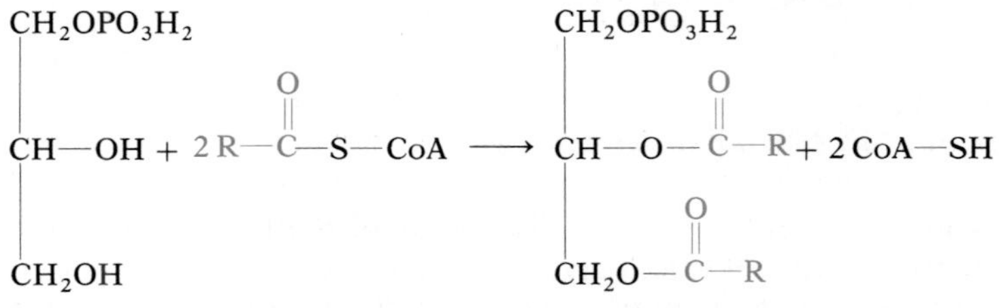

$$
\begin{array}{l}
CH_2OPO_3H_2 \\
| \\
CH—OH \quad + \; 2\,R—\overset{\displaystyle O}{\overset{\|}{C}}—S—CoA \quad \longrightarrow \\
| \\
CH_2OH
\end{array}
\qquad
\begin{array}{l}
CH_2OPO_3H_2 \\
| \\
CH—O—\overset{\displaystyle O}{\overset{\|}{C}}—R \; + \; 2\,CoA—SH \\
| \\
CH_2O—\overset{\displaystyle O}{\overset{\|}{C}}—R
\end{array}
$$

Glycerol
phosphate

Phosphatidic
acid

The enzyme phosphatase removes the phosphate group, leaving a diglyceride as a product.

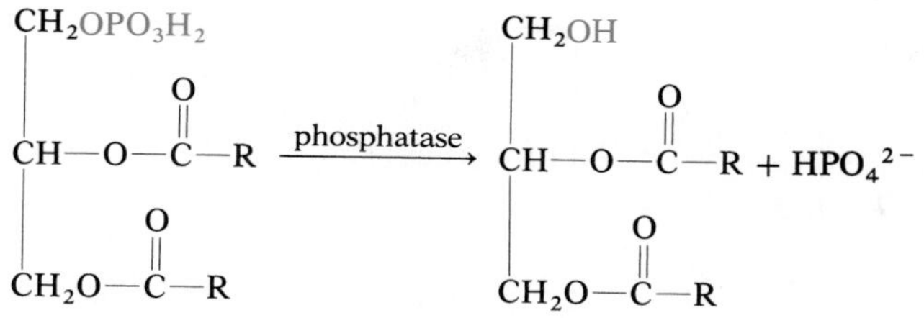

$$
\begin{array}{c}
CH_2OPO_3H_2 \\
| \\
CH-O-\overset{\overset{O}{\|}}{C}-R \\
| \\
CH_2O-\overset{\overset{O}{\|}}{C}-R
\end{array}
\xrightarrow{\text{phosphatase}}
\begin{array}{c}
CH_2OH \\
| \\
CH-O-\overset{\overset{O}{\|}}{C}-R \;+\; HPO_4{}^{2-} \\
| \\
CH_2O-\overset{\overset{O}{\|}}{C}-R
\end{array}
$$

A diglyceride

The diglyceride then reacts with the coenzyme A derivative of another fatty acid, giving the triglyceride, or fat.

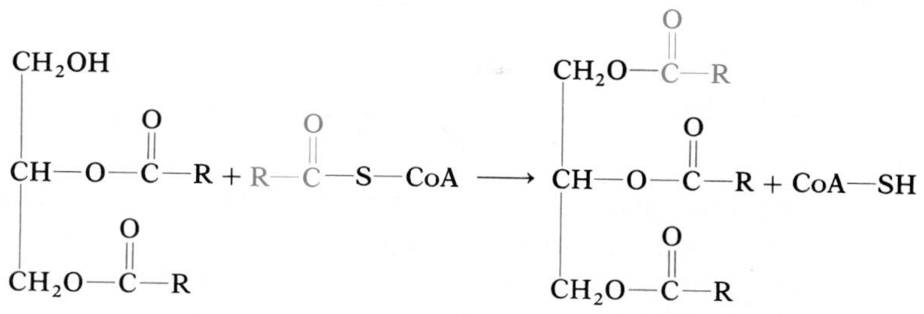

$$
\begin{array}{c}
CH_2OH \\
| \\
CH-O-\overset{\overset{O}{\|}}{C}-R \\
| \\
CH_2O-\overset{\overset{O}{\|}}{C}-R
\end{array}
\;+\; R-\overset{\overset{O}{\|}}{C}-S-CoA
\longrightarrow
\begin{array}{c}
CH_2O-\overset{\overset{O}{\|}}{C}-R \\
| \\
CH-O-\overset{\overset{O}{\|}}{C}-R \;+\; CoA-SH \\
| \\
CH_2O-\overset{\overset{O}{\|}}{C}-R
\end{array}
$$

A triglyceride (fat)

These fats, like any others, can be stored in the adipose tissue and used as an energy reserve.

28.6 BIOSYNTHESIS OF PHOSPHOLIPIDS

A phospholipid generally has a polar end and a nonpolar end. Phospholipids often serve to bridge the gap between the less polar fats and the highly polar aqueous body fluids. Phospholipids are involved in electron transport systems and in the movement of fats and fatty acids through the walls of the digestive tract and from the liver and fat depots to other tissues.

When eaten, phospholipids are probably digested in much the same way as other lipids. The constituent parts are then reassembled to make new phospholipids. Intermediates from fat metabolism or other sources can also be used in the synthesis of phospholipids.

Let us suppose that we want to make phosphatidylethanolamine, a cephalin (Chapter 21).

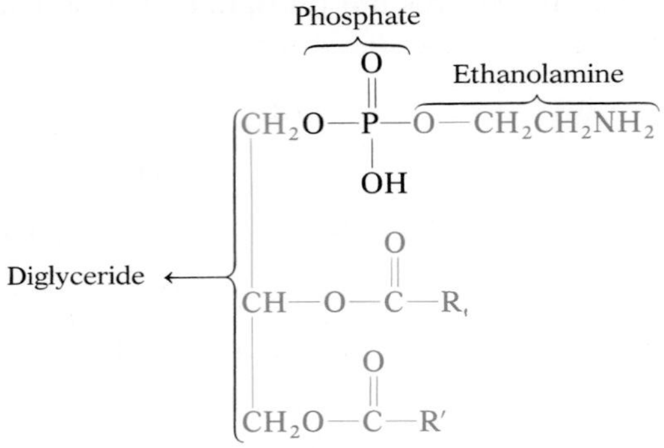

$$
\text{Diglyceride} \leftarrow
\begin{cases}
\overset{\overbrace{}^{\text{Phosphate}}}{CH_2O}-\underset{OH}{\overset{\overset{O}{\|}}{P}}-O\overbrace{-CH_2CH_2NH_2}^{\text{Ethanolamine}}\\[2em]
CH-O-\overset{\overset{O}{\|}}{C}-R,\\[2em]
CH_2O-\overset{\overset{O}{\|}}{C}-R'
\end{cases}
$$

Somehow we must combine a diglyceride with a unit of phosphate and a unit of ethanolamine. In higher animals, the alcohol group of ethanolamine is first phosphorylated.

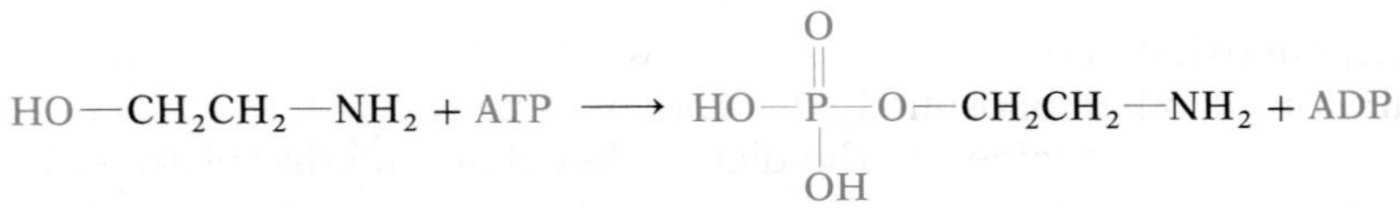

$$
HO-CH_2CH_2-NH_2 + ATP \longrightarrow HO-\underset{OH}{\overset{\overset{O}{\|}}{P}}-O-CH_2CH_2-NH_2 + ADP
$$

Then the ethanolamine phosphate is coupled to a cytidine nucleotide. The cytidine nucleotide acts as a carrier or activator; it picks up one molecule and ultimately transfers it to another molecule. In this case, cytidine triphosphate (CTP) reacts with the ethanolamine unit.

$$
CTP + \text{phosphoethanolamine} \longrightarrow CDP-\text{ethanolamine} + P_2O_7^{4-}
$$

CDP then transfers the ethanolamine unit with one phosphate unit to a diglyceride molecule.

$$
CDP-\text{ethanolamine} + \text{diglyceride} \longrightarrow \text{phosphatidylethanolamine} + CMP
$$

The phosphatidylethanolamine can be biomethylated to form phosphatidylcholine, a lecithin (Chapter 21).

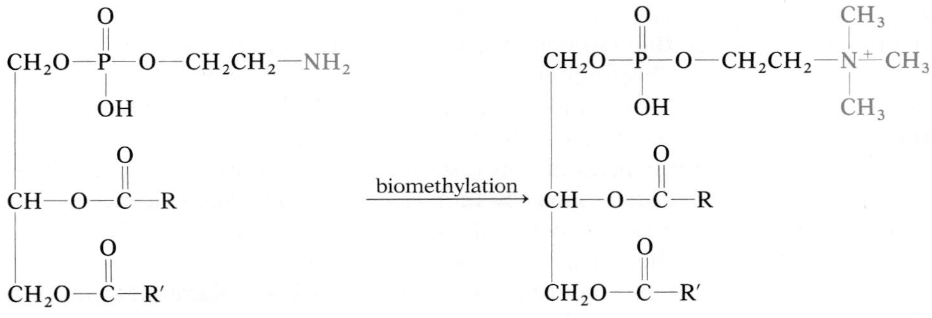

$$
\begin{array}{l}
CH_2O-\underset{OH}{\overset{\overset{O}{\|}}{P}}-O-CH_2CH_2-NH_2\\[2em]
CH-O-\overset{\overset{O}{\|}}{C}-R\\[2em]
CH_2O-\overset{\overset{O}{\|}}{C}-R'
\end{array}
\qquad\xrightarrow{\text{biomethylation}}\qquad
\begin{array}{l}
CH_2O-\underset{OH}{\overset{\overset{O}{\|}}{P}}-O-CH_2CH_2-\overset{CH_3}{\underset{CH_3}{N^+CH_3}}\\[2em]
CH-O-\overset{\overset{O}{\|}}{C}-R\\[2em]
CH_2O-\overset{\overset{O}{\|}}{C}-R'
\end{array}
$$

Phosphatidylethanolamine   Phosphatidylcholine

Or the phosphatidylethanolamine can exchange its ethanolamine unit, under the influence of an enzyme, for a serine unit.

$$
\begin{array}{ccc}
\underset{\text{Phosphatidylethanolamine}}{
\begin{array}{l}
\text{CH}_2\text{O}-\overset{\displaystyle \text{O}}{\overset{\|}{\text{P}}}-\text{OCH}_2\text{CH}_2-\text{NH}_2 \\
\quad\quad\; | \\
\quad\quad\text{OH} \\
| \quad\quad\;\; \overset{\displaystyle \text{O}}{\overset{\|}{}} \\
\text{CH}-\text{O}-\text{C}-\text{R} \\
| \quad\quad\;\; \overset{\displaystyle \text{O}}{\overset{\|}{}} \\
\text{CH}_2\text{O}-\text{C}-\text{R}'
\end{array}}
& +\; \underset{\text{Serine}}{\text{HO}-\text{CH}_2\text{CH}-\text{NH}_2}
& \longrightarrow
\end{array}
$$

(structural equation)

Phosphatidylethanolamine + HO—CH₂CH(COOH)—NH₂ (Serine) ⟶ Phosphatidylserine + HO—CH₂CH₂—NH₂ (Ethanolamine)

The product, **phosphatidylserine**, is another cephalin.

Phosphatidylethanolamine does not always serve as an intermediate in the preparation of other phospholipids. The cytidine nucleotide carrier can pick up a choline unit (obtained in the diet) rather than an ethanolamine unit and transfer it directly to the diglyceride molecule.

$$
\underset{\text{Choline}}{\text{HO}-\text{CH}_2\text{CH}_2-\overset{\overset{\displaystyle \text{CH}_3}{|}}{\underset{\underset{\displaystyle \text{CH}_3}{|}}{\text{N}^{+}}}-\text{CH}_3} + \text{ATP} \longrightarrow \underset{\text{Phosphocholine}}{\text{HO}-\overset{\overset{\displaystyle \text{O}}{\|}}{\underset{\underset{\displaystyle \text{OH}}{|}}{\text{P}}}-\text{O}-\text{CH}_2\text{CH}_2-\overset{\overset{\displaystyle \text{CH}_3}{|}}{\underset{\underset{\displaystyle \text{CH}_3}{|}}{\text{N}^{+}}}-\text{CH}_3} + \text{ADP}
$$

$$\text{Phosphocholine} + \text{CTP} \longrightarrow \text{CDP}\text{—choline} + \text{P}_2\text{O}_7{}^{4-}$$

$$\text{CDP}\text{—choline} + \text{diglyceride} \longrightarrow \text{phosphatidylcholine} + \text{CMP}$$

Figure 28.8 shows syntheses and interconversions of the phospholipids.

28.7 KETONE BODIES

Two related compounds, acetoacetic acid and β-hydroxybutyric acid, are intermediates in the fatty acid spiral (Section 28.3). A third compound, acetone, is formed by the decarboxylation of acetoacetic acid. Collectively, these three compounds are called the **ketone bodies** (even though one is *not* a ketone). The relationship among the three is shown in Figure 28.9.

The ketone bodies are formed in the liver and are normal components of blood. The two acids are used as energy sources by resting muscles. The kidneys excrete about 20 mg of ketone bodies each day. Normally, blood levels

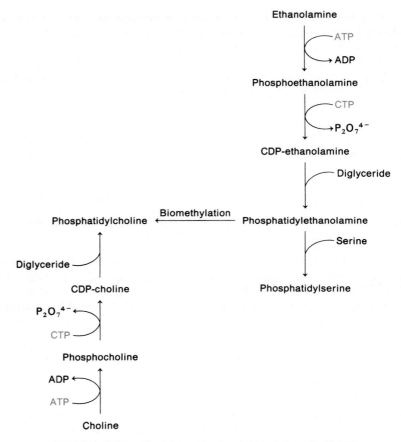

FIGURE 28.8   *The biosynthesis of some phospholipids.*

are maintained at about 1 mg per 100 mL. Large concentrations are present in the blood, however, during starvation and in certain carbohydrate metabolism disorders, such as diabetes mellitus. This high concentration of ketones in the blood, called **ketosis,** also results from certain diets that severely restrict carbohydrate intake.

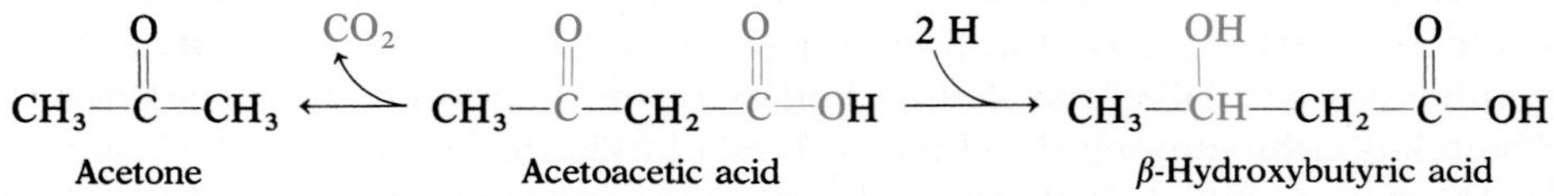

$$\text{CH}_3\text{—}\overset{\overset{\text{O}}{\|}}{\text{C}}\text{—CH}_3 \xleftarrow{\text{CO}_2} \text{CH}_3\text{—}\overset{\overset{\text{O}}{\|}}{\text{C}}\text{—CH}_2\text{—}\overset{\overset{\text{O}}{\|}}{\text{C}}\text{—OH} \xrightarrow{\text{2 H}} \text{CH}_3\text{—}\overset{\overset{\text{OH}}{|}}{\text{CH}}\text{—CH}_2\text{—}\overset{\overset{\text{O}}{\|}}{\text{C}}\text{—OH}$$

Acetone   Acetoacetic acid   β-Hydroxybutyric acid

FIGURE 28.9   *The three ketone bodies.*

During ketosis, the odor of acetone may be detected on the person's breath. Uncontrolled ketosis leads to acidosis. Severe ketosis can be fatal.

In uncontrolled diabetes, glucose is not taken into cells efficiently, so fats become the alternative energy source. Similar changes occur during starvation and in dieters who take in too little carbohydrate. In each case, the supply of oxaloacetic acid (which comes from the carboxylation of pyruvic acid) is diminished. There is then a shortage of oxaloacetic acid to react with acetyl coenzyme A from the fatty acid spiral. The surplus acetyl coenzyme A undergoes a self-condensation (much like the aldol condensation in Chapter 16) to form acetoacetyl coenzyme A.

$$2\ CH_3\overset{\displaystyle O}{\overset{\|}{C}}-S-CoA \longrightarrow CH_3\overset{\displaystyle O}{\overset{\|}{C}}-CH_2-\overset{\displaystyle O}{\overset{\|}{C}}-S-CoA + CoA-SH$$

It is from acetoacetyl coenzyme A that the three ketone bodies are formed. Acetoacetic acid is formed in two steps.

$$CH_3\overset{\displaystyle O}{\overset{\|}{C}}-CH_2-\overset{\displaystyle O}{\overset{\|}{C}}-S-CoA + H_2O \longrightarrow \longrightarrow CH_3\overset{\displaystyle O}{\overset{\|}{C}}-CH_2-\overset{\displaystyle O}{\overset{\|}{C}}-OH + CoA-SH$$

In one subsequent reaction, acetoacetic acid is reduced to β-hydroxybutyric acid. In another, it is decarboxylated to form acetone. These reactions are shown in Figure 28.9.

28.8 ACIDOSIS: UNCONTROLLED DIABETES

Two of the ketone bodies are carboxylic acids. Continued production of these acids in amounts larger than the kidneys can excrete leads to **acidosis.** More acids are produced than the blood buffers can handle. The pH of the blood drops. Bicarbonate ions are used up as the buffers act to maintain constant pH. Normal bicarbonate concentration ranges from 24 to 31 meq/L. In mild acidosis, the level drops to around 18 meq/L. At about 13 meq/L, moderate acidosis occurs. Less than 10 meq/L of bicarbonate in blood results in severe acidosis. Coma sets in at about 6 or 7 meq/L in adults.

Acidic blood cannot transport oxygen very well. In moderate to severe acidosis, "air hunger" sets in. Breathing becomes labored and very painful. The body also loses fluids as the kidneys eliminate large quantities of water trying to get rid of the acids. The person becomes dehydrated. The short oxygen supply and dehydration lead to depression. Even mild acidosis leads to lethargy, loss of appetite, and a generally run-down feeling. Untreated patients may go into diabetic coma. At that point, prompt treatment is necessary if the patient's life is to be saved.

FIGURE 28.10
While much of the world suffers from starvation, many people in the industrialized nations show the effects of overindulgence in the "good life." (Top photo © Reuters/Bettmann Newsphotos. Bottom photo courtesy Stock, Boston; © George W. Gardner.)

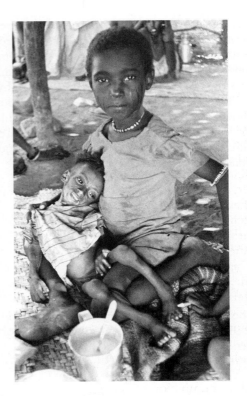

28.9 THE CHEMISTRY OF STARVATION

When the body is totally deprived of food, whether voluntarily or involuntarily, the condition is known as **starvation** (Figure 28.10). During total fasting, the glycogen stores are rapidly depleted. The body calls on its fat reserves. Fat is first obtained from around the kidneys and heart. Then it is removed from other locations. Ultimately even the bone marrow (which is also a fat storage depot) is depleted, and it becomes red and jellylike rather than white and firm. In the early stages of a total fast, body protein is also metabolized at a relatively rapid rate. The preferred energy source for brain cells is glucose. If none is available in the diet, the cells will make it from amino acids. (This process, called gluconeogenesis, is discussed in Chapter 29.) Fasting humans get 8–10% of their energy from metabolism of protein. The nitrogenous metabolites from protein catabolism must be excreted through the kidneys. This requires large volumes of water. Starving people often die of dehydration.

After several weeks, the rate of protein breakdown slows considerably as the brain adjusts to using fatty acid metabolites for its energy source. When no more fat reserves remain, the body must again draw heavily on structural protein for its energy requirements. The emaciated appearance of a starving individual is due to depleted muscle protein.

Starvation is seldom the sole cause of death. Weakened by starvation, a person succumbs to disease. Even "minor" diseases such as chicken pox and measles become life-threatening disorders. Barring disease, starvation alone will lead to death from circulatory failure as the heart muscle becomes too weak to pump blood.

28.10 OBESITY

One of the anomalies of our time is that while much of the world suffers from hunger and malnutrition, the major dietary problem in industrialized or developed nations is **obesity.** We eat too much, and our bodies store the excess fat in our fat depots. In rare cases, obesity is due to glandular malfunctions. The overwhelming majority of obese people are that way, though, because they eat more food than their bodies use as fuel. Adults need food primarily for the energy it can supply. Amounts eaten in excess of immediate energy requirements are stored as fat.

Extra weight puts an extra load on the heart. Not only must the body move a greater bulk around, but the heart must supply blood to more tissue.

There are many diets available for reducing one's weight. And there are many best-selling books enriching authors who recognize the public's interest in weight reduction. All successful reducing diets, though—regardless of the gimmick—have one thing in common: food intake is reduced to less than the amount burned for the production of energy. If you eat more food than you need, you gain weight. If you eat less food than you need, you lose weight. And that's the way it is.

Obesity is a major factor in diseases of the heart and circulatory system. Too much food, whether taken in as fat, carbohydrate, or protein, leads to the storage of fat. And it is this fat that shortens our life spans, killing us slowly through increased stress on the heart and arteries.

28.11 CHOLESTEROL AND SOME OTHER STEROIDS

Steroids are lipids, but they are usually distinguished from the fats or phospholipids because of their unique structure (Chapter 21). Fats, for example, are esters, and they can be readily hydrolyzed. Steroids are nonsaponifiable lipids; they cannot be hydrolyzed.

Steroids have been the objects of intensive research, and much has been learned about their biosynthesis and their metabolism in general. Yet much is still to be learned. The precise mechanism by which cholesterol and other lipids are deposited on arterial walls to produce the condition called arteriosclerosis is still not known. Nor is it known exactly why diets rich in saturated fats produce high blood cholesterol levels.

What is known? The details of cholesterol biosynthesis have been worked out. The synthetic pathway is so long, however, that we shall simply list a few of the more important intermediates. These are given in Figure 28.11. We also know that bile acids and steroid hormones are synthesized from cholesterol. The details of these interconversions will not concern us here. An outline of the path of interconversion is given in Figure 28.12.

Steroids play so many interesting roles within the body (as sex hormones, for example, or as the main ingredients of gallstones) that they will certainly continue to be objects of chemical and medical research for many years.

28.12 LIPID STORAGE DISEASES

We have considered the metabolism of relatively simple lipids. The breakdown of far more complex lipids, particularly the compounds called glycolipids, will not be considered in detail here. We shall, however, take a brief look at the consequences of defects in these metabolic pathways.

There are several diseases that result from the faulty metabolism of complex lipids. The diseases are genetic in origin, caused by defective genes that produce defective enzymes. Because these enzymes are responsible for breaking down the lipids, their failure to work properly results in an abnormal accumulation of the compounds. The resulting disorders are referred to as **lipid**

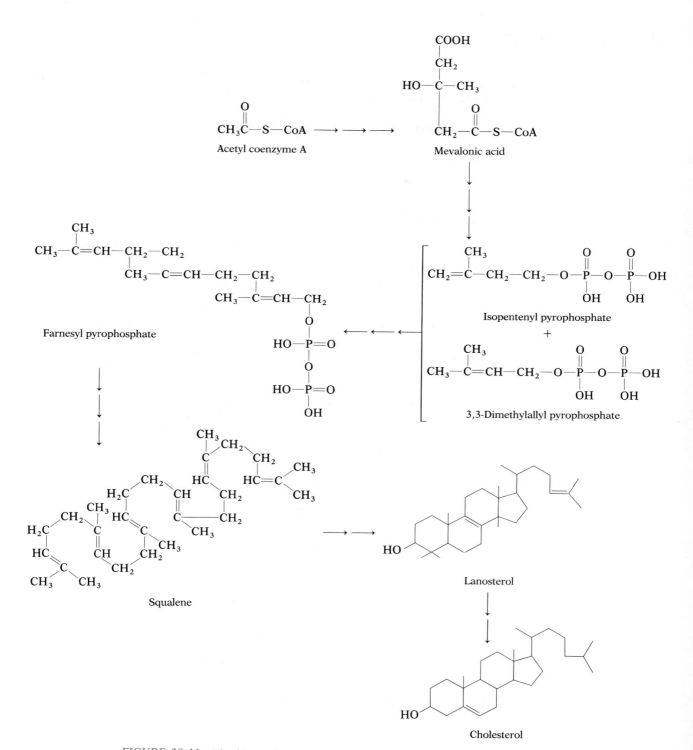

FIGURE 28.11 *The biosynthesis of cholesterol. The basic steroid skeleton can be constructed from acetyl coenzyme A.*

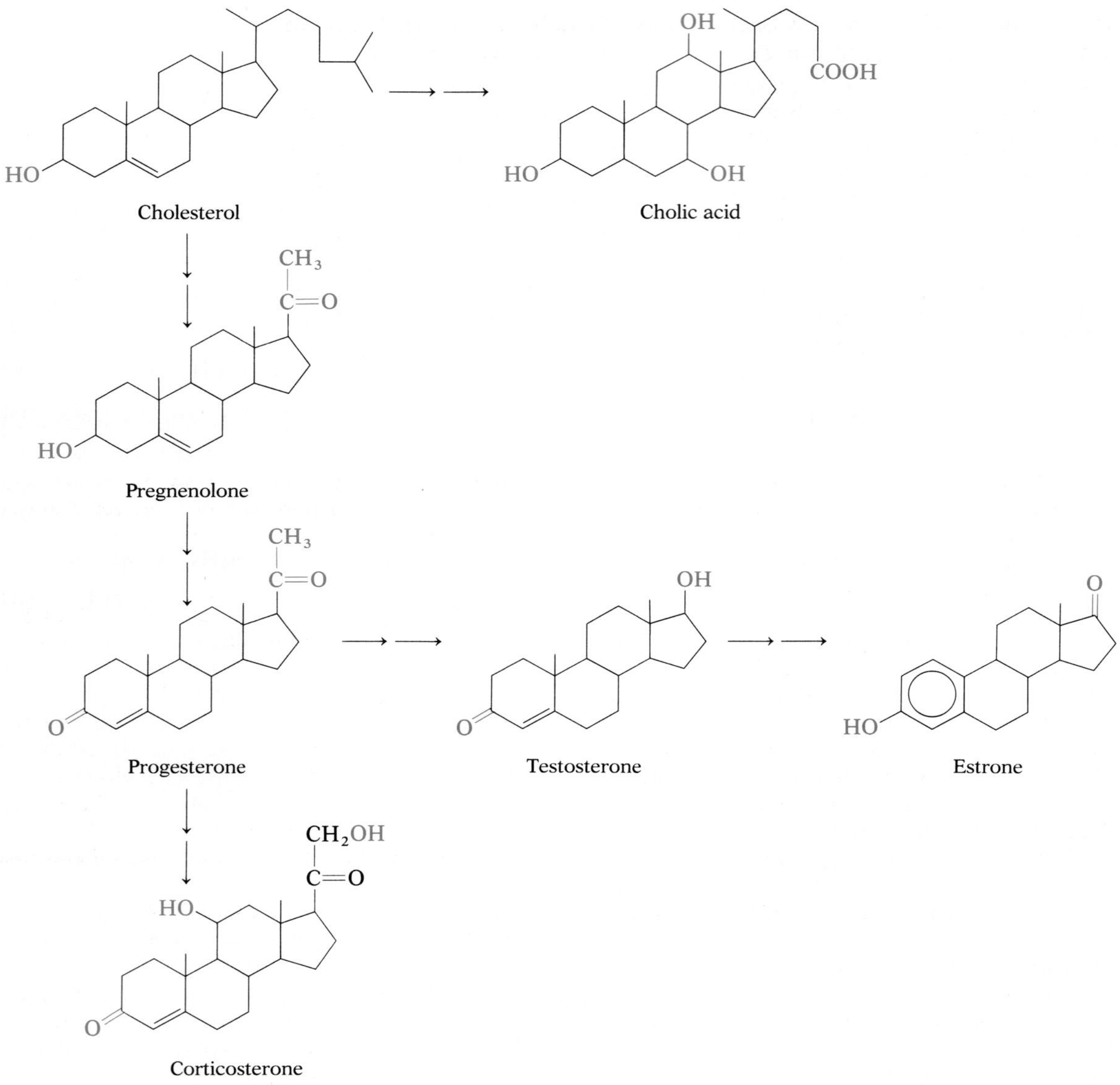

FIGURE 28.12 *The conversion of cholesterol to other biologically important steroids. Cholic acid is a bile acid. Progesterone is the pregnancy hormone. Corticosterone is an adrenal cortical steroid.*

storage diseases and include Tay-Sachs disease, Gaucher's disease, Niemann-Pick disease, and others. The genetic nature of several of these diseases is evident from the very restricted populations they affect. For example, Tay-Sachs disease is found almost exclusively among Jews of Eastern European extraction.

The glycolipids are components of brain and nerve tissue. The abnormal accumulation of these compounds frequently results in mental retardation and other brain disorders and in paralysis and other neurological problems.

The enzymes that are responsible for lipid storage diseases can be named. The juncture at which the metabolic pathways go awry can be pinpointed. Unfortunately, however, no cure for these diseases has yet been developed. At present, genetic counseling of prospective parents who carry the defective gene is the only approach to control of the diseases.

PROBLEMS

1. Define these terms:
   a. adipose tissue
   b. fat depot
   c. beta oxidation
   d. ketone bodies
   e. ketosis
   f. acidosis
2. How are lipoproteins classified?
3. What kind of molecules are transported by VLDL?
4. What do LDL transport? To where?
5. What do HDL transport? To where?
6. Which of the lipoproteins is associated with cardiovascular disease?
7. How might people increase their HDL levels?
8. If lipids are insoluble in water, how is it possible for them to be transported in the blood?
9. What is the end product of the fatty acid spiral?
10. What three roles does acetyl coenzyme A play in metabolism?
11. In what physical form are triglycerides stored in adipose tissue?
12. What types of reactions are involved in fatty acid metabolism?
13. What types of reactions are involved in fatty acid synthesis?
14. How many molecules of acetyl coenzyme A are produced in the metabolism of one molecule of palmitic acid (hexadecanoic acid)?
15. How many molecules of acetyl coenzyme A are produced in the metabolism of decanoic acid?
16. How many times does the spiral reaction sequence occur in the metabolism of a molecule of myristic acid (tetradecanoic acid)?

17. How many times does the spiral reaction sequence occur in the metabolism of a molecule of decanoic acid?
18. How and where is $FADH_2$ from the fatty acid spiral converted back to FAD?
19. What happens to the glycerol formed when mobilized fats are hydrolyzed?
20. Why do most naturally occurring fatty acids have an even number of carbon atoms?
21. What three-carbon group is condensed with the growing fatty acid chain during fatty acid synthesis?
22. Why does the chain length of a growing fatty acid increase by only two carbons, even though it is condensed with a three-carbon molecule?
23. What are three functions of fats (adipose tissue) in the body?
24. What is the principal cause of obesity?
25. What is the role of cytidine triphosphate in phospholipid synthesis?
26. What common steroid serves as a starting material for steroid hormone synthesis in the body?
27. Compare and contrast glycogen and adipose tissue as reserve energy sources.
28. Which pair of carbon atoms are added last in the biosynthesis of stearic acid, C-1 and C-2 or C-17 and C-18?
29. Why does a deficiency of carbohydrates in the diet lead to the formation of ketone bodies?
30. During a fast, what energy reserves are used first?

31. What energy reserves supply the major part of the body's needs during a fast?

32. During starvation, what is the final source of energy called on by the body?

33. Why does acidosis result from ketosis?

34. Why does starvation result in acidosis?

35. How does lack of insulin lead to acidosis?

36. Use equations to show how acetone and β-hydroxy-butyric acid are formed from acetoacetic acid.

37. Comment on the claim by some health faddists that "fasting cleanses the body."

38. How many molecules of $FADH_2$ are formed during the complete oxidation of one molecule of palmitic acid?

39. How many molecules of NADH are formed during the complete oxidation of one molecule of palmitic acid?

40. How many molecules of ATP are formed during the complete oxidation of one molecule of palmitic acid? (Refer to Problems 14, 38, and 39.)

41. How many molecules of ATP are formed during the complete oxidation of one molecule of decanoic acid? (Hint: Use your calculations in Problem 40 as a guide.)

42. What is the origin of lipid storage diseases?

43. What type of tissue is particularly susceptible to the effects of lipid storage diseases?

44. Name two lipid storage diseases.

Protein Metabolism

Consider green plants. If proper amounts of nutrients—nitrates, phosphates, sulfur compounds, and so on—are available, green plants can make all the amino acids they need. From these the plants put together all their necessary protein. We poor animals are not quite so versatile. We too can put together all the proteins we require, but only if we eat properly. There are eight amino acids that we need for protein synthesis but cannot make. These **essential amino acids** (Chapter 22) must be included in our diet. Perhaps if we turn green with envy . . . .

Proteins are the very stuff of our existence. They make up structural tissues such as muscles and tendons. The all-important enzymes are proteins. Some hormones are proteins. Our bodies must be able to make all of the myriad proteins that we require from those that we eat. The process of building large molecules such as proteins for growth, replacement, and repair is called **anabolism.** Proteins in excess of those needed for anabolism can be torn down and used for energy. The process of degrading proteins, fats, and carbohydrates is called **catabolism.**

Proteins—we build them, we destroy them. As long as all is in a beautiful dynamic balance, we are healthy. That's life.

29.1 THE AMINO ACID POOL

Carbohydrates can be stored in the liver and muscles as glycogen. Fats can be placed on reserve in the fat depots. For proteins, however, there are no comparable storage facilities. Proteins are digested in the small intestine, and the resulting amino acids are absorbed directly into the bloodstream. Here they join a circulating pool of amino acids.

Members of the **amino acid pool** are there only on temporary assignment. New members are constantly added to the pool, and old ones are regularly withdrawn. The body can supply the pool by synthesizing amino acids from scratch or by breaking down tissue protein to obtain the constituent amino acids. It drains the pool to obtain raw materials for new protein synthesis or for the synthesis of other nitrogen-containing compounds such as heme. It can also use the amino acids as a source of energy. The nitrogen stored in the pool cannot be recycled within the body endlessly. Each day some of the amino acids are catabolized, and the nitrogen is eliminated from the body as urea. This represents a net drain of material from the pool. It is to compensate for this drain that we require proteins in our diet.

Ingested proteins satisfy two needs. First, they provide nitrogen to replace that which has been eliminated from the body. Although the nitrogen comes in as part of specific amino acid residues, the body has a considerable capacity for shifting that nitrogen around (Section 29.4). In this respect, then, the ingested protein acts simply as a source of nitrogen.

In addition to serving as a general source of nitrogen, dietary protein must also supply some specific amino acids. These are the essential amino acids. The body cannot put together the carbon skeletons of these compounds. Yet amino acids with these skeletons are required for protein synthesis, among other things. By eating complete proteins (those that contain essential amino acid

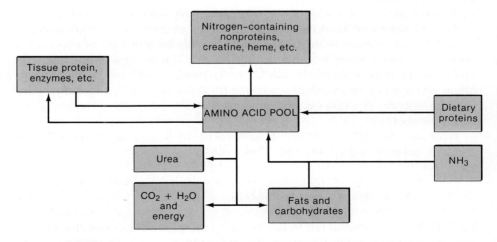

FIGURE 29.1   The amino acid pool and protein metabolism.

residues), we see to it that the amino acid pool is regularly resupplied with the amino acids we cannot synthesize ourselves.

The relationship between the amino acid pool and other aspects of protein metabolism is shown in Figure 29.1.

29.2 TURNOVER: NITROGEN BALANCE

Just as amino acids are only temporary participants in the amino acid pool, so, too, are specific protein molecules only temporary residents in the body. Our bodies continually take in proteins and break them down into their constituent amino acids. The liver and other body tissues take these amino acids and incorporate them into new proteins. Structural proteins, enzymes, and proteinlike hormones are continually being degraded and rebuilt. Our proteins are in a state of dynamic equilibrium as "old" ones are hydrolyzed and "new" ones are synthesized.

The amount of protein synthesized or degraded per unit time gives the **turnover rate.** The rate is different for proteins in different kinds of body tissue. **Turnover times** represent the average residence time for a protein molecule in the tissues. Turnover times are usually expressed as half-lives. A half-life is the time interval in which one-half of the protein molecules in a given tissue have been replaced. Liver proteins and those in the blood plasma have rapid turnovers, with half-lives of from 2 to 10 days. Muscle proteins have a more stable employment, with half-lives of about 6 months. Some collagen molecules may hang around for 3 years.

Protein synthesis is carried on continuously in the body. Its control by the genetic code and an outline of its mechanism are given in Chapter 23. Protein synthesis is limited by the availability of each of the eight essential amino acids. For example, if a given protein requires 4 units of phenylalanine per molecule and 40 are available, only 10 protein molecules can be made, even though there may be enough of all the other amino acids to make a million protein molecules. Within these limits, proteins are synthesized as needed by our bodies.

The state of the protein equilibrium within the body is usually evaluated by measurement of the amount of nitrogen being taken into and excreted from the body. As long as the amount of nitrogen taken in is equal to the amount of nitrogen excreted, a person is said to be in **nitrogen balance.** During growth or substantial repair of body tissues, more nitrogen is taken in than is excreted. Growing children and adults who are recovering from starvation or wasting illness are in such a state of **positive nitrogen balance.** Starvation and debilitating disease may result in a **negative nitrogen balance;** less nitrogen is ingested than is excreted. Diets lacking sufficient quantities of one or more of the essential amino acids also result in the excretion of more nitrogen than is taken in.

29.3 AMINO ACID METABOLISM

If amino acids are taken from the pool for something other than protein synthesis, then the elimination of the amino group is frequently the first step in the metabolic pathway. There is a set of reversible reactions that serves as a link between amino acid catabolism and synthesis. As they proceed in one direction, these reactions result in the removal of the amino group of an amino acid. The reverse reactions provide a means for attachment of an amino group to form a new amino acid. There are two reactions in the set. They are called, respectively, *oxidative deamination* and *transamination.* We shall consider first the catabolism of amino acids by way of these reactions.

Some of the proteins we eat are used for the production of energy. Protein eaten in excess of that needed for energy or for growth, replacement, and repair of tissue is converted to glycogen or to fat for storage. First the proteins are hydrolyzed to amino acids. These enter the amino acid pool. If they are to be burned for energy or processed for storage as fat or glycogen, the amino acids must be converted to intermediates of the Krebs tricarboxylic acid cycle or to pyruvic acid or acetyl coenzyme A. In all cases, the amino acid must first be stripped of its amino group.

Transamination is an exchange of functional groups between amino compounds and keto compounds. The α-amino group of any one of 11 amino acids—alanine, arginine, asparagine, aspartic acid, cysteine, isoleucine, lysine, phenylalanine, tryptophan, tyrosine, and valine—can be removed by transamination. The amino group is transferred to one of three α-keto compounds—pyruvic acid, α-ketoglutaric acid, or oxaloacetic acid. The following equation illustrates the general reaction:

$$\begin{array}{cccc} COOH & COOH & COOH & COOH \\ | & | & | & | \\ CH-NH_2 + & C=O & \longrightarrow & C=O & + CH-NH_2 \\ | & | & | & | \\ R & R' & R & R' \end{array}$$

The three possible receptor compounds are converted to alanine, glutamic acid, and aspartic acid, respectively. These reactions are catalyzed by three aminotransferase enzymes. The enzymes, named for the product amino acid, are alanine transaminase, glutamate transaminase, and aspartate transaminase.

Two of the product amino acids, alanine and aspartic acid, are ultimately converted to the third, glutamic acid, through transamination reactions.

$$\begin{array}{cccc} COOH & COOH & COOH & COOH \\ | & | & | & | \\ CH-NH_2 + & C=O & \longrightarrow & C=O & + CH-NH_2 \\ | & | & | & | \\ CH_3 & CH_2 & CH_3 & CH_2 \\ & | & & | \\ & CH_2 & & CH_2 \\ & | & & | \\ & COOH & & COOH \end{array}$$

$$
\begin{array}{ccc}
\underset{\substack{|\\ \text{CH}-\text{NH}_2 \\ |\\ \text{CH}_2 \\ |\\ \text{COOH}}}{\text{COOH}}
+
\underset{\substack{|\\ \text{C}=\text{O} \\ |\\ \text{CH}_2 \\ |\\ \text{CH}_2 \\ |\\ \text{COOH}}}{\text{COOH}}
\longrightarrow
\underset{\substack{|\\ \text{C}=\text{O} \\ |\\ \text{CH}_2 \\ |\\ \text{COOH}}}{\text{COOH}}
+
\underset{\substack{|\\ \text{CH}-\text{NH}_2 \\ |\\ \text{CH}_2 \\ |\\ \text{CH}_2 \\ |\\ \text{COOH}}}{\text{COOH}}
\end{array}
$$

These reactions are catalyzed by glutamic-pyruvic transaminase (GPT) and glutamic-oxaloacetic transaminase (GOT).

Note that whatever the pathway, the final receptor for the amino group from several different amino acids is α-ketoglutaric acid and the final product amino acid is glutamic acid.

The reverse of these reactions can occur. In this case, glutamic acid acts as the (almost) universal donor of its amino group to keto acids (Figure 29.2).

In **oxidative deamination,** the amino acid is converted to a keto acid, and the amino group is released as an ammonium ion. The ammonium ion is converted to urea (Section 29.5) and excreted.

Biochemists have isolated many enzymes that will catalyze the oxidative deamination of a variety of amino acids. However, the only compound for which this is a significant metabolic route is glutamic acid.

$$
\underset{\text{Glutamic acid}}{\underset{\substack{|\\ \text{CH}-\text{NH}_2 \\ |\\ \text{CH}_2 \\ |\\ \text{CH}_2 \\ |\\ \text{COOH}}}{\text{COOH}}}
+ \text{NAD}^+ + \text{H}_2\text{O} \longrightarrow
\underset{\alpha\text{-Ketoglutaric acid}}{\underset{\substack{|\\ \text{C}=\text{O} \\ |\\ \text{CH}_2 \\ |\\ \text{CH}_2 \\ |\\ \text{COOH}}}{\text{COOH}}}
+ \text{NADH} + \text{NH}_4{}^+
$$

Through a combination of transamination and oxidative deamination, the amino group of amino acids is converted to ammonium ions. Figure 29.3 shows that glutamic acid merely serves as an intermediate in this process. The nitrogen released by glutamic acid as ammonia can originate in any amino acid.

If the amino acid that transfers its amino group to α-ketoglutaric acid is alanine, the keto acid product of the transamination is pyruvic acid.

$$
\underset{\text{Alanine}}{\underset{\substack{|\\ \text{CH}-\text{NH}_2 \\ |\\ \text{CH}_3}}{\text{COOH}}}
\xrightarrow{\text{transamination}}
\underset{\text{Pyruvic acid}}{\underset{\substack{|\\ \text{C}=\text{O} \\ |\\ \text{CH}_3}}{\text{COOH}}}
$$

Pyruvic acid can be converted to acetyl coenzyme A. The acetyl coenzyme A can be degraded through the tricarboxylic acid cycle to produce energy or can be used for fatty acid synthesis.

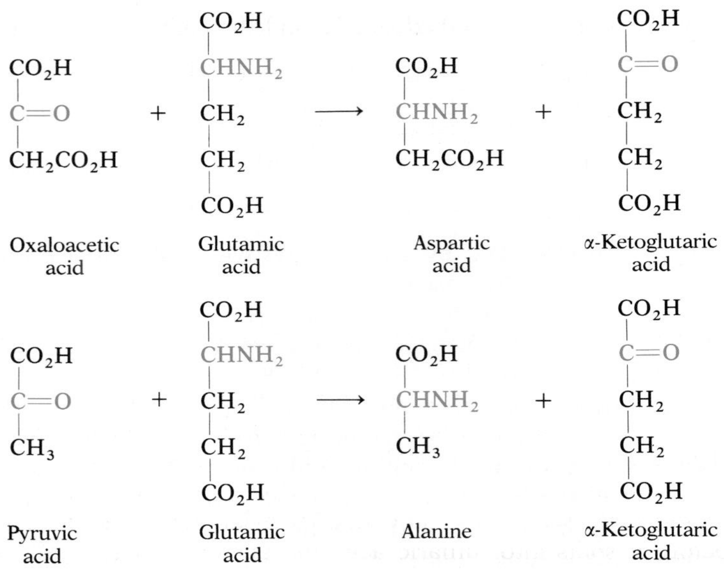

$$
\begin{array}{ccccccc}
\begin{array}{c}
\text{CO}_2\text{H} \\
| \\
\text{C}{=}\text{O} \\
| \\
\text{CH}_2\text{CO}_2\text{H}
\end{array}
& + &
\begin{array}{c}
\text{CO}_2\text{H} \\
| \\
\text{CHNH}_2 \\
| \\
\text{CH}_2 \\
| \\
\text{CH}_2 \\
| \\
\text{CO}_2\text{H}
\end{array}
& \longrightarrow &
\begin{array}{c}
\text{CO}_2\text{H} \\
| \\
\text{CHNH}_2 \\
| \\
\text{CH}_2\text{CO}_2\text{H}
\end{array}
& + &
\begin{array}{c}
\text{CO}_2\text{H} \\
| \\
\text{C}{=}\text{O} \\
| \\
\text{CH}_2 \\
| \\
\text{CH}_2 \\
| \\
\text{CO}_2\text{H}
\end{array}
\\[2mm]
\text{Oxaloacetic} & & \text{Glutamic} & & \text{Aspartic} & & \text{α-Ketoglutaric} \\
\text{acid} & & \text{acid} & & \text{acid} & & \text{acid}
\end{array}
$$

$$
\begin{array}{ccccccc}
\begin{array}{c}
\text{CO}_2\text{H} \\
| \\
\text{C}{=}\text{O} \\
| \\
\text{CH}_3
\end{array}
& + &
\begin{array}{c}
\text{CO}_2\text{H} \\
| \\
\text{CHNH}_2 \\
| \\
\text{CH}_2 \\
| \\
\text{CH}_2 \\
| \\
\text{CO}_2\text{H}
\end{array}
& \longrightarrow &
\begin{array}{c}
\text{CO}_2\text{H} \\
| \\
\text{CHNH}_2 \\
| \\
\text{CH}_3
\end{array}
& + &
\begin{array}{c}
\text{CO}_2\text{H} \\
| \\
\text{C}{=}\text{O} \\
| \\
\text{CH}_2 \\
| \\
\text{CH}_2 \\
| \\
\text{CO}_2\text{H}
\end{array}
\\[2mm]
\text{Pyruvic} & & \text{Glutamic} & & \text{Alanine} & & \text{α-Ketoglutaric} \\
\text{acid} & & \text{acid} & & & & \text{acid}
\end{array}
$$

FIGURE 29.2 *Glutamic acid serves as a "universal" donor of amino groups to α-keto acids.*

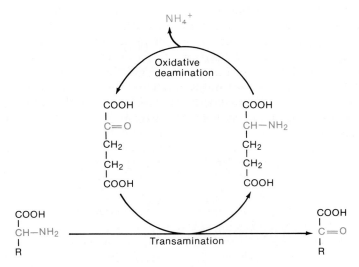

FIGURE 29.3 *α-Ketoglutaric acid and glutamic acid as intermediates in the deamination of amino acids.*

Aspartic acid is converted to oxaloacetic acid through transamination.

$$
\begin{array}{ccc}
\text{COOH} & & \text{COOH} \\
| & & | \\
\text{CH—NH}_2 & \xrightarrow{\text{transamination}} & \text{C}=\text{O} \\
| & & | \\
\text{CH}_2 & & \text{CH}_2 \\
| & & | \\
\text{COOH} & & \text{COOH} \\
\text{Aspartic acid} & & \text{Oxaloacetic acid}
\end{array}
$$

Oxaloacetic acid is a tricarboxylic acid cycle intermediate and, as such, represents a raw material for the synthesis of glucose or glycogen.

All of the amino acids can make their way from the amino acid pool to the Krebs tricarboxylic acid cycle. Each individual amino acid has a unique pathway. Most of the amino acids lose their amino groups through transamination. Each α-keto acid thus formed then follows its own special pathway. Some of these pathways are quite devious. Phenylalanine undergoes a six-step reaction before it splits into fumaric acid and acetoacetic acid. The fumaric acid is an intermediate in the tricarboxylic acid cycle, but acetoacetic acid must be converted to acetoacetyl coenzyme A and then to acetyl coenzyme A before it enters the cycle.

Eventually the 20 or more individual pathways leading from the amino acid pool converge into five routes that join the tricarboxylic acid cycle. The various paths are summarized in Figure 29.4.

If carbohydrates are lacking in the diet (through voluntary restriction or starvation) or if glucose is not getting into cells (as in diabetes), the body responds by converting proteins to carbohydrates through gluconeogenesis.

Gluconeogenesis is the synthesis of glucose from glycerol (obtained from the hydrolysis of fats) or from some amino acids. The pathway is diagrammed in Figure 29.5. The amino acids may come from the diet. In case of fasting or dieting, they come from the breakdown of proteins in muscle tissue. The amino acids are converted by way of pyruvic acid or oxaloacetic acid. The net reaction for gluconeogenesis is

$$
\text{2 pyruvic acid} + \text{4 ATP} + \text{2 GTP} + \text{2 NADH} + \text{6 H}_2\text{O} \longrightarrow
$$
$$
\text{glucose} + \text{4 ADP} + \text{2 GDP} + \text{6 HPO}_4^- + \text{2 NAD}^+
$$

Note that the process requires six high-energy phosphate molecules for each glucose molecule formed. Gluconeogenesis can occur only at the expense of ATP from other energy-producing reactions.

Gluconeogenesis satisfies some needs. The brain gets its preferred fuel, glucose, this way when carbohydrate is lacking in the diet. Without gluconeogenesis, we would be at risk of brain death when deprived of food even for one day. Gluconeogenesis may also cause some problems. For example, the

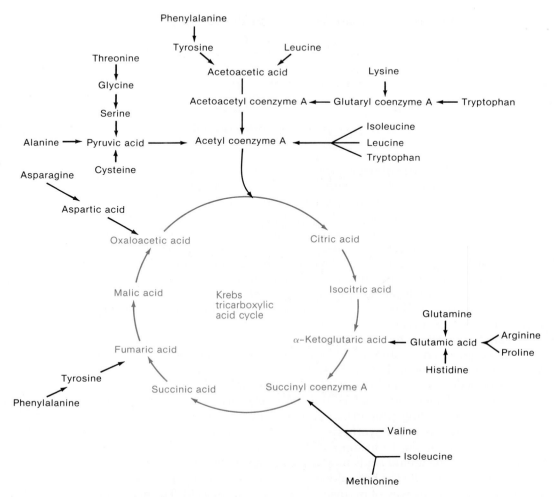

FIGURE 29.4 *Pathways of amino acid catabolism. All paths eventually lead to the Krebs cycle.*

buildup of keto acids that occurs as proteins are catabolized and converted to carbohydrates can result in ketosis or acidosis (Chapter 28).

Since amino acids can also be converted to fatty acids (through acetyl coenzyme A), proteins can make you fat—if you eat too many of them.

The amino acid pool can supply you with almost anything—proteins, fats, carbohydrates, energy, and even a tan (the coloring matter in your skin starts out as phenylalanine, an amino acid).

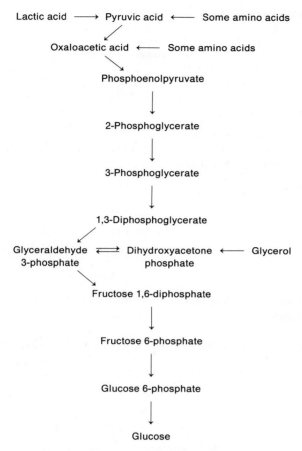

FIGURE 29.5 *The gluconeogenesis pathway.*

29.4 AMINO ACID SYNTHESIS

Dietary proteins supply amino acids to the amino acid pool. Catabolism of tissue protein also provides amino acids for the pool. One other way to get amino acids is to synthesize them. The liver is particularly good at that.

Transamination provides a route from available α-keto acids to amino acids. In Section 29.3, we considered the synthesis of alanine and aspartic acid in this manner (see Figure 29.2). Recall that glutamic acid serves as a source of amino groups for the synthesis of other amino acids.

$$
\begin{array}{c}
\underset{\substack{R}}{\overset{\substack{COOH \\ | \\ C=O \\ |}}{}} \; + \; \underset{\substack{| \\ CH_2 \\ | \\ CH_2 \\ | \\ COOH}}{\overset{\substack{COOH \\ | \\ CH-NH_2}}{}} \; \rightleftharpoons \; \underset{\substack{R}}{\overset{\substack{COOH \\ | \\ CH-NH_2 \\ | }}{}} \; + \; \underset{\substack{| \\ CH_2 \\ | \\ CH_2 \\ | \\ COOH}}{\overset{\substack{COOH \\ | \\ C=O}}{}}
\end{array}
$$

An α-keto acid   Glutamic acid   An amino acid   α-Ketoglutaric acid

The α-ketoglutaric acid product can pick up ammonia through a reversal of the oxidative deamination reaction.

$$
\underset{\substack{| \\ COOH}}{\overset{\substack{COOH \\ | \\ C=O \\ | \\ CH_2 \\ | \\ CH_2}}{}} + NADH + H^+ + NH_3 \rightleftharpoons \underset{\substack{| \\ COOH}}{\overset{\substack{COOH \\ | \\ CH-NH_2 \\ | \\ CH_2 \\ | \\ CH_2}}{}} + NAD^+ + H_2O
$$

The two reactions combined incorporate ammonia into a variety of amino acids, with glutamic acid acting as intermediate. This two-step process is summarized in Figure 29.6.

The body can't synthesize the essential amino acids, because it cannot make the corresponding α-keto acids. In contrast, keto acids such as pyruvic acid,

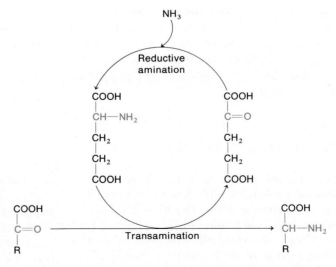

FIGURE 29.6 *Glutamic acid and α-ketoglutaric acid as intermediates in the synthesis of amino acids. Compare with Figure 29.3.*

oxaloacetic acid, or α-ketoglutaric acid are available from carbohydrate metabolism. Thus, the corresponding amino acids, alanine, aspartic acid, and glutamic acid, are readily synthesized within the body.

The transamination reaction also permits the body to adjust the composition of the amino acid pool to fit its needs. The nitrogen from an amino acid present in excess of need can transfer to a keto acid to provide more of an amino acid in low supply. All of this can only be done within the limitations imposed by the availability of the keto acid with the proper carbon skeleton.

Several nonessential amino acids are made by processes other than transamination. Serine and glycine are made from phosphoglyceric acid. Cysteine is made from methionine, an essential amino acid. Glutamine and proline are synthesized from glutamic acid. The reaction pathways for these amino acids are somewhat complicated and need not concern us here.

There is a relatively simple conversion that is of considerable importance to the health of some individuals. Tyrosine is made by the hydroxylation of the essential amino acid phenylalanine.

$$\text{C}_6\text{H}_5\text{—CH}_2\text{CH—COOH} + \text{NADPH} + \text{H}^+ + \text{O}_2 \xrightarrow{\text{phenylalanine hydroxylase}}$$
$$\overset{|}{\text{NH}_2}$$

$$\text{HO—C}_6\text{H}_4\text{—CH}_2\text{CH—COOH} + \text{NADP}^+ + \text{H}_2\text{O}$$
$$\overset{|}{\text{NH}_2}$$

A lack of phenylalanine hydroxylase, the enzyme that catalyzes the conversion of phenylalanine to tyrosine, results in a condition known as *phenylketonuria* (PKU). This disease was one of the first of the genetic disorders to be thoroughly investigated from the standpoint of biochemistry. If the enzyme is lacking, the phenylalanine is metabolized via an alternate route; it enters into a transamination reaction.

$$\text{C}_6\text{H}_5\text{—CH}_2\text{—CH—COOH} \xrightarrow{\text{transamination}} \text{C}_6\text{H}_5\text{—CH}_2\text{—C—COOH}$$
$$\overset{|}{\text{NH}_2} \qquad\qquad\qquad\qquad\qquad \overset{\|}{\text{O}}$$

Phenylpyruvic acid

The resulting keto acid, phenylpyruvic acid, accumulates in the blood and is eliminated in the urine. (*Phenylketonuria* refers to the appearance of a phenyl ketone in the urine.) Clinically, the disease is characterized by severe mental retardation. This is one case where chemical research has had very practical benefits. Many hospitals now routinely screen newborn infants for phenylketonuria. If affected infants are provided with diets low in phenylalanine, mental retardation can be prevented.

Another inborn error in the metabolism of tyrosine leads to albinism. Tyrosine serves as a precursor for the melanins, the pigments that color the skin, hair, and eyes. A defective enzyme prevents the occurrence of one of the reactions necessary for this conversion. The lack of pigmentation characteristic of albinism is the result.

Before we leave the subject of amino acid synthesis, let's take a closer look at a particularly important transamination reaction. Glutamic acid and oxaloacetic acid react to form α-ketoglutaric acid and aspartic acid (Section 29.3), in a reaction catalyzed by an enzyme called GOT. The heart muscle is particularly rich in GOT. A sharp rise in the concentration of GOT in the blood (where it is referred to as serum GOT or SGOT) is an indication of a myocardial infarction. This condition is caused by a clot in a coronary artery that interferes with the blood supply to the heart muscle. The resulting deterioration of heart muscle is accompanied by the release of GOT into the blood. The situation here is similar to that mentioned in Chapter 27 with respect to the enzyme creatine kinase (CK). CK is involved in energy transformations associated with muscle contraction. An increase in the serum concentration of CK may also accompany the heart muscle deterioration associated with a myocardial infarction.

The synthesis of aspartic acid through this transamination reaction is very important to animals who excrete nitrogen in the form of urea. Aspartic acid is needed for the synthesis of urea. We shall consider this synthesis in detail in the next section.

29.5 THE UREA CYCLE

The Krebs cycle was named for Hans Krebs, its discoverer. Krebs also worked out the **urea cycle,** or, as it is sometimes called, the **ornithine cycle.** The urea cycle describes the reactions by which the nitrogen released in amino acid metabolism is converted to the compound urea. These reactions occur in the liver. The *net* reaction of the cycle involves the synthesis of urea from ammonia and carbon dioxide.

$$2\,NH_3 + CO_2 \longrightarrow H_2N-\overset{\displaystyle O}{\overset{\displaystyle \|}{C}}-NH_2 + H_2O$$

The stepwise process of the urea cycle is summarized in Figure 29.7. The ammonia molecules shown in the net reaction enter the cycle in disguise, using two routes. One route brings the ammonia in as the amino group of an amino acid. The second route requires the input of a carbon dioxide molecule as well as an ammonia molecule. The ammonia is provided through the oxidative deamination of amino acids (Section 29.3). Let's focus on the second route.

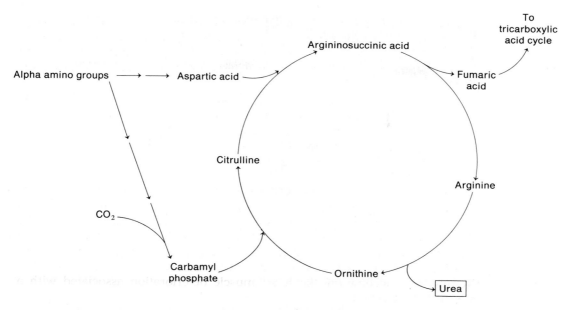

FIGURE 29.7 *The urea cycle.*

Carbon dioxide combines with ammonia in the presence of ATP to form carbamyl phosphate.

$$NH_3 + CO_2 + 2\,ATP \longrightarrow H_2N-\overset{\overset{\displaystyle O}{\|}}{C}-O-\overset{\overset{\displaystyle O^-}{\|}}{\underset{\underset{\displaystyle O}{\|}}{P}}-O^- + 2\,ADP + HPO_4{}^{2-}$$

Carbamyl phosphate

Carbamyl phosphate is a high-energy compound. It reacts with a basic amino acid named ornithine to form citrulline.

$$H_2N-\overset{\overset{\displaystyle O}{\|}}{C}-O-\overset{\overset{\displaystyle O^-}{\|}}{\underset{\underset{\displaystyle O}{\|}}{P}}-O^- + H_2N-(CH_2)_3-\underset{\underset{\displaystyle NH_2}{|}}{CH}-COOH \longrightarrow$$

Ornithine

$$H_2N-\overset{\overset{\displaystyle O}{\|}}{C}-NH-(CH_2)_3-\underset{\underset{\displaystyle NH_2}{|}}{CH}-COOH + HPO_4{}^{2-}$$

Citrulline

The citrulline molecule next reacts with aspartic acid. Aspartic acid represents the second entry route to the cycle. The amino groups of other alpha amino acids can be transferred to aspartic acid by a series of transamination reactions (Section 29.4). The aspartic acid and the previously formed citrulline combine to produce argininosuccinic acid.

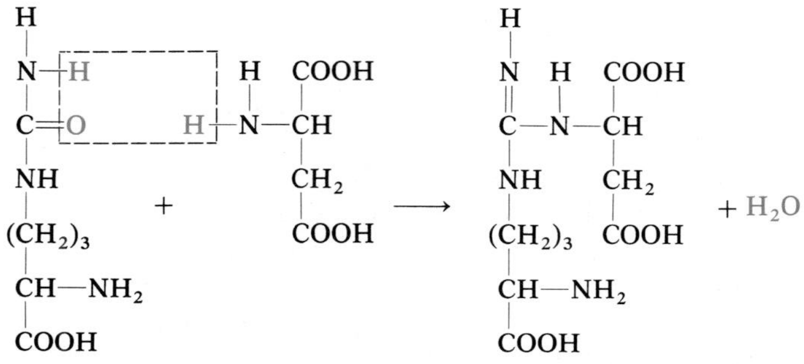

Citrulline   Argininosuccinic acid

The carbon and both nitrogens required for the formation of urea have now entered the cycle.

The argininosuccinic acid splits out fumaric acid, leaving arginine.

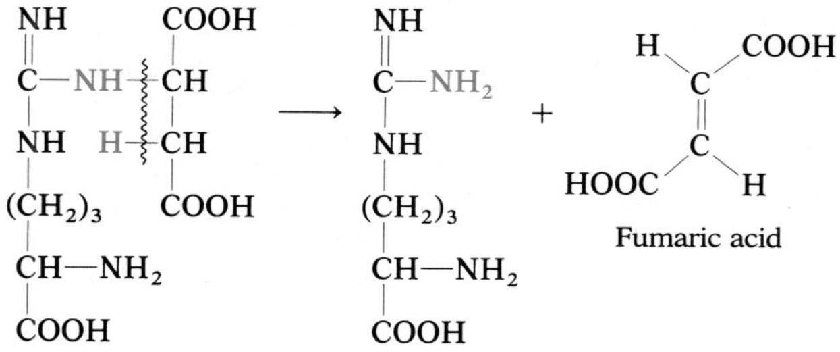

Arginine   Fumaric acid

In essence, the fumaric acid is what's left of the aspartic acid after it completes its delivery of an ammonia molecule to the cycle. The fumaric acid enters Krebs's tricarboxylic acid cycle. Arginine is hydrolyzed to urea and ornithine.

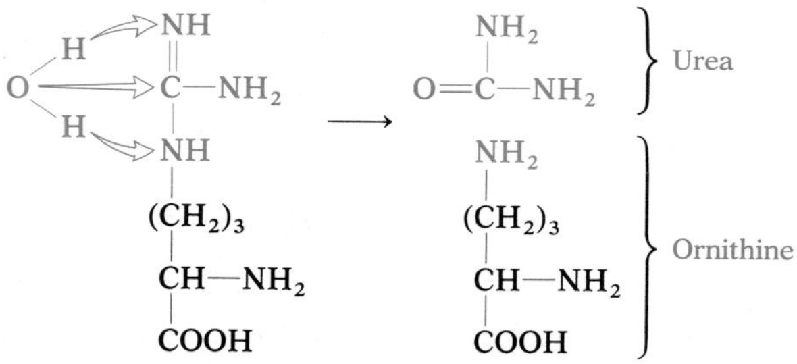

Urea

Ornithine

The urea is transported to the kidneys and excreted in urine (Chapter 26). Ornithine starts another turn of the cycle by reacting with a fresh molecule of carbamyl phosphate. In human beings, over 80% of the nitrogen from protein catabolism is excreted as urea.

29.6 NUCLEOPROTEIN METABOLISM

Nucleoproteins are almost inevitable constituents of our diets. Both animal and plant food sources are cellular in nature. The cells contain nuclei; nuclei contain nucleoprotein. During digestion, the protein portion is removed from the nucleic acid part of the nucleoprotein. The protein is hydrolyzed to its constituent amino acids, and these are metabolized as described in previous sections. The nucleic acids are also hydrolyzed during digestion. They are broken down first to nucleotides by enzymes called nucleases. The nucleotides are further cleaved by nucleotidases to nucleosides and phosphates. The nucleosides are absorbed and then split by nucleosidases in the tissues, forming pentose sugars and purine and pyrimidine bases. These reactions are summarized in Figure 29.8.

The sugars—deoxyribose and ribose—are metabolized by the usual carbohydrate pathways (Chapter 27). Phosphates can be used in the formation of new phosphate compounds, or they may be excreted in the urine. The purines and pyrimidines are metabolized by different pathways. Let's look at purine metabolism first.

Adenine is first converted to hypoxanthine. This involves the substitution of a hydroxyl group for adenine's amino group (Figure 29.9). The hypoxanthine is again hydroxylated on the six-membered ring to give xanthine. That compound is then hydroxylated on the five-membered ring to give the trihydroxy compound uric acid.

Guanine is converted directly to xanthine by replacement of its amino group by a hydroxyl group, and the xanthine is then converted to uric acid. Each of these steps is catalyzed by a specific enzyme. Uric acid is generally excreted in the urine.

The painful metabolic disorder called **gout** results from the deposition of uric acid salts in cartilage. It is far more common in men than in women. Uric acid salts are also deposited in the form of kidney and bladder stones.

$$\text{Nucleic acids} \xrightarrow{\text{nucleases}} \text{Nucleotides} \xrightarrow{\text{nucleotidases}} \text{Nucleosides} + HPO_4^{2-}$$

$$\text{Purine bases} + \text{Pentose sugars} + \text{Pyrimidine bases} \xleftarrow{\text{nucleosidases}}$$

FIGURE 29.8 *The hydrolysis of nucleic acids. For structures, see Chapter 23.*

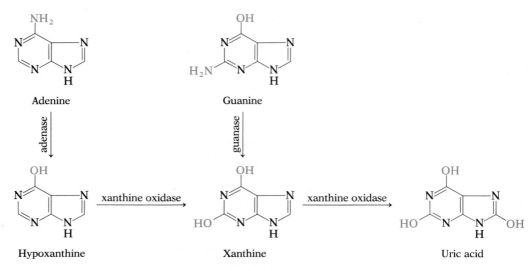

FIGURE 29.9 *The conversion of purine bases to uric acid.*

It is of interest that mammals other than humans and apes convert uric acid into allantoin.

Allantoin is more soluble in water than uric acid and is more readily excreted. Birds and terrestrial reptiles excrete large quantities of uric acid. They make it not only from purines but from the nitrogen in amino acids. Excretion of protein nitrogen in the form of semisolid uric acid suspensions helps these organisms conserve vital water. The excreta of some snakes is nearly pure uric acid.

Whereas purines wind up as uric acid, pyrimidine bases are converted to products that can be oxidized to carbon dioxide, water, and urea. Cytosine and uracil, through a series of enzyme-catalyzed reactions, wind up as β-alanine. Thymine is converted to β-aminoisobutyric acid. These conversions are summarized in Figure 29.10.

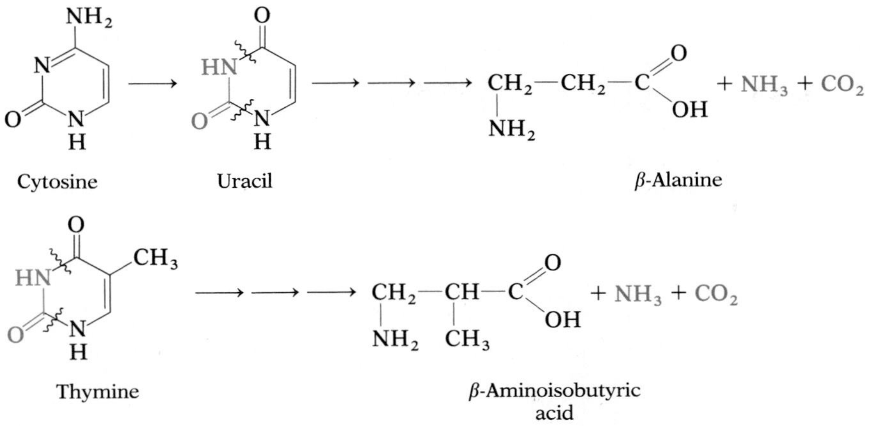

FIGURE 29.10   *The metabolism of pyrimidine bases.*

Nucleic acids are not necessary components of our diets, even though they are nearly always present. Our bodies can synthesize all the component parts of nucleic acids, including the purine and pyrimidine bases. These bases can be made from amino acids and other metabolites. The process is quite complicated, and we will not consider it here.

29.7 A SUMMARY OF METABOLISM

At several places in this chapter and Chapters 27 and 28, we have mentioned how one metabolic pathway is connected to another. Figure 29.11 summarizes the interconnections of carbohydrate, lipid, and protein metabolism. Note that energy can be obtained from any of the three classes of compounds. Note also that the classes can be interconverted by way of common intermediate metabolites.

And there you have it. Chemistry from protons to proteins. Carbohydrate, lipid, and protein metabolism are all tied together. Indeed, in a living organism, everything is connected to everything else. Life is one huge, complicated set of chemical reactions—and, of course, a lot more. The whole of life is certainly much more than the sum of a set of chemical reactions.

We hope that we have enriched your life by helping you to learn something of the basis of chemistry and of the many ways chemistry touches your life every day. And we wish for you the proper reward for the many hours you have spent studying chemistry: the joy of success in your chosen profession.

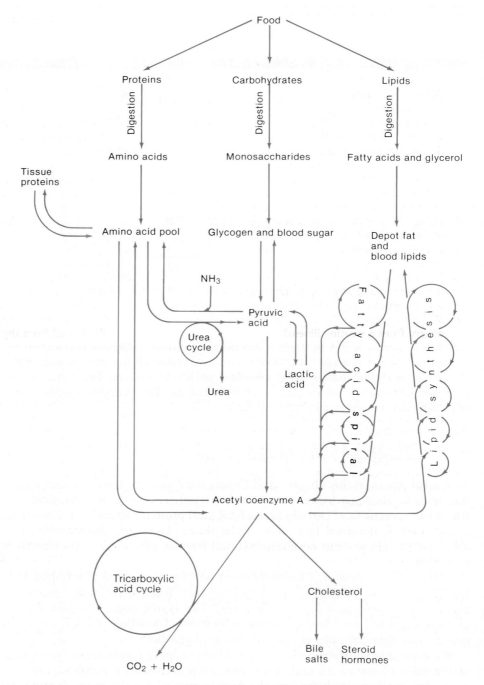

FIGURE 29.11 The interconnections of carbohydrate, lipid, and protein metabolism.

I

The International System of Measurement

Measurement is discussed in Chapter 1. Conversions within a system of measurement and between systems are discussed in Special Topic A. Further discussion and additional tables are provided here.

The standard unit of length in the International System of Measurement is the **meter.** This distance was once meant to be 0.000 000 1 of the Earth's quadrant, that is, of the distance from the North Pole to the equator measured along a meridian. The quadrant was difficult to measure accurately. Consequently, for many years the meter was defined as the distance between two etched lines on a metal bar (made of a platinum-iridium alloy) kept in the International Bureau of Weights and Measures at Sèvres, France. Today, the meter is defined even more precisely as being the distance that light travels in a vacuum during 1/299 792 458 of a second.

The primary unit of mass is the **kilogram** (1 kg = 1000 g). It is based on a standard platinum-iridium bar kept at the International Bureau of Weights and Measures. The **gram** is a more convenient unit for many chemical operations.

The basic SI unit of volume is the cubic meter. The unit more frequently employed in chemistry, however, is the cubic decimeter, which is often called a liter.

$$1 \, dm^3 = 1 \, L = 0.001 \, m^3$$

All other SI units of length, mass, and volume are derived from these basic units.

TABLE I.1   Some SI Prefixes and Their Relationship to the Basic Units

Prefix	Abbreviation	Connotation
Pico-	p	$0.000\ 000\ 000\ 001 \times$ (or $10^{-12} \times$)
Nano-	n	$0.000\ 000\ 001 \times$ (or $10^{-9} \times$)
Micro-	μ	$0.000\ 001 \times$ (or $10^{-6} \times$)
Milli-	m	$0.001 \times$ (or $10^{-3} \times$)
Centi-	c	$0.01 \times$ (or $10^{-2} \times$)
Deci-	d	$0.1 \times$ (or $10^{-1} \times$)
Deka-	da	$10 \times$ (or $10^{1} \times$)
Hecto-	h	$100 \times$ (or $10^{2} \times$)
Kilo-	k	$1\ 000 \times$ (or $10^{3} \times$)
Mega-	M	$1\ 000\ 000 \times$ (or $10^{6} \times$)
Giga-	G	$1\ 000\ 000\ 000 \times$ (or $10^{9} \times$)
Tera-	T	$1\ 000\ 000\ 000\ 000 \times$ (or $10^{12} \times$)

TABLE I.2   Some Metric Units of Length

1 kilometer (km) = 1000 meters (m)
1 meter (m) = 100 centimeters (cm)
1 centimeter (cm) = 10 millimeters (mm)
1 millimeter (mm) = 1000 micrometers (μm)

TABLE I.3   Some Metric Units of Mass

1 kilogram (kg) = 1000 grams (g)
1 gram (g) = 1000 milligrams (mg)
1 milligram (mg) = 1000 micrograms (μg)

TABLE I.4   Some Metric Units of Volume

1 liter (L) = 1000 milliliters (mL)
1 milliliter (mL) = 1000 microliters (μL)
1 milliliter (mL) = 1 cubic centimeter (cm^3)

TABLE I.5 *Some Common Metric Conversions*

Length
1 mile (mi) = 1.61 kilometers (km) 1 yard (yd) = 0.914 meter (m) 1 inch (in.) = 2.54 centimeters (cm)
Mass
1 pound (lb) = 454 grams (g) 1 ounce (oz) = 28.4 grams (g) 1 pound (lb) = 0.454 kilogram (kg) 1 grain (gr) = 0.0648 gram (g) 1 carat (car) = 200 milligrams (mg)
Volume
1 U.S. quart (qt) = 0.946 liter (L) 1 U.S. pint (pt) = 0.473 liter (L) 1 fluid ounce (fl oz) = 29.6 milliliters (mL) 1 gallon (gal) = 3.78 liters (L)

TABLE I.6 *Some Conversion Units for Pressure*

1 millimeter of mercury (mm Hg) = 1 torr 1 atmosphere (atm) = 760 millimeters of mercury (mm Hg) = 760 torr 1 atmosphere (atm) = 29.9 inches of mercury (in. Hg) = 14.7 pounds per square inch (lb/in.^{2}) = 101 kPa

TABLE I.7 *Some Temperature Equivalents**

Phenomenon	Fahrenheit	Celsius
Absolute zero	−459.69 °F	−273.16 °C
Nitrogen boils/liquefies	−320.4 °F	−195.8 °C
Carbon dioxide solidifies/sublimes	−109.3 °F	−78.5 °C
Bitter cold night, northern Minnesota	−40 °F	−40 °C
Cold night, Indiana	0 °F	−18 °C
Water freezes/ice melts	32 °F	0 °C
Pleasant room temperature	72 °F	22 °C
Body temperature	98.6 °F	37.0 °C
Very hot day	100 °F	38 °C
Water boils/steam condenses	212 °F	100 °C
Temperature for baking biscuits	450 °F	232 °C

* Keep these equations in mind:

$$°F = \frac{9}{5}(°C) + 32$$

$$°C = (°F - 32)\frac{5}{9}$$

TABLE I.8   *Some Conversion Units for Energy*

$$
\begin{aligned}
1 \text{ calorie (cal)} &= 4.184 \text{ joules (J)} \\
1 \text{ British thermal unit (Btu)} &= 1053 \text{ joules (J)} \\
&= 252 \text{ calories (cal)} \\
1 \text{ food ``Calorie''} &= 1 \text{ kilocalorie (kcal)} \\
&= 1000 \text{ calories (cal)} \\
&= 4184 \text{ joules (J)}
\end{aligned}
$$

II

Exponential Notation

S cientists often use numbers that are so large—or so small—that they boggle the mind. For example, light travels at 30 000 000 000 cm/s. There are 602 300 000 000 000 000 000 000 carbon atoms in 12 g of carbon. On the small side, the diameter of an atom is about 0.000 000 000 1 m. The diameter of an atomic nucleus is about 0.000 000 000 000 001 m.

It is obviously difficult to keep track of the zeros in such quantities. Scientists find it convenient to express such numbers as **powers of ten.** Tables II.1 and II.2 contain partial lists of such numbers.

The speed of light is usually expressed as 3×10^{10} (i.e., $3 \times 10 \times 10 \times 10 \times 10 \times 10 \times 10 \times 10 \times 10 \times 10$) cm/s. The mass of an atom of cesium (Cs) is expressed as 2.21×10^{-22} g, that is, as

$$2.21 \times \frac{1}{10\ 000\ 000\ 000\ 000\ 000\ 000\ 000}\ \text{g}$$

TABLE II.1 *Positive Powers of Ten*

$$10^{0} = 1$$
$$10^{1} = 10$$
$$10^{2} = 10 \times 10 = 100$$
$$10^{3} = 10 \times 10 \times 10 = 1000$$
$$10^{4} = 10 \times 10 \times 10 \times 10 = 10\ 000$$
$$10^{5} = 10 \times 10 \times 10 \times 10 \times 10 = 100\ 000$$
$$10^{6} = 10 \times 10 \times 10 \times 10 \times 10 \times 10 = 1\ 000\ 000$$
$$\vdots$$
$$10^{23} = 100\ 000\ 000\ 000\ 000\ 000\ 000\ 000$$

TABLE II.2   Negative Powers of Ten

$$10^{-1} = 1/10 = 0.1$$
$$10^{-2} = 1/100 = 0.01$$
$$10^{-3} = 1/1000 = 0.001$$
$$10^{-4} = 1/10\ 000 = 0.000\ 1$$
$$10^{-5} = 1/100\ 000 = 0.000\ 01$$
$$10^{-6} = 1/1\ 000\ 000 = 0.000\ 001$$
$$|$$
$$10^{-13} = 1/10\ 000\ 000\ 000\ 000 = 0.000\ 000\ 000\ 000\ 1$$

Numbers such as 10^6 are called exponential numbers, where 10 is the **base** and 6 is the **exponent.** Numbers in the form 6.02×10^{23} are said to be written in **scientific notation.**

Exponential numbers are often used in calculations. The most common operations are multiplication and division. Two rules must be followed: (1) to **multiply** exponentials, **add** the exponents, and (2) to **divide** exponentials, **subtract** the exponents. These rules can be stated algebraically as

$$(x^a)(x^b) = x^{a+b}$$

$$\frac{x^a}{x^b} = x^{a-b}$$

Some examples follow:

$$(10^6)(10^4) = 10^{6+4} = 10^{10}$$
$$(10^6)(10^{-4}) = 10^{6+(-4)} = 10^{6-4} = 10^2$$
$$(10^{-5})(10^2) = 10^{(-5)+2} = 10^{-5+2} = 10^{-3}$$
$$(10^{-7})(10^{-3}) = 10^{(-7)+(-3)} = 10^{-7-3} = 10^{-10}$$
$$\frac{10^{14}}{10^6} = 10^{14-6} = 10^8$$
$$\frac{10^6}{10^{23}} = 10^{6-23} = 10^{-17}$$
$$\frac{10^{-10}}{10^{-6}} = 10^{(-10)-(-6)} = 10^{-10+6} = 10^{-4}$$
$$\frac{10^3}{10^{-2}} = 10^{3-(-2)} = 10^{3+2} = 10^5$$
$$\frac{10^{-8}}{10^4} = 10^{(-8)-4} = 10^{-12}$$
$$\frac{10^7}{10^7} = 10^{7-7} = 10^0 = 1$$

Problems involving both a coefficient (a numerical part) and an exponential are solved by multiplying (or dividing) coefficients and exponentials separately.

To what is the following expression equivalent?

$$(1.2 \times 10^5)(2.0 \times 10^9)$$

First, multiply the coefficients.

$$1.2 \times 2.0 = 2.4$$

Then multiply the exponentials.

$$10^5 \times 10^9 = 10^{5+9} = 10^{14}$$

The complete answer is

$$2.4 \times 10^{14}$$

example
II.1

To what is the following expression equivalent?

$$\frac{(8.0 \times 10^{11})}{(1.6 \times 10^4)}$$

First, divide the coefficients.

$$\frac{8.0}{1.6} = 5.0$$

Then divide the exponentials.

$$\frac{10^{11}}{10^4} = 10^{11-4} = 10^7$$

The answer is

$$5.0 \times 10^7$$

example
II.2

Give an equivalent for the following expression.

$$\frac{(1.2 \times 10^{14})}{(4.0 \times 10^6)}$$

It is convenient, before carrying out the division, to rewrite the dividend (the numerator) so that the coefficient is larger than that of the divisor (the denominator).

$$1.2 \times 10^{14} = 12 \times 10^{13}$$

example
II.3

Note that the coefficient was made larger by a factor of 10 and the exponential was made smaller by a factor of 10. The quantity as a whole is unchanged. Now divide.

$$\frac{12 \times 10^{13}}{4.0 \times 10^6} = 3.0 \times 10^7$$

example II.4   Give an equivalent for the following expression.

$$\frac{(3 \times 10^7)(8 \times 10^{-3})}{(6 \times 10^2)(2 \times 10^{-1})}$$

In problems such as this, you can carry out the multiplications specified in the numerator and in the denominator separately and then divide the resulting numbers.

$$(3 \times 10^7)(8 \times 10^{-3}) = 24 \times 10^4$$
$$(6 \times 10^2)(2 \times 10^{-1}) = 12 \times 10^1$$
$$\frac{24 \times 10^4}{12 \times 10^1} = 2 \times 10^3$$

The multiplications and divisions in problems like this can be carried out in any convenient order.

There is only one other mathematical function involving exponentials that is of importance to us. What happens when you raise an exponential to a power? You just multiply the exponent by the power. To illustrate:

$$(10^3)^3 = 10^9$$
$$(10^{-2})^4 = 10^{-8}$$
$$(10^{-5})^{-3} = 10^{15}$$

If the exponential is combined with a coefficient, the two parts of the number are dealt with separately, as in the following example:

$$(2 \times 10^3)^2 = 2^2 \times (10^3)^2 = 4 \times 10^6$$

For a further discussion of—and more practice with—exponential numbers, see one of the following references:

1. Goldish, Dorothy M., *Basic Mathematics for Beginning Chemistry*, 3rd ed., New York: Macmillan, 1983. Chapter 3 covers exponential notation.
2. Loebel, Arnold B., *Chemical Problem Solving by Dimensional Analysis*, 2nd ed., Boston: Houghton Mifflin, 1978. Chapter 1, section A, deals with exponential notation.

PROBLEMS

1. Express each of the following in scientific notation:

   a. 0.000 01

   b. 10 000 000

   c. 0.0034

   d. 0.000 010 7

   e. 4 500 000 000

   f. 406 000

   g. 0.02

   h. 124×10^3

2. Carry out the following operations. Express the answers in scientific notation.

   a. $(4.5 \times 10^{13})(1.9 \times 10^{-5})$

   b. $(6.2 \times 10^{-5})(4.1 \times 10^{-12})$

   c. $(2.1 \times 10^{-6})^2$

   d. $\dfrac{(4.6 \times 10^{-12})}{(2.1 \times 10^3)}$

   e. $\dfrac{(9.3 \times 10^9)}{(3.7 \times 10^{-7})}$

   f. $\dfrac{(2.1 \times 10^5)}{(9.8 \times 10^7)}$

   g. $\dfrac{(4.3 \times 10^{-7})}{(7.6 \times 10^{22})}$

(Answers are provided in Appendix IV.)

III

Significant Figures

U nlike counting, measurement is never exact. You can *count* exactly 10 people in a room. If you asked each of those 10 people to *measure* the length of the room to the nearest 0.01 m, however, the values they determine are likely to differ slightly. Table III.1 presents such a set of measurements.

Note that all 10 students agree on the first 3 digits of the measurement; differences occur in the fourth digit. Which values are correct? Actually, all are accurate within the accepted range of uncertainty for this physical measurement. The accuracy of measurement depends on the type of measuring instrument and the skill and care of the person making the measurement. Measured values are usually recorded with the last digit regarded as uncertain. The data in Table III.1 allow us to state that the length of the room is

TABLE III.1   *A Set of Measurements of the Length of a Room*

Student	Length (m)	Student	Length (m)
1	14.14	6	14.14
2	14.15	7	14.17
3	14.17	8	14.17
4	14.14	9	14.16
5	14.16	10	14.17

between 14.1 m and 14.2 m, but we are not sure of the fourth digit. The measurements in the table have four *significant figures*, which means that the first three are known with confidence and the fourth conveys an approximate value. **Significant figures** include all digits known with certainty plus one uncertain digit.

In any properly reported measurement, all nonzero digits are significant. The zero presents problems, however, because it can be used in two ways: to position the decimal point or to indicate a measured value. For zeros, follow these rules.

1. A zero between two other digits is always significant.
   Examples: The number 1107 contains four significant figures.
            The number 50.002 contains five significant figures.
2. Zeros to the left of *all* nonzero digits are not significant.
   Examples: The number 0.000 163 has three significant figures.
            The number 0.068 01 has four significant figures.
3. Zeros that are *both* to the right of the decimal point *and* to the right of nonzero digits are significant.
   Examples: The number 0.2000 has four significant figures.
            The number 0.050 120 has five significant figures.
            The number 802.760 has six significant figures.
4. Zeros in numbers such as 40 000 (that is, zeros to the right of *all* nonzero digits in a number that is written without a decimal point) may or may not be significant. Without more information, we simply do not know whether 40 000 was measured to the nearest unit or ten or hundred or thousand or ten-thousand. To avoid this confusion, scientists use exponential notation (Appendix II) for writing numbers. In exponential notation, 40 000 would be recorded as 4×10^4 or 4.0×10^4 or 4.0000×10^4 to indicate one, two, and five significant figures, respectively.

ADDITION OR SUBTRACTION

In addition or subtraction, the result should contain no more digits to the right of the decimal point than the quantity that has the least digits to the right of the decimal point. Align the quantities to be added or subtracted on the decimal point, then perform the operation, assuming blank spaces are zeros. Determine the correct number of digits after the decimal point in the answer and round off to this number. In rounding off, you should increase the last significant figure by one if the following digit is 5 through 9.

Add the following numbers: 49.146, 72.13, 5.9432. Align the numbers on the decimal point and carry out the addition.

example
III.1

$$
\begin{array}{r}
49.146 \\
72.13 \\
5.9432 \\
\hline
127.2192
\end{array}
$$

The quantity with the least digits after the decimal point is 72.13. The answer should have only two digits after the decimal point. Since the third digit after the decimal point is 9, the second digit after the decimal point should be rounded up to 2. Correct answer: 127.22

Perform the following addition:

example
III.2

$$
\begin{array}{r}
744 \\
2.6 \\
14.812 \\
\hline
761.412
\end{array} \longrightarrow 761
$$

The first quantity has no digits to the right of the decimal point (which is understood to be at the right of the digits). The answer must therefore be rounded so that it, too, contains no digits to the right of the decimal point.

Perform the indicated operation.

example
III.3

$$
\begin{array}{r}
71.124\ 96 \\
-\ 9.143 \\
\hline
61.981\ 96
\end{array} \longrightarrow 61.982
$$

Since the second quantity has only three digits to the right of the decimal point, so must the answer.

MULTIPLICATION AND DIVISION

In multiplication and division, our answers can have no more significant figures than the factor that has the least number of significant figures. In these operations, the *position* of the decimal point makes no difference.

example
III.4

Multiply 10.4 by 3.1416.

$$10.4 \times 3.1416 = 32.672\,64 \longrightarrow 32.7$$

The answer has only three significant figures because the first term has only three.

example
III.5

Divide 5.973 by 3.0.

$$\frac{5.973}{3.0} = 1.991 \longrightarrow 2.0$$

The answer has only two significant figures because the divisor has only two.

EXACT VALUES

Some quantities are not measured but are defined. A kilometer is defined as 1000 meters: 1 km = 1000 m. Similarly, 1 foot can be defined as 12 inches: 1 ft = 12 in. The "1 km" should not be regarded as containing one significant figure; nor should "12 in." be considered to have two significant figures. In fact, these values can be considered to have an infinite number of significant figures (1.000 000 000 000 000 0 . . .) or, more correctly, to be *exact*. Such defined values are frequently used as conversion factors in problems (Chapter 1). When you are determining the number of significant figures for the answer to a problem, you should ignore such exact values. Use only the measured quantities in the problem to determine the number of significant figures in the answer.

PROBLEMS

Perform the indicated operations and give answers with the proper number of significant figures.

1. **a.** 48.2 m + 3.82 m + 48.4394 m
   b. 151 g + 2.39 g + 0.0124 g
   c. 15.436 mL + 9.1 mL + 105 mL
2. **a.** 100.53 cm − 46.1 cm
   b. 451 g − 15.46 g
   c. 19.71 L − 10.4 L
3. **a.** 73 m × 1.340 m × 0.41 m
   b. 0.137 cm × 1.43 cm
   c. 3.146 cm × 5.4 cm
4. **a.** $\dfrac{5.179\ \text{g}}{4.6\ \text{mL}}$ **b.** $\dfrac{4561\ \text{g}}{3.1\ \text{mol}}$ **c.** $\dfrac{40.00\ \text{g}}{3.2\ \text{mL}}$
5. $\dfrac{1.426\ \text{mL} \times 373\ \text{K}}{204\ \text{K}}$

(Answers are provided in Appendix IV.)

Answers to Selected Problems

Answers are provided for all odd-numbered numerical problems and for other selected odd-numbered problems.

Chapter 1

5. yes **7.** no **9.** steam **11.** sprinter
13. automobile **15.** diver on 10-m platform
17. roller coaster at the top of the hill
21. a. cm **b.** kg **c.** dL **23. a.** m **b.** lb **c.** gal
25. a. 50 000 m **b.** 0.25 m
27. a. 10 dL **b.** 0.20 dL **29. a.** 15 g **b.** 86 mg
31. a. 1500 mL **b.** 0.018 L
33. (coldest) 0 K < 0 °F < 0 °C (hottest) **35.** 100 °C
37. a. 2750 cal **b.** 740 cal **39.** 1000 cal
41. 27 g **43.** 1.1 g/cm^3 **45.** 1.6 g/mL
47. 680 g **49.** 1.05 g/mL **51.** 39.5 g
53. 0.73 g/mL

Special Topic A

1. a. 3400 m **b.** 5.70 m **c.** 0.121 m **d.** 91.4 m
3. a. 1290 g **b.** 1.575 g **c.** 421 mg **d.** 2.55 m
  e. 18.3 mm **f.** 4220 mL
5. 141 kg **7.** 317 lb **11.** 0.325 g; 0.0114 oz
13. 7.5 mi/min **15.** 72 km/hr

17. 299 000 000 m/s **19.** $2.09/gal **21.** 750 lb
23. 23 mi/gal **25.** 633 g

Chapter 2

3. in agreement with Dalton's theory: a; not in agreement: b, c
11. carbon monoxide and carbon dioxide (many other answers are possible)
17. Check your answers in Table 2.1.
19. elements: a, b, d, f, h; compounds: c, e, g, i
21. physical change: a, d, e; chemical change: b, c, f
31. A and B are not isotopes. A and C are not isotopes. A and D are isotopes of neon (atomic number 10). B and C are isotopes of sodium (atomic number 11).
33. a. 2 **b.** 19 **c.** 17 **d.** 8 **e.** 12 **f.** 16
35. a. 20 **b.** 9 **c.** 4 **d.** 11 **e.** 18 **f.** 7
37. 32
39. An electron has dropped from a higher to a lower energy level.
43. a. 4 **b.** 8 **c.** 7 **d.** 3 **e.** 2
51. d **53.** e **55.** 20 g H$_2$; 18 t water; 9 t water
57. 4 **59.** 11 g; definite proportions

A-15

Special Topic B

1. Check your answers in Table B.1.
3. **a.** $6s^2$ **b.** $4s^24p^5$ **c.** $5s^1$ **d.** $4s^24p^4$ **e.** $4s^24p^3$
   f. $5s^25p^2$
5. **a.** Ne   **b.** Si   **c.** Na   **d.** Be   **e.** N   **f.** Al
9. maximum of 2 electrons in *any* orbital; maximum of 6 electrons in a *p* sublevel

Chapter 3

5. isotopes: b   **11.** $^{24}_{11}Na$
13. $^{209}_{82}Pb \longrightarrow {}_{-1}^{\ 0}e + {}^{209}_{83}Bi$ **15.** 26 s
19. treatment of malignancy: patient A; imaging: patient B
27. $^{18}_{8}O$ **29.** atomic number 105; mass number 258
33. $^{3}_{2}He$ **35.** $^{30}_{15}P$ **37.** about $2\frac{2}{3}$ hours
41. medical X rays

Special Topic C

3. The like (positively) charged particles repel one another.
7. $^{30}_{14}Si + {}^{1}_{0}n \rightarrow {}^{31}_{14}Si;\ {}^{31}_{14}Si \rightarrow {}_{-1}^{\ 0}e + {}^{31}_{15}P$
11. mass defect or binding energy
13. 224 years (8 half-life periods)

Chapter 4

5. **a.** 3+   **b.** 2−   **c.** 1+   **d.** 1−
7. cations: a, c; anions: b, d
9. **a.** NH_4^+ **b.** HCO_3^- **c.** PO_4^{3-}
11. **a.** carbonate ion   **b.** hydrogen phosphate ion
   c. nitrate ion   **d.** hydroxide ion
13. **a.** $MgSO_4$ **b.** KNO_3 **c.** NaCN   **d.** CaC_2O_4
15. **a.** sodium bromide   **b.** calcium chloride
   c. aluminum oxide
17. **a.** magnesium acetate   **b.** aluminum acetate
   c. ammonium phosphate   **d.** ammonium oxalate
   e. sodium hydrogen phosphate
   f. calcium dihydrogen phosphate
   g. magnesium hydrogen carbonate
   h. calcium hydrogen sulfate   **i.** ammonium nitrite
19. **a.** VIIA, VIA, VA   **b.** $H:\ddot{X}:,\ :\ddot{Y}:H,\ H:\ddot{Z}:H$
                                  H   H

   c. $Na^+\ :\ddot{X}:^-;\ 2\,Na^+\ :\ddot{Y}:^{2-}$

25. **a.** N   **b.** Cl   **c.** F   **27.** polar covalent
29. No. Sodium chloride consists of sodium ions and chloride ions.
31. **a.** 1   **b.** 1   **c.** 4   **d.** 3   **e.** 2   **f.** 1
33. **a.** ionic   **b.** polar covalent   **c.** nonpolar covalent
   d. polar covalent   **e.** ionic   **f.** nonpolar covalent
   g. ionic   **h.** ionic   **i.** polar covalent
35. **a.** bent   **b.** tetrahedral   **c.** linear   **d.** planar

Chapter 5

3. **a.** 4   **b.** 9
5. balanced: a, d, e; not balanced: b, c
7. **a.** not balanced   **b.** balanced
9. **a.** $Zn + 2\,HCl \longrightarrow ZnCl_2 + H_2$
   b. $2\,H_2S + O_2 \longrightarrow 2\,H_2O + 2\,S$
   c. $Al_2(SO_4)_3 + 6\,NaOH \longrightarrow 2\,Al(OH)_3 + 3\,Na_2SO_4$
   d. $Zn(OH)_2 + 2\,HNO_3 \longrightarrow Zn(NO_3)_2 + 2\,H_2O$
   e. $3\,NH_4OH + H_3PO_4 \longrightarrow (NH_4)_3PO_4 + 3\,H_2O$
11. **a.** 16 amu   **b.** 84 amu   **c.** 352 amu
13. **a.** 164 amu   **b.** 58 amu   **c.** 132 amu
15. **a.** 0.10 mol   **b.** 0.001 00 mol   **c.** 10.0 mol
   d. 0.0300 mol
17. **a.** 0.016 g   **b.** 504 g   **c.** 14 100 g
19. **a.** 246 g   **b.** 340 g   **c.** 33 g
21. **a.** 1.0 mol   **b.** 1.0 mol
23. **a.** 0.8 mol   **b.** 50 mol   **c.** 5.0 mol   **d.** 0.86 mol
25. **a.** 32 g   **b.** 220 g
27. **a.** 2.5 mol   **b.** 0.075 mol
31. exothermic   **33.** increases the rate
35. decreases the rate   **37.** increases the rate
39. **a.** Equilibrium shifts to the left.
   b. Equilibrium shifts to the right.
   c. Equilibrium shifts to the right.
41. Equilibrium shifts to the right.
45. 56 g CaO   **47.** 250 g HNO_3 **49.** 160 kg O_2
51. 2090 g Fe   **53.** 0.50 g H_2
55. 7.9×10^{10} g; 4.0×10^8 mol   **57.** 1.31 g Au

Chapter 6

5. **a.** decrease in volume   **b.** decrease in volume
   c. increase in volume
7. The temperature is decreasing.
9. Assuming the temperature in both containers is the same, the pressure is the same.
11. **a.** 1500 torr   **b.** 380 torr   **c.** 300 Pa

13. a. 150 lb/in.^{2} **b.** 14 cm H_2O **c.** 0.10 atm
15. 1000 mL **17. a.** 9000 L **b.** 19 hr
21. 10 mL **23.** 3.3 L **25.** 2470 torr
27. 0.48 m^3 **29.** 750 cc **31.** 0.95 atm
33. 708 torr **35.** 95% **37.** 22 L **39.** 19 L
41. 1.2 atm **43.** 390 K **45.** 7.3 × 10^8 mol
47. 67% **51.** 280 s

Chapter 7

5. N_2 (weakest), HCl, H_2O (strongest)
7. a. dispersion forces **b.** dipolar forces
  c. hydrogen bonding
13. a. H_2S (lowest), H_2Se, H_2Te (highest)
  b. O_2 (lowest), CO, H_2O (highest)
  c. CCl_4 (lowest), CBr_4, CI_4 (highest)
15. 7200 cal/mol **17.** 96.8 cal **21.** 44.5 cal
23. 7400 cal/mol **25.** ethanol **27.** 72 kcal
29. 16 kcal **33.** ethane **35.** Xe **37.** 180 kcal

Chapter 8

1. a. $CO_2 + H_2O$ **b.** CO_2 **c.** SO_2 **d.** $CO_2 + SO_2$
  e. NO **f.** $CO_2 + H_2O$
3. oxidation: only d **5.** oxidized
7. a. 0 **b.** +6 **c.** −2 **d.** −1
9. a. +4 **b.** +2 **c.** +6
17. a. −3 **b.** 0 **c.** +3 **d.** +5 **e.** +5 **f.** +1
  g. −3 **h.** +5 **i.** +3 **j.** +1 **k.** +5 **l.** +5
  m. +5 **n.** +3 **o.** +5 **p.** +5 **q.** +5 **r.** +3
  s. −3 **t.** −3 **u.** +3 **v.** +5
19. neither oxidized nor reduced **21.** reduced
23. +7 **27.** 11 000 L of air; 2300 L of O_2
29. redox reactions: b, c

Chapter 9

5. soluble in water: a, c, d, g, h, i, j, k
7. saturated **11.** 4.0% by mass
13. 5.0% by mass
15. Dissolve 0.50 kg of NaCl in 4.5 kg of water.
17. 1 ppb < 1 ppm < 1 mg% < 1%
  (lowest concentration → highest concentration)
19. a. Dilute 0.1 mL of drug to a total volume of
    100 mL.
  b. Dilute 0.0005 L (0.5 mL) of drug to a total
    volume of 1 L.

21. 2.0 M **23.** 3.0 M
25. Dissolve 5.35 g of NH_4Cl in water to give a total
  of 100 mL of solution.
29. a. 2 **b.** 1 **c.** 3 **d.** 2 **e.** 3
31. a. 0.50 mol **b.** 1.0 mol **c.** 0.33 mol
  d. 0.50 mol **e.** 0.33 mol
35. a. 0.1 M $NaHCO_3$ **b.** 1 M NaCl **c.** 1 M $CaCl_2$
  d. 3 M glucose
41. no **45.** 1.0 g
47. a. 0.05% by mass **b.** 500 ppm **c.** 0.50 g/kg
  d. 0.010 M

Chapter 10

9. OH^- **13.** hydroselenic acid
15. phosphorous acid **17.** benzoic acid
19. $HAsO_3$; AsO_3^-; $KAsO_3$

21.

	Stronger Base	Weaker Base
a.	F^-	Br^-
b.	CN^-	F^-
c.	IO_3^-	$H_2PO_4^-$
d.	$HCOO^-$	CCl_3COO^-
e.	NO_3^-	ClO_3^-

23. a. $NaHCO_3 + HNO_3 \longrightarrow NaNO_3 + H_2O + CO_2$
  b. $CaCO_3 + 2\,HBr \longrightarrow CaBr_2 + H_2O + CO_2$
  c. $H_2SO_4 + K_2CO_3 \longrightarrow K_2SO_4 + H_2O + CO_2$
25. $H_3O^+ + OH^- \longrightarrow 2\,H_2O$
27. $HCO_3^- + H_3O^+ \longrightarrow 2\,H_2O + CO_2$
29. a. $Mg + 2\,HCl \longrightarrow MgCl_2 + H_2$
  b. $2\,Al + 6\,HCl \longrightarrow 2\,AlCl_3 + 3\,H_2$
  c. $Mg + H_2SO_4 \longrightarrow MgSO_4 + H_2$
  d. $Zn + H_2SO_4 \longrightarrow ZnSO_4 + H_2$
  e. $2\,Al + 3\,H_2SO_4 \longrightarrow Al_2(SO_4)_3 + 3\,H_2$
33. Aniline is a weak base.

Chapter 11

3. a. 6 eq **b.** 0.5 eq **c.** 10 eq
5. a. 2 mol **b.** 0.1 mol **c.** 1.5 mol
7. a. 0.0200 eq **b.** 1.5 eq **c.** 0.25 eq
9. a. yes **b.** no **c.** yes
11. a. 0.5 N **b.** 1 N **c.** 1 N **d.** 0.4 N **e.** 2 N
13. a. 2.0 L **b.** 1.4 mL
17. $NH_4^+ + H_2O \longrightarrow NH_3 + H_3O^+$
19. basic **21.** 0.13 N **23.** 10.0 mL
25. 0.121 N **27. a.** 2 **b.** 3 **29. a.** 1 **b.** 5

31. 27a. 12   27b. 11   28a. 10   28b. 9   28c. 13
33. a. basic   **b.** acidic   **c.** neutral   **d.** acidic
39. decrease   **41. a.** 2.5   **b.** 6.3   **c.** 9.1
43. 1.06   **45.** 7.34   **47.** 11.70

Special Topic D

1. a. $HBO_2 \rightleftharpoons H^+ + BO_2^-$
   b. $HClO_2 \rightleftharpoons H^+ + ClO_2^-$
   c. $HC_9H_7O_4 \rightleftharpoons H^+ + C_9H_7O_4^-$
   d. $H_2Se \rightleftharpoons H^+ + HSe^-$
3. 4×10^{-7}
7. a. 1.3×10^{-4} **b.** 1.2×10^{-3} **c.** 1.8×10^{-5}
9. a. 7.7×10^{-11} **b.** 8.3×10^{-12} **c.** 5.6×10^{-10}
11. a. 6.7×10^{-4} **b.** 6.5×10^{-3} **c.** 6.5×10^{-6}
13. a. 1.5×10^{-11} **b.** 1.5×10^{-12} **c.** 1.5×10^{-9}
15. a. 6.2×10^{-10} **b.** 6.6×10^{-4} **c.** 6.6×10^{-5}
17. 1.8×10^{-5}; 5.6×10^{-10} **19.** 10   **21.** 9.21
23. 4.18   **25.** 4.34

Chapter 12

5. a. K and Br_2 **b.** Li and Cl_2 **c.** Al and O_2
7. $HI + H_2O \longrightarrow H_3O^+ + I^-$
13. a. base   **b.** acid   **c.** salt   **17.** yes   **19.** yes
21. $CH_3COOAg + HNO_3 \longrightarrow CH_3COOH + AgNO_3$
33. 15 t

Chapter 13

9. Be   **11.** Al   **13.** $\cdot \dot{Ga} \cdot$; Ga^{3+}
15. $2\,Al_2O_3 \longrightarrow 4\,Al + 3\,O_2$
19. $CaO + H_2O \longrightarrow Ca^{2+} + 2\,OH^-$
21. a. $4\,Li + O_2 \longrightarrow 2\,Li_2O$
   b. $2\,Ca + O_2 \longrightarrow 2\,CaO$
   c. $S + O_2 \longrightarrow SO_2$
   d. $CaO + H_2O \longrightarrow Ca(OH)_2$
   e. $SO_2 + H_2O \longrightarrow H_2SO_3$
   f. $Ca + S \longrightarrow CaS$
39. Na^+ and K^+ **41.** chlorophyll
43. water containing Ca^{2+}, Mg^{2+}, or Fe^{2+}

Chapter 14

3. organic: a, c, d; inorganic: b, e, f   **5.** organic
7. a. same compound   **b.** same compound
   c. isomers   **d.** same compound   **e.** isomers
11. saturated: b, d; unsaturated: a, c

17. a. C_6H_{10} **b.** C_4H_8
19. a. para   **b.** ortho   **c.** meta
27. a. 2-methyl-1-pentene   **b.** 2,5-dimethyl-2-hexene
   c. *cis*-3-hexene
29. a. methylcyclopropane   **b.** cyclobutene
31. a. $2\,H_2$, Ni   **b.** H_2O, H^+ **c.** H_2O, H^+ **d.** Cl_2
33. a, b, c

Special Topic E

1. an alkyl group
3. $CH_3CH_2CH_2Br$ and $CH_3CHBrCH_3$
5. a. $CF_3CF_2CF_3$ **b.** $CF_3CF{=}CF_2$
7. a. CH_2Cl_2 **b.** $CHCl_3$ **c.** CCl_4
9. a. $Cl_2C{=}CHCH_3$ **b.** $CH_3\underset{\underset{Cl}{|}}{C}HCH_2\underset{\underset{CH_3}{|}}{C}HCH_2CH_2CH_3$

11. a. 2,4,6-trichloroheptane
   b. 3-chloro-2-methylhexane
23. 30 g   **25.** 300 μg

Chapter 15

3. 1-pentanol, 2-pentanol, 3-pentanol,
   3-methyl-1-butanol, 3-methyl-2-butanol,
   2-methyl-2-butanol, 2-methyl-1-butanol,
   and 2,2-dimethyl-1-propanol
5. butyl methyl ether, isobutyl methyl ether,
   sec-butyl methyl ether, *tert*-butyl methyl ether,
   ethyl propyl ether, and ethyl isopropyl ether
7. a. dipropyl ether   **b.** *sec*-butyl ethyl ether
21. a. oxidation   **b.** dehydration   **c.** oxidation
   d. hydration   **e.** dehydration
27. methanol, ethanol, 1-propanol
29. diethyl ether, 1-butanol, propylene glycol
31. pentane, diethyl ether, propylene glycol
35. a. H_2O, H^+ **b.** H_2O, H^+ **c.** H_2SO_4, heat
   d. $K_2Cr_2O_7$, H^+ **e.** $K_2Cr_2O_7$, H^+
   f. H_2SO_4, heat
39. a. $CH_2{=}CHCH_3$ **b.** $CH_2{=}CH_2$
   d. $CH_2{=}CHCH_2CH_3$ or $CH_3CH{=}CHCH_3$
45. 23 t

Chapter 16

1. valeraldehyde (pentanal), β-methylbutyraldehyde
   (3-methylbutanal), α-methylbutyraldehyde
   (2-methylbutanal), dimethylpropionaldehyde
   (dimethylpropanal)

3. a. benzaldehyde **b.** propionaldehyde (propanal)
c. β,β-dimethylbutyraldehyde (3,3-dimethylbutanal)
5. a. *o*-chlorobenzaldehyde (2-chlorobenzaldehyde)
b. *sec*-butyl methyl ketone (3-methyl-2-pentanone)
9. propanol **11.** acetaldehyde
15. a, c, d can be distinguished by Tollens's reagent
17. Phenylhydrazine gives an orange precipitate
with 2-pentanone.
19. $CH_3CH=CHOH$ **25.** alkene, alcohol, ketone
27. a, d **29.** c, e

Chapter 17

7. a. $CH_3CHClCOOH$ **b.** $CH_3CHBrCH_2COOH$
c. $CH_3CHFCH_2CH_2COOH$
9. a. sodium benzoate
b. calcium propionate (calcium propanoate)
c. ammonium acetate (ammonium ethanoate)
d. zinc butyrate (zinc butanoate)
e. calcium phthalate
17. a. $CH_3CH_2CH_2OH$ **b.** $HOCH_2CH_2OH$
c. CH_3OH **d.** $(CH_3)_2CHCH_2CH_2OH$
21. a. benzamide **b.** N-isopropylacetamide
c. acetamide **d.** *p*-chlorobenzamide
e. malonamide **f.** N,N-dimethylpropionamide
g. butyranilide
23. butyric acid **25.** butyric acid
27. acetic acid **29.** methyl acetate
45. a. $K_2Cr_2O_7, H^+$ **b.** $K_2Cr_2O_7, H^+$ **c.** NaOH
47. a. NH_3 **b.** $SOCl_2$ **c.** H_2O, H^+ **d.** H_2O, OH^-
49. a. $H_2NCH_2CH_2COO^-Na^+$
b. $H_3N^+CH_2CH_2CH_2COOH$
51. a. carboxylic acid **b.** alcohol, ketone
c. alcohol, ketone **d.** aldehyde, alcohol
e. aldehyde, ether **f.** aldehyde, ether **g.** ester
h. ether, ketone **i.** ester **j.** ester
k. ketone, alcohol **l.** alcohol, aldehyde
55. The ester is hydrolyzed to butyric acid.

Special Topic F

7. isobutyl **9. a.** CH_3OH, H^+ **b.** NaOH
11. 3

Chapter 18

1. butylamine (1°); *sec*-butylamine (1°); isobutylamine
(1°); *tert*-butylamine (1°); methylpropylamine (2°);

methylisopropylamine (2°); diethylamine (2°);
dimethylethylamine (3°)
3. a. 1° alcohol **b.** 1° amine **c.** 2° alcohol
d. 1° amine **e.** ether **f.** 2° amine **g.** 2° amine
h. ether **i.** 3° amine **j.** 3° alcohol **k.** 1° amine
l. phenol
7. $H_2NCH_2CH_2OH$
11. a. propylamine **b.** ethylmethylamine
c. triethylamine
13. diethylammonium bromide
15. a. propylamine **b.** 2-amino-4-methylpentane
17. butylamine **19.** propylamine
21. propylamine **23.** It reacts with HCl to form a salt.
31. none **39. a.** NH_3 **b.** NH_3 **c.** H_2, NH_3, Ni
41. a. $NaNO_2$ and HCl **b.** H_2, NH_3, Ni **c.** heat
45. acidic: c; basic: a; neutral: b, d

Special Topic G

3. tryptophan → serotonin and
tyrosine → norepinephrine
9. ketamine and phencyclidine
19. phenol, amide, alcohol, amine
23. NH_4^+ is the stronger acid; $C_{17}H_{25}N$ is the stronger
base.

Special Topic H

3. a. propanethiol **b.** 2-methyl-1-butanethiol
c. 3-hexanethiol
7. a. CH_3SSCH_3 **b.** $CH_3CH_2CH_2SSCH_2CH_2CH_3$
c. $(CH_3)_3CSSC(CH_3)_3$
9. $CH_3CH_2SSCH_2CH_3$
21. a. H_2SO_4, SO_3 **b.** NaOH

Chapter 19

7. a. . . . $CH_2CHCH_2CHCH_2CH$ . . .

with Cl, Cl, Cl substituents

b. . . . $CF_2CF_2CF_2CF_2CF_2CF_2$ . . .

13. . . . structure showing benzene ring with two C=O groups, —OCH₂—, cyclohexane ring, —CH₂O— . . .

15. . . .

$$\overset{O}{\underset{\|}{C}}—\bigcirc—\overset{O}{\underset{\|}{C}}—NH—\bigcirc—NH \ldots$$

21. . . . CH_2CH⎯⎯CH_2CH⎯⎯CH_2CH⎯⎯CH_2CH . . .
  $\underset{OCOCH_3}{|}\quad \underset{OCOCH_3}{|}\quad \underset{OCOCH_3}{|}\quad \underset{OCOCH_3}{|}$

23. . . . CH_2CH⎯CH_2CH⎯CH_2CH⎯CH_2CH . . .
  (with phenyl groups)

25. (see bottom of page)

27. . . . $OCH_2\overset{O}{\underset{\|}{C}}CH_2\overset{O}{\underset{\|}{C}}CH_2\overset{O}{\underset{\|}{C}}CH_2\overset{O}{\underset{\|}{C}}$ . . .

29. The rubber in an automobile tire is more extensively cross-linked.

Special Topic I

  3. yes, yes   **9.** a, d
11. a. no **b.** no (meso) **c.** yes
15. 2-Pentanol, 2-methyl-1-butanol, and 3-methyl-2-butanol have chiral carbon atoms.
19. 8   **21.** yes, no, no

Chapter 20

  3. a. D **b.** L **c.** L **d.** L
  5. a. hexose **b.** pentose **c.** hexose **d.** triose
13. a. beta **b.** alpha **c.** beta **d.** alpha
15. d   **17.** glucose   **19.** glucose
21. glucose and fructose   **23.** galactose and glucose
25. yes, alpha   **33. a.** $Ag(NH_3)_2{}^+$ **b.** OH^-
37. galactose and fructose
43. Xylulose is a 2-ketopentose, L-form.   **45.** L

Chapter 21

  3. a. 6 **b.** 18 **c.** 18 **d.** 16 **e.** 18 **f.** 8
11. triolein   **13.** liquid margarine
23. saponifiable: a and c; nonsaponifiable: b and d

25. a and c
27. All have both polar and nonpolar ends.
31. glycerol: a, b, and d; sphingosine: f; neither: c, e, and g
37. a. H_2, Ni **b.** OH^- **c.** heat   **49.** c

Chapter 22

11. the L-family
17. Strict vegetarian diet often is lacking in one or more essential amino acids, vitamin B_{12}, calcium, iron, and riboflavin.
19. alpha helix
21. salt bridges, hydrogen bonds, disulfide linkages, and hydrophobic interactions
23. fibrous: b, c, e; globular: a, d, f   **25.** globular
27. secondary and tertiary   **35.** 20   **37.** proline
43. inside: a, e; outside: b, c, d, f
47. GABA is a neurotransmitter. It is not found in proteins.

Chapter 23

  7. ribose: a, b, and d; deoxyribose: c, e, and f
  9. pyrimidine: a; purine: b
11. a. nucleoside **b.** pyrimidine **c.** DNA
17. hydrogen bonding
19. a. uracil **b.** cytosine **c.** adenine **d.** guanine
21. Each daughter DNA molecule contains one of the parent DNA strands.
25. DNA and messenger RNA   **27.** messenger RNA
29. . . . T-A-A-G-C . . .   **31.** . . . A-G-G-C-T-A . . .
33. a. A-A-C **b.** C-U-U **c.** A-G-G **d.** G-U-G
35. a. Asn **b.** Leu **c.** Arg **d.** Val
37. Asn-Glu-Ser (The codons on mRNA would be A-A-U-G-A-G-A-G-U.)
47. uracil and deoxyribose
49. adenosine diphosphate (ADP)

Chapter 24

  3. urease   **5.** no, only the *rate* of the reaction
13. No. Coenzymes are organic.
15. The rate is doubled.

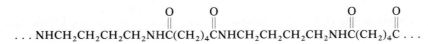
. . . $NHCH_2CH_2CH_2CH_2NH\overset{O}{\underset{\|}{C}}(CH_2)_4\overset{O}{\underset{\|}{C}}NHCH_2CH_2CH_2CH_2NH\overset{O}{\underset{\|}{C}}(CH_2)_4\overset{O}{\underset{\|}{C}}$ . . .

17. less active at normal body temperature; even less active at 40 °C
35. prothrombin   39. an isomerase
41. a lyase (more specifically, a decarboxylase)
43. It increases the rate.
45. around 2, the pH of the stomach
47. by saturating the active sites, making them unavailable for the oxidation of methanol

Chapter 25

5. The lightning-induced reaction is endergonic. The firefly reaction is exergonic.
11. **a.** -2.8 kcal/mol **b.** -3.5 kcal/mol
    c. -3.2 kcal/mol **d.** -5.1 kcal/mol
15. a net synthesis   17. a net consumption
19. At the end of the respiratory chain, oxygen (O_2) is reduced to water (H_2O).
21. pyruvic acid $\longrightarrow$ FAD $\longrightarrow$ cytochromes $\longrightarrow$ O_2
23. aerobic oxidation of glucose   27. carbohydrates
29. anaerobic
37. pheasants: Type IIB fibers; great blue heron: Type I fibers
43. ADP, ribose, and nicotinamide (a B vitamin)
45. the ADP part   47. acetyl coenzyme A
49. $\Delta G° = -5.5$ kcal/mol

Special Topic J

3. See Table J.1.   5. See Table J.1.
9. water-soluble: b and c; fat-soluble: a and d
11. fat-soluble   15. no   21. nicotinamide
23. pantothenic acid

Special Topic K

3. See Table K.1.   5. the pituitary gland
7. See Table K.1.   9. steroids
11. the ethynyl group
13. alcohol, aldehyde, alkene, ketone
15. alkene, ketone

Chapter 26

3. Ca^{2+} 9. Na^+, K^+, Ca^{2+}, Mg^{2+}
11. **a.** 1 eq **b.** 3 eq **c.** 4 eq

13. **a.** 2.0 mol **b.** 0.025 mol **c.** 110 mmol
15. $Cl^- = 120$ meq/L; $Ca^{2+} = 1.3$ meq/L; $Na^+ = 140$ meq/L
25. lung obstruction
27. production of acids during metabolism
35. gamma globulins or antibodies
37. globulins   55. diffusion   57. lower

Special Topic L

5. HCO_3^- 9. cholesterol
11. fatty acids and glycerol
33. No. It would be destroyed by digestive enzymes.
35. Monosaccharides are absorbed directly. Disaccharides and polysaccharides end up as monosaccharides.
37. The starch is partially converted to glucose by salivary amylase.

Chapter 27

11. glycogen   15. lactic acid   21. NAD^+ and FAD
23. 2   25. anhydride
27. The anhydride in Problem 25 is a high-energy phosphate.
29. the respiratory chain
37. Fructose is phosphorylated in the liver. It may enter the glycolysis pathway directly or be interconverted with other monosaccharides.

Chapter 28

3. triglycerides   5. cholesterol (to the liver)
7. exercise   9. acetyl coenzyme A
11. as fat globules
13. condensation, followed by reduction, dehydration, and another condensation
15. 5   17. 4
19. It is phosphorylated in the liver to glycerol phosphate, then oxidized to dihydroxyacetone phosphate which enters the Embden-Meyerhof glycolytic pathway.
21. the malonyl group (as malonyl coenzyme A)
25. It acts as a carrier molecule.   31. fats

33. Two of the three ketone bodies are also acids.
39. 7   **41.** 129   **43.** brain and nerve tissue

Chapter 29

3. no
7. A negative nitrogen balance develops as more nitrogen is excreted than taken in.
9. alanine   **11.** phenylalanine   **15.** glycerol
19. muscle, collagen   **25.** pyruvic acid
29. α-ketoglutaric acid   **31.** urea   **33.** uric acid
35. arginine and citrulline   **37.** phenylalanine
39. phenylketonuria   **41.** albinism
43. indication of a myocardial infarction
45. The precursor thymine occurs in DNA but not in RNA.

Appendix II

1. a. 1×10^{-5} **b.** 1×10^{7} **c.** 3.4×10^{-3}
   d. 1.07×10^{-5} **e.** 4.5×10^{9} **f.** 4.06×10^{5}
   g. 2×10^{-2} **h.** 1.24×10^{5}
2. a. 8.6×10^{8} **b.** 2.5×10^{-16} **c.** 4.4×10^{-12}
   d. 2.2×10^{-15} **e.** 2.5×10^{16} **f.** 2.1×10^{-3}
   g. 5.7×10^{-30}

Appendix III

1. a. 100.5 m   **b.** 153 g   **c.** 130 mL (1.3×10^{2} mL)
2. a. 54.4 cm   **b.** 436 g   **c.** 9.3 L
3. a. 40 m^3 **b.** 0.196 cm^2 **c.** 17 cm^2
4. a. 1.1 g/mL   **b.** 1500 g/mol   **c.** 13 g/ml
5. 2.61 mL

Index

References to definitions of terms are printed in italic. References to tables have a *t* following the number.

SYMBOLS AND NAMES FOR SOME SIMPLE IONS

Group	Element	Name of Ion	Symbol for Ion
IA	Hydrogen	Hydrogen ion*	H^+
	Lithium	Lithium ion	Li^+
	Sodium	Sodium ion	Na^+
	Potassium	Potassium ion	K^+
IIA	Magnesium	Magnesium ion	Mg^{2+}
	Calcium	Calcium ion	Ca^{2+}
IIIA	Aluminum	Aluminum ion	Al^{3+}
VA	Nitrogen	Nitride ion	N^{3-}
VIA	Oxygen	Oxide ion	O^{2-}
	Sulfur	Sulfide ion	S^{2-}
VIIA	Chlorine	Chloride ion	Cl^-
	Bromine	Bromide ion	Br^-
	Iodine	Iodide ion	I^-
IB	Copper	Copper(I) ion (cuprous ion)	Cu^+
		Copper(II) ion (cupric ion)	Cu^{2+}
	Silver	Silver ion	Ag^+
IIB	Zinc	Zinc ion	Zn^{2+}
VIIIB	Iron	Iron(II) ion (ferrous ion)	Fe^{2+}
		Iron(III) ion (ferric ion)	Fe^{3+}

* Does not exist independently in aqueous solution.

SOME COMMON POLYATOMIC IONS

Charge	Name	Formula
1+	Ammonium ion	NH_4^+
	Hydronium ion	H_3O^+
1−	Hydrogen carbonate (bicarbonate) ion	HCO_3^-
	Hydrogen sulfate (bisulfate) ion	HSO_4^-
	Acetate ion	$CH_3CO_2^-$ (or $C_2H_3O_2^-$)
	Nitrite ion	NO_2^-
	Nitrate ion	NO_3^-
	Cyanide ion	CN^-
	Hydroxide ion	OH^-
	Dihydrogen phosphate ion	$H_2PO_4^-$
	Permanganate ion	MnO_4^-
2−	Carbonate ion	CO_3^{2-}
	Sulfate ion	SO_4^{2-}
	Monohydrogen phosphate ion	HPO_4^{2-}
	Oxalate ion	$C_2O_4^{2-}$
	Dichromate ion	$Cr_2O_7^{2-}$
3−	Phosphate ion	PO_4^{3-}